Technology Innovation for the Circular Economy

Scrivener Publishing
100 Cummings Center, Suite 541J
Beverly, MA 01915-6106

Publishers at Scrivener
Martin Scrivener (martin@scrivenerpublishing.com)
Phillip Carmical (pcarmical@scrivenerpublishing.com)

Technology Innovation for the Circular Economy

Recycling, Remanufacturing, Design, Systems Analysis and Logistics

Edited by

Nabil Nasr

REMADE Institute, Rochester, New York, USA

Scrivener
Publishing

WILEY

Wiley Global Headquarters
111 River Street, Hoboken, NJ 07030, USA

For details of our global editorial offices, customer services, and more information about Wiley products visit us at www.wiley.com.

Limit of Liability/Disclaimer of Warranty
While the publisher and authors have used their best efforts in preparing this work, they make no representations or warranties with respect to the accuracy or completeness of the contents of this work and specifically disclaim all warranties, including without limitation any implied warranties of merchant-ability or fitness for a particular purpose. No warranty may be created or extended by sales representatives, written sales materials, or promotional statements for this work. The fact that an organization, website, or product is referred to in this work as a citation and/or potential source of further information does not mean that the publisher and authors endorse the information or services the organization, website, or product may provide or recommendations it may make. This work is sold with the understanding that the publisher is not engaged in rendering professional services. The advice and strategies contained herein may not be suitable for your situation. You should consult with a specialist where appropriate. Neither the publisher nor authors shall be liable for any loss of profit or any other commercial damages, including but not limited to special, incidental, consequential, or other damages. Further, readers should be aware that websites listed in this work may have changed or disappeared between when this work was written and when it is read.

Library of Congress Cataloging-in-Publication Data

ISBN 978-1-394-21426-6

Cover images: Pixabay.Com
Cover design by Russell Richardson

Set in size of 11pt and Minion Pro by Manila Typesetting Company, Makati, Philippines

Printed in the USA

10 9 8 7 6 5 4 3 2 1

Contents

Part 4: Systems Analysis   197

Part 6: Chemical Recycling   407

Preface

The consensus that we need a circular economy is gaining speed. Decision-makers in government and industry increasingly see the immediate value that circularity can bring to the manufacturing sector, while addressing some of today's greatest global challenges. At the same time, academic research continues to investigate the tough technological and logistical questions that need to be answered for a circular economy to become reality.

At the REMADE Institute, my colleagues and I wanted to capture this gathering energy by bringing together the best research and innovation looking to solve circular economy implementations challenges. The result was the first-ever REMADE Circular Economy Tech Summit and Conference, which took place in March 2023 at the National Academy of Sciences building in Washington, D.C.

Attracting over 300 attendees, the scientific research, business models and logistics conference featured nearly 60 presentations of original research, as well as keynotes and plenaries from visionary thought leaders. The conference was held in partnership with the Ellen MacArthur Foundation, one of the most influential advocates for a circular economy at work today, and supported by the U.S. Department of Energy (DOE).

The following book compiles the peer-reviewed papers that were presented over the course of the conference. These materials cover in-depth areas of circular economy design, planning, business models, and enabling technologies.

The REMADE Institute is a public-private partnership, national institute, that focuses on the acceleration of circular economy implementations in the United States. REMADE is one of six U.S. manufacturing institutes that operate under the DOE's Advanced Materials and Manufacturing Technologies Office (*AMMTO*). REMADE is a consortium of 170 members (90 industry, 42 universities, 32 trade associations, and six national labs). It addresses knowledge gaps that, once overcome, can lead to faster adoption of circular economy practices. The 2023 conference included REMADE members, as well as non-members, and included international researchers from many countries including Japan, Germany, France, UK, and Ireland.

First, it is important to understand why circular economy has gained so much attention over the last decade and what role it can play in reducing the environmental footprint of industrial development. The future should be circular, and in the manufacturing economy this means that economic growth will come to depend more on extracting value from existing materials than securing new virgin material supplies. Remanufacturing, refurbishment, reuse, and recycling are processes that need more advancement and expansion to achieve this goal. In addition, our design methods have to evolve to ensure that these processes are effective and economical through design. In many industries, future survival will require transformative redevelopment of business structures to create more inherently regenerative models — an idea that forward-thinking groups like the Ellen MacArthur Foundation are adamantly promoting.

Some of the greatest opportunities for innovation in the circular economy are in remanufacturing, refurbishment, reuse, and recycling. Critical to its growth, however, are developments in product design approaches and the manufacturing business model that are often met with challenges in the current, largely linear economies of today's global manufacturing chains. Beyond the technical and logistical, these processes also meet with both market and policy barriers that stifle its growth across the industrial economy. To combat these challenges, significant investments in technology research, both public and private, highlight the importance of and opportunities for innovation in pursuit of a more resilient, circular economy.

This book consists of 56 chapters in 10 parts covering broad areas of research and applications in the circular economy area. The first four parts explore the system level work related to circular economy approaches, models and advancements including the use of artificial intelligence (AI) and machine learning to guide implementation, as well as design for circularity approaches. Mechanical and chemical recycling technologies follow, highlighting some of the most advanced research in those areas. Next, Innovation in remanufacturing are addressed with descriptions of some of the most advanced work in this field. This is followed by tire remanufacturing and recycling, highlighting innovative technologies in addressing the volume of end-of-use tires. Pathways to net-zero emissions in manufacturing of materials concludes the book, with a focus on industrial decarbonization.

I would like to acknowledge the contribution of the conference Organizing Committee, as well as the Program Committee and its members who reviewed the papers. This book would not have been possible without their contributions. I also would like to acknowledge the following members of the REMADE Institute team (Magdi Azer, Michelle Schlafer, Ed Daniels, Megan Connor Murphy, Steve Remmler, Carrie Degláns, John Kreckel, Jared Ratzel, Michele Gibson, Bonnie Schiffmaker, Sarah Beisheim, Mike Haselkorn) for their support in every aspect of the planning and execution of the conference. Special thanks go to the REMADE Institute CTO, Magdi Azer, who was instrumental in leading and orchestrating the papers' peer reviews and finalization. Additions thanks go to Carrie Degláns from REMADE who assisted in the tracking and organization of submitted papers.

We also highly appreciate the dedicated support and valuable assistance rendered by Martin Scrivener and the Scrivener Publishing team during the publication of this book.

Part 1
CIRCULAR ECONOMY

Standards as Enablers for a Circular Economy

K.C. Morris[1]*, Vincenzo Ferrero[1], Buddhika Hapuwatte[2], Noah Last[3] and Nehika Mathur[1]

[1]National Institute of Standards and Technology, Gaithersburg, USA
[2]National Institute of Standards and Technology, Gaithersburg, USA; University of Maryland, College Park, USA
[3]National Institute of Standards and Technology, Gaithersburg, USA; Georgetown University, Washington DC, USA

Abstract

A successful transition to a circular economy (CE) will require global participation, but the path to that transition will follow many unique routes depending on local situations. The transition must have rigorous technical underpinnings and well-conceived social interventions. In addition to solid technical foundations, consensus will add legitimacy to new and revised business practices thereby reducing the risk in their adoption. Standards created by voluntary consensus bodies are uniquely positioned to serve these purposes. In these bodies, stakeholders from a broad spectrum of society (industry, academia, government) come together to define solutions for unique circumstances and communicate them through published standards. Standards are developed by systematically determining the scope of the work, agreeing on terminology, and building on that foundation to create detailed specifications. Early engagement in the standards bodies can position stakeholders to be leaders in the path that lies ahead.

This chapter reviews several efforts to coordinate industries to facilitate the adoption of circular practices and technologies, highlights opportunities for further development, and discusses the role of consensus-based standards in these efforts. It highlights two recent international standards activities supporting the transition to a CE in the International Organization for Standardization (ISO) and ASTM International, followed by an example of a carbon savings measure designed to encourage more reuse of materials. While the initial standards efforts are underway, greater participation will be needed to complete the necessary agreements to establish a successful CE. ISO Technical Committee 323 on Circular economy is developing a set of standards including terminology, fundamental principles, metrics, product circularity data sheet definitions, and documenting business models and industrial case studies. These standards will support the UN Sustainable Development Goals and hence are applicable across many levels of economic and infrastructural development. ASTM International, in contrast, focuses on specialized technical standards to be used to operationalize changes in existing practices. The ASTM Committee E60 Sustainability produces standards for operationalizing sustainability in practice and supports the work of other committees to pursue sustainability objectives. E60 recently developed a roadmap for standards to foster a CE of manufacturing materials and is initiating new work in this area. The chapter concludes with an example for incentivizing broader stakeholder

---

*Corresponding author: kcm@nist.gov

Nabil Nasr (ed.) Technology Innovation for the Circular Economy: Recycling, Remanufacturing, Design, Systems Analysis and Logistics, (3–16) © 2024 Scrivener Publishing LLC

participation in the transition to a CE through metrics for calculating carbon avoidance and highlights the need for standards to support the approach.

Keywords: Circular economy, standards, ASTM, ISO, carbon avoidance, sustainable manufacturing, smart manufacturing

1.1   Introduction

Facilitating the transition from a linear (take-make-use-dispose) economy to a circular one requires that we reevaluate our relationship with products, the materials and processes we use to make them, and our attitudes about them once they reach the end of their useful life. This requires that we see both the trees and the forest–i.e., take a systems-level perspective in which we zoom in on individual product life cycle stages while understanding how the materials and information flowing in and out of those stages influences the entire product life cycle. Due to its role in the transformation of materials into products and the generation of economic activity, the manufacturing sector is a major stakeholder in the transition to a circular economy (CE). In fact, the CE shows great promise for manufacturers to fulfill sustainability goals in part because they can view it through the lenses of both sustainability and economics. In terms of sustainability, a CE promotes the efficient use/reuse and equitable allocation of resources. If successfully implemented, a CE will reduce our reliance on the extraction of non-renewable resources [1], decrease environmental damage from resource extraction [2], and promote manufacturing and better waste management [3]. From a purely economic lens, a CE for materials and products challenges us to build a hyper-efficient closed-loop economic system in which waste and products at their end-of-life are seen as a resource instead of a burden [4, 5].

The transition to this more sustainable and economically attractive system of material use and product creation has already begun, but it remains a patchwork of initiatives and policies [6, 7]. To solve this patchwork problem, broad coordination is needed among local and federal governments, international governing bodies, manufacturers themselves, financial institutions, and consumers. Standards are a key tool for creating this coordination. First, foundational standards create a consensus around, for example, terminology (e.g., an agreed upon definition of a CE), practice methods (e.g., establishing best practices for measuring, predicting, and reducing environmental impacts.), and reporting standards (e.g., the Greenhouse Gas Protocol). Standards also build trust among consumers to have faith that products are designed for circularity, made of post-consumer materials, and can satisfy claims of environmental quality [8]. In addition, because the improvements can be difficult to implement, especially for small- and medium-sized firms, standards can decrease the barriers that organizations face when adapting and improving their practices [9]. Finally, while most standards are voluntary, meaning firms can choose to adhere to them, the fact that they are developed through consensus by a diversity of relevant stakeholders means that they are often widely implemented in practice and can be contractually relied on and used to develop and adhere to regulations [10].

Standards may be key to coordinating stakeholders in the transition to a CE, but identifying the standards that are needed and determining the best way to create them is a daunting task. A successful CE will require three categories of overlapping standards:

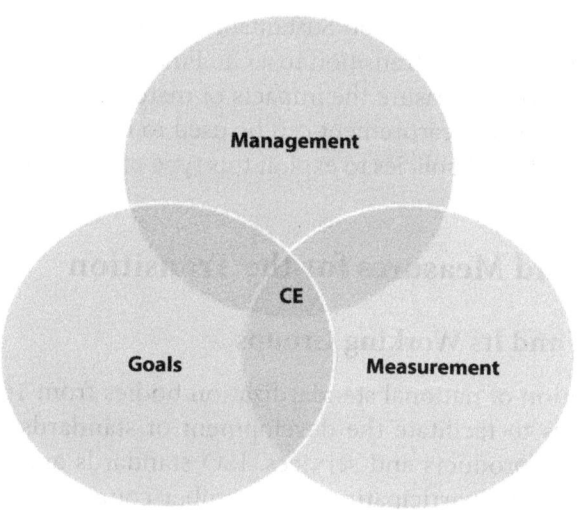

Figure 1.1 Types of standards needed to transition to a circular economy (CE).

1) shared goals, 2) management standards, and 3) measurement standards (Figure 1.1) [6, 9]. Shared goals involve initiatives that direct broad standards efforts, like the UN's Sustainable Development Goals and Sustainability Accounting Standards Board (SASB) guidance, which incorporates metrics for 77 different industries. Management standards specify how organizations should manage themselves and their supply chains, for instance, to reduce negative impacts on the environment and human health and safety; the International Organization for Standardization (ISO) gathers representatives from different countries to create international standards that help bring consistency and quality to management practices worldwide. Finally, the measurement category involves standards for material quality/ performance, communicating technical specifications, testing, and process improvements; ASTM International contributes heavily to this area.

Along with broad standards initiatives, circular business models are needed to transition to a CE [11–13]. Companies are increasingly making ESG (environmental, social, governance) commitments in response to consumer and shareholder demands, regulations, and planning for longevity. However, the types and scopes of commitments being made need to be balanced across a circular system that extends beyond the efforts and interests of individual organizations. Mechanisms are needed to incentivize companies to make their own operations more circular, and in doing so strengthen the larger value chains. For instance, large organizations are seeking means to account for the impacts of their supply chains in addition to their own individual contributions. From there, they are reporting these impacts, reduction goals, and progress towards those goals through reporting standards (e.g., SASB; the Global Reporting Initiative). These industries and their consumers, shareholders, and governments are increasingly desiring metrics for measuring progress towards these goals that are traceable and transparent.

This chapter reviews three initiatives–two standards efforts and one metric-based approach to incentivize firms–to coordinate manufacturers to operationalize circularity and further the transition to a CE. The first is ISO Technical Committee (TC) 323 on Circular economy, a technical committee established in 2018 to work on management standards. The second is

ASTM International's Committee E60 on Sustainability, which is creating operational measurement standards to support the transition to a CE. Finally, we describe a research effort to calculate carbon avoidance to measure the impacts of material reuse. We show how metrics such as the carbon avoidance measurement can be used to incentivize participation in a CE and the need for standards and policies to exploit this type of measurement.

1.2 Standards and Measures for the Transition

1.2.1 ISO TC 323 and Its Working Groups

ISO is a global federation of national standardization bodies from 167 member countries. Its primary objective is to facilitate the development of standards to ensure the quality, safety, and efficiency of products and services. ISO standards are agreements developed through consensus with the participating ISO member countries and thus help maintain current and enable new trade links and fair practices between nations.

Over the years the ISO has developed a suite of management standards for supporting manufacturing practices. Several ISO standards are frequently deployed to address sustainable manufacturing goals, such as improving production quality (ISO 9000 series), quantifying environmental impacts (ISO 14000 series), and optimizing the performance of energy systems (ISO 50001 and ISO 20140 series) [9, 14]. The concept of CE builds on these goals of sustainable manufacturing while extending the scope from the internal operations of a business to support the integration of materials back into the economy at the end of their initial use.

Implementing a CE requires a systems-level perspective which enables materials to retain the most value in reentering the system (i.e. the economy) after their initial use. Realizing that the current ISO suite of standards lack a dedicated set of standards on CE and its implementation, ISO Technical Committee (TC) 323 on Circular economy was established in 2018. The scope of ISO TC 323 extends beyond the manufacturing sector. Currently numerous organizations around the world have implemented sustainability practices independent of any standardized framework, thus limiting the overall benefit across product and/or system value chains. The ISO TC 323 standards aim to provide a global vision to organizations (public and private) and countries (developed and developing) across the world to collectively and systematically transition towards a CE model.

Recognizing that the CE needs to be implemented collectively by different regions and countries around the world, at this stage the ISO TC 323 standards are being drafted as foundational standards as opposed to operational standards as they may be implemented to support a broad range of circumstances. Each working group (WG) in TC 323 remains cognizant of the diverse challenges developing and developed countries face in terms of transitioning to a new economic paradigm such as a CE. This awareness is particularly important given the highly interconnected nature of global supply chains. Thus, the ISO TC 323 aims to promote harmonization and interoperability to enable a CE across the world.

Under ISO procedures, technical committees (TCs) produce standards as is illustrated in Figure 1.2. Each country can contribute to the development of the standard but only receives a single vote in the final approval process. ISO TC 323 comprises the five working groups (WGs) as shown in Table 1.1 each producing an initial standard as indicated. The WGs

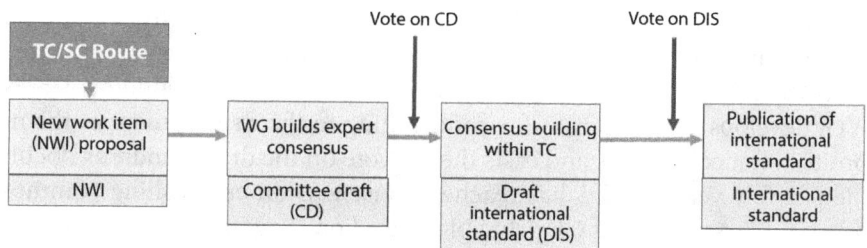

Figure 1.2 The stages of development for an ISO standard, including each step in the process and deliverables that emerge from each step (adapted from [33]).

Table 1.1 An overview of standards development across ISO TC 323's five working groups.

Working group	Standard	Scope
1	ISO/ DIS 59004 Circular Economy - Terminology, Principles and Guidance for implementation	Identify CE-related terms and develop technically sound definitions [34].
2	ISO/DIS 59010 Circular Economy – Guidance on the transition of business models and value networks	Collate business case studies that are implementing circular economy by identifying a framework for organizations to develop circular business models [35].
3	ISO/DIS 59020 Circular Economy - Measuring and assessing circularity	Identify metrics to measure and quantify circular economy (value-based, LCA-based, etc.) and work to harmonize the most relevant metrics [36].
4	ISO/DTR 59031 Circular economy – Performance-based approach – Analysis of case studies ISO/DTR 59032.2 Circular Economy – Review of business model implementation	Collate data on developing a performance-based approach for circular economy (Economy of Functionality and Cooperation) via the identification of relevant definitions and concepts that will help analyze and evaluate case studies from a triple-bottom line perspective [37, 38].
5	ISO/WD 59040.2 Circular Economy - Product circularity data sheet (PCDS)	Develop a PCDS, a standardized document that enables the digital exchange of data related to circularity characteristics of products across supply chains to facilitate data related to circularity characteristics to support standardization and transparency [39].

produce committee draft (CD) standards documents. In the US, these five WGs are mirrored in the ANSI-appointed technical advisory group, i.e., the US TAG. The US TAG for ISO 323 is coordinated by ASTM's E60 Sustainability Committee. The TAG appoints US experts to the ISO WG, develops US consensus for contributing to the development of the new standards, submits ballot comments, and casts the US vote on the draft standards documents.

Over the last few years, WG 1 has reached consensus on establishing common terminology and an understanding of CE principles based on a rigorous consensus process. The ISO 59004 has passed through the working group and is now at the Draft International Standard (DIS) stage; from there it will proceed to final balloting, in which the standard is evaluated for technical robustness and prepared for publication (Figure 1.2). Since its initiation, experts participating in the WG 1 have debated extensively in the context of CE principles, fine tuning definitions and understanding the notion of material recovery and formulating guidance for resource management actions (ISO/ DIS 59004). The WG has deemed this document technically complete as indicated by approval of the committee draft (CD) document and final publication of ISO 59004 is expected in 2024. WG 2 focused on developing a compendium of CE use cases with the aim of creating a framework that helps businesses adopt CE practices. The WG 2 standard, 59010, is also approved and is at the DIS stage. WG 3 has developed DIS 59020 that identifies a set of metrics to assist in quantifying and measuring a CE. At this point the metrics identified are mostly restricted to physical material flows, emissions, and water. Given the uncertainty around data associated with computing sustainability metrics, this harmonization for both data and methodologies is needed. While most of the definitions in ISO 59010 are informative some are normative, meaning that they will be required to be in conformance with the standard. WG 4 produces draft technical reports (DTR 59031 and DTR 59032) that focus on understanding specific case studies. Finally, WG 5 is undertaking the development of a Product Circularity Data Sheet (PCDS), 59040. The document from WG 4 and WG 5 have not yet gained committee approval and are at the working draft stage.

1.2.2 ASTM Committee E60

ASTM International is another international standards body that creates consensus-based technical standards to support many of the goods that society relies on. Its Committee E60 on Sustainability addresses sustainability goals and is beginning to include the transition to a CE. While the ISO TC 323 working groups are developing standards that lay the foundation for a CE generally (through principles, definitions, business guidelines, and product datasheets), ASTM seeks to integrate circularity and sustainability into existing engineering, manufacturing, and sector-specific practices. E60 technical subcommittees cover a variety of sectors, including building and construction (E60.01), healthcare sustainability (E60.42), and manufacturing (E60.13). As such, E60 has the expertise to drive refined sector-specific CE standards either within the committee or by supporting other committees. Whereas ISO TC 323 specifies best practices for transitioning to a CE through macro-level management, ASTM E60 contributes the operational expertise to, for example, principle and performance standards (Figure 1.1). E60's technical expertise on sustainability and CE supports other ASTM committees and, as mentioned earlier, formulates U.S. positions for ISO 323 via the ISO TAG. E60 members collaborate with and participate directly in other committees across ASTM to support coordination across the sustainability efforts of these groups (e.g., Committee D20 on Plastics; Committee E54 on Homeland Security Applications).

In April 2022, ASTM E60 hosted a workshop to identify drivers and barriers that manufacturers are experiencing when implementing circularity into their business practices. Information was gathered from a pre-workshop survey (N=260), speaker presentations, and roundtable discussions among participants and speakers from across the manufacturing sector. The two major takeaways from this effort were a list of six categories of standards needs (Table 1.2) and a proposed roadmap to fill those needs. One category of standards addresses foundational needs, such as consensus around terminology and reporting towards circularity goals. The other categories identify needs from a systems perspective, to transition through the entire product life cycle (from cradle to cradle as opposed to cradle to grave). The categories include support for addressing system-wide needs, front-end design, manufacturing production, back-end recovery, and recycling. While recycling can be viewed as a back-end recovery process, its critical role in transitioning generated enough interest and momentum among the survey respondents and workshop participants to create its own category. In the US, participation in moving towards circularity is being driven by both government and corporations responding to consumer demands. For example, part one of the EPA's National Recycling Strategy explicitly strives for a CE [15]. An example from the corporate side is the US Plastics Pact, which is an agreement among large plastics value chain participants to use circularity principles to reduce, redesign, and more

Table 1.2 Categories of standards needed for manufacturers to transition to a circular economy, as identified from a 2022 survey and workshop hosted by ASTM International.

Category	Focus Areas
Foundational Circular Economy Standards	• Definitions/terminology • Corporate Benchmarking & Reporting • Life Cycle Assessment & Inventories
System Support Standards	• Systems Thinking • Traceability & Digital Records • Labeling
Front-End Design Standards	• Design for Circularity (general) • Design for Material Circularity • Product Design for Recovery
Manufacturing Production Standards	• Supply Chain • Process Improvements • Secondary Materials Marketplace
Back-End Recovery Standards	• General • Use, Reuse, & Repurpose • Repair • Refurbish & Remanufacture
Recycling Standards	• Collection for Recycling • Sorting • Recycling • Recycled Content

responsibly source plastic packaging [16]. Standards can support these efforts by improving collection infrastructure and strengthening the market for sorting and recycling technologies making them more accessible, especially to small- and medium-sized recyclers.

ISO 323 will address some of the identified standards and ASTM Committee E60 is well positioned to address others. The roadmap from the CE workshop proposes a strategy for the committee and ASTM more broadly to move forward. Because ASTM standards development relies on volunteers, its members decide which standards are developed and when. This roadmap recommends that standards development efforts be paired with a) education and training to build the expertise needed to deploy the standards into practice as the CE transition progresses and b) research to validate approaches and enable technological, infrastructural, economic, and social progress towards new standards. The new standards themselves may begin as guidelines, then progress to best practices addressing testing and certification as experience and case studies increase.

As an outcome of the workshop, ASTM E60.13 introduced a new work item (NWI) titled *A Guide to Principles for Circular Product Design*. This NWI introduces a list of generalized design principles to apply during product design to support reintroduction of products and/or their constituent materials into the CE. These principles are created to address operationalizing CE concepts in increments to support the maturation of the CE. A goal for the work is to provide guidance for manufacturers to engage in the transition toward the CE with fewer barriers to entry and for other stakeholders, such as recyclers, to rely on consistent practices across a variety of products. This NWI also serves as a means for coordinating CE standards across ASTM and within different and specific sectors.

The NWI is structured around a primary set of general circular design principles and will feature sector-specific appendices. The appendices have unique drafting teams that include members outside of E60.13. These drafting teams work to create complementary sector-specific principles and guidelines. The NWI guidelines are defined as recommendations to operationalize circular design principles.

Overall, the *Principles of Circular Product Design* NWI encompasses all tenants of E60 by introducing a generalized CE standard that also addresses sector-specificity and enlists members from outside of ASTM E60. Once published, the *Principles of Circular Product Design* could form the basis for a certification program for circularity factors like water consumption and waste generation. Best practices and certifications are important as they improve trustworthiness among consumers and other stakeholders, ensure regulatory compliance, support the development of markets, and incentivize the use of standardized metrics [17, 18].

1.2.3 Metrics as Incentivization

Introduction to carbon accounting

Greenhouse gas emissions (GHG) are a common metric for the environmental impacts of products, processes, and corporations. The metrics and activities to reduce emissions are frequently called "carbon" metrics or "carbon" reduction activities because the GHG emissions are typically measured as *carbon dioxide equivalent (CO_2e)*. Regulations on corporate GHG emissions are getting stricter. Certain regions (e.g., California, European Union) have even implemented carbon taxes and *cap-and-trade* programs [19]. Many corporations have also set their own voluntary carbon reduction goals, including *net-zero* targets, spurring the growth of voluntary carbon offset credits. Such practices have heightened the industry's interest in carbon reduction activities and methods to evaluate the carbon emissions of their actions. Given the interest and value of "carbon-related" metrics, we are exploring the potential of carbon metrics to incentivize material recovery practices in manufacturing, and therefore, facilitate the broader CE.

Material recovery activities and carbon accounting
The demand for extraction of *virgin* material from the environment (i.e., *primary* material) and the landfilling or incineration of material at the end-of-use (EoU) can be minimized through effective implementation of CE activities such as *reuse, remanufacture*, and *recycle* [20–22]. Recovery activities are typically less resource intensive than the production of virgin materials and brand-new products due to fewer processing steps; the former therefore has the potential to curtail environmental impacts including carbon emissions [21, 23]. Furthermore, if these activities lead to *reduction-* or *avoidance*-based *carbon offsets*, those bring additional benefits for the stakeholders involved [19].

A secondary materials marketplace (SMM) is where a manufacturer can list their excess or discarded materials, so that other companies can obtain the recoverable materials. SMMs facilitate the interaction between businesses avoiding (or reducing) material wastage and the need for primary materials production. Therefore, SMMs can be a significant component in manufacturing-related carbon emissions reduction. Yet, our work with one of the SMMs operating in multiple regions of the United States has identified that SMMs are largely underutilized. A recent ASTM workshop conducted on CE [24] also established the lack of standards and incentives as part of the reasons for overall paucity in recovery practices in the US industry.

If the carbon reduction or avoidance of SMM activities may be quantified and attributed, businesses' interest in carbon accounting can stimulate the use of SMMs. Furthermore, that ability to quantitatively realize the carbon benefits of recovery activities can drive the decisions to design products and processes which better employ material recovery activities, including utilization of SMMs, forming positive reinforcements.

Available guidelines and their limitationss
In terms of quantifying the carbon credits and offsets of specific projects and activities, guidance and standards are available in the form of literature such as GHG Protocol [25] and Clean Development Mechanism (CDM) methodologies [26]. Many carbon regulations and voluntary marketplaces also use the GHG Protocol as their basis. While the guidelines such as GHG Protocol are extremely useful to identify and verify the activities that may claim carbon offset credits, the allocation of those credits between the stakeholders is left to the discretion of users of these guidelines [25, 26]. This can be especially challenging in SMMs and other such materials recovery activities as those involve multiple stakeholders. For example, SMM transactions involve at least two stakeholders—a *seller* and a *buyer*. An *equitable allocation* of avoided emissions and/or potential carbon offsets can incentivize each stakeholder's participation and also avert any *double counting errors* [27] of the carbon avoidance. We are working with industry representatives to develop a framework for accounting and allocating the carbon reductions and voluntary offset credits related to SMM activities.

A potential solution based on allocation approaches of LCA
The life cycle assessment (LCA) literature provides *allocation approaches* [28, 29] for material recovery applications and have potential to be a basis for voluntary carbon offsets allocation. Ekvall *et al.* [28] describe a good allocation approach as one that is equitable as well as can incentivize reduction of environmental impacts–including carbon emissions. Schrijvers *et al.* [29] and [30] discuss allocation approaches in detail and their potential to incentivize material recovery. For example, the *End-of-life recycling approach* can be applied when the recovered secondary material has a higher demand than the supply. In this approach, the product system providing the recovered material is allocated the benefits (i.e., avoidance of producing virgin material) of recycling. Therefore, in principle, it motivates the producers of (easily) recoverable material. If the recovered secondary material has a lower demand (e.g., a by-product that is typically considered a "waste"), then the *Cut-off allocation approach* can be applied. Here the environmental burden of primary material extraction is fully allocated to the first product system which produced the material (i.e., the user of recovered material receives the recovered material without any environmental burden related to extraction). This approach motivates the use of recovered secondary materials. LCA literature also includes other approaches such as *Market-price-based substitution approach* which allocates a higher percentage of avoidance credits producer if the market price of secondary material is closer to the price of virgin material it substituted [29].

Once the total carbon offset is calculated according to a guideline such as the GHG Protocol, taking a similar approach in principle to LCA allocations, the carbon offset can be allocated between the stakeholders considering the demand for the secondary material. If a specific secondary material has a lower market value, a higher portion of the offset could be credited to the user of the recovered material creating an incentive to utilize the recovered material. Similarly, for higher-valued secondary materials, more offset can be credited to the producer of the recovered material. A detailed discussion of a framework for such allocation is discussed in an upcoming publication [31]. Ultimately, these allocations must be set using contractual agreements between stakeholders of the carbon avoidance activities.

Need for application-specific standards
Building consensus-based standards within specific industries will help establish frameworks to base the carbon-related agreements between stakeholders. Forming material- or industry-specific standards streamline such stakeholder agreements and help more manufacturers utilize material recovery activities–including through SMMs. A listing of existing standards to guide application-specific sustainability practices is available from ASTM [32].

Given the complexities in value chains and carbon offset calculations, the following are a few crucial ways material and industry-specific standards can facilitate this process, by providing guidance on

- defining scope: identify the types of activities and stakeholders that need to be considered in carbon offsets calculations
- selecting the "baseline scenarios" and evaluating the "additionality tests" and other qualifications for carbon offsets (described in WBCSD and WRI [25])
- specifying industry-specific *common* practices as market conditions and recovery practices change over time

- allocation practices for the transfer of material ownership between multiple stakeholders within the recovery value chain (e.g., intermediaries between the supply and consumption of the recovered material)
- estimating and accounting of GHG emissions for special industry practices (e.g., backhauling)

Since many of these considerations are application-specific, any general framework developed must be further adapted to the specific sector or application. That requires contribution from stakeholders of all levels of each sector. Such standards will provide consistency and transparency in the system leading to more credibility to society and engagement of stakeholders.

1.3   Conclusions & Recommendations

A CE is a promising path to move toward a more sustainable economic system. The world's environmental crises–including climate, biodiversity, pollution, and finite natural resources–require both immediate solutions and more fundamental systemic changes, which the CE approach attempts. A circular approach to resource use is largely viewed as a strong basis for better managing our worldwide resources to reduce environmental harm and reverse the course towards future environmental degradation. Standards are foundational to this transition across nations and economic sectors. Standards for a CE will help

- design more durable and efficient products and services for the future
- assess the life cycle impacts of products before they are produced
- assure that future materials derived from non-virgin sources retain the characteristics needed to support the integrity of new products
- address the source of waste and pollution problems streamlining materials back into the economy
- provide verifiable baselines for evaluating future improvements to manufacturing processes
- rapidly deploy new technologies and scale up their use
- enable automation of global systems for recovery and traceability of goods and materials
- create transparency and accountability across organizations and industrial practices through accurate and comparable metrics
- effectively plan for the reintroduction of materials into the economy
- efficiently handle waste through automation in collection, sorting, and recovery practices
- motivate stakeholders to participate in new socio-economic systems increasing synergies

The standards needs to transition to a CE are great. The standards development process is a means by which stakeholders come together to align their different points-of-view towards common solutions. Standards setting can be a slow process as different stakeholders must come to common understandings not only of terminology but also agree on the principles by which they are willing and able to move forward. Standards setting is also a living process. Standards change over time with more experience and greater needs. If we as a society are to complete this transition to a CE, the time to start this process is now. We can lay the foundations for more rigorous standards in the future while setting the course on which those conversations will be built.

As new technologies and solutions are developed through fundamental research, the goal- and management-level standardization efforts can effectively guide and interface new approaches, technologies and methods. The measurement-level standards provide a means to introduce new technologies and transparency to evaluate their effectiveness. Standards can also establish a basis for incentivizing participation in CE activities as discussed in the example provided above. Therefore, standards will be key to the adoption and widespread deployment of a CE. The creation of standards requires participation across a range of stakeholders to work towards solutions that meet the common good.

Acknowledgements

The authors would like to acknowledge the many participants of the ISO 323 and ASTM E60 standards bodies whose hours of dedication and hard work are reflected here. Kelsea Schumacher, Amy Costello, and Maya Reslan are acknowledged for their significant roles in leading the ASTM workshop.

References

1. Nadeem, S.P., Garza-Reyes, J.A., Anosike, A.I., Kumar, V., Coalescing the Lean and Circular Economy (pp. 1–12). Michigan: IEOM Society, 2019.
2. van Loon, P., Delagarde, C., Van Wassenhove, L.N., The role of second-hand markets in circular business: A simple model for leasing versus selling consumer products. *Int. J. Prod. Res.*, 56(1–2), 960–973, 2017.
3. Halog, A., Anieke, S., A Review of Circular Economy Studies in Developed Countries and Its Potential Adoption in Developing Countries. *Circular Economy Sust.*, 1(1), 209–230, 2021.
4. Schmitt, T., Wolf, C., Lennerfors, T.T., Okwir, S., Beyond "Leanear" production: A multi-level approach for achieving circularity in a lean manufacturing context. *J. Clean. Prod.*, 318, 2021.
5. Webster, K., The Circular Economy: A Wealth of Flows, Ellen MacArthur Foundation Publishing, 2016.
6. Reslan, M., Last, N., Mathur, N., Morris, K.C., Ferrero, V.; Circular Economy: A Product Life Cycle Perspective on Engineering and Manuf. Practices. *Procedia CIRP*, 105, 851–858, 2022.
7. Hapuwatte, B.M., Mathur, N., Last, N., Ferrero, V., Reslan, M., Morris, K.C., Optimizing Product Life Cycle Systems for Manufacturing in a Circular Economy. In: Kohl, H., Seliger, G., Dietrich, F. (eds) *Manufacturing Driving Circular Economy. GCSM 2022.* Lecture Notes in Mechanical Engineering, 419-427, 2023.
8. Flynn, A., Hacking, N., Setting standards for a circular economy: A challenge too far for neoliberal environmental governance? *J. Clean. Prod.*, 212, 1256–1267, 2019
9. Escoto, X., Gebrehewot, D., Morris, K.C., Refocusing the barriers to sustainability for small and medium-sized manufacturers. *J. Clean. Prod.*, 338, 130589, 2022.

10. Association of Equipment Manufacturers, Regulations vs Standards: Clearing Up the Confusion, https://www.aem.org/news/regulations-vs-standards-clearing-up-the-confusion, 2021

11. Agrawal, V.V., Atasu, A., Van Wassenhove, L. N., OM Forum—New Opportunities for Operations Management Research in Sustainability. *Manuf. Serv. Op.*, 21(1), 1–12, 2019.

12. Hopkinson, P., Zils, M., Hawkins, P., Roper, S., Managing a Complex Global Circular Economy Business Model: Opportunities and Challenges. *Calif. Manage. Rev.*, 60(3), 71–94, 2018.

13. Romero, D., Rossi, M. Towards Circular Lean Product-Service Systems. *Procedia CIRP*, 64, 13–18, 2017.

14. Brundage, M. P., Bernstein, W. Z., Hoffenson, S., Chang, Q., Nishi, H., Kliks, T., and Morris, K. C. (2018). Analyzing environmental sustainability methods for use earlier in the product lifecycle. *Journal of Cleaner Production*, 187, 877–892. https://doi.org/10.1016/j.jclepro.2018.03.187

15. US Environmental Protection Agency, National Recycling Strategy: Part One of a Series on Building a Circular Economy for All, 2021.

16. US Plastics Pact, Plastics Pact: Roadmap to 2025, US Plastics Pact, https://usplasticspact.org/roadmap/, 2021.

17. European Commission, A European Strategy for Plastics in a Circular Economy.

18. Tassey, G., Standardization in technology-based markets. *Res. Policy*, 29(4–5), 587–602, 2000.

19. World Bank, State and Trends of Carbon Pricing 2022 (State and Trends of Carbon Pricing), World Bank, 2022.

20. Ellen MacArthur Foundation, Towards the circular economy Vol. 1: An economic and business rationale for an accelerated transition [Non-Profit], Ellen MacArthur Foundation (EMF), 2013

21. Hapuwatte, B.M., Jawahir, I.S., Closed-loop sustainable product design for circular economy. *Journal of Industrial Ecology*, 25(6), 1430–1446, 2021.

22. Kirchherr, J., Reike, D., Hekkert, M., Conceptualizing the circular economy: An analysis of 114 definitions. *Resour. Conserv. Recy.*, 127, 221–232, 2017.

23. Yang, M., Chen, L., Wang, J., Msigwa, G., Osman, A.I., Fawzy, S., Rooney, D.W., Yap, P.-S., Circular economy strategies for combating climate change and other environmental issues. *Environ. Chem. Lett.*, 2022.

24. Schumacher, K., Last, N., Morris, K., Costello, A., Hapuwatte, B., Mathur, N., Ferrero, V., Reslan, M., Fostering a Circular Economy of Manuf. Materials Workshop Report, ASTM International, 2023.

25. WBCSD and WRI, The greenhouse gas protocol: The GHG protocol for project accounting. World Business Council for Sustainable Development (WBCSD); World Resources Institute (WRI), 2005.

26. UNFCC, Clean Development Mechanism: Methodology Booklet, pp.286, United Nations Framework Convention on Climate Change (UNFCC), 2021.

27. WBCSD and WRI, The Greenhouse Gas Protocol. A Corporate Accounting and Reporting Standard. World Business Council for Sustainable Development (WBCSD); World Resources Institute (WRI), 2004.

28. Ekvall, T., Albertsson, G.S., Jelse, K., Modeling recycling in life cycle assessment, IVL Svenska Miljöinstitutet, http://urn.kb.se/resolve?urn=urn:nbn:se:ivl:diva-27, 2020

29. Schrijvers, D. L., Loubet, P., Sonnemann, G., Developing a systematic framework for consistent allocation in LCA. *Int. J. Life. Cycle Ass.*, 21(7), 976–993, 2016.

30. EC-JRC, International Reference Life Cycle Data System (ILCD) Handbook: General guide for life cycle assessment: Detailed guidance (EUR 24708 EN), European Commission-Joint Research Centre (EC-JRC), https://op.europa.eu/en/publication-detail/-/publication/325e9630-8447-4b96-b668-5291d913898e/language-en, 2010

31. Hapuwatte, B.M., Mathur, N., Morris, K., Emissions avoidance quantification and allocation framework for secondary materials marketplaces supporting the circular economy. (In press) *ASME 2023 18th International Manuf. Sci. Eng. Conf.*, 2023.

32. ASTM International, ASTM Sustainability Reference Database, https://www.astm.org/get-involved/technical-committees/other-programs-services/sustainability-reference.html, 2022.

33. ISO, The different types of ISO publications, https://www.iso.org/deliverables-all.html, 2023.

34. Circular Economy Terminology, Principles and Guidance for implementation, ISO TC 323/DIS 59004, 2023.

35. Circular Economy Guidance on the transition of business models and value networks, ISO TC 323/DIS 59010, 2023.

36. Circular Economy Measuring and assessing circularity, ISO TC 323/DIS 59020, 2023.

37. Circular Economy Terminology, Principles & Guidance for implementation, ISO TC 323/DIS 59004, 2023.

38. Circular Economy Review of business model implementation, ISO TC 323/DIS 59032.2, 2023.

39. Circular Economy Product circularity data sheet (PCDS), ISO TC 323/DIS 59040.2, 2023.

2

Circularity Index: Performance Assessment of a Low-Carbon and Circular Economy

Luis Gabriel Carmona¹*, Kai Whiting² and Jonathan Cullen¹

¹University of Cambridge, Cambridge, UK
²Université Catholique de Louvain, Louvain, Belgium

Abstract

The circular economy (CE) is defined as a sustainable economic system in which socioeconomic development is sustained via the reduction of non-renewable resource extraction and material recirculation. In practice, however, the additional resource inputs and losses involved with various CE initiatives, including recycling, can often trump many of the environmental gains associated with the concept. The purpose of this chapter is to update Cullen's [1] circularity index approach and its two quantitative indicators to achieve a more comprehensive circularity assessment, in terms of quantity and quality. In particular, we use these indicators to assess the degree of (1) reuse and repurposing; (2) decarbonisation; and (3) resource use intensity (the amount of material required to provide a unit of service), relative to a system entirely reliant on virgin material extraction. The resulting circularity index (CI) value is derived from an aggregation of the two indicators and is measured from 0 (zero circularity) to 1 (100 percent circularity). We applied the CI to UK car-based passenger mobility with respect to carbon emissions and material cycles. We compared an internal combustion engine vehicle (ICEV) to a battery electric vehicle (BEV). The results show that in 2015 the average UK ICEV's CI value was 0.01 and a BEV's CI value was 0.1. The introduction of low-carbon strategies such as electrification, resource efficiency, and shared-mobility improves the CI value to 0.5. This suggests that if UK car-based passenger transport is to approach full circularity additional innovations, both social and technological, will be required. It is our view that the CI approach can facilitate the operationalisation of CE strategies assessment and environmental performance at the micro (e.g. product, company), meso (e.g. sectorial) and macro (e.g. national/global) levels. It also holds relevance for the development of public policies and corporate strategies that promote evidence-based sustainable solutions.

Keywords: Circularity, decarbonisation, eco-efficiency, material flows and stocks, metrics, resource efficiency, sustainable resource use

2.1 Introduction

The goal of the Circular Economy (CE) is to encourage a sustainable and regenerative cycle of production and consumption. CE is a set of sustainable policies and strategies that target

*Corresponding author: lgc29@cantab.ac.uk

Nabil Nasr (ed.) *Technology Innovation for the Circular Economy: Recycling, Remanufacturing, Design, Systems Analysis and Logistics*, (17–28) © 2024 Scrivener Publishing LLC

reduced material intake and waste generation, in order to lessen economic and environmental costs [2, 3]. CE has been widely supported in recent years due to its potential to decouple socioeconomic development from natural resource extraction and use. It is gaining popularity in many different countries, sectors, and organisations [4]. Several strategies have been proposed to facilitate the shift from a linear economy to a circular one. These include eco-design, resource efficiency measures, waste R's hierarchy (e.g. reduce-reuse-recycle), industrial symbiosis, etc. [5].

One way to quantify the degree of a system's resource circularity is through a circularity index (CI). The resulting value is a number, usually ranging from 0 (zero circularity) to 1 (100 percent circularity). The higher the number, the more sustainable the system is deemed to be. Examples of CIs include the Material Circularity Indicator - MCI [6], the Circular Economy Index – CEI [7], the Global Circularity Metric [8] and Cullen's [1] Circularity Index approach. Most CIs focus on circularity quantity, that is to say the degree to which a product or system has closed material loops. It is typically measured as the proportion of materials that comes from reused or recycled sources relative to the total amount of material required.

It is our view that a circularity index should not be solely focused on aspects related to *quantity* but expanded to encompass other characteristics such as resource *quality*. Although far from common, circularity quality in quantitative terms is a measure of environmental performance, with a higher quality reflecting a lower dependency on virgin materials or fossil fuels, for example [9, 10]. Cullen [1] proposed a CI approach that accounts for material quality using energy intensity for material production as a proxy. He used this approach to calculate energy requirements for primary and secondary produced materials.

With the intention of further aligning Cullen's [1] CI to the principles of the CE, the aim of this present chapter is to propose a modification of his approach. We do this to better reflect material inflow and outflow patterns, and to capture a system's lifecycle environmental impacts, relative to a reference system or baseline (Section 2.2). The updated approach is applied to a case study on UK car-based mobility. We calculate the circularity of UK-registered cars in terms of their material cycles and lifecycle carbon emissions. The two types of car categories considered were the "internal combustion engine vehicle – ICEV" and the "battery electric vehicle – BEV". In addition, and for the BEV, we considered the common sustainability strategy of lightweighting via material substitution and its impact on the car's circularity. We also explore aspects of Grubler *et al.* [11]'s low energy and material demand scenario and the challenges society may face on the adoption of their suggested targets and transport transition ideals (Section 2.3). We then summarise our findings and offer brief recommendations (Section 2.4).

2.2 Circularity Index Approach

The Circularity Index approach as proposed by Cullen [1] calculates material recirculation in both terms of quantity and quality. For the latter, Cullen calculates the ratio of energy needed for material recovery relative to that needed for primary material production. The index's indicators alpha (α) and beta (β) are shown in equations 2.1 and 2.2, respectively. The overall CI value is obtained by multiplying the result of alpha with that of beta, as shown in Equation 2.3. The parameter alpha is a value between 0 and 1 (referred to in Graedel *et al.* [12] and Fellner and Lederer [13] as the 'recycling rate'). Beta is necessarily less than 1 (and can be negative). The CI value (Equation 2.3) is between 0 (zero circularity) and 1 (full circularity).

$$\alpha = \frac{recovered\ EOL\ material}{total\ material\ demand} \tag{2.1}$$

$$\beta = 1 - \frac{energy\ required\ to\ recover\ material}{energy\ required\ to\ primary\ production} \tag{2.2}$$

$$CI = \alpha \bullet \beta \tag{2.3}$$

Figure 2.1 shows the estimates of alpha, beta, and the overall CI values for five energy-intensive materials (aluminium, concrete, paper, plastic, and steel). It also suggests that there is a long way to go before any one global material production stream reaches anywhere near the ideal state of circularity (CI = 1).

Cullen [1] incorporated quality metrics in an attempt to mitigate the potential negative effects of growing energy use in the pursuit of resource circularity. He argues that there are *practical* limits to how much an economy can be circular due to factors linked to resource availability, the reliance on non-renewable energy sources, and technical/technological constraints. Existing infrastructure often relies on non-renewable sources of energy, which can severely undermine circular economy progress. Fundamentally, even with optimal conditions, there are also thermodynamic considerations that will prevent 100 percent closed loops: there is simply a hard physical limit to just how circular any economy can be, as dictated by the Second Law. This limit is referred to as the *theoretical* limit.

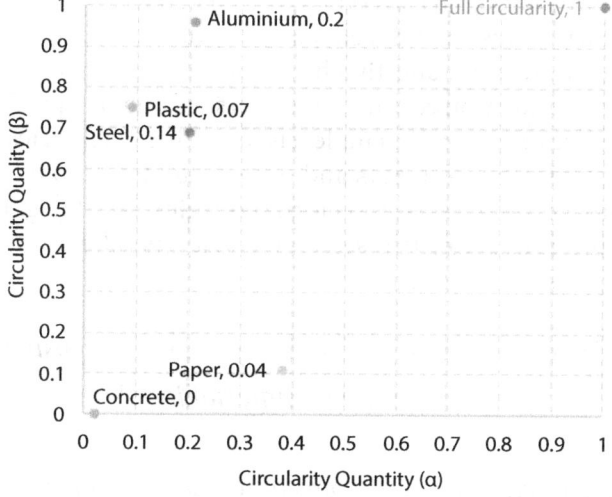

Figure 2.1 α, β and CI of five materials. Note: The number next to the material label corresponds to the overall CI value (obtained via α x β).

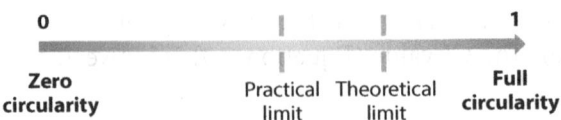

Figure 2.2 CI degrees.

Figure 2.2 depicts the circularity index range from 0 (zero circularity) to 1 (100 percent circularity) and the imposed limits (benchmarks). While we must acknowledge the practical and theoretical limits of our systems, if we value improved levels of sustainability and consider circularity to be an important measure of sustainability, then it is worth considering how we might close the gap between where the system is now and full circularity.

Cullen's [1] approach, while a step in the right direction, especially when it comes to more comprehensive quantitative circularity assessments, overlooked the following issues: (1) his circularity quantity value (as derived from the alpha indicator) is obtained via the assumption that all reused or recycled materials that leave a system will be reincorporated eventually back into the same system. In reality, some materials never re-enter and are instead used in lower-value applications (e.g. secondary steel from the car industry is used in buildings and not vehicles because of its deteriorated quality); (2) his circularity quality value (beta indicator) captures the energy associated with the material inputs of specific processes within the system but fails to account for the system's entire life cycle and its environmental impacts. This means that his original approach is one dimensional.

2.2.1 A More Comprehensive CI Approach

2.2.1.1 Alpha Indicator: Circularity Quantity

We modified Cullen's [1] alpha indicator to reflect the fact that some material flows will never re-enter a system once they have left. The practical implication of this is a reduction in the total quantity of these flows and, consequently, a lowering of the theoretical circularity quantity value. It is worth noting that secondary metal contamination, due to the proliferation of alloys, is a primary cause of reduced *quantities* of resource inflows (affecting alpha) and is detrimental to flow quality (affecting beta). Appropriate measurement of the impact of secondary metal contamination requires an understanding of a product lifecycle including metal recycling processes, contaminant levels, recycling efficiencies, and energy use.

As different streams of material inputs and outputs are likely to have different levels of circularity, particularly at the meso (e.g. sectorial) or micro level (e.g. product, company), our alpha indicator is a product of two separate equations (alpha inflows and alpha outflows), as shown in Equations 2.4 to 2.6.

$$\alpha_{inflow} = \frac{LC\ circular\ inflow\ (from\ recycling, repurposing, reuse)}{total\ LC\ material\ inflow} \tag{2.4}$$

$$\alpha_{outflow} = \frac{LC\ circular\ outflow\ (sent\ to\ recycling, repurposing, reuse)}{total\ LC\ material\ outflow} \tag{2.5}$$

$$\alpha = \alpha_{inflow} \bullet \alpha_{outflow} \qquad (2.6)$$

The scope of macro, meso, and micro levels of analysis refers to the level of granularity or detail at which circularity is being measured [14]. At the macro level, the scope of analysis is broad and covers the entire economy or a large geographical area, such as a country or region. At the meso level, the scope of analysis is more focused and covers specific sectors or industries. At the micro level, the scope of analysis is highly detailed and covers specific companies or products.

2.2.1.2   Beta Equation: Circularity Quality

We modified Cullen's beta equation to capture the various environmental impacts produced throughout a system's lifecycle relative to a reference system, as captured by the denominator in Equation 2.7. In the case study provided in Section 2.3, the value corresponds to carbon emissions, but the equation can be applied to other impacts, including acidification and toxicity, etc. This allows for a multidimensional perspective, and quantification, of a system's environmental performance throughout its lifecycle.

$$\beta = 1 - \frac{life\ cycle\ emissions\ of\ circular\ system}{life\ cycle\ emissions\ of\ linear\ system\ (baseline)} \qquad (2.7)$$

Unlike, the alpha indicator (which measures quantity), the beta (because it assesses quality) must be calculated with a reference baseline, and in this respect the beta value is relative. An appropriate reference point for the beta CI value may be the carbon emissions created per unit of service when only virgin materials are consumed in its provision. Such a system would be completely "linear", in the sense that the sequences of the processes involved would be unidirectional (Figure 2.3a). Any positive delineation from this baseline is thus a measure of increased circularity, as made manifest by an increase in the proportion of secondary materials that are reincorporated back into the system (Figures 2.3b). It may also result from a decrease in the energy required per unit of service. In this regard, the beta CI value could be a useful mechanism to evaluate the resource sustainability of light-weighting, lifetime extension, repurposing and sustainable lifestyle transition practices (see description in Carmona *et al.* [15]).

The beta CI value is affected by the system's energy mix, with the degree of influence dependent on the subject at hand. If the subject of the beta CI value is carbon dioxide emissions, for example, a system with a high proportion of renewable energy source inputs will have a higher beta CI value compared to the entirely linear system baseline, where all energy production requires fossil fuel combustion. In Cullen's [1] rendition of the CI approach, there was no way of distinguishing the environmental impact of 1 kJ of energy consumed, when the origin of that kJ was produced by a wind turbine relative to another kJ produced by a gas turbine. Our modified beta CI indicator supports a more holistic environmental performance assessment, while Cullen's original beta indicator only measured energy consumption, as a proxy for environmental sustainability.

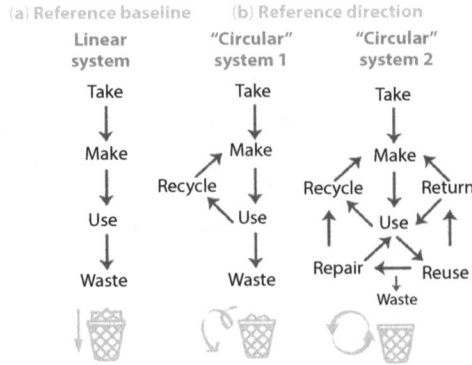

Figure 2.3 A graphical depiction of different types of systems (a) Reference baseline (b) Reference direction.

The flexibility of the new beta CI indicator means that it can also calculate the environmental performance of service units, relative to passenger mobility, visual comfort, shelter and nutrition, for example (see Carmona *et al.* [16]; Whiting *et al.* [17] for more information on resource services). The service approach is increasingly recognised as a useful way of evaluating resource production and consumption because it offers additional insights for advocates of the circular economy, especially when it comes to resource decoupling [18, 19]. A service approach is thought to be helpful when measuring resource efficiency because it focuses on societal activity and is linked to the material aspects of human needs and wellbeing rather than economic output or physical products [20, 21]. In this regard, a service approach is also more aligned with CE principles, given that circular economy thinking is not primarily concerned with maximised production or GDP.

2.2.1.3 Alignment with CE Principles

It is our view that the development of the alpha and beta indicators more readily aligns Cullen's [1] circularity index with the Ellen MacArthur Foundation [22] principles. The latter are fundamental to the advancement of the circular economy. Table 2.1 shows how we tied our updated metrics to the ideals promoted by the EMF, and thus the philosophical underpinning of the circular economy in general.

Table 2.1 Summary of the CI framework for a sustainable low-carbon and circular economy.

EMF principles	CI measuring aspects	CI indicator
Keep products and materials in use	EOL recovery, reuse, and repurposing	Alpha
Phase out waste and pollution	Decarbonisation	Beta
Decouple resource use to allow for the regeneration of natural systems	Service efficiency (unit of service provided relative to resource input)	Beta

2.3   Case Study: UK Car-Based Passenger Mobility

The transportation sector is critical to socioeconomic development because it enables the mass movement of people and products. It is a key matter for the circular economy because it is one of the biggest consumers of energy and materials. It is also responsible for 37 percent of CO_2 emissions from end use sectors globally [23]. The UK transport sector was selected due to ease of data accessibility and its quality. Due to the prevalence of car-based mobility, we focused on conventional (internal combustion engine vehicle – ICEV) and electric cars (battery electric vehicle - BEV) as opposed to public transport options. The vehicle data was obtained from Geyer [24], World Auto Steel [25] and Rodrigues *et al.* [26]. The data scope was vehicle's design, manufacturing, use phase, and end-of-life processes. The timespan for the data was restricted to a single year: 2015.

Figure 2.4 shows a summary of the average ICEV and BEV material composition in the UK and their respective alpha circularity index (CI) value. A large difference between the input alpha value (0.22) and output alpha value (0.91) is observed, which indicates that it is correct to assume that not all material outflows are reintegrated back into the same system eventually. In reality, these outflows are of low quality and are typically downcycled. If they are reincorporated back into the system, they are first mixed with virgin materials.

Figure 2.5 shows the results obtained for the carbon emission lifecycles for an average UK-registered ICEV and BEV. It also shows their respective beta CI values relative to carbon emissions. It states the reference baseline beta CI value for the average ICEV and BEV carbon emissions respectively. In both cases, the baseline represents a vehicle manufactured with entirely virgin material, that is reliant on fossil fuels (for the BEV the electricity is provided via coal) and disposed of via landfill. The significant variation between the beta values for each vehicle occurs during the operational phase. This suggests that electric vehicles, while more circular than ICEVs, in terms of carbon dioxide, will not close loops sufficiently to advance circularity beyond 0.1 (Figure 2.7a). This is primarily because there is no evidence to suggest that electric vehicles result in an increased circularity with regards to material cycles (Figure 2.4).

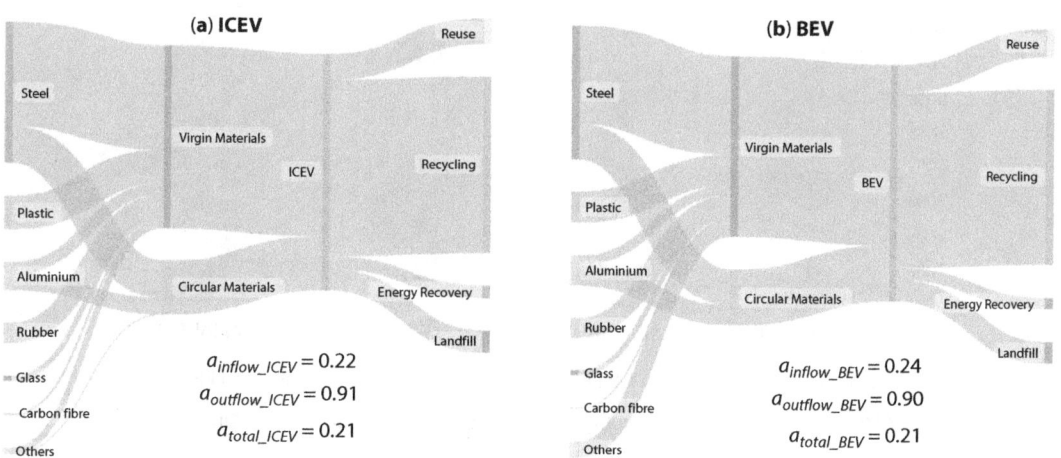

Figure 2.4 Material composition and alpha values (a) ICEV (b) BEV.

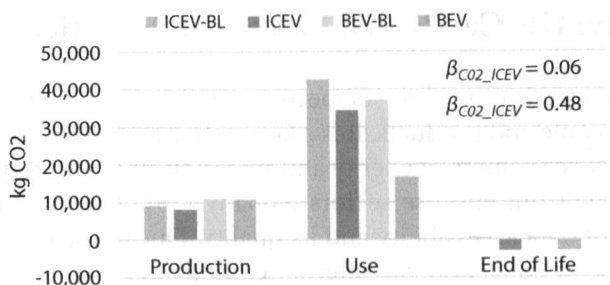

Figure 2.5 ICEV and BEV lifecycle carbon emissions. Note: BL – baseline.

In addition, we simulated three distinct lightweighting scenarios for the average BEV relative to the aforementioned reference baseline. The three scenarios (Figure 2.6) all involved the *substitution* of a proportion of conventional steel (with an eye on maintaining car performance) in the vehicle body with one of the following: (a) advanced high strength steel (AHSS); (b) aluminium and (c) aluminium and carbon fibre.

As demonstrated by Figure 2.7b, the introduction of lightweighting relative to the reference baseline has a negligible effect on the overall CI value (0.09 is the best performance). In the aluminium-carbon fibre scenario, lightweighting has a negative effect on environmental performance relative to the BEV baseline. This is because carbon fibre manufacturing is highly energy intensive while recycling is not currently practiced for this material.

In summary, lightweighting via material substitution, as opposed to lighting via absolute material reduction in mass terms, is not a good strategy for the advancement of circularity, in terms of carbon emissions. Rodrigues *et al.* [26] came to similar conclusions regarding the benefits of lightweighting via substitution in the car industry when it came to service provision. They discovered that the practice doesn't improve unit efficiency per passenger kilometre. In addition, lightweighting in general carries the risk of further secondary metal

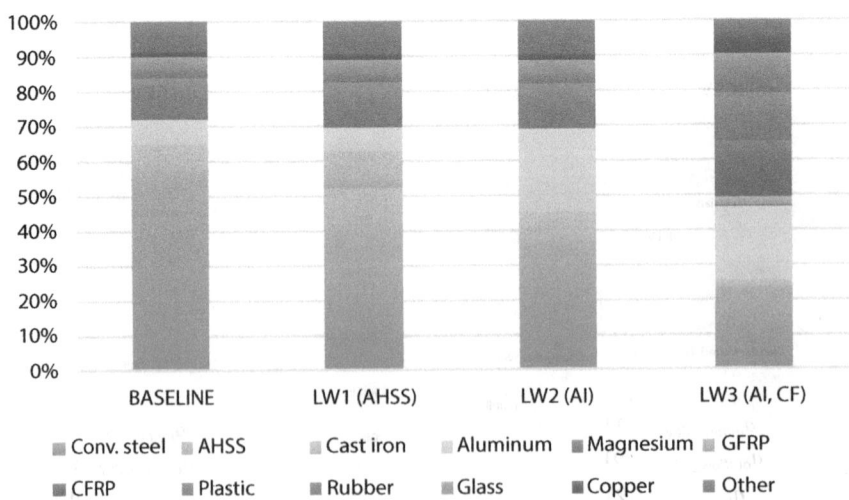

Figure 2.6 Material composition of BEV car body baseline and lightweighting scenarios. Note: AL – aluminum, CF- carbon fiber.

contamination and higher energy intensity for recycling and recovery, due to the reliance on alloying. This practice negatively affects both the alpha and beta values, reducing the overall circularity of the system. This is thus indicative of the futile nature of employing this type of lightweighting to close material loops.

Grubler *et al.* [11]'s low energy and material demand (LEMD) scenario indicates that substantial improvements can be made when it comes to the development and maintenance of a low-carbon circular economy, only if resource consumption is halved by 2050. This is something they believe is possible in vehicle manufacture and operation if:

- Industry dematerialises by 89 percent and increases material efficiency by 72 percent.
- Vehicle fuel efficiency improves by 50 percent, which for the UK requires going from 1.4 to 0.7 MJ/pkm.
- Stock efficiency increases by 40 percent, which means reaching 25,383 pkm per UK-registered vehicle.

Under the LEMD scenario, with the aforementioned targets in mind, the overall CI value (combined alpha and beta) reaches 0.48 relative to the reference baseline (Figure 2.7b).

The results obtained for the overall CI value are promising, but as Grubler *et al.* [11] make clear, the transition to a more circular economy requires a low energy and low material demand. They argue that this necessitates 100 percent electrification of currently motorised transport. However, this carries its own issues due to the availability and environmental impact of lithium and rare earth extraction and disposal (see Valero and Valero [27]). Grubler *et al.* [11] also state the need for policies that transition people away from their car and onto public forms of transport and non-motorised options. However, current UK infrastructure, lifestyle choices and expectations have locked most people into car ownership [28, 29]. Dematerialisation is also a key aspect in the transition to a low-carbon circular economy. While it may be easier for the car industry to dematerialise, if the right

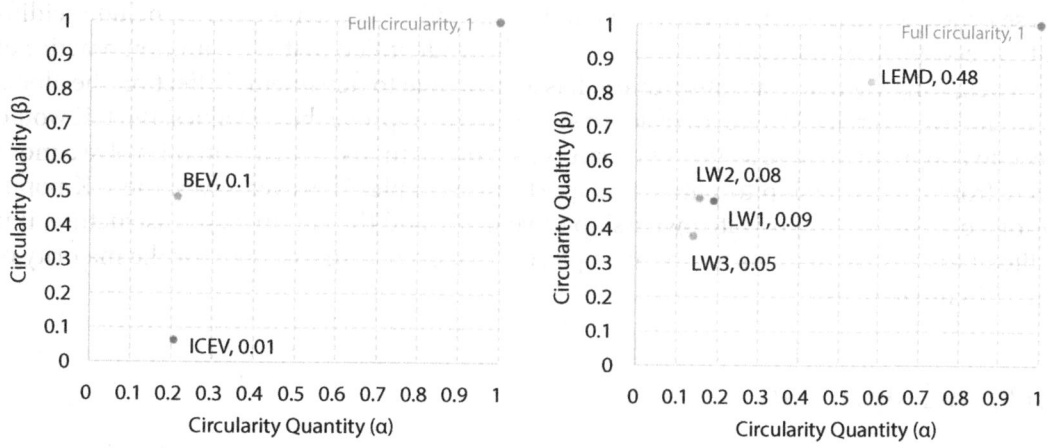

Figure 2.7 Alpha, beta and CI for: (a) ICEV and BEV current state (b) BEV lightweighting (LW) and low energy and material demand scenarios (LEMD) Note: The number next to the label corresponds to the overall CI value (obtained via α x β).

incentives are provided, industry leaders must be careful not to simply transmaterialise (switch out one material for another). They must also be aware of the trade-offs that become apparent when trying to optimise energy savings relative to material savings and vice versa (see Whiting *et al.* [30]).

2.4 Conclusions & Recommendations

In this chapter, we improved Cullen's [1] circularity index approach by expanding his alpha and beta indicators to better reflect the dynamics of material inflows and outflows and the environmental impacts across the life cycle of a system. The beta indicator is now more robust, as it can be used to examine a range of environmental impacts, not just energy consumption. Finally, we have tied Cullen's circularity index approach to the philosophical underpinnings of the circular economy, as first proposed by the Ellen MacArthur Foundation [23]. We hope that this will increase the adoption of this index among those stakeholders interested in quantitatively evaluating the circularity of their system.

System circularity is measured by a wide range of stakeholders, including businesses, governments, and academics. The CI updated in this chapter can communicate to these stakeholders the degree of circularity at various stages (design, operation, decommission) and scopes (macro, meso, micro). If done properly, the results can guide management and operational decisions, without the need for excessive or extra expertise and training. Indeed, they would only be asked to consider three values (alpha, beta, and overall CI).

With regard to our case study, our results suggest that the UK car sector is far from circular when it comes to closing the carbon emission and material cycles. However, there are some advancements that could be made in our progress towards circularity in terms of carbon emissions. The main improvement in the car sector will come from a transition towards fully BEV fleets, although this has a negligible impact on material cycle circularity. That said, a low-carbon circular economy is not without it challenges, nor environmental impacts, given the toxicity of rare earth mining and disposal, not to mention the logistics of having a much-expanded supply chain of rare earths and lithium. To better understand these issues, the research here can be expanded beyond carbon emissions to include acidification, eco-toxicity, mineral depletion, etc. We have identified that the common practice of lightweighting via material substitution, has a negligible to detrimental effect on the alpha, beta, and the overall CI value. In order to significantly improve the CI values, the UK would need to adopt much more stringent decarbonisation methods, which would involve a move away from car ownership to public transport, for example. However, given the UK population's dependency on private ownership of vehicles, and the country's infrastructure, it is difficult to imagine that this will be a simple, easy, or even desirable path for the majority of the UK's inhabitants.

Acknowledgements

L.G.C. acknowledges the financial support of the European Union's Horizon 2020 research and innovation programme under the Marie Sklodowska-Curie grant agreement

No 101027892. K.W. acknowledges the financial support of the Fonds de la Recherche Scientifique – FNRS under Grant No CR40001149.

References

1. Cullen, J. M.: Circular Economy: Theoretical Benchmark or Perpetual Motion Machine?, *Journal of Industrial Ecology*, 21, 483–486, 2017.
2. Blomsma, F. and Brennan, G.: The emergence of circular economy: a new framing around prolonging resource productivity, *Journal of Industrial Ecology*, 21, 603–614, 2017.
3. Homrich, A. S., Galvão, G., Abadia, L. G., and Carvalho, M. M.: The circular economy umbrella: Trends and gaps on integrating pathways, *Journal of Cleaner Production*, 175, 525–543, 2018.
4. Pauliuk, S.: Critical appraisal of the circular economy standard BS 8001: 2017 and a dashboard of quantitative system indicators for its implementation in organizations, *Resources, Conservation and Recycling*, 129, 81–92, 2018.
5. Corona, B., Shen, L., Reike, D., Carreón, J. R., and Worrell, E.: Towards sustainable development through the circular economy—A review and critical assessment on current circularity metrics, *Resources, Conservation and Recycling*, 151, 104498, 2019.
6. EMF: Circularity Indicators: Methodology, Ellen MacArthur Foundation, Cowes, UK, 2019.
7. Di Maio, F. and Rem, P. C.: A robust indicator for promoting circular economy through recycling, *Journal of Environmental Protection*, 6, 1095, 2015.
8. De Wit, M., Hoogzaad, J., Ramkumar, S., Friedl, H., and Douma, A.: The Circularity Gap Report: An analysis of the circular state of the global economy, Circle Economy, Amsterdam, 2018.
9. Desing, H., Braun, G., and Hischier, R.: Resource pressure–a circular design method, *Resources, Conservation and Recycling*, 164, 105179, 2021.
10. Hummen, T. and Sudheshwar, A.: Fitness of product and service design for closed-loop material recycling: A framework and indicator, *Resources, Conservation and Recycling*, 190, 106661, 2023.
11. Grubler, A., Wilson, C., Bento, N., Boza-Kiss, B., Krey, V., McCollum, D. L., Rao, N. D., Riahi, K., Rogelj, J., and De Stercke, S.: A low energy demand scenario for meeting the 1.5 C target and sustainable development goals without negative emission technologies, *Nature Energy*, 3, 515–527, 2018.
12. Graedel, T. E., Allwood, J., Birat, J.-P., Buchert, M., Hagelüken, C., Reck, B. K., Sibley, S. F., and Sonnemann, G.: What do we know about metal recycling rates?, *Journal of Industrial Ecology*, 15, 355–366, 2011.
13. Fellner, J. and Lederer, J.: Recycling rate–The only practical metric for a circular economy?, *Waste Management*, 113, 319–320, 2020.
14. Alaerts, L., Van Acker, K., Rousseau, S., De Jaeger, S., Moraga, G., Dewulf, J., De Meester, S., Van Passel, S., Compernolle, T., and Bachus, K.: Towards a more direct policy feedback in circular economy monitoring via a societal needs perspective, *Resources, Conservation and Recycling*, 149, 363–371, 2019.
15. Carmona, L. G., Whiting, K., and Cullen, J.: A stock-flow-service nexus vision of the low carbon economy, *Energy Reports*, 8, 565–575, 2022.
16. Carmona, L. G., Whiting, K., Carrasco, A., Sousa, T., and Domingos, T.: Material Services with Both Eyes Wide Open, Sustainability, 9, 1508, 2017.
17. Whiting, K., Carmona, L. G., and Carrasco, A.: The resource service cascade: A conceptual framework for the integration of ecosystem, energy and material services, *Environmental Development*, 41, 100647, 2022.

18. Haberl, H., Wiedenhofer, D., Virág, D., Kalt, G., Plank, B., Brockway, P., Fishman, T., Hausknost, D., Krausmann, F. P., Leon-Gruchalski, B., Mayer, A., Pichler, M., Schaffartzik, A., Sousa, T., Streeck, J., and Creutzig, F.: A systematic review of the evidence on decoupling of GDP, resource use and GHG emissions, part II: synthesizing the insights, *Environmental Research Letters*, 2020.

19. Steinberger, J. K. and Roberts, J. T.: From constraint to sufficiency: The decoupling of energy and carbon from human needs 1975–2005, *Ecological Economics*, 70, 425–433, 2010.

20. Whiting, K., Carmona, L. G., and Carrasco, A.: Material and Energy Services, Human Needs and Wellbeing, in: Environmental Sustainability and Economy, edited by: Pardo Martinez, C. I. and Cotte Poveda, A., Elsevier, Cambridge, MA, US, 275–296, 2021.

21. Kalt, G., Wiedenhofer, D., Görg, C., and Haberl, H.: Conceptualizing energy services: A review of energy and well-being along the Energy Service Cascade, *Energy Research & Social Science*, 53, 47–58, 2019.

22. EMF: Towards the Circular Economy Vol. 1, Ellen MacArthur Foundation, Cowes, UK, 2013.

23. IEA: Transport: Improving the sustainability of passenger and freight transport, International Energy Agency, Paris, France, 2022.

24. Geyer, R.: Life Cycle Energy and Greenhouse Gas (GHG) Assessments of Automotive Material Substitution User Guide for Version 5 of the UCSB Automotive Energy and GHG Model, University of California at Santa Barbara, CA, on behalf of WorldAutoSteel, 2017.

25. World Auto Steel: Future Steel Vehicle - Results and Reports & Cost Model, World Auto Steel, Brussels, Belgium, 2012.

26. Rodrigues, B., Carmona, L. G., Whiting, K., and Sousa, T.: Resource efficiency for UK cars from 1960 to 2015: From stocks and flows to service provision, *Environmental Development*, 100676, 2021.

27. Valero, A. and Valero, A.: Thanatia: The Destiny of the Earth's Mineral Resources: a Thermodynamic Cradle-to-cradle Assessment, *World Scientific*, Singapore, 2014.

28. Mattioli, G.: "Forced car ownership" in the UK and Germany: socio-spatial patterns and potential economic stress impacts, *Social Inclusion*, 5, 147–160, 2017.

29. Mattioli, G.: Where sustainable transport and social exclusion meet: Households without cars and car dependence in Great Britain, *Journal of Environmental Policy & Planning*, 16, 379–400, 2014.

30. Whiting, K., Carmona, L. G., Brand-Correa, L. I., and Simpson, E.: Illumination as a material service: A comparison between Ancient Rome and early 19th century London, *Ecological Economics*, 169C, 106502, 2020.

3

Biodegradable Polymers For Circular Economy Transitions—Challenges and Opportunities

Koushik Ghosh* and Brad H. Jones

Sandia National Laboratories, 1515 Eubank SE, Albuquerque, New Mexico, United States

Abstract

Unprecedented material influx into modern consumer life without a thoughtful consideration of sources and after-life is unsustainable and requires a close scrutiny of the way we produce, consume, and discard materials. The successful circular economy transitions require synergy of technologies slowly making inroads in the material value chain without disrupting the convenience, value, and versatility of plastics. Biodegradable materials are often thought to decompose in the natural soil, compost, or marine environments in an environmentally benign manner overcoming the concerns of environmental persistence. Despite their biomimetic appeal, zero-waste promises, and decade-long presence in the marketplace, these classes of polymers are still far from realizing their potential. Here, we discuss these apparent challenges through a lifecycle perspective (Figure 3.1)—where we will trace the flow of polymers through the traditional plastics value-chain and identify the research gaps across the entire lifecycle. A system-level analysis through material value chain perspective allows us to identify the leakage routes of plastics into the environment along with the time-scale requirements often ignored or overlooked with a fragmented outlook.

We will revisit commercial biodegradable plastics through established database analysis and contrast it against the desired state of the zero-waste-focused circular economy. To bridge the gap, we will present a holistic framework inclusive of all the stakeholders across the traditional value chain and new post-consumer value chain where producers, consumers, and waste management facilities play a critical role in enabling timely circular economy transitions. Along these lines we also identify critical research needs concerning the structure and properties of biodegradable plastics, testing standards, application development, and waste management.

Keywords: Biodegradable, compostable, polymers, plastics, persistence, bioplastics, circular economy, zero-waste

3.1 Introduction

Global-scale plastic pollution didn't happen when humans started replacing animal-based or natural materials with cheap, lightweight fossil-fuel-based alternatives. It happened after unrestrained growth, irresponsible waste management, and a cultural acceptance of

*Corresponding author: kghosh@sandia.gov

Nabil Nasr (ed.) *Technology Innovation for the Circular Economy: Recycling, Remanufacturing, Design, Systems Analysis and Logistics*, (29–42) © 2024 Scrivener Publishing LLC

exchanging convenience through growing plastics use, not regarding their EOL management. While no one can argue against the value plastics bring to our daily life, we do understand this current trend is unsustainable as long as recycling rates are inefficient [1–3] or a linear economy viewpoint [4–7]. Between 1980 and 2019, annual plastic use tripled, with 380 million tons of plastics produced in 2019 [8]. COVID-19 pandemic has magnified the problem with urgent demands for single-use-plastics [9, 10]. In addition to the necessary increase of single-use plastics for personal protective equipment, some governments and businesses have delayed or scrapped plastic bag and packaging bans. With less than 10% plastics being recycled in the USA [2, 3]. the inevitable fate of plastics is landfills or incineration [11]. Both methods of current waste management can lead to further environmental pollution and climate change concerns where the leakage of plastics in water or soils and their further intake by animals or plants can have more serious unintended consequences [12–15]. The scrutiny of plastic accumulation in the ecosystem is beyond aesthetics [8, 12, 13] and has been discussed in the media, academia, industry, and policy literature [16–18].

The concept of biodegradable polymers spring from the idea of keeping the convenience, value, and functionality of the non-biodegradable alternatives yet providing an added benefit of environmental non-persistence in water, soil, compost, or marine medium. Biodegradable polymers borrow the idea from nature, where most natural materials provide functional value throughout their life cycle. After use, they leave no trace through the environmental decay process. However, this is a significantly challenging task where synthetic products are expected to survive the entire value chain (Figure 3.1) without losing quality but degrade completely in the natural or contained environment as soon as the

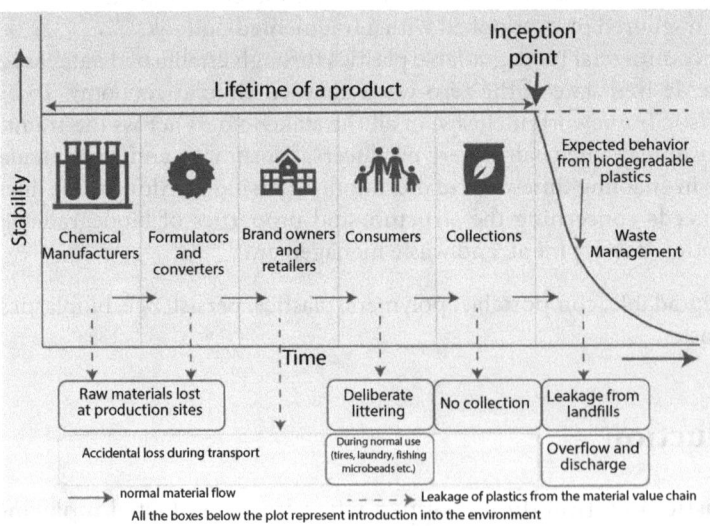

Figure 3.1 Material flow perspective, borrowing the value chain concept from the marketing literature, establishes the challenges and opportunities for biodegradable plastics where they need to survive the entire value chain and yet expected to decompose in a timely manner when they are discarded into the environment after the post-consumer use. Plastics often leak into the environment and create microplastics and nanoplastics that are difficult to track. Figure is adapted from Ghosh, K. and B.H. Jones, *Roadmap to Biodegradable Plastics—Current State and Research Needs*. ACS Sustainable Chemistry & Engineering, 2021. 9(18): p. 6170-6187.

consumer disposes of them. No wonder, despite the immense appeal of these materials or several decades of their presence in the marketplace, these biodegradable polymers, often in the form of different consumer-end products are subject to lots of scrutiny, confusing terminologies, high prices, poor performances, or inconsistent quality standards.

Here, we revisit the perspective [19] we shared around the roadmap of biodegradable plastics while reaffirming the challenges and research opportunities from a material value-chain point of view (Figure 3.1). A material value chain is a concept introduced in the marketing literature [20] that aptly captures the system-level complexity of plastic pollution. This viewpoint enables us to recognize the interconnectedness of the different parts and cautions us against the perils of ignoring one part of the value chain by labeling it "someone else's problem." At the same time, it reemphasizes the importance of an extended value chain, a key requisite for successful circular economy transitions. Further, this is a basis for tracking material, energy, and money flow and a good visual map to minimize waste.

3.2   Clarification of Confusing Terminologies

One of the major criticisms against biodegradable polymers is confusing terminologies [21] used by several stakeholders. For instance, biodegradable polymers are different from biobased polymers. Biobased polymers deal with the source of the raw feedstocks while biodegradable polymers deal with the end-of-life fate. While these terms are used interchangeably in trade literature or marketing claims, the difference between these two value propositions must be clarified. Biodegradable polymers can be sourced from fossil-fuel based feedstocks and biobased products may not be biodegradable. For instance, bio-based durable polyethylene (bioPE) is sourced from biomass derivatives but is not claimed to be biodegradable, polybutylene succinate (PBS) is typically fossil-based yet biodegradable.

The other source of confusion comes from the loose use of the term biodegradability without mentioning the medium. Environmental erosion could be a source of disintegration of plastic articles, but for complete mineralization to occur, microbes must be present in those mediums. Biodegradation is a subset of degradation involving mineralization by microorganisms primarily to CO_2, H_2O, and CH_4, which are the final products of aerobic or anaerobic degradation. Compost is one possible biodegradation medium along with water, soil, and marine (Figure 3.2). The rate of biodegradation differs significantly based on the test medium.

Biodegradation can be mediated in manufactured environments like compost or anaerobic digesters, or it can occur spontaneously without human intervention in natural water, soil, or marine medium. While one assumes the biodegradable claims should complement responsible waste management and must be limited to composts or anaerobic digesters, the term biodegradable often invokes the image of environmental non-persistence and promotes littering.

3.3   Structures, and Application-Space of Biodegradable Polymers

Generally, the structures of biodegradable polymers can be either aliphatic polyesters or the derivatives of natural biodegradable polymers like cellulose or starch. Major biodegradable polymers include polylactic acid (PLA); polyhydroxyalkanoates (PHAs), including

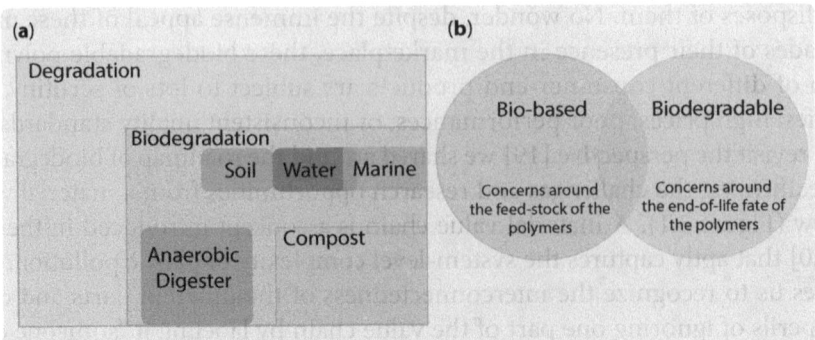

Figure 3.2 (a) Difference between the degradation, biodegradation and degradation in compost. (b) Distinction between the biobased and biodegradable polymers. Figure is adapted from Ghosh, K. and B.H. Jones, Roadmap to Biodegradable Plastics—Current State and Research Needs. *ACS Sustainable Chemistry & Engineering*, 2021. 9(18): p. 6170-6187.

polyhydroxybutyrate (PHB) and related copolymers; cellulose diacetate; regenerated cellulose; copolyesters such as polybutylene adipate terephthalate (PBAT) and polybutylene succinate adipate (PBSA); polybutylene succinate (PBS); polycaprolactone (PCL); polyglycolic acid (PGA); and starch compounds (mixtures of starch and other biodegradable polymers such as PBAT or PLA). The most understood biodegradation mechanisms are the environmentally mediated hydrolytic degradation process followed by or parallel with the microbial mineralization process. While hydrolysis can be surface- or bulk-driven [22] with water attacking the ester backbone or side chains of the derivatives, the microbial process can be convoluted with the water content, oxygen level, and the nature of feedstocks can play critical roles in determining the microbial diversity.

According to the S&P Global report [23], Northeast Asia—Mainland China, Japan, South Korea, and Taiwan—has the highest consumption growth rate: greater than 37%/year on average during 2020-25. For comparison, the predicted consumption growth rate for the world is almost 29% during the forecast period. The expected consumption growth rates in other regions are average of 5-7 %/year.

Commercial biodegradable plastics reach consumers through a certification (or labeling) by certifying bodies like DIN Certco, Biodegradable Products Institute (BPI), TÜV Austria, and others. Certification of the marketplace products is generally a sign of the credibility of sustainability claims and aims to clarify the confusion around terminologies in easily understandable logos. Generally, the certification process follows three steps:

1. The brand owner recognizes the brand to certify and which certification to obtain based on the customer's needs.
2. A third-party testing laboratory (OWS, AIMPLAS, Innovhub SSI, etc.) conducts the test following the prescribed test protocols by the International Organization for Standardization (ISO), American Society for Testing and Materials (ASTM), European Committee for Standardization (CEN), and others.
3. The certifying body reviews the data from the third-party testing lab and decides on certification. Different market reports and our analysis (Table 3.1)

Table 3.1 Commercial biodegradable products as listed in the TÜV Austria certification database [27] (as of 01/20/23).

Certified Products as listed in the TÜV Austria database	No. of Certified Products	Main Constituent	Representative forms
OK Biodegradable Marine	67	PHA and their derivatives, cellulose	Films, nonwoven fibers, granulates, ink, colorants, and adhesives
OK Biodegradable Soil	123	Regenerated cellulose, cellulose acetate, PHA	Nonwoven fibers, paper, films, plastic additives
OK Biodegradable Water	52	Regenerated cellulose, polymer blends	Granulates, film, biomaterials, and additives
OK Compost Industrial	2890	PLA, PBAT, PHA, Regenerated cellulose and cellulose acetate	Granulates, films, bags/sack trade, trays and plates, additives
OK Compost Home	1508	PBAT, PHA, polymer blends, Regenerated cellulose and cellulose acetate	Granulates, films, bags, additives,

of certified biodegradable products (BPI and TUV Austria certified products) in the marketplace confirm that food packaging and bags are the major end uses for biodegradable polymers. Within the food packaging category, the largest application is shopping bags; other applications include produce bags and food service ware—cold cups, drinking straws, utensils, and clamshell containers. Compost bags—bin liners and garbage bags—constitute the second largest end-use category. Although shopping and produce bags often have a second life as organic waste bags, they are excluded from this category.

Regulations and corporate initiatives are the main drivers of demand for biodegradable polymers. In this context, "regulation" usually translates to "ban. "National or regional bans on plastic shopping bags, notably in Europe and Mainland China, drive demand for biodegradable polymers in those regions.[23] Produce bags, drinking straws, utensils, and other single-use plastic products are targets of other bans. Regulations prohibiting food and yard waste in landfills support the consumption of compost bags made of biodegradable polymers.

Biodegradable polymers have their place when we realize the economics of sorting mixed waste plastics [24], the performance degradation of repeated mechanical recycling [25], the unsustainability of landfill environments [26], or the inevitable leakage of polymers [4] into the environment. However, some of the enduring questions remain. Why do the compostable plastics end up in landfills? Are biodegradable plastics promoting an alternative to contained waste management practices? How do these formulations fit to the existing practices and processes?

3.4 Knowledge Gaps and Research Needs

Biodegradable polymers have a commercial presence, but the scale and controversies around them force us to probe deep into the heart of the issue. As we have pointed out several shortcomings of biodegradable polymers over performance, greenwashing [28–31], and littering concerns, we identify the knowledge gaps (Table 3.2) according to the four key functions—structure and properties, testing, application development, and waste management—highlighting the interdependence in different parts of the value chain. This system level and holistic outlook help us to emphasize the need for multidisciplinary research involving multiple stakeholders involving researchers, industry, media, government agencies, nongovernmental organizations, and the general public. In this discussion, we focus on the extended value chain for enabling the collection, sorting, and engineered waste management that often remains at the forefront of the discussion in the circular

Table 3.2 Knowledge gaps (similar to the perspective [19]) around four key functions of biodegradable plastics.

Structure and properties	Can we develop a model for the lifetime prediction of biodegradable polymers? How do we access the alternative feedstocks that will match the scale and cost of fossil-fuel- based raw materials? What other alternative materials and processes can we think of that will allow us to expand the structural diversity? Can we engineer design principles to degrade?
Testing of biodegradation	What are the test methods that can verify the biodegradability claims? How do we address the gap between lab and field tests? What are our options if we need test methods for uncontrolled marine and soil mediums? Can we predict or accelerate biodegradation tests to reduce the length of standard tests?
Application Development	Is your product designed with end-of-life-specific applications in mind? While developing new formulations, do we understand the tradeoff between performance and degradability? Can the new product or formulations fit the existing process? What are the impacts of different additives used in product development in the ecology?
Waste management	Do we have enough contained waste management capabilities to collect and treat discarded biodegradable plastics? Why will the composters take the discarded plastics? Can we develop engineered waste management based on fundamental biotechnology knowledge? What are the incentives for biodegradable plastics? Is it used as permission to litter?

economy but is neglected while developing biodegradable polymers. Biodegradable value propositions often invoke the promise of non-persistence but should not be used as a substitute for littering. For biodegradable polymers to thrive, we need engineered waste management, not depending merely on the composters.

3.4.1   Structure and Properties

Most of the fundamental knowledge of commercial biodegradable polymers is translated from their applications in biomedical applications [32, 33]. However, we have a limited understanding of how these polymers degrade in real environments, precluding us from having a lifetime prediction [34]. Nor do we know about the source of alternative feedstocks while competing with the scale of fossil fuel-based materials [35–37]. We must be cognizant of not competing with agricultural feedstocks or other food sources. Instead, we must focus on converting waste [38]—industrial solid waste or CO2— to raw materials- to avert a resource security catastrophe. More ambitious but practical endeavors can address this using alternative materials with scalable processing techniques or engineered degradation. Engineered degradation [39, 40] can have an enticing prospect. However, the unsuccessful and premature introduction of oxo-degradable plastics [41] into the marketplace cautions us against the fragmented outlook without considering the ecological fate of degraded minor constituents.

3.4.2   Testing of Biodegradation

Testing biodegradable materials against the proper ASTM or ISO test methods is an essential criterion often unique to these classes of materials. We recommend adhering to the standard test development protocols [42] before claiming biodegradability. Also, we recognize the need for new test methods in open and uncontrolled medium such as soils or marine and the lack of predictability [43] given the lengthy timescales of some of these tests [44]. Most importantly, the gap between laboratory and field tests is crucial [45], especially when the microbial crosstalk between the polymeric substrates and the environment differs significantly [46].

3.4.3   Application Development

When we discuss plastics, we often fail to recognize that plastics are products-a well-chosen mixture of different additives [47] to meet the processing and final application needs. Often to meet the application requirements, significant polymer modification [48] or additive package [47, 49] is needed and that may hinder the biodegradability [50]. This performance-degradability trade-off again brings back the need for multidisciplinary research early in application development. If the polymers cannot be processed with standard compounding, injection molding, blow molding, or several other film processing techniques, it is extremely difficult to see their successful transition. Most of the aliphatic polyesters are inherently susceptible to thermal and hydrolytic degradation [51] during processing. Similarly, the need for ecologically conscious formulations [52] or deep knowledge of the end application [53] are critical.

3.4.4   Waste Management

Possibly, the most overlooked aspect of biodegradable product development is the consideration of waste management. The current infrastructure of waste management in the form of compost and anaerobic digesters [54] is often inadequate to handle the zero-waste demands [55]. Also, we must seriously consider the misalignment of goals between traditional composting practices and compost as a medium for waste management. Compost is often a logical path for plastics contaminated with food waste. But do composters need plastics in their compost feedstocks? Composting is a delicate balancing of carbon-to-nitrogen ratio, water, and oxygen contents [56]. How will the process be impacted if we suddenly demand polymeric formulations of different thicknesses dominate the feedstocks? Put simply, if we suddenly ask about all the biodegradable water bottles as a potential feedstock for compost, how big will your compost piles need to be? Also, for economic purposes, composters target shorter compost cycles [56] with optimized feedstocks. They have little incentive to incorporate plastics which will inevitably prolong the process. This is where we emphasize the need for engineered waste management with a deep knowledge of biotechnology. We know very little about the anaerobic biodegradation [57] of commercial plastics, but the environments in marine sediments or different animal guts are mostly anaerobic. We know the enzymes can be engineered [58] to degrade polymeric materials, but can that fundamental knowledge be translated for a scalable and accelerated process of bioremediation? Further, we know little about how microbial degradation will be impacted by more than one food source coming in the form of multiple plastics.

Plastics, irrespective of their nature, do leak into the environment. Unfortunately, because of the confusion it carries with their ambiguous terminologies, biodegradable plastics have a much greater risk of being leaked into the environment [59]. Most often, without proper composting, anaerobic digester, or even a lack of knowledge of the consumers encourage littering. Because of their size, microplastics or nano plastics can act as a vector for pollutants [60] or impede the assimilation of microbes [61]. We still do not know the consequences of this problem, but we can always improve our ability to track the plastics in the environment.

3.5   Biodegradable Plastics and Circular Economy Transitions

A short-term focused produce-use-discard based linear economy necessitates the need for shift of perspectives. Implementation of circular economy is not straightforward with logistics, technical, and financial issues [62, 63]. Additionally, circularity automatically does not guarantee sustainability [64]. Sustainability is a system-level concept, and it can only be measured with a comprehensive accounting for energy flow, material flow, and money flow at each step of the value chain [10, 64]. Twelve principles of green chemistry [65], Material flow analysis, and life cycle analysis (LCA) [66] are some of the tools often used to capture sustainability at a systems level. The concept of circular economy, initially introduced by Ellen McArthur Foundation [5], brings discrete tools and methods of sustainability under one umbrella, encouraging the minimization of waste in any form-material, energy, or financial by innovation. The idea of biodegradable plastics is not a "silver bullet." Their true environmental impact must be considered rigorously as nondegradable counterparts [67, 68].

Biodegradable plastics are not an excuse for littering; therefore, the successful transition and integration [6, 69] to the circular economy requires an extended value chain (Figure 3.3)

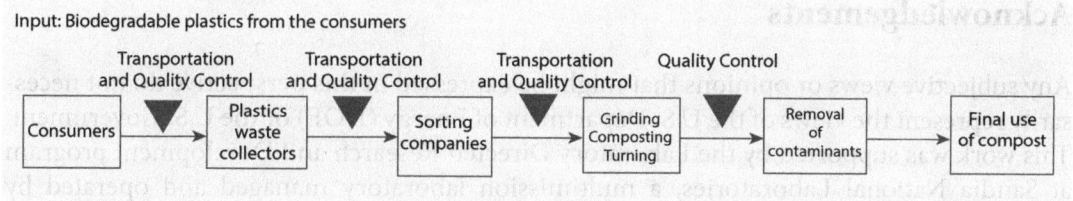

Figure 3.3 The concept of extended value chain to enable circular economy transitions. Discarded plastic materials need to be collected and mixed with other feedstocks to generate a quality compost in a reasonable timeframe. Composting process can be replaced by anaerobic digesters or other engineered waste managements. To make profit, composters need to optimize the process based on feedstocks, time, or other operating costs. By creating an alternative or extended value chain circular economy shifts the job from production and manufacturing to collection and sorting.

and a proper infrastructure for collection, sorting, grinding, and, eventually, waste treatment. Biodegradable plastics may avoid sorting, but grinding does help [70] in expediting the process in a controlled medium. The World Economic Forum (WEF), together with the Ellen MacArthur Foundation and McKinsey & Company, is stressing the importance of government-backed infrastructures and science-based policy initiatives for a circular plastics economy [71]. As part of the regulatory framework, the European Union (EU) plans to revise and address outstanding concerns against existing standards (such as EN 13432) to account for more realistic biodegradation testing [72].

3.6   Conclusions and Recommendations

According to an estimate by European Bioplastics, the global production of bioplastics is about 2.11 million tonnes compared to 335 million tonnes of the overall global production of plastics in 2017 [73]. There is room to grow for biodegradable plastics, but as outlined in the previous section, the path is complex and intertwined with challenges. Often, we are misguided when biodegradable polymers are introduced as the panacea of current plastics pollution, but as we attempt to convey the successful transition to a circular economy requires a drastic change in the process and outlook of stakeholders. For example, most often biodegradable plastics often represented as a drop-in replacement for non-degradable single-use plastics and become an excuse to litter. As we have illustrated here, the successful transition to biodegradable plastics do require infrastructure, capabilities, initiatives—often in the form of extended value chain, an integral part of circular economy transition.

In this perspective, we investigate the prospect of biodegradable plastics in meeting the demands of circular economy transitions. We recommend research needs in the form of 16 key knowledge gaps in all technology readiness levels (TRLs) with an emphasis on the interrelatedness of the problem. By utilizing these questions as guidelines, we hope to create an inclusive dialog or have a right conversation among stakeholders with conflicting priorities. At this critical juncture of environmental crisis, what we choose today will make a transformative impact in the future. Rather than engaging social and political discussions around consumption habits or dematerialization initiatives, we remain hopeful that we will reach the goal of non-persistence if we choose the right application or right end-of-life scenarios with biodegradable plastics.

Acknowledgements

Any subjective views or opinions that might be expressed in this perspective do not necessarily represent the views of the U.S. Department of Energy (DOE) or the U.S. Government. This work was supported by the Laboratory Directed Research and Development program at Sandia National Laboratories, a multimission laboratory managed and operated by National Technology and Engineering Solutions of Sandia, LLC, a wholly owned subsidiary of Honeywell International, Inc., for the U.S. Department of Energy's National Nuclear Security Administration under Contract DE-NA-0003525.

References

1. Geyer, R., J.R. Jambeck, and K.L. Law, Production, use, and fate of all plastics ever made. *Sci Adv*, 2017. 3(7): p. e1700782.
2. Heller, M.C., M.H. Mazor, and G.A. Keoleian, Plastics in the US: toward a material flow characterization of production, markets and end of life. *Environmental Research Letters*, 2020. 15(9): p. 094034.
3. Di, J., B.K. Reck, A. Miatto, and T.E. Graedel, United States plastics: Large flows, short lifetimes, and negligible recycling. *Resources, Conservation and Recycling*, 2021. 167: p. 105440.
4. Kosuth, M., S.A. Mason, and E.V. Wattenberg, Anthropogenic contamination of tap water, beer, and sea salt. *Plos One*, 2018. 13(4).
5. MacArthur, D.E., Beyond plastic waste. *Science*, 2017. 358(6365): p. 843-843.
6. Bucknall, D.G., Plastics as a materials system in a circular economy. *Philosophical Transactions of the Royal Society a-Mathematical Physical and Engineering Sciences*, 2020. 378(2176).
7. Zheng, J.J. and S. Suh, Strategies to reduce the global carbon footprint of plastics. *Nature Climate Change*, 2019. 9(5): p. 374-+.
8. Chen, Y., A.K. Awasthi, F. Wei, Q. Tan, and J. Li, Single-use plastics: Production, usage, disposal, and adverse impacts. *Science of The Total Environment*, 2021. 752: p. 141772.
9. Hale, R.C. and B. Song, Single-Use Plastics and COVID-19: Scientific Evidence and Environmental Regulations. *Environmental Science & Technology*, 2020. 54(12): p. 7034-7036.
10. Klemes, J.J., Y. Van Fan, R.R. Tan, and P. Jiang, Minimising the present and future plastic waste, energy and environmental footprints related to COVID-19. *Renewable & Sustainable Energy Reviews*, 2020. 127.
11. Zhang, F., Y. Zhao, D. Wang, M. Yan, J. Zhang, P. Zhang, T. Ding, L. Chen, and C. Chen, Current technologies for plastic waste treatment: A review. *Journal of Cleaner Production*, 2021. 282: p. 124523.
12. Amaral-Zettler, L.A., E.R. Zettler, and T.J. Mincer, Ecology of the plastisphere. *Nature Reviews Microbiology*, 2020. 18(3): p. 139-151.
13. Shen, M.C., W. Huang, M. Chen, B. Song, G.M. Zeng, and Y.X. Zhang, (Micro)plastic crisis: Un-ignorable contribution to global greenhouse gas emissions and climate change. *Journal of Cleaner Production*, 2020. 254.
14. Zhang, B., X. Yang, L. Chen, J. Chao, J. Teng, and Q. Wang, Microplastics in soils: a review of possible sources, analytical methods and ecological impacts. *Journal of Chemical Technology & Biotechnology*, 2020. 95(8): p. 2052-2068.
15. Center for International Environmental Law (CIEL), *Plastic & climate. The hidden costs of a plastic planet*. 2018.
16. US Congress, *Break Free From Plastic Pollution Act*. 2020. p. 1-127.

17. Moshood, T.D., G. Nawanir, F. Mahmud, F. Mohamad, M.H. Ahmad, and A.A. Ghani, Expanding Policy for Biodegradable Plastic Products and Market Dynamics of Bio-Based Plastics: Challenges and Opportunities. *Sustainability*, 2021. 13(11).

18. Fadeeva, Z. and R. Van Berkel, 'Unlocking circular economy for prevention of marine plastic pollution: An exploration of G20 policy and initiatives. *Journal of Environmental Management*, 2021. 277.

19. Ghosh, K. and B.H. Jones, Roadmap to Biodegradable Plastics—Current State and Research Needs. *ACS Sustainable Chemistry & Engineering*, 2021. 9(18): p. 6170-6187.

20. Porter, M.E., The value chain and competitive advantage. *Understanding Business Processes*, 2001. 2: p. 50-66.

21. Zumstein, M.T., R. Narayan, H.-P.E. Kohler, K. McNeill, and M. Sander, *Dos and Do Nots When Assessing the Biodegradation of Plastics*. Environmental Science & Technology, 2019. 53(17): p. 9967- 9969.

22. Burkersroda, F.v., L. Schedl, and A. Göpferich, Why degradable polymers undergo surface erosion or bulk erosion. *Biomaterials*, 2002. 23(21): p. 4221-4231.

23. S&P Global, *S&P Global's Chemical Economics Handbook – Biodegradable Polymers*. 2021.

24. Gundupalli, S.P., S. Hait, and A. Thakur, A review on automated sorting of source-separated municipal solid waste for recycling. *Waste Management*, 2017. 60: p. 56-74.

25. Schyns, Z.O.G. and M.P. Shaver, Mechanical Recycling of Packaging Plastics: A Review. *Macromolecular Rapid Communications*, 2020. **n/a**(n/a): p. 2000415.

26. Themelis, N.J. and P.A. Ulloa, Methane generation in landfills. *Renewable Energy*, 2007. 32(7): p. 1243- 1257.

27. Austria, T. [cited 2023 January 20th]; Available from: https://www.tuv-at.be/green-marks/certified- products/.

28. Zhu, J. and C. Wang, Biodegradable plastics: Green hope or greenwashing? *Marine Pollution Bulletin*, 2020. 161: p. 111774.

29. Shen, M., B. Song, G. Zeng, Y. Zhang, W. Huang, X. Wen, and W. Tang, Are biodegradable plastics a promising solution to solve the global plastic pollution. *Environ. Pollut. (Oxford, U. K.)*, 2020. 263(Part_A): p. 114469.

30. Nazareth, M., M.R.C. Marques, M.C.A. Leite, and Í.B. Castro, Commercial plastics claiming biodegradable status: Is this also accurate for marine environments? *Journal of Hazardous Materials*, 2019. 366: p. 714-722.

31. Narancic, T. and K.E. O'Connor, Plastic waste as a global challenge: are biodegradable plastics the answer to the plastic waste problem? *Microbiology (London, U. K.)*, 2019. 165(2): p. 129-137.

32. Tian, H., Z. Tang, X. Zhuang, X. Chen, and X. Jing, Biodegradable synthetic polymers: Preparation, functionalization and biomedical application. *Progress in Polymer Science*, 2012. 37(2): p. 237-280.

33. Park, J.H., M.L. Ye, and K. Park, Biodegradable polymers for microencapsulation of drugs. *Molecules*, 2005. 10(1): p. 146-161.

34. Laycock, B., M. Nikolić, J.M. Colwell, E. Gauthier, P. Halley, S. Bottle, and G. George, Lifetime prediction of biodegradable polymers. *Progress in Polymer Science*, 2017. 71: p. 144-189.

35. Huo, J. and B.H. Shanks, Bioprivileged Molecules: Integrating Biological and Chemical Catalysis for Biomass Conversion. *Annual Review of Chemical and Biomolecular Engineering*, 2020. 11(1): p. 63-85.

36. Sheldon, R.A., Chemicals from renewable biomass: A renaissance in carbohydrate chemistry. *Current Opinion in Green and Sustainable Chemistry*, 2018. 14: p. 89-95.

37. Gandini, A., Polymers from Renewable Resources: A Challenge for the Future of Macromolecular Materials. *Macromolecules (Washington, DC, U. S.)*, 2008. 41(24): p. 9491-9504.

38. Grignard, B., S. Gennen, C. Jerome, A.W. Kleij, and C. Detrembleur, Advances in the use of CO2 as a renewable feedstock for the synthesis of polymers. *Chemical Society Reviews*, 2019. 48(16): p. 4466-4514.

39. Albertsson, A.-C. and M. Hakkarainen, Designed to degrade. *Science*, 2017. **358**(6365): p. 872.

40. Robert B. Wilson, J.S.J., *Hydrolytically degradable olefin copolymers*. 2001 US6534610B1, SRI International Inc.

41. Vazquez, Y.V., J.A. Ressia, M.L. Cerrada, S.E. Barbosa, and E.M. Vallés, Prodegradant Additives Effect onto Comercial Polyolefins. *Journal of Polymers and the Environment*, 2019. 27(3): p. 464-471.

42. De Wilde, B., Biodegradation testing protocols. *ACS Symp. Ser.*, 2012. 1114(Degradable Polymers and Materials): p. 33-43.

43. Abi-Akl, R., E. Ledieu, T.N. Enke, O.X. Cordero, and T. Cohen, Physics-based prediction of biopolymer degradation. *Soft Matter*, 2019. 15(20): p. 4098-4108.

44. Lucas, N., C. Bienaime, C. Belloy, M. Queneudec, F. Silvestre, and J.-E. Nava-Saucedo, Polymer biodegradation: Mechanisms and estimation techniques – A review. *Chemosphere*, 2008. 73(4): p. 429-442.

45. Rudnik, E., Compostable polymer materials. *Compostable Polymer Materials*. 2019. 1-410.

46. Oberbeckmann, S., A.M. Osborn, and M.B. Duhaime, Microbes on a Bottle: Substrate, Season and Geography Influence Community Composition of Microbes Colonizing Marine Plastic Debris. *PLoS One*, 2016. **11**(8): p. e0159289.

47. Hahladakis, J.N., C.A. Velis, R. Weber, E. Iacovidou, and P. Purnell, An overview of chemical additives present in plastics: Migration, release, fate and environmental impact during their use, disposal and recycling. *J Hazard Mater*, 2018. 344: p. 179-199.

48. Anderson, K.S., K.M. Schreck, and M.A. Hillmyer, Toughening Polylactide. *Polym. Rev. (Philadelphia, PA, U. S.)*, 2008. **48**(1): p. 85-108.

49. Zhao, X., K. Cornish, and Y. Vodovotz, Narrowing the Gap for Bioplastic Use in Food Packaging: An Update. *Environ. Sci. Technol.*, 2020. 54(8): p. 4712-4732.

50. Zenkiewicz, M., R. Malinowski, P. Rytlewski, A. Richert, W. Sikorska, and K. Krasowska, Some composting and biodegradation effects of physically or chemically crosslinked poly(lactic acid). *Polym. Test.*, 2012. 31(1): p. 83-92.

51. Signori, F., M.-B. Coltelli, and S. Bronco, Thermal degradation of poly (lactic acid)(PLA) and poly (butylene adipate-co-terephthalate)(PBAT) and their blends upon melt processing. *Polymer Degradation and Stability*, 2009. **94**(1): p. 74-82.

52. Zimmermann, L., A. Dombrowski, C. Volker, and M. Wagner, Are bioplastics and plant-based materials safer than conventional plastics? In vitro toxicity and chemical composition. *Environment International*, 2020. 145.

53. Narancic, T., S. Verstichel, S. Reddy Chaganti, L. Morales-Gamez, S.T. Kenny, B. De Wilde, R. Babu Padamati, and K.E. O'Connor, Biodegradable Plastic Blends Create New Possibilities for End-of-Life Management of Plastics but They Are Not a Panacea for Plastic Pollution. *Environ. Sci. Technol.*, 2018. 52(18): p. 10441-10452.

54. Wan, S., L. Sun, Y. Douieb, J. Sun, and W. Luo, Anaerobic digestion of municipal solid waste composed of food waste, wastepaper, and plastic in a single-stage system: Performance and microbial community structure characterization. *Bioresource Technology*, 2013. 146: p. 619-627.

55. Hottle, T.A., M.M. Bilec, N.R. Brown, and A.E. Landis, Toward zero waste: Composting and recycling for sustainable venue based events. *Waste Management*, 2015. **38**: p. 86-94.

56. Awasthi, S.K., S. Sarsaiya, M.K. Awasthi, T. Liu, J. Zhao, S. Kumar, and Z. Zhang, Changes in global trends in food waste composting: Research challenges and opportunities. *Bioresour Technol*, 2020. **299**: p. 122555.

57. Quecholac-Pina, X., M.D.C. Hernandez-Berriel, M.D.C. Manon-Salas, R.M. Espinosa-Valdemar, and A. Vazquez-Morillas, Degradation of Plastics under Anaerobic Conditions: A Short Review. *Polymers (Basel)*, 2020. 12(1).

58. Purohit, J., A. Chattopadhyay, and B. Teli, Metagenomic exploration of plastic degrading microbes for biotechnological application. *Curr. Genomics*, 2020. 21(4): p. 253-270.

59. Degli Innocenti, F. and T. Breton, Intrinsic Biodegradability of Plastics and Ecological Risk in the Case of Leakage. *ACS Sustainable Chem. Eng.*, 2020. 8(25): p. 9239-9249.

60. Ziccardi, L.M., A. Edgington, K. Hentz, K.J. Kulacki, and S.K. Driscoll, Microplastics as vectors for bioaccumulation of hydrophobic organic chemicals in the marine environment: A state-of-the-science review. *Environmental Toxicology and Chemistry*, 2016. 35(7): p. 1667-1676.

61. Shen, M., G. Zeng, Y. Zhang, X. Wen, B. Song, and W. Tang, Can biotechnology strategies effectively manage environmental (micro)plastics? *Sci. Total Environ.*, 2019. 697: p. 134200.

62. Schirmeister, C.G. and R. Mulhaupt, Closing the Carbon Loop in the Circular Plastics Economy. *Macromolecular Rapid Communications*, 2022. 43(13).

63. Syberg, K., M.B. Nielsen, L.P.W. Clausen, G. van Calster, A. van Wezel, C. Rochman, A.A. Koelmans, R. Cronin, S. Pahl, and S.F. Hansen, Regulation of plastic from a circular economy perspective. *Current Opinion in Green and Sustainable Chemistry*, 2021. 29.

64. Blum, N.U., M. Haupt, and C.R. Bening, Why "Circular" doesn't always mean "Sustainable". *Resources Conservation and Recycling*, 2020. 162.

65. Anastas, P.T. and J.C. Warner, Principles of green chemistry. *Green chemistry: Theory and practice*, 1998. 29.

66. Hellweg, S. and L. Milà i Canals, Emerging approaches, challenges and opportunities in life cycle assessment. *Science*, 2014. 344(6188): p. 1109.

67. Walker, S. and R. Rothman, Life cycle assessment of bio-based and fossil-based plastic: A review. *Journal of Cleaner Production*, 2020. 261.

68. Yates, M.R. and C.Y. Barlow, Life cycle assessments of biodegradable, commercial biopolymers-A critical review. *Resources Conservation and Recycling*, 2013. 78: p. 54-66.

69. Schneiderman, D.K. and M.A. Hillmyer, 50th Anniversary Perspective: There Is a Great Future in Sustainable Polymers. *Macromolecules (Washington, DC, U. S.)*, 2017. 50(10): p. 3733-3749.

70. Jones, D., Grinding waste away. *Nature*, 2000. 403(6772): p. 847-847.

71. World Economic Forum, E.M.F., McKinsey & Company,, *The new plastics economy: rethinking the future of plastics* 2016.

72. European Commission, *The European Green Deal* 2019.

73. European Bioplastics, *Bioplastics Market Development Update* in *European Bioplastics, Berlin, Germany*. 2019.

4

Evaluating Nationwide Supply Chain for Circularity of PET and Olefin Plastics

Tasmin Hossain[1]*, Damon S. Hartley[1], Utkarsh S. Chaudhari[2], David R. Shonnard[2], Anne T. Johnson[3] and Yingqian Lin[1]

[1]Operations Research and Analysis, Idaho National Laboratory, Idaho Falls, USA
[2]Department of Chemical Engineering, Michigan Technological University, Houghton, Michigan, USA
[3]Resource Recycling Systems, Ann Arbor, Michigan, USA

Abstract

PET (Polyethylene Terephthalate, #1) and olefin plastics including HDPE (High Density Polyethylene, #2), LDPE/LLDPE (Low Density/Linear Low-Density Polyethylene, #4) and PP (Polypropylene, #5) together comprise nearly 80% of the U.S. plastic market. Due to their high market share and extensive application in packaging these polymers have better potential for circularity than other polymer types. To understand the potential of future scenarios with higher recycling rates, supply chain scenarios for PET and olefin plastic packaging need to be analyzed with increased availability and collection of plastics. Currently there exists a knowledge gap to understand the circular supply chain on a national level. Performing a nationwide analysis introduces certain challenges such as variability in the plastic mix of the recycling stream in different municipalities, lack of collection of some types of plastics, regional differences in cost and estimating a national supply curve based on state-level access rate and participation rate. Currently, a limited number of Plastic Reclaimers recycle the nation's collected plastic involving transportation over long distances surpassing state boundaries. Modeling a nationwide scenario instead of regional/state based scenario will facilitate transportation beyond state boundaries for the development of a circular economy. A Mixed-integer Linear Programming model was developed, to identify the optimal location and capacity of the Material Recovery Facilities (MRFs) nationwide subject to maximizing the profit margin of the industrial entities within the model. We considered three different scenarios including the base case scenario (S1) collecting 2.27 million metric tons/year as well as two additional scenarios with available plastic supply of 1.8 times (S2) and 2.2 times (S3) of the base case scenario. The model identified 166, 275 and 319 counties as potential MRF locations for scenarios S1, S2 and S3 respectively. Compared to the existing number of counties having MRFs in the US, the model results indicated a reduction of 38% of MRFs for S1. For S2 and S3, the number of counties with MRFs increased 3% and 20% respectively compared to counties currently with MRFs. The average profit remains constant between $180-$181/ton regardless of the increased plastic collection.

Keywords: Recycling, circular economy, PET, plastics, supply chain, MRF

---

*Corresponding author: tasmin.hossain@inl.gov

Nabil Nasr (ed.) Technology Innovation for the Circular Economy: Recycling, Remanufacturing, Design, Systems Analysis and Logistics, (43–54) © 2024 Scrivener Publishing LLC

4.1 Introduction

Plastics are one of the fastest growing material streams in the U.S. As the focus on circular economy grows, there is a great interest in understanding the potential of a circular economy for plastics. With the advancement in mechanical and chemical plastic recycling technologies, the closed-loop circular economy holds a promising future. However, the current plastic economy mostly relies on the linear supply chain with "take-make-use-dispose" system incurring a loss of $80-$120 billion annual for the global plastics that ends up in landfill [1, 2]. About 250 million metric tons (mt) of postconsumer plastic waste were generated in 2018, out of which only 50 million mt have been recycled, increasing the risk of negative environmental impact due to virgin plastic generation from fossil fuel and plastic disposal in land and sea [3]. Estimation suggests that there are around 150 million tons of plastic in the ocean today [1]. To reduce the fossil fuel extraction as well as to mitigate the environmental impact of plastic disposal, the circular economy for plastics needs to be incorporated in the current plastic supply chain, thereby increasing the recycling rate of postconsumer plastics while reducing the production of virgin plastics.

In 2018, the US generated 28 million tons of waste PET (Polyethylene Terephthalate, #1) and olefin plastics including HDPE (High Density Polyethylene, #2), LDPE/LLDPE (Low Density/Linear Low-Density Polyethylene, #4) and PP (Polypropylene, #5) comprising 80% of the total plastic waste in municipal solid waste [4]. These resins have better potential for circularity compared to other polymer types. However, the U.S. currently has a plastics recycling rate with national average of only 8.7% [4]. To achieve higher recycling rates, future recovered supply chain scenarios for PET and olefin plastics need to be analyzed with increased collection as well as recycling rates. Utilizing the circular economy for the plastic packaging could reflect earnings of $2-$4 billion per year in the U.S. [5]. With the growing consumption of plastic, the unrecovered plastic will continue to grow as well. Demand in plastic packaging could increase 35% by 2040 [5]. However, increasing the current low recycling rate will require solving numerous challenges related to collection and sorting mechanisms, access to recycling, collection and processing cost, consumer behavior, supply and demand coordination as well as competing price of virgin plastic [5–7]. Designing a supply chain for plastic recycling involves allocating postconsumer recyclables to Material Recovery Facilities (MRF) for sorting and processing and transfer of the processed material from MRFs to Reclaimer Facilities for recycling. Determining the optimal facility size and location of MRFs and Reclaimers while managing the supply-demand coordination is imperative to handle the logistical challenges of a plastic circular economy. Several studies have focused on material flow analysis (MFA), life cycle assessment (LCA), and scenario analysis of the circular economy to understand the material flow and GHG emission of the process [8–12]. Studies have also addressed current challenges such as policy regulations, collection and sorting, production as well as material recovery rates in Europe [13–15]. There have been very few studies in the US involving the plastic supply chain management. Heller *et al.* [16] analyzed the plastic material supply for different resins in the US concluding that packaging was the major consumption. Thakker, & Bakshi [17] analyzed circular economy value chain for paper and plastic carrier bags in the US. Moreover, studies involving the facility location and optimization for cost and profit analysis in the US have rarely been conducted with only a few studies focusing on analyzing the current infrastructure location and capacity [6, 10]. The novelty of our study lies in identifying the research gap in optimal economic conditions of the plastic supply chain in the US considering national scenario for plastic collection, sorting and transportation. The objective is to develop an optimization model i.e., a Mixed-integer Linear Programming model to identify the optimal location and capacity of the Material Recovery Facilities (MRF) nationwide that would benefit the collection of the plastic waste while maximizing the profit margin of the stakeholders for a sustainable circular economy. Introducing MILP instead of integer linear programming (LP) facilitates the use of both integer and continuous decision variables in the model increasing the computational efficiencies. We considered different scenarios that differ in the national average recycling rate estimating total collected plastic nationwide, as well as cost and profit analysis to supply PET and olefin plastics to the MRFs nationwide.

4.2 Methods

4.2.1 Model Input

This study considers a county-based scenario for developing plastic supply curve data for curbside recycling. It is assumed that each county is a collection point and the available plastic is located at the centroid of each county. According to the most updated MSW generation data from EPA, 86.68 kg per capita/year of PET, LDPE, HDPE and PP are generated in the US [4]. Using the 2018 per capita generation and the US population for 2022, the county-based plastic generation data has been developed for 2022 (Figure 4.1). The state collection rates have been estimated using the Eunomia report for 50 States of Recycling [18]. Estimated state recycling rates from Eunomia has been adjusted to reflect the reduced recycling rates of United States Environmental Protection Agency (USEPA) for PET, LDPE, HDPE and PP [4]. The estimated state collection rates for this study varied between 0.5% to 28.69% where 16 states had a collection rate higher than 11%. The county-based supply

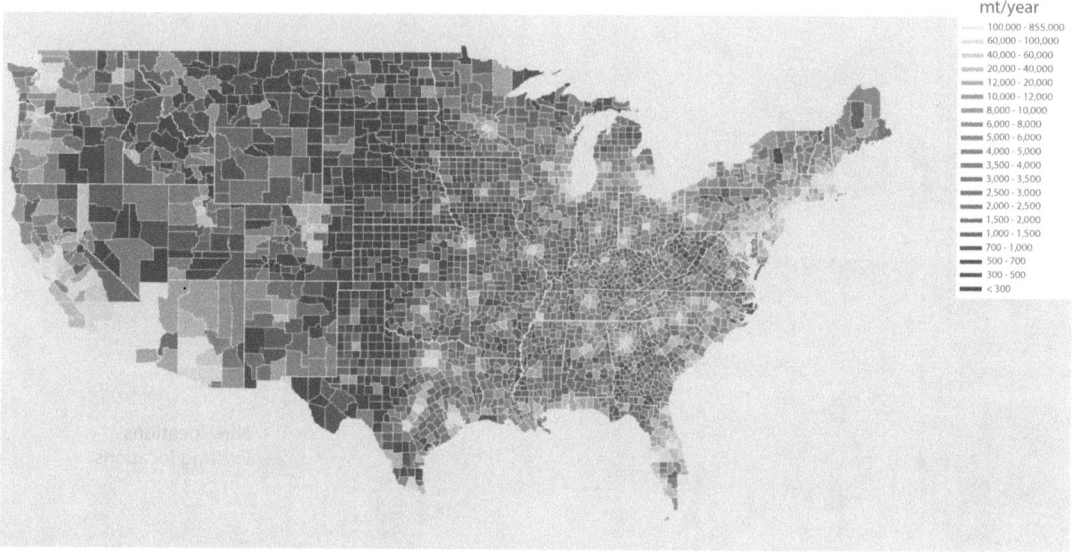

Figure 4.1 Plastic (PET, HDPE, LDPE, PP) generated per county in 2022.

(plastic available for collection) for the model was calculated using the state collection rates and county-based plastic generation.

Currently, there are 355 residential MRFs located in the US in 266 counties [19]. The candidate MRF locations were selected by combining the existing MRF locations with a random set of points on a 100-mile grid. A total of 530 counties have been considered as MRF candidates for the model. The centroid of each county has been identified as a unique candidate location (Figure 4.2). The capacity ranges for the MRFs have been considered to vary from 20,000 up to 220,000 mt/year with 10,000 mt capacity increment. MRFs having a capacity less than 90,000 mt/year were considered smaller MRF working on a single shift while bigger MRFs were assumed to be working on a double shift. The collection and transportation cost for picking up the single stream (SS) recyclables using a compactor truck has been estimated using a linear equation having a distance-fixed cost of $10.792/mt and a distance-variable cost of $0.7193/mt-mile (Eq. 4.1) [20].

$$y = 0.7193x + 10.792 \qquad (4.1)$$

Table 4.1 lists all the data assumptions for the MRF processing required as input to the model. According to the Resource Recycling Systems (RRS) database, the cost to build a 78,000-88,000 mt/year capacity MRF can vary from $40-$50/mt. Depending on the RRS data, the capital costs for varying sizes of MRFs were scaled using the rule of six-tenth [21]. The estimated capital cost can be as high as $92/mt for smaller MRFs with 20,000 mt/year capacity while they can be $35/mt for bigger MRFs with 220,000 mt/year capacity.

4.2.2 Model Formulation

The Mixed-integer Linear Programming (MILP) model presented in this study identifies the optimal location and capacity of the MRFs to maximize the profit while collecting the plastic waste for sorting and prepare them for recycling for the entire system. The decision

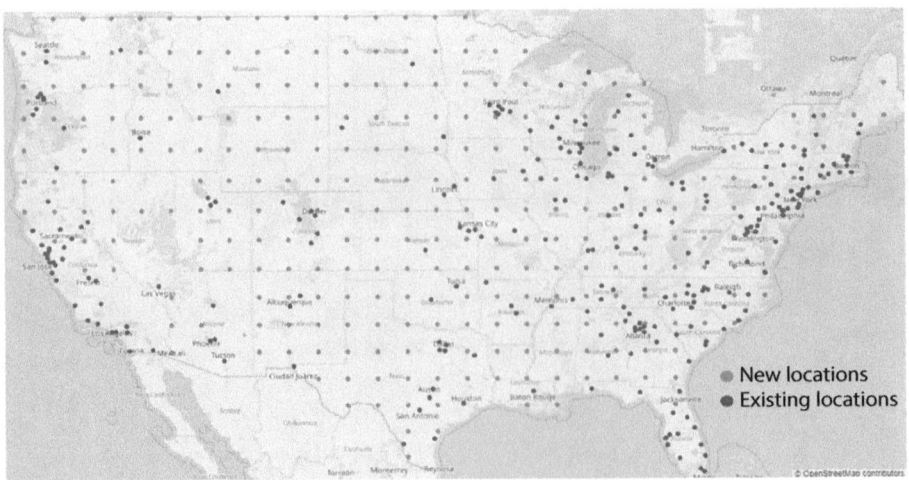

Figure 4.2 Candidate locations for the MRF.

Table 4.1 MRF data assumptions.

Capacity Ranges	20,000 – 220,000 mt/year	
Single shift hours	8 hours/day	
Double shift hours	16 hours/day	
Plastic mix in recyclables	PET	2.8%
	Natural HDPE	0.8%
	Colored HDPE	0.7%
	LDPE, PP	0.3%
Plastic bale blended price	$431.46/mt	
Operating cost/Processing cost for plastics	$135/mt	
Tipping fees	$90/mt	
Plastic contamination	30%	
Plastic at MRF (including residue)	10% of the MRF processing capacity	
MRF capacity utilization	80%	
Profit sharing for Bale Revenue	85%-15% between MRF and Municipalities	

network in Figure 4.3 explains the flow of plastics in the supply chain between different stages. Each collection point (i) is a unique county assuming the total available plastic is aggregated at the centroid of that county. The PET and olefin plastics are collected using garbage trucks as single stream (SS) recyclables and then transferred to the MRF located at county centroids (j). For the single stream recyclables, a maximum collection distance of 100 miles is assumed in the model. The collected recyclables are assumed to go through sorting to create relevant plastic commodity bales which are then ready to be shipped to reclaimers for recycling. Table 4.2 presents the data sets, parameters and decision variables used in the MILP formulation.

The profit at the MRF is calculated using the revenue and costs related to the MRF (Eq. 4.2). The revenue comes from the sale of commodity plastic bales and the processing fees charged to customers. Costs include the capital cost to build the MRF, operating costs to process plastic, collection or transportation costs and cost of residue disposal.

$Profit_{MRF}$ = (*Revenue from Bales + Tipping fees charged*)
  – Capital and operating cost at MRF – Transportation from household
    to MRF – Residue disposal cost (4.2)

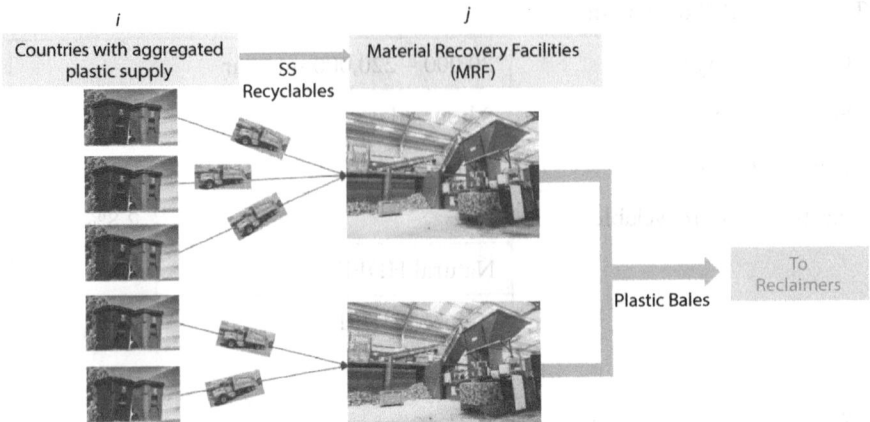

Figure 4.3 Decision network of the supply chain.

The objective function of the model is to maximize the profit at the MRF while meeting the required set of constraints (Table 4.3),

$$\max \left(\sum_{i \in I} \sum_{j \in J} X_{ij} * ((1 - C_M) * S_b * 0.85 + T) \right) - \left(\sum_{j \in J} \sum_{c \in C} L_{jc} * C_c * P_m * (C_{cm} + C_{om}) \right)$$

$$- \sum_{i \in I} \sum_{j \in J} X_{ij} * t_{ij} - \sum_{i \in I} \sum_{j \in J} X_{ij} * C_M * D \qquad (4.3)$$

Where,

$$t_{ij} = d_{ij} * C_v + C_f \qquad (4.4)$$

Constraint (1) ensures that the amount of SS recyclables collected from each location is not more than the available plastic at that location. Constraint (2) sets a minimum utilization of the capacity of the MRF facilities meaning if a facility is built with capacity C_c, a minimum $U\%$ of that capacity needs to be utilized. Constraint (3) sets the maximum limit that can be shipped to a MRF facility which will be bound by the capacity. Constraint (4) ensures that only one MRF is built in each county and it will represent the aggregated capacity for that county. Constraint (5) is the non-negativity constraint and constraint (6) is the binary constraint on the decision variables.

4.3   Results and Discussion

Three different scenario runs were performed including S1, S2 and S3. Scenario S1 is the base case scenario collecting 2.27 million mt/year. S2 collects 1.8 times the base case

Table 4.2 Data sets, parameters, and decision variables.

Sets	
I	Set of counties for SS recyclables pickup
J	Set of candidate locations for the MRF
$C = \{C_c, j \text{ in } J\}$	Set of capacities for MRF
$DP = \{d_{ij}, i \text{ in } I, j \text{ in } J\}$	Set of distances between location i and location j within a 100-mile radius
$S = \{S_i, i \text{ in } I\}$	Set of available SS recyclables at location i
$Tm = \{t_{ij}, i \text{ in } I, j \text{ in } J\}$	Transportation cost of SS recyclables between i and j using truck
Parameters	
C_{cm}	Capital cost/ton at MRF
C_{om}	Operating cost/ton at MRF
C_f	Truck fixed cost/ton for SS recyclables
C_v	Truck variable cost/ton-mile for SS recyclables
S_b	Market/selling price of plastic bales
P_c	Percent of SS recyclables collected using curbside recycling
P_m	Percent of plastics at MRF
U	Utilization for MRFs
D	Residue disposal cost at MRF
C_M	Contamination rate at MRF
T	Tipping fees charged by MRF to customers
F_M	Average annual profit per ton at MRF
Decision Variables	
X_{ij}	Weight of SS recyclables collected from location i and shipped to j
L_{jc}	1 if MRF is built at location j with capacity c; 0 otherwise

scenario and S3 collects 2.2 times the base case scenario. The increased collection assumes the future scenarios where postconsumer plastic availability increases with increased demand. The MILP model was developed using Python and CPLEX solver for an error gap of less than 3%. The results for the different scenarios in this study are presented in Table 4.4. The base case scenario identified 166 counties for MRF locations which is around 38% less than the current number of counties having MRFs in the US [19]. Compared to the base case scenario, total plastic collected for processing increased 78% and 118% for S2 and S3 respectively. Whereas the number of counties having MRFs increased only 3% and 20%

Table 4.3 Model constraints.

Constrain no.	Mathematical formulation
1	$\sum_{j \in J} X_{ij} \leq P_c * S_i; \forall\, i \in I$
2	$\sum_{i \in I} X_{ij} \geq U * (\sum_{c=C}(L_{jc} * C_c * P_m)); \forall\, j \in J$
3	$\sum_{i \in I} X_{ij} \leq (\sum_{c=C}(L_{jc} * C_c * P_m)); \forall\, j \in J$
4	$\sum_{c \in C} L_{jc} \leq 1; \forall\, j \in J$
5	$X_{ij} > 0; \forall\, i \in I, j \in J$
6	$L_{jc} \in \{1,0\}; \forall\, j \in J, c \in C$

Table 4.4 Scenario results.

Scenario	Plastic collected for processing (million mt/year)	Number of MRF	Average annual profit ($/ton)
S1	2.27	166	180.73
S2	4.04	275	180.42
S3	4.95	319	181.70

compared to the current MRF county numbers [19]. Collecting more plastic material with a lower number of MRFs indicates bigger size facilities to realize the benefit of economies of scale. As the availability of plastic increases, the total number of MRFs identified by the model also increases to allocate more material. However, the average profit remains constant between $180-$181/ton regardless of the increased plastic collection.

Figure 4.4 presents the nationwide location and capacity of the MRFs identified by the model for the three different scenarios. The radius of the circle indicates the capacity of the MRF while the color of the circle indicates whether the facility is in a county with an existing MRF facility or at a new location. As evident from the findings of this study, more than 65% of the MRF locations selected by the model is in a county with an existing MRF reducing the need to build a new facility. California, New York, Pennsylvania, Michigan, Texas, New Jersey, Washington, Wisconsin, Massachusetts, Maryland and Florida contributed to more than 60% of the total plastic collected for all three scenarios.

To analyze the capacity of the MRFs selected by the model in different scenarios, the capacities have been divided into five bins between 20,000 to 220,000 mt/year with each bin having a range of 40,000 mt/year (Figure 4.5). Around 39%, 52% and 58% of the MRFs have high capacity beyond 180,000 mt/year for S1, S2 and S3. In all three scenarios, MRFs with capacities of 60,000 – 180,000 mt/year were less than 30% of the total capacity. MRFs having capacities less than 60,000 mt/year and capacities greater than 180,000 mt/year contributed to 80% of the total capacity for the scenarios with increased recycling rates. As the recycling rate increases for these plastics in the future, high capacity MRFs will be more economical and sort more plastic material while smaller MRFs can access areas having lower collection rates.

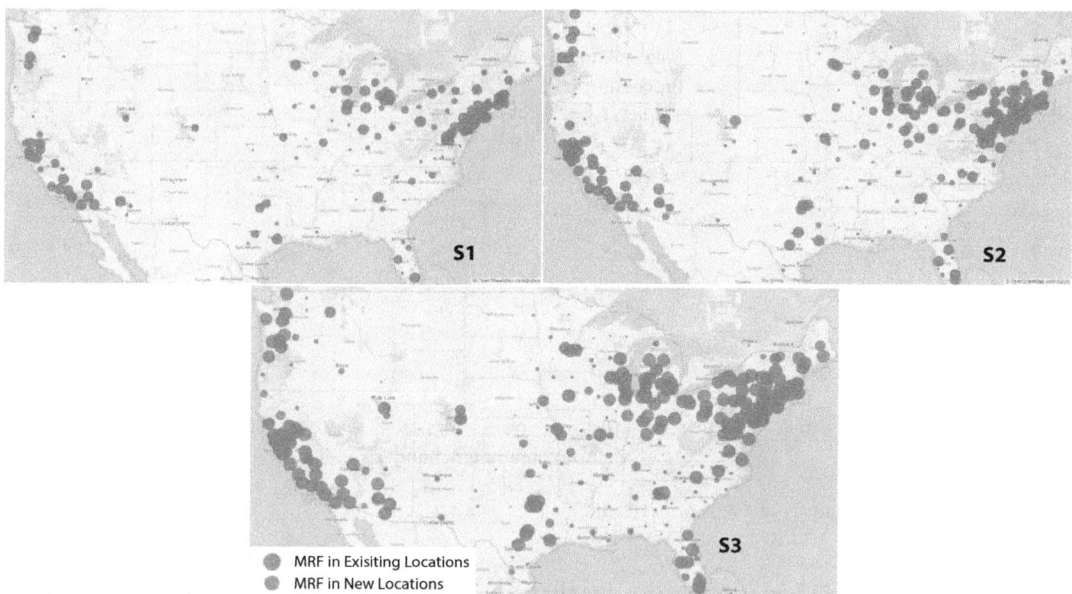

Figure 4.4 MRF locations in the US selected by the model for the three scenarios: S1, S2 and S3.

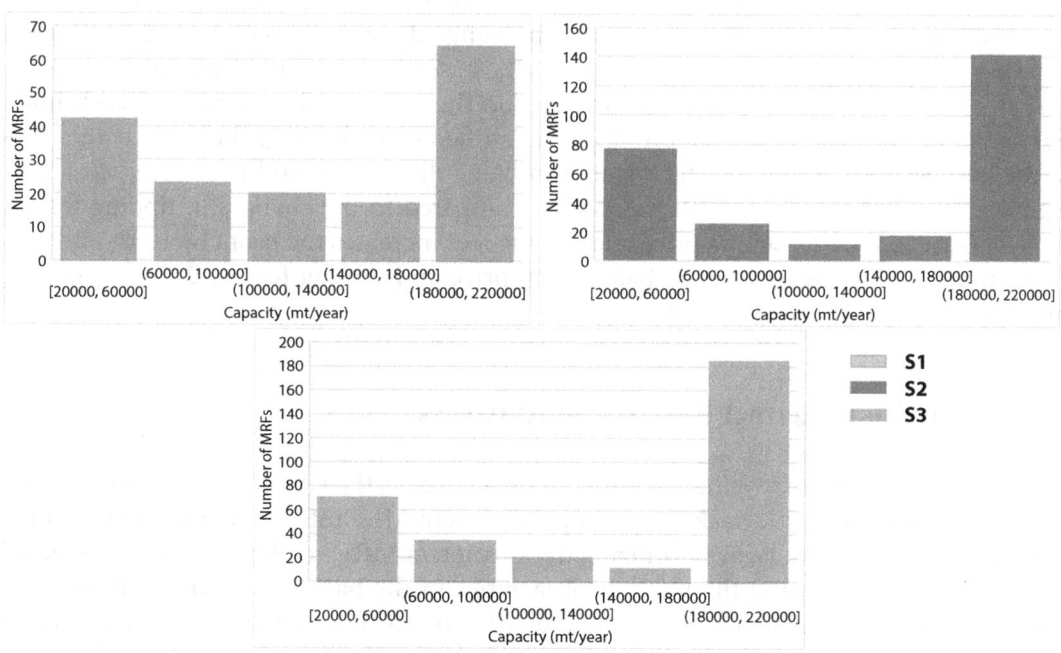

Figure 4.5 Capacities of the MRF selected by the model for the three scenarios: S1, S2 and S3.

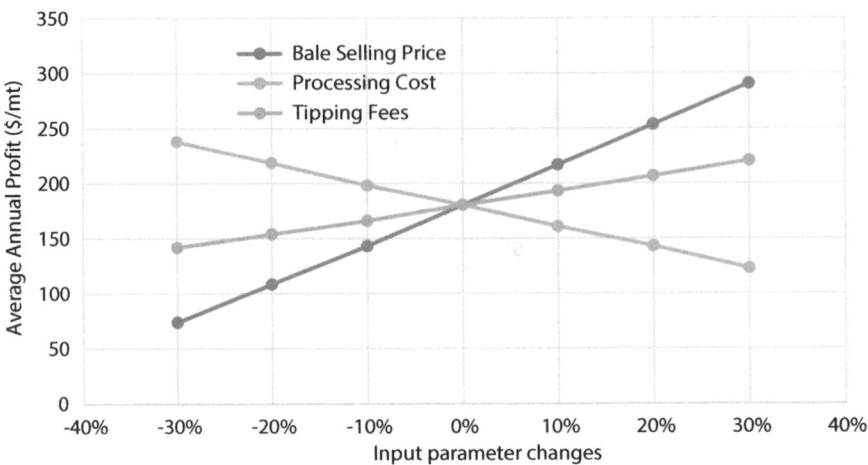

Figure 4.6 Sensitivity analysis of the model results.

The economies of scale benefits are greater than the additional cost of transport to the larger scale MRFs.

To understand the impact of different input parameters on the model results, sensitivity analysis was conducted for different input parameters such as bale selling price, processing cost and tipping fees as these parameters can change depending on supply chain market conditions. The S1 scenario was considered as the base case and the parameter values were varied ±30% from their base values (Table 4.1). The effect of the changes of each input parameter on the average annual profit was quantified while keeping the other parameters at the base level (Figure 4.6). The changes in bale selling price and processing cost have significant impact on the annual profitability compared to the changes in tipping fees. A 30% increase in the bale selling price and tipping fees increases the profit by 100% and 22% respectively. While decreasing the bale selling price and tipping fees by 30% reduces the profit by 60% and 21% respectively.

4.4   Conclusions and Recommendations

With a national average recycling rate of 8.7% for all plastic, the US collects around 1.96 million mt of primarily plastic packaging using the current MRF infrastructure [4]. According to the findings of this study, by optimizing the location of MRFs, collection can be increased by 14% without increasing the number of facilities in the base case. If the availability of postconsumer plastic is increased to 15% and 20%, collection can increase 100% and 150% respectively compared to the current plastic collection. High capacity MRFs will play a significant role in future plastic supply chain scenarios as they benefit from economies of scale with increased recycling rates. With the collection of more material, the total profit for the stakeholders will increase. However, the average profit per ton will not increase beyond a certain limit, $181/ton for the scenarios considered in this study. This is due to the high collection and processing cost of the MRF regardless of sorting more material. Although increased recycling rates will facilitate bigger MRFs which are more economical, the reduction in the average capital cost is not enough to offset the high logistics cost. Further work is needed to model a circular plastic supply chain, incorporating the supply of plastic bales to reclaimers for processing into recycled plastic materials. Future work will include finding the optimal location and capacity of both the MRF and Reclaimers with using truck and rail transportation for plastic bales. The results of the model are highly dependent on the input parameters as well as the available plastic for collection. Current modeling framework does not consider the impact of uncertain disruptions within the supply chain. Updating the current model to a stochastic framework can be the appropriate method to include the buffer for uncertainties of the supply chain.

Acknowledgements

This work was funded by the REMADE institute and the U.S. Department of Energy's Office of Energy Efficiency and Renewable Energy (EERE) under the Advanced Manufacturing Office Award Number DE-EE0007897. The views expressed in this publication do not necessarily represent the views of the DOE or the U.S. Government. The U.S. Government retains and the publisher, by accepting the article for publication, acknowledges that the U.S. Government retains a nonexclusive, paid-up, irrevocable, worldwide license to publish or reproduce the published form of this work, or allow others to do so, for U.S. Government purposes. We acknowledge the help from people from RRS Inc. including Michael Timpane, Kerry Sandford and Sean Duffy for their help with the data collection. We would also like to thank Alejandra Peralta, Robert Handler, Barbara Reck, Vicki S. Thompson, Daniel Kulas and David Atkins for their valuable insights and comments towards the development of this study.

References

1. Ellen MacArthur Foundation (EMF), The New Plastics Economy: Rethinking the Future of Plastics & Catalysing action. https://ellenmacarthurfoundation.org/the-new-plastics-economy-rethinking-the-future-of-plastics-and-catalysing, 2017.
2. Chaudhari, U. S., Johnson, A. T., Reck, B. K., Handler, R. M., Thompson, V. S., Hartley, D. S., ... & Shonnard, D., Material Flow Analysis and Life Cycle Assessment of Polyethylene Terephthalate and Polyolefin Plastics Supply Chains in the United States. *ACS Sustainable Chemistry & Engineering*, 10(39), 13145-13155, 2022.
3. Lindner, C., & Beylage, H., Global Plastics Flow 2018. Converstion Market & Strategy GmbH, 2020.
4. USEPA, Advancing Sustainable Materials Management: 2018 Tables and Figures. https://www.epa.gov/sites/default/files/2021-01/documents/2018_tables_and_figures_dec_2020_fnl_508.pdf, December 2020.
5. Hundertmark, T., Prieto, M., Ryba, A., Simons, T. J., & Wallach, J., Accelerating plastic recovery in the United States. McKinsey & Company, December 2019.
6. Waste Management (WM), WM Report on Recycling. https://sustainability.wm.com/downloads/WM_Report_on_Recycling.pdf, March 2020
7. Northeast Recycling Council (NERC), Report in Blended Commodity Values – EPA Regions 1,2 and 3. https://nerc.org/documents/NERC, September 2020.

8. Starreveld, P. F., & Van Ierland, E. C., Recycling of plastics: A materials balance optimisation model. *Environmental and Resource Economics*, 4(3), 251-264, 1994.

9. Westin, A. L., Kalmykova, Y., Rosado, L., Oliveira, F., Laurenti, R., & Rydberg, T., Combining material flow analysis with life cycle assessment to identify environmental hotspots of urban consumption. *Journal of Cleaner Production*, 226, 526-539, 2019.

10. Closed Loop Partners (CLP). US and Canada Recycling Infrastructure and Plastic Waste Map. https://www.closedlooppartners.com/research/us-and-canada-recycling-infrastructure-and-plastic-waste-map/?preview=true, November 2020.

11. Lonca, G., Lesage, P., Majeau-Bettez, G., Bernard, S., & Margni, M., Assessing scaling effects of circular economy strategies: A case study on plastic bottle closed-loop recycling in the USA PET market. *Resources, Conservation and Recycling*, 162, 105013, 2020.

12. Di, J., Reck, B. K., Miatto, A., & Graedel, T. E., United States plastics: Large flows, short lifetimes, and negligible recycling. *Resources, Conservation and Recycling*, 167, 105440, 2021

13. Salmenperä, H. (2021). Different pathways to a recycling society–Comparison of the transitions in Austria, Sweden and Finland. *Journal of Cleaner Production*, 292, 125986.

14. Mueller, W. (2013). The effectiveness of recycling policy options: waste diversion or just diversions?. *Waste Management, 33*(3), 508-518.

15. Paletta, A., Leal Filho, W., Balogun, A. L., Foschi, E., & Bonoli, A. (2019). Barriers and challenges to plastics valorisation in the context of a circular economy: Case studies from Italy. *Journal of Cleaner Production*, 241, 118149.

16. Heller, M. C., Mazor, M. H., & Keoleian, G. A. (2020). Plastics in the US: toward a material flow characterization of production, markets and end of life. *Environmental Research Letters*, 15(9), 094034.

17. Thakker, V., & Bakshi, B. R. (2021). Designing Value Chains of Plastic and Paper Carrier Bags for a Sustainable and Circular Economy. *ACS Sustainable Chemistry & Engineering*, 9(49), 16687-16698.

18. Eunomia, The 50 States of Recycling. https://www.ball.com/50-States-of-Recycling-Eunomia-Report, 2021.

19. The Recycling Partnership, Map of Commingled Residential MRFs in the U.S. Residential MRFs - The Recycling Partnership, 2023.

20. Lin Y., Severson M., Nguyen R., Johnson A., King C., Coddington B., Hu H., & Madden B., Economic and environmental feasibility of recycling flexible plastic packaging from single stream collection. *Resources, Conservation and Recycling*, 2022. Under revision.

21. Guthrie, K. M., Capital cost estimation. *Chem. Eng.*, 24, 114-142, 1969.

5

NextCycle: Building Robust Circular Economies Through Partnership and Innovation

Juri Freeman

Resource Recycling Systems Inc., RRS, Boulder, Colorado, USA

Abstract

NextCycle is an innovative public-private partnership model operating across several US states that successfully advances local and regional circular economies. The authors outline the NextCycle design and value proposition and describe how the project has grown from a regional initiative to a multimillion-dollar program with scores of partners collaborating to support projects that drive transformation to circularity. The article share the lessons learned in identifying gaps, developing the partnerships, and managing the challenges tracks, helping readers and conference attendees advance circularity in their regions.

NextCycle is designed to accelerate projects that will improve a state or regions' circular economy to an investment- ready status. NextCycle partners with states, brands, investors and others to recruit and develop a pipeline of high quality, vetted, investments with the potential to create local and regional circular economies. The project leverages public sector grants with private sector investments to increase the impacts of the investments. For example, NextCycle Michigan partners with nearly 100 different investors and brands to provide matching grant funds for teams applying to grants, and Closed Loop Partners has committed millions in investments in the states that have funded NextCycle programs.

NextCycle is centered around 'Innovation Challenges'. Each challenge is specifically designed to recruit and support teams as they develop investment-ready initiatives to fill 'gaps' in the regional circular economy. Participants can be entrepreneurs or start-ups, small businesses, established corporations, universities, non-profits, or a collaboration of entities. By providing technical and business support and nurturing projects that incorporate waste prevention, repair, reuse, recycling, and/or composting models, the program helps develop equitable local economies while reducing waste, keeping materials in use longer and regenerating natural systems. To date, NextCycle has helped nearly 150 teams and tracked over $700M of direct and leveraged investments in circularity projects.

Keywords: Circular economy, recycling, end markets, accelerator, funding

Email: jfreeman@recycle.com

Nabil Nasr (ed.) Technology Innovation for the Circular Economy: Recycling, Remanufacturing, Design, Systems Analysis and Logistics, (55–62) © 2024 Scrivener Publishing LLC

5.1   Introduction

NextCycle is a public private partnership initiative designed to accelerate projects that improve a state or regions' circular economy. NextCycle partners with states, brands, investors and others to recruit and develop a pipeline of high quality, vetted, investments with the potential to create local and regional circularity. The project leverages public sector grants with private sector investments to increase the impacts of the investments. Since its inception in 2018, the project has grown from a regional initiative to a multimillion-dollar program with scores of partners collaborating to support projects that drive transformation to circularity across the US.

The early successes of the initiative demonstrate that the concepts and ideas behind NextCycle can be replicated across North America and beyond to catalyze circularity. NextCycle fills the need for entities in the circular economy by helping to establish value chain partnerships and attract impact investment funding that does not necessarily require a high multiple return on investment. For funding partners, NextCycle identifies and supports investable businesses that fill the gaps in a circular economy. For participants, NextCycle provides the technical resources and pathways to funding to grow their businesses. The concepts behind NextCycle can be modified to meet city, regional, or state needs, and can be adopted by local agencies to meet their material management goals.

5.2   The NextCycle Concept

5.2.1   How It Started

NextCycle originated as a collaborative effort between the Colorado Department of Public Health and Environment (CDPHE) and Resource Recycling Systems Inc. (RRS, www.recycle.com). More than 15 years ago, the Colorado legislature passed the Recycling Resources Economic Opportunity (RREO) Act of 2007. The act established a grant program within the Colorado Department of Public Health and Environment with the statutory goal of providing 'funding that promotes economic development through the management of materials that would otherwise be landfilled'. RREO is funded through a $0.14/cubic yard surcharge on landfill tipping fees and provides approximately $3.0M to $4.0M in grant funding annually[1].

Over the next decade, RREO built a strong record of successfully supporting capital investments in material recovery infrastructure, collection, and engagement. Despite these investment, the state continued to struggle to significantly increase diversion rates, create jobs, or grow regional circularity. In 2018, the state's overall MSW recycling rate hovered just over 17% which was well shy of the goal of 45% by 2036. In trying to understand why this was so, research conducted by RRS indicated that the state had done well to support parts of the recovery ecosystem, but there remained significant gaps in circularity. Specifically, analysis found that additional or expanded end markets, both local and regional, were needed to create truly circular systems in Colorado. NextCycle was proposed as a way to fill the gaps in the circular supply chain.

---

[1]https://cdphe.colorado.gov/recycling-grants

Through NextCycle Colorado, CDPHE was able to recruit teams that could grow end markets in the state, prepare the teams for future investments, and leverage the state's grant dollars with private sector investments to achieve success. Since its inception, NextCycle Colorado alone has leveraged $2.5M of state grants investments in participating businesses and non-profits to enable those entities to secure more than $70M in tracked project financing from outside sources. Colorado projects range from a non-profit exclusively hiring hard-to-employ people to disassemble and recycle mattresses to a company commercializing a novel technology to recover high value composites.

5.2.2   The NextCycle Model

The overall concept of NextCycle is simple. It starts with the identification of system gaps followed by the recruitment of teams that can fill the gaps. Selected teams are provided with direct consulting and mentorship through the NextCycle accelerator program, also known as 'challenge tracks'. Teams that have the highest potential to fill the identified gaps receive investments in the form of state grants and / or private sector funding including debt financing and equity. Teams are also connected with the project partners to foster long term growth. The replicable elements of the basic NextCycle concepts are provided below.

> **Data Based Gap Analysis**: NextCycle starts with a data driven assessment of the state or regional gaps to circularity. The analysis begins at the point of material manufacturing or import, and looks at the flows of the material through the recovery value chain, ending with reuse, upcycling, remanufacturing, or as an input into a new material. The NextCycle gap analysis assesses 1) where recycled inputs could be substituted for virgin materials in local manufacturing, 2) which materials cannot easily be sold as commodities in the regional marketplace, and 3) the existing collection and processing infrastructure. In some locations, the gap analyses include a focus on local job creation and equity to identify areas for investment that can support state or local environmental justice goals. By keeping valuable commodities in a community, localities can create job opportunities resulting in local economic benefits. The data collected through the gap analysis is used to identify the key intervention points to improve regional and state circularity; these include waste reduction and reuse, as well as material recovery, recycling, composting, and remanufacturing.
>
> **Marketing and Recruitment**: Integral to the success of the NextCycle model is building a strong pipeline of potentially investable projects. This is driven through extensive marketing and team recruitment. Teams can be made up of businesses, entrepreneurs, universities, non-profit entities, local governments and other entities. A team can consist of an individual, a single organization or may include any number of cross- sector entities and the definition of a team can be adjusted depending on the needs and focuses of the program. In each of the existing program models, recruitment extends beyond the boarders of the region, and is designed to pull in prospective projects and teams from around the globe. The caveat, of course, is that while a team may be headquartered out of region, any potential projects must be based in the

targeted region. This allows NextCycle to provide economic and job creation benefits to the funding state.

Project Funding: In the current NextCycle models, project funding is provided through state agencies. The states partner with Resource Recycling Systems (RRS) to design and manage the program. While two of the three models are linked to state grant funding (Colorado and Michigan), partnerships with other private sector funders for teams means that states do not necessarily need to provide grant funding to teams. However, one of the distinguishing concepts of the NextCycle program is the ability of state grants to focus on supporting people and planet first, followed by a reduced emphasis on profits. For a NextCycle team to succeed revenues must exceed expenses, thus allowing the project to maintain financial viability and be sustainable economically. However, unlike other accelerator programs, by partnering with state's NextCycle does not require returns for investors that may not be viable for either larger scale infrastructure such as organics composting or small scale sustainability projects like a non-profit sewing cooperative that focus on people and planet, not profit. Additionally, as a state funded program, NextCycle does not take an equity stake, require a convertible note, or otherwise charge teams that wish to participate. This distinguishes the NextCycle model from others in the space, increasing equitable access to resources to support wider diversity of teams and projects.

Renew Partnership Portal: One of novel concepts that has bolstered the growth of NextCycle, especially in Michigan, is the incorporation of the partnership portal. The portal creates a pathway for outside investors, companies, and others with an interest in circularity to become actively involved in the project. For example, Closed Loop Partners – New York based investment firm comprised of venture capital, growth equity, private equity, and project finance as well as an innovation center focused on building the circular economy – has pledged up to $5.0M in flexible financing per project in each of the active NextCycle states. Each state, through its NextCycle initiative, will identify projects that develop recovery infrastructure solutions for post-consumer recyclable materials with a focus on polyethylene terephthalate (PET) and aluminum, optimize innovative collection systems for polyethylene (PE) and polypropylene (PP), and divert materials from landfills back into the supply chain. Over the next three years of partnership, Closed Loop Partners will closely collaborate with the various NextCycle initiatives, identifying investable opportunities that advance collective circularity goals. In Michigan alone, NextCycle includes 122 active partners and has supported 73 partners projects. Project partners have contributed over $90M in planned or in progress investments in Michigan that support circularity.

Challenge Tracks: Each of the NextCycle initiatives includes various challenge tracks. Each of the challenge tracks are based specifically on the gaps and opportunities identified through data analytics. The opportunities identified through the gap analysis form the basis of the challenge track aims. For example, NextCycle Michigan includes a challenge track focused on expanding use of recycled content in road and path construction and NextCycle

Washington includes an upstream challenge track that recruits teams and projects that emphasize waste prevention, reuse, and repair. The challenge track is the conduit for teams in the program to access technical support ranging from economic analysis, pitch development, marketing assistance, and business planning to partner networking and feedstock analysis. Technical support varies by team and may even include activities such as designing a logo or website, conducted laboratory testing on materials, or activities as basic as determining if a team should be set up as an LLC, partnership, or not for profit entity. One-on-one technical support is provided to teams throughout the challenge track. In addition, each track also provides a multi-day bootcamp event for all participating teams. Mentors, industry experts, local governmental partners and others provide interactive workshop sessions at the bootcamps on a range of subjects including but not limited to funding, grant writing, business models, transportation and logistics, partnerships, permitting, and marketing.

Technical Advisory Committee: NextCycle work closely with a cohort of experts uniquely situated to provide mentorship and support to teams in the program. The group of experts, referred to as the Technical Advisory Committee (TAC), help to hone the recruitment process, review team applications, and choose the teams that are accepted into the program. Once teams are accepted, the advisors continue to remain engaged by meeting one-on-one with teams, participating in the bootcamp, and connecting eligible teams with additional project partners.

Pitch Competition: Each of the challenge tracks culminates in a judged pitch competition. Depending on the challenge track and the state, the competition may be closed-door or open to the public. During the competition, teams have a limited amount of time to clearly articulate their value proposition, identify the market need and opportunity, and verbalize an ask. Depending on the team, the ask may range from a state grants or equity investment to project partners or support in securing a site location for their anticipated project. Additionally, depending on the track and the location, the winning team is awarded a 'no strings' monetary ward to help advance their project.

5.2.3 Project Outcomes

Since its inception in Colorado in 2018, NextCycle has expanded to include programs in Washington and Michigan. Each of the initiatives is designed to specifically address the gaps in state circularity, and thus, while there are certainly overlapping activities and resources, the programs differ by location. Across the three active states NextCycle participants have secured over $8M in state grant funding. The state grant investments have been leveraged through the program's partners, networking, bootcamps, and other NextCycle support activities to generate over $750M in tracked project investments. Table 5.1 summarizes the three program offerings and their outcomes.

Table 5.2 below presents examples of past NextCycle challenge tracks and the teams supported through the program. The table is not an exhaustive list of all tracks and participants

Table 5.1 NextCycle program offerings and outcomes.

State	Colorado	Michigan	Washington
Year Established	2018	2020	2022
Funder	Colorado Department of Public Health and Environment (CDPHE)	Michigan Department of Environment, Great Lakes, and Energy (EGLE)	King County's Solid Waste Division
Website	www.nextcyclecolorado.com	www.nextcyclemichigan.com	www.nextcyclewashington.com
Initiative Goals and Focus	Solid Waste Diversion goals of 35% by 2026 and 45% by 2036 End Market Development	Recycling rate goals of 30% by 2025, then on to 45% Expanding circularity across supply chain; DEI/EJ; climate stability	Expand the state's circular economy. Meet zero waste goals – representing approximately 70% landfill diversion. Create a more equitable investment landscape
Circularity Modes	Entities using recovered commodities as inputs in their manufacturing process and reuse	Reuse, recycling and composting access, collection, processing and end market development, anaerobic digestion, waste reduction	Waste reduction, reuse, recycling and organic recovery access, collection, processing and end market development
Associated Grants	Recycling Resources Economic Opportunity (RREO) program. Front Range Waste Diversion (FRWD)	NextCycle MI MICROS, Renew Michigan Infrastructure, Market Development, partner investment	SEED Grants, Industrial Symbiosis Grants, Re+ Circular Economy Grants
Number of Ongoing and Complete Projects	34 Teams Accepted	59 Teams Accepted Over 50 MICROS Projects 133 applicants 122 Partners	14 Teams accepted into the Accelerator. 41 Seed Grants selected
Diversity, Equity, Inclusion metrics	50% of 2023 cohort is WMBE	42% of participants representing underserved community or business in first year	Over 60% are Women and/or BIPOC owned businesses
Estimated Investment Public | private	$10,000 in NextCycle Awards $2.5M in state grants $72.7 tracked private investment	$81,600 in NextCycle Awards $5.3M in State of MI EGLE Grants to NextCycle Network $689M Tracked investment	$410,000 in Seed Grants Over $7.5 million in estimated funding raised by teams in the Accelerator Tracks

Table 5.2 Examples of NextCycle challenge tracks and past participants.

Challenge track	Participant examples
End Market Development	• Vartega – A technology development company specializing in the carbon-fiber-reinforced plastic recycling process • Timber Age – Cross laminated timber based walls using standing dead timber • Trash Panda – 100% recycled content disc golf discs • Sana Packaging – Pioneer sustainable packaging brand in the cannabis industry in the USA
Food, Liquid, and Organics Waste	• Savormetrics – Food waste prevention solution through QA/QC using biochemical sensors and AI • Wormies – Scaled vermicompost operation • Urban Ashes – Modular urban wood recovery and utilization • Unlimited Recycling – Organics collection and processing services
Recycling Innovation and Technologies	• Goodwill Associates – Textile recycling technologies through regional hub and spoke systems • SCOPS – Heat sealable coating technology for paper to replace single use plastic packaging • Glacier – Advanced robotic sortation for material recovery facilities • Nextiles – Building insulation materials from recycled auto-manufacturing waste • Birch Biosciences – Recycling non-bottle PET using enzymes to depolymerize plastics
Upstream	• DeliverZero – Service that allows customers to utilize reusable containers for food delivery • GeerGarage – Online company that facilitates peer-to-peer rental of outdoor gear • Industrial Sewing and Innovation Center – Production model to upcycle textile fabric into felted fabric

and is included to demonstrate the wide array of projects that have been supported though NextCycle programming.

5.3 Conclusions & Recommendations

Creating robust circular economies requires innovative ideas and businesses, partnerships that include the public and private sectors, and significant financial and resource investments. NextCycle utilizes a data driven and approach to clearly identify the gaps to circularity, foster innovation, create partnership networks, and leverage state grant funding with private sector investments to close the loop. States, regions, or other entities considering adopting a NextCycle style approach for increasing circularity should consider including the following key elements in their program design:

1) Data analysis to identify system gaps, needs, and opportunities, evaluate full material system from point of generation to remanufacture, including collection, processing, end markets, and other system elements.

2) NextCycle is only as strong as the pipeline of teams that participate in the program. Thus, the program must include dedicated marketing activities, pre-application networking and support, and clear provision of benefits to participating teams.

3) Establishment of partnerships with local experts to support team recruitment, review applications, and provide on-going mentorship to accepted teams.

4) A viable pathway for funding projects that can transform the regional system, if possible, include impact funding, state grants, and other patient capital that does not require a quick, high multiple, return on investment.

Acknowledgements

The author would like to acknowledge the collaboration and support for NextCycle from the state project funders, partners, and staff including:

Cascadia Consulting Group https://www.cascadiaconsulting.com/
Centrepolis Accelerator at Lawrence Technological University https://www.centrepolisaccelerator.com/
Colorado Department of Public Health and Environment (CDPHE), Recycling Resources Economic Opportunity grant fund https://cdphe.colorado.gov/sustainability-programs/recycling-grants-support/recycling-resources- economic-opportunity
King County's Solid Waste Division https://kingcounty.gov/depts/dnrp/solid-waste.aspx
Michigan Department of Environment, Great Lakes, And Energy (EGLE) https://www.michigan.gov/egle Michigan Recycling Coalition https://michiganrecycles.org/
Resource Recycling Systems NextCycle project team including; Frey, Jim; Hesterman, Bryce; Higgins, Keira; Seltzer, Elisa; Wiebe, Meghan, and the many other RRS staff that have collaborated to create and manage NextCycle around the US https://recycle.com/
Seattle Public Utilities https://www.seattle.gov/utilities
START Consulting https://www.startsustainability.com/our-team
Traversal https://traversal.design/
Washington Department of Ecology https://ecology.wa.gov/
Washington Recycling Development Center https://ecology.wa.gov/Waste-Toxics/Reducing-recycling- waste/Strategic-policy-and-planning/Recycling-Development-Center
Washington State Department of Commerce https://www.commerce.wa.gov/

6

My So-Called Trash: Evaluating the Recovery Potential of Textiles in New York City Residential Refuse

Sarah Coulter[1]*, Constanza Gomez[2], Agustina Mir[2] and Janel Twogood[1]

[1]Accelerating Circularity, Inc. Campbell Hall, New York, USA
[2]Sortile, Inc., New York, USA

Abstract

In its most recent, publicly available waste characterization study (2017), the New York City Department of Sanitation (DSNY) estimated that roughly 6% of NYC's residential solid waste is textile, but little is known about this material beyond its gross volume. Accelerating Circularity's previous research suggests that up to 80% of the textiles in municipal solid waste (MSW) are potentially recoverable as feedstocks for recycled fibers. If correct, this implies that roughly 300 thousand pounds of recyclable textiles are lost from NYC households each year (based on 2019 residential refuse collection volumes). We will perform fieldwork to test these assumptions and synthesize research results with ongoing, real-world trials of circular textile-to-textile (T2T) systems with a goal of illuminating the potential recovery value of textile in MSW.

In its Fall 2022 Waste Characterization Study, DSNY will set aside the textile fraction of sampled MSW for further analysis by our research team. Each item will be hand-sorted and visually assessed to identify category, color, and the presence of T2T contaminants (also known as "disruptors" or "irritants") such as plastic and metal trims. Fiber identification will be aided by near-infrared spectroscopy. This type of material assessment has typically been conducted on textiles that have already been diverted to the secondhand materials market and have established, if not realized, market value. Our study focuses on those materials that have been disposed of as refuse to assess potential recovery value. To the extent possible, our study methods will align with common practice, however, because of the unique characteristics of this study (materials have been extracted from commingled refuse), evaluation will exclude any assessment of item condition (and implicit wearability/resale value). Rather, the collected data will be used to evaluate fractions potentially suitable for T2T recycling and/or downcycling (wipers, shoddy). The paper will map these fractions onto learnings from Accelerating Circularity's ongoing circular product System Trials and its insights on the T2T potential of low-value secondhand textile grades to envision future diversion pathways in fully realized, commercial T2T circular systems.

Keywords: Textile, municipal solid waste, textile-to-textile recycling, waste characterization study, new york city, post-consumer, NIR

*Corresponding author: sarah@acceleratingcircularity.org

Nabil Nasr (ed.) Technology Innovation for the Circular Economy: Recycling, Remanufacturing, Design, Systems Analysis and Logistics, (63–74) © 2024 Scrivener Publishing LLC

6.1   Introduction

6.1.1   Background

Textile waste is a global problem. We estimate that at least 110 million tons of post-consumer textiles are wasted annually, approximately 30% of garment production is never sold at retail, and more than 15% of all cutting waste is not recycled. Despite recent trends in the secondhand textile market, the US still sends nearly 17 million tons per year to landfill [1]. Diverting these textiles from incineration and landfill has the potential to lower the textile industry's GHG emissions as well as reduce the use of virgin materials while helping brands and retailers meet their public commitments to reduce resource consumption and incorporate recycled materials into their products.

While municipal waste characterization studies generally include a textile sort category, they do not identify fiber type, fabric construction, or other material characteristics of the textiles. What little textile waste characterization data is available has primarily been focused on understanding postconsumer textile that is already in the secondhand market, rather than material that would be destined for landfill. Our previous research exposed this gap in understanding which textiles currently going to landfill have the potential to be diverted to recycling streams but suggests that up to 80% is potentially or readily recyclable (Figure 6.1).

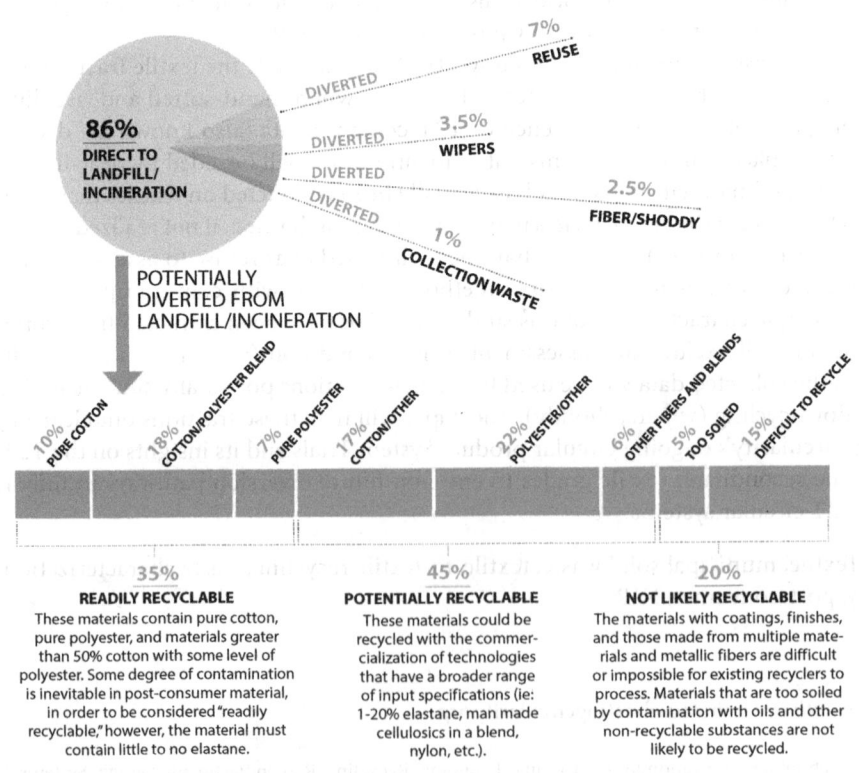

Figure 6.1 Textile landfill diversion potential to recycling [2].

These estimates are therefore extrapolated from assumptions about the characteristics of textiles in municipal solid waste that are not based on primary data.

In its most recent, publicly available waste characterization study [8], the New York City Department of Sanitation (DSNY) estimated that nearly 6% of NYC's residential solid waste is textile, but little is known about this material beyond its gross volume. If ACP's hypothesis about the recyclability of this material is correct – and based on a 2022 total collection of 3,321,856 tons year [3] – we could reasonably infer that the environmental, social, and economic value of roughly 150 thousand tons of recyclable textiles is lost from NYC households each year:

MSW Collected (3,321,856T) x Textile (5.8%) x Recycling Potential (.8) = Textile Value Wasted (154,134T)

DSNY is conducting an updated waste characterization study comprising three seasonal sorts between fall 2022 and summer 2023. The agency agreed to set aside the textile fraction of sampled MSW for further analysis by Accelerating Circularity's research team to begin to remediate the data gaps presented here and test assumptions about the potential recyclability of landfilled textile.

Each item will be hand-sorted and visually assessed to identify category, color, and the presence of T2T contaminants (also known as "disruptors" or "irritants") such as plastic and metal trims. Fiber identification will be aided by near-infrared spectroscopy. This type of material assessment has typically been conducted on textiles that have already been diverted to the secondhand materials market and have established, if not realized, market value. Our study focuses on those materials that have been disposed of as refuse to assess potential recovery value. To the extent possible, our study methods will align with common practice, however, because of the unique characteristics of this study (materials have been extracted from commingled refuse), evaluation will exclude any assessment of item condition (and implicit wearability/resale value). Rather, the collected data will be used to evaluate fractions potentially suitable for T2T recycling and/or downcycling (wipers, shoddy). The study team will map these fractions onto learnings from Accelerating Circularity's ongoing circular product System Trials and its insights on the T2T potential of low-value secondhand textile grades to envision future diversion pathways in fully realized, commercial T2T circular systems.

6.1.2   Project Goal

Field-test desk research on the characteristics of post-consumer spent textiles in MSW:
- Volume of post-consumer textile collected in the study period.
- Key material characteristics, including garment/item type, fabric, fiber content, color, and
- Percent of textiles potentially recoverable for:
  - Appropriate feedstocks for mature and emerging recycling processes:
    - Mechanical cotton, polyester
    - Chemical cotton, polyester, blended textiles
  - Wipes/Shoddy

6.2   Textile Sub-Sort of DSNY Waste Characterization Study, Fall 2022 Season

6.2.1   Methodology

Primary MSW Sort

In its Fall 2022 sort, conducted between October 24 and December 18, 2022, DSNY collected approximately eight hundred samples of refuse and recycling, totaling approximately 40,000 pounds. Accelerating Circularity collected the set aside textile samples the week of December 12, 2022. It was transferred to a waste transfer station in Elizabeth, New Jersey operated by WM. The material was sorted by a team of six researchers: four Accelerating Circularity staff, two Sortile staff running item scans, with general, safety, and logistics support provided by WM. At the conclusion of the subsort, the material was/will be disposed of by WM. The research methodology for this study is founded in the methodology of the larger Waste Characterization Study conducted by DSNY. The samples collected in the DSNY's primary sort include curbside residential refuse and recycling collections, school collections, and street bins throughout the five boroughs of New York City [4].

Textile Subsort

The textile subsort operation was adapted from practices such as those documented by FFG in their Sorting for Circularity Europe project [5]. Project-specific adjustments include:

Sampling method: the sorting studies outlined by FFG are strictly time- and location-based and limited to textiles *already diverted from the waste stream*. Our study is based on textiles pulled from the municipal solid waste stream as sampled by DSNY as part of a comprehensive waste characterization study.

The sampled materials considered in-scope for this study are the fraction characterized in the primary sort as "textile-clothing," which were set aside by the primary sorting agency (DSNY) for the purpose of this study. This excludes mattresses, footwear, accessories, and home textiles. We consider all textiles assessed in this study **non- rewearable** by nature of having been disposed of as refuse and commingled with mixed municipal waste.

Fiber Identification

Accelerating Circularity identified fiber composition using AI-enabled near-infrared spectroscopy developed and provided by Sortile, Inc., validated with care-and-content label information, where available. Sortile leverages machine learning integrated with a tabletop scanning device to efficiently identify fibers. Sorters can identify the fiber composition of items in less than a second, improving productivity over manual sortation over 5x. Sortile's current materials library allows identification of 6 mono-materials and one blend. The company conducts extensive fieldwork to collect data from different post-consumer sources to constantly improve the accuracy of the algorithms that drive the fiber ID prediction. All the models are tested with a curated and validated sample set, and all the devices go through a live testing cycle prior to shipping to make sure that target accuracy rates are being met.

Item-level Material Assessment

Each item sorted was identified by fiber, color, and category as detailed in the data collection application (Figure 6.2).

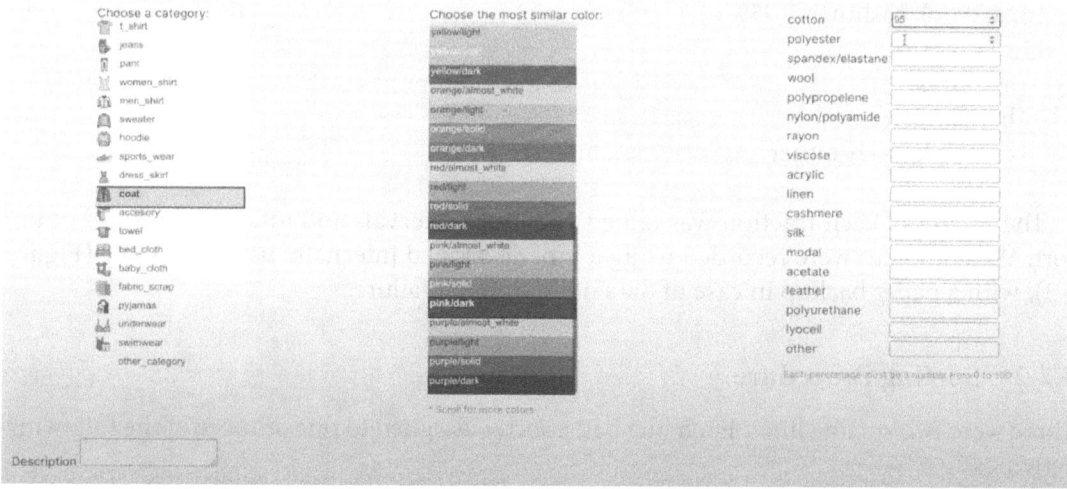

Figure 6.2 Data collection dashboard [6].

The basic Data Collection Dashboard was modified to include four categories of textile-to-textile disruptors:

1 = **Metal** (fastener, rivet, underwire, eyelet, charm)
2 = **Plastic** (reflective strip, button, zipper, epaulet, collar support, pearl, foam, fastener)
3 = **Fabric** (elastic, string, ribbon, patch, pompon, inset/yoke, embroidery, pocket)
4 = **Other** (leather, fur, pendant, print, wood, sequin, lurex)

After documenting category, color, fiber and disruptors, each item was placed in a fraction to be weighed. Only the outermost layer of multi-layer items are identified by the NIR spectrometer, so the utility of fiber ID for those items – as well as their suitability for textile-to-textile recycling – is extremely limited. Therefore, sorters were instructed to note multi-layer items and set them aside in a fraction designated "other." The material fractions assigned were:

1. Acrylic (95%+)
2. Cotton (95%+), White, Woven
3. Cotton (95%+), White, Knit
4. Cotton (95%+), Color, Woven
5. Cotton (95%+), Color, Knit
6. Cotton/Poly Blend (>50%Cotton)
7. Leather (95%+)
8. Linen, hemp, other plant fibers (95%+)
9. MMCFs (95%+) - ex: rayon, viscose, modal, acetate, lyocell
10. Nylon
11. Other blends
12. Other synthetics - ex: polyurethane, polypropylene
13. Polyester (95%+)
14. Poly/Cotton Blend (>50%Polyester)
15. Silk (95%+)

16. Wool, cashmere (95%+)
17. OTHER:
    a. Bras
    b. Multi-layer
    c. Unknown/Other

The weight of each fraction was tallied at regular intervals and totaled at the end of the sort. Weight tallies were recorded using a tool developed internally using AirTable (Figure 6.3), with a paper backup in case of data or equipment failure.

6.2.2 Sorting Procedure

There were two sorting lines. Each line had a sorter assigned to one or more of the following activities:

1. Staging – Sorters opened bags and pre-sorted items based on a visual assessment of color and item type to ensure consistent and efficient flow of material to be sorted.

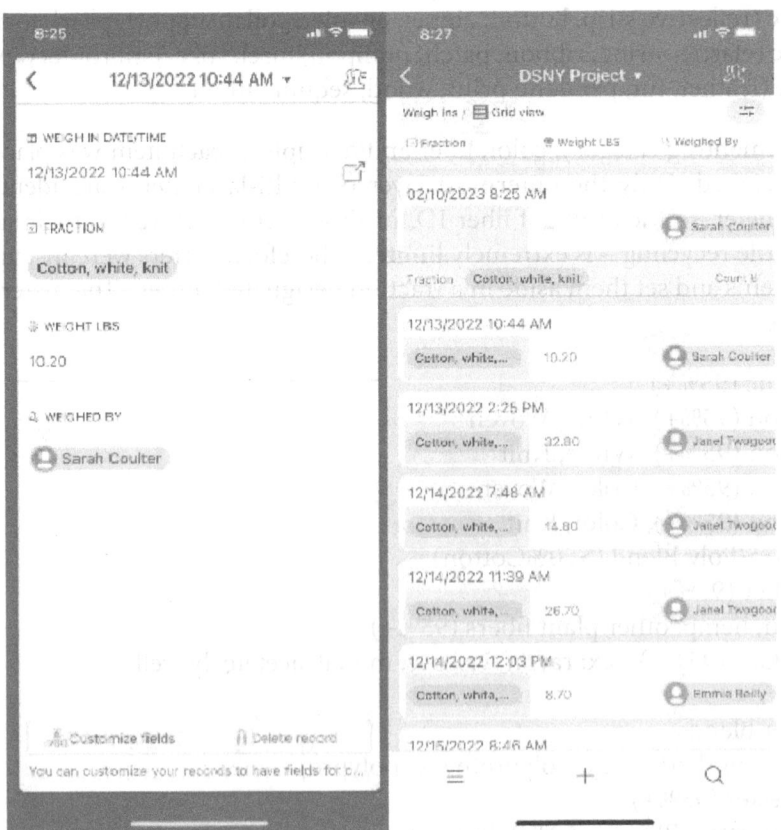

Figure 6.3 Weigh-in dashboard.

2. Preliminary assessment – Sorters picked an item to be scanned; located the care and content label for fiber content validation, if available; and visually assessed item type, color, and disruptors. They informed the scanning operator of the preliminary assessment characteristics when presenting it for scan.
3. Scanning – Each item was scanned, and item fiber content, category, color, and disruptors were recorded in the data collection dashboard.
    1. For multi-layer items, the primary layer was scanned and the fiber content recorded. If a care and content label was available, the fiber content of the other layers was recorded in the free field.
4. Sorting – After scanning, sorters assigned each item to the correct fraction and called for a weigh-in when adequate volume was aggregated (roughly 32 liters, the volume of containers used for the weigh-ins).
5. Disposal – After being weighed, the material was placed over the barrier from the sorting area to the main facility work area for final disposal.

Personnel rotated activities as needed per live conditions.

6.2.3 Safety

The study materials and work site conditions required safety precautions that are not typically taken in a conventional, secondhand textile sorting workflow. The research team received daily safety briefings and was required to use personal protective equipment suitable for the industrial setting. The work site was partially open to the outdoors, so warm clothing, hand warmers, moisture protection, and warming breaks were required.

6.2.4 Results

The study took place over roughly 4 working days comprising approximately 12 hours of active sorting. Based on our knowledge of the typical fiber content and construction of socks, women's underwear, hats, and gloves, we assumed these items to be mostly unsuitable for T2T applications. Therefore, these items were tallied and weighed but not scanned to avoid overburdening the NIR sorting lines. The total item count was 6,567 (total weight 2118.4 lbs.), for a total sorting throughput of 547.25 items (176.5 lbs.) per hour. With seven sorters, average throughput per sorter-hour was 78 items (25.2 lbs.) per hour.

6.2.5 Fractions

We found that nearly 56% of the sorted textile would have been readily recyclable in a textile-to-textile stream (as defined in our earlier research), and an additional 7.5% potentially recyclable as recycling technologies commercialize over the next 3 – 5 years, for total recycling potential of about 63% of sorted textile. This far exceeds the estimates extrapolated from earlier desk research for "readily recyclable" materials [2]. The cotton fraction was further sorted into color and fabric sub-fractions due to the narrower feedstock specifications of mechanical cotton recycling for T2T applications. Denim was assigned a separate fraction as it has multiple remanufacturing and recycling pathways and is therefore in high demand throughout the secondhand materials market. Although some brands have take-back programs for bras, they were assigned to the "not recyclable" category due to their high concentration of blended fibers, multi-layer construction and high ratio of trims to fabric that result in low yield from pre-processing, and limited take-back availability.

Recycling Potential of Sorted Textile

Fraction	Weight (lbs.)	% Total	Readily recyclable	Potentially recyclable	Not recyclable
Cotton, color, knit	276.6	13.1%	55.6%		
Polyester	206.7	9.8%			
Denim	161.2	7.6%			
Cotton, color, woven	158.5	7.5%			
Cotton, white, knit	126	5.9%			
Cotton, white, woven	37.9	1.8%			
Wool, cashmere	19.5	0.9%			
Cotton/Polyester (Cotton-rich)	191.7	9.0%			
Polyester/Cotton (Poly-rich)	96.4	4.6%		7.5%	
Manmade Cellulosic	35.4	1.7%			
Acrylic	16.2	0.8%			
Nylon	6.3	0.3%			
Other plant fibers	3.9	0.2%			
Silk	1.6	0.1%			
Other blends	254.7	12.0%			36.8%
Multi-layer	138.8	6.6%			
Socks	130.4	6.2%			
Unknown/other	98.6	4.7%			
Underwear (W)	61.8	2.9%			
Bras	29.6	1.4%			
Miscellaneous	23.9	1.1%			
Hats	19.5	0.9%			
Other synthetics	13.3	0.6%			
Gloves	9.9	0.5%			
Total	**2,118.4**	**100%**			

6.2.6 Item Level Data

Item type is not usually a critical consideration for T2T recycling, but it provides a useful filter for obviously non- recyclable items (most outerwear and accessories, for example) and can also serve as a useful proxy for fabric construction, which can be difficult for sorters to determine and is relevant to some mechanical recycling processes:

Product Category

Item type	Count	% Total
Socks*	2,341	35.65%
T-shirt	1,143	17.41%
Underwear (W)*	630	9.59%
Baby	406	6.18%
Pants/shorts	309	4.71%
Bras*	190	2.89%
Underwear (M)	168	2.56%
Sportswear	158	2.41%
Other	149	2.27%
Hoodie	129	1.96%
Sweater	129	1.96%
Jeans	118	1.80%
Hats*	114	1.74%
Dress/skirt	110	1.68%
Gloves*	108	1.64%
Shirt, men	107	1.63%
Coat	88	1.34%
Linens	54	0.82%
Pajamas	50	0.76%
Swimwear	40	0.61%
Towel	18	0.27%
Fabric scrap	8	0.12%
Total	*6,567*	*100%*
Excluded from NIR scan	*(2,341)*	
Items Assessed for Recyclability	**3184**	

Textile-to-textile recycling streams have varying thresholds for contamination from non-target fibers. This contamination tolerance impacts not only an item's overall recycling potential, but also the target recycling technology. For example, a thermomechanical polyester recycler may accept a small percentage of elastane contamination but have no tolerance for cellulosic fiber contamination. Similarly, a chemical cellulosic recycler may have zero tolerance for non-cellulosic fibers, whereas a mechanical cotton recycler may tolerate minimal amounts of spandex or polyester contamination. Our research on more than one hundred recycler specifications indicates that 100% and 95% purity thresholds are useful in determining the appropriate recycling stream [7]:

Fiber Content, Assessed Items

	100% Threshold	95% Threshold	100% Threshold	95% Threshold
Cotton	1148	1317	36.06%	41.36%
Polyester	349	394	10.96%	12.37%
Wool	21	23	0.66%	0.72%
MMCF	58	74	1.82%	2.32%
Acrylic	19	19	0.60%	0.60%
PolyCotton (Poly+)	130	130	4.08%	4.08%
PolyCotton (Cotton+)	368	368	11.56%	11.56%
Total PolyCotton	498	498	15.64%	15.64%
Other	593	361	18.62%	11.34%
Total	**3184**	**3184**		

Similarly, mechanical cotton recycling streams require color sortation, and certain color chemistries are not allowed in various chemical recycling streams. Therefore, it is useful to identify item color in a sortation exercise even though it may not directly impact an item's overall recycling potential:

Color, Assessed Items

Color	Count	% Total
White	687	21.58%
Blue	685	21.51%
Gray	521	16.36%
Black	510	16.02%

Pink	212	6.66%
Green	154	4.84%
Red	145	4.55%
Beige	89	2.80%
Purple	67	2.10%
Orange	50	1.57%
Yellow	40	1.26%
Brown	24	0.75%
Total	**3184**	

6.3   Conclusions & Recommendations

This research is a first step in remediating a significant gap in understanding textile in the waste stream (as opposed to the secondhand market, which also has significant data gaps). As the need to better-manage used textiles becomes increasingly urgent and jurisdictions begin to regulate the handling of these textiles, it is important to understand the potential diversion pathways available. We recommend that municipalities and others commissioning waste characterization studies include more robust classification of the textile fraction to inform policy decisions related to textile waste diversion. These categories should, at minimum, include fiber screening for 100% cotton, polyester, and poly-cotton blends, although we recommend a full assortment of fractions in our "readily" and "potentially" recyclable categories. Although our sub-sort required only 84 total active sorter-hours, it did require two additional set-up/break-down days, appropriate transportation and facilities, and additional third-party support. It could have been much more time-, labor-, and cost-efficient if it had been integrated into the primary sort.

We did not set out to assess re-wearability, but many of the items were plainly wearable, including like-new children's clothing, workwear uniforms in good condition, and new-with-tags apparel. This points to a need for consumer education and more ready access to the secondhand markets (including donation, freecycling, repair/remanufacturing, and resale). Of those items that were not in re-wearable condition, many were so because they were soiled, wet, and/or moldy *because of* having been discarded as trash. A significant portion of the remaining non-rewearable items appear to have been discarded due to heavy soiling, which in many cases (with appropriate laundering) would not exclude an item from T2T streams.

Overall, we found that most items classified as "textile-apparel" in DSNY's Fall 2022 waste characterization sort had high recyclability potential in commercial or near-commercial textile-to-textile processes. To validate these results and account for seasonality, we will continue to collect data through DSNY's Spring and Summer 2023 waste characterization sorts and will update our results accordingly.

Acknowledgements

Carmelo Freda, Senior Manager, Policy and Planning, Bureau of Recycling and Sustainability, NYC Department of Sanitation

Katherine Kitchener, Chief of Staff, Bureau of Recycling and Sustainability, New York City Department of Sanitation

Raymond Randall, Senior Manager, Textiles, WM Sustainability Growth Solutions Jose De Leon, Sr Associate, WM Advisory Services

James Van Woert, Area Director of Disposal Operations, WM Greater Mid-Atlantic Market Area

References

1. US EPA. (2019). *Facts and Figures about Materials, Waste and Recycling*. Retrieved from United States Environmental Protection Agency Web site: US Environmental Protection Agency, https://www.epa.gov/facts-and-figures-about-materials-waste-and-recycling/guide-facts-and-figures-report- about#Materials_and_Products, 2019
2. Accelerating Circularity, Inc. (2020). *Research and Mapping Report*. New York.
3. New York City Department of Sanitation. (2022). *Annual Reports for DSNY & Non-DSNY Collections*. Retrieved from NYC Web site: https://www1.nyc.gov/assets/dsny/site/resources/statistics/annual-dsny-non-dsny- collection
4. New York City Department of Sanitation. (2022). *Annual Reports for DSNY & Non-DSNY Collections*. Retrieved from NYC Web site: https://www1.nyc.gov/assets/dsny/site/resources/statistics/annual-dsny-non-dsny- collection
5. Freda, C. (2022, January 12). Senior Manager, Policy and Planning, Bureau of Recycling and Sustainability, DSNY. (S. Coulter, Interviewer)
6. Fashion for Good. (2022, 9). *Reports*. Retrieved from Fashion for Good: chrome-extension://efaidnbmnnnibpcajpcglclefindmkaj/https://reports.fashionforgood.com/wp-content/uploads/2022/09/Sorting-For-Circularity-Europe_Sorters-Handbook_Fashion-for-Good.pdf
7. Sortile, Inc. (2022, December 1). Data Collection Dashboard. 2022, New York, United States.
8. New York City Department of Sanitation. (2017). 2017 NYC Residential, School, and NYCHA Waste Characterization Study. Retrieved from NYC Web site: dsny.cityofnewyork.us

When is it Profitable to Make a Product Sustainable? Insights from a Decision-Support Tool

Karan Bhuwalka[1]*, Jessica Sonner[1], Lisa Lin[1], Mirjam Ambrosius[2] and A. E. Hosoi[1]

[1]Massachusetts Institute of Technology, Cambridge, Massachusetts, MA, USA
[2]adidas AG, Adi-Dassler-Strasse, Herzogenaurach, Germany

Abstract

Meeting corporate sustainability goals requires policies and business models that incentivize sustainable product design. We create a simple modeling tool that examines the conditions in which it is profitable for companies to sell more sustainable versions of a product. Users can input variables that relate to consumer demand (e.g. proportion of consumers willing to pay the price premium for a sustainable product), policy (e.g. carbon tax) and costs of production including the added costs associated with incorporating sustainable materials and practices. To demonstrate the utility of our tool, we input values for sneaker production and identify the demand and policy conditions under which sustainable sneaker design leads to increased profit. Policymakers and companies can use this as a decision-support tool to promote sustainable design for various ranges of products.

Keywords: Product design, policy, sustainability, demand, shoes, apparel, sneakers

7.1 Introduction

With sustainability initiatives becoming increasingly important in order to reduce environmental impacts across various industries, it is essential that decision-makers are equipped with the tools and knowledge necessary to implement effective sustainability strategies. Different factors might play a role when deciding upon an efficient sustainability strategy: consumers' preferences, cost to manufacture sustainable products, and incentives or penalties like e.g. carbon taxes introduced by policymakers. With strong interdependencies between these factors, it is often hard to quantify the exact direct and indirect effects of making a more or less sustainable product. To facilitate decision-making, we developed a tool that takes all these factors into account and allows us to calculate the impact of introducing a sustainable alternative to a product on profitability. Hence, this tool can help decision-makers in industry to assess future scenarios with different regulatory set-ups and changing consumer preferences. Furthermore, our tool can help policymakers to predict

Corresponding author: bhuwalka@mit.edu

Nabil Nasr (ed.) Technology Innovation for the Circular Economy: Recycling, Remanufacturing, Design, Systems Analysis and Logistics, (75–94) © 2024 Scrivener Publishing LLC

the effects of a new regulatory framework for incentivizing sustainable behaviour. The tool is especially useful in providing decision-support for a wide range of products to help companies prioritize which products to make more sustainable and set cost targets for sustainable manufacturing. Since the fashion industry is a large source of carbon emissions (10% globally [1]), we specifically apply our tool to study the impact of changing a sneaker's design from using virgin to recycled polyester in the shoe upper. We are then able to identify regulatory and demand scenarios under which such a substitution would be desirable to a wide range of manufacturers.

7.1.1   Literature Review

As interest in sustainable design grows, there is a growing need to understand how companies' profitability is impacted as they switch towards sustainable products. Taticchi *et al.* review 394 papers relating to decision support tools for sustainable supply chain management and argue that the economic aspects of industrial sustainability have been neglected and reiterate that understanding and measuring the costs of implementing sustainability practices should be high on the future research agenda [2]. Making products sustainable can be costly, as it may need more expensive materials or changing manufacturing processes. In order to ensure that companies prioritize sustainable product development, it's imperative to understand how sustainability can be made profitable. It is specifically important to understand the impact of consumer demand and policies on profitability so that companies can form strategies that encourage green innovation. While many companies have embedded sustainability goals within their 'key performace indicators' (KPIs), they need decision tools to prioritize trade-offs between sustainability, product quality and profit margins [3]. In a recent review of papers on green innovation, Khanra *et al.* argue that large multinational companies are implementing strategies that encourage green innovations to achieve sustainable development, which is often evaluated using a 'triple bottom line' approach [4]. They go on to argue that profitability from green products is a serious concern for firms and that researchers should aid the development of green products by suggesting ways for firms to reconceive products and markets. They also highlight the need for research that provides useful inputs for policymakers to promote innovations in green product design. This need for understanding the interplay of economics, product design and policymaking is echoed by Rao and Holt who argue that "if green supply chain management practices are to be fully, a demonstrable link between such measures and improving economic performance and competitiveness is necessary" [5].

Many recent papers in the management science and operations research literature have focused on addressing this need to understand the interplay of consumer demand, manufacturing costs and policies on product greening. The goal of these analyses is to derive the conditions under which producers are incentivized to switch to greener products by building models in which producers maximize profits based on costs, consumer demand, market structure and government regulations. The modelling approaches used by these papers to provide insights into the economics of product greening can be classified as either i) game theory models or ii) optimization approaches that identify a market equilibrium. Game theory models typically study the competition between two firms selling differentiated products in a market where consumers are sensitive to the price and the 'greenness' of the products. These models are used to derive insights for policymaking and corporate strategy. For example, Ling *et al.* [6] study the impact of government subsidies in a duopoly while Liu *et al.* [7] study the impact of government intervention on green innovation. In certain game theory papers, the 'players' are not competing firms but rather actors in a supply chain. For example, Zhu and He [8] study how supply chain structures and the type of market competition impact the greenness of products. Similarly, Zhang and Zhang [9] model optimal pricing and greening decisions to demonstrate the conditions under which the manufacturer should engage in remanufacturing.

Other researchers study firm pricing strategy and investment decisions by using equilibrium modelling. In these models, firms make pricing and investment decisions which maximize their profits under different market conditions. Much of the research in this area applies these models to study the impact of government policy on green investment. In early work, Chen (2001) [10] analyses a single (monopolist) producer's strategic decision regarding their price and greenness, under different environmental standards. Bi *et al.* [11] study the impact of policies in a market with two products that differ in manufacturing costs, selling prices and amount of emission per product. In their model, firms determine the selling price after taking into account consumer demand and costs. Li (2022) [12] develops a dynamic control model to study the impact of taxes and demand on firm pricing strategies and green investment in different cases. They use a linear demand function and different tax levels to identify equilibrium conditions under monopoly, duopoly and social planning cases. Similarly, Ghosh *et al.* [13] build a model that derives the optimal level of pricing and 'level of greening' for a firm. They consider linear demand functions as a function of price and consumer willingness to pay extra for greener products, as well as additional costs of making a product greener. They then use this model to investigate the conditions under which a firm is incentivized to make its products more sustainable and how government taxes and subsidies impact these conditions.

Other research uses these models to specifically focus on the impact of consumer behavior on green investment and product pricing. For example, Ling and Xu [14] study product pricing in a duopoly after considering consumer premiums for green products. Liu *et al.* [15] study pricing strategies as a function of consumer sensitivity to environmental attributes. More recently, Jiang *et al.* [16] built a model that analyses consumer fairness when differentiated pricing is used for green products. In most of the optimization papers mentioned above, firms typically choose a 'level of greening' based on costs and consumer sensitivity. However, other models explicitly consider a 'green' and a 'brown' product with different attributes. In such models, optimization is used to identify when a firm is incentivized to introduce a green product. Mukherjee and Carvalho [17] determine equilibrium decisions for pricing and greening investments in a supply chain with a green and brown product. Zhang *et al.* [18] analyse production costs to study whether a green or brown product is optimal for a manufacturer. Similarly, Yenipazarli and Vakharia [19] analyse whether green and brown products can co-exist in a market.

Overall, the literature highlights the importance of understanding the interplay between consumer preferences, government regulations, and firm strategies to promote green product development. In this paper, we build on this literature and present an equilibrium model that evaluates company profitability when transitioning towards sustainable production as a function of manufacturing costs, consumer demand, and policy. While our goal of studying the interplay of demand, manufacturing costs and taxation is similar to the other models, we combine useful model attributes from different approaches. For example, we model linear consumer demand for green products similarly to Li (2022) and Ghosh *et al*. However, instead of using their approach of firms selecting a 'level of greening', we model firms as having a choice to introduce a 'green' product similar to Zhang *et al*. and Yenipazarli and Vakharia. We believe such a framework where producers have to select whether to introduce a new, 'greener' product is more useful for product designers who often have to choose between a set of design options and do not have control over a conceptual parameter such as 'level of greening'. However, neither Zhang *et al*. nor Yenipazarli and Vakharia explicitly analyse the impact of government taxation levels on the decision to introduce a green product (Zhang *et al*. discuss this qualitatively). By combining various aspects of different modeling approaches in this space, we are able to create a tool that explicitly studies whether it is profitable to introduce a green product under varying conditions of demand, cost and taxation.

Although there are some papers that apply such models to real cases such as agricultural products (Perlman *et al*. [20]), most of the research papers discussed above present general theoretical results for the conditions where greening is optimal. Each model makes different assumptions and has certain advantages in analysing the scenario they focus on. While these models present useful theoretical insights across different scenarios and market structures, stakeholders (designers, policymakers, etc) are typically not provided with a way to apply the models to help make decisions. In this paper, we present an equilibrium model along with an easy-to-use tool and demonstrate its applicability in product design decisions as well as policymaking. We both describe the theoretical results for when manufacturing a green product is more profitable than a non-green one and also apply it to a real-world case study of sneaker production under different government policies. Moreover, rather than just describing the model with mathematical equations, we also present an easy-to-use web app which allows stakeholders to analyze product greening decisions. This tool can be further used by policymakers to incentivize green innovation or by corporations to set company-level or industry-level strategies by identifying manufacturing cost targets, profitable product greening opportunities.

7.2   Main Content of the Chapter

7.2.1   Methods

We model producer choice as a profit optimisation problem in which a producer is a price-setter i.e. the producer decides the price such that they maximise profits given certain input conditions regarding product costs and consumer demand for their product. The goal of the model is to identify the range of conditions in which manufacturing a sustainable version of the product produces greater profits relative to manufacturing a purely non-sustainable version. The input parameters to the model can be thought of as influencing either a) consumer demand or b) production costs.

For illustration purposes, here we consider consumer demand to be a linearly decreasing function of product price (slope and intercept are inputs), although measured or more general forms of the demand curve can easily be incorporated into the framework. We assume that consumer substitution decisions are buried within the demand curve that the firm observes i.e. the curve tells us how a change in price will impact product sales for that company alone. Stated differently, the amount consumers are willing to buy under different prices already incorporates their utility given the characteristics and prices of other competitors' products. In this model, for simplicity, we are assuming that the price-setting does not significantly impact the market i.e. other companies are not impacted enough to respond to this decision by changing their own prices, (if other market actors responded by changing prices for their products, the firm's observed demand curve would change). The price-setting assumption is also relevant for cases when the firm is effectively a monopoly or when the product in question is sufficiently differentiated from other products and can be treated as having its own market.

A fraction of consumers (f) are willing to pay a price premium (m) to buy a more sustainable product where m represents the excess price e.g. if consumers are willing to pay 10% more for a sustainable product, $m = 0.1$. The cost of production, c, along with the additional

Table 7.1 Input parameter values for the baseline model run. Since different products vary significantly in their values for demand and cost parameters, we use reasonable baseline values for our model but study the impact of varying all the parameters on the results for the profitability of sustainable designs in this paper.

Input	Description	Value
c'	Cost of non-sustainable production. For the baseline results (Figure 7.1), the cost input is chosen to be less than half of p_m which is the price at which no consumption occurs	0.9
Δ	Excess Cost to make a product sustainable, normalized as a proportion of the non-sustainable product cost c. Default value of 0.2 corresponds to a baseline in which making a sustainable product is 20% more expensive than making a non-sustainable product	0.2
t	Tax on the non-sustainable product. This represents any legislation that increases the cost of non-sustainable products. The value is input is as a proportion of the non-sustainable product cost c. Default value is 0. A value of t=0.1 means that the tax increases the cost of producing a non-sustainable product by 10%	0
m	Excess price charged for sustainable product. This is the premium that consumers are willing to pay for sustainable product (in baseline, sustainable products are 10% more expensive than non-sustainable options)	0.1
f	Proportion of consumers that purchase the sustainable product (In baseline, 10% of customers are willing to pay the 10% price premium and buy the sustainable product)	0.1
p_m	Price at which no consumption occurs. This parameter is a measure of the slope of the demand-curve. A higher value means that consumers are less price sensitive, a lower value means consumers are more price sensitive	2
r_m	The quantity consumers are willing to purchase if the price was zero. This represents the intercept of the linear demand function.	1

cost (Δ) it takes to make a sustainable version of the product – which, as with m, represents the excess cost as a proportion of the non-sustainable product cost – are inputs to the model. The values for f and m should be informed based on an environmental assessment of the two product options. If, in reality, the two product options have a similar environmental impact, consumers will not be willing to pay a premium for the product that is branded as 'greener'. However, if there is a significant difference in their environmental impacts, more consumers may be willing to pay a larger premium. Finally, the model includes a tax (t) that can be levied on the non-sustainable product so policymakers can use the tool to study the impact of policies on manufacturers' decisions to make sustainable products. The input value for the tax (t) also requires an environmental assessment of the products, as the level of taxation aims to internalize the environmental costs in the life-cycle of a product. A full list of input parameters, along with illustrative, approximate values for the running shoe industry, is given in Table 7.1. In the following we will explore various parameter combinations and "baseline" values will refer to the representative values in Table 7.1.

7.2.1.1   Production Costs

Using the inputs above, the cost of producing a non-sustainable product is $c(1 + t)$, while the cost of producing a sustainable product is $c(1 + \Delta)$. Here we assume that the tax is only levied on non-sustainable products. In cases where there is a tax on the sustainable version of the product as well, we can incorporate the common taxation into the shoe production cost c.

7.2.1.2   Demand Function

The amount of consumption is r', which, for simplicity, we take to be a linear function of shoe price, p' i.e. as price increases, consumption decreases linearly:

$$r' = r_m(1 - \frac{p'}{p_m})$$

where p_m is the price above which which no consumption occurs and r_m is the total units of consumption as the price asymptotes to zero. (Note that it is straightforward to incorporate more complicated $r(p)$ relationships but here we illustrate the method with a linear model). The sustainable product is sold at a premium price $p(1 + m)$ to a fraction of consumers, f who purchase the sustainable version at that premium price. The relationship between f and m represents the consumers' willingness to pay extra for sustainable products. In reality, f and m are strongly correlated (lower premiums mean more consumers are willing to pay it). However, for now we will treat them as independent inputs and can then evaluate scenarios that lie on realistic points of the f-m tradeoff curve.

We can normalize the demand function such that $r = \frac{r'}{r_m}$ and $p = \frac{p'}{p_m}$ so that the equation $\frac{r'}{r_m} = (1 - \frac{p'}{p_m})$ can be rewritten as $r = 1 - p$. We can similarly normalize the costs of production c' such that $c = \frac{c'}{p_m}$.

7.2.1.3   Price Optimization

We can write the (normalized) profit (P) that a producer receives as a product of the normalized sales volume (r) and the net revenue per unit of products sold. The net revenue per unit of product is a function of the difference between the product price p and the (normalized) manufacturing cost c. We can say, in general:

$$P = r\,(ap - bc)$$

where a, b and c are transformed versions of the model inputs. Note that we can easily convert the normalized profit P to absolute profit P':

$$P' = Pr_m p_m = r\,r_m p_m\,(ap - bc) = r'\,(ap' - bc')$$

When the producer only sells the 'non-sustainable' product, $a = 1$ and $b = (1 + t)$ so that the profit equation becomes $P_{NS} = r\,(p - c(1 + t))$. In the case where the producer sells a mix of sustainable and non-sustainable products (with f representing the proportion of sustainable products sold at a price premium of m), $a = (1 - f) + f(1 + m)$ and $b = (1 - f)(1 + t) + f(1 + \Delta)$.

We now optimize over price to maximize the profits P, such that $\dfrac{\partial P}{\partial p} = 0$. The price at which P is maximised is given as a function of inputs by the expression:

$$p_{opt} = \frac{1}{2}\left(1 + \frac{bc}{a}\right)$$

In the general case with a mix of sustainable and non-sustainable products we can write the absolute price as follows (by substituting in a and b)

$$p'_{opt} = \frac{p_m}{2} + \frac{c_{avg}}{2(1 + mf)}$$

where c_{avg} is the average cost of producing a mix of sustainable and non-sustainable products (weighted by their production fractions) i.e. $c_{avg} = c'\,[f(1 + \Delta) + (1 - f)(1 + t)]$.

In this model, the optimum price is a linear function of consumer price-sensitivity (measured by p_m). If consumers are more price sensitive, p_m is smaller and the optimal strategy for the manufacturer is to lower the price of the goods. This is especially true for products that have a large number of substitutes, because consumers can easily substitute a different product and hence p_m is small. The optimized price is also a linear function of the average cost of production. A larger cost of making a product sustainable is passed on to the consumer as an increase in price, but all of the additional cost is not passed on to the consumer (since the effect of Δ is weighted by willingness of consumers to pay premiums $\dfrac{f}{2(1 + mf)}$).

If more consumers are willing to pay larger premiums (mf is larger), p_{opt} is reduced. The price paid by consumers for the sustainable product, $p_{opt} * (1 + m)$ therefore increases sub-linearly when increasing m because producers are optimizing for total profit and reducing the price can increase total sales. The change in p_{opt} when changing m $\left(\dfrac{\delta p_{opt}}{\delta m} \right)$ is a function of the shape of the demand curve (given by p_m).

It is worth noting that this rebound effect is likely to be small in most real cases as the values for m and f are both <1 so the product mf is likely to be small. When mf is close to 0 the price defaults to being the average between the max-price a consumer is willing to pay and the weighted cost (same as the non-sustainable case). How much price changes as a function of the premium consumers are willing to pay depends on the slope of the demand curve. If the demand curve is inelastic, p_{opt} will change less with a change in the premium that consumers are willing to pay. If the demand is very elastic, p_{opt} will decrease with an increase in the premium so that sales are increased.

We can substitute the optimum price to the profit expressions above to obtain an expression for the maximum profit possible given certain input parameters:

$$P_{max} = \frac{a}{2}(1 - \frac{bc}{a})^2$$

We can obtain the maximum profit for the non-sustainable production case ($P'_{max,NS}$) as well as the sustainable mix case ($P'_{max,S}$) by substituting in the appropriate values for the parameters (a,b) and de-normalizing. It is more profitable to make a sustainable product when $P'_{max,S} > P'_{max,NS}$ i.e. the profit-maximizing condition for the sustainable product mix is larger than the non-sustainable product. This can also be written as $\dfrac{P_{max,S}}{P_{max,NS}} > 1$.

After solving the equations, we get the ratio:

$$\frac{P'_{max,S}}{P'_{max,NS}} = \frac{[p_m(1+mf) - c(1+f + (1-f)t)]^2}{(1+mf)(p_m - c(1+t))^2}$$

If this ratio is larger than one it is more profitable to make sustainable products; the value of the ratio informs how much more profitable it is to also sell sustainable products. We call this ratio the 'sustainability ratio'.

7.2.2   Results

Figure 7.1 shows the sustainability ratio as a function of changing two input parameters (f, Δ) with all other parameters given the baseline values as per Table 7.1. The grey contour shows the line at which the profit from selling a sustainable product is equivalent to the profit from selling a non-sustainable product. This condition occurs at a (almost) fixed value of the extra cost Δ as a function of f, suggesting that there is a threshold excess cost such that selling sustainable and non-sustainable shoes are equivalent. This threshold value of excess cost is given by the following equation:

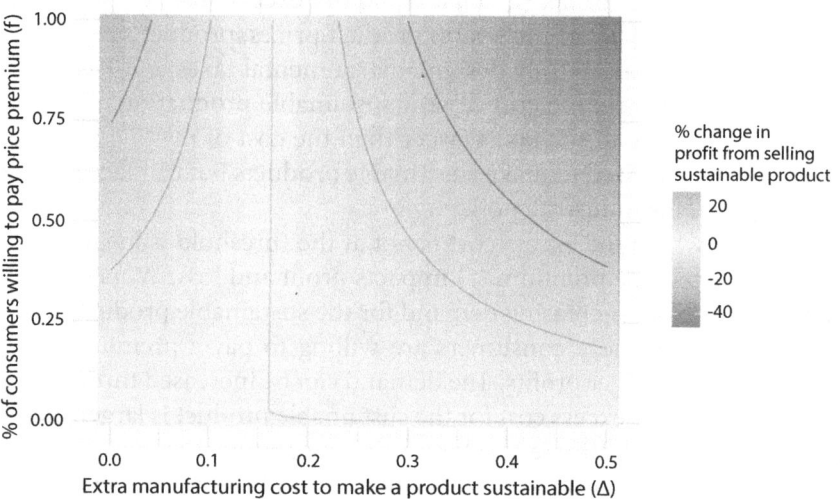

Figure 7.1 The colored regions represent the change in profit from selling a sustainable product (red is a decrease in profit, blue is an increase in profit i.e. blue represents regions of the phase space where manufacturing a sustainable option is more profitable). All other input variables are assigned values based on Table 7.1. In this case, it is assumed that the (non-sustainable) product cost is 90% of the price and there is no tax on the non-sustainable product. The vertical axis represents the proportion of consumers willing to pay the 10% price premium on a sustainable product. The horizontal axis represents the extra cost to make a product sustainable (either through design changes or using recycled/bio-based material). The grey contour represents the line at which profits are identical. The other colored contours represent the -20%, -10%, +10% and +20% profit conditions.

$$threshold \cong t + \frac{m}{2}\left(1+t+\frac{1}{c}\right)$$

When the excess cost of producing a sustainable product is below this threshold, it is profitable to introduce the product to the market. It is interesting that threshold cost is independent of the proportion of consumers willing to pay the premium (f does not matter). Designers and manufacturers who want to make greener products should set this threshold value as a cost target. They can then determine strategies to help meet this cost target such as using cheaper feedstocks, increasing process efficiency or improving product design. Alternatively, given an excess cost of production (Δ) and a taxation level (t), producers can set the premium for sustainable products (m) in a way that they will not result in a loss by selling a sustainable version of the product. Once the premium is set, the proportion of consumers willing to pay that additional premium is a function of the demand curve. The larger the value of f is for that given price premium, the more substitution occurs and environmental outcomes are improved. If the premium (m) is very large due to a high $\Delta_{threshold}$, it is possible that no consumers will be willing to pay the premium and buy the sustainable product ($f=0$).

If the model is being used by policymakers interested in promoting green innovation, they can select a level of taxation such that manufacturers increase their profits by switching to a sustainable strategy. For consumers to bear no additional premium for the sustainable product (i.e. m=0), the level of tax (t) must equal the excess cost Δ. In this scenario, our model can calculate the impact of the tax on product prices, product sales and firm profitability. Policymakers should carefully design environmental taxes and Extended Producer Responsibility polices to make the cost of non-sustainable production as close to the sustainable product as possible. If the tax is larger than the cost of making a product sustainable, producers are incentivised to make sustainable products but can face significant losses which may be undesirable industrial policy.

In all scenarios in which the excess cost is not at the threshold value, the proportion of consumers willing to pay the premium (f) impacts profit and loss. When the excess cost is cheaper than the threshold, increasing demand for the sustainable product increases profit. A larger value of f means more consumers are willing to pay a premium for sustainable goods, thereby leading to larger profits. The demand can be increased through marketing or other campaigns. When the excess cost for the sustainable product is larger than the threshold value, increasing demand increases losses (because more consumers buy sustainable goods that are more costly to produce). Under these conditions when there is high demand but costs are too large, manufacturers are disincentivized to offer a sustainable product to the market.

In reality, manufacturers might respond to these high costs (and low profitability) by increasing the margin charged on sustainable products. While we held the margin constant in Figure 7.1, we demonstrate the impact of increasing the premium in Figure 7.2. (Figure 7.2 also shows the impact of adding taxation which is discussed later). The central-right heatmap in Figure 7.2 corresponds to the previously discussed Figure 7.1. We can see from the right-hand column of Figure 7.2 that as we increase the price premium, the excess cost threshold at which there is no profit change shifts further to the right. The implication of this is that if costs of making a product sustainable are high, charging higher premiums can make it profitable to be sustainable. It is important to note that increasing the premium has a trade-off where it may reduce the percentage of consumers willing to pay the higher premium (f). While we treat these two parameters as independent, we can evaluate the combinations of f and m that lead to increasing profits from selling sustainable products.

Given a particular cost of making a product sustainable, the model is, therefore, able to inform the premium that must be charged to ensure that there is no change in profit. However, if the premium is too high, the proportion of consumers that may be willing to pay it may become too small for there to be any significant impact of creating a sustainable product. In the extreme case when f=0, no sustainable products are sold so the change in profit is zero.

In situations where costs cannot be sufficiently reduced to make sustainability profitable, policies such as taxes can act to influence the manufacturers' decisions. In this analysis assume that taxes only apply to the non-sustainable product. (Taxes that are applied equivalently to the sustainable and non-sustainable product can be absorbed in the base cost, c.) In Figure 7.2, we can see the effect of changing taxation along the horizontal grid. When taxes are increased, the area where sustainable products are more profitable is larger. The most profitable condition is when there is taxation combined with high demand (high premium, as well as a large proportion of consumers willing to pay the premium). If the demand is low, taxation is necessary to make it profitable to manufacture sustainable products.

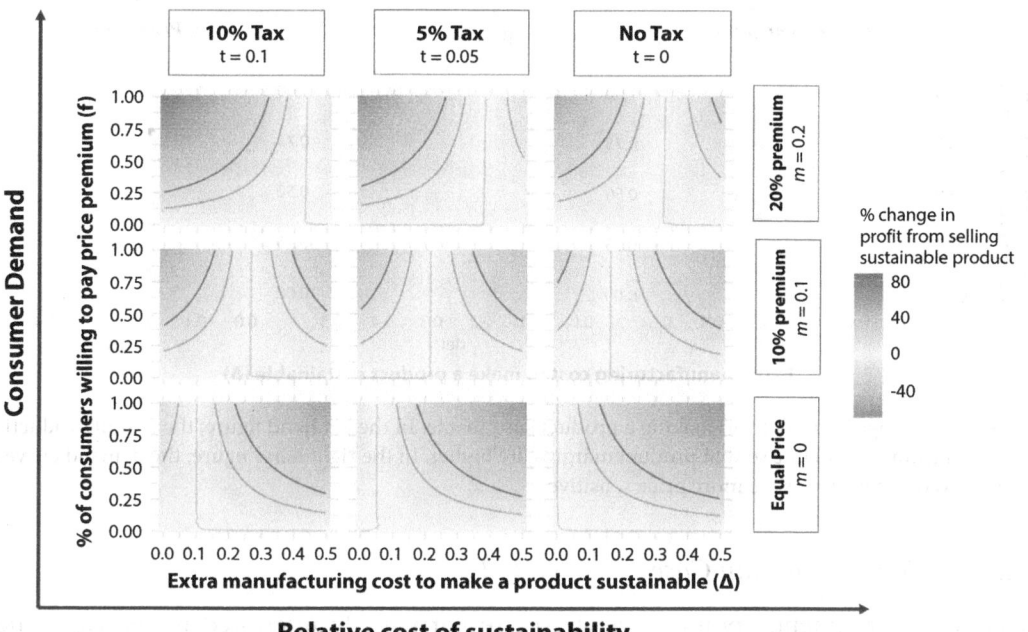

Figure 7.2 The colored regions represent the change in profit from selling a sustainable product (red is a decrease in profit, blue is an increase in profit). In this case, it is assumed that the (non-sustainable) product cost is 90% of the price. The horizontal grid axis represents 3 levels of taxation (0, 5%, 10% of cost) for the non-sustainable product option applied to the cost of manufacturing. Within each grid cell, the horizontal axis represents the additional cost (in %) to make a product sustainable. The vertical grid axis represents three levels of price premium (m; 0,10%,20%) for the sustainable product. Within each grid cell, the vertical axis represents the % of consumers willing to pay the price premium (f).

7.2.2.1 Impact of Product Characteristics and Market Structure

In Figure 7.2, we did not vary two parameters that relate to product characteristics: product margins and substitutability. When products have more substitutes, the consumers are more price sensitive and the slope of the demand curve changes. In Figure 7.3, we see the impact of changing these product characteristics on the sustainability ratio. On the left-hand side, we see the impact of reducing the product cost (i.e. the products have a higher margin). In this scenario, the contours move further apart suggesting that the impact of cost changes and demand changes on profit are reduced. On the other hand, when there are more substitutes available for a product (and margins are low), the contours are closer together, suggesting that the impact of cost changes is magnified. Margins can also be lower when there is increased market competition with large number of players in the market. In scenarios with more product substitutes and higher competition, it may be more challenging to transition towards sustainable production as the impact of higher costs leads to even greater losses due to reduced demand. In these cases, reducing the costs of sustainable products is vital to increasing sustainability.

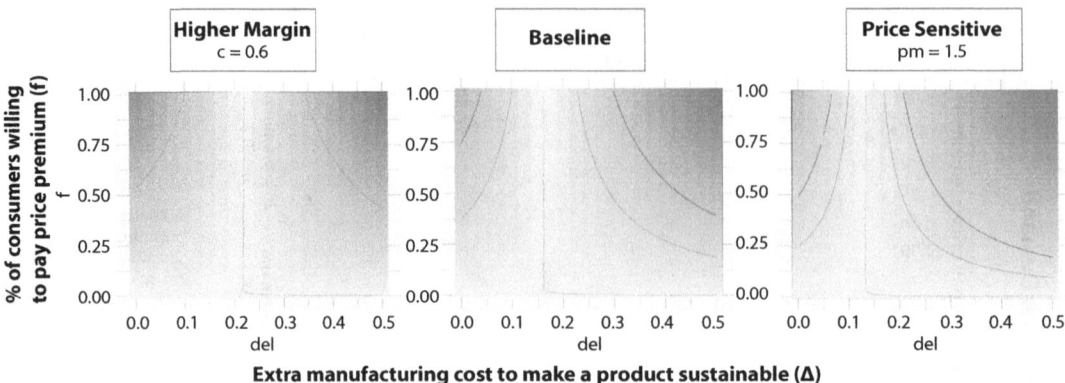

Figure 7.3 Change in profit from making a product sustainable. In the left hand figure, the cost of production is reduced from the baseline so that product margins are higher. In the right hand figure, the demand curve is changed so that consumers are more price sensitive.

7.2.2.2   Full Substitution Case i.e. when f=1

From an environmental point of view, we are particularly interested in the case where all consumers buy the sustainable product i.e. f=1 and full substitution occurs from the non-sustainable to sustainable case. Figure 7.4 demonstrates how profits are impacted in this scenario as a function of the premium and the cost. Considering the first case when there is no taxation, we find a linear relationship between the margin and the cost. Charging a higher premium, while keeping costs low can increase profits (if all consumers are willing to pay the higher premium). In reality, it is unlikely that all consumers will continue to buy

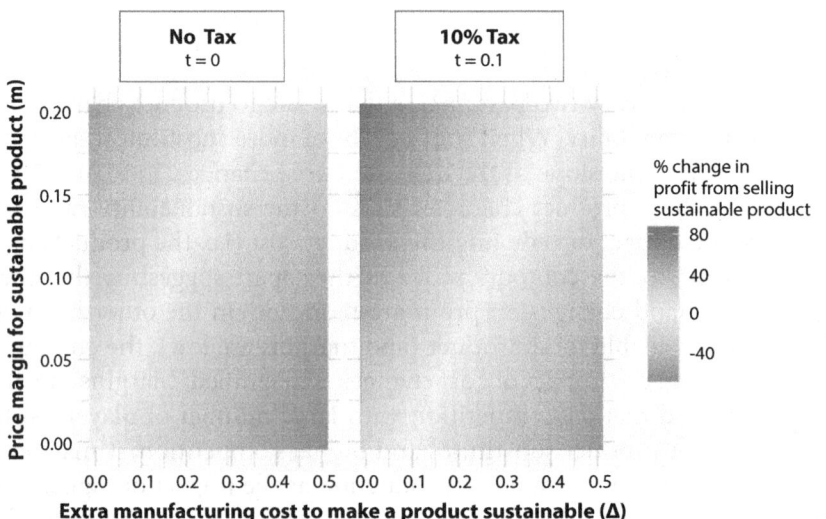

Figure 7.4 In this figure, we show the profit from a full substitution to a sustainable version of the product (f=1). On the left-hand side, we assume no taxation and see the trade-off between price margin and excess cost to make a product sustainable. For the product to be sustainable, all consumers have to be willing to pay a premium that is higher than the excess cost. In the right-hind side figure, we apply a 10% tax on the non-sustainable product and observe that it creates a larger area for sustainable production.

the sustainable product while the price premium is increased. A response to this could be to increase taxation on the non-sustainable product, as shown in the Figure 7.4. Increasing the tax increases the region where it is profitable to be sustainable such that profits may increase even when consumers are charged no price premium. In this situation, the tax has to be larger than the extra cost to make products sustainable. This large tax will then incentivise all manufacturers to switch to making sustainable products even without a price premium. The tax could lead to higher costs to consumers as manufacturers may increase the price premium to further drive up profit. The extent to which they can increase the premium depends on the shape of the consumer demand curve (as consumers may not be willing to buy the product at high premiums).

7.2.3 Case Study on Sneakers: Full Substitution of Virgin Polyester Upper to Recycled Polyester

Our goal in this section is to demonstrate the utility of our model by comparing the profitability of two shoe designs: one which uses more sustainable materials and one which uses less sustainable materials. Life cycle assessments (LCAs) determine the environmental impacts of the two designs and we input values for the costs of production and consumer demand. We are specifically looking at a case where there is consumer demand for more sustainable products. A report by Simon-Kucher & Partners surveyed 10,000 people and found that a third of consumers may be willing to pay a 25% premium on goods that are known to be more sustainable [21]. For the non-sustainable shoe option, we apply three levels of carbon taxes to study how policymakers could use this model to study the effect of policies on producers' decisions to make sustainable products. All parameter inputs for this case study are described in Table 7.2.

For our case study, we assume that it costs a company (the producer) $48 to make a pair of shoes (including manufacturing cost, freight, overhead, marketing etc) that it sells to retailers for a variable price that retailers sell to end-consumers for $100. From the producers' point-of-view, the buyer is the retailer and they make decisions based on the retailers' demand curve. We assume that the retailer has a $44 operating cost so can pay up to a *maximum* of $56 to buy the shoe-pair from the producer. These cost estimates are rough averages reported by 'solereview' that are taken from adidas' and Footlockers' 2015 income statement (and a few additional assumptions) [22]. We assume that an average pair of shoes is 674kg in weight with virgin polyurethane and polyester being the main components [23, 24]. We assume, based on LCAs, the net emissions impact of the shoe pair is 12kg CO2e per pair [23]. For the sustainable shoe option, the materials assumed are recycled polyester and bio-based sugarcane, with manufacturing in factories which use 100% renewable energy. The materials use and manufacturing assumptions are based on Allbirds' 'Futurecraft.Footprint' shoe which has an estimated emissions impact of ~3kgCO2 per shoe pair [24]. We assume that the cost of producing the sustainable shoe is 30% higher than the non-sustainable option. This assumption is for demonstration purposes as actual cost data is typically proprietary. Companies using this tool for their own products can input actual cost numbers for a more realistic estimate.

Table 7.2 Input values for model parameters used for sneakers case study.

Input	Description	Value
c'	Cost of non-sustainable production in USD (absolute value)	48
Δ	Excess Cost to make a product sustainable. Input is normalized as a proportion of the non-sustainable product cost c.	0.3
t	Tax on the non-sustainable product. We assume 3 scenarios corresponding to a \$0/tCO2e, \$50/tCO2e and \$100/tCO2e carbon tax. We assume that the carbon tax applies only to the difference in emissions from using a recycled vs virgin polyester upper. We then normalize the carbon tax by the cost of production c	0, 0.02, 0.04
m	We assume an average margin that environmentally-sensitive consumers are willing to pay based on market reports.	0.25
f	The proportion of consumers buying the greener shoe is assumed based on market reports that a third of consumers may be willing to pay extra for sustainable products. Because of a difference in price, the total consumption can still reduce based on the demand curve.	.33
p_m	Price at which no consumption occurs.	56

We assume three levels of carbon taxation, \$0/tCO2e, \$100/tCO2e and \$200/tCO2e, and study the impact of these policies on profitability and decision-making. For simplicity, we assume that the carbon tax only applies to the difference in emissions between the two sneaker designs of \$9kg CO2/pair. The carbon tax in the 3 scenarios would be \$0, \$0.9 and \$0.18 (or 0%, 2%, and 4% of costs). In the baseline scenario, 400 million pairs of sneakers are assumed to be sold in the US [25]. As carbon taxes apply on the non-sustainable shoe or additional costs incurred for the sustainable shoe, prices rise and sales drop (Results for case studies are displayed in Table 7.3). In the case with no taxation, the producers' profits reduce by almost 10% if they offer a sustainable shoe, as the costs are greater than the margin consumers are willing to pay. In the absence of policies, producers would not be incentivised to switch to recycled materials unless the materials cost reduced or demand was higher and consumers were willing to pay a premium. When there is a \$100/t Carbon Tax, the profits in the case that only non-sustainable shoes are sold reduces due to the tax from \$1.6 billion to \$1.24 billion. The additional tax raises the price of shoes and decreases shoe sales. In this scenario, the profits are still slightly larger than they would be if a sustainable shoe was introduced to the market at a 25% premium and purchased by a third of the consumers. However, when a \$200/t Carbon Tax is applied, it is now more profitable for the producers to sell the sustainable shoe.

It should be noted that the reason it is *relatively* more profitable to be sustainable is that the profits from using the non-sustainable reduce due to the taxation. Some of the tax revenue can be used to support companies that are making sustainable decisions via investment in recycling infrastructure that can reduce materials costs. Even though carbon taxes incentivize sustainable shoe production, the reduction in industry profits can have adverse effects on other policy goals such as jobs. Policymakers can use this tool to determine the optimal

Table 7.3 Results for 3 taxation scenarios examined for sustainable sneaker production.

	Tax= $0/t		Tax= $100/t		Tax= $200/t	
Sustainable Shoe Sold?	**No**	**Yes**	**No**	**Yes**	**No**	**Yes**
Price (S)	-	$65.45	-	$65.83	-	$66.2
Price (NS)	$52	$52	$52.48	$52.48	$52.96	$52.96
Profit/pair (S)	-	$3.05	-	$3.43	-	$3.8
Profit/pair (NS)	$4	$4	$3.52	$3.52	$3.04	$3.04
Quantity Sold	400 million	400 million	352 million	352 million	304 million	304 million
Total Profit	$1.6 billion	$1.47 billion	$1.24 billion	$1.22 billion	$0.93 billion	$0.99 billion
Tax Collected	0	0	$317 million	$209 million	$547 million	$361 million

trade-off point that incentivizes sustainable production without a significant impact on employment. Conversely, producers can use this model to determine a cost target such that it is more profitable to sell the sustainable shoe given a tax level and consumer demand. They can also explore ways that would increase the premiums consumers would be willing to pay. To study other scenarios and policies, we have created a web tool that allows users to input values and see which product options are more sustainable. Use the tool at https://bhuwalka.shinyapps.io/app-1/.

7.3   Conclusions & Recommendations

Through our model, we find that the profitability of sustainable products is determined by a complex interplay of various factors including production costs, consumer demand and willingness to pay a premium, competition, and government regulations and incentives. The developed tool can help in quantifying the effects of these factors in different scenarios and identify important levers for policymakers as well as private companies. Policymakers can implement taxes and penalties, and support the development and upscaling of sustainable materials and processes to decrease the additional cost that companies pay. Also, they can educate and engage consumers to increase their willingness to pay for sustainable products. Private companies, on the other hand, can invest in marketing efforts to change consumers' willingness to pay, or in innovation to achieve technological advancements that will minimize additional costs of sustainability. As our calculations show, the interaction between the different levers strongly influences the effectiveness of an action. In scenarios where demand is very elastic, it is important to focus efforts on decreasing additional costs of sustainability by investing in more efficient technologies and supply chains. At the same time, policy can focus on financial levers such as subsidizing more sustainable products or penalizing a product with higher environmental impacts.

In this paper, we provided a simple model that is applicable to a wide range of products and industries so that the model is easy to use for corporate decision-makers and policy-makers alike. The model can also be combined with other decision-support tools such as life cycle assessments (LCAs). LCAs can determine the relative environmental impacts of two differentiated products and then the inputs for the consumer demand variables can be set based on the results of the LCA. For example, if the two products are found to have a large difference in emissions impacts, a greater proportion of consumers may be willing to pay a premium. Given certain demand conditions, the model can help producers determine cost targets that manufacturers need to meet so that the green product becomes more viable than the non-green option. To help determine this target, we presented a formula for the threshold cost at which it is equally profitable to manufacture the green product (as a function of demand and policy variables). Designers and manufacturers can evaluate the relative ease of meeting the cost target to identify product greening opportunities. Once product greening opportunities are identified via our model, a more detailed analysis can be undertaken to create a concrete product development plan. This also helps firms conduct longer-term planning as policy changes take place: a cost target may seem difficult to reach at present but may be possible when there is a subsidy provided (or a tax implemented on the non-green option).

Moreover, the model helps policymakers identify tax (or subsidy) levels to incentivize green design. For example, in the case study for sneakers, we demonstrated how much carbon tax would be needed to make a shoe made with recycled polyester more profitable than one made with virgin polyester. While it is straightforward to calculate a tax level that simply makes the two product costs equal, the contribution of our model is the endogenous inclusion of demand-side variables that help identify the impact of this tax on product sales and industry profits. The results can help make policy design more collaborative and identify policies that promote green design without significantly harming industry profitability (which has negative impacts on factors such as employment and wages).

Future work should focus on collecting real-world empirical data on consumer demand and production costs. Once this data is used to calibrate our model, it can help to determine where firms should focus their efforts in order to maximize sustainability e.g. increasing demand or reducing costs. Furthermore, efforts should be directed in modelling some of the functional forms more realistically. The demand curve is very simplified in our example and can be re-formulated and calibrated with historical demand data. It is specifically important to collect information on consumer demand for green products and willingness to pay for sustainable goods. The current academic literature on how much consumers are willing to pay for sustainable products is mixed, as echoed by Polyportis et al. who review consumer preferences for products made out of recycled material [26]. For example, while Herrmann et al took a mixed methods approach and found that people are willing to pay more for sustainable food packaging [27], De Marchi et al. demonstrated that consumers are willing to pay more for bio-based options than bottles made out of virgin PET but less for recycled PET bottles [28]. Similarly, while Pretner et al. finds that consumers are willing to pay less for circular garments compared to traditional ones [29], Magnier et al. find that a subset of consumers are willing to pay more for clothing made form ocean plastic [30]. Consumer preferences also vary across consumer demographics and geography. Better understanding of consumer preferences will be key to improving the real-world applicability of the equilibrium model we present in this paper.

Efforts should also be focused on understanding the profitability under different market structures and accounting for game-theoretic competition between firms. For example, under perfect competition, the producer is a 'price taker' and prices are considered endogenous. In this scenario, the optimization decision of the producer is to decide what quantity of the sustainable good they will produce given their cost structure and externally determined prices. We will demonstrate alternative models under different market structures in future work. Another major assumption in our simplified model is that we do not endogenously consider the co-dependence of the input parameters. For example, the cost of production is assumed as constant in this model but can be modified to be a function of the quantity of production and therefore influence in price.

While we input various parameters independently and study the combinations of their values that lead to profitability, it is important to note that they are often very correlated. For example, the size of the price premium (f) will influence the number of consumers willing to buy the product (m). In interpreting the results of our work, it is important to keep in mind that it is unrealistic that policymakers or producers can increase f but not affect m. Moreover, it could be plausible that taxes or subsidies directly or indirectly influence the excess cost of producing sustainable goods, which can be incorporated by modeling modelling Δ as a function of t. Of course, it is important to note that also other factors that cannot be quantified as easily play a role and should therefore be considered when deciding about the profitability of a sustainable product. Some examples are brand reputation and customer loyalty, marketing efforts, and product life cycle. A thorough analysis of these factors is necessary to make informed decisions about implementing new products or policies.

References

1. "The impact of textile production and waste on the environment (infographic) | News | European Parliament." https://www.europarl.europa.eu/news/en/headlines/society/20201208STO93327/the-impact-of-textile-production-and-waste-on-the-environment-infographic (accessed Jan. 31, 2023).
2. P. Taticchi, P. Garengo, S. S. Nudurupati, F. Tonelli, and R. Pasqualino, "A review of decision-support tools and performance measurement and sustainable supply chain management," *International Journal of Production Research*, vol. 53, no. 21, pp. 6473–6494, Nov. 2015.
3. M. Moktadir, Y. Mahmud, A. Banaitis, T. Sarder, and M. Khan, "Key Performance Indicators for Adopting Sustainability Practices in Footwear Supply Chains," *E a M: Ekonomie a Management*, vol. 24, pp. 197–213, Mar. 2021.
4. S. Khanra, P. Kaur, R. P. Joseph, A. Malik, and A. Dhir, "A resource-based view of green innovation as a strategic firm resource: Present status and future directions," *Business Strategy and the Environment*, vol. 31, no. 4, pp. 1395–1413, 2022.
5. P. Rao and D. Holt, "Do green supply chains lead to competitiveness and economic performance?," *International Journal of Operations & Production Management*, vol. 25, no. 9, pp. 898–916, Jan. 2005.
6. Y. Ling, J. Xu, and M. A. Ülkü, "A game-theoretic analysis of the impact of government subsidy on optimal product greening and pricing decisions in a duopolistic market," *Journal of Cleaner Production*, vol. 338, p. 130028, Mar. 2022.

7. L. Liu, Z. Wang, J. Xu, and Z. Zhang, "Green baton: how government interventions advance green technological innovation," *Environ Dev Sustain*, Jul. 2022.

8. W. Zhu and Y. He, "Green product design in supply chains under competition," *European Journal of Operational Research*, vol. 258, no. 1, pp. 165–180, Apr. 2017.

9. Y. Zhang and W. Zhang, "Optimal pricing and greening decisions in a supply chain when considering market segmentation," *Ann Oper Res*, Mar. 2022.

10. C. Chen, "Design for the Environment: A Quality-Based Model for Green Product Development," *Management Science*, vol. 47, no. 2, pp. 250–263, Feb. 2001.

11. G. Bi, M. Jin, L. Ling, and F. Yang, "Environmental subsidy and the choice of green technology in the presence of green consumers," *Ann Oper Res*, vol. 255, no. 1, pp. 547–568, Aug. 2017.

12. D. Li, "Dynamic optimal control of firms' green innovation investment and pricing strategies with environmental awareness and emission tax," *Managerial and Decision Economics*, vol. 43, no. 4, pp. 920–932, 2022.

13. D. Ghosh, J. Shah, and S. Swami, "Product greening and pricing strategies of firms under green sensitive consumer demand and environmental regulations," *Ann Oper Res*, vol. 290, no. 1, pp. 491–520, Jul. 2020.

14. Y. Ling and J. Xu, "Price and greenness competition between duopoly firms considering consumer premium payments," *Environ Dev Sustain*, vol. 23, no. 3, pp. 3853–3880, Mar. 2021.

15. K. Liu, W. Li, E. Cao, and Y. Lan, "A behaviour-based pricing model of the green product supply chain," *Environ Sci Pollut Res*, vol. 28, no. 46, pp. 65923–65934, Dec. 2021.

16. Y. Jiang, X. Ji, J. Wu, and W. Lu, "Behavior-based pricing and consumer fairness concerns with green product design," *Ann Oper Res*, Feb. 2023.

17. A. Mukherjee and M. Carvalho, "Dynamic decision making in a mixed market under cooperation: Towards sustainability," *International Journal of Production Economics*, vol. 241, p. 108270, Nov. 2021.

18. Q. Zhang, Q. Zhao, X. Zhao, and L. Tang, "On the introduction of green product to a market with environmentally conscious consumers," *Computers & Industrial Engineering*, vol. 139, p. 106190, Jan. 2020.

19. A. Yenipazarli and A. Vakharia, "Pricing, market coverage and capacity: Can green and brown products co-exist?," *European Journal of Operational Research*, vol. 242, no. 1, pp. 304–315, Apr. 2015.

20. Y. Perlman, Y. Ozinci, and S. Westrich, "Pricing decisions in a dual supply chain of organic and conventional agricultural products," *Ann Oper Res*, vol. 314, no. 2, pp. 601–616, Jul. 2022.

21. "Simon-Kucher_Global_Sustainability_Study_2021.pdf." Accessed: Jan. 29, 2023. [Online]. Available: https://www.simon-kucher.com/sites/default/files/studies/Simon-Kucher_Global_Sustainability_Study_2021.pdf

22. solereview, "What does it cost to make a running shoe?," *https://www.solereview.com/*, May 22, 2016. https://www.solereview.com/what-does-it-cost-to-make-a-running-shoe/ (accessed Jan. 29, 2023).

23. L. Cheah *et al.*, "Manufacturing-focused emissions reductions in footwear production," *SSRN*, Dec. 2012, Accessed: Jan. 29, 2023. [Online]. Available: https://dspace.mit.edu/handle/1721.1/102070

24. "Carbon Footprint Calculator & Tools | Manually Calculate Carbon Footprint, For Students, Businesses," *Allbirds*. https://www.allbirds.com/pages/carbon-footprint-calculator (accessed Jan. 29, 2023).

25. "Sneakers - United States | Statista Market Forecast," *Statista*. https://www.statista.com/outlook/cmo/footwear/sneakers/united-states (accessed Jan. 29, 2023).

26. A. Polyportis, R. Mugge, and L. Magnier, "Consumer acceptance of products made from recycled materials: A scoping review," *Resources, Conservation and Recycling*, vol. 186, p. 106533, Nov. 2022.

27. C. Herrmann, S. Rhein, and K. F. Sträter, "Consumers' sustainability-related perception of and willingness-to-pay for food packaging alternatives," *Resources, Conservation and Recycling*, vol. 181, p. 106219, Jun. 2022.

28. E. De Marchi, S. Pigliafreddo, A. Banterle, M. Parolini, and A. Cavaliere, "Plastic packaging goes sustainable: An analysis of consumer preferences for plastic water bottles," *Environmental Science & Policy*, vol. 114, pp. 305–311, Dec. 2020.

29. G. Pretner, N. Darnall, F. Testa, and F. Iraldo, "Are consumers willing to pay for circular products? The role of recycled and second-hand attributes, messaging, and third-party certification," *Resources, Conservation and Recycling*, vol. 175, p. 105888, 2021.

30. L. Magnier, R. Mugge, and J. Schoormans, "Turning ocean garbage into products–Consumers' evaluations of products made of recycled ocean plastic," *Journal of Cleaner Production*, vol. 215, pp. 84–98, 2019.

8

Clean Energy Technologies, Critical Materials, and the Potential for Remanufacture

T.E. Graedel

Yale University, New Haven, CT, USA

Abstract

Planet Earth is in the early stages of a transition to "clean" (non-fossil fuel) energy production, with wind and solar power, battery technology, and nuclear power all envisioning rapid growth. Accordingly, energy transmission from point of generation to point of use must evolve dramatically as well. Concurrently, most major technologies (buildings, heavy equipment, mining, transport) are evolving into possessing continuous performance monitoring. All of these technologies are enabled by critical materials, whose potential for significantly increased production from virgin ores appears problematic. As a consequence, a significant opportunity for greatly enhanced remanufacturing of energy- and technology-related equipment will occur during the decades from now until at least mid-century. This discussion explores the materials involved in clean energy and product technologies, the potential of these technologies for remanufacture and reuse, and the apparent need for a strong national focus on creating a near-circular economy if clean energy technologies are to be functional over time.

Keywords: Critical materials, remanufacture, solar power, wind power

8.1 Introduction

Materials science and its associated technologies are in a period of rapid evolution, leading to demands for a very wide range of material diversity across much of the product spectrum of today and tomorrow. Common examples of this trend include smart phones (now containing at least seventy different elements [1]) and automobiles (now containing ~76 different elements [2]). Indeed, the modern automobile is now so complex that it is sometimes compared to a spaceship. In many instances, the elements contained in these varied products are not employed in pure form, but rather in complex forms such as mixtures or alloys.

Many of the elements of modern technology are designated as critical by countries or regions that made such an assessment [Australia [3]; Canada [4], European Union [5], Japan [6, 7]. A "periodic table" illustration of that situation is shown in Figure 8.1. This broad concern indicates that there is a significant level of doubt over the long-term availability of these elements in amounts sufficient to enable the full realization of their potential

Email: thomas.graedel@yale.edu

Nabil Nasr (ed.) Technology Innovation for the Circular Economy: Recycling, Remanufacturing, Design, Systems Analysis and Logistics, (95–100) © 2024 Scrivener Publishing LLC

benefits. As a consequence, products utilizing these elements increasingly need to be recovered and then remanufactured as necessary or recycled as not, so as to avoid the senseless approach of investing energy and effort in their acquisition and employment and then discarding them after a single use.

The situations just described seem likely to have major influence upon the remanufacturing and recycling industries. At a stage of technological development when materials diversity in products is increasingly great, little thought is being given to the implication of these trends for remanufacturing and recycling. This paper explores the evolving technology in some detail, from the perspective that REMADE should mount a significant effort toward preparing to deal with tomorrow's products rather than focusing largely on the industrial products of yesterday and today.

8.2   Modern Examples of Materials Complexity, and Their Implications

For more than a century, the energy employed to enable a wide variety of technologies and to create more desirable living conditions has been supplied by fossil fuels. Now, however, businesses and government entities are in a vigorous effort to transform the technological world. And, at a time when climate concerns are widespread and intense, the focus of energy provisioning is on clean energy technologies. The enabling technologies are, however, dependent on an appropriate set of raw materials, supplies of a number of those materials are under pressure worldwide, and a substantial majority of all elements have been designated as "critical" by governments worldwide [3–7]. When these designations are plotted on the periodic table (Figure 8.1), it is easy to see that most non-radioactive elements have been so designated.

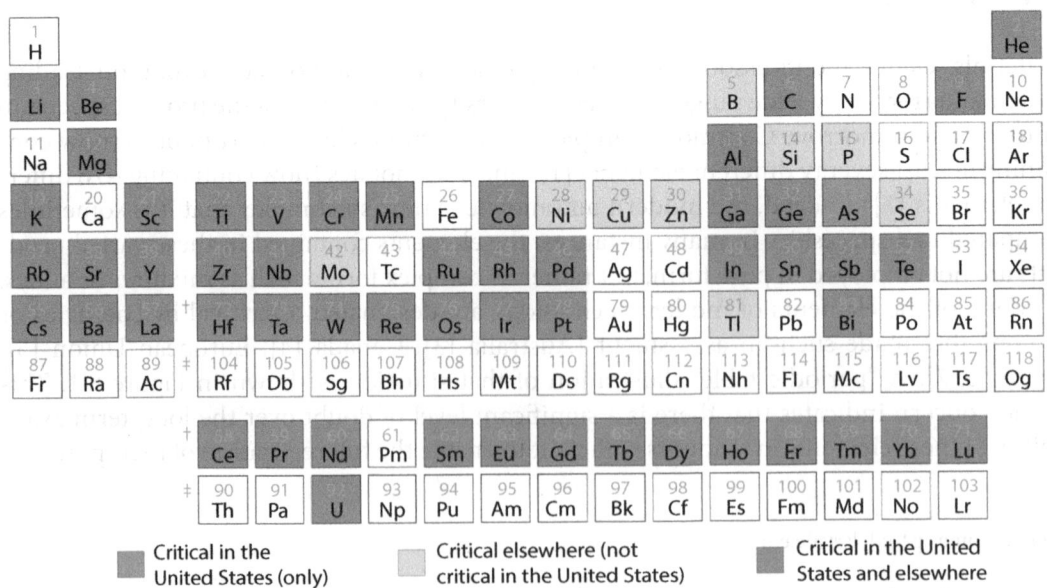

Figure 8.1 Current designations of critical materials by national and international governments.

An important realization from critical materials and clean energy perspectives is that new clean energy technologies are rapidly being deployed in the United States and elsewhere. This use brings with it an increasingly heightened demand for a variety of critical (and a few non-critical) materials. Table 8.1 provides typical critical element lists of these technologies.

A periodic table of materials in clean technology applications (Figure 8.2) shows that more than twenty elements are required as demand for these technologies accelerates.

Energy-related technologies and their materials requirements must compete for resources with the accelerations in demand for materials to serve other expanding technologies. The

Table 8.1 Elements in clean energy technologies.

Energy technology	Elements typically included	References
Wind power	Neodymium, praseodymium, terbium, dysprosium	[8] (revised 2017)
Solar power	Lithium, silver, cadmium, tellurium,	[9]
Battery power	Manganese, cobalt, nickel	Nitta *et al.* [10]; [11]
Nuclear power	Uranium, thorium, iron, chromium, nickel	[12]
Energy transmission	Aluminum, magnesium, silicon, vanadium, iron, copper, zinc, cadmium, niobium	[13], [14]

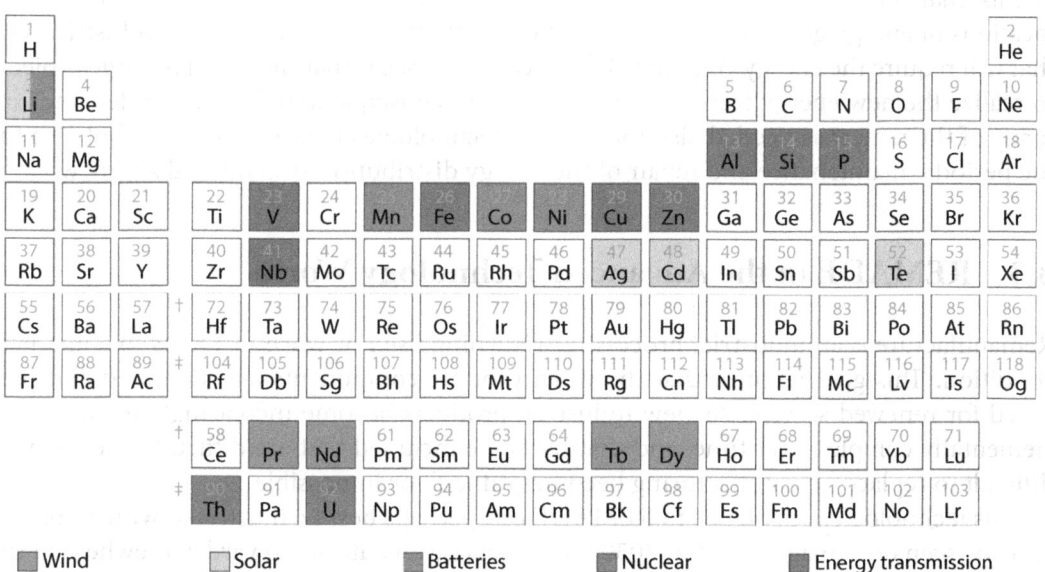

Figure 8.2 Elements in clean energy applications.

All options

Figure 8.3 The National Renewable Energy Laboratory 2035 "all options" energy transmission scenario [21]. Transmission lines – gray; land-based wind – blue; photovoltaics – red; offshore wind – black.

list of these technologies could be very long, but surely includes electric vehicles for transport of people and goods [15], agricultural machinery [16], construction machinery [17, 18]; and mining equipment [19, 20].

A related issue as clean energy technologies expand is that their spatial locations will generally not be where energy has traditionally been generated (often near ports, highways, and rail lines) to enable the distribution of fossil fuels to power plants. Rather, they will be located where natural sources of energy are abundant: regions with dependable winds, those with high solar intensities and minimal cloud cover, ocean margins, etc. The result will be a dramatic enhancement in the construction of new energy transmission systems. A study of this challenge by the U.S. National Renewable Energy Laboratory predicts the major locations of energy generation by 2035 and of the locations of populations, industries, etc. that will require the energy (Figure 8.3). The consequence is that there will be large projects to realize the new energy transmission vision, with consequential competing demands on many of the materials needed also for the new technologies mentioned above, but also for the periodic maintenance and repair of the energy distribution systems, old and new.

8.3   REMADE in the Advanced Technology World

Remanufacture is an industrial process for returning used or worn products to a like-new condition. This goal requires that the worn product components are identified and prepared for renewed service. As new industrial products become increasingly intricate and elementally complex over time, and that parts are included that are difficult to identify or difficult to replace, remanufacturing becomes difficult or impossible.

This technological challenge to REMADE can perhaps best be illustrated with a not-un-realistic scenario. Imagine that in 2030 you supervise an auto scrap yard somewhere in the United States. The vehicles that you receive may be any or all of the following:

2005-era vehicles with cast aluminum engine blocks (Al 7Si 0.4Mg 0.2Cu 0.2Fe 0.1Zn).

2015-era vehicles with high-strength-low-alloy components (Fe 0.1Re 0.1V).

2017-era vehicles with substantial electronics buried within.

2020-era vehicles with extensive use of alloy steel and other metallurgically-complex components.

2025-vehicles with electric engines (Li Ni Mn Co Al etc.).

Can you create or maintain the capability of identifying, remanufacturing, and recycling the products of such a mix of technologies? At a minimum, doing so will require extensive disassembly, precise component identification with sophisticated equipment, removal and replacement of non-functioning but highly complex components, and reinstalling and evaluating the performance of the electronic operating systems. Advanced analytical training and the acquisition of associated equipment will almost certainly be involved.

To summarize, the remanufacturing/recycling sector of the U.S. economy is rapidly approaching a major challenge: that of dealing with more and more complex and challenging products and of retaining as much of product function and value as possible. The transition is likely to require significantly upgraded analytical capability and well educated and trained personnel. To accept this challenge is to play a vital role in enhancing the technical and manufacturing capabilities of the country, while to refuse the challenge will be to relegate materials technologies to other countries and regions that accept it. REMADE's future is either a much more highly technological one, or it is one of ignoring the handwriting on the wall and becoming moribund.

References

1. Rohrig, B., Smartphones: Smart Chemistry, ChemMatters, April/May 2015, 10-12, 2015, https://www.acs.org/education/resources/highschool/chemmatters/past-issues/archive-2014-2015/smartphones.html.

2. Bhuwalka, K., Field, F.R. III, De Kleine, R.D., Kim, H.C., Wallington, T.J., and Kirchain, R.E., Characterizing the changes in material use due to vehicle electrification, Environmental Science & Technology, 55, 10097-10107, 2021.

3. Australian Government, Australian Critical Mineral Prospectus, p. 172, 2020.

4. Government of Canada, Canada's Critical Minerals List 2021, https://nrcan.gc.ca/sites/nrcan/files/mineralsmetals/pdf/Critical_Minerals_List_2021-EN.pdf, 2021.

5. Blengini, G.A., et al., European Commission, Study on the EU's list of Critical Raw Materials, ISBN-978-92-76-21056-6, Brussels, Belgium, 2022.

6. Nakano, J., The Geopolitics of Critical Minerals Supply Chains, Center for Strategic and International Studies, Washington, D.C., 2021.

7. U.S. Geological Survey, U.S. Geological Survey Releases 2022 List of Critical Minerals, https://www.usgs.gov/news/national-news-release/us-geological-survey-releases-2022-list-of-critical-minerals, 2022.

8. Moore, C., M. Hand, M. Bolinger, J. Rand, D. Heimiller, and J. Ho, 2015 cost of green energy, NREL/TP-6A20-66861, Revised 2017.

9. Hudedmani, M.G., Soppimath, V., and Jambotkar, C., A study of materials for solar PV technology and challenges, European Journal of Applied Engineering and Scientific Research, 5 (1), 1-13, 2017.

10. Nitta, N., Wu, F., Lee, J.T., and Lushin, G., Li-ion battery materials: present and future, Materials Today, 18 (5), 2014.

11. Friedemann, A., Peak energy and resources, climate change, and the preservation of knowledge, https://energyskeptic.com/2021/battery-materials-rare-declining, 2022.

12. Maziasz, P.J., and J.T. Busby, Properties of austenitic stainless steels for nuclear reactor applications, Comprehensive Nuclear Materials, 2, 267-283, 2012.

13. Electrical4U, Materials used for transmission line conductor, https://www.electrical4u.com/materials-used-for-transmission-line-conductor/, 2020.

14. Somavanshi, C., Materials used in transmission lines and their specifications, https://www.Cselectricalandelectronics.commaterials-used-in-transmission-lines-and-their specifications/, 2022.

15. Field, F.R. III, T.J. Wallington, M. Everson, and R.E. Kirchain, Strategic materials in the automobile: A comprehensive assessment of strategic and minor metals use in passenger cars and light trucks, Environmental Science & Technology, 51, 14436-14444, 2017.

16. Crummett, D., Ag prepares for electric-powered future, https://www.farm-equipment.com/articles/20408-ag-prepares-for-electric-powered-future, 2022, accessed Jan. 16, 2023.

17. Volvo CE, Reviews of electric construction equipment by early adopters, https://volvoceblog.com/benefits-of-early-electric-construction-equipment-adopters/, 2021, accessed Jan. 23, 2023.

18. Conexpo Con/Agg, More electric-powered vehicle options than ever, https://Conexpoconagg.com/news/electricpowered-construction, accessed Jan. 25, 2023.

19. Chakravorty, A., Underground robots: How robotics is changing the mining industry, Eos, 100, https://doi.org/10.1029/2019EO121687, 2019.

20. Lugana, J., Rock-solid communications for the underground mine, Global Mining Review, 5 (4), 30-32, 2022.

21. National Renewable Energy Laboratory, Examining Supply-Side Options to Achieve 100% Clean Energy by 2025, 161 pp. Golden, CO, 2022.

Part 2

ENABLING A CIRCULAR ECONOMY THROUGH AI & MACHINE LEARNING

Part 2

9

Towards Eliminating Recycling Confusion: Mixed Plastics and Electronics Case Study

Amin Sarafraz[1]*, Nicholas Alvarez[1], Jonas Toussaint[2], Felipe Rangel[1], Lamar Giggetts[2] and Shawn Wilborne[2]

[1]University of Miami, Miami, FL, USA
[2]Lidvizion, Miami, FL, USA

Abstract

Current waste and recycling programs are suffering from consumer confusion, which causes contamination. To prevent contamination, we developed an Application Programming Interface (API) that uses object recognition and the consumer's location information to identify recyclable materials and inform the consumer. Our API can be used to educate consumers on how to properly handle their materials according to local laws and materials accepted in specific collection programs. The API delivers proper materials handling responses combining computer vision and database analytics. As an example use-case for our API, we collected a custom dataset of different mixed plastics and electronics and fine-tuned several existing object recognition models using our dataset. We assess the performance of each model on our custom dataset with different metrics. We describe all the components of our API, which provides a blueprint for deploying existing state-of-the-art computer vision models in a commercial application.

Keywords: Recycling, machine learning, artificial intelligence, convolutional neural networks

9.1 Introduction

Recycling confusion leads to contamination, which causes 68% of all recyclables to end up in landfills, incinerators, or the environment [12]. This demonstrates a considerable need for a tool to reduce confusion by identifying and matching the type of material with a specific collection program. Regardless of their location, residents must be able to receive simple, immediate, and accurate insights to ensure proper recycling behavior. Currently, recycling educators lack an efficient and cost-effective way of eliminating contamination from the waste and recycling streams. Current methods include feet-on-the-street inspections and tagging, rejecting contaminated carts, issuing direct mailers and bill inserts, tabling, billboards, bus-wraps, and digital media outreach [12]. However, these solutions lose momentum, are costly, and can only target a sample of the overall population, while still not delivering specific insights at the source; the point where a person decides on how

---

Corresponding author: a.sarafraz@umiami.ed

Nabil Nasr (ed.) Technology Innovation for the Circular Economy: Recycling, Remanufacturing, Design, Systems Analysis and Logistics, (103–114) © 2024 Scrivener Publishing LLC

to handle a material before placing it in their receptacle. This is further compounded by the abundance of misinformation spread through social media, television, and other advertisements about what is considered recyclable. These issues, coupled with the finite amount of consumer attention to recycling, without an immediate response when a person is confused at the receptacle, people will continue to contaminate waste and recycling streams, be confused, and continue to lose faith in our recycling systems.

Improper recycling can be hazardous and even deadly. Improper disposal of electronics, metals, chemicals, and other household hazardous waste can spark fires that destroy valuable recyclable materials, ignite collection trucks, and cause explosions at facilities. According to a report [3] published by the Environmental Protection Agency, in 2019 alone, 343 fires were reported at waste and recycling facilities in the United States and Canada, causing 49 injuries and two deaths. The report's author believed that this number of fires was underreported and that that were more than 1,800 facility fires in 2019. It is paramount to identify these materials before they enter either our recycling or waste stream, to ensure we create safe workplaces, and protect lives with proper material handling at the source of contamination, upstream when a consumer makes a decision on how to handle a material.

Residents currently have no simple way to know what is recyclable at a specific receptacle, especially considering the various receptacles available to consumers, including, but not limited to, curbside carts and bins, public receptacles, dumpsters, open-top containers, waste and recycling chutes in multifamily buildings, and other collection mechanisms.

This paper presents an Application Programming Interface (API) that could be used in solutions to address the recycling confusion (see section 9.3). We also introduce a new dataset for recycling materials called UM-LV Recycling dataset (see section 9.4). The API can immediately recognize different recyclable objects in images and send proper responses based on the consumer's location information. It could be used as a micro-service to develop smartphone apps or web applications. Our solution proposes that a person can scan an object using a mobile device and our API will provide a response based on their specific recycling program. It is envisioned that our solution must be robust enough to identify numerous household materials, and be flexible enough to adapt to various recycling programs. Thus, we will test the performance of multiple state-of-the-art object recognition models on our recycling dataset (see sections 9.5 and 9.6).

The technical barriers we are addressing in this paper are focused on developing a scalable API and using state-of-the-art object recognition models to identify mixed plastics and electronics at the source as an example case study.

9.2 Related Work

Currently, computer vision is used to identify materials on a conveyor belt for debris and litter in the material recovery facilities (MRFs) [22], but to the best of our knowledge, not on residential devices for immediate recycling clarity and at the point where contamination and confusion occur, the source.

Similar to this study is a work where researchers created a smartphone application to identify various types of recycled waste, including plastic, glass, metal, and garbage [7]. They used a deep convolutional neural network to train on the TrashNet dataset [11]. They obtained 82% test accuracy using transfer learning and fine-tuning pre-trained models.

In addition, other works have researched waste classification using ResNet-50 [1], Support Vector Machines (SVM) [13, 1], and multiple convolutional neural network architectures [16] to automate trash sorting tasks. Melinte *et al.* [10] used the TrashNet dataset to fine tune several types of Single Shot Detectors (SSD) and Regional Proposal Networks (RPN) for deployment on an autonomous ground-based robot system for waste collection. Wu *et al.* [21] have deployed a variant of the YOLO model [14] on robots to perform real-time garbage classification.

Although we were could not identify a comprehensive dataset of images for recycling materials, there is a commonly used dataset in waste classification called TrashNet [10, 7, 13, 5]. Researchers have also used Pascal Visual Object Classes [10], ImageNet [1], and the COCO dataset [10]. Other researchers created their datasets from scratch, see [22, 9].

9.3   Object Recognition API

Figure 9.1 shows the flowchart of our API. Although we used Amazon Web Services (AWS) for deploying our API, other cloud companies provide similar services.

9.3.1   Back-End Verification

Back-end verification begins with the initiation of the front-end camera. The user captures and transmits the image via API to the backend. To validate the API call, the proper headers need to be passed and the authentication key needs to be passed along with the request. Once authenticated, a Lambda function triggers model inference, which determines the type of material in the image. Then this response from the model is returned to the application and passed to the MongoDB query. The query includes the type and subcategory of the item in the image. This query searches the database for the specific item to determine

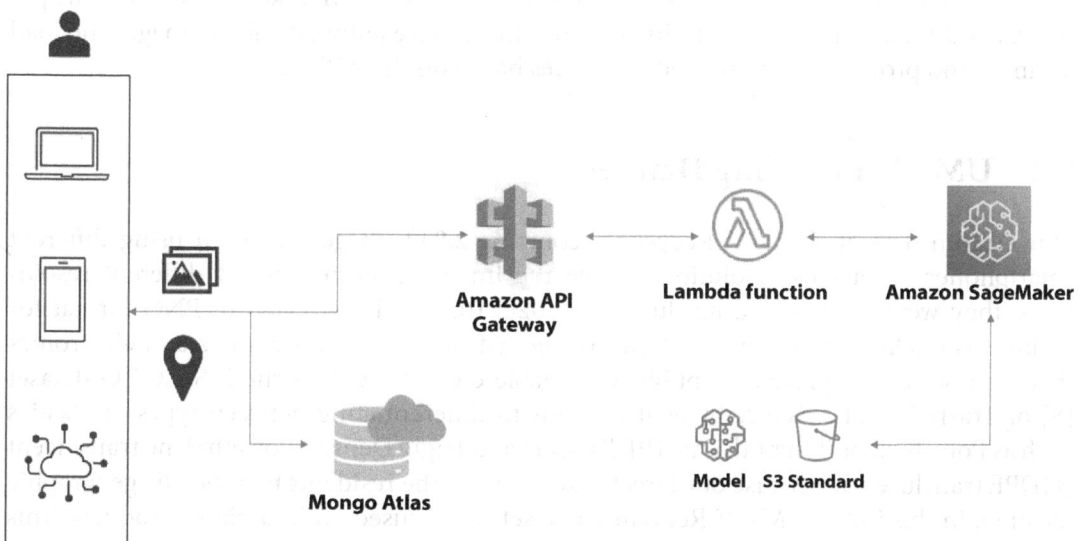

Figure 9.1 The flowchart of our API's software architecture.

if it is accepted in the recycling program. If the query is empty or returns a non-recyclable object, the object fails verification and the user interface displays a nonrecyclable message. Otherwise, the object is verified and information on how to recycle it is displayed.

9.3.2   Software Architecture

Data is stored and retrieved from MongoDB, a viral NoSQL database. Mongo Atlas is a cloud implementation that allows us to use the same database across multiple applications. The cloud implementation allows us to integrate the database in virtually any environment or application we choose.

This project's software architecture consisted of three main components, a simple user interface, a database of recycling laws, and an object recognition model. Starting from the user interface, the components that build the display are created using React, a JavaScript library. This prevalent library provides many supporting libraries and the flexibility needed to easily integrate the different components necessary to provide real-time results from the model. The application allows a user to press a button to begin inference. Once pressed, the image is transmitted via API to the object recognition model until the object is detected. Based on the object recognition model response and conditions set in the database, a specific response is displayed to the user and gives the user information on how to recycle and disposed of the item properly. If an item is not detected, then it continues searching for an item until the user stops the detection by clicking a button.

We store the object recognition models in a S3 bucket on AWS. AWS provides a service called Amazon API Gateway, that connects the application and the model. This setup allows the model to access multiple interfaces.

Amazon API Gateway is a comprehensive AWS service that handles the API's creation, testing, and deployment. This service also provides authentication and authorization services. The service will generate an API key based on the usage plan selected. These usage plans can be created directly from the API Amazon Gateway console. The API key is used to authorize each call, and the marketplace will track and use that key to throttle calls per second and total calls per month. In addition, the service automatically manages the load balances and provides more or fewer instances based on the API traffic.

9.4   UM-LV Recycling Dataset

Our custom dataset, UM-LV Recycling, contains 2,842 images captured using different smartphones at various resolutions. Since the images collected are of different resolutions, they were resized to a resolution of 1024x1024 and converted to PNG format for its lossless quality. The primary interest of the dataset focuses on plastics and electronics that are not already present in publicly available datasets such as the MS COCO dataset [8] or TrashNet [11]. Especially as it pertains to differentiating between types of plastics such as Polyethylene Terephthalate (PET Clear) and High-Density Polyethylene translucent (HDPE translucent). Because our target audience was the residents in a specific geographic location, in this initial UM-LV Recycling dataset, we focused on household products. This dataset has two super categories: plastics and electronics. For the plastics super category, we chose two out of seven types of plastics, namely HDPE Translucent (hereafter HDPE) and PET Clear (hereafter PET). We chose HDPE (recycling code #2) and PET (recycling code #1) because plastics with recycling codes #3 to #7 have very low or no market value and flexible plastics (films, bags, and shrink-wrap) disrupt operations [2]. We aim to target the high value plastics with the best chance of meeting a market demand. For the electronics super category, we focused on four kinds of electronics, namely batteries, gamepads, remotes, and calculators. Table 9.1 shows the list of all different categories along with the number of images in each category.

Objects of various shapes, sizes, and brands at multiple angles were collected to encourage variation in the dataset. The images were collected in both indoor and outdoor environments. The outdoor images provided different lighting depending on the time of day and numerous backgrounds, including concrete, sand, grass, and different vegetation.

The boundaries of objects in all images were traced to create masks. Unlike bounding boxes, masks capture the target object's geometry at a pixel-by-pixel level. We used GIMP (www.gimp.org), a free and open source graphics editor software, to create image masks. Utilizing several image layers within GIMP, the target items were differentiated using different RGB color values. For images with multiple target items, a soft limit of seven layers was set per image. Each layer contained a corresponding target item with an assigned unique color in that image. The RGB color values chosen were red (250, 0, 0) for the first item, green (0, 255, 0) for second item, blue (0, 0, 255) for the third item, yellow (255, 255, 0) for the fourth item, fuchsia (255, 0, 255) for the fifth item, cyan (0, 255, 255) for the sixth item, and orange (255, 125, 0) for the seventh item. For example, if there were three objects in the frame, red, green, and blue would be used to differentiate the three items in the frame with the background color black (0, 0, 0). Once the masks were completed, they were "cleaned" through a script to remove any anti-aliasing fragments between the target items' masks and the black background such that the masked images only contain the aforementioned RGB values.

In addition, we utilized a labeling software called LabelImg [20] to create bounding boxes around an item using the Pascal VOC [4] format. Every image has an accompanying XML file with all the information in Pascal VOC format. Maximizing the tightness of a bounding box around a target item and minimizing the overlap of the boxes were especially important

Table 9.1 The list of all categories in our dataset and the number of corresponding images in each category.

Super category	Category	Count
Electronics	Battery	499
Electronics	Calculator	449
Electronics	Gamepad	349
Electronics	Remote	501
Plastic	HDPE Translucent	547
Plastic	PET Clear	497
Total Number of Images		2,842

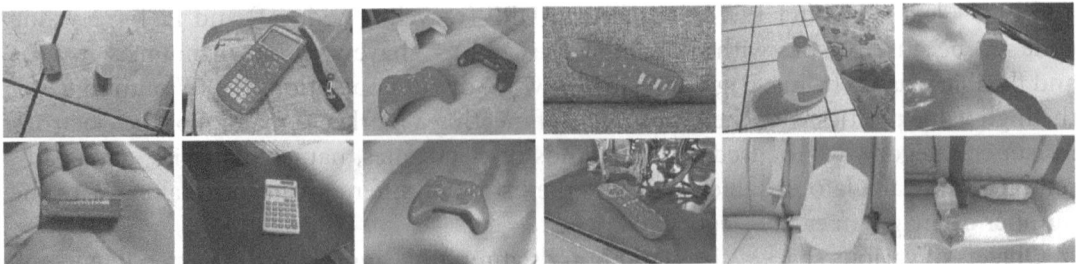

Figure 9.2 Sample masked images from UM-LV recycling dataset.

to reduce noise in the training images. Although masking an image is more precise than utilizing a bounding box, the boxes took significantly less time to create, especially with multisided target items.

Samples of images for different categories in our dataset are shown in Figure 9.2. The dataset is available for research purposes upon request, and we plan to make it publicly available in the near future.

9.5   Object Recognition Models

We selected four object recognition models (Faster R-CNN [15], MobileNetV2 [17], ResNet-50 [6], and EfficientDet [19]) to compare their respective performances. The four models chosen are expected to perform differently due to their fundamental differences.

Faster R-CNN has become a staple in computer vision due its history as an evolution of one of the most powerful techniques. However that comes with the tradeoff of a much larger model size. It uses multiple steps to detect objects in an image, first by dividing the image into region proposals, classified as either background or a trained object class.

The other three models are different from the Faster R-CNN in how they perform inference. The ResNet-50 model is a single-shot object detection, which means it is a single-stage multi-object detector. Once trained, a single-stage detector is expected to be faster but slightly less accurate than a two-stage one.

Similar to ResNet-50, MobileNetV2 is a one-stage object detection model which has gained popularity for its lean network and novel depth-wise separable convolutions. It is a model commonly deployed on low compute devices such as mobile with high accuracy performance. Both ResNet-50 and MobileNetV2 have roughly the same pipeline for training and testing and only vary slightly in the setup of the network internally.

EfficientDet is a convolutional neural network that builds upon the EfficientNet [18] model. Due to its scalability and efficiency, as the name suggests, it uses up to 9x less computation than the prior state-of-the art models [19].

The models were initialized with weights from the pre-trained models on the TensorFlow model zoo. For training, all four models used the same dataset with six categories. We randomly split the dataset into 80% training and 20% testing images. All four models were trained and tested using the same data split. We used several metrics to assess the performance of each model. The first metric was the intersection over union ratios, which measure the model's predicted bounding boxes' intersections with the ground truths and can even be easily visualized over the image to gather an intuitive grasp of the model's performance. The remaining performance metrics, i.e. average precision, recall rate, and F1-score can be extracted from confusion matrices. We selected the best performing model based on these metrics and exported the final weights as an inference graph to be deployed in our API.

9.6   Results

We calculated multiple metrics for each model including the Recall, F1-scores, Average Precision (APrecision), and Mean Average Precision (mAP). For all metrics, we used an Intersection Over Union (IOU) of 0.5 to match the COCO and PascalVOC metrics, which both use mAP at 0.5 IOU as their primary metric.

All four models were trained on a local machine, which consisted of an AMD Ryzen 5800x 8c16t, NVIDIA RTX 2070 Super 8GB VRAM, and 32 GB DRAM DDR4 clocked at 3200MHZ. A different image resolution was chosen to train each model. Our goal was to choose the highest possible image resolution that could be used to train each model on our local machine.

For Faster R-CNN, we had to reduce the training batch size to 2 and image resolution to 800x1333. This led to the model having the worst performance of the four models with mAP of 83.21%, and average F1-Score of 0.72. Table 9.2 shows all the metrics per category for the Faster R-CNN model. Among all categories, the PET Clear did not meet the minimum requirement in our use case (85% in AP for each category).

MobileNetV2 was one of our best performing models. We can attribute its accuracy to its flexibility with object placement and size in the image, as well as the slightly lower resolution of 320x320 reducing the amount of unnecessary information that may cause excessive false positives. It measured as the second-best model, with mAP of 96.15%, and average F1-score of 0.88. Table 9.3 shows all the metrics per category for the MobileNetV2 model. The model outperformed the minimum requirement of 85% AP in all categories.

The single-shot version of ResNet-50 was used as a middle point between Faster R-CNN and the other single-shot models, since it uses the same backbone architecture as Faster R-CNN, but as a single stage detection. At 640x640 resolution, the model produced results with mAP of 92.36%, and average F1-score of 0.84. Table 9.4 shows all the metrics per category for the ResNet-50 model. Although ResNet-50 model performed much better than the Faster R-CNN in PET Clear category, it still did not meet our minimum requirement in AP.

Lastly, EfficientDet, at 512x512 resolution produced mAP of 96.35%, and average F1-score of 0.89. The performance differences between the EfficientDet and MobileNetV2 are negligible. Table 9.5 shows all the metrics per category for the EfficientDet model. Similar to the MobileNetV2 model, EfficientDet outperformed our minimum requirement in AP in all categories.

Table 9.6 summarizes the overall precision, recall, and F1-score at 0.5 IOU for all four models. Considering all the metrics, the EfficientDet was the best performing model on our recycling dataset. Sample results for different categories for all four models are shown in Figure 9.3.

Table 9.2 Performance metrics for Faster R-CNN model.

Categories	APrecision	Recall	F1-score
Battery	86.2%	79.1%	79.5%
Calculator	98.4%	96.7%	93.0%
Gamepad	92.8%	97.1%	69.7%
Remote	86.0%	60.8%	73.8%
HDPE Translucent	88.9%	80.3%	78.0%
PET Clear	47.0%	73.1%	40.6%
mAP: 83.21%			

Table 9.3 Performance metrics for MobileNetV2 model.

Categories	APrecision	Recall	F1-score
Battery	96.0%	87.8%	88.2%
Calculator	99.9%	88.9%	89.9%
Gamepad	94.2%	81.6%	85.7%
Remote	88.9%	79.7%	79.0%
HDPE Translucent	99.9%	96.6%	96.2%
PET Clear	97.9%	89.8%	90.2%
mAP: 96.15%			

Table 9.4 Performance metrics for ResNet-50 model.

Categories	APrecision	Recall	F1-score
Battery	90.6%	86.1%	80.2%
Calculator	98.0%	92.2%	93.8%
Gamepad	91.8%	79.6%	85.9%
Remote	92.5%	61.4%	74.0%
HDPE Translucent	98.2%	89.7%	92.5%
PET Clear	83.2%	74.1%	77.3%
mAP: 92.36%			

Table 9.5 Performance metrics for EfficientDet model.

Categories	APrecision	Recall	F1-score
Battery	94.7%	85.2%	84.8%
Calculator	99.5%	92.2%	96.0%
Gamepad	95.9%	89.3%	88.5%
Remote	93.5%	88.0%	85.8%
HDPE Translucent	99.2%	94.9%	93.7%
PET Clear	95.1%	80.6%	87.0%
mAP: 96.35%			

Table 9.6 Overall metrics for all four models.

Metrics	Faster R-CNN	MobileNetV2	ResNet-50	EfficientDet
Overall Precision @0.5IOU	0.606	0.883	0.881	0.894
Overall Recall @0.5IOU	0.792	0.870	0.790	0.883
Overall F1-Score @0.5IOU	0.687	0.876	0.833	0.889
Overall Precision @0.5IOU	0.606	0.883	0.881	0.894

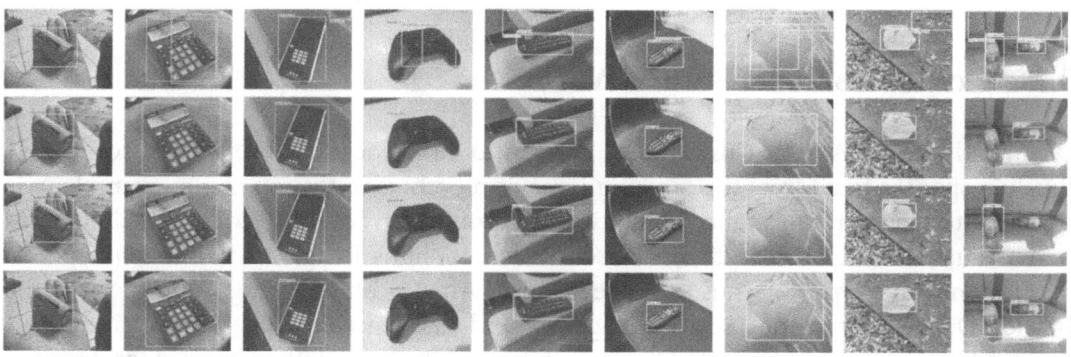

Figure 9.3 Sample results for different categories for all four models. From top row to bottom row: 1) Faster R-CNN, 2) MobileNetV2, 3) ResNet-50, 4) EfficientDet.

9.6.1 Potential Impacts

Potential economic impacts: With the widespread deployment of this solution, it is estimated that if it could increase the curbside recycling participation rate by 15%, there is the potential to save approximately 4,540,002 metric tons of CO2 from the extraction and processing of raw materials. For this calculation, we used the Recycling Partnership's tool built from the Environmental Protection Agency's WARM Model adjusted for residential recyclables [12]. We entered 69,800,000 for the number of households with access to curb-side recycling, 80% for the number of households currently receiving service and an increase in participation from 65-80% to factor in eliminating confusion and increasing participation. At 65% there was a carbon emission savings 19,673,339 metric tons of CO_2, whereas when the participation is increased to 80%, the savings were 24,213,341. Thus, if our tool can increase recycling participation by 15% it would save 4,540,002 metric tons of CO_2 per year.

Potential negative impacts: The negative impacts of this technology include an inaccurate object recognition model, API security, accessibility, and the various terminology and definitions used in recycling programs. The object recognition model issues can cause contamination if a model returns an incorrect response. For example, if a model responds that a 'battery' is an 'aluminum can', the consumer would place a battery in their recycling cart, which could potentially cause a fire or other adverse downstream effects. Regarding API security, if images or responses are transmitted unencrypted, a bad actor could intercept this, leading to privacy concerns. Accessibility issues are focused on the interface and are often caused by low contrast of text, missing alt text on images, uncontrolled time-outs, and responsiveness of an interface. Lastly, the terminology and definitions of materials change based on regions and even within communities. For instance, one location might call an 'aluminum can' a 'pop can', which could both be confused with a 'tin or steel can', which are different materials (aluminum versus steel), or provide specific restrictions on types of materials (some places do not accept greasy pizza boxes, while other programs do). However, this type of classification can cause definition issues and cause incorrect responses to the consumer because these broad definitions used in recycling programs can conflict with model responses and the specific dataset used to train it. These negative impacts should be addressed before any practical, real-world deployment of the technology occurs.

9.7   Conclusion and Future Work

We introduced an API that uses object recognition and location information to provide timely responses to address the growing issue of recycling confusion. As an example case study, we tested the performance of multiple state-of- the-art object recognition algorithms on our custom UM-LV Recycling dataset, which contains images of two major types of plastics and four types of consumer electronics. Among object recognition models, the EfficientDet performed the best on average on our dataset based on different metrics. In our case study, this model was selected to be used in the API that will then be used for developing smartphone apps, web apps, or in IoT devices. Our API is flexible and can be extended to be adapted to the needs of different communities. We plan to extend our current dataset and create a comprehensive image dataset for recycling. We are in the process of developing an educational web app as well as a smartphone app that uses our API. The apps will be tested and used in our community to increase awareness about recycling right and decrease recycling confusion at the source.

Acknowledgements

This research was conducted with support from the ReMade Institute with funding from SMIA and the Office of Energy Efficiency and Renewable Energy within the U.S. Department of Energy, Number DE-EE0007897, in collaboration with Lid Vizion, and the resources of the University of Miami Institute for Data Science and Computing.

This material is based upon work supported by the U.S. Department of Energy's Office of Energy Efficiency and Renewable Energy (EERE) under the Advanced Manufacturing Office Award Number DE-EE0007897 awarded to the REMADE Institute, a division of Sustainable Manufacturing Innovation Alliance Corp. This report was prepared as an account of work sponsored by an agency of the United States Government. Neither the United States Government nor any agency thereof, nor any of their employees, makes any warranty, express or implied, or assumes any legal liability or responsibility for the accuracy, completeness, or usefulness of any information, apparatus, product, or process disclosed, or represents that its use would not infringe privately owned rights. Reference herein to any specific commercial product, process, or service by trade name, trademark, manufacturer, or otherwise does not necessarily constitute or imply its endorsement, recommendation, or favoring by the United States Government or any agency thereof. The views and opinions of authors expressed herein do not necessarily state or reflect those of the United States Government or any agency thereof.

References

1. Olugboja Adedeji and Zenghui Wang, Intelligent waste classification system using deep learning convolutional neural network. Procedia Manufacturing, 35:607–612, 2019. *The 2nd International Conference on Sustainable Materials Processing and Manufacturing, SMPM 2019, 8-10 March 2019, Sun City, South Africa.*

2. Carlos Alberto Correa, Marcio Adilson De Oliveira, Christiane Jacinto, and Giulliana Mondelli, Challenges to reducing post-consumer plastic rejects from the msw selective collection at two mrfs in são paulo city, brazil. *Journal of Material Cycles and Waste Management,* 24(3):1140–1155, 2022.

3. EPA. An analysis of lithium-ion battery fires in waste management and recycling. Technical report, Environmental Protection Agency, Office of Resource Conservation and Recovery, July 2021.

4. M. Everingham, S. M. A. Eslami, L. Van Gool, C. K. I. Williams, J. Winn, and A. Zisserman, The pascal visual object classes challenge: A retrospective. *International Journal of Computer Vision,* 111(1):98–136, Jan. 2015.

5. Ali Usman Gondal, Muhammad Imran Sadiq, Tariq Ali, Muhammad Irfan, Ahmad Shaf, Muhammad Aamir, Muhammad Shoaib, Adam Glowacz, Ryszard Tadeusiewicz, and Eliasz Kantoch, Real time multipurpose smart waste classification model for efficient recycling in smart cities using multilayer convolutional neural network and perceptron. *Sensors (Basel),* 21(14), Jul 2021.

6. Kaiming He, X. Zhang, Shaoqing Ren, and Jian Sun, Deep residual learning for image recognition. *2016 IEEE Conference on Computer Vision and Pattern Recognition (CVPR),* pages 770–778, 2016.

7. Pratima Kandel, Computer vision for recycling. *Student Research Submissions 379*, University of Mary Washington, 2020.

8. Tsung-Yi Lin, Michael Maire, Serge J. Belongie, James Hays, Pietro Perona, Deva Ramanan, Piotr Dollár, and C. Lawrence Zitnick, Microsoft coco: Common objects in context. In *ECCV*, 2014.

9. Weisheng Lu, Junjie Chen, and Fan Xue, Using computer vision to recognize composition of construction waste mixtures: A semantic segmentation approach. *Resources, Conservation and Recycling*, 178:106022, 2022.

10. Daniel Octavian Melinte, Ana-Maria Travediu, and Dan N. Dumitriu, Deep convolutional neural networks object detector for real-time waste identification. *Applied Sciences, 10(20)*, 2020.

11. Gary Thung Mindy Yang, Classification of trash for recyclability status. *Technical report*, Stanford University, 2016.

12. Scott Mouw, Lily Schwartz, , Sherry Yarkosky, Elizabeth Biser, Ali Blandina, Joe Bontempo, Anthony Brickner, Jackie Caserta, Sarah Dearman, Dylan de Thomas, Allison Francis, Keefe Harrison, Samantha Kappalman, Cody Marshall, Asami Tanimoto, Rob Taylor, Laura Thompson, and Aditi Varma, State of curbside recycling report, 2020.

13. Nadish Ramsurrun, Geerish Suddul, Sandhya Armoogum, and Ravi Foogooa, Recyclable waste classification using computer vision and deep learning. In 2021 Zooming Innovation in Consumer Technologies Conference (ZINC), pages 11–15, 2021.

14. Joseph Redmon, Santosh Kumar Divvala, Ross B. Girshick, and Ali Farhadi, You only look once: Unified, real- time object detection. *2016 IEEE Conference on Computer Vision and Pattern Recognition (CVPR)*, pages 779– 788, 2016.

15. Shaoqing Ren, Kaiming He, Ross B. Girshick, and Jian Sun, Faster r-cnn: Towards real-time object detection with region proposal networks. *IEEE Transactions on Pattern Analysis and Machine Intelligence*, 39:1137–1149, 2015.

16. Victoria Ruiz, Ángel Sánchez, José F. Vélez, and Bogdan Raducanu, Automatic image-based waste classification. In José Manuel Ferrández Vicente, José Ramón Álvarez-Sánchez, Félix de la Paz López, Javier Toledo Moreo, and Hojjat Adeli, editors, From *Bioinspired Systems and Biomedical Applications to Machine Learning*, pages 422–431, Cham, 2019. Springer International Publishing.

17. Mark Sandler, Andrew G. Howard, Menglong Zhu, Andrey Zhmoginov, and Liang-Chieh Chen, Mobilenetv2: Inverted residuals and linear bottlenecks. *2018 IEEE/CVF Conference on Computer Vision and Pattern Recognition*, pages 4510–4520, 2018.

18. Mingxing Tan and Quoc V. Le, Efficientnet: Rethinking model scaling for convolutional neural networks. ArXiv, abs/1905.11946, 2019.

19. Mingxing Tan, Ruoming Pang, and Quoc V. Le. Efficientdet: Scalable and efficient object detection. *2020 IEEE/CVF Conference on Computer Vision and Pattern Recognition (CVPR)*, pages 10778–10787, 2020.

20. Tzutalin. Labelimg. git code. https://github.com/tzutalin/labelimg, 2015.

21. Xuyun Zhang, Yuezhong Wu, Xuehao Shen, Qiang Liu, Falong Xiao, and Changyun Li, A garbage detection and classification method based on visual scene understanding in the home environment. *Complexity*, 2021:1055604, 2021.

22. Dimitris Ziouzios, Nikolaos Baras, Vasileios Balafas, Minas Dasygenis, and Adam Stimoniaris. Intelligent and real-time detection and classification algorithm for recycled materials using convolutional neural networks. *Recycling, 7(1)*, 2022.

10

Identification and Separation of E-Waste Components Using Modified Image Recognition Model Based on Advanced Deep Learning Tools

Rahulkumar Sunil Singh, Subbu Venkata Satyasri Harsha Pathapati, Michael L. Free and Prashant K Sarswat*

Department of Materials Science and Engineering, University of Utah, Salt Lake City, Utah, USA

Abstract

Globally, the use of electrical and electronic equipment has increased tremendously in recent years as a result of technological advancements, planned obsolescence, new methods of data storage, and increased affordability. Because of this, e-waste has emerged as the fastest-growing waste stream (~ 57.4 million tons for year 2021), making its proper management a top priority for the economy, environment, and even human health. The main challenges in managing e-waste are the collection, sorting, and product complexity, low energy density, emissions, and recycling economics. Efficient e-waste recycling provides a way to recover valuable materials to combat the shortage of various primary resources, reduce hazardous materials in the environment, and support the economy. In order to increase recycling efficiency, a holistic approach that includes the development of innovative methods for sorting and screening is required. The separation must be effective to prevent material loss that cannot be recovered. The printed circuit boards (PCBs) account for 8–14% of the total volume of e-waste but 48% of the financial value, highlighting the significance of separation.

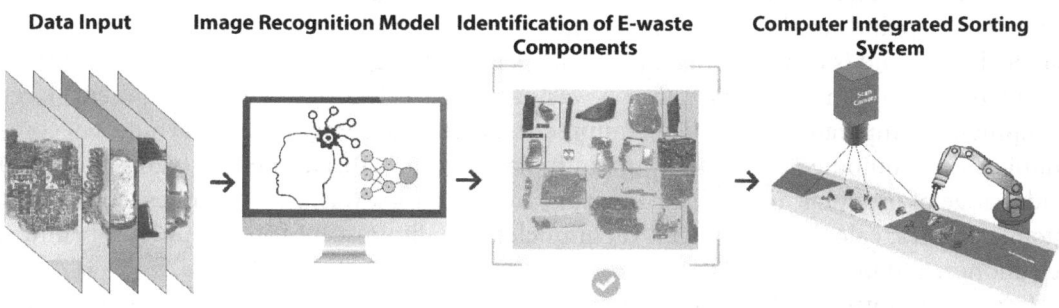

*Corresponding author: u0403179@utah.edu

Nabil Nasr (ed.) Technology Innovation for the Circular Economy: Recycling, Remanufacturing, Design, Systems Analysis and Logistics, (115–128) © 2024 Scrivener Publishing LLC

The mechanical shredding or crushing of the electronics provides a mixture of e-waste components. Identification and sorting of these generated components are increasingly difficult. Labor-intensive manual separation is effective but not sustainable. Use of a deep learning-based image recognition model, combined with an appropriate integrated computer sorting system can enable the process to be sustainable, automated, and economical. The proposed method primarily focuses on identifying and categorizing the e-waste components into PCBs, Steel, Copper, Aluminum, and Polymers. Experimental results show that the model accurately identifies the images of e-waste components. Overall, the study not only aims to improve the efficiency in the identification of e-waste components through deep learning but also provides an approach for its separation without human intervention to make e-waste recycling an efficient and safer process.

Keywords: E-waste, recycling, computer vision, object detection, image identification, YOLO, automation

10.1 Introduction

E-waste, or electronic waste, refers to any electronic appliance or device that has reached the end of its useful life, including items such as smartphones, computers, televisions, and household appliances. E-waste is a growing problem, as the number of electronic devices being produced and discarded is increasing rapidly. The primary reason behind this is the rapid pace of technological advancement and consumer demand for newer and more advanced electronic devices. The amount of e-waste generated globally in 2019 is 53.6 million metric tons with a tremendous growth of 21% in the last half-decade [1, 2], and expected to reach 120 million metric tons by 2050 [3]. In 2019, only 17% of e-waste is recycled, with most of the e-waste ending up in landfills or burnt in the open environment [1, 4], where it is often handled in a way that is harmful to both the environment and human health [5]. In terms of geography, Asia is the largest generator of e-waste, with China, the United States, and India being the top three countries [6]. The e-waste recycling scenario varies greatly between countries due to a range of factors such as economic development, population size, infrastructure, and regulations. In 2020, the global e-waste recycling market was valued at $49.88 billion and is expected to be three times by end of 2028 [7]. The Directive 2012/19/EU of the European Union classifies e-waste into six categories: small equipment, large equipment, Temperature-exchange equipment, Screen equipment, Small IT and telecommunications equipment, and lighting equipment (Figure 10.1). These categories cover a wide range of products, including televisions, refrigerators, washing machines, computers, mobile phones, calculators, routers, vacuum cleaners, fluorescent tubes, cameras, printers, telephones, and power tools [8].

It is worth noting that e-waste is a complex mixture of materials, and the component shares can vary widely depending on the type and age of the equipment, brand, and country of origin. However, in general, PCBs, plastics, and metals are three of the major components found in e-waste [9, 10]. Printed circuit boards (PCBs) typically contain a variety of valuable metals, such as gold, silver, copper, and palladium, as well as other materials like ceramics and glass fibers [11, 12]. In TVs, computers, and cell phones, PCBs are typically 30% plastic or polymer, 40% metals, and 30% ceramics [13]. Precious

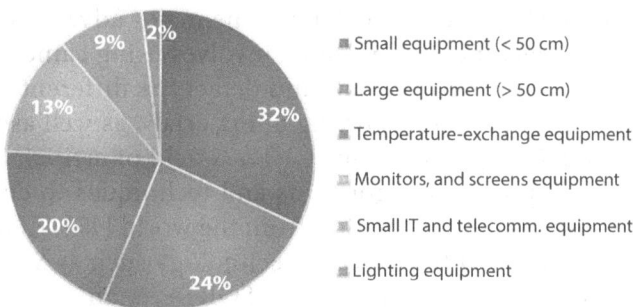

Figure 10.1 The classification of e-waste into six categories and their percentage share (based on ref. [7]).

metals (platinum, gold, palladium, and silver) and critical metals (lithium, indium, tin, tantalum, and rare earth elements) are also found in e-waste, along with useful metals like copper, bulky materials like iron and aluminum, and recyclable plastics [13]. Copper makes up 4% of the total amount of metal in e-waste, but in particular materials, specifically PCBs, its concentration is higher than 20%. As a result, 80–400 pounds of copper, which has a market value of $220–1100, per ton of conventional e-waste, can be found in it. E-waste contains a sizeable amount of gold and other precious metals ($1-$300/ton depending on the origin) [13]. Apart from the valuable materials, E-waste also contains potentially harmful chemicals and pollutants that can lead to environmental pollution and health hazards. Improper disposals of e-waste, such as through incineration or land-filling, can release toxic chemicals into the air and water [14]. Therefore, e-waste recycling is crucial for not only achieving economic benefits, but also for protecting the environment, conserving resources, saving energy, and fulfilling social responsibility. The main challenges associated with e-waste management are collection, sorting, product complexity, low energy density, emissions, and recycling economics [15, 16]. There is a need for an alternative strategy to recover more economic value from e-waste through improved metal recovery and facilitate more excellent opportunities to differentiate individual plastics constituents from e-waste, thus offering great financial and environmental incentives to recycle more e-waste. As can be seen that the annual increase in e-waste is alarming in most parts of the world. Thus, appropriate recycling and reuse strategies are essential. After the mechanical processing, such as shredding and crushing, the identification and sorting of a mixture of e-waste components becomes increasingly difficult. The development of a more efficient sorting method is required to increase recycling efficiency as it can prevent material loss.

Labor-intensive manual separation can result in a significant recovery, but it cannot be considered an economically sustainable solution. The presence of toxic materials, particularly heavy metals like lead, mercury, and cadmium can cause severe health issues, which makes this a more challenging task for human operators. Many of those problems can be addressed effectively with the help of object detection techniques. Optical sorting technology is a method of separating materials based on their physical properties, such as color, shape, and size. One of the major advances in optical sorting technology is the use of machine vision, which involves using cameras and image processing algorithms to analyze the materials and detect the different components. This has led to more accurate and efficient sorting of materials, as well as the ability to sort materials at a faster rate. Computer vision (CV) is a field of artificial intelligence (AI) that deals with the development of algorithms and techniques to enable machines to interpret and understand visual information from the world [17]. In recent years, there have been several advances in optical sorting technology such as Sesotec VARISORT+ and NRT SpydIR with MAX-AI. Sesotec VARISORT+ is an advanced optical sorting technology developed by Sesotec GmbH. Sensotec is a German-based company that uses a combination of imaging and sensor technologies such as color, shape, and near-infrared spectroscopy to sort materials based on their physical and chemical properties. It can sort a wide range of materials, including plastics, metals, paper, and glass. NRT SpydIR with MAX-AI is another advanced optical sorting technology that combines the power of non-destructive testing (NRT) and artificial intelligence (AI) to sort a wide range of materials, including metals, plastics, and ceramics based on their physical and chemical properties [18].

In the context of waste separation, CV has been used to analyze images and videos of waste materials and to detect and classify different types of waste including municipal solid waste [19] and medical waste [20]. The use of CV could also potentially help with the upstream/downstream or source sorting of e-waste at recycling companies [17, 21]. Therefore, training a CV model can provide identification of e-waste components, and it has the ability to automatically classify them at faster rate. This can lead to improved efficiency, automation, adaptability, and real-time monitoring of the e-waste, which can provide economic benefits and a reduction in environmental impact. Despite the bright future, the function of CV in waste separation had long been constrained and remained quite minor [17]. The lengthy manual process needed for feature handcrafting and the early stage's low resilience are to blame for the slow progress [17].

The limitations of conventional CV algorithms can be addressed by deep learning (DL) approaches in two ways. First, using DL's strength in feature extraction, tedious feature handcrafting can be avoided because big data can be used to automatically learn the visual features shown by various trash types [17]. Real-time object identification is very important in computer vision systems. Numerous real-time object detection methods have been developed with the quick advancement of artificial intelligence technology. These object detection algorithms can be roughly divided into two types: Two-Stage and One-stage algorithms [22]. Two-Stage object detection algorithms consist of two stages: the first stage is to generate a set of region proposals, and the second stage is to classify the objects within the proposals. Examples of Two-Stage object detection algorithms include Faster R-CNN [23], R-FCN [24], and Mask R-CNN [25]. These algorithms typically have a higher accuracy but are also computationally more expensive. On the other hand, One-stage object detection algorithms perform object detection in a single stage. These algorithms directly predict the bounding boxes and class scores for objects in an image. One-stage algorithms are faster and more efficient but may not be as accurate as the Two-Stage algorithms. Examples of One-stage object detection algorithms include YOLO [26], SSD [27], and RetinaNet [28]. YOLO (You Only Look Once) is a popular real-time object detection system, presents a balance between speed and accuracy, with a smaller model size and a faster inference time compared to other algorithms [29]. YOLO uses a single neural network to perform object detection, which means it is less complex and easier to train and deploy. Researchers have updated the YOLO series, which creates the high-performance anchor-free detector known as YOLOX, with certain practical improvements [30]. With the use of certain modern sophisticated detection approaches, such as the decoupled head, anchor free, and enhanced label assignment strategy, YOLOX outperforms its competitors across all model sizes in terms of the trade-off between speed and accuracy. It is impressive that they improved the YOLOv3 architecture to 47.3% average precision (AP) on COCO, surpassing the current best practice by 3.0% AP. YOLOv3 is still one of the most extensively used detectors in industry due to its wide compatibility. YOLOv7 (You Only Look Once version 7) is the latest version, released in 2021 and has been met with significant interest from the computer vision community [31].

In this study, the purpose is to investigate the use of the YOLOv7 object detection algorithm as a tool for the identification and sorting of e-waste components. The proposed research will focus on evaluating the YOLOv7 algorithm on our self-prepared dataset of e-waste images to determine its accuracy in identifying and sorting various type of components like plastic, PCB, copper, aluminum, and steel. The research will also investigate the potential for using the YOLOv7 algorithm in a real-world waste sorting setting and the possibility of integrating it into a robotic system for sorting of e-waste components. The results of this study will provide valuable information on the feasibility of using advanced object detection algorithms in the field of e-waste management and could lead to the development of more efficient, accurate, and sustainable e-waste sorting solutions. Furthermore, it will pave the way to the implementation of robotic systems for waste sorting.

10.2   Materials & Methods

The electronic wastes were supplied by Sunnking Inc, an electronic recycling business in Brockport, New York, and included laptop, hard-drive, user scrap, flat-panel scrap, and motherboard junk. Using an electric shaker, the parts were broken into smaller pieces, which were then screened and sieved. Digital photographs were captured using DSLR (Canon) Camera. For identification and sorting of e-waste components, first, we examined our materials using TensorFlow version 2.8.0 based model. In this sets of simulations, we have utilized categorical cross entropy which computes the cross-entropy loss between the labels and predictions. After the success of this, we have used YOLOv7 for real-time object detection. YOLOv7 model was used by training the model on a dataset of waste images labeled with the corresponding class of e-waste, such as PCB, plastic, copper, aluminum, and steel. The model was the used to detect and classify different types of waste in real-time images, thus making it a suitable solution for waste sorting.

YOLOv7 is a state-of-the-art real-time object detection model that is widely used for various computer vision tasks, such as image classification and object detection. YOLOv7 is known for its high accuracy and fast inference speed, making it suitable for real-time applications, such as waste sorting. It works by using a single convolutional neural network

(CNN) to detect objects in an image by dividing the image into a grid of cells and using each cell to predict multiple bounding boxes and class probabilities. The model makes a prediction for each box, indicating whether or not an object is present in that box, and if so, which class of object it is. YOLOv7 is trained on large datasets to learn the features of different objects and to generalize well to new images. It uses anchor boxes to handle different aspect ratios and scales of the objects in the image. It also uses techniques like multi-scale prediction and feature-up sampling to improve the model's performance. One of the key improvements in YOLOv7 is its increased accuracy. This is achieved through several changes, including the use of a new backbone network and the implementation of a new loss function. The backbone network used in YOLOv7 is a variant of the EfficientNet architecture, which is highly effective in several image classification tasks. Another important aspect of YOLOv7 is its ability to detect objects at multiple scales. This is achieved through the use of multiple branches in the network, each responsible for detecting objects at a different scale. This allows the system to be more robust to variations in object size and also improves its ability to detect small objects. In addition to these technical improvements, YOLOv7 also includes a number of practical features that make it more user-friendly. For example, it includes a built-in data augmentation system that allows users to easily apply various types of transformations to their training data. It also includes support for multiple GPUs, making it possible to train large models on powerful hardware. Overall, YOLOv7 is a powerful and versatile object detection system that offers a number of benefits over its predecessors. Its improved accuracy, ability to detect objects at multiple scales and user-friendly features make it an attractive choice for a wide range of applications [31].

The first step in the methodology was to collect a large dataset of images of e-waste components obtained after shredding. The total of 400 images was captured that includes 80 images of each e-waste component (PCB, plastic, copper, aluminum, and steel). Single class and multiclass e-waste components data were captured. The collected images were then preprocessed to prepare them for the training process. This included resizing the images to a common size, removing noise and artifacts, and creating a dataset of labeled images. The images were labeled manually with boundary boxes around the different components of e-waste. Further, the dataset was split into a training set and a test set, with a ratio of 80:20. Finally, the labeled images of the prepared dataset were used to the YOLOv7 model and fined-tuned to improve the performance. The trained model was then evaluated using several metrics, such as precision, recall, and F1-score. The model was tested on a dataset of a variety of e-waste images and the results were analyzed to investigate the possibility of implementation in a real-time system using a robotic arm in identifying and sorting different components of e-waste.

10.3 Results & Discussion

In this study, we aimed to develop an object detection model based on YOLOv7 for the identification and sorting of e-waste components. However, the first TensorFlow version 2.8.0-based model was developed and examined. Subsequently, a YOLO-based model was utilized. The model was trained and tested on a dataset of e-waste images and evaluated its performance in terms of precision, recall, and F1 score. For initial sets of simulations, a total of ~ 400 images were chosen (belonging to 5 classes namely: 'Aluminum', 'Copper', 'PCB',

'Plastics', and 'Steel'). The performance of the TensorFlow version 2.8.0 model was evaluated on a test dataset of e-waste components, and the test output plot of the model on PCB, Plastic, and Copper is shown in Figure 10.2. Typical step size was ~ 280 ms. It was observed that for a relatively small number of epochs (~10), the model accuracy was ~ 0.98 for training (~ 0.81 for validation). The loss for such an epoch was less than 0.1 (~ 0.7 for validation). The validation loss occasionally exceeds the training loss, and in the present, a case such a scenario was observed. This could mean that the model is not fitting the data well for a small number of epochs. When a model is unable to effectively represent the training set of data, underfitting occurs and large errors are produced. For 50 epochs, the model accuracy was ~ 0.72 both for validation and training data with the mostly overlapped curve. The loss was also similar ~ 0.8, both for training and validation datasets with significant overlapping. However, for greater numbers of epochs overfitting is noted as training loss is approaching ~ 0.1 whereas validation loss was approaching ~ 0.6 with fluctuation. The model (with ~ 50 epochs) was examined on different sets of images (Figure 10.2). We have tested images of different e-waste components and noticed that the prediction probability was greater than 0.95 with correct prediction. The prediction probability for other options was significantly lower as can be seen in the images. Only in a few cases steel was identified as aluminum.

A confusion matrix is a table that is used to show the performance of a classification model. It shows the number of correct and incorrect predictions made by the model for each class. The rows of the matrix represent the actual classes, while the columns represent the predicted classes. The diagonal cells of the matrix represent the number of correct predictions, while the off-diagonal cells represent the number of incorrect predictions. It can be seen in Figure 10.3 that the model is able to classify the majority of e-waste components correctly.

Precision, recall, and F1 score are commonly used performance evaluation metrics to identify the overall accuracy of a trained model for a given dataset. Precision is a metric that

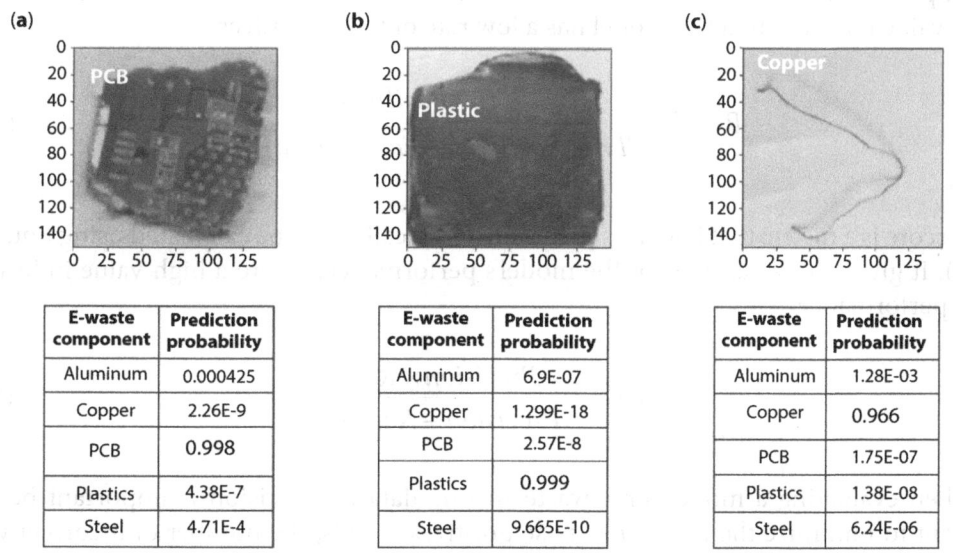

Figure 10.2 The test result of TensorFlow version 2.8.0 on different components (PCB, plastic, copper) of the e-waste.

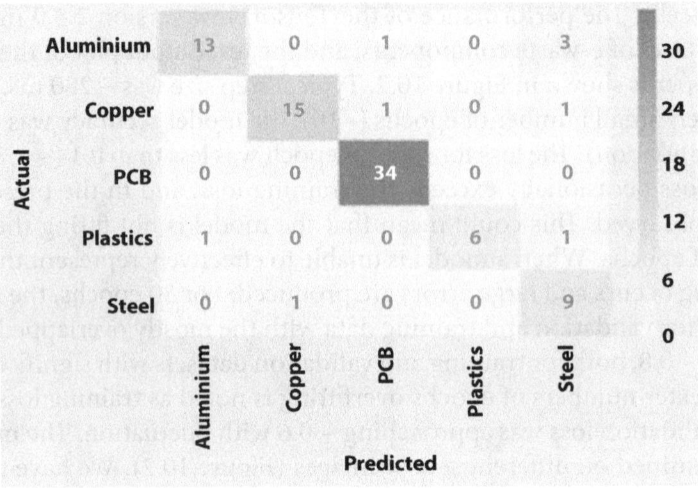

Figure 10.3 Confusion matrix of TensorFlow-based model for 50 epochs.

calculates the proportion of true positive detections out of all positive detections made by the model. It can be calculated using Equation (10.1). A high precision value indicates that the model has a low rate of false positives.

$$Precision = \frac{True\ Positives}{True\ Positives + False\ Positives} \tag{10.1}$$

The recall is a metric that calculates the proportion of true positive detections out of all actual positive examples in the dataset. It can be calculated using Equation (10.2). A high recall value indicates that the model has a low rate of false negatives.

$$Recall = \frac{True\ Positives}{True\ Positives + False\ Negatives} \tag{10.2}$$

F1 score is a metric that balances precision and recall, it can be calculated using Equation (10.3). It gives an overall idea of the model's performance, where a high value indicates a good performance.

$$F1 = \frac{2 \times Precision \times Recall}{Precision + Recall} \tag{10.3}$$

When evaluating a model for a waste sorting dataset, precision is important because we want to minimize the number of false positives, that is, the number of incorrect waste classifications. The recall is also important because we want to minimize the number of false negatives, that is, the number of waste items that are not correctly classified. The F1 score gives an overall idea of the model's performance, where a high value indicates a good performance [32].

Figures 10.4 and 10.5 give the mean precision (mAP), precision, and recall values of the YOLOv7 model for a single class and multi-class, respectively.

The prediction of the YOLOv7 model in an image that contains single and multiple e-waste components is shown in Figure 10.6. The visual representation of the results shows

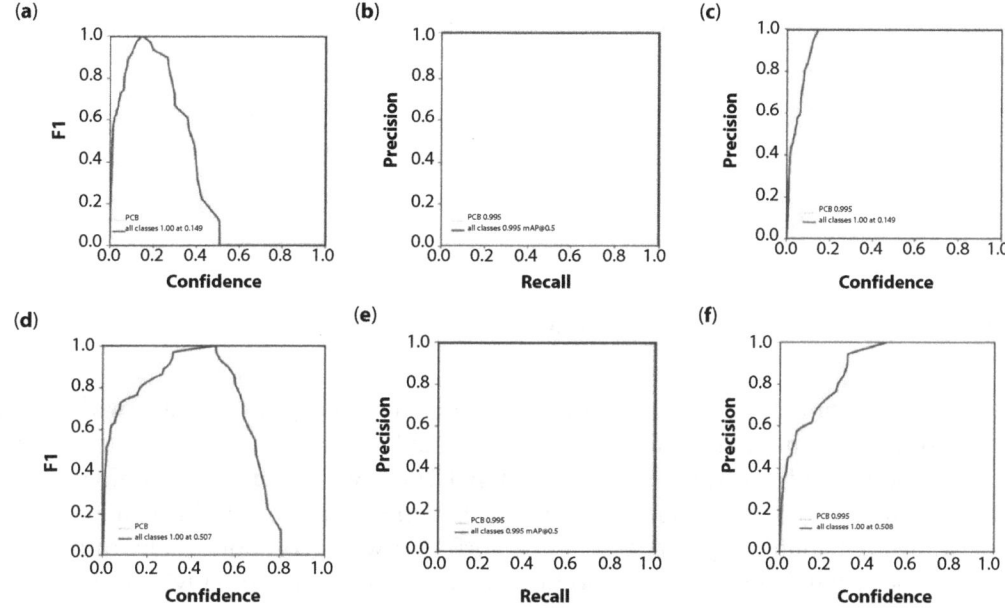

Figure 10.4 F1, PR, and P curves for single class: (a)-(c): batch size: 4; (d)-(f) batch size: 8.

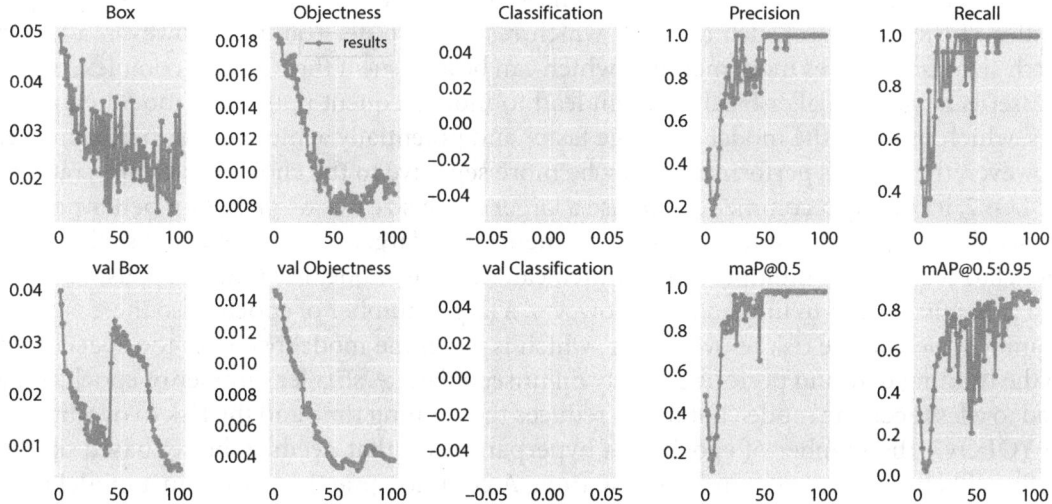

Figure 10.5 Precision, recall, and mAP curves in YOLOv7 for multiclass.

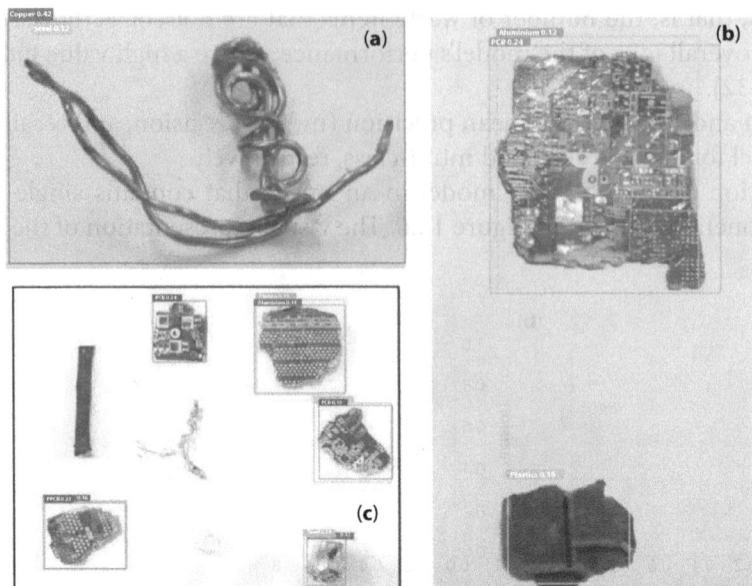

Figure 10.6 (a) single (b, c) multiple object prediction by the YOLOv7 model.

that the model accurately detects and classifies different components of e-waste with bounding boxes drawn around them. The results showed that YOLOv7 achieved good accuracy in sorting components of e-waste. It was observed that YOLOv7 was able to accurately identify and classify the e-waste components, which is a challenging task for traditional methods.

This study contributes to the field of waste sorting e-waste components by introducing a new method using deep learning. Additionally, we showed that YOLOv7 can be trained on a relatively small dataset and still achieve high accuracy, which is important for practical applications in waste sorting. However, our study also had some limitations. The used dataset was limited in terms of the number of classes and the diversity of the objects.

A larger batch size allows the model to make better use of the parallel processing capabilities of the hardware, such as GPU, which results in faster training. However, a larger batch size also requires more memory, which can be an issue if the model is complex, or the dataset is large. A smaller batch size can lead to more frequent updates of model parameters, which can help the model converge faster and potentially achieve better performance. However, the model's performance may be more sensitive to the choice of learning rate. In YOLOv7, it's highly recommended to use a larger batch size like 64 or 128 for better performance, but it's also depended on the GPU memory. A larger number of epochs allows the model to see the data multiple times, which can help the model learn more robust features and generalize better to unseen data. However, a larger number of epochs also increases the training time and the risk of overfitting, which is when the model becomes too specialized to the training data and performs poorly on unseen data. A smaller number of epochs may lead to a less accurate model, but it also reduces the training time and the risk of overfitting. In YOLOv7, the number of epochs is a hyperparameter that needs to be set based on the trade-off between accuracy and training time. A common practice is to train a model for a small number of epochs, evaluate its performance on a validation set, and then increase the number of epochs if the model's performance can be improved further.

It is worth noting that the batch size and a number of epochs are the important parameters that need to be fine-tuned to achieve optimal performance, other parameters like the learning rate, and optimizer also play an important role in the training process. Overall, our study demonstrates the potential of using YOLO and TensorFlow-based models for e-waste sorting and highlights the importance of considering deep learning methods. It is also important to have a diverse and high-quality dataset, as well as to properly evaluate the performance of the model to ensure it is able to accurately classify different types of waste in real-world scenarios. Future research directions include using other deep learning methods, such as SSD and RetinaNet to compare their performance with YOLOv7 for e-waste sorting. Finally, the model has the potential to be deployed in computer-assisted automated sorting systems in e-waste recycling facilities, which would make the process more efficient, automated, and sustainable.

10.4   Conclusions

This study aimed to develop an object detection technique based on a deep learning model (i.e., YOLOv7) for the identification and sorting of e-waste components. The e-waste dataset was composed of 400 images of different components, including PCB, plastic, copper, aluminum, and steel. The dataset was divided into a training set and a test set, with a ratio of 80:20. The model was trained using the self-prepared dataset and fine-tuned it using a variety of techniques such as adjusting the threshold, non-maxima suppression, and other parameters. The trained model was then evaluated on the test dataset. The results show that our model achieved a precision of 0.95, a recall of 0.87, and F1 score of 0.91 on the test dataset. The visual representation of the results shows that the model accurately detects and classifies different components of e-waste with bounding boxes drawn around them. A confusion matrix also shows that the model was able to classify the majority of e-waste components correctly, with the highest accuracy for PCB. These results demonstrate the potential of TensorFlow and YOLOv7 as effective tools for e-waste sorting. However, the model performance on the classification of aluminum and steel needs improvement. In future work, we plan to test the model on a larger dataset of e-waste components and explore the use of other object detection algorithms.

In conclusion, this study has investigated a new method for e-waste separation using promising model TensorFlow version 2.8.0 and YOLOv7, achieving high precision, recall, and F1 score on a dataset of e-waste images in preliminary investigations. The proposed method has the potential to significantly improve the efficiency of e-waste separation.

Acknowledgment

The University of Utah's research facilities helped this project along, and the authors are appreciative of the support received.

References

1. Shahabuddin, M., *et al.*, A review of the recent development, challenges, and opportunities of electronic waste (e-waste). *International Journal of Environmental Science and Technology*, 2022.

2. Murthy, V. and S. Ramakrishna A Review on Global E-Waste Management: Urban Mining towards a Sustainable Future and Circular Economy. *Sustainability*, 2022. **14**.

3. Gaidajis, G., K. Angelakoglou, and D. Aktsoglou, E-waste: environmental problems and current management. *Journal of Engineering Science and Technology Review*, 2010. **3**(1): p. 193-199.

4. Siddiqua, A., *et al.* E-Device Purchase and Disposal Behaviours in the UAE: An Exploratory Study. *Sustainability*, 2022. **14**.

5. Abalansa, S., *et al.*, Electronic waste, an environmental problem exported to developing countries: the GOOD, the BAD and the UGLY. *Sustainability*, 2021. **13**(9): p. 5302.

6. Forti, V., *et al.*, Quantities, flows and the circular economy potential. *The Global E-waste Monitor*, 2020. **2020**: p. 13-15.

7. Ruiz, A., *Latest Global E-Waste Statistics And What They Tell Us.* 2022: Fredericksburg, TX 78624.

8. Kumar, A. and D.C.R. Espinosa, Electronic waste: recycling and reprocessing for a sustainable future. 2022.

9. Krishna, K., Effective electronic waste management and recycling process involving formal and non-formal sectors. *International Journal of Physical Sciences*, 2009. **4**(13): p. 893-905.

10. Tembhare, S.P., *et al.*, E-waste recycling practices: a review on environmental concerns, remediation and technological developments with a focus on printed circuit boards. *Environment, Development and Sustainability*, 2021: p. 1-83.

11. Arshadi, M., S. Yaghmaei, and S.M. Mousavi, Content evaluation of different waste PCBs to enhance basic metals recycling. *Resources, Conservation and Recycling*, 2018. **139**: p. 298-306.

12. Kaya, M., Recovery of metals and nonmetals from waste printed circuit boards (PCBs) by physical recycling techniques. *Energy Technology 2017*, 2017: p. 433-451.

13. Murali, A., *et al.*, Determination of metallic and polymeric contents in electronic waste materials and evaluation of their hydrometallurgical recovery potential. *International Journal of Environmental Science and Technology*, 2021: p. 1-14.

14. Alam, T., *et al.*, E-Waste Recycling Technologies: An Overview, Challenges and Future Perspectives. *Paradigm Shift in E-Waste Management*, 2022: p. 143-176.

15. Cucchiella, F., *et al.*, *Recycling of WEEEs:* An economic assessment of present and future e-waste streams. *Renewable and Sustainable Energy Reviews*, 2015. **51**: p. 263-272.

16. Tanskanen, P., Management and recycling of electronic waste. *Acta Materialia*, 2013. **61**(3): p. 1001-1011.

17. Lu, W. and J. Chen, *Computer vision for solid waste sorting:* A critical review of academic research. *Waste Management*, 2022. **142**: p. 29-43.

18. Barker, K., The latest optical sorters for recycling are fast, acccurate and versatile. 2021: Vancouver.

19. Lin, K., *et al.*, MSWNet: A visual deep machine learning method adopting transfer learning based upon ResNet 50 for municipal solid waste sorting. *Frontiers of Environmental Science & Engineering*, 2023. **17**(6): p. 77.

20. Mehendale, N., *et al.*, Computer vision based medical waste separator. Available at SSRN 3857802, 2021.

21. Nowakowski, P. and T. Pamuła, Application of deep learning object classifier to improve e-waste collection planning. *Waste Management*, 2020. **109**: p. 1-9.

22. Yang, Z., A YOLOv7 based visual detection of waste. 2022.

23. Girshick, R., *et al.* Rich feature hierarchies for accurate object detection and semantic segmentation.
24. Dai, J., *et al.*, R-fcn: Object detection via region-based fully convolutional networks. *Advances in Neural Information Processing Systems*, 2016. **29**.
25. He, K., *et al.* Mask r-cnn.
26. Redmon, J., *et al.* You only look once: Unified, real-time object detection.
27. Liu, W., *et al.* Ssd: Single shot multibox detector. Springer.
28. Alhasanat, M.N., *et al.*, RetinaNet-based Approach for Object Detection and Distance Estimation in an Image. 2021.
29. Tan, L., *et al.*, Comparison of RetinaNet, SSD, and YOLO v3 for real-time pill identification. *BMC Medical Informatics and Decision Making*, 2021. **21**(1): p. 1-11.
30. Ge, Z., *et al.*, Yolox: Exceeding yolo series in 2021. arXiv preprint arXiv:2107.08430, 2021.
31. Wang, C.-Y., A. Bochkovskiy, and H.-Y.M. Liao, YOLOv7: Trainable bag-of-freebies sets new state-of-the-art for real-time object detectors. arXiv preprint arXiv:2207.02696, 2022.
32. Chen, J., *et al.*, A Multiscale Lightweight and Efficient Model Based on YOLOv7: Applied to Citrus Orchard. *Plants*, 2022. **11**(23): p. 3260.

11

Enhanced Processing of Aluminum Scrap at End-of-Life via Artificial Intelligence & Smart Sensing

Sean McCoy Langan[1], Emily Molstad[2*], Ben Longo[2], Caleb Ralphs[2], Robert De Saro[3], Diran Apelian[4] and Sean Kelly[1]

[1]Solvus Global, LLC, Worcester, MA, USA
[2]Solvus Global, LLC and Valis Insights, Worcester, MA, USA
[3]Energy Research Company, Plainfield, NJ, USA
[4]University of California, Irvine, CA, USA

Abstract

Sustainability initiatives such as lightweighting and electrification have shifted demand for aluminum away from cast alloys for components such as internal combustion engines to higher-value cast and wrought alloys. As aluminum-intensive vehicles reach their end of life, and international trade trends toward increased quality restrictions for scrap, there is a growing need to transition metal recycling away from unsustainable practices, such as downcycling, toward processes that retain material value for use in high quality applications. Domestic scrap processors and consumers need next generation technology, powered by advanced sensors and artificial intelligence, that enables data-driven sorting and scrap blending capabilities to produce high-quality scrap packages from end-of-life automotive scrap. This will enable an increase in aluminum production with a decrease in emissions. The successful upcycling of aluminum scrap requires three major technical capabilities: (1) optimal scrap processing (i.e., intelligent sorting), (2) optimal scrap melting, and (3) a feedback loop to ensure processing is adjusted based on melt results. This work will deliver a scrap melting control tool that integrates with the necessary sensing capabilities to create a feedback loop that drives optimal scrap recycling decision-making. Previous work, focused on optimal scrap sorting, will additionally be leveraged to ensure all information is considered in this decision-making process. To accomplish this, a smart sensor suite that includes a rapid scrap characterization quality assessment tool is being developed. This provides baseline solid state materials data that is coupled with the output from an *in-situ* laser-induced breakdown spectroscopy sensor, providing analysis of melt composition in real time; both sensing tools are integrated with a novel process control software. The completion of this effort will demonstrate increased consumption of post-consumer scrap by at least 30% at the commercial scale to produce value-added aluminum alloys. This paper covers subject background, work to date on this project, as well as future work to be performed.

Keywords: Artificial intelligence, aluminum, aluminum recycling, circular economy, scrap, twitch, VALI-MELT, quality control

Corresponding author: emily.molstad@valisinsights.com

Nabil Nasr (ed.) Technology Innovation for the Circular Economy: Recycling, Remanufacturing, Design, Systems Analysis and Logistics, (129–142) © 2024 Scrivener Publishing LLC

11.1   Introduction

The aluminum industry stands at the precipice of unprecedented change dictated by competing external factors. The International Energy Agency has allotted the industry a budget of 250 Mt CO_2-e per year by 2050 as a part of its efforts to limit global warming to 2°C [1]. This will require a dramatic drop in emissions, as the industry currently generates 1.1 Gt CO_2-e every year [1]. Further complicating matters, demand for aluminum is projected to increase 80% by that time [1]. Multiple elements are driving this spike in demand. As lightweighting and electrification become essential for the automotive industry, some of steel's dominant market share is being replaced by aluminum, due to its strength and lighter weight [2]. This increase in demand is exacerbated by the need not just for more aluminum, but a growing necessity for more "advanced", high-value cast and wrought aluminum alloys. All these factors are summarized in Figure 11.1. "Business as usual" will no longer suffice, as these competing pressures have created a perilous imbalance in the aluminum industry that needs to be addressed through bold, transformative technologies to achieve society's economic, environmental, and social goals [3].

One important solution for this increase in demand and corresponding decrease in emission targets is upcycling of aluminum scrap as a part of a circular economy. For this, at its end of life, aluminum scrap is used to create aluminum alloys of equal or greater value, which is environmentally preferable to using purely "virgin" metal sources [4, 5]. However, approximately 60-80% high-value aluminum alloys found in auto-shred aluminum are downcycled [6]. This trend is a significant hurdle in the effort towards a more circular economy, but one that has been dictated by technological readiness [7]. Valuable material precursors such as 5xxx, 6xxx, and low-copper-containing alloy scrap are currently being blended into lower-value cast products. This leads to resource processing inefficiencies [2], 3, 6–12] and financial losses, as seen in Figure 11.2, as more virgin materials are needed,

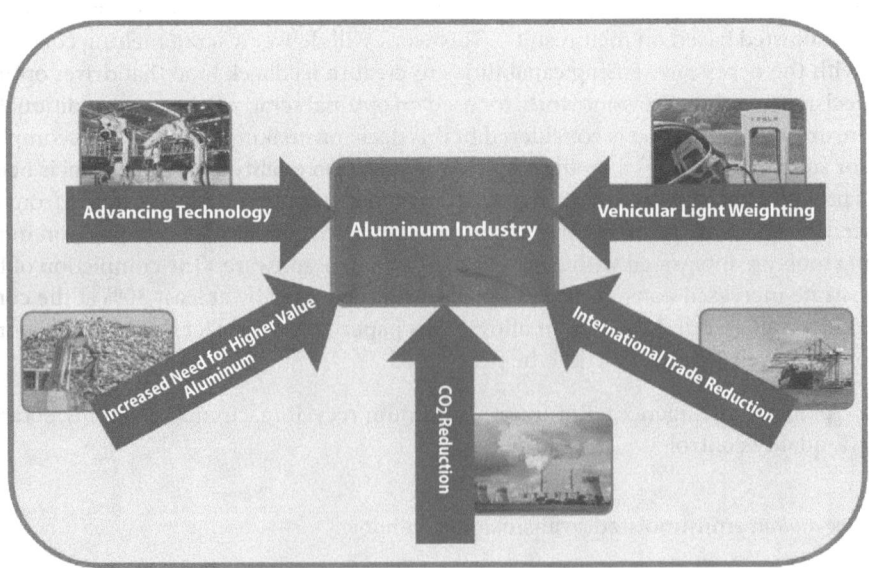

Figure 11.1 Pressures on the aluminum industry.

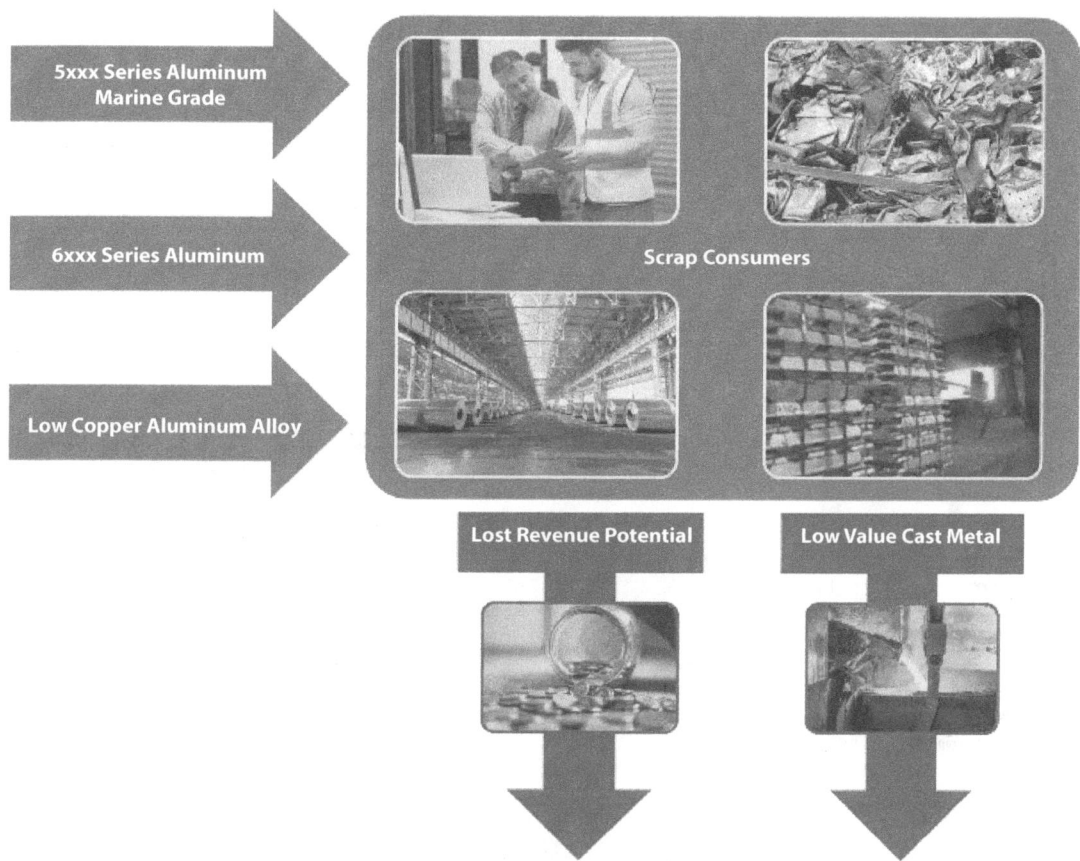

Figure 11.2 Scrap processing and its products.

and scrap consumers lose out on the opportunity to maximize their profits on higher-value materials. For the industry to hit its emission targets, upcycling, not downcycling, must become the accepted paradigm, and indeed, "business as usual."

Aluminum is used in a variety of ways, and so, the scrap sources are similarly diverse. In 2020, aluminum end-use was as follows: 23% for transportation, 25% for construction, 8% for packaging, 9% for foil stock, 12% for electrical, 6% for consumer durables, 11% for machinery and equipment, and 6% for other uses [13]. As shown in Figure 11.3, end-of-life vehicles (ELVs) are first processed by dismantlers, where certain high-scrap-value components, components that can be re-used, and parts of the vehicle that would be dangerous to shred are removed. These vehicles are often flattened, then sent to scrap processing yards and shredded; the shredded scraps are then sorted by type. Ferrous scrap, high in iron-based metals, is typically first separated from the shreds via magnetic separation, leaving auto-shred residual (ASR). This ASR then undergoes air separation to segment out "light material" and is further refined with eddy current separators to give Zorba, i.e., non-ferrous auto shred. This is often exported, but what is not exported is separated via density separation into two fractions: Zebra, the "heavy" portion of Zorba containing stainless steel, copper alloys, zinc alloys, etc.; and Twitch, the "light" portion with a composition of 90-98% aluminum alloys, as well as magnesium along with steel and plastic contaminants [6].

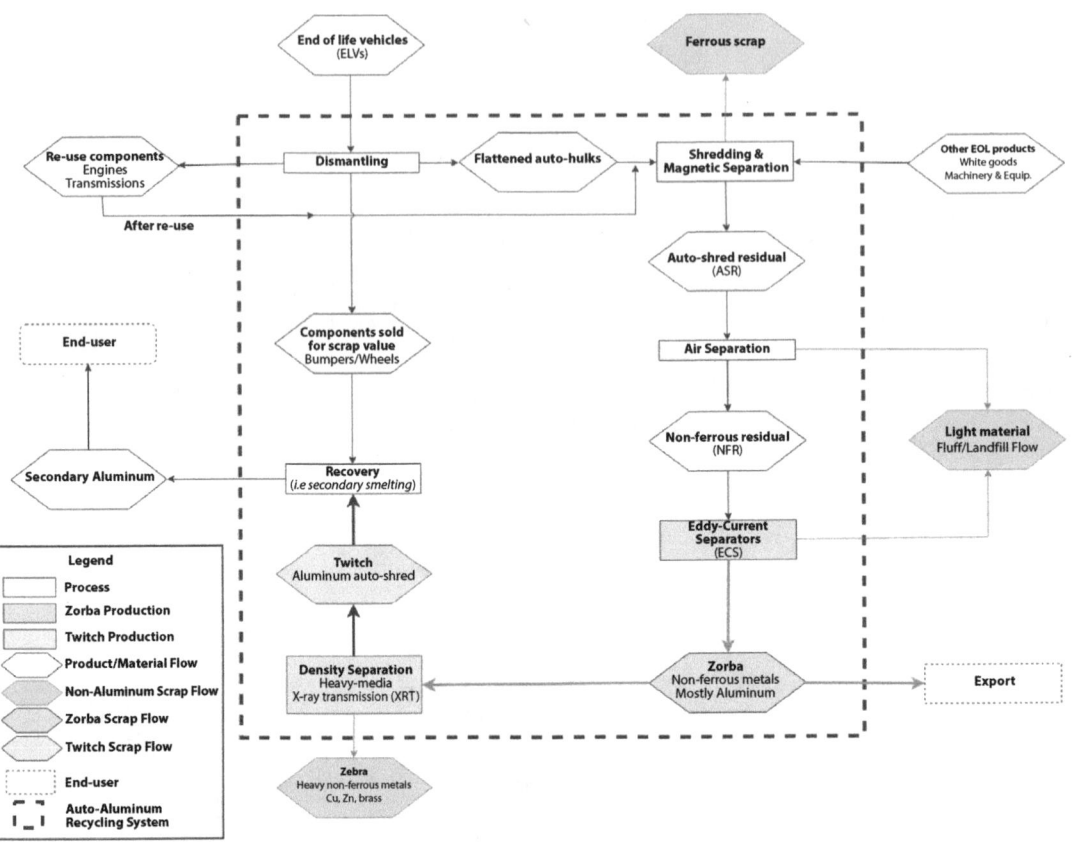

Figure 11.3 ELV processing flow.

This is currently the end of the scrap processing line for aluminum auto-shred. Twitch is then charged in bulk by secondary aluminum producers in the US to form die-cast alloys with wide compositional specifications (e.g., *A380/A319*). This juncture is where the paradigm can shift to upcycling Twitch through sortation into varying alloy classes leading to increased profit and sustainable practice by closing the recycling loop [3, 6, 10, 14–20].

To best identify the optimal sorting criteria to upcycle Twitch and best prepare it for recovery (i.e., melt processing), the contents of the scrap product must be known, in addition to a few key quality metrics relating to contamination levels. The compositional make up of Twitch affects end-use application and sorting criteria management; the contamination evaluation of Twitch affects melt processing and yield. Carbon-containing coatings such as paints and oils increase the complexity of the auto-shred recycling process. Direct remelting of auto-shred scrap without proper cleaning creates gaseous emissions at higher volumes and unwanted chemical reactions, including the formation of aluminum carbide and other intermetallic compounds that result in increased melt loss due to further oxidative reactions [21, 22].

At a minimum, the steps taken by industry to characterize shredded scrap include: (i) extracting, at random, a representative sample (>50lbs per sample, [23]) of the larger scrap stream from the production line and/or final product output to determine various material properties and (ii) evaluating the bulk composition of scrap including quantifying the various metals, aluminum forms (i.e., cast vs. wrought) and free plastic/rubber contamination in the mix. This analysis is primarily driven by the high-level guidelines set in the Institute for Scrap Recycling Industries' (ISRI) Scrap Specification Circular [24].

These guidelines set by ISRI have been instrumental in buyer-seller arrangements historically but do not provide the granularity required to ensure that the scrap properties critical to upcycling are known by both the scrap processor and scrap consumer. The list of required properties for evaluation are discussed in Table 11.1. The third column specifies whether the material property is likely or unlikely to be evaluated, as determined by characterization efforts in the industry. It is difficult to say for certain if specific analytical techniques are practiced facility-to-facility and company-to-company, as each scrap processor approaches quality control differently and there is not a standardized procedure or requirement for this type of analysis industry wide. For high-caliber upcycled materials to be produced from this Twitch scrap, high-caliber quality control is necessary. Sorted scrap can vary in quality, cleanliness, and consistency between scrap processors.

Table 11.1 Scrap properties.

Property*	Description	Common practice?
Bulk composition (C)	Analysis of base metal composition (i.e. aluminum, copper, brass, stainless steel, etc.)	Likely
Major form/category type (Y)	Determination of form/category type (i.e. cast, wrought, other metal, polymer, mixed metal, mixed material)	Likely
Mass (C&Y)	Quantifying the weight of scrap pieces so weighted calculations can be made	Likely
Size & Volume (Y)	Quantifying the dimensions of the scrap piece for surface area estimation and volume for surface-area-to-volume ratio	Likely
Elemental Composition (C)	Compositional analysis to determine elemental distribution/alloy family	Unlikely
Bulk contamination (C&Y)	Identification of non-base metal/alloy metallic contamination (i.e. steel bolt in aluminum cast)	Likely
Organic contamination (Y)	Identification of non-metallic contamination (i.e. rubber connection hoses, paint etc.)	Unlikely

*(C) = affects composition and (Y) = affects melt yield.

11.2 Results and Discussion

11.2.1 Twitch Characterization

11.2.1.1 Compositional Evaluation

In a previous study [6], various Twitch samples were characterized to set baseline quality values for composition and contamination levels. Bulk Twitch composition was determined by melting five representative samples from two different scrap processing locations in an induction furnace and casting three optical emission spectroscopy samples per each melt for compositional analysis. A handheld x-ray fluorescence (XRF) analyzer in conjunction with an electrical conductivity probe were used to determine the alloy/alloy family distribution for these same five samples before melting. Table 11.2 displays average bulk Twitch composition. Two major takeaways from this data are (1) the variability of (±) the composition from a single facility over time and (2) the dissimilarity of this composition from A380/A319 specifications, the secondary alloys typically produced with Twitch. For reference, the silicon content specification for A380 is 7.5 – 9.5wt.% and A319 is 5.5 – 6.5wt.%. The copper content specification for both A380 and A319 is 3.0 – 4.0wt.%. Table 11.3 displays the distribution of aluminum alloys and alloy series in Twitch. It is clear by the breakdown presented that there is significant opportunity to extract value by sorting out high-value 5xxx-series, 6xxx-series, and 356-cast alloys (>30wt.% of total mixture).

11.2.1.2 Contamination Evaluation

Additionally, from this prior work, a visual hand-sort determined the major bulk contaminants in the mixture including compounded metals (e.g., aluminum cast with a steel bolt), free plastics/rubbers, mixed materials (e.g., compounded aluminum tubing and rubber), and other non-aluminum metals. It was found that Twitch can also contain ~1 – 2.5% compounded metals, ~1 – 2% free plastic and rubber, ~1% mixed materials, and ~1% other metals (primarily magnesium alloys). Further, the organic content adhered to auto-shred

Table 11.2 Bulk twitch compositional analysis.

Alloying/Matrix elements	Sample set 1	Sample set 2
Si	3.8 ± 0.75%	4.8 ± 0.94%
Fe	0.62 ± 0.11%	0.68 ± 0.06%
Cu	1.2 ± 0.34%	1.4 ± 0.36%
Mn	0.22 ± 0.05%	0.23 ± 0.05%
Mg	1.9 ± 1.2%	1.9 ± 0.54%
Zn	0.95 ± 0.38%	1.0 ± 0.41%
Other	0.13 ± 0.05%	0.15 ± 0.03%
Al	91 ± 2.2%	90 ± 1.8%

Table 11.3 Twitch aluminum alloy distribution.

Alloy/Alloy family	Twitch 1	Twitch 2
319	5%	7%
356	3%	3%
380	16%	31%
390	1%	3%
413	4%	6%
2xxx Series	0%	0%
3xxx Series	1%	1%
4xxx Series	0%	0%
5xxxx Series	8%	8%
6xxx Series	26%	20%
7xxx Series	2%	1%
Other (Size-Restricted)	34%	20%

Table 11.4 Twitch de-coating results.

Sample	Mass burn-off [wt.%]	Ultrasonic cleaning [wt.%]	Total de-coated [wt.%]
Twitch 1 – 1	0.87%	0.30%	1.17%
Twitch 1 – 2	0.46%	0.48%	0.94%
Twitch 1 – 3	0.29%	0.15%	0.44%
Twitch 1 – 4	0.38%	0.19%	0.57%
Twitch 1 – 5	0.31%	0.16%	0.47%
Average:	0.46%	0.26%	0.61%
Twitch 2 – 1	0.54%	0.21%	0.75%
Twitch 2 – 2	0.96%	0.11%	1.07%
Twitch 2 – 3	0.41%	0.25%	0.66%
Twitch 2 – 4	0.36%	0.24%	0.60%
Twitch 2 – 4	0.92%	0.17%	1.09%
Average:	0.64%	0.19%	0.83%

aluminum was evaluated. The amount of adhered organic content (i.e., paint, oil, lubricants, etc.) was quantified using a mass balance approach on five randomly selected subsamples from each processing site (T1-1 – T1-5, T2-1 – T2-5). Each sample was initially weighed and then placed in a box furnace at 425°C for 1h. This temperature was selected as it is comfortably below the thermal oxidation temperature for aluminum when heated in air [25] and is within the decomposition temperature range for these adhered organics, also in air, as determined by thermogravimetric analysis (TGA) [6]. After cooling, each sample was weighed and ultrasonically cleaned for 60 minutes to remove any residual char. The amount of organic contamination de-coated after each step and in total is shown in Table 11.4 as a weight percent. The average amount of surface contamination removed was 0.6wt.% and 0.8wt.% for site 1 and site 2, respectively, with relatively significant variation. In summary, contaminated scrap particulates and free contaminants totaled between 4.6 – 7.3 wt.%; this is a significant fraction of Twitch that can negatively affect both composition control and melt recovery yields.

While it is always important to monitor melt composition during metal production, it becomes vital when using scrap as a charge material. Due to contamination concerns, as well as the inherent variability of scrap lots, there is a greater chance that melt composition will fall out of specified limits and will require modification. Monitoring of the melt is generally done by removing small samples of molten metal and analyzing it with techniques such as optical emission spectroscopy (OES). While this process functions, the sample preparation can be time consuming, and the act of retrieving the sample from the melt is dangerous to the operator. Furthermore, this must be done at precise intervals, which can be difficult for workers in a production environment. Because of this, numerous "active" methods for melt evaluation, as described in [26], are being developed to monitor the melt in its liquid form, including laser-induced breakdown spectroscopy (LIBS), Raman spectroscopy, laser-induced fluorescence (LIF), and diffuse reflectance spectroscopy (DRS). These technologies can eliminate the necessity of frequent samplings of the melt by a person and replace them with continuous or near-continuous monitoring of the melt's composition.

11.2.2 Next Generation Technology: VALI-Melt

Providing technological solutions that enable Twitch upcycling is a generational opportunity, but not one that can be met by advancements in scrap-sorting capabilities alone. Sensor-based sorting system (SBS) manufacturers are making strides in advancing the technology (e.g., LIBS) needed to sort fragmented aluminum scrap packages into increased-value products; however, it is known that some related system performance metrics (e.g., material throughput, sort accuracy, sort resolution, reduced costs) must be improved to achieve industry-wide adoption. Additionally, technologies that focus on quality control and real-time decision-making must be developed and operate in concert with sorting systems to effectively upcycle shredded aluminum scrap. VALI-Melt will do exactly this – connect the material property output from a scrap quality assessment tool (SQA) with the compositional data from an in-melt LIBS analyzer to create closed-loop quality assurance and process monitoring to aid upcycling of auto-shred scrap, as seen in Figure 11.4. The development and validation of these supporting systems that will span the aluminum recycling industry from scrap processing to scrap consumption facilities are the focus of the REMADE-funded transformational program discussed here.

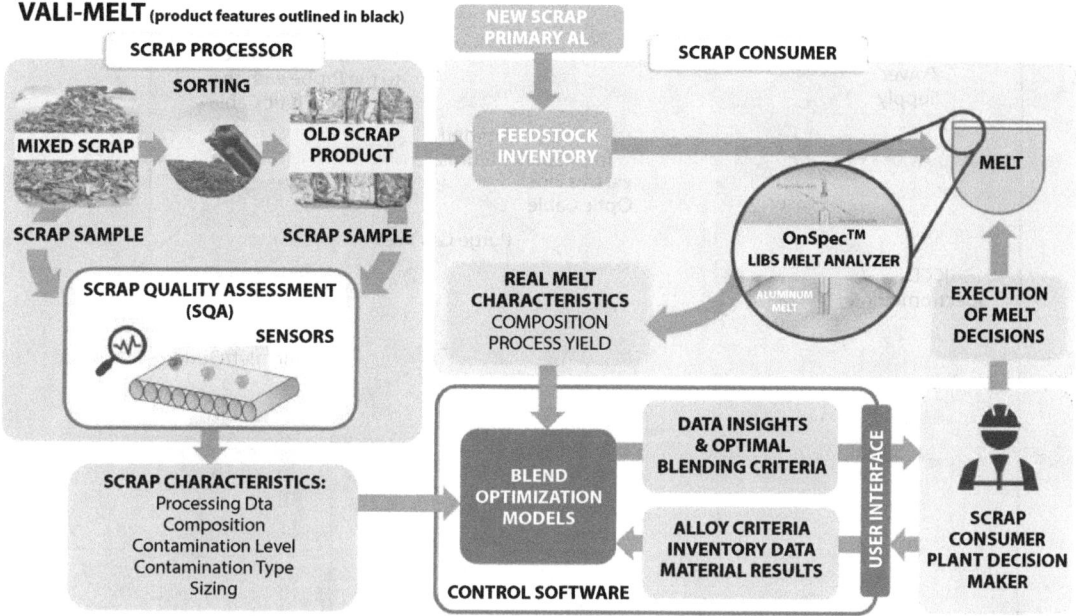

Figure 11.4 VALI-melt: closed loop quality assurance.

11.2.3 Quality Control

11.2.3.1 Solid Scrap Quality Control

The SQA system demonstrated from this program will provide baseline solid-state material data while limiting additional personnel involvement to reduce cost. As discussed, and as shown in Table 11.1, detailed characterization efforts are required to gauge scrap quality for reporting to scrap consumers for melt processing preparation, as well as to review scrap processing efficacy and where changes and enhancements need to be made. However, knowing scrap processors work within tight operational margins, the additional costs and time needed to achieve such detailed analysis must be addressed. Representative scrap sample sets are fed piece-by-piece to the SQA system and are analyzed by off-the-shelf sensors. This sensor data feeds proprietary algorithms to evaluate the necessary properties that effect composition and melt yield in real time. The data is automatically reported to the scrap processor and customer (i.e., scrap consumer) for quality assurance and seamless linkage to melt quality data. The SQA's analytical capability target is critical alloy composition evaluation within ±5wt.%, mass within ±5%, and surface/bulk contamination within ±10%.

11.2.3.2 Molten Metal Quality Control

The second quality-control feature to be optimized within VALI-Melt during this program is the OnSpec™ in-melt LIBS sensor. This sensor provides real-time melt composition data, replacing the need for inefficient sample analysis via offline analysis methods. Streamlined and immediate feedback to metallurgists and quality control managers at melt facilities allows for expedited decision-making and adjustments to ensure the melt meets compositional specifications of the

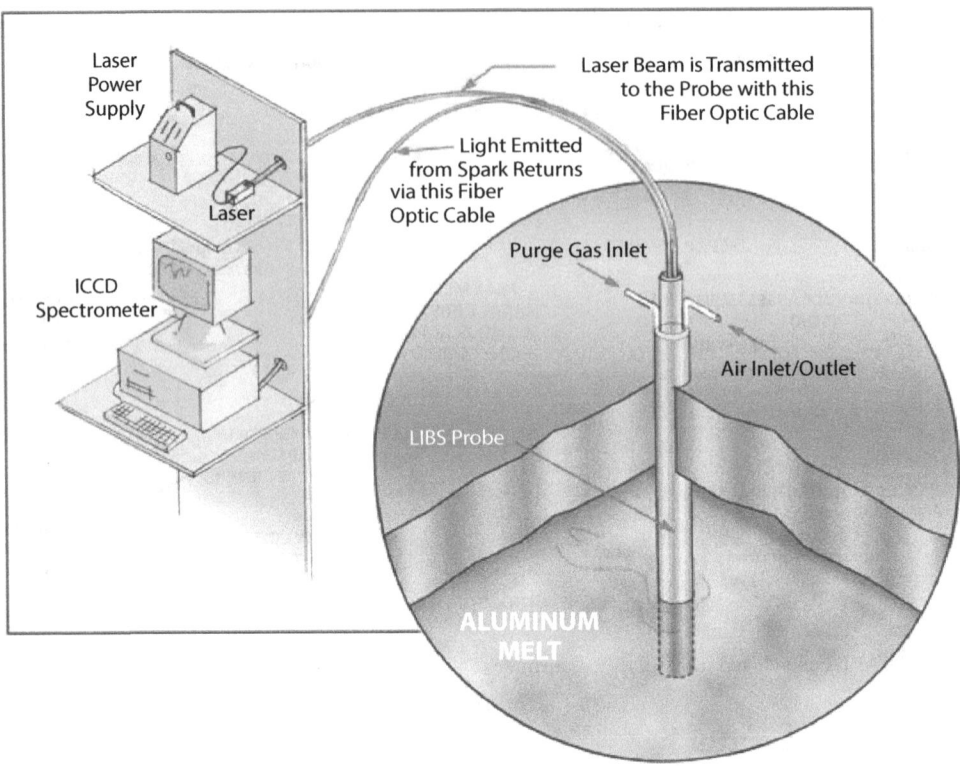

Figure 11.5 In-melt LIBS sensor.

aluminum alloy being produced. This *in-situ* sensor can analyze all critical alloying and metallic impurities related to aluminum including silicon, copper, iron, magnesium, manganese, zinc, etc. The general functionality of the OnSpec™ is highlighted in Figure 11.5.

11.2.4   Optimal Decision Making

As discussed, connecting scrap property data from processing facilities with an immediate feedback loop from melt pool analysis creates a system that permits immediate processing adjustments to be made across the scrap metal recycling industry. The VALI-Melt product provides critical insights into how scrap feedstock material directly impacts melt composition and yield, with the ability to make adjustment recommendations in real time. This connection and immediate feedback will provide the data necessary to optimize material blending, accounting for scrap composition and contamination levels. Resource blending can scale from in-house recipes through to market availability as increasingly more scrap processors bring SBS systems online with varying composition and quality levels. Figure 11.6 displays an example output of resource blending. Actual primary and scrap resource compositions and availabilities are inputted to an optimization model that presents options for how to meet demand and compositional requirements for various automotive alloys. Such a tool will enable scrap consumers to quickly scan the commodity market and determine what scrap resources are available and can be used to meet demand, so long as detailed compositional data is available. VALI-Melt via the SQA tool offers what is needed to provide this insight.

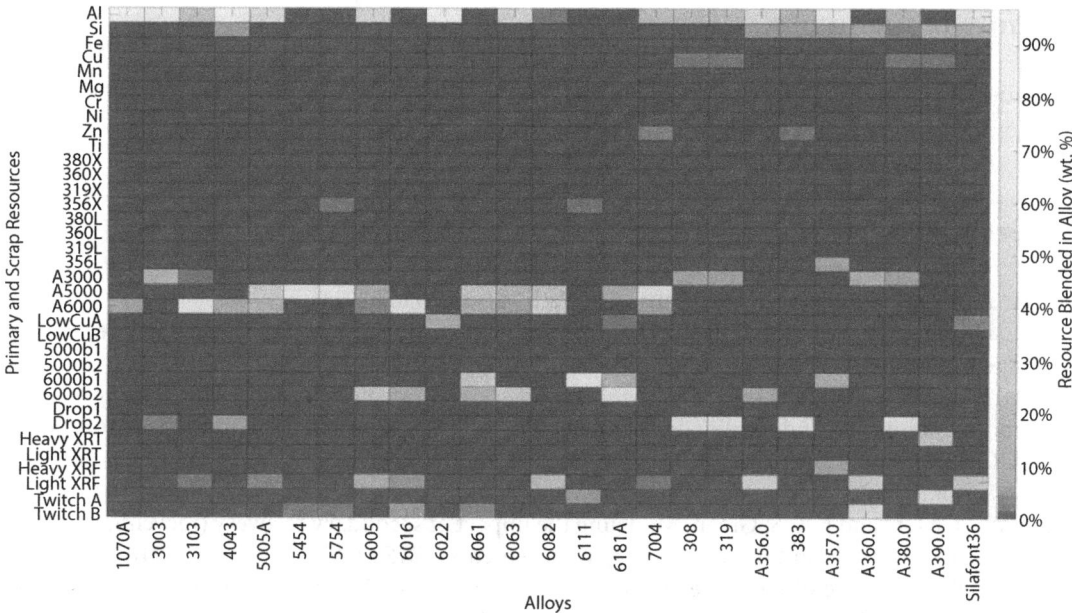

Figure 11.6 Commodity blending possibilities to form common automotive alloys from primary metal and scrap resources.

11.3   Conclusions & Recommendations

This study examined the complex, dynamic situation that the aluminum industry is in due to the simultaneous needs to meet increased demand and to reduce emissions, the historical efforts at scrap characterization, and the solution the authors are pursuing to solve these problems. The pressures that the aluminum industry is facing, including the need to drastically reduce emissions while substantially increasing production, and the need for higher value aluminum alloys, are forcing a dramatic change in direction for the industry. The environmental benefits of utilizing aluminum scrap cannot only be realized for low value cast alloys such as A380 and A319 but must also be applied to high-value wrought and cast alloys as well.

Understanding the nature of aluminum scrap is tremendously important for a successful transition from downcycling to upcycling. It has previously been found that Twitch could be better suited for high-value alloys rather than lower- value cast due to its composition. Twitch has been found to be high in 5xxx, 6xxx, and 356-cast alloys, showing its vast potential for upcycling. Understanding the contamination in Twitch is similarly important. This contamination can affect melt yield and purity. These factors show the need for the wholistic, closed-loop quality control system being developed through this work. It will gather information on both the composition and contamination levels of scrap being fed to the melt, and, in concert with an in-melt LIBS detector and a novel blending model, will allow for optimal decision making. Future works will describe the progress of the technology and explore more deeply how efforts like this can solve the aluminum industry's difficulties.

Acknowledgements

The authors would like to thank The REMADE Institute for support through 21-01-RR-5102, and Nicole Elizabeth Boyson for her assistance and advice on this chapter.

References

1. IAI, "Aluminium Sector Greenhouse Gas Pathways to 2050," 2021.
2. L. Cheah, J. Heywood, and R. Kirchain, Aluminum stock and flows in U.S. passenger vehicles and implications for energy use. *J. Ind. Ecol.*, 13, 5, pp. 718–734, 2009.
3. A. N. Løvik, R. Modaresi, and D. B. Müller, Long-term strategies for increased recycling of automotive aluminum and its alloying elements. *Environ. Sci. Technol.*, 48, 8, pp. 4257–4265, 2014.
4. J. Cui and H. J. Roven, Recycling of automotive aluminum. *Trans. Nonferrous Met. Soc. China (English Ed.)*, 20, 11, pp. 2057–2063, 2010.
5. W. T. Choate and J. A. S. Green, Modeling the impact of secondary recovery (recycling) on U. S. Aluminum supply and nominal energy requirements. *TMS Light Met.*, pp. 913–918, 2004.
6. S. Kelly, Recycling of Passenger Vehicles: A Framework for Upcycling and Required Enabling Technologies. Worcester Polytechnic Institute, 2018.
7. G. Gaustad, E. Olivetti, and R. Kirchain, Toward sustainable material usage: Evaluating the importance of market motivated agency in modeling material flows. *Environ. Sci. Technol.*, 45, 9, pp. 4110–4117, 2011.
8. W. Q. Chen, Recycling rates of aluminum in the United States. *J. Ind. Ecol.*, 17, 6, pp. 926–938, 2013.
9. W. Q. Chen and T. E. Graedel, Dynamic analysis of aluminum stocks and flows in the United States: 1900-2009. *Ecol. Econ.*, 81, pp. 92–102, 2012.
10. G. Gaustad, E. Olivetti, and R. Kirchain, Improving aluminum recycling: A survey of sorting and impurity removal technologies. *Resour. Conserv. Recycl.*, 58, pp. 79–87, 2012.
11. H. Hatayama, H. Yamada, I. Daigo, Y. Matsuno, and Y. Adachi, Dynamic substance flow analysis of aluminum and its alloying elements. *Mater. Trans.*, 48, 9, pp. 2518–2524, 2007.
12. R. Modaresi, A. N. Løvik, and D. B. Müller, Component- and Alloy-Specific Modeling for Evaluating Aluminum Recycling Strategies for Vehicles. *JOM*, 66, 11, pp. 2262–2271.
13. Norsk Hydro ASA, Annual Report. Oslo, 2021. [Online]. Available: https://www.hydro.com/globalassets/download- center/investor-downloads/ar20/annual-report-2020-new.pdf, 2020.
14. H. Hatayama, I. Daigo, Y. Matsuno, and Y. Adachi, Evolution of aluminum recycling initiated by the introduction of next-generation vehicles and scrap sorting technology. *Resour. Conserv. Recycl.*, 66, pp. 8–14, 2012.
15. S. Koyanaka and K. Kobayashi, Automatic sorting of lightweight metal scrap by sensing apparent density and three- dimensional shape. *Resour. Conserv. Recycl.*, 54, 9, pp. 571–578, 2010.
16. S. Koyanaka, K. Kobayashi, Y. Yamamoto, M. Kimura, and K. Rokucho, Elemental analysis of lightweight metal scraps recovered by an automatic sorting technique combining a weight meter and a laser 3D shape-detection system. *Resour. Conserv. Recycl.*, 75, pp. 63–69, 2013.
17. M. B. Mesina, T. P. R. de Jong, and W. L. Dalmijn, Automatic sorting of scrap metals with a combined electromagnetic and dual energy X-ray transmission sensor. *Int. J. Miner. Process.*, 82, 4, pp. 222–232, 2007.

18. S. Merk, C. Scholz, S. Florek, and D. Mory, Increased identification rate of scrap metal using Laser Induced Breakdown Spectroscopy Echelle spectra. *Spectrochim. Acta - Part B At. Spectrosc.*, 112, pp. 10–15, 2015.
19. T. Takezawa, M. Uemoto, and K. Itoh, Combination of X-ray transmission and eddy-current testing for the closed-loop recycling of aluminum alloys. *J. Mater. Cycles Waste Manag.*, 17, 1, pp. 84–90, 2015.
20. P. Werheit, C. Fricke-Begemann, M. Gesing, and R. Noll, Fast single piece identification with a 3D scanning LIBS for aluminium cast and wrought alloys recycling. *J. Anal. At. Spectrom.*, 26, 11, pp. 2166–2174, 2011.
21. J. Grayson, Reducing melt loss and dross generation. *Light Met. Age*, 75, 1, pp. 32–35, 2017.
22. A. Kvithyld, C. E. M. Meskers, S. Gaal, M. Reuter, and T. A. Engh, Recycling light metals: Optimal thermal de-coating. *JOM*, 60, 8, pp. 47–51, 2008.
23. V. Kevorkjjan, The recycling of standard quality wrought aluminum alloys from low-grade contaminated scrap. *JOM*, 62, 8. pp. 37–42, 2010.
24. ISRI, Scrap Specifications Circular. 2021.
25. L. Shih, Teng-Shih; Zin, Thermally-Formed Oxide on Aluminum and Magnesium. *Mater. Trans. JIM*, 47, pp. 1347– 1353, 2006.
26. A. K. Myakalwar, C. Sandoval, M. Velásquez, D. Sbarbaro, B. Sepúlveda, and J. Yáñez, LIBS as a spectral sensor for monitoring metallic molten phase in metallurgical applications—A review. *Minerals*, 11, 10, 2021.

12

Deep Learning for Defect Detection in Inspection

Mohammad Mohammadzadeh[1], Pallavi Dubey[1], Elif Elcin Gunay[2]*, John K. Jackman[1], Gül E. Okudan Kremer[3] and Paul A. Kremer[4]

[1]Iowa State University/Industrial, Manufacturing and Systems Engineering, Ames, IA, United States
[2]Sakarya University/Department of Industrial Engineering, Sakarya, Turkey
[3]The University of Dayton, School of Engineering, Dayton, OH, United States
[4]Iowa State University/Civil, Construction and Environmental Engineering, Ames, IA, United States

Abstract

Remanufacturing is the process of restoring a used product to the condition that satisfies the original performance specifications. One of the most critical stages of this process is inspection, which assesses the returned item's condition. Practically, the inspection process is carried out with manual or automated systems. Manual systems are generally expensive due to labor-intensive and time-consuming workloads. Additionally, they depend highly on the inspector's experience and psychophysical abilities, e.g., morning shift vs. night shift. With recent advancements in automated inspection enabled by machine vision and machine learning techniques, high inspection accuracy is possible without significant economic barriers. In this paper, we implement three object detection algorithms, YOLO v5s (small), Faster R-CNN ResNet101, and MobileNet-SSD with FPN, to classify and locate the defects encountered during remanufacturing. Using an industry-originated data set, the overall performance analysis and comparison results of three algorithms are presented for average accuracy metric. These initial results provide insights about the capability of the algorithms in defect detection as an alternative to human-operated systems and to understand which algorithm performs better under which circumstances. Overall, our findings will contribute to the design of future studies focusing on improvements in automated inspection technology.

Primary and Secondary Topics: Innovation of Remanufacturing Technologies, Remanufacturing, and Material Reuse

Keywords: Defect detection, inspection, deep learning, computer vision, YOLO v5s, Faster R-CNN ResNet101, Mobile-Net and SSD

**Corresponding author*: ekabeloglu@sakarya.edu.tr

Nabil Nasr (ed.) Technology Innovation for the Circular Economy: Recycling, Remanufacturing, Design, Systems Analysis and Logistics, (143–156) © 2024 Scrivener Publishing LLC

12.1 Introduction

Remanufacturing helps reduce material consumption, energy use, and waste by retaining the value of extracted and refined raw materials. Thus, it is a foundational element supporting circular economy implementations. Remanufacturing has multiple stages, such as cleaning, inspection, and reconditioning, to return a used core product to a condition that meets the specification requirements of a new product [1]. As one of the important stages, inspection assesses the core condition to determine the size and location of defects and deviations from design specifications [2]. In fact, quality inspection is critical to remanufacturing objectives associated with end customer satisfaction, the elimination of rework originating from poor inspection and earlier decision making related to the viability of remanufacture of a given core component.

Inspection can be performed in any of three forms: manual, semi-automated, or automated. Manual inspection is labor-intensive and time-consuming for remanufacturing, where 100 % inspection is necessary to gain a second user's confidence and to increase the manufacturer's profitability [3]. Additionally, variability in inspector experience, inspector skills, time of inspection (e.g., night shift vs. morning shift inspection periods) or inspection location impacts the Type I and Type II error rates [4–7]. In response to concerns associated with manual inspection quality, semi-automated and automated inspection tools are merged to assist with the inspection process. Mital *et al.* [2] reported that semi-automated systems, for which both human inspectors and machine vision systems are in the loop, can detect defects in less time, with fewer errors (approximately 47% less), and with less variation. As tools and techniques evolve, fully automated systems employing machine learning (ML) techniques and machine vision systems can supplant human inspectors for defect detection in remanufacturing processes. With advancements in machine vision, it is possible to detect many defects of interest with sufficient resolution and accuracy without increasing inspection time unduly. In today's manufacturing industry automation plays a significant role and inspection and quality control using image processing techniques for defect detection is now common [8]. Deep learning-based inspection systems have become more prevalent. These systems include stacked, multi-layer, neural networks to extract the features that help describe the data patterns.

As a deep learning architecture class, the convolutional neural network (CNN) is inspired by the visual perception of living things [9] and has been successfully applied in image recognition, object detection, and localization [10]. In general, CNN-based object detection algorithms are classified as one- stage and two-stage object detection models. One-stage models do not require any preliminary steps for object detection; thus, they provide quick predictions that are capable of real-time responses. In contrast, two-stage object detection models employ preliminary stages with important image regions being detected first followed by object classification. Faster regions with convolutional neural networks (Faster R-CNN ResNet101), one of the high- accuracy, two-stage object detector models, divides the region proposal and classification steps into two stages to locate and classify objects. Despite achieving higher prediction performance in terms of accuracy as compared to other approaches, Faster R-CNN ResNet101 needs improvement in prediction speed [11]. Consequently, one-stage detector models have been proposed as alternative approaches. For instance, MobileNet- Single shot multi- box detector (SSD) with feature pyramid network (FPN) is a lightweight, one-stage object detector model that has been developed to yield quick response with less computational power demands [12]. The MobileNet- SSD with FPN approach reduces the number of model parameters thanks to its depth-wise separable convolutional neural network structure. As a result, it has been found to provide faster predictions with lower memory requirements [11]. The SSD support in MobileNet helps extract features from different feature maps to gain strong information from an image [13-15] with the intention of detecting multi-scale objects with increased accuracy. YOLO v5s, on the other hand, is a one-stage detector model that belongs to the You Only Look Once (YOLO) family of computer vision models [16] and is known for processing images multiple times faster than the above-mentioned models by making inline transformations to the base training data, by providing a wider range of semantic variation and by carefully calculating the loss functions to maximize the objective of mean average precision (mAP).

This study investigates the performance of three algorithms Faster-RCNN ResNet101, MobileNet-SSD with FPN, and YOLO v5s, which provide utility for defect detection because of their accuracy and quick response. The effectiveness comparison of these algorithms is evaluated for accuracy using datasets drawn from images of observed defects on the metal surfaces of real components being remanufactured. Section 12.2 introduces related work in this field, informing the methodology outlined in Section 12.3. Section 12.4 presents an initial assessment of the prediction performance of the three algorithms. Section 12.5 concludes the paper and discusses the limitations and extensions of the present study.

12.2   Literature Review

Classic machine learning methods for image recognition rely on conventional image processing and machine learning algorithms [17]. Conventional image processing techniques attempt to retrieve significant data characteristics from photos. Such attributes are then used to train an algorithm for machine learning to detect trends in photos containing target objects. Three common ML approaches exist involving: supervised methods, semi-supervised methods, and unsupervised methods [17]. For supervised approaches, the user needs to label data with their class information and use labeled data for model training. Hence, supervised methods require a person to be able to recognize and classify defects. Unsupervised methods utilize data without labels and attempt to anticipate defects on their own. Semi-supervised methods utilize labeled and unlabeled data and combine supervised and unsupervised training approaches. In this paper, we employ supervised methods for defect detection because competent persons are available to label images with defects of interest to train the models.

Object detectors based on deep learning can be one-stage or two-stage detectors. One-stage detectors like YOLO v1 [16], SSD [13], YOLO v2 [18] search for the existence of an item at predetermined locations. One-stage detectors are most often employed as a dense grid of rectangular shapes, with the model determining whether to include a certain object. Two-stage detectors including R-CNN [19], Fast-RCNN [20], and Faster-RCNN [21] execute a region proposal method before classifying objects. This way, the number of locations to be categorized is reduced; however, the total complexity increases because the region proposal requires additional processing. In comparison to one-stage detectors, two-stage detectors are often slower but more precise. In recent times this disparity in precision has shrunk with the current state-of-the-art methods, such as YOLO v5 [22]. Detection accuracy is usually the key factor when selecting a model. If a large amount of data must be processed, however, it is advantageous for the selected algorithm to be fast.

12.3   Methodology

This section briefly describes three algorithms used to detect cracks, mainly selected due to their superiority in providing quick and accurate predictions. Section 12.3.1 gives information about dataset preparation, while Sections 12.3.2-12.3.4 present the structure of object detection models Faster R-CNN, MobileNet-SSD with FPN, and YOLO v5s, respectively. Section 12.3.5 explains data pre-processing and experimental setting. Last, Section 12.3.6 presents the performance metrics used to evaluate the model.

12.3.1   Dataset Preparation

With guidance from our project partner, John Deere Reman, certain defects of interest were selected for study in this project. Specifically, we focus on detecting cracks in cast iron metallurgy that must be identified in the remanufacture of cylinder heads. For this investigation, we used two approaches to acquire images for use in the training and testing of the selected ML algorithms for this study. The first, more controlled approach in a laboratory environment at Iowa State uses real cylinder heads provided by John Deere Reman with defects of interest present. For this approach, automated cobot and gantry systems shown in Figure 12.1 that have been developed are used to mount and move the camera used. A Basler ace 3088-16 gm area scan camera outfitted with an Edmund Optics 8 mm HP series lens is used for in-laboratory image acquisition. A working distance of 100 mm and f/4 aperture setting with a lens orientation normal to the cylinder head surface was used for all images acquired

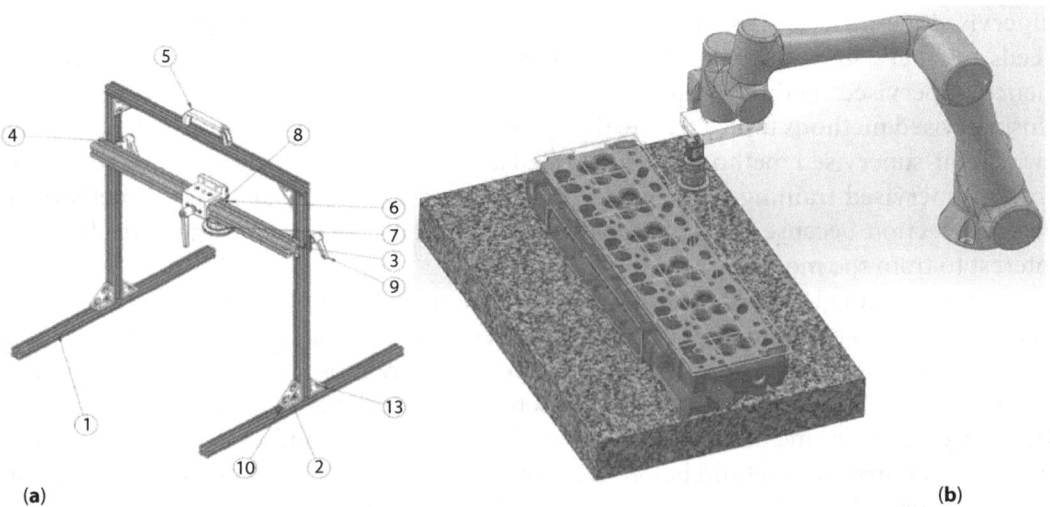

(a)   (b)

Figure 12.1 Example image acquisition apparatus employed in this study: (a) standalone area scan camera gantry rig; (b) cobot-mounted area scan camera.

for this investigation with this C-Mount lens. The camera has a Sony IMX178LLJ-C 1/1.8" format sensor providing 3088 by 2064 HxV pixel resolution. Images of the cylinder head were taken at the mid-point of each grid box location shown in Figure 12.1b. Although optimum control of multiple factors used in the acquisition of images, which includes lighting, is an area of research in this investigation [23], the images used in this investigation only employed ambient lighting with a fixed exposure value setting for the camera.

The second, less-controlled approach in a remanufacturing shop environment uses a simple gantry system shown in Figure 12.1a to acquire images using a conventional, integrated lens SLR camera with 2592 by 1944 pixel resolution for this investigation with less control than the laboratory environment over the aforementioned factors, which are known to affect the quality of acquired images. This second approach was necessary for this investigation to provide a sufficient number of supplemental defect images to support model training efforts. These ambient light images acquired in two distinct settings are merged to create a single, curated data set of images. From the images acquired for this study, the inspection team curated 110 crack images with a total of 121 cracks present in the image set.

Image labeling, annotation of the defects and their location on the images, was manually conducted collaboratively by three inspectors using "labelimg" which is an open-source labeling software. Each inspector labeled the same image sequentially to decrease the error and bias in labeling, meaning that the second inspector could see his or her predecessor's labeling and decide whether it should be changed. Additionally, meetings were conducted when there was a disagreement between inspectors to reach consensus. The labelling process continued until all inspectors agreed on the type and location of each defect.

12.3.2   Faster R-CNN

Faster R-CNN may utilize the convolutional feature maps employed by region-based detectors, such as its predecessors R-CNN and Fast R-CNN, for region proposals that include objects. This is performed by Region Proposal network (RPN) proposed by Ren *et al.* [21]. RPN is developed by adding an additional convolutional layer to the feature extraction network (pre-trained CNN), transforming it into a fully convolutional network that can be trained from the ground up for creating detection proposals [21]. Faster R-CNN includes the following components: 1) Region proposal network (RPN), which creates object proposals and ratings at each place concurrently; and 2) Fast-RCNN detector for predicting the true object category. An image serves as the input for the convolutional layers shared by both RPN and detector. RPN takes feature maps from convolutional layers as input and creates rectangular item suggestions together with confidence scores for each proposal as output. After receiving the region suggestions from RPN, the desired prediction results are generated using Region of Interest (ROI) pooling, an upstream classifier, and a bounding box regressor. This RPN approach allows one to avoid computationally expensive methods such as Selective Search. Using anchor boxes, the RPN creates region suggestions with a broad spectrum of sizes and aspect ratios [21].

12.3.3   MobileNet-SSD with FPN

MobileNet employs depth-separable convolution layers to achieve faster predictions. Depth separable layers almost mimic the operation of typical convolutional layers; however, they split filtering and feature integration processes into two separate layers. In other words, depth-separable convolution layers utilize a separate layer for filtering and another separate layer for merging. First, a single filter is applied to each input channel. Then, a linear combination of the output of the depth-based layers is created by using 1×1 convolutions in point-based convolution [11]. This approach minimizes the size of the model and reduces computational power demands. Therefore, they are much faster compared to the same depth CNN architecture.

Integration of SSD and FPN to the MobileNet architecture provides the following advantages. SSD combines region proposal and classification in a single stage to regress bounding boxes. Because it uses default, fixed size, bounding boxes with different aspect ratios, it significantly reduces the space of bounding box proposals; thus, it provides fast predictions [13]. FPN supports the general structure in merging multi-scale feature maps gathered from different convolutional layers to enrich the semantic information [14]. For instance, small-scale feature maps are utilized for large object detection, whereas large-scale feature maps are employed for small objects. Integration of different scale feature maps boosts model performance in detecting multi-scale objects accurately. Consequently, FPN improves the model's performance in detecting multi-scale objects, and SSD satisfies quick response needs. The model's loss function is the weighted sum of localization loss, calculated through Smooth L1, and the confidence loss, calculated by focal loss [24].

12.3.4   YOLO v5s

YOLO (You Only Look Once) is a popular object detection model developed by Joseph Redmon and Ali Farhadi at the University of Washington and is known for its high accuracy and speed. YOLO v5 is an extension of YOLO v3 [25] and comes in five different sizes; nano, small, medium, large, and extra-large, of CNN depth used in the model in ascending order [22]. We have used YOLO v5s (small) and the model generates weights with only 14MB, 7.25 M model parameters and 16.5 billion floating-point operations per second (GFLOPs) which leads to fast inferencing speeds allowing quick simulation, a desired feature in automated defect detection in real-time applications. However recent versions of YOLO v5, for instance YOLO v7 demands 37.2 million parameters requires higher number of model parameters, leading to YOLO v7 taking more than twice as much time to train as YOLO v5 with marginally improved performance (less than 3%) [26]. The YOLO vision models architecture comprises of three main components, known as backbone, neck, and head. Backbone extracts features from the images at the minutest level of detail giving it to the neck, where a series of these features are mixed and combined to forward to the head for predicting the classes and the bounding boxes. The backbone of the model is Cross Stage Partial Network (CSPNet) [27], and as a model neck, Path Aggregation Network (PANet) developed by Liu et al. [28] was used. The neck of the model also generates feature pyramids that consist of feature maps of different scales (similar to Mobile-Net SSD with FPN), which enhances the capability of the model to detect objects of various sizes.

YOLO v5s performs online augmentation of three types, scaling, color space adjustments and mosaic augmentation to provide a greater and richer variation while training the model [22]. Mosaic augmentation is especially useful for the small object detection, as it addresses the issue where small objects are not as accurately detected as large objects during the transfer learning when performed using COCO (Common Objects in Context) dataset [29, 30]. The auto learning bounding box anchor feature introduced in YOLO v3 is used in YOLO v5 as well, that customizes distribution of bounding box sizes and locations based on custom dataset rather than preset COCO dataset. This helps improve precision in localizing the object, especially for odd sized objects. YOLO v5 can also be easily implemented on a mobile device given its easy installation, fast training, and intuitive data file system structure.

12.3.5   Data Pre-Processing and Experimental Settings

Working with a small dataset can cause low prediction performance since a single wrong prediction greatly influences overall accuracy. Therefore, we created five different partitions of the original dataset, as shown in Figure 12.2, keeping the 80%-20% train-test ratio. In each partition, we kept 20% of defects as the test set and used the remaining 80% of defects for the training set. Then, we trained five models and calculated the mean accuracy of all five testing sets to compare three algorithms.

For MobileNet-SSD with FPN and Faster R-CNN ResNet101, pre-trained base models on the COCO dataset inside TensorFlow Object Detection API (application program interface) are employed. For Faster R-CNN ResNet101, batch size, learning rate, and total steps are set to 4, 0.04, and 13000, respectively. For MobileNet-SSD with FPN, the batch size is 8, the learning rate is 0.04, and the number of steps is 2500. The rest of the parameters are based on default values provided in the TensorFlow pipeline for this model. YOLO v5s model was trained with a batch size of 16 for 200 epochs. For YOLO, with a standard GPU of P100, it took almost 4 hours to train each model, including its evaluation on the test dataset. The transfer learning is performed using COCO dataset while training YOLO v5s with the industrial dataset. Google Colab Pro was used to run all the codes written in PyTorch. For a more detailed explanation of the architecture and loss functions used, please refer to Jocher *et al.* [31].

12.3.6   Performance Metric

The initial focus of the study was the accurate detection of cracks in a given image. In this regard, we used the accuracy and mean average accuracy metrics to evaluate the performance of Faster R-CNN ResNet101, MobileNet- SSD with FPN, and YOLO v5s. For the accuracy of each set, we divided the number of true positives (TP) by the ground truth. Higher values of accuracy represent more accurate defect detection. The mean accuracy is the average accuracy of all sets.

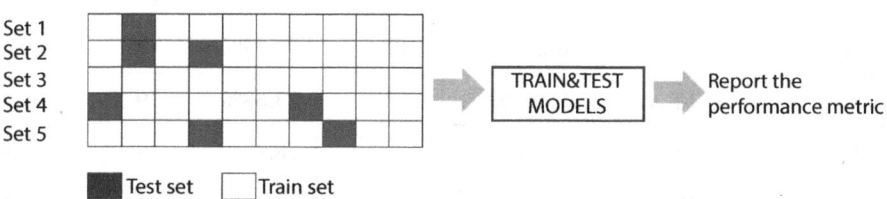

Figure 12.2 Random sampling procedure.

12.4   Results

This section presents the performance comparison of three ML algorithms, MobileNet-SSD with FPN, Faster R- CNN ResNet101, and YOLO v5s, in detecting cracks on cylinder head surfaces.

The accuracy of the Faster R-CNN ResNet101, MobileNet-SSD with FPN, and YOLO v5s for crack images are presented in Table 12.1. The results indicate that YOLO v5s outperforms the other two models with a mean accuracy of 96.39%. There were a few miss detections for YOLO v5s in the first two experimental sets, and 100% accuracy was achieved in the last three sets. Faster R-CNN ResNet101 achieved 80.62% mean accuracy and MobileNet-SSD with FPN obtained the poorest accuracy, 59.79%. For MobileNet-SSD with FPN and Faster R-CNN ResNet101, lower accuracy was related to missed detections. This occurred due to the variation in the shape of the cracks. There were a few unusual type cracks, which were bigger than most of the crack images fed into the model. These unusual cracks caused difficulties in extracting the features for the models.

Figures 12.3-12.4 present sample prediction results of MobileNet-SSD with FPN. In Figure 12.3(a), MobileNet-SSD with FPN can recognize small cracks in sharp and high-resolution images acquired in the laboratory setting at 62% confidence. However, the model missed detecting bigger cracks in blur images collected in the factory setting when there is a long distance between the camera and the object, as in Figure 12.3(b).

In Figure 12.4(a), cracks in sharp images with low contrast could be detected by MobileNet-SSD with FPN above 65% confidence. On the other hand, unusual crack images that are relatively bigger and do not resemble the rest of the crack images were missed by MobileNet-SSD with FPN, as in Figure 12.4(b).

Sample prediction results for Faster R-CNN ResNet101 are demonstrated in Figure 12.5. In Figure 12.5, Faster R- CNN ResNet101 detected small cracks at high confidence, above 94%. However, similar to MobileNet-SSD with FPN, Faster R-CNN ResNet101 often failed to detect unusual defects, as exemplified in Figure 12.5(b). In Figure 12.5(b), Faster R-CNN ResNet101 could detect the crack on the left; however, could not detect the one on the right.

Table 12.1 Accuracy comparison of three models.

SETS	Total Labels (Ground truth)	MobileNet-SSD with FPN		Faster R-CNN ResNet101		YOLO v5s	
		TP	Accuracy	TP	Accuracy	TP	Accuracy
Set 1	16	8	50.00%	16	100.00%	14	87.50%
Set 2	18	12	66.67%	14	77.77%	17	94.44%
Set 3	19	14	73.68%	14	73.68%	19	100.00%
Set 4	19	12	63.16%	15	78.94%	19	100.00%
Set 5	22	10	45.45%	16	72.7%	22	100.00%
		Average	59.79%	Average	80.62%	Average	96.39%

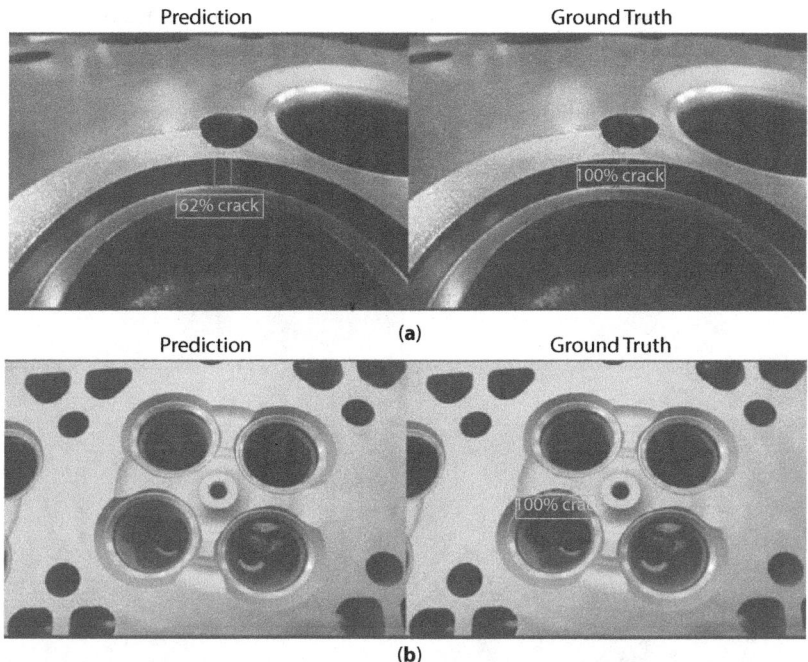

Figure 12.3 MobileNet-SSD with FPN performance in blur vs. sharp images.

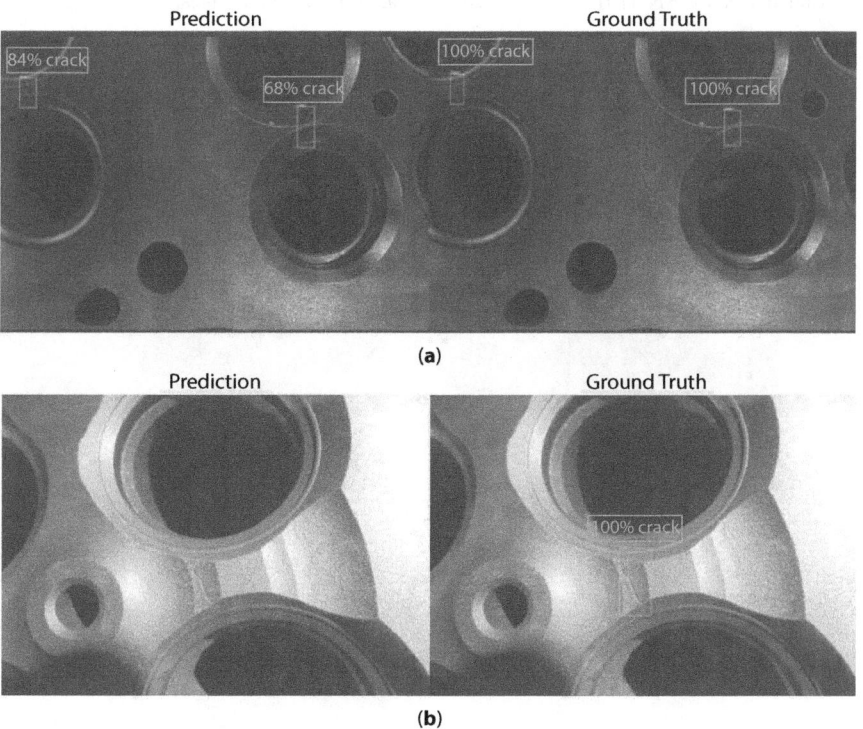

Figure 12.4 MobileNet-SSD with FPN performance in sharp images.

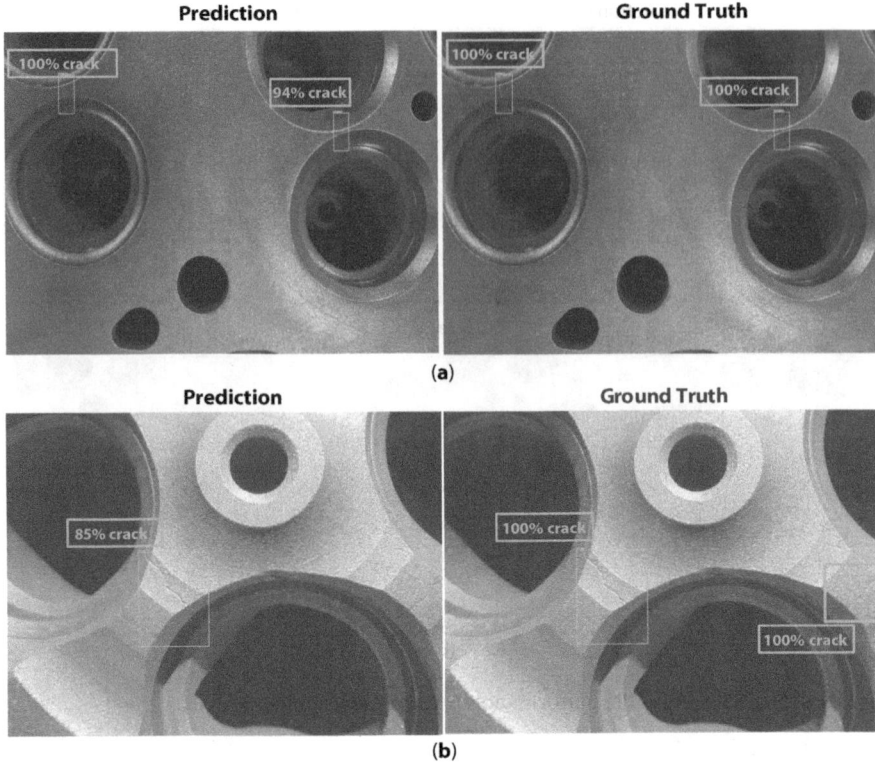

Figure 12.5 Faster R-CNN ResNet101 performance in usual and unusual images.

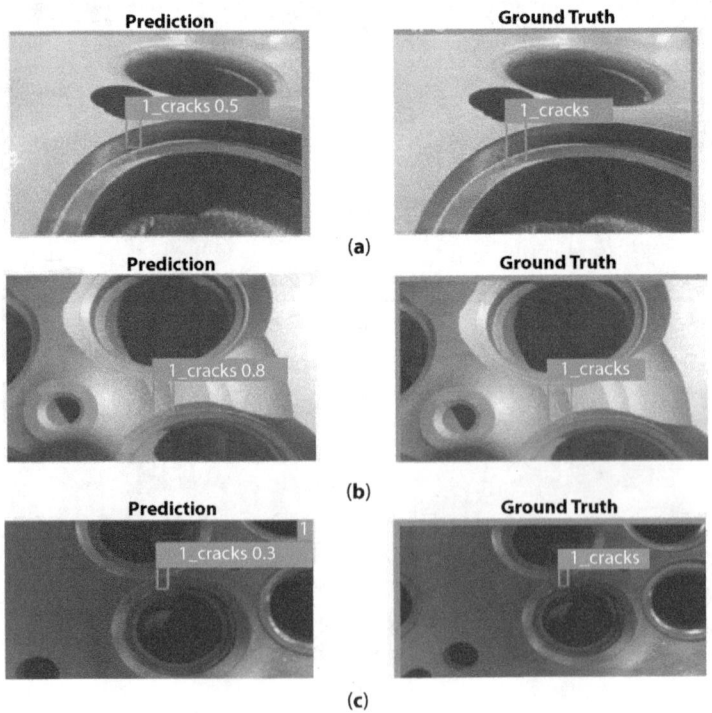

Figure 12.6 YOLO v5s predictions vs. ground truth.

Figure 12.6 shows various prediction results for YOLO v5s. The model could detect the small size cracks even in the blurred image at 50% confidence, as seen in Figure 12.6(a). As opposed to MobileNet and Faster R-CNN ResNet101, YOLO v5s could detect unusual defects at high confidence (80%), as seen in Figure 12.6(b). This can be due to the customization of the anchor boxes per the industrial dataset in YOLO v5s. YOLO v5s was triggered to create custom anchor boxes for objects and introduced the idea of learning anchor boxes based on the distribution of bounding boxes in the custom dataset with K-means and genetic algorithms. When there is a high disparity between the distribution of bounding box sizes and the preset bounding box anchors in the COCO dataset, the customization process may significantly impact the accurate detection of defects. Figure 12.6(c) exemplifies the circumstance that even a human inspector might have difficulty locating the defect; however, YOLO v5s could accurately detect the crack at 80% confidence.

12.5   Conclusion and Future Work

In this study, we evaluated the performance of three deep learning object detection models, Faster R-CNN ResNet101, MobileNet-SSD with FPN, and YOLO v5s using industry data to detect cracks on the surface of cylinder heads. YOLO v5s outperforms crack detection with an average accuracy of 96.39% for cracks. Faster R-CNN ResNet101 is superior to the MobileNet-SSD with FPN an average accuracy of 80.62%. Detection of unusual cracks was difficult for three algorithms, especially for MobileNet-SSD with FPN. However, despite the high variation in crack shapes, YOLO v5s achieved higher accuracy in detecting unusual cracks. These results clearly demonstrate that YOLO v5s is capable of practical crack detection applications.

In deep learning, the performance of the algorithms usually improves with a high volume of data set. Therefore, we plan to apply data augmentation techniques in a future study to increase the total number of images contained within a dataset. Additionally, we aim to investigate the impacts of factors such as light, camera-object distance, and resolution on defect detection by comparing two data collection platforms (the laboratory setting, where factors are in control, versus the factory setting, where there is high noise in data collection).

Acknowledgements

This material is based upon the work supported by the U.S. Department of Energy's Office of Energy Efficiency and Renewable Energy (EERE) under the Advanced Manufacturing Office Award Number DEEE0007897–RM05.

References

1. Paterson, D.A., Ijomah, W.L. and Windmill, J.F., End-of-life decision tool with emphasis on remanufacturing. *Journal of Cleaner Production*, 148, 653-664, 2017.
2. Mital, A., Govindaraju, M. and Subramani, B., A comparison between manual and hybrid methods in parts inspection. *Integrated Manufacturing Systems*, 9(6), 344-349, 1998.

3. Errington, M. and Childe, S.J., A business process model of inspection in remanufacturing. *Journal of Remanufacturing*, 3(1), 1-22, 2013.

4. Hammond, R., Amezquita, T. and Bras, B., Issues in the automotive parts remanufacturing industry: a discussion of results from surveys performed among remanufacturers. *Engineering Design and Automation*, 4, 27-46, 1998.

5. Kujawińska, A. and Vogt, K., Human factors in visual quality control. *Management and Production Engineering Review*, 6, 25-31, 2015.

6. Drury, C.G., Karwan, M.H. and Vanderwarker, D.R., The two-inspector problem. *IIE Transactions*, 18(2), 174-181, 1986.

7. Dubey, P., Jackman, J., Kremer, G.E., Kremer, P. (2023). A Probabilistic Model to Estimate Automated and Manual Visual Inspection Errors. *The Human-Data-Technology Nexus*. FAIM 2022. Lecture Notes in Mechanical Engineering. Springer, Cham. https://doi.org/10.1007/978-3-031-17629-6_72

8. Cha, Y.J., Choi, W. and Büyüköztürk, O., Deep learning-based crack damage detection using convolutional neural networks. *Computer-Aided Civil and Infrastructure Engineering*, 32(5), 361-378, 2017.

9. LeCun, Y., Bengio, Y. and Hinton, G., Deep learning. *Nature 521*, 436–444 (2015). https://doi.org/10.1038/nature14539

10. He, K., Zhang, X., Ren, S. and Sun, J., Spatial pyramid pooling in deep convolutional networks for visual recognition. *IEEE Transactions on Pattern Analysis and Machine Intelligence*, 37(9), 1904-1916, 2015.

11. Howard, A.G., Zhu, M., Chen, B., Kalenichenko, D., Wang, W., Weyand, T., Andreetto, M. and Adam, H., Mobilenets: Efficient convolutional neural networks for mobile vision applications. arXiv preprint arXiv:1704.04861, 2017.

12. Meng, J., Jiang, P., Wang, J., Wang, K., A MobileNet-SSD Model with FPN for Waste Detection. *Journal of Electrical Engineering & Technology*, 17(2), 1425-1431, 2022.

13. Liu, W., Anguelov, D., Erhan, D., Szegedy, C., Reed, S., Fu, C.Y. and Berg, A.C., SSD: Single shot multibox detector. In *European conference on computer vision*, 21-37, Springer, Cham., October 2016.

14. Lin, T.Y., Dollár, P., Girshick, R., He, K., Hariharan, B. and Belongie, S., Feature pyramid networks for object detection. In *Proceedings of the IEEE Conference on Computer Vision and Pattern Recognition*, 2117-2125, 2017.

15. Dubey, P., Günay, E.E., Jackman, J., Kremer, G.E., Kremer, P. (2023). Deep Learning-Powered Visual Inspection Using SSD Mobile Net V1 with FPN. *The Human-Data-Technology Nexus*. FAIM 2022. Lecture Notes in Mechanical Engineering. Springer, Cham. https://doi.org/10.1007/978-3-031-17629-6_78

16. Redmon, J., Divvala, S., Girshick, R. and Farhadi, A., You only look once: Unified, real-time object detection. In *Proceedings of the IEEE Conference on Computer Vision and Pattern Recognition*, 779-788, 2016.

17. Baumgartl, H., Tomas, J., Buettner, R. and Merkel, M., A deep learning-based model for defect detection in laser-powder bed fusion using *in-situ* thermographic monitoring. *Progress in Additive Manufacturing*, 5(3), 277-285, 2020.

18. Redmon, J. and Farhadi, A., YOLO9000: Better, Faster, Stronger. In *Proceedings of the IEEE Conference on Computer Vision and Pattern Recognition*, 7263-7271, 2017.

19. Girshick, R., Donahue, J., Darrell, T. and Malik, J. Rich feature hierarchies for accurate object detection and semantic segmentation. In *Proceedings of the IEEE Conference on Computer Vision and Pattern Recognition*, 580-587, 2014.

20. Girshick, R., Fast R-CNN., *IEEE Int. Conf. Comput. Vis. Santiago*, Chile, 7-13, December 2015.

21. Ren, S., He, K., Girshick, R., and Sun, J., Faster R-CNN: Towards realtime object detection with region proposal networks, in *Proc. NIPS*, 91–99, 2015.

22. Jocher G., Chaurasia A., Stoken A., Borovec J., *et al.* (2022). ultralytics/yolov5: v7.0 - YOLOv5 SOTA Realtime Instance Segmentation (v7.0). Zenodo. https://doi.org/10.5281/zenodo.7347926

23. Kremer, P., White, B., Kremer, G., Jackman, J., Gunay, E., Dubey, P., Ahmed, N., Blamey, S., Barger, J. and Chong, J., A Ruggedness Test Approach to the Design of An Automated Defect Inspection System. *REMADE Circular Economy Tech Summit and Conference,* March 2023.

24. Lin, T.Y., Dollár, P., Girshick, R., He, K., Hariharan, B. and Belongie, S., Feature pyramid networks for object detection. In *Proceedings of the IEEE Conference on Computer Vision and Pattern Recognition*, 2117-2125, 2017.

25. Lin, T.Y., Goyal, P., Girshick, R., He, K. and Dollár, P., Focal loss for dense object detection. In *Proceedings of the IEEE International Conference on Computer Vision*, 2980-2988, 2017.

26. Lin, T.Y., Maire, M., Belongie, S., Hays, J., Perona, P., Ramanan, D., Dollár, P. and Zitnick, C.L., Microsoft coco: Common objects in context. In *European Conference on Computer Vision* 740-755, Springer, Cham., September 2014.

27. Redmon, J. and Farhadi, A., YOLO9000: Better, Faster, Stronger. *In Proceedings of the IEEE Conference on Computer Vision and Pattern Recognition*, 7263-7271, 2017.

28. Redmon, J. and Farhadi, A., Yolov3: An incremental improvement. arXiv preprint arXiv:1804.02767, 2018.

29. Wang C. Y., Bochkovskiy A., and Liao H. Y. M. YOLOv7: Trainable bag-of-freebies sets new state-of-the-art for real-time object detectors Institute of Information Science, Academia Sinica, Taiwan, arXiv:2207.02696

30. Kisantal, M., Wojna, Z., Murawski, J., Naruniec, J. and Cho, K., Augmentation for small object detection. arXiv preprint arXiv:1902.07296, 2019.

31. Jocher, G., Chaurasia, A., Stoken, A., Borovec, J., Kwon, Y., Michael, K., and Fang, J., "ultralytics/yolov5: v6.2-yolov5 classification models, apple m1, reproducibility, clearml and deci. ai integrations." Zenodo. org, 2022.

Part 3

DESIGN FOR CIRCULARITY

13

Calculator for Sustainable Tradeoff Optimization in Multi-Generational Product Family Development Considering Re-X Performances

Michael Saidani[1]*, Xinyang Liu[1], Dylan Huey[1], Harrison Kim[1†], Pingfeng Wang[1], Atefeh Anisi[2], Gul Kremer[3], Andrew Greenlee[4] and Troy Shannon[4]

[1]University of Illinois Urbana-Champaign, Champaign, IL, USA
[2]Iowa State University, Ames, IA, USA
[3]University of Dayton, Dayton, OH, USA
[4]Deere and Company, Moline, IL, USA

Abstract

Design for Re-X (reuse, remanufacturing, recycling) in a multi-generational product family setup is a challenging task that requires new and more advanced, quantitative, and user-friendly tools to ensure environmental savings and economic profitability for product makers. Through this REMADE-funded project, state-of-the-art Re-X, reliability, and lifecycle-based models are combined within a new design tool in order to generate and compare designs for Re-X alternatives. This newly developed Calculator aims to assess the potential of reliability-informed analysis of long-term benefits/costs of Re-X in product family design to enable substantial reductions in embodied energy and carbon emissions. To do so, three integrated modules are working together: (i) the LCA/LCC module; (ii) the reliability module; and (iii), the Re-X policy optimization module. Notably, the lifecycle-based module allows testing and comparing the impact of different material alternatives, part designs, and sub-assemblies. In addition, the Re-X and reliability analysis modules enable to compare and optimize different designs based on, e.g., material composition, the percentage of remanufactured components, the energy consumption, or the remaining useful life of critical components. In particular, for the reliability module, the inputs are the following: the shape/scale from reliability info or estimated using warranty claims, the warranty length (provided from the product or part information), the total number of new and used parts, and the weight (initial weight of product/subassembly being evaluated). The outputs enable the user to estimate the additional units for warranty claims, the reuse rate, and the adjusted weight. The Re-X policy optimization module includes: the bi-objective function (as of now, minimize cost and energy consumption), the Pareto front exploration, the impact of component reliability, the impact of warranty length, the impact of return rate, and the impact of remanufacturing cost. This new Re-X Calculator is being tested, fine-tuned, and validated based on two complementary product family categories (utility tractor product family design problem and mobile phone design and recovery problem) provided by the industrial partners, John Deere and Global Electronics Council.

**Corresponding author*: msaidani@illinois.edu
†*Corresponding author*: hmkim@illinois.edu

Nabil Nasr (ed.) Technology Innovation for the Circular Economy: Recycling, Remanufacturing, Design, Systems Analysis and Logistics, (159–170) © 2024 Scrivener Publishing LLC

Keywords: Circular economy, remanufacturing, life cycle assessment, product sustainability, product reliability, design performance, impact quantification, computation tool

13.1   Introduction

13.1.1   Context and Motivations

Measuring the potential ecological savings and quantifying financial benefits associated with product recovery and remanufacturing is a challenge due to variable market conditions and many uncertainties at the end-of-use of product systems [1]. The lack of design tools that can quantify both the environmental savings potential and financial benefits of design for Re-X solutions, as well as their impact on component reliability, during the product design and development (PDD) process is a significant barrier to overcoming this challenge in the context of product family design [1]. For instance, Table 13.1 illustrates how different models and tools available in the literature are only considering a subset of the following elements: (i) Re-X strategies, (ii) reliability analysis, (iii) life cycle impact assessment, (iv) economic profitability, and (v) product family design optimization.

To make robust design for Re-X decisions considering post-design uncertainties and maximizing the benefits of Re-X for a family of products both from financial and environmental perspectives, new design theory and tools that concurrently consider the interdependencies of different design principles, their impacts on long-term techno-economic

Table 13.1 Illustration of the contributions and limitations of related models and tools available in the literature.

References	Re-X strategies	Reliability analysis	Environmental impact	Economic profitability	Product family optimization
[2]	x	x		x	
[3]				x	x
[4]	x	x		x	
[5]	x		x	x	
[6]	x		x		x
[7]		x		x	
[8]			x	x	x
[9]	x	x	x		
[10]	x		x		
Newly developed Calculator	x	x	x	x	x

'x' indicates that this aspect is included.

benefits/costs for Re-X options, and uncertainties at the product operating stage (e.g., market demands) must be created.

This task is motivated by the fact that existing tools for reliability analysis lack the ability to consider Re-X options after the occurrence of product failures or replacements prior to failures. There is a need for the integration of Re-X options with existing reliability analysis tools to maximize the Re-X benefits. In this line, Figure 13.1 illustrates the workflow and the building blocks of the newly developed Calculator.

13.1.2   Research Objectives and Industrial Relevance

The objectives are to develop fundamental models and new design tools with capabilities of generating and comparing designs for Re-X alternatives considering economic profitability and environmental impact savings. The specifics of the research objectives are to (i) identify design for reliability process factors that are interdependent with Re-X options, thus establish models for the interdependencies, (ii) integrate these interdependence models with existing reliability analysis tools so that new analysis tools could take into account Re-X options in design for reliability, (iii) create a decision support system for the optimization of product family design considering reliability and Re-X options concurrently under the objectives of environmental impact saving and economic profitability, and (iv) take into account the uncertainties resulted from post design activities so that robust design tradeoff decisions can be made.

First, in this paper, new knowledge is created for a better understanding of the interdependencies between product reliability and Re-X options in the context of product family design under uncertainty. Then, a newly developed Calculator, to be readily used by designers, material engineers, and product leads, is detailed and showcased for the analysis of longer-term techno-economic benefits/costs tradeoffs, therefore, maximizing Re-X benefits at the product design phase. Overall, this project aims to contribute to the REMADE goals of: (i) developing technologies capable of achieving cost parity for secondary materials,

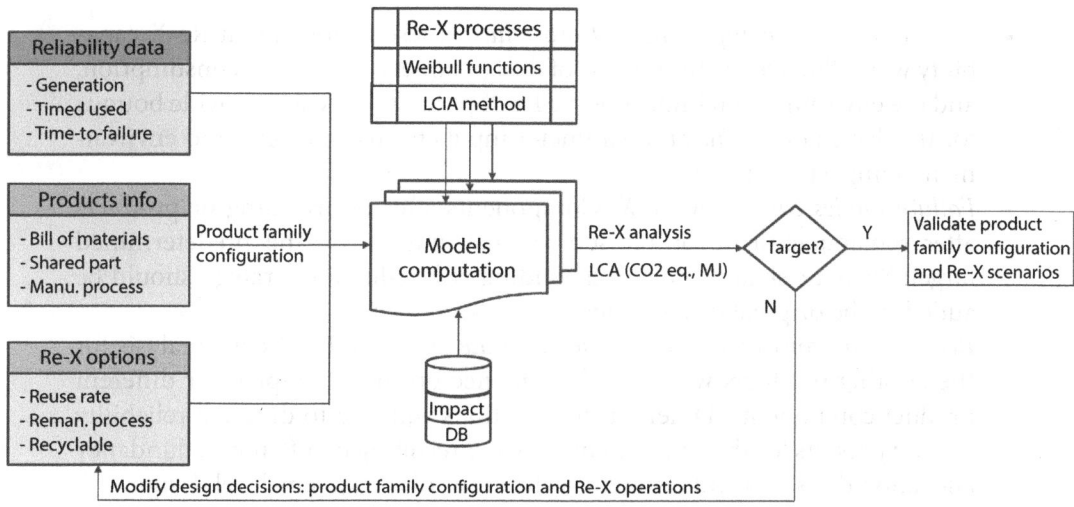

Figure 13.1 Workflow and building blocks of the proposed calculator.

(ii) developing manufacturing processes that enable cross-industry reuse of recycled feedstock, and (iii) developing technologies capable of achieving a reduction in primary material feedstock and reduction in greenhouse gases.

13.2   Design for Reliability Process Review and Re-X Interdependence Identification

Reliability is generally treated as a performance target to be met through product design, whereas the costs and benefits of Re-X after product failure or maintenance replacement are not systematically considered. As such, the identification of the interdependent factors would enable the development of models that help designers take into account the Re-X options in the product family design process. Here, a thorough review of the design for the reliability process is performed with the aim of identifying key interdependence factors between reliability and Re-X options. This task starts with the well-established design for reliability processes without any consideration of Re-X options as the baseline model and then incorporates Re-X options into the design for reliability process models.

Six critical steps exist in the design for the reliability process [11, 12], from identification of the reliability requirements and product goals to process control. In this task, the project team reviewed the purposes and methods in each step to see how Re-X options could make a difference in the process [13–15]. In practice, the interdependence models between reliability and Re-X options will be used for the development of more advanced reliability analysis and design tools (e.g., stand-alone software that can be run in Excel; add-on module to Excel) considering Re-X options.

While Table 13.2 summarizes all the potential interdependencies between reliability and Re-X options, the research team further analyzed the key factors that can be modeled to incorporate the Re-X conditions and decisions into the design for the reliability process. Below is the discussion of key interdependence factors and how we plan to model and integrate them into the Calculator:

- *Re-X technology, equipment, and procedure*: These factors about Re-X capability will influence the fulfillment of design targets, the energy consumption, and the environmental impact of the Re-X process. They will provide bounds for the design decisions and parameter inputs for the financial and environmental impact evaluation.
- *Failure modes caused by Re-X*: Components after refurbishing or products after remanufacturing or reassembly may not satisfy the predetermined target. Such a risk and the corresponding risk reduction strategy should be added to the original design stage.
- *Re-X options for different components in the products*: Field data analysis for the existing products will provide guidance on the Re-X plan for different product components. Different Re-X options will lead to different reliability recovery results for the components, further resulting in different redundancy allocation decisions and different realizations of the system-level design.

Table 13.2 Summary of the interdependence factors between design for reliability and design for Re-X.

DfR Process	Interdependence Factors	Notes
1-Identify (Identify reliability requirements, product goals and translation from customer requirements to functional requirements.)	Reliability target for the Re-X result	Whether to refurbish an assembly or remanufacture a product, the target of the Re-X results will influence the design.
	Environmental conditions (e.g., environmental evaluation target)	Metrics selected to evaluate the environmental impact should be determined and will influence the design and verification process.
	Re-X technology, equipment, and procedure	Re-X capabilities will influence the component-level reliability and the redundancy allocation to satisfy the system-level target.
2-Design (Understand what changes are brought to the current design, identify and prioritize the key reliability risk items)	Failure modes caused by Re-X	Additional manufacturing steps in Re-X (e.g., inspection, cleaning, recovery) require supplementary failure modes analysis.
	Quality of returned cores	Quality of returned cores will influence the Re-X result and the environmental impact.
	Property of recycled materials	The material property will influence how a design can satisfy the reliability target.
	Supply of Re-X resources	Supplier, quantity, and lead time of the Re-X resources will add potential risk in the Re-X process.
3-Analysis (A rough estimation of product reliability)	Re-X options for different components in the products	What procedure is to be applied to a certain component or assembly to which degree is a decision to be analyzed.
	Cost, energy consumption and emission of different Re-X procedures	These facts of the Re-X procedures should be provided for the system-level estimation.
	Maintenance activities or options during the product service life	Maintenance activities will influence the time-dependent reliability and the expected warranty cost during the product service life.

(Continued)

Table 13.2 Summary of the interdependence factors between design for reliability and design for Re-X. (*Continued*)

DfR Process	Interdependence Factors	Notes
4-Verify (An iterative process where different types of tests are performed, the results are analyzed, and design changes are made)	Test on effects of different maintenance activities	Testing if the impact of maintenance activities will verify the correctness or improve the post-design modeling method.
	Different allocations of product and product family	Re-X may influence the module-sharing or component-sharing decisions.
5-Validate and 6-Control (Ensure the manufacturing process does not deviate from the specifications, continuous monitoring and field data analysis)	Warranty policy and records	Different warranty policies may be selected for new products and remade products
	Transportation and distribution of materials, components, returned cores and products	The reverse supply chain due to Re-X will influence the control process.

- *Maintenance activities and warranty policies during the product service life*: One of the critical Re-X options is repair during the service life, which will influence the degradation process, the condition of remanufacturing resources, the lifecycle cost, and environmental impact. Thus, a new Weibull analysis tool enabling the Re-X data to be incorporated into the Weibull analysis process for system reliability analysis needs to be developed.

13.3　Integrated Tool for Quantifying Reliability and Re-X Performances During Product Design

13.3.1　Working Principles of the Reliability, Re-X, LCA and LCC Modules

Figure 13.2 shows the detailed working flow and links between the reliability, Re-X and lifecycle modules of the proposed Calculator to assess the potential of reliability-informed analysis of long-term benefits/costs of Re-X in product family design to enable reductions in embodied energy and carbon emissions.

In the first step, through the reliability module, lifetime and warranty data are used to execute the warranty process simulation in order to generate the required Re-X quantities for maintaining and/or remanufacturing a given product at a parts and components level. Then, using the Re-X policy optimization module, the users can input the remaining useful

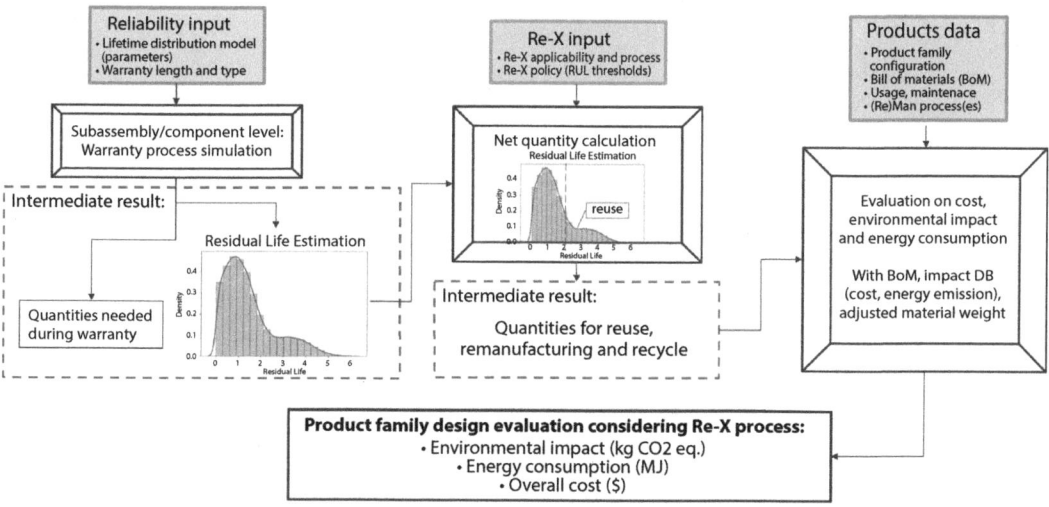

Figure 13.2 Detailed workflow of the calculator, with links between the modules.

life (RUL) threshold, and simultaneously use Re-X and product data inputs to calculate the costs, the environmental impacts and the energy consumption of multi-lifecycle products, through the life cycle assessment (LCA) and life cycle costing (LCC) modules.

In all, coupling Re-X and reliability analyses with environmental and economic life cycle assessment capabilities, enables the user to compare and optimize different designs, based on, e.g., material composition, the percentage of recycled/remanufactured components, the energy consumption, or the RUL of critical components. For instance, Figure 13.3 is a snapshot of the reliability module. The inputs are the following: (i) shape/scale (from reliability info or estimated using warranty claims); (ii) warranty length (provided from a product or part information); (iii) total number of parts (number of both new and used parts); (iv) used parts (proportion of parts that were not new); and, (i) weight (initial weight of product/subassembly being evaluated). The outputs are the following: (i) additional units for warranty claims; (ii) reuse rate; and, (iii) adjusted weight.

For illustrative purposes, the method has been tested on a benchmark problem, where we optimize the Re-X thresholds for one product component by considering the demand and usage process uncertainty. A fixed Re-X policy is obtained based on a static demand profile, while the optimized policy is obtained by considering the changed demand levels in future stages, as illustrated in Figure 13.4.

13.3.2   Application

A real-world industrial sub-assembly – the axle illustrated in Figure 13.5 – that can be assembled and operational on different heavy-duty agricultural machines has been used first to demonstrate the capabilities and practicality of the lifecycle and Re-X modules of the developed Calculator. While the illustrative case study showcased in this paper focused on a single product/component design/redesign, the final version of the Calculator will be applied in the context of product family optimization. In fact, this axle is a key and common component of John Deere tractors, loaders, dozers, and excavators. This axle is made of

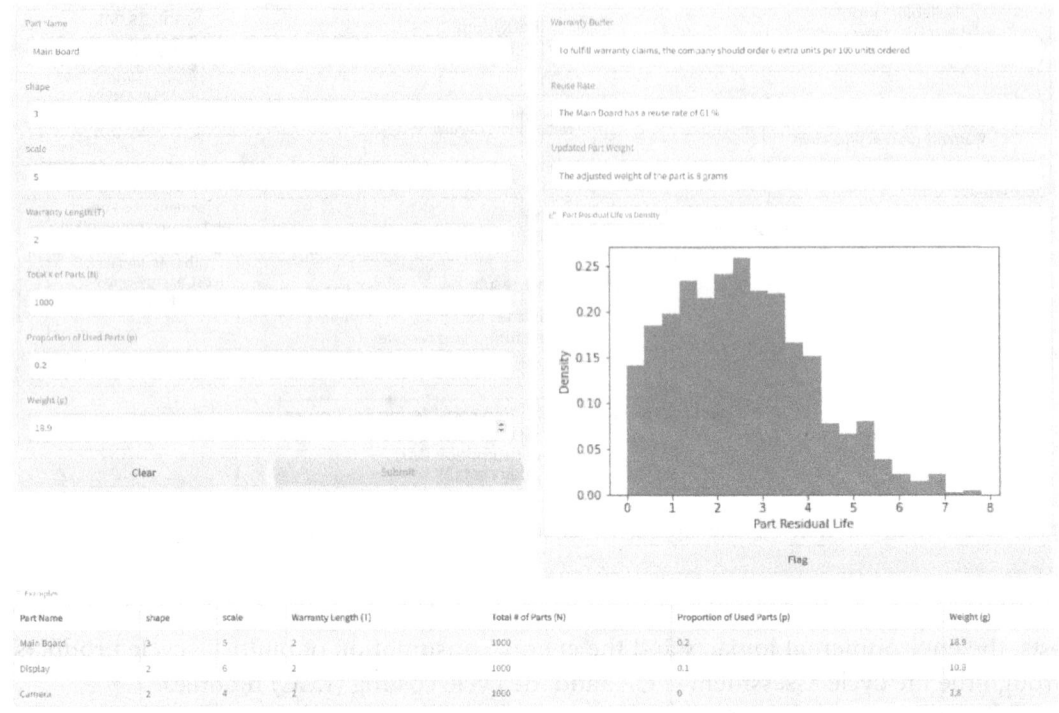

Figure 13.3 Reliability module.

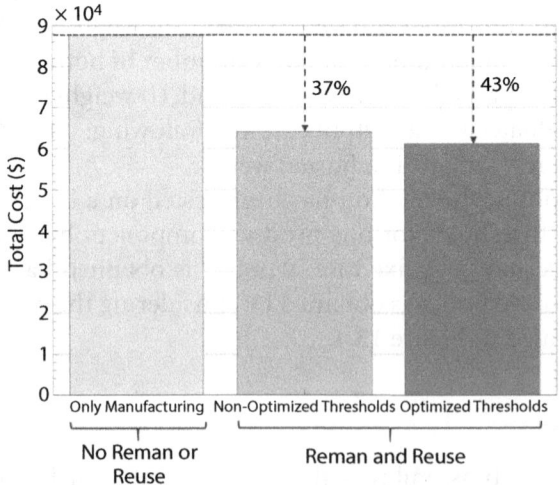

Figure 13.4 Example of results: manufacturing only vs (non-)optimized threshold for Re-X options.

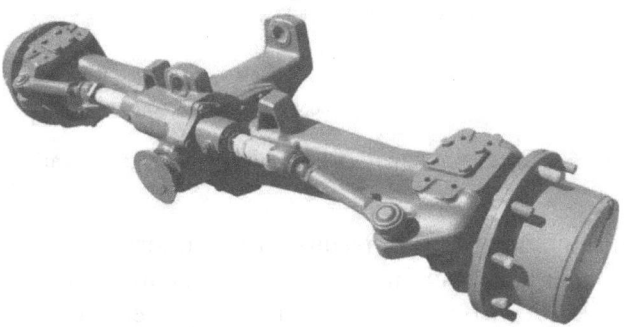

Figure 13.5 Axle.

more than 100 unique parts for a total of approximately 1,200 kg. Interestingly, around 50% of the parts can be reused/remanufactured. This corresponds to the parts with a reuse rate superior to 0, as shown in Table 13.3. The exact numbers are not displayed in this paper for confidentiality reasons. Also, note that while accurate lifecycle information – such as the bill of materials, manufacturing processes, replacement and reuse rates – have been used

Table 13.3 Illustration of carbon footprint savings on reused/remanufactured parts.

Parts #	Carbon intensity of material extraction	Carbon intensity of manufacturing process	Total carbon footprint for brand new (kg CO2 eq.)	Replacement rate	Reuse rate	Carbon footprint savings (kg CO2 eq.)
19M###1	2.7	1.4	4.1	0.1	0.9	2.6
19M###2	3.1	1.9	5.0	0.1	0.9	2.5

Replacement implies brand-new parts; Reuse implies cleaning and remanufacturing.
Note that all the above numbers are slightly altered for confidentiality purposes.

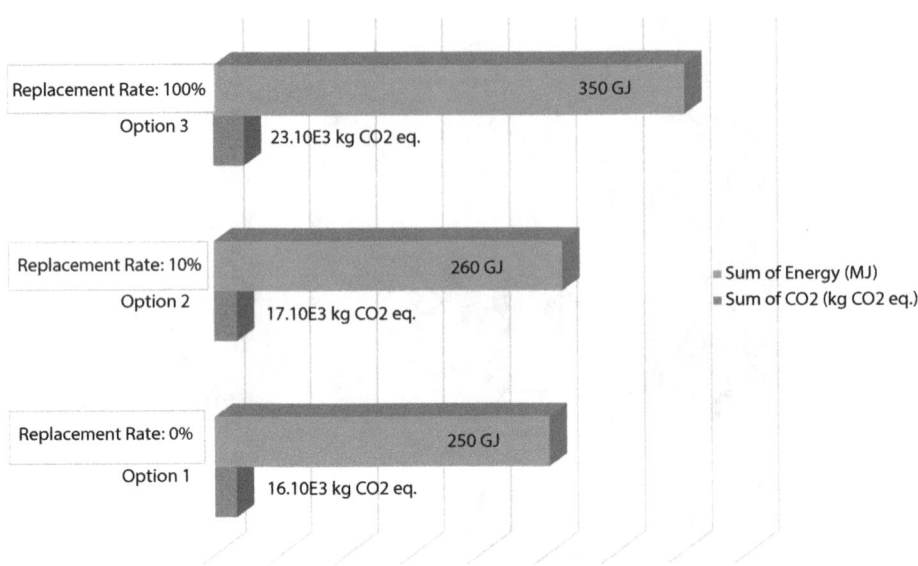

Figure 13.6 Comparison of different Re-X scenarios and associated energy and carbon impact.

to ensure the relevance of the generated results for the industrial partner, the results show-cased in the present paper have been modified for confidentiality reasons.

While the Calculator allows one to quickly quantify the environmental savings (energy consumption and carbon footprint) of a remanufactured product with its actual reuse rates, it also has the capability to compare different Re-X scenarios, e.g., when replacement rates as altered, as illustrated in Figure 13.6. It also allows the user to rapidly identify which sub-assemblies, parts, components, or materials have the highest impact (e.g., case differential, gear) to further research Re-X optimization on these critical items.

13.4   Conclusion and Perspectives

The proposed product family design platform ultimately aims to facilitate robust design decision-making that enables cost-effective Re-X at end-of-life. Such improvements will come from: (i) reliability-informed analysis of long-term benefits/costs of Re-X in product family design (i.e., better quantifying the long-term values of Re-X at the design stage), and (ii) robust design decision-making on long-term techno-economic tradeoffs (i.e., promoting Re-X through product family design integrations). The next step, as the end-state technology validation phase, is to experiment and validate the newly developed Calculator that jointly considers product reliability and Re-X options with compact utility tractor product family design problem and recovery problem provided by the industry partner John Deere.

Acknowledgments

The authors would like to thank Dr. Mike Haselkorn, senior project manager at the REMADE Institute, and Prof. Bert Bras, Professor of Mechanical Engineering at the Georgia Institute of Technology, for their continuous support and constructive feedback all along this project. This material is partially based upon the work supported by Deere and Company. Any opinions, findings, conclusions, or recommendations expressed in this publication are those of the authors and do not necessarily reflect the views of the sponsor.

References

1. Liu, X., Wang, P., & Kim, H. (2022). Multi-Generational Product Family Design for Reliability and Environmental Sustainability, *Proceedings of the IISE Annual Conference & Expo 2022*, K. Ellis, W. Ferrell, J. Knapp, eds.
2. Jiang, Z., Zhou, T., Zhang, H., Wang, Y., Cao, H., & Tian, G. (2016). Reliability and cost optimization for remanufacturing process planning. Journal of cleaner production, 135, 1602-1610.
3. Ma, J., & Kim, H. M. (2016). Product family architecture design with predictive, data-driven product family design method. *Research in Engineering Design*, 27(1), 5-21.
4. Yang, S. S., Nasr, N., Ong, S. K., & Nee, A. Y. C. (2016). A holistic decision support tool for remanufacturing: end-of-life (EOL) strategy planning. *Advances in Manufacturing*, 4, 189-201.
5. Saidani, M., Kim, H., Yannou, B., Leroy, Y., & Cluzel, F. (2019). Framing product circularity performance for optimized green profit. In *International Design Engineering Technical Conferences and Computers and Information in Engineering Conference* (Vol. 59223, p. V004T05A022). American Society of Mechanical Engineers.
6. Kim, J., & Kim, H. M. (2020). Impact of generational commonality of short life cycle products in manufacturing and remanufacturing processes. *Journal of Mechanical Design*, 142(12).
7. Gao, F., Cui, S., & Cohen, M. (2021). Performance, Reliability, or Time-to-Market? Innovative Product Development and the Impact of Government Regulation. *Production and Operations Management*, 30(1), 253-275.
8. Kim, J., Saidani, M., & Kim, H. M. (2021). Designing an optimal modular-based product family under intellectual property and sustainability considerations. *Journal of Mechanical Design*, 143(11).
9. Li, M., Nemani, V. P., Liu, J., Lee, M. A., Ahmed, N., Kremer, G. E., & Hu, C. (2021). Reliability-Informed Life Cycle Warranty Cost and Life Cycle Analysis of Newly Manufactured and Remanufactured Units. *Journal of Mechanical Design*, 143(11).
10. Khan, M. A. A., Cárdenas-Barrón, L. E., Treviño-Garza, G., & Céspedes-Mota, A. (2023). Optimal circular economy index policy in a production system with carbon emissions. *Expert Systems with Applications*, 212, 118684.
11. Kapur, K. C., & Pecht, M. (2014). Reliability engineering (Vol. 86). John Wiley & Sons.
12. Crowe, D., & Feinberg, A. (Eds.). (2017). Design for reliability. CRC press.
13. Charter, M., & Gray, C. (2008). Remanufacturing and product design. *International Journal of Product Development*, 6(3-4), 375-392.
14. Nasr, N., Hilton, B., & German, R. (2011). A framework for sustainable production and a strategic approach to a key enabler: remanufacturing. In *Advances in sustainable manufacturing: Proceedings of the 8th Global Conference on Sustainable Manufacturing* (pp. 191-196). Springer Berlin Heidelberg.
15. Bras, B. (2015). Design for remanufacturing processes. *Mechanical engineers' handbook*, 301-328.

14

A Practical Methodology for Developing and Prioritizing Remanufacturing Design Rules

Brian Hilton

Golisano Institute for Sustainability, Rochester Institute of Technology, Rochester, NY, United States

Abstract

A circular economy is designed to maintain a product's resources at their highest value for as long as possible. Value is retained insofar that the product's performance continues to meet user needs. When the product fails, its performance degrades, or the user needs change, it is removed from service. Traditionally, a product at the end of its life is either thrown away or, if possible, recycled. However, studies have shown that the best industrial method for capturing the most value in the product and returning it to a like-new or better condition is remanufacturing. Unfortunately, the potential of remanufacturing as a value-recovery strategy is limited by a product's original design. Current design and engineering tools do not include the inherent knowledge—design for remanufacturing (DfReman) rules—that would allow a designer to create a product that can be easily and consistently remanufactured.

This paper presents research into DfReman and how the research was used to create a framework. First, the paper explains how existing knowledge was collected through an extensive literature review, industry survey, and interviews with remanufacturing practitioners. Data collection yielded over 1,500 DfReman guidelines with significant repetition. The paper then shows how these guidelines were combined and classified into high-level design principles using an affinity-diagramming process. Since each principle will not always apply to every product; the third task was to build a structure that would enable a design team to select the principles that best apply to their specific product. This structure was developed by customizing a Reliability Centered Maintenance (RCM) process and organizing the design principles into a usable framework. This paper concludes by discussing plans to use the framework to develop DfReman tools.

Keywords: Remanufacturing, design, principles, guidelines, framework, RCM

14.1 Introduction

Our planet is currently supporting a record-level 8 billion people, whose consumption and economic activity have accelerated resource consumption and emissions to land, air, and sea to unprecedented levels. Consumption has decreased the available material stocks and emissions have significantly degraded the environment. The need to improve the efficiency of

Email: Brian.Hilton@rit.edu

Nabil Nasr (ed.) Technology Innovation for the Circular Economy: Recycling, Remanufacturing, Design, Systems Analysis and Logistics, (171–182) © 2024 Scrivener Publishing LLC

material resources and to modify our economic activity is clear. Among many proposed solutions, the circular economy is a promising option for transitioning industrial economies towards longer-term sustainable economic systems. [1]

Expanding the remanufacturing industry, a vital component of the circular economy [2], is considered a means of reducing the material consumption and other impacts associated with modern products. [3] By definition, remanufacturing "is a comprehensive and rigorous industrial process by which a previously sold, leased, used, worn, remanufactured, or non-functional product or part is returned to a like-new, same-as-when-new, or better-than-when-new condition from both a quality and performance perspective, through a controlled, reproducible, and sustainable process." [4] Remanufacturing can also lower the cost to the consumer, reduce energy use and greenhouse gas emissions by 79–89 percent, and reduce material consumption by 80–98 percent. [1] A remanufactured product will therefore meet the market performance need at a significantly reduced cost and environmental impact.

Despite the significant and well-recognized benefits of remanufactured products, the remanufacturing industry only accounts for approximately 2 percent of manufacturing's share of U.S. gross domestic product (GDP). [5] Remanufacturing can be a powerful driver for change; however, to capitalize on this change, the industry must solve many of the barriers that are restricting its growth.

In May-June of 2016, the Golisano Institute for Sustainability (GIS) at Rochester Institute of Technology (RIT) surveyed the remanufacturing industry to identify the challenges facing the remanufacturing industry as well as the cross-cutting technologies that can advance the transition to a circular economy. [6] The survey was distributed through industry associations, trade organizations, and existing networks of manufacturing sector contacts, accruing over 120 responses. Half of all survey respondents stated that products generally lack specific features that enable remanufacturing, and over half of this group suggested that this deficit is a major barrier to increasing their remanufacturing business.

Nearly half of the survey respondents identified design for remanufacturing (DfReman) as the most effective way to make a product easier to remanufacture. Current DfReman is limited to an individual company's own knowledge and experience; no definitive standardized set of practices exist that companies can leverage. [7] Additionally, traditional engineering and design tools (enterprise resource planning, product life-cycle management, computer-aided design, cost models, etc.) have no remanufacturing-specific content. DfReman was therefore identified as one of eleven core research themes in the Technology Roadmap for Remanufacturing in the Circular Economy. [7]

This paper intends to contribute new knowledge to improve the remanufacturability of products through more intentional design practices (the essence of DfReman). It is organized as follows: In "Design Principles," We discuss the collection of publicly available remanufacturing design guidelines, and the process used to aggregate and format design guidelines into design principles. Next, in "Design Framework for Remanufacturing," We introduce reliability-centered maintenance (RCM) as a process that can be customized for DfReman, showing how remanufacturing design principles can be aligned with the RCM framework. Finally, in "Conclusions and Recommendations," We summarize these research findings before offering a brief overview of the proposed next steps towards developing design tools that can improve remanufacturing methodology and expand its application.

14.2 Design Principles

Terms like "design principle," "design guideline," and "design heuristic" can be—and often are—used interchangeably in scientific literature. [8] Through an extensive review and consolidation process, the authors [8] define a guideline as a

> context-dependent directive, based on extensive experience and/or empirical evidence, which provides design process direction to increase the chance of reaching a successful solution, and a heuristic as a context-dependent directive, based on intuition, tacit knowledge, or experiential understanding, which provides design-process direction to increase the chance of reaching a satisfactory but not necessarily optimal solution.

To develop the remanufacturing-focused principles discussed in this paper, DfReman guidelines and heuristics were collected from literature through an industry survey and direct interviews with design and remanufacturing practitioners. This collection was then organized and grouped using an affinity-diagramming process and developed into high-level DfReman principles using a function-analysis system technique (FAST) diagramming process. These principles were defined following [8], who state that a principle is a "fundamental rule or law, derived inductively from extensive experience and/or empirical evidence, which provides design process guidance to increase the chance of reaching a successful solution."

14.2.1 Design Guideline and Heuristic Collection

Design guidelines were first collected by conducting a comprehensive study of published literature on remanufacturing design and related topics. The central focus of the literature review was to map out prior research that has been performed in the field of DfReman and identify design guidelines that could potentially be integrated into a DfReman tool. Scientific papers were drawn from the electronic databases Scopus, Semantic Scholar, Science Direct, Web of Science, and Compendex, as well as academic journals known to cover the subject matter, such as the *Journal of Mechanical Design*, the *Journal of Remanufacturing*, and the *Journal of Cleaner Production*.

The following search terms were used:

1. Initial searches used a combination of terms: "remanufactur*" AND "design" AND "guidelines."
2. Search terms were then expanded to areas related to DfReman strategies, including "design for" AND "quality," "modularity," "disassembly," "durability," "viability," "upgradeability," "maintenance," "restoration," "cleaning," "assessment," and "inspection."
3. Other high-level strategies were searched, such as "design for" AND "sustainability," "end-of-life," "circular processes," "the environment," or "eco-design."

Articles, conference papers, theses, and book chapters were considered without any limitation regarding year or source. Only documents in English where considered. This exhaustive literature search was carried out from December 2019 to February 2020.

Given the large amount of papers retrieved, the collection of papers was narrowed down by reviewing the abstracts to determine the relevance of the paper to new product design. This reduced the number of articles from thousands to only 298. Full copies of these articles were then added to Zotero reference management software, and each article was read to determine if they contained specific guidelines to assist a product designer. The majority of these articles addressed other domains of design and remanufacturing, such as mathematical models, cost models, frameworks, case studies, or theoretical application; however, just 79 articles were identified with specific designer guidelines. Finally, through snowballing, 22 more articles were added, increasing the article count to 101 papers. Each guideline was extracted from every paper and built into a large pivot table to enable further data manipulation. Every guideline was represented with no regard for duplication or common theme. This process yielded over 1,500 design guidelines with significant repetition.

To supplement the guidelines from literature, design heuristics were collected from practitioners by surveying the members of the Remanufacturing Industries Council (RIC), a strategic alliance of global manufacturers and remanufacturers, businesses, academia, and industry-specific trade associations supporting the remanufacturing industry. The survey aimed to identify design features that currently make remanufacturing a challenge as well as guidelines or heuristics that could maximize part recovery during remanufacturing. Additionally, design heuristics were collected through a series of virtual interviews with an original equipment manufacturer's (OEM) internal design and remanufacturing teams. Overall, multiple heuristics were identified from both the industry survey and the practitioner interviews that demonstrated the significant value of going beyond published data and engaging directly with design and remanufacturing practitioners.

14.2.2   Sorting Design Guidelines and Heuristics

Many of the guidelines and heuristics discovered during the literature study were insightful, but they did not all fit the following research objectives:

1. Improve remanufacturing through design practices.
2. Address design issues in industry sectors in which OEM companies have a large remanufacturing presence.
3. Integrate design guidelines into design engineering tools.
4. Advance remanufacturing as a distinct resource-recovery methodology.

Because of this, any guidelines that did not meet these research goals were removed according to the following sort criteria:

- To meet the first objective above, specific guidelines for OEMs were differentiated from general design guidelines meant to improve aftermarket remanufacturing performance.
- For the second objective, any guidelines for designing parts made with inexpensive, lightweight non-metal materials were removed. Focus was instead placed on guidelines created for high-value, heavy mechanical equipment

(mining and construction, agricultural, trucking, automotive, large industrial and medical equipment, etc.), since most OEMs with remanufacturing
capabilities produce these kinds of products.

- The third objective led the research team to distinguish guidelines specific to
  designers and engineers from those with the larger business team in mind,
  such as managers, logistics, marketing, or other stakeholders.
- To support the fourth objective, guidelines specifically for remanufacturing
  were isolated from those addressing recycling or other sustainable design
  strategies.

The next step was to organize and group the over 1,500 relevant design guidelines using
an affinity-diagramming process. Affinity diagramming is one of the seven management
and planning tools of quality function deployment (QFD). It is used to organize large
amounts of data into themes based on their relationships. [9] Guidelines were grouped into
common themes based on their relationships, and then a top-level guideline was created
to represent the theme. The top-level guideline was developed using a common descriptive
functional format:

1. verb-noun-intended effect (or objective or performance standard)

These guidelines were formed using action verbs, avoiding passive verbs such as "ensure,"
"allow," "use," or "provide." They were constructed from the designer perspective and focus
on positive actions such as "detachably join parts," versus negative actions such as "avoid
permanent connections."

Finally, a steering committee comprised of design and remanufacturing practitioners in
the target industry sectors was used to rank order the top-level guidelines. Consequently,
some principles found in scientific literature were deemed by the committee to have a low
relevance to the targeted heavy-equipment industry sectors. Principles such as "emotion-
ally connect the customer with the product to assure the product endures" or "automate
assembly and disassembly to improve part to part consistency" are examples of principles
that were determined to be of lower priority to the target sectors, but that may apply to
consumer products, appliances, or other sectors that produce consumer-facing products in
high volume.

Of note, the committee's ranking of guideline priorities was in sharp contrast to an observation found in the scientific literature review. Sixty-two percent of the research papers
focused on simplifying the disassembly process in terms of time and cost by standardizing
and minimizing fasteners, sequencing, or improving ergonomic handling. Published scientific literature shows a concern with improving the recovery of low-cost systems, but guidelines addressing this tend to be lowly rated by remanufacturing practitioners in the target
sectors. This may be because, in most cases, the value recovered in a "core"—a used part
or component recovered for remanufacturing—far outweighs the effort to streamline the
disassembly process. With this in mind, the rapid disassembly guidelines are aspirational,
attempting to move design towards circularity even in the low-cost product categories.
However, very few designers and manufacturers of low-cost products also remanufacture
those products currently.

14.2.3   Developing Design Principles

As stated earlier, a circular economy is designed to maintain a product's resources at their highest value for as long as possible. A primary function of a product designed for the circular economy and remanufacturing is therefore to maintain the required performance beyond the initial use for the full product life cycle.

A value-engineering diagramming process based on function-analysis system technique (FAST) was used to link multi-life-cycle performance to the top-level design guidelines. The FAST diagram represents the logical relationships between different functions, answering "how" and "why" questions. [10] The process puts higher order functions on the left of the diagram to describe what is being accomplished. Lower order functions go on the right of the diagram to describe how they are being accomplished through heuristics and design guidelines. For this work, we define the far-left high-order functions as design principles. This process not only linked the design principles and hundreds of generic guidelines found in literature, it also helped define how they can be implemented. In the end, seven principles have significant relevance to heavy-equipment industry sectors. The relationships between an example principle, a few top-level guidelines, and a few guidelines from literature are illustrated in Table 14.1.

14.3   DfReman Framework

The sorting and data-aggregation activities resulted in an unorganized list of principles, top-level guidelines, and design guidelines without an easy way to apply each to a specific part design. In general, principles and guidelines apply to specific failure modes, and therefore not all principles and guidelines will apply to all parts or all failure modes. With this

Table 14.1   Design principles and top level DfReman guidelines.

Principle 1: Prevent failure modes from occurring.
1.1 Top Level DfReman Guideline: Adjust geometric features through design practices so stresses are below the fatigue limit to protect parts from overstress
1.1.1 Literature Guideline: Minimize sharp corners
1.1.2 Literature Guideline: Apply generous radii to reduce stress concentrations
1.1.3 Literature Guideline: Avoid sudden changes in geometry cross-section
1.1.4 Literature Guideline: Maintain a uniform wall thickness
1.1.5 Literature Guideline: Optimize the shape of holes or other stress concentrations
1.2 Top Level DfReman Guideline: Select durable materials to last the full life cycle
1.3 Top Level DfReman Guideline: Future-proof the design to be tolerant of changing regulatory standards (e.g. efficiency, materials, refrigerants, GWP, etc).
1.4 Top Level DfReman Guideline: Design high quality components to reduce failure and increase core value and likelihood of recovery and future use.

in mind, a structure was developed to enable a design team to select the DfReman principles and guidelines that best apply to a specific product. For this, we turned to the RCM process. The following subsections describe the methods used to build a customized RCM framework.

14.3.1   Reliability-Centered Maintenance (RCM)

RCM is a systematic process for preserving a system's function and performance by identifying functional failures, failure modes, and failure effects, as well as selecting and applying effective corrective measures or maintenance tasks. RCM is used to [11, 12]

- preserve a desired level of system or equipment functionality;
- as a life-cycle management tool and should be applied from design through end-of-life;
- manage the consequences of failure, not prevent all failures; and
- identify the most applicable and effective maintenance task or other logical action.

These uses of RCM make it a suitable framework around which to organize DfReman principles and guidelines.

The RCM process, explained in the technical standard SAE JA1012, [13] asks the following seven questions:

- Q1: Functions - What is the item supposed to do and what are its associated performance standards?
- Q2: Functional Failures - In what ways can it fail to provide the required functions?
- Q3: Failure Modes - What are the events that cause each failure?
- Q4: Failure Effects - What happens when each failure occurs?
- Q5: Failure Consequences - In what way does each failure matter?
- Q6: Proactive Tasks - What can be performed to prevent the consequences of the failure?
- Q7: Default Actions - What must be done if a suitable preventive task cannot be found?

14.3.2   Aligning DfReman Principles with a Customized RCM Framework

The first five questions of RCM focus on developing a failure mode effects and criticality analysis (FMECA). Applying the RCM process to a product to be remanufactured, the first question in RCM asks users to define what the product is supposed to do. A primary function for a circular product designed for remanufacturing is to *maintain the required performance beyond the initial use for the full product life cycle.*

The second question looks at the different ways the product is unable to perform a specific function to a desired level of performance. The functional failure considers various failed states, such as full or partial loss of performance, for the same function. In general, a product's life fails to meet the full life cycle because 1) it ceases to function, 2) the function degrades to a point that it no longer meets the customer requirements, or 3) the market needs change and the performance does not meet the new market requirements. [14]

The third question evaluates the different events that cause the function to fail, degrade, or not meet the market requirements. Known product failure modes include degradation (adhesive wear, abrasive wear, galling, rolling contact fatigue, fretting, erosion, corrosion, oxidation, stress cracking, creep, yield, etc), damage, obsolescence, contamination, and human error, among others.

After fully developing the product FMECA through Questions 4–5, the second part of an RCM analysis is to apply a structured decision process or "RCM logic," asking a defined set of questions to determine the appropriate mitigating tasks for identified failure modes. [13] Question 6 asks if any proactive tasks can be performed to prevent the consequences of failure. Finally, Question 7 asks what else can be done, such as a design change, if a proactive task cannot be performed.

Answering the last two questions is how the process is customized to evaluate a remanufacturable design. The RCM analysis is typically performed on fielded equipment and therefore the process logic prioritizes maintenance or restorative tasks over redesign tasks. If appropriate restorative tasks cannot be defined, then the logic moves to design changes as a last resort because changing a design and retrofitting fielded equipment can be cost prohibited. However, when evaluating design principles and guidelines for a new product early in the development cycle and yet to be manufactured, the order of the RCM logic is reversed and Question 7 should come before Question 6. Design changes which in traditional RCM were a last resort are now prioritized.

In the logic for Question 7, RCM recognizes three major categories of default actions: redesign or one-time change (OTC), failure-finding tasks (FF), and run-to-failure (RTF). For remanufacturing, we focused on the redesign action and then built logic to prevent the consequences of failure. First, can the designer prevent the failure mode from happening? Are there design features that would enable the product to last the full life cycle? Next, if the failure mode cannot be prevented, can it be delayed until after the product reaches it full useful life? Finally, if it cannot be prevented or delayed, can you move the failure so that it happens on a less important, less costly part to preserve the function of the high value part? This RCM redesign logic aligns with the DfReman principles focused on improving the initial design.

1. Principle 1: Prevent failure modes from occurring.
2. Principle 2: Minimize failure modes to an acceptable level.
3. Principle 3: Relocate failure modes to a less impactful location.

If the failure mode cannot be prevented or designed out of the product, we then look to additional DfReman principles to restore the failure mode through remanufacturing processes. In the logic for RCM, Question 6 divides the restoration tasks into three categories: on-condition tasks, scheduled-restoration tasks, and scheduled-discard tasks. On-condition tasks, also known as condition-based maintenance (CBM) or predictive maintenance (PM), use equipment or personnel to monitor and detect specific symptoms of a potential failure. These tasks are called "on condition" because the items that are monitored are left in service *on the condition* that they continue to meet performance requirements. If monitoring

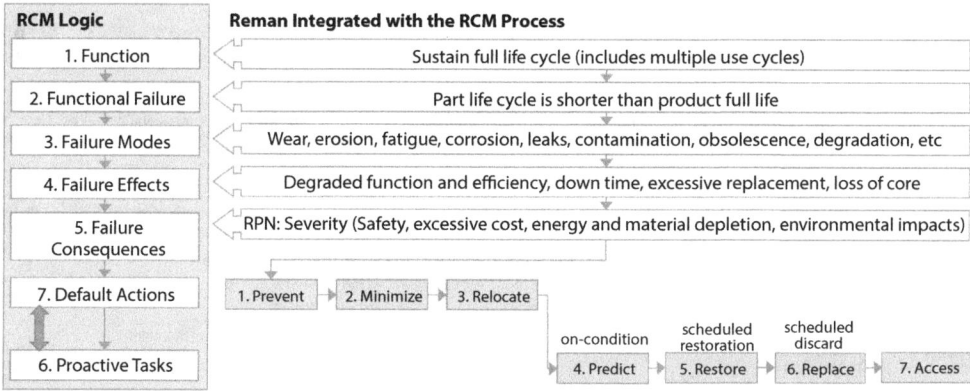

Figure 14.1 DfReman principles integrated with a customized RCM framework.

reveals a potential failure condition, corrective action must be taken. The on-condition tasks align with the following DfReman principle:

4. Principle 4: Detect or predict failure modes before the onset of significant damage.

RCM scheduled-restoration tasks involve preventatively restoring or remanufacturing a part or overhauling an assembly at or before a specified age limit, regardless of its condition at the time. The remanufacturing industry restores part features through cleaning, machining, additive manufacturing, coatings, or inserts among other processes. The scheduled restoration tasks align with the following DfReman principle:

5. Principle 5: Restore failure modes back to a like new condition.

Similarly, RCM scheduled-discard tasks involve discarding an item at or before a specified life limit, regardless of its condition at the time. Parts typically discarded can be filters, items that wear out like brake pads or bearings, consumables, technology like processing chips, or aesthetic features that become obsolete over time. The scheduled discard tasks align with the following DfReman principle:

6. Principle 6: Replace failed or obsoleted parts.

Additionally, to implement Principles 4–6, the part being inspected, restored, or replaced must be accessible. Accessibility is therefore the last principle.

7. Principle 7: Gain access to parts needing to be restored or replaced.

Figure 14.1 shows how the DfReman principles align within the modified RCM structure.

14.3.3   DfReman Principles and Top-Level Guidelines

With the principles aligned with the RCM process logic, the results of the FAST-diagramming process can now be used to link the design principles to the top design guidelines and subsequent guidelines and heuristics from literature and practitioners as illustrated in Figure 14.2. This framework now allows a design engineer to identify failure modes that limit the

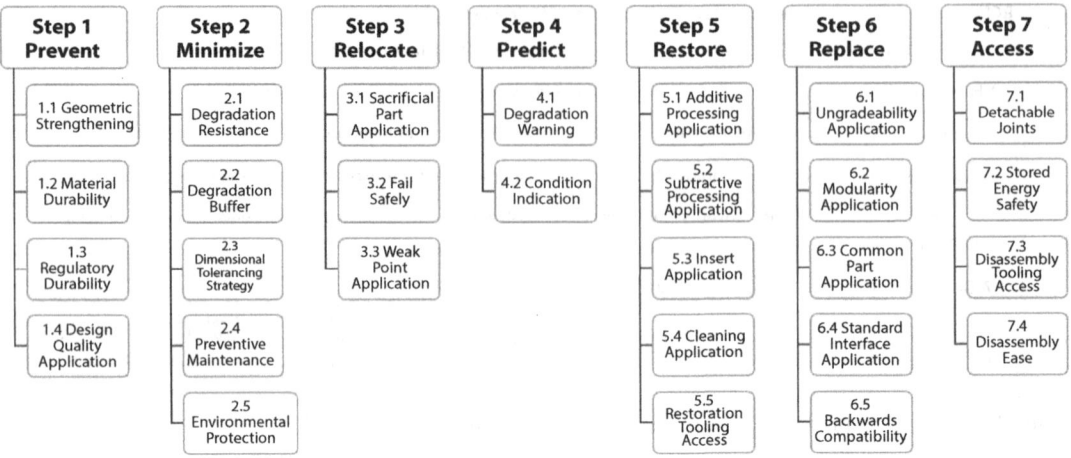

Figure 14.2 Top level DfReman guidelines and principles within the RCM framework.

life of their product, and then work through various design guidelines in the priority of the design principles.

14.4   Conclusions and Recommendations

Remanufacturing can be a powerful tool to transition industrial economies towards longer-term sustainable economic systems. But to capitalize on this change, the industry must solve many of the barriers that are restricting its growth. One of these barriers is the knowledge gap on how to design products for remanufacturing.

Significant knowledge exists in literature and among remanufacturing practitioners, as the collection of publicly- available remanufacturing design guidelines and heuristics documented herein shows. This paper has aimed to show how this store of knowledge can be used to aggregate design guidelines and format them into design principles to improve remanufacturing as a methodology. These principles were further arranged into a framework by aligning with a customized RCM process, first by outlining that a remanufacturing primary function is to maintain the required performance beyond the initial use for the full product life cycle, and then working through the RCM logic to identify actions and tasks to address life-cycle-limiting failure modes.

The next step of this research will be to provide DfReman information to engineers in usable design tools. One such tool being developed leverages the design guidelines, principles, and RCM framework as the basis for a DfReman computer-aided design (CAD) plugin to allow the design engineer to check parts for remanufacturing. Many of the design guidelines lend themselves to prescriptive design rules.

Consider this illustrative example: An engine head gasket surface is known to fail and shorten the engine life by the failure modes of pitting or eroding with use. Applying Principle 5 of restoring failure modes back to a like-new condition, a few design guidelines address sealing-surface failure modes. One such guideline from remanufacturing practitioners is to machine the degraded surface to restore the flatness or surface-finish specifications.

However, this remanufacturing process is limited if there is not enough material thickness on the original surface available to machine and still maintain the design specifications. A design rule can therefore be applied by a CAD plugin to check a sealing-surface thickness against a prescriptive threshold thickness required for a machining working allowance. If the threshold is not met, then the surface can be flagged in CAD for non-conformance.

Future work will be to define the design guidelines as design rules and automate a design rule checker, as well as develop a plugin for specific CAD software packages.

Acknowledgements

This material expands on the original research found in [15] and is based upon work supported by the U.S. Department of Energy's Office of Energy Efficiency and Renewable Energy (EERE) under the Advanced Manufacturing Office Award Number DE-EE0007897 awarded to the REMADE Institute, a division of Sustainable Manufacturing Innovation Alliance Corp. This report was prepared as an account of work sponsored by an agency of the United States Government. Neither the United States Government nor any agency thereof, nor any of their employees, makes any warranty, express or implied, or assumes any legal liability or responsibility for the accuracy, completeness, or usefulness of any information, apparatus, product, or process disclosed, or represents that its use would not infringe privately owned rights. Reference herein to any specific commercial product, process, or service by trade name, trademark, manufacturer, or otherwise does not necessarily constitute or imply its endorsement, recommendation, or favoring by the United States Government or any agency thereof. The views and opinions of authors expressed herein do not necessarily state or reflect those of the United States Government or any agency thereof.

In addition, RIT received funding, in part, from New York State Empire State Development under grant #AC118 that supported the work on this project. The views expressed herein do not necessarily represent the views of New York State.

I would also like to thank the industry stakeholder committee for their review of the guideline priorities including: Brandon Baker (RIT), Yeshwant Bhoskar (BorgWarner), Tyrone Ellis (Trane Technologies), Curt Graham (Caterpillar), Michelle Hayes (RIC), Murthy Kotike (Trane Technologies), Patrick Layman (ZF Group), Allen Luccitti (RIT), Swathy Ramaswamy (Trane Technologies), Kristi Sisak (RIT), Chris Stickling (Caterpillar), Jeff Stukenborg (ZF Group), Jeff Sutherland (Caterpillar), Michael Thurston (RIT), and Mark Walluk (RIT).

References

1. Nasr, N., Russell, J., Bringezu, S., Hellweg, S., Hilton, B., Kreiss, C., and von Gries, N., Re-defining Value – The Manufacturing Revolution. Remanufacturing, Refurbishment, Repair and Direct Reuse in the Circular Economy. A Report of the International Resource Panel. United Nations Environment Programme, Nairobi, Kenya, 2018
2. Singhal, D., Tripathy, S., Jena, S., Remanufacturing for the circular economy: Study and evaluation of critical factors, Resources, Conservation and Recycling, Volume 156, 2020, 104681, https://doi.org/10.1016/j.resconrec.2020.104681.

3. Hilton, B. and Thurston, M., Design for Remanufacturing, in Nasr, N. (ed.) Remanufacturing in the Circular Economy: Operations, Engineering and Logistics. New Jersey: John Wiley & Sons, pp 137-168, 2019.

4. ANSI, Specifications for the process of remanufacturing, RIC001.2-2021, https://webstore.ansi.org/standards/ansi/ansiric0012021, 2021.

5. USITC, Remanufactured Goods: an Overview of the U.S. And Global Industries, Markets and Trade, United States International Trade Commission publication, 2012, 4356

6. Nasr, N., Hilton, B., and Parnell, K., Remanufacturing Industry Survey Analysis, Sept 2016, https://remancouncil.org/wp-content/uploads/_pda/2020/05/RIT-Reman-industry-roadmap-workshop-survey-analysis-9-24-16.pdf

7. Nasr, N. and Hilton, B., Technology Roadmap for Remanufacturing in the Circular Economy, U.S. Department of Commerce Report, Web. May 2017. https://www.rit.edu/sustainability-institute/public/Reman_Roadmap_2017.pdf

8. Fu, K. K., Yang, M. C., and Wood, K. L., Design Principles: Literature Review, Analysis, and Future Directions, ASME J. Mech. Des., 138(10), p. 101103 (13 pages), 2016.

9. Tague, N., Quality Toolbox (2nd Edition). American Society for Quality (ASQ), 2005. Retrieved from https://app.knovel.com/hotlink/toc/id:kpQTE00001/quality-toolbox-2nd-edition/quality-toolbox-2nd-edition

10. Fox, J., Quality through design: The key to successful product delivery, McGraw-Hill Book Company, 1993, 9780077077815.

11. Nowlan, S., and Heap, H., Reliability–Centered Maintenance, Department of Defense, Report No. AD-A066579, Washington D.C., 1978.

12. U.S Department of Defense (DOD), Reliability Centered Maintenance (RCM), Manual, DoDM 4151.22-M, 2011.

13. SAE, A Guide to the Reliability-Centered Maintenance (RCM) Standard, JA1012, 2002, The Engineering Society for Advancing Mobility Land, Sea. Air, and Space, Warrendale, PA.

14. Nasr, N. and Hilton, B., Design for Remanufacturing In: CIRP International Conference on Life Cycle Engineering, LCE 2008: 15th CIRP International Conference on Life Cycle Engineering: Conference Proceedings. Sydney, N.S.W.: CIRP, 2008: 19-22.

15. Hilton, B., Design for remanufacturing. Report, DOE-RIT-0007897-3, United States, https://doi.org/10.2172/1876418. https://www.osti.gov/servlets/purl/1876418, 2021.

15

Recyclability Feedback for Part Assemblies in Computer-Aided Design Software

Bert Bras* and Richard Lootens

George W. Woodruff School of Mechanical Engineering, Georgia Institute of Technology, Atlanta, GA, USA

Abstract

Currently, it is not possible to evaluate recyclability of product assemblies in a commercial Computer-Aided Design (CAD) system. Some tools already provide feedback about individual plastic recyclability based on limited data sets. Feedback on individual materials, however, does not address recyclability feedback on assemblies and products made from mixed plastics and/or metals that are in the process of being designed using the various CAD software. In this paper, we will highlight the development of a prototypical tool that evaluates a product assembly's recyclability as a plug-in for a commercial CAD system. The goal of this project is to provide feedback to designers about recyclability of mixed plastic assemblies and products while they are being designed directly in the CAD software. In order to test the feasibility of providing recyclability assessments and feedback to designers, the CAD systems electronic Bill of Materials (eBOM) data was used to identify the different materials present in the assembly/product, as well as their masses based on the parts model volume representations, and design for recycling rules obtained from select sources were integrated. Specifically, the plastics recycling guidelines from the Association of Plastic Recyclers (APR) were analyzed and coded into a plug-in module that interfaces with a commercial CAD system. One limitation of the APR guidelines is the limited number of plastics covered in the guidelines. To extend the usability, the capability of selecting other recyclability guidelines was added. Specifically, recyclability assessments based on material compatibility matrices such as found in a previous version of the German VDI 2243 – Design for Recycling standard and in the UNEP "Metal Recycling - Opportunities, Limits, Infrastructure" publication. Assembly recyclability evaluations based on these guidelines provide feedback to designers and CAD users with respect to areas of concern where some incompatibility may be present (warning), parts and materials that will definitely prevent recycling of an assembly (dangers/incompatibilities), as well as materials present that are not covered by the recycling guideline. A summary shows what parts and materials are compatible versus parts and materials that may need to be changed to improve overall recyclability. Once changes are made, a new recyclability assessment can be made by a simple click of the "evaluate" button in the software. To further boost usability, the tool also includes an assessment of embodied energy and carbon emissions based on the REMADE Calculator. Together, this allows both a quick assessment of potential recyclability bottlenecks as well as insight into overall embodied energy and carbon of the product assembly or part at hand.

**Corresponding author:* bert.bras@me.gatech.edu

Nabil Nasr (ed.) Technology Innovation for the Circular Economy: Recycling, Remanufacturing, Design, Systems Analysis and Logistics, (183–196) © 2024 Scrivener Publishing LLC

As will be shown, the tool is useful, but is hampered by a low number of materials for which compatibility information exists. As a next step, we would urge the recycling industry to provide more data that allows codifying compatibility and recyclability of mixed material combinations for different recycling processes.

Keywords: Design for recycling, recyclability assessments, plastics recycling, material compatibility

15.1 Introduction

Product design is a process in which product attributes such as cost, performance, manufacturability, safety, and consumer appeal are considered together. Modern design is characterized by an increase in information because, in addition to manufacturing, life cycle concerns and even recycling are receiving increased attention.

The importance of incorporating both recyclability and sustainability considerations in the design phase of products cannot be understated. According to the Ellen MacArthur Foundation, the current consumption pattern of consumer goods annually sends over \$2.6 trillion of goods to landfills and incineration plants [1]. This same report explained that shifting to a circular model could generate up to \$700 billion in economic opportunity. For a more circular economy to become a reality, product designers and engineers have to begin to consider how their product will be processed once the use phase of its lifetime has come to an end. While Design for Recycling (DFR) guidelines been developed for many years or even decades to help designers in this area (e.g., [2–4]), they are often presented in an inaccessible format and require considerable effort, time, or expertise to be particularly effective. Furthermore, changing collection infrastructure, recycling technology and material demands ensure that DFR guidelines differ over time and locality. Nevertheless, most agree that reductions in the number of different materials, additives, adhesives, etc. in product assemblies are universally preferable.

Although it is relatively easy to determine the recyclability of specific single materials, it is currently not possible toeffectively evaluate the recyclability of mixed material assemblies using traditional CAD/CAM systems such as SolidWorks, Autodesk Inventor, Fusion 360, etc. Software modules such as Moldflow give feedback on individual material recyclability, however an assessment of mixed material combinations is not given. A multi-material assessment can be critical for determining the actual recyclability of a product, as two readily recyclable materials can be joined in such a way as to make both of them, together, unrecyclable.

This paper presents the development of a prototype CAD software plug-in to automatically evaluate the recyclability of a product design consisting of multiple parts using a number of available DFR guidelines. This evaluation is presented to the designer with the goal of changing their design decisions to improve a product's recyclability. This problem is not new and the need for better CAD integration was already noted in the 1990s [5–8]. Our goal is to (finally) produce software that will allow designers who are not experts in eco-design to get accurate feedback as they are designing a product.

15.2   Current State of Design for Recycling (DFR) Integration

A number of tools have been developed in academia, but many have not been widely used or accepted by industry. The ideal state is that major CAD vendors include Design for Re-X tools in their product suite.

15.2.1   CAD Vendor Developments

Autodesk's Moldflow 2017 software was one of the few commercial engineering software packages that had a recyclability rating built in. The software displays the recyclability rating of the family of plastics to which the selected material belongs based on publicly available data from PlasticsEurope. The feedback to the designer, however, is based on a single material part. No automated feedback was available for an assembly composed of multiple separately molded parts.

Dassault Systems offers a variety of engineering software products. Catia and Solidworks are perhaps its best-known CAD software products. Solidworks includes a "Solidworks Sustainability" module that includes the GaBi LCA Environmental Database and has various features that include screening-level Life Cycle Assessment, sustainability report generations, and environmental impact dashboard, among others. The SOLIDWORKS Sustainability module does not include explicit Design for Recyclability assessments unless it is part of the Life-Cycle Analysis info. AS is, the software does not provide explicit feedback with respect to actual recyclability of plastic assemblies.

Siemens offers a wide variety of industrial software products. However, Siemens NX, Team Center, and SolidEdge are probably the best-known Siemens CAD software systems for design engineers. They are part of Siemens Product Life Cycle Management (PLM) suite of software tools. The NX and SolidEdge CAD systems do not include software modules that explicitly address Re-X assessments. Teamcenter also does not include modules that perform Re-X assessments, but Teamcenter is a PLM system that is customizable with various add-ons. For example, per Siemens' environmental compliance and sustainability needs webpage, it includes features to address environmental compliance and sustainability needs.

15.2.2   Industry Organizations/Third Party Developments

There are a number of industry organizations that have developed design for recycling guidelines worth noting. Many, however, focus on packaging materials, which are not necessarily applicable for designers of mixed material product assemblies. For example, Plastics Recyclers Europe (PRE) created RecyClass™ with the aim of improving the design of packaging so that it is easily recyclable into high-quality recyclates which can then be used in a new plastic product. Part of RecyClass are Recyclability Evaluation Protocols, Design for Recycling Guidelines, and the RecyClass Online Tool based on the Design for Recycling Guidelines developed by The European PET Bottle Platform (EPBP) and other RecyClass activities. The tool tests the recyclability of a plastic packaging.

To enable circular thinking the Ellen MacArthur Foundation is also developing tools that can be used at various stages in developing products and/or business strategies. ResCoM, which stands for Resource Conservative Manufacturing, has developed a collection of methodologies and tools for the implementation of closed-loop manufacturing systems.

The tool set includes a Reman Design checklist, circularity calculator, etc., but does not include a product recyclability analysis or evaluator. In the US, the Sustainable Packaging Coalition also offers various resources, including a "Design for Recycled Content Guide". This guide provides information and guidelines on using recycled plastics (incl. PET, HDPE, PP, PS, PE), paper (incl. paperboard and corrugate), glass, aluminum and steel. As a spin-off, Trayak is marketing the COMPASS (Comparative Packaging Assessment) LCA tool. COMPASS was originally developed by the Sustainable Packaging Coalition (SPC) in a collaborative process. The tool was created to incorporate environmental performance criteria into the concept development and material selection stages of package design.

The Association of Plastic Recyclers (APR) has created a Design Guide for Plastics Recyclability [9] that is regularly updated to it accurately reflects the operations and technology in use by today's North American plastics recycling infrastructure. The APR Design Guide specifically addresses plastic packaging, but the principles can be applied to all potentially recycled plastic items. Although not a computer-based tool, the APR guide is viewed by many as one of the best resources for designers seeking to understand how to improve plastic packaging and product recyclability in North America. The APR defines an item as recyclable when the following conditions are met:

- At least 60% of consumers or communities have access to a collection system that accepts the item
- The item is most likely sorted correctly into a market-ready bale of a particular plastic meeting industry standard specifications, through commonly used material recovery systems, including single- and dual stream MRFs, PRF's, systems that handle deposit system containers, grocery store rigid plastic and film collection systems.
- The item can be further processed through a typical recycling process cost effectively into a postconsumer plastic feedstock suitable for use in identifiable new products.

This definition is particularly useful as it considers the accessibility and likelihood of correct sorting, in addition to the product's ability to be processed into plastic feedstock. The fundamentals and scope of the APR Guide align well with providing design feedback, as the feature categorization system in the APR guide (preferred, requires testing, detrimental, or non-recyclable) provides a clear way to assign point values that help quantify design improvements.

15.3 Our Approach for a DFR Evaluator CAD Plug-In

This paper is not focused on identifying or evaluating the validity of specific Design for Recycling (DFR) guidelines, but on showing how existing DFR guidelines can be incorporated in CAD software. The approach taken was to find guidelines that were developed by experts (like the APR) and code them into a plug-in so that feedback could be given based on data retrieved from a given 3D model. The prototype "DFR Evaluator" tool is a software plug-in for evaluating the recyclability of a product assembly in a CAD system. This tool simplifies recyclability design by scoring a CAD assembly's recyclability-relevant features, based on a user-selected DFR standard/guideline and returning intelligent feedback for design improvements. Many designers would probably make their product more recyclable if they only knew how, but are often daunted by the time needed to understand or implement recyclability guidelines. This tool aims to surmount that obstacle by providing a clean, quick, and easy-to-understand assessment of their product's recyclability. A prototype plug-in was developed for Autodesk's Fusion 360 CAD software for providing design for recycling feedback for assemblies based on different types of guidelines or standards, namely:

- The APR Design Guide
- Compatibility Matrix Feedback
- Embodied energy/Carbon Dioxide Equivalent Feedback

These different types of guidelines were chosen to demonstrate how different assessments can (and have to) be made because guidelines can vary by region, type of product (for example, packaging versus bulk products), post-consumer vs post-industrial recycling, or material. For example, one set of DFR guidelines may focus on plastic recycling while another may focus on metal recycling, which are typically sorted by entirely different recycling plants for processing. All DFR standards should give an indication of improvement for the assembly. Having feedback based on multiple DFR standards is a good way to evaluate the recyclability of a product from different angles.

The first step towards CAD integration was to digitize the APR design guide as a set of flowcharts to make the logic and scoring method clear for each APR plastic. These logical flowcharts underlie the DFR evaluation process and allow for a computer-based comparison between different types of plastics. The scoring method is a penalty-based scheme based on APR's "preferred" (0 points), "requires testing" (2 points), "detrimental" (3 points), or "detrimental"/"non-recyclable" (4 points) feature categorization system. Thus, the design goal is to reduce the number of penalty points as much as possible.

In addition to the APR Design Guide, compatibility matrices have been used in design for recycling to quickly determine if two materials are compatible based on various criteria. These matrices provide highly simplified representations to what extent the lack of separation and thus mixing of two material will degrade material properties and/or market demand of the combined recycled material. In Figure 15.1, two examples of compatibility matrices are shown, which have been incorporated in the DFR plug-in. These two matrices cover different materials and are likely based on different assumptions with respect to recycling technologies and material properties and markets and thus could cover compatibility in different domains.

The prototype software framework allows for quick addition of future matrices that can be used for a myriad of materials/feedback applications. The logic for the addition of these matrices is straightforward. The entire matrix is decomposed into a list of dictionaries and each dictionary corresponds to a material in the matrix and contains a key-value pair for each material it has a conflict with and another key-value pair for each material it is entirely incompatible with. The final key-value pair in the dictionary is a list of all the components that are made of the specified material. When the evaluation is conducted, the list of dictionaries is looped through in a way that returns all relevant data, including parts that are incompatible with each other, parts that are compatible with each other and materials that are not covered by the specified matrix. Having developed and implemented

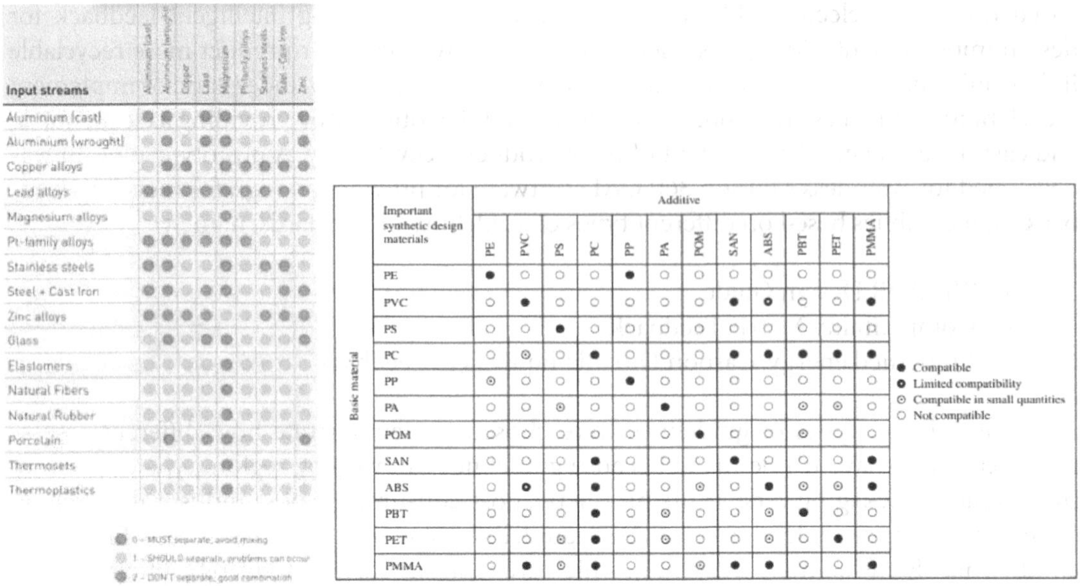

Figure 15.1 Material compatibility matrices: Left: UNEP metals compatibility matrix [10]; Right: 1993 VDI 2243 plastics compatibility matrix [4].

this framework into the software, it is easy to add many more matrices that cover different materials or consider different criteria of compatibility.

A third type of feedback provided is the product assembly's Embodied Energy (EE) and Carbon Dioxide Equivalent (CO2e), which are two metrics used in quantifying a product's environmental impact. These two metrics were selected to be included for design feedback because they are widely used and data for many materials are available. Embodied energy is the energy required by all processing associated with converting a material into a manufactured product – from mining to final product delivery. Feedback for this metric is given in terms of kiloJoule (kJ). Carbon dioxide equivalent (CO2e) is like embodied energy but represents the equivalent amount of CO2 generated by the same span of the lifecycle of a product (from mining to final product delivery). Feedback for this metric is given in terms of kg CO2e. The final feedback 'score' is simply based on the mass of the object being scored and its material. Each material has a factor associated with it which is multiplied by the mass of the associated object and that is how its embodied energy or CO2e is calculated. Sources like the REMADE Impact Calculator and [11] present these factors for a multitude of materials, some of which have been used in the CAD plug-in. These factors can easily be changed in the software as they can vary for the same material based on a multitude of reasons such as location, manufacturing processes used, and even material extraction processes used. Whenever an evaluation is executed, the material and mass data for each component of an assembly is pulled from Fusion 360 and multiplied by its associated factor.

15.4 DFR Evaluator Plug-In Demonstration

In Figure 15.2, a soap bottle assembly with associated parts and materials is shown that will be used for demonstrating the plug-in: In Figure 15.3, the DFR Evaluator plug-in is shown after installation in the Fusion 360 utilities tab. When the user clicks on it, the DFR Evaluator will immediately pull data from the CAD model and convert the data to a JSON format. The converted JSON file can be exported using a "File – Export Model Data" option

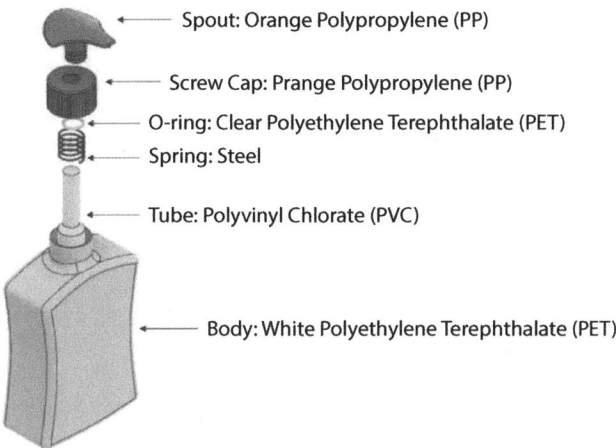

Spout: Orange Polypropylene (PP)

Screw Cap: Prange Polypropylene (PP)

O-ring: Clear Polyethylene Terephthalate (PET)

Spring: Steel

Tube: Polyvinyl Chlorate (PVC)

Body: White Polyethylene Terephthalate (PET)

Figure 15.2 Exploded assembly view of mixed materials soap rottle assembly.

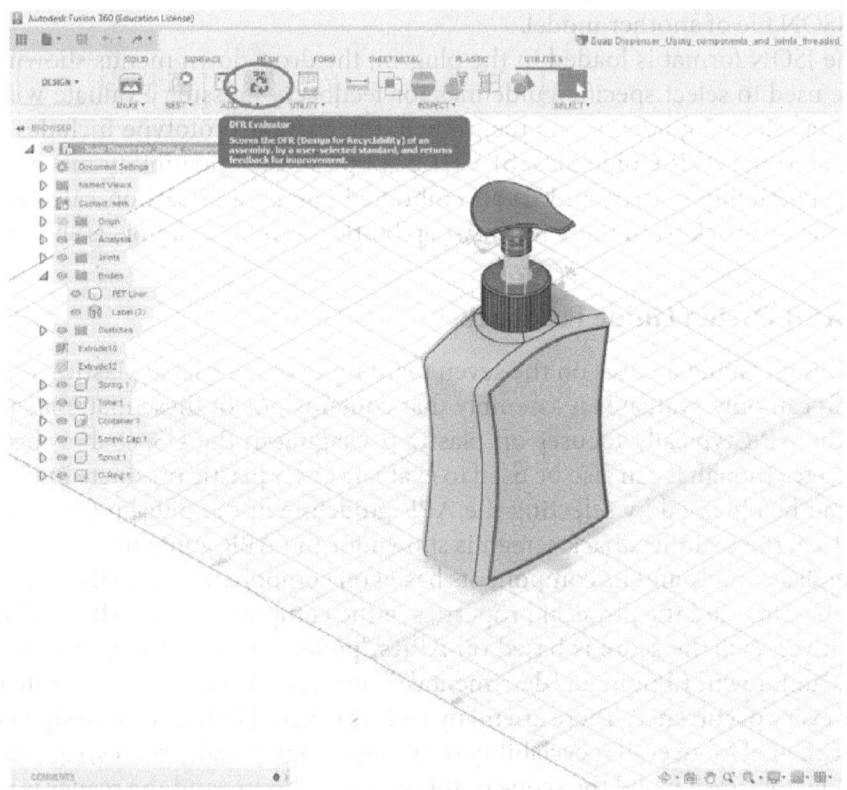

Figure 15.3 DFR evaluator plug-in in fusion 360.

Figure 15.4 DFR evaluator main screen and guideline selection menu.

in the plug-in's File menu. The user can also import model data external to Fusion 360 by loading a JSON file of another model.

After the JSON format is loaded in the plug-in, the drop-down menus shown in Figure 15.4 can be used to select specific guidelines for feedback. Pressing Evaluate will start the analysis for a selected guideline. As mentioned, the current prototype includes an implementation of a) the APR Guidelines, b) VDI 2243 plastics compatibility matrix, c) UNEP material compatibility matrix, and d) an embodied energy and carbon calculator. In the following, the feedback from these for the soap bottle assembly example is shown.

15.4.1 APR Design Guide Feedback

The APR Design Guide focuses on the seven main types of recyclable plastics, PET, HDPE, PP, etc., and can only evaluate an assembly that contains one of these materials in bulk. In addition, the APR typically focuses on plastic packaging in the US, such as bottles, soap dispensers, etc., though it can also be used to evaluate other plastic products in the US. APR feedback can be obtained by selecting the APR guideline in the plug-in (see Figure 15.4). In Figure 15.5, the main feedback screen is shown for the APR guidelines. A score is given for the overall assembly and its components based on component materials, their combinations, attachments, and the physical properties of the components, as well as color, melting point, additives, etc. The score is based on APR's "preferred" (0 points), "requires testing" (2 points), "unknown" (3 points), "detrimental"/"non-recyclable" (4 points) categorization system for every occurrence. There are many factors covered in the APR Design Guide that can be detrimental for overall recyclability resulting in high penalty scores, e.g., 99 in Figure 15.5 (these factors are beyond the scope of this paper to discuss and the reader is referred to the comprehensive APR Design Guide for details). Hovering over the feedback gives a help

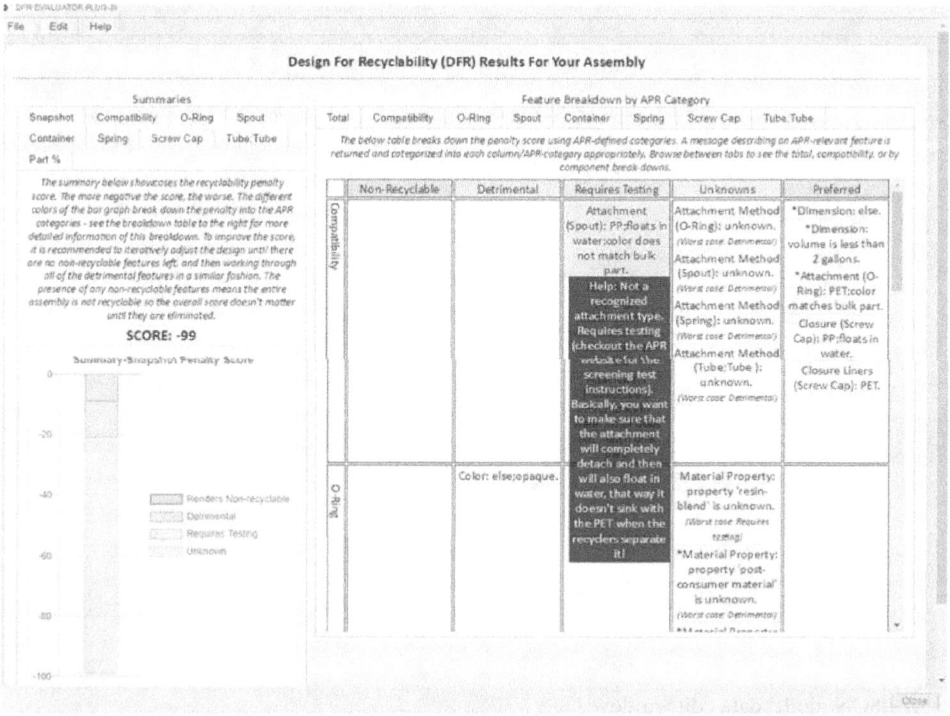

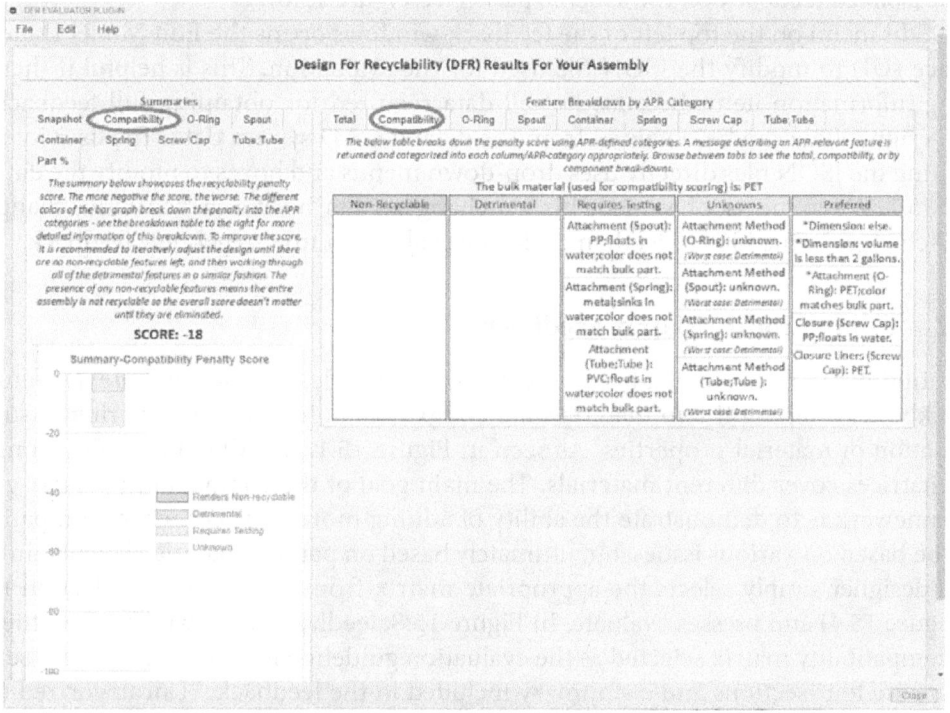

Figure 15.5 APR design guide DFR feedback – main screen.

Figure 15.6 APR design guide DFR feedback – part details screen.

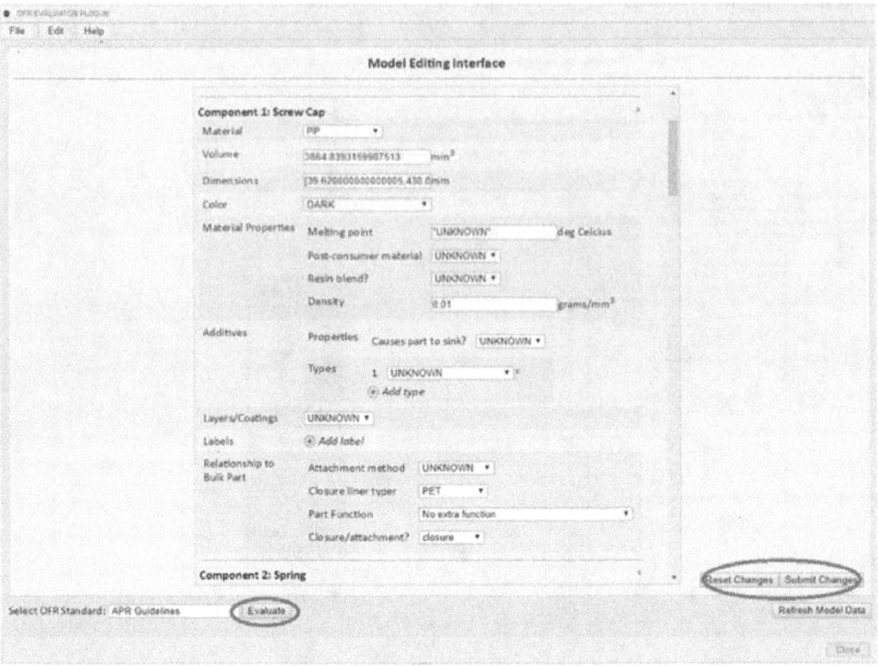

Figure 15.7 JSON model data edit window.

message that provide additional design suggestions. Clicking on the tabs with part names provide isolated feedback on the specific parts (see Figure 15.6).

The Edit menu on the top left of the feedback windows opens the Edit Model Data User Interface (UI) to modify the JSON file used for the evaluation. This is helpful if there are missing information items because not all data required for obtaining full feedback per APR Design Guide can be obtained from the default CAD system data. The user interface for editing the JSON files directly uses drop-down menus and boxes to modify the data, see Figure 15.7. When the user has made his/her changes/edits, he/she clicks Submit changes to submit changes to the JSON file, then clicks the evaluate button to get new feedback.

15.4.2   Compatibility Matrix Feedback

Compatibility matrices entail a broad spectrum of possible feedback and are a useful way of quickly determining if two materials are compatible based on various criteria, such as degradation of material properties. As seen in Figure 15.1, the VDI and UNEP compatibility matrices cover different materials. The main goal of the compatibility matrix guideline framework is to demonstrate the ability of adding more matrices where compatibility could be based on various issues, but ultimately based on pairs of materials. To obtain feedback a designer simply selects the appropriate matrix from the guideline selection menu (see Figure 15.4) and presses evaluate. In Figure 15.8, feedback is shown based on the VDI 2243 compatibility matrix selected as the evaluation guideline for the Soap Bottle assembly.

There are four sections and a summary included in the feedback. "Uncategorized materials" lists all materials and their component names from Fusion360 that are not included in

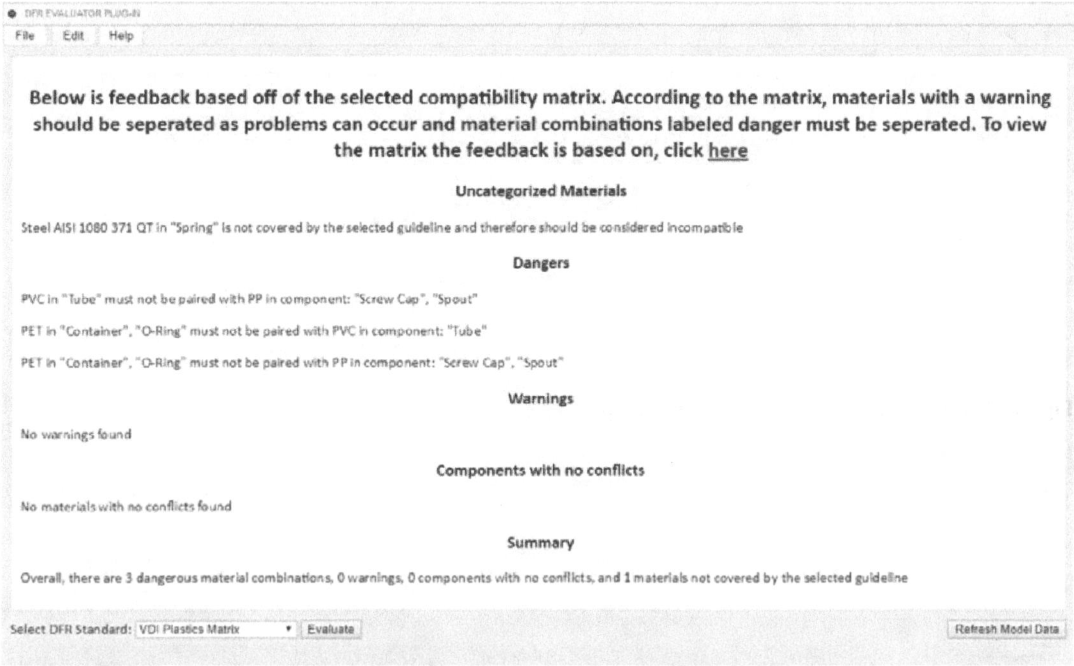

Figure 15.8 Compatibility matrix feedback example.

the currently selected matrix. It is recommended that one treats uncategorized materials as incompatible unless there is data to the contrary. "Dangers" lists components that are not compatible, and "Warnings" lists components that are conditionally compatible. Finally, all components with no conflicts are listed and a summary is provided.

15.4.3   Embodied Energy/Carbon Dioxide Equivalent

The feedback generated by selecting embodied energy or embodied carbon dioxide equivalent in the guideline selection menu is not a measure of the assembly's recyclability per se, but more a direct quantification of the environmental impact of the assembly. Embodied energy is the energy required by all processing associated with converting a material into a manufactured product, typically from mining to final product delivery. Feedback for this metric is given in terms of kJ. Carbon dioxide equivalent (CO2e) is like embodied energy but represents the equivalent amount of CO2 generated during the same span of the lifecycle of a product (from mining to final product delivery). Feedback for this metric is given in terms of kg CO2e.

To received feedback, simply select either embodied energy or carbon dioxide equivalent from the drop-down menu as shown in Figure 15.4. After selecting Evaluate, a feedback table will appear for embodied energy or carbon dioxide equivalent, dependent on the selection (see Figure 15.9 and Figure 15.10, respectively). For a given design, the goal would be to reduce the total for the assembly while still meeting all required design specifications. It is worth noting that these values can change for a given material based on many factors, so check that the values being used to calculate these are accurate for your location/manufacturing processes being considered. Additions of other metrics like these

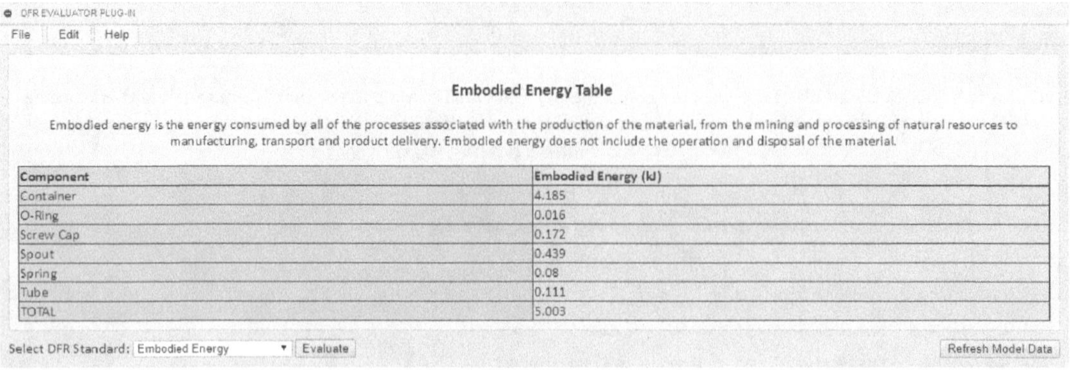

Figure 15.9 Embodied energy evaluation example.

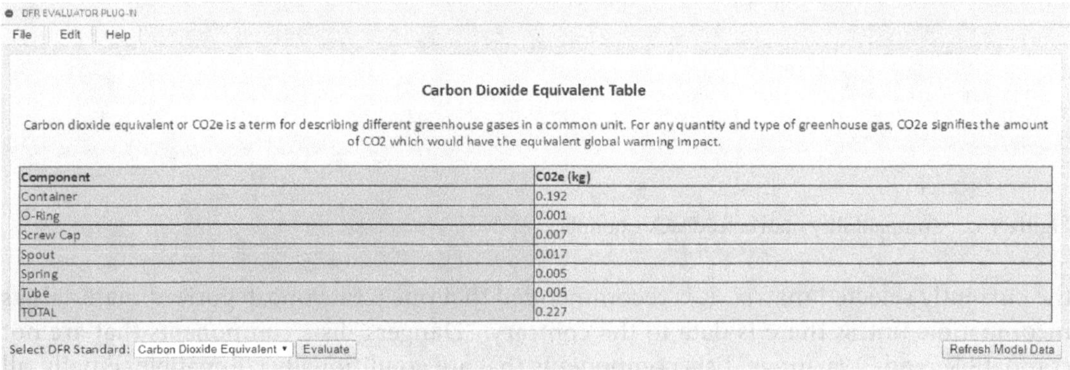

Figure 15.10 Embodied carbon dioxide equivalent example.

for feedback is relatively simple because they require simply multiplying the mass by a factor that is predetermined.

15.5   Conclusions & Recommendations

In this chapter, a prototype DFR CAD plug-in was presented which includes multiple feedback modes covering different types of DFR guidelines and materials while requiring no or minimal additional user input. Different DFR feedback frameworks were developed and implemented to allow for future expansion with additional guidelines that follow similar guideline styles. Recommendations for future work are:

Continued Expansion of DFR guidelines: While the current plug-in has achieved the goal of giving feedback with limited additional user input, there is still potential for further expansion. One of the largest hurdles of implementing complex guidelines like the APR in a CAD plug-in software is that many simplifying assumptions must be made if user input is to be reduced. Careful consideration must be taken here in what assumptions are made so that the feedback will not be heavily scrutinized down the line. Given that the APR guidelines cover only 7 materials, more guidelines addressing mixed material assemblies are needed for a larger variety of materials.

Improvement of user experience: Currently, the feedback of the compatibility matrix guidelines is rudimentary. Not much time was dedicated to improving user experience overall, but several features may easily be added, such as a way to print feedback for a given guideline. More in depth improvements may include a way to combine all applicable guidelines into one comprehensive 'recyclability report'.

Continued Improvement of data collection functionalities: If data collection functionalities can be improved, then more complex features can be identified directly from a CAD model, and more complex guidelines can be added while lessening the need for simplifying assumptions. However, a larger issue may be that many designers do not add the level of detail required to definitively identify certain features like different types of welds/bonds, as many designers simply mate two surfaces together without further input of how they are mated together. Perhaps the need for some user input is unavoidable for recyclability assessments in the design process. Nevertheless, opportunities for intelligent feedback from CAD data exist, but material guidelines need to be expanded.

Acknowledgements

We gratefully acknowledge the encouragement and support from Autodesk, Brook Byers, and the Woodruff School of Mechanical Engineering, as well as the work from Sophie Lehtikoski, Sarah Kate Carpenter and Jason Lee.

References

1. World Economic Forum, Ellen MacArthur Foundation, McKinsey & Company, *The New Plastics Economy - Rethinking the future of plastics*, Ellen McArthhur Foundation, 2016.
2. E. Masanet, R. Auer, D. Tsuda, T. Barillot, A. Baynes, An assessment and prioritization of "design for recycling" guidelines for plastic components, *Conference Record 2002 IEEE International Symposium on Electronics and the Environment (Cat. No.02CH37273)*, 2002, pp. 5-10.
3. S.L. Coulter, B.A. Bras, G. Winslow, S. Yester, Designing for Material Separation: Lessons from the Automotive Recycling, *Journal of Mechanical Design* 120(3) (1998) 501-509.
4. VDI, VDI 2243 - Konstruieren Recyclinggerechter Technischer Produkte (Designing Technical Products for ease of Recycling), VDI-Gesellschaft Entwicklung Konstruktion Vertrieb, Germany, 1993.
5. U. Kalyan-Seshu, B.A. Bras, Integrating DFX Tools with Computer-Aided Design Systems, 1998 ASME Design Automation Conference, ASME Design Technical Conferences and Computers in Engineering Conference, ASME, Atlanta, Georgia, 1998, pp. Paper no. 98-DETC/DAC-5621.
6. B. Bras, Incorporating Environmental Issues in Product Realization, United Nations Industry and Environment 20(1-2) (1997) 7-13.
7. D.W. Rosen, B.A. Bras, S.L. Hassenzahl, P.J. Newcomb, T. Yu, Computer-Aided Design for the Life-Cycle, *Journal of Intelligent Manufacturing* 7 (1996) 145-160.
8. A. Kriwet, E. Zussman, G. Seliger, Systematic integration of design-for-recycling into product design, *International Journal of Production Economics* 38(1) (1995) 15-22.
9. APR, APR Design® Guide for Plastics Recyclability, 2018. https://plasticsrecycling.org/apr-design-guide. 2022.

10. United Nations Environment Programme International Resource Panel, Metal Recycling: Opportunities, Limits, Infrastructure, UNEP Publications, 2019.
11. M.F. Ashby, Materials and the Environment: Eco-informed Material Choice, 2nd ed., Butterworth-Heinemann 2013.

Part 4
SYSTEMS ANALYSIS

16

Preliminary Work Towards A Cross Lifecycle Design Tool for Increased High-Quality Metal Recycling

Daniel R. Cooper[1]*, Aya Hamid[2], Seyed M. Heidari[1], Alissa Tsai[1] and Yongxian Zhu[1]

[1]*George G. Brown Laboratory, Mechanical Engineering Department, University of Michigan, Ann Arbor, USA*
[2]*Research and Innovation Center, Ford Motor Company, Dearborn, MI, United States*

Abstract

The embodied energy of vehicles is growing as energy-intensive materials such as aluminum auto body sheet (ABS) are used to deliver improved performance. This presents an opportunity for recyclers to shift towards high-value recycling into wrought alloys and for car makers to increase the end-of-life (EOL) recycled content of their sheet, reducing their material costs and energy burden. However, the current system cannot effectively recycle the aluminum and steel sheets. Shredded and contaminated EOL metal (e.g., mixed aluminum alloys with steel rivets and mixed steel alloys with embedded copper wiring) is often exported, downcycled to castings, or recycled as rebar. In this paper, we will discuss preliminary work on developing a design tool that couples the effects of vehicle design, recycling system practices and technologies, and alloy compositional tolerances, under different ecosystem scenarios from 2020-2050. Ultimately, design and R&D recommendations will be made using the tool to test the effect of cross-lifecycle design changes on the *system metrics*, which include cumulative energy demand (primary energy), greenhouse gas emissions, and primary metal demand associated with automotive sheet metal production.

Keywords: Vehicle design, recycling systems, alloy design, system optimization

16.1 Introduction

The embodied energy of vehicles is growing as energy-intensive materials such as aluminum auto body sheet (ABS) and advanced high strength steels are used to deliver improved performance [1–3]. This presents an opportunity for recyclers to shift towards high-value recycling into wrought alloys and for car makers to increase the end-of-life (EOL) recycled content of their sheet (on average, currently $\approx$0% and $\approx$14% for U.S. aluminum and steel auto body sheets respectively), reducing their material costs and energy burden [4]. However, the current system cannot effectively recycle the aluminum [4–6] and steel sheets [7, 8]. Shredded and contaminated EOL metal (e.g., mixed aluminum alloys with steel rivets

---

**Corresponding author*: drcooper@umich.edu

Nabil Nasr (ed.) Technology Innovation for the Circular Economy: Recycling, Remanufacturing, Design, Systems Analysis and Logistics, (199–210) © 2024 Scrivener Publishing LLC

and mixed steel alloys with embedded copper wiring) is often exported, downcycled to castings, or recycled as rebar [4, 7–9]. Aluminum is classed as a critical material because the U.S. is 100% import reliant on bauxite [10]; yet, 40-70% of U.S. aluminum auto shred is exported due to its residual content [11]. Furthermore, a shift to electric vehicles (EVs) will see greater high-quality sheet demand (e.g., aluminum ABS and Advanced High Strength Steels (AHSS) for light weighting), a doubling of vehicle copper wiring that acts as an EOL contaminant [8, 12], and the potential loss of traditional residual sinks (vehicle castings). Combined with trade tensions [13], then without action we could be facing lower future (down-)recycling rates [7–9].

This conference article summarizes the early work in a DOE REMADE Institute project, where the goal of the study is to increase automotive aluminum and steel sheet metal EOL (post-consumer) recycled contents; thus, reducing vehicle embodied energy and primary feedstock consumption. The objectives are to produce a new analytical *design for recycling* tool tailored for automotive metal sheet, and to generate new knowledge on how EOL sheet recycling is affected by vehicle design (e.g., alloy specification), recycling system infrastructure (e.g., deployment of emerging separation processes), and sheet manufacturing process decisions (e.g., temperature profiles informed by new Integrated Computational Materials Engineering (ICME) tools). The intention of this conference article is to present the approach and the preliminary methods being used and results being generated. The work presented is the result of an ongoing collaboration between the University of Michigan, The Aluminum Association (AA), Argonne National Laboratory, Ford Motor Company, Institute of Scrap Recycling Industries (ISRI), Light Metal Consultants, and Novelis.

The technical approach is a design tool that couples the effects of vehicle design, recycling system practices and technologies, and alloy compositional tolerances, under different eco-system scenarios from 2020-2050 (e.g., continued ICEV dominance vs. rapid deployment of EVs). Many elements of the tool will be integrated into the Argonne National Laboratory (ANL) GREET model, which is already widely used across the car industry to quantify environmental impacts [14–17]. *System metrics* calculated by the design tool will include sheet metal EOL recycled contents, closed and open-loop recycling rates, and the reduction in primary energy demand, greenhouse gas (GHG) emissions, and primary feedstock consumption. The design tool is being validated against today's recycling system, and the results of case studies based on industry feedback.

Within the design tool, we are building sub-models for vehicle design, recycling systems, and sheet manufacturing that are then combined using a linear programming model that minimizes the primary energy (or other *system metric* objective function) needed to meet new alloy demand while satisfying compositional and scrap supply constraints. Design and future R&D recommendations will be made using the tool to test the effect of cross-lifecycle design changes on the *system metrics*. Figure 16.1 summarizes the cross life cycle design decisions being considered in the project and the tools being used to evaluate them.

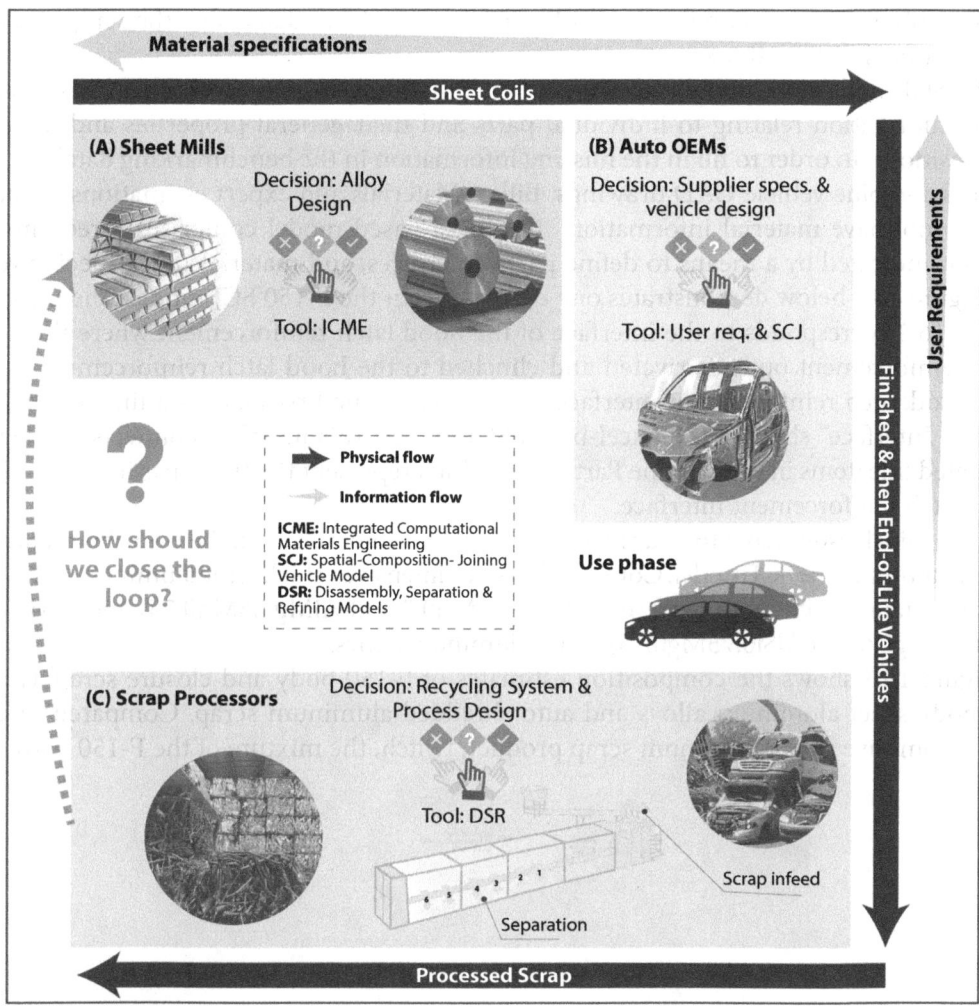

Figure 16.1 The project approach is to evaluate the effect of alloy, vehicle, and recycling system design decisions on the potential for closed-loop recycling of automotive sheet.

16.2   A Quantitative Recycling Pertinent Model of Vehicle Design

A new **Spatial-Composition-Joining (SCJ)** model has been developed that quantitatively defines the recycling pertinent aspects of a vehicle's design. We have initially focused on four common aluminum ABS alloys (high & low Mg 5xxx and high & low Cu 6xxx), and mild steel, bake hardenable, HSLA, and AHSS steel sheets. The Excel-based Spatial Composition Joining (SCJ) model has been developed to represent the current designs (the alloys, the joints, the coatings, and how they are spatially distributed) used in an aluminum-intensive Ford F-150 and a steel-intensive vehicle Ford Escape. The material breakdown quantified by the SCJ models will be used as an input for the material flow analysis to estimate the demand, stock and scrap of aluminum and steel ABS alloys in the automotive industry. The vehicle part alloys, joining and spatial distribution information will be fed into the disassembly, separation & refining (DSR) processes model to determine the scrap size and compositional distributions.

The SCJ models are populated using a benchmarking database which consists of tear-down information relating to individual parts and their general properties and material compositions. In order to fill in the missing information in the benchmarking database, the authors combine vehicle CAD drawings, Bill of Materials and expert estimations for parts that do not have material information. The Excel-based model comprises three important tabs indicated by a means to define parts, interfaces, and materials for a specific vehicle. Figure 16.2 below demonstrates one example from the F-150 SCJ model. The image in Figure 16.2 corresponds to the interface of the hood latch reinforcement where the hood latch reinforcement outer is riveted and clinched to the hood latch reinforcement inner. This hood latch reinforcement interface is recorded in the first table as a line item in the "Define Interface" sheet of the Excel-based SCJ model. This interface connects to the two separated line items in the "Define Part" sheet which represent the two parts involved in the hood latch reinforcement interface.

As an initial result, Table 16.1 summarizes the material breakdown of the F-150 body and closure according to the SCJ model. Compared to steel-intensive vehicles, the aluminum-intensive F-150 contains much more 5xxx (e.g., AlMg3.2Mn0.5 and AlMg4.6Mn0.35) and 6xxx (e.g., AlSi0.75Mg0.6 and AlSi0.75Mg0.75) series aluminum alloys.

Figure 16.3 shows the composition estimates of F-150 body and closure scrap, typical autobody sheet aluminum alloys and auto shredded aluminum scrap. Compared to current automotive shred aluminum scrap product, twitch, the mixture of the F-150 body and

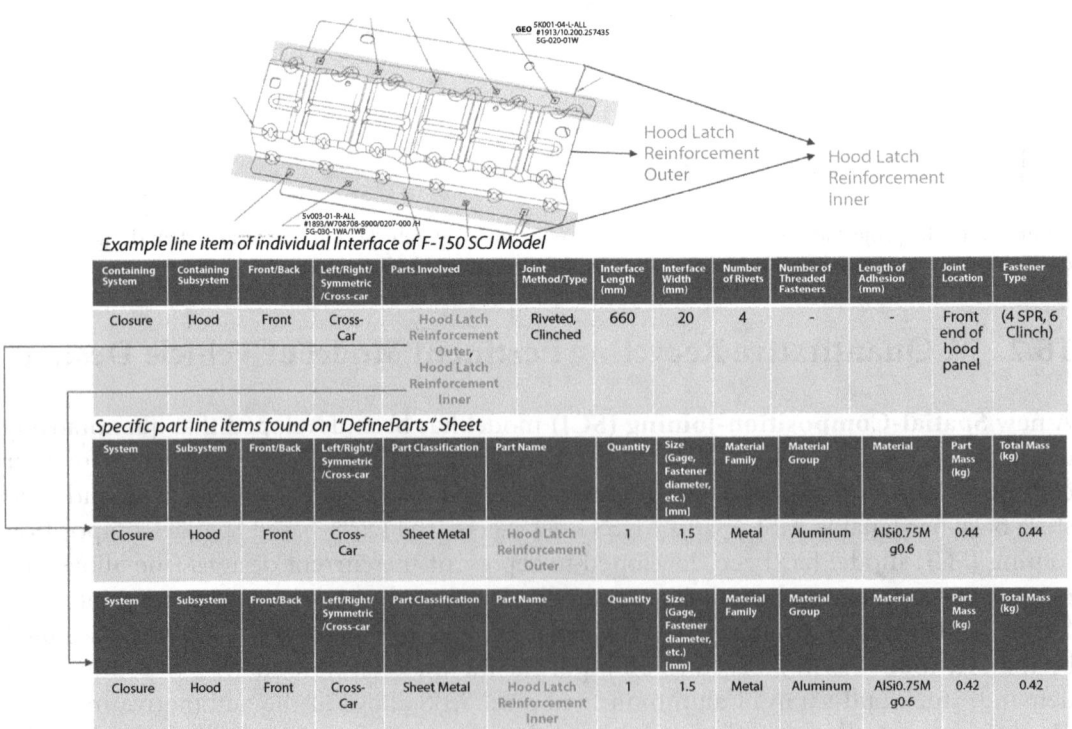

Figure 16.2 Example of how parts and interfaces are defined in the SCJ model.

Table 16.1 Material breakdown of the F-150 body and closures.

List of materials	Category	Mass by materials (kg)	Mass percentage (%)
Motor Material	Mixed materials	3.6	1%
Other Electronic	Mixed materials	1.5	0%
Speaker Material	Mixed materials	0.9	0%
Harness	Mixed materials	31.4	6%
AlMg0.675Si0.4	Aluminum Extrusion	2.4	0%
AlMg1Si0.6	Aluminum Extrusion	7.5	1%
AlMg3.2Mn0.5	Low Mg 5xxx Aluminum Sheet	31.3	6%
AlMg4.6Mn0.35	High Mg 5xxx Aluminum Sheet	20.1	4%
AlSi0.75Mg0.6	Low Cu 6xxx Aluminum Sheet	94.1	18%
AlSi0.75Mg0.75	High Cu 6xxx Aluminum Sheet	92.7	18%
AlSi1Mg0.9	Aluminum Extrusion	13.2	3%
AlSi1Mn0.35	Aluminum Extrusion	5.5	1%
Zinc	Other metals	3.8	1%
FeC0.12Mn1.5	HSLA steel sheet	164.0	32%
FeC0.13Mn0.6	Mild steel sheet/Steel Extrusion	35.4	7%
FeC0.1Mn0.5	Mild steel sheet	2.7	1%
FeC0.25Mn1.4	AHSS steel sheet	4.6	1%
FeC0.37Mn0.76	Fastener Steel	4.1	1%
Total aluminum and steel		521.3	100%

closure scrap will likely have much lower Si content while higher Mg content. Currently, the majority of automotive aluminum scrap is recycled into cast aluminum alloys. However, recycling the mixed F-150 body and closure scrap into cast aluminum alloys is both technically difficult and economically inefficient due to the high Mg content in the scrap and high Mg price. Figure 16.3 also shows that it is difficult to close the loop and recycle the mixed F-150 body and closure scrap into any of the four auto body sheet alloys used in the F-150 because of the high Mg, Cu, and Fe content in the scrap.

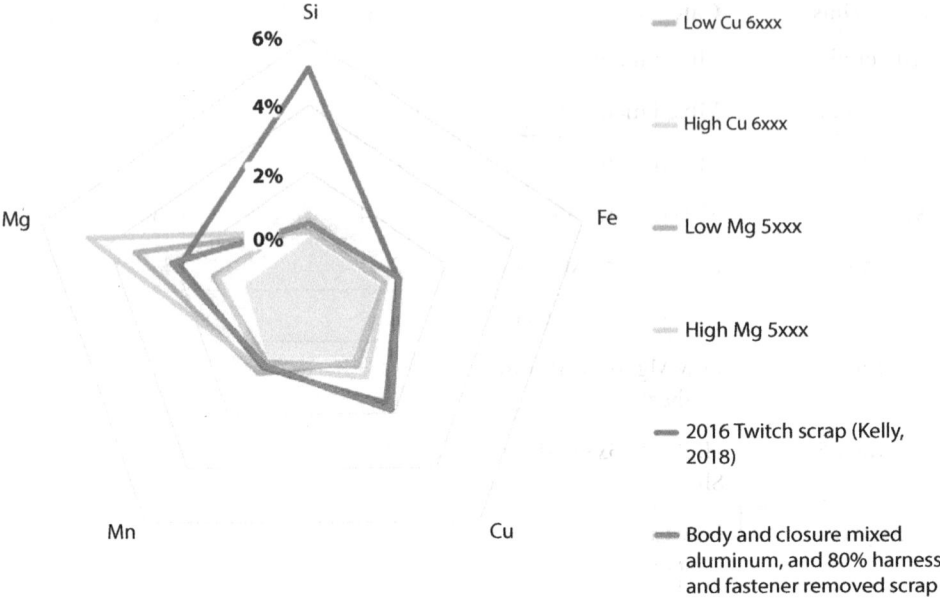

Figure 16.3 Composition estimates of specifically F-150 body and closure scrap (from the new SCJ model), typical autobody sheet aluminum alloys (from Ford reports [18, 19], 2014 & 2020), and typical auto shredded aluminum scrap from [20].

16.3   Modeling Existing and Emerging Recycling Systems and Processes

In order to predict how the EOL metal scrap is handled in the current and future recycling system, we first model the scrap stream available for U.S. recyclers and the alloy demand across the automotive sector and the rest of U.S. industry. This model is being completed using a new dynamic material flow analysis (DMFA) of U.S. steel and aluminum flows. Concurrently, we are developing a disassembly, separation & refining (DSR) processes model to calculate the change in chemistry between inflowing scrap streams and outgoing material streams for different recycling processes and systems. The DSR processes studied as part of this modeling work includes end-of-life vehicle disassembly (e.g., removal of aluminum wheels and catalytic converters), separation (e.g., shredding of mixed EOL vehicles, old appliances and furniture into 10-100 mm pieces and sorting by material compositions), and refining (e.g., removal of tramp/contamination elements through fluxing, electrolysis, fractional crystallization, etc.). To construct the model, we have designed a survey to collect information from U.S. shredders about the size and composition of the steel and aluminum scrap processed at U.S. shredding facilities. This survey is being combined with an extensive review of the academic and gray literature, physics-based extrapolations, and semi-structured interviews with industry experts to validate the DSR model.

As an example of our modeling approach, Figure 16.4 shows the process flow through a typical U.S. shredding facility and Figure 16.5 shows a question from our survey in order to

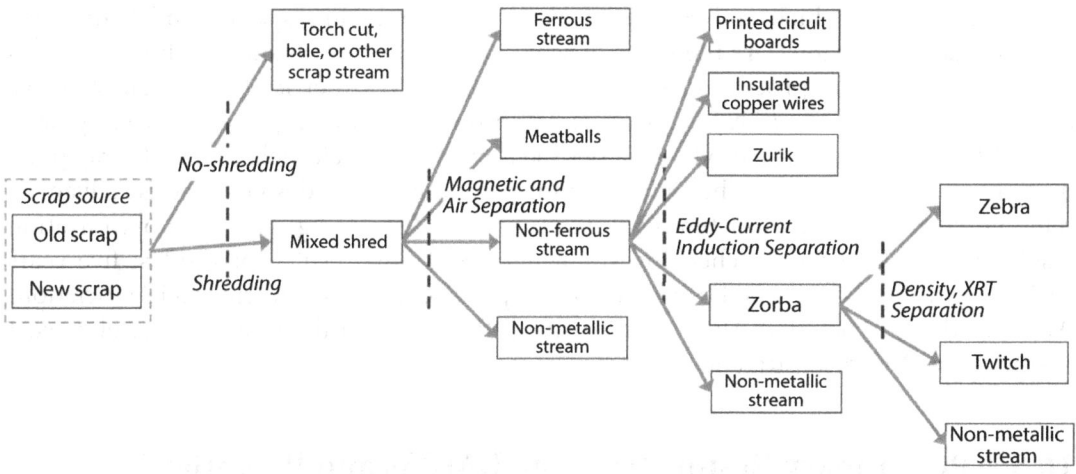

Figure 16.4 Conceptual diagram of typical material flow through a U.S. shredding facility.

3.2 For each scrap stream, what is the breakdown of its compositions by mass? (Please note that the sum of each column should add up to 100%.)

	Ferrous stream	Meatball	Non-ferrous stream	Non-metallic stream
Ferrous content	0 %	0 %	0 %	0 %
Copper content	0 %	0 %	0 %	0 %
Non-ferrous content	0 %	0 %	0 %	0 %
Non-metallic content	0 %	0 %	0 %	0 %
Other content, please give one or more examples below	0 %	0 %	0 %	0 %
Total	%	%	%	%

Figure 16.5 Example of a question from our shredder survey to determine the composition and contaminant levels in the main material streams post shredding, magnetic and air separation (second dashlane in Figure 16.4).

model the separation process of the DSR model. For the example shown in Figure 16.5, the expert elicitation is used to determine the composition and contaminant levels in the main material streams post shredding.

16.4 Optimizing the Cross Lifecycle Supply Chain for Maximized Recycling

A linear programming model will be developed to simulate alloy production by minimizing an objective function (user-chosen *system metric*; e.g., energy) while meeting the volumetric and compositional industry demands for steel and aluminum alloys. The optimization is a function of the vehicle and material design, and the recycling system model. The optimization design variables are the flows of scrap and primary feedstocks to different alloys via different recycling processes. It calculates all the system metrics for a given objective function. For the *system level analysis*, this optimization will be repeated for each year of the 2020-2050 analysis. The model includes the impacts and melt/yield losses of scrap purification processes, which can be significant for recycling system designs that use advanced refining technology; e.g., low-temperature electrolysis (11 MJ/kg) [21]. The optimization results will be validated by checking simulated metal flow results for the last five years against the reality as discerned from statistical reports (e.g., USGS, World Steel Association/ AISI and the International Aluminum Institute/The AA) and the internal data of project partners; e.g., Novelis, ISRI, and The AA.

16.5   Preliminary Results from the DMFAs and Potential Environmental Benefits

Figure 16.6 shows the baseline projected metal sheet demand (demand for sheet embedded in vehicles, not demand for sheet coils) and scrap generation in the U.S. automotive industry for automotive aluminum (Figure 16.6a) and steel (Figure 16.6b) sheet by 2050. These graphs were constructed from preliminary DMFA results constructed by combining vehicle compositions (i.e., the SCJ models shown in Section 16.2) with vehicle sales projections from EIA, 2020 [22], and expected vehicle lifespan distribution from Liao *et al.*, 2023 [23].

Automotive aluminum sheet demand was about one million metric tons (Mt) in 2022 and is expected to reach around 1.2 Mt by 2050 under the baseline scenario. As mass production of aluminum-intensive vehicles is relatively recent (i.e., with the aluminum-bodied Ford F-150 in 2015), then the current quantities of aluminum sheet scrap available are small relative to the sheet demand; however, there is a wave of high-quality aluminum sheet scrap that will become available for recycling in the 2030s [24].

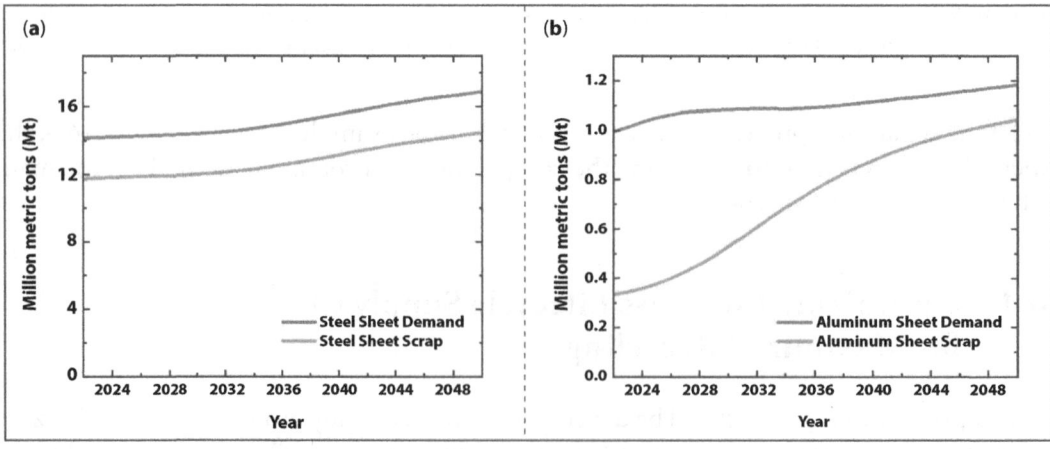

Figure 16.6 The baseline projection of (embedded) sheet demand and end-of-life sheet scrap generation for (a) steel and (b) aluminum for the U.S. automotive industry from 2022 to 2050.

Automotive steel sheet demand was about 14 Mt for 2022 and is expected to reach around 17 Mt by 2050. The automotive sheet scrap generation in 2022 was about 0.3 Mt for aluminum and 12 Mt for steel. It is projected that the scrap generation for aluminum will increase by about 200% by 2050 and reach more than 1 Mt. This is about 80% of the projected aluminum sheet demand in the automotive industry in 2050. The scrap generation for steel is estimated to be in the range of 80% to 85% of steel demand throughout the period.

Figure 16.7 illustrates preliminary results on the GHG emissions attributable to the production of U.S. automotive aluminum and steel sheet products for four scenarios from 2022 to 2050. The baseline represents the current industry situation (including the current grid emissions intensity) continuing to 2050. For example, for aluminum sheets, the End-of-Life Closed Loop Recycling Rate (EOL-RR) of automotive sheet products will remain at zero, and there will be no further electricity decarbonization. For the steel sheet, the EOL-RR will continue to be around 30% without electricity decarbonization. For this baseline, it is projected that the carbon footprint of sheet products will increase from about 12 Mt CO_{2eq} in 2022 to 15 Mt CO_{2eq} in 2050 for aluminum and from 53 Mt CO_{2eq} in 2022 to more than 60 Mt CO_{2eq} in 2050 for steel.

Scenario 1 in Figure 16.7 shows the amount of GHG emissions if the EOL-RR remains unchanged and electricity decarbonization reaches 80% by 2050 (based on the 2022 National Renewable Energy Laboratory (NREL) Standard Scenarios Report [25]). The electricity emission factors are 421 gCO_2/kWh in 2022 and 84 gCO_2/kWh in 2050. It is worth mentioning that we consider electricity decarbonization for aluminum processing in the US only as aluminum is imported from Canada, where renewable energy is already the primary source of aluminum production. It is estimated that the GHG emissions of aluminum sheet products will increase from 12 Mt CO_{2eq} in 2022 to about 14 Mt CO_{2eq} in 2050. In scenario 1, the GHG emissions of steel sheet products decrease from 53 Mt CO_{2eq} in 2022 to 46 Mt CO_{2eq} in 2050.

Scenario 16.2 in Figure 16.7 represents an 80% improvement in EOL-RR by 2050 and no electricity decarbonization. Under this scenario, the GHG emissions of aluminum sheet

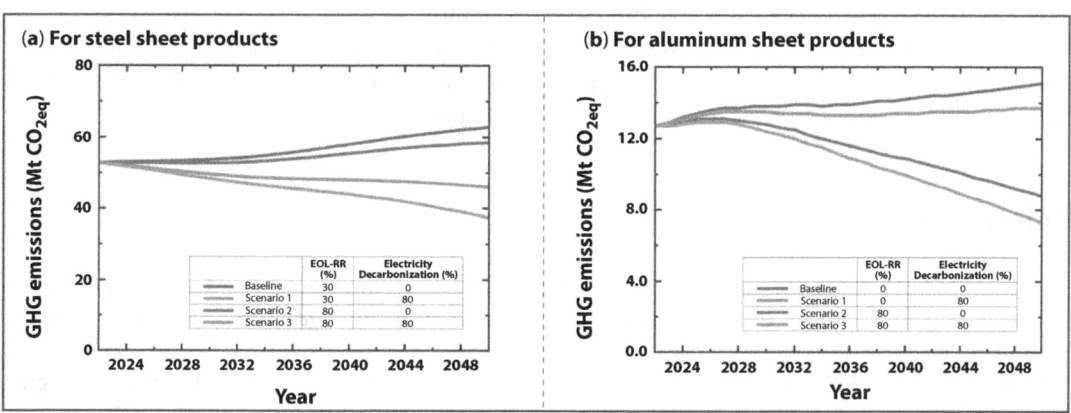

Figure 16.7 Preliminary results for the greenhouse gas (GHG) emissions of the U.S. automotive sheet products for (a) steel and (b) aluminum for baseline and Scenario 1-3 from 2022 to 2050. EOL-RR: Closed-loop end-of-life sheet recycling rate.

products are expected to decrease from about 12 Mt CO_{2eq} in 2022 to 8.8 Mt CO_{2eq} in 2050. For steel, GHG emissions are estimated to increase from 53 Mt CO_{2eq} to 59 Mt CO_{2eq} in 2050.

In Scenario 3, EOL-RR improvement and electricity decarbonization is assumed to increase to 80% by 2050. The aluminum sheet products' GHG emissions decrease from 12 Mt CO_{2eq} in 2022 to 7.3 Mt CO_{2eq} in 2050. The GHG emissions of steel sheet products are expected to drop from 53 Mt CO_{2eq} in 2022 to 37 Mt CO_{2eq} in 2050.

16.6   Conclusions & Recommendations

The development of the *design for recycling* tool, coupled with the recycling-focused ICME tool, will enable development of alloy, vehicle, and recycling systems that maximize the recycled content of the feedstock, while maintaining or improving the process efficiency and product performance of the resulting materials. This creates value for stakeholders throughout the sheet life cycle, reducing material costs and the energy burdens for car makers and sheet manufacturers and shifting recyclers towards producing more profitable wrought alloys. After project completion, there will be the opportunity to further develop the tool to include other vehicle material systems such as plastics, composites, glass, and electronics. Longer term ($\approx$5 years), a cross-lifecycle mathematical design optimization will become feasible with a greater understanding of the design space boundaries.

Acknowledgements

The authors were funded by the U.S. Department of Energy's REMADE Manufacturing USA Institute via the Sustainable Manufacturing Innovation Alliance, the University of Michigan, Ford Motor Company, Light Metal Consultants, Novelis, the Institute of Scrap Recycling Industries (ISRI), and the Aluminum Association (AA). The authors thank our collaborators across the above organizations, including but not limited to Gregory Keoleian, Jarod Kelly, George Luckey, Daniel Freiberg, Constantin Chirac, Robert De Kleine, Chul Kim, Laurent Chappuis, Jinlong Marshall Wang, and Dean Kanelos. The statements in this paper reflect the understanding of the current authors only and any mistakes are solely of the current authors rather than the wider project team.

References

1. Kim, H.J., McMillan, C., Keoleian, G.A., Skerlos, S.J., Greenhouse Gas emissions payback for lightweighted vehicles using aluminum and high-strength steel. *Journal of Industrial Ecology*, 14(6), pp.929-946, 2010.
2. Kim, H.C., Wallington, T.J., Life-cycle energy and greenhouse gas emission benefits of light-weighting in automobiles: review and harmonization. *Environmental Science & Technology*, 47(12), pp.6089-6097, 2013.

3. Sun, X., Meng, F., Liu, J., McKechnie, J., Yang, J., Life cycle energy use and greenhouse gas emission of lightweight vehicle–A body-in-white design. *Journal of Cleaner Production*, 220, pp.1-8, 2019.

4. Zhu, Y., Chappuis, L.B., De Kleine, R., Kim, H.C., Wallington, T.J., Luckey, G., Cooper, D.R. The coming wave of aluminum sheet scrap from vehicle recycling in the United States. *Resources, Conservation and Recycling*, 164: 105208, 2021.

5. Bertram, M., Ramkumar, S., Rechberger, H., Rombach, G., Bayliss, C., Martchek, K.J., Müller, D.B., Liu, G., A regionally-linked, dynamic material flow modelling tool for rolled, extruded and cast aluminium products. *Resources, Conservation and Recycling*, 125, pp.48-69, 2017.

6. Løvik, A.N., Modaresi, R., Müller, D.B., Long-term strategies for increased recycling of automotive aluminum and its alloying elements. *Environmental Science & Technology*, 48(8), pp.4257-4265, 2014.

7. Cooper, D.R., Ryan, N.A., Syndergaard, K., Zhu, Y., The potential for material circularity and independence in the US steel sector. *Journal of Industrial Ecology*, 2020.

8. Daehn, K.E., Cabrera Serrenho, A., Allwood, J.M., How will copper contamination constrain future global steel recycling?. *Environmental Science & Technology*, 51(11), pp.6599-6606, 2017.

9. Zhu, Y., Cooper, D.R., An optimal reverse material supply chain for US aluminum scrap. *Procedia CIRP*, 80, pp.677-682, 2019.

10. Congressional Research Services, Critical Minerals and U.S. Public Policy (CRS Report No. R45810), 2019.

11. Kelly, S., Apelian, D., Automotive aluminum recycling at end of life: a grave-to-gate analysis. *Center for Resource Recovery and Recycling (CR3), Metal Processing Institute,* Worcester Polytechnic Institute, 2016.

12. Copper Facts., Copper the Metal. Retrieved from https://www.copper.org/education/cfacts/facts- print.html, 2020.

13. Trump, D., Proclamation on Adjusting Imports of Derivative Aluminum Articles and Derivative Steel Articles into the United States. White House Press Office, 2020.

14. NHTSA, Draft Environmental Impact Statement for The Safer Affordable Fuel-Efficient (SAFE) Vehicles Rule for Model Year 2021–2026 Passenger Cars and Light Trucks, 2018.

15. Obnamia, J.A., Dias, G.M., MacLean, H.L., Saville, B.A., Comparison of US Midwest corn stover ethanol greenhouse gas emissions from GREET and GHGenius. *Applied Energy*, 235, pp.591- 601, 2019.

16. Kelly, J.C., Sullivan, J.L., Burnham, A., Elgowainy, A., Impacts of vehicle weight reduction via material substitution on life-cycle greenhouse gas emissions. *Environmental Science & Technology*, 49(20), pp.12535-12542, 2015.

17. Woertz, I.C., Benemann, J.R., Du, N., Unnasch, S., Mendola, D., Mitchell, B.G., Lundquist, T.J., Life cycle GHG emissions from microalgal biodiesel–a CA-GREET model. *Environmental Science & Technology*, 48(11), pp.6060-6068, 2014.

18. Ford, Engineering material specification. Aluminum alloy, sheet, heat treatable, enhanced hemming, 2014.

19. Ford, Engineering material specification. Aluminum alloy, sheet, heat treatable, structural, high strength, thick gage, 2020.

20. Kelly, S.M., Recycling of Passenger Vehicles: A framework for upcycling and required enabling technologies. 2018.

21. Kamavaram, V., Mantha, D., Reddy, R.G., Electrorefining of aluminum alloy in ionic liquids at low temperatures. *Journal of Mining and Metallurgy B: Metallurgy*, 39(12), pp.43-58, 2003.

22. U.S. EIA. Annual Energy Outlook. Table 38. *Light-Duty Vehicle Sales by Technology Type*, https://www.eia.gov/outlooks/aeo/tables_side.php, 2021

23. Liao, J., Zhu, Y., Cooper, R. D., Reducing emissions by using products more intensively. *Journal of Industrial Ecology (under review),* 2023.

24. Zhu, Y., Chappuis, L.B., De Kleine, R., Kim, H.C., Wallington, T.J., Luckey, G. and Cooper, D.R., The coming wave of aluminum sheet scrap from vehicle recycling in the United States. *Resources, Conservation and Recycling,* 164, p.105208, 2021.

25. Gagnon, P., Brown, M., Steinberg, D., Brown, P., 2022 Standard Scenarios Report: A U. S. Electricity Sector Outlook. *National Renewable Enerrgy Laboratory (NREL),* 2022.

17

Assessing the Status Quo of U.S. Steel Circularity and Decarbonization Options

Barbara K. Reck[1*], Yongxian Zhu[2], Shahana Althaf[1,3] and Daniel R. Cooper[2]

[1]Yale School of the Environment, New Haven, CT, USA
[2]Department of Mechanical Engineering, University of Michigan, Ann Arbor, MI, USA
[3]Aligned Incentives, LLC, Middleton, MA, USA

Abstract

The iron and steel sector is highly energy- and emissions intensive, accounting for 8% of global final energy use and 7% of global direct energy-related CO_2 emissions. In steel production, most emissions are generated when steel is produced from primary raw materials, but 60% to 80% of energy can be saved when steel is produced from scrap in electric arc furnaces (EAFs). This study provides a comprehensive update of the U.S. steel cycle for year 2017, demonstrating that the U.S. already uses much higher shares of scrap in steel production than the world average, with detailed information on primary and secondary steel production and the manufacture of steel into intermediate products and finished goods. Options to further reduce the embodied energy of steel in the U.S. will be discussed and include increasing secondary production through more efficient collection and separation, reducing the energy intensity of primary production through a shift from blast furnaces to direct reduced iron (DRI) technology, and a shift towards low-carbon electricity sources for DRI and EAF furnaces.

Keywords: Global steel cycle 2017, material flow analysis, scrap, embodied energy, direct reduced iron (DRI), electric arc furnace (EAF), material efficiency, low-carbon electricity

17.1 Introduction

With growing concerns about climate change, decarbonizing emission-intensive industries like steel is necessary to meet ambitious climate goals set by the 2016 Paris agreement and the U.S. climate targets for 2030 of reducing greenhouse gas emissions by 50-52% compared to 2005 levels [1]. To identify the most impactful mitigation strategies requires a good understanding of the existing steel production routes as well as the steel demand by its key end-use sectors. Such information can be provided by regularly updating existing detailed material flow studies since production and trade patterns can change substantially within just a few years. This study provides such an update for the U.S. steel cycle in 2017, covering the entire life cycle from production to manufacturing, end use, and end-of-life management.

Corresponding author: barbara.reck@yale.edu

Nabil Nasr (ed.) Technology Innovation for the Circular Economy: Recycling, Remanufacturing, Design, Systems Analysis and Logistics, (211–222) © 2024 Scrivener Publishing LLC

Having such information allows to assess the process efficiencies across all life stages, illustrating where and how large the potential for improvement is. To allow putting the U.S. steel economy into the global context we also provide an update of the global steel cycle in 2017.

The earliest existing study showed the U.S. steel cycle in year 2000 [2], providing a good overview on the main flows but lacking the details of later studies. The global steel supply chain from steelmaking to end-use in 2008 provides a much more detailed map of the different production options and their raw material sourcing as well as information on the type of intermediate steel products used by the respective end-use sectors [3]. The most detailed study on the U.S. material flows of steel is also the most recent, covering year 2014 [4]. This study uses the same methodology and framework as Zhu *et al.* [4] (i.e., the detailed spreadsheets published in the respective appendices) and also uses mostly similar data sources to update the material cycles to 2017. The present study also offers a brief discussion on options to further reduce the carbon impact caused by the production, manufacturing, and use of steel products.

The iron and steel sector is highly energy- and emissions intensive, responsible for 8% of global final energy use, 7% of global direct energy-related CO_2 emissions [5], and 20% of total global direct industrial CO_2-eq emissions in 2019 [6]. In the United States, the iron and steel sector is the 4th most energy-intensive manufacturing industry, only surpassed by the cement & lime, paper, and bulk chemical feedstock industries [7]. Broadly speaking, opportunities to reduce the carbon impact of steel can be grouped into improvements at the level of steel production, material efficiency measures at all levels of the steel life cycle, and demand reduction.

In steel production, the energy requirements and corresponding GHG emissions differ greatly depending on the steel being produced from primary (virgin) or secondary (scrap) raw materials. Most emissions in steel production are generated when steel is produced from primary raw materials in the blast furnace/basic oxygen furnace (BF/BOF) route: this process requires carbon as a reducing agent, provided in the form of coal-derived coke and resulting in CO_2 emissions, a process that also requires energy to heat the melt. In contrast, secondary steel production uses scrap as its key raw material in an electricity-powered electric arc furnace (EAF) that requires 60% to 80% less energy than the BF/BOF route [5, 8] as it does not entail the coke-requiring chemical reduction step. These substantial energy savings translate into cost savings for steel producers, together with the lower price of using scrap rather than pig iron as raw material input and explain why the use of scrap as input material is already very high, leaving only modest room for improvement. This is particularly true in a mature economy like the U.S., where steel has been produced and used for many decades, providing a reliable stream of old scrap. As a result, with 63% the U.S. has one of the world's highest shares of (secondary) EAF steel production [4].

17.2   Methodology

This study assesses the flows of carbon steel (also called mild steel) in 2017, for simplicity referred to as 'steel' from here on. The construction of the U.S. steel cycle follows the same approach as [4] and is described in detail in their article. The main data sources were USGS [9, 10], World Steel Association [11], and UN Comtrade [12] for trade flows. The construction of the global steel cycle is explained and follows the approach by Cullen *et al.* [3], with the main data sources having been the World Steel Association [11, 13] and BIR [14].

17.3   Results and Discussion

17.3.1   The 2017 U.S. Steel Cycle and Global Context

The U.S. steel cycle in 2017

The U.S. steel supply chain for 2017 is shown as a Sankey diagram in Figure 17.1. It starts with steel production on the left and continues with the production of intermediate and finished products and their final end use on the right. Import and export flows are shown at the top of the diagram (light and dark pink flows, respectively, plus blue exports of finished goods). Yield losses are shown in light grey and assumed to be recycled as new scrap (bottom flow). The raw materials for steel production (shown as grey flows) are iron ore (top left), old and purchased scrap (middle flow, left; purchased includes imported scrap that can be either old or new scrap), and new scrap.

Data reconciliation. A common challenge in material flow analysis is the balancing of in- and output flows of a given process since different data sources for the respective flows can lead to various degrees of inconsistency. Here, we use data reconciliation to address this challenge. We reconcile the U.S. steel flow network by minimizing the sum of square residuals between final assigned values of the generated values by our steel flow network and the USGS [9, 10] data. The optimization was implemented with Matlab's nonlinear 'fmincon' algorithm using the 'interior-point' method and setting as initial values of the variables the balanced U.S. steel cycle in 2014 in Zhu *et al.* [4], where the approach including the functions of the optimization algorithm is further described. The optimization converges after 1100 iterations when mass balance has been reached and the objective value has been reduced by 85%.

In 2017, the U.S. produced 84.6 million metric tons (MMT) of steel (indicated as "liquid steel" in Figure 17.1, medium/light blue flows), of which 32% (26.9 MMT) were produced in Basic Oxygen Furnaces (BOF), 67% (56.4 MMT) in Electric Arc Furnaces (EAF), and the remainder as castings (ingot casting, steel product casting, foundry iron casting). Slabs, billet, and bloom were then transformed into different forms of sheet, plate, bar, and sections for use in the main end use sectors construction, automotive, machinery and defense, energy, and steel products. Overall, the U.S. imported more steel products than it exported, with the main imports being finished goods (59 MMT) and to a lesser extent intermediate products (27 MMT). The largest exports were in the form of finished goods (35 MMT), followed by intermediate products (8 MMT) and iron ore (5 MT). Steel's largest end use sector is construction (35 MMT), with two thirds used for buildings (23 MMT) and one third for infrastructure (12 MMT). The automotive sector is the second largest end use sector for steel (28 MMT), with most steel used in cars (25MMT). 14 MMT steel were used for Machinery, 13 MMT in the Energy sector, and the remainder for Defense and Steel Products.

Table 17.1 provides an overview on the raw materials used in steelmaking. It shows that pig iron is the predominant input material for BOFs (82% pig iron) while it is scrap for EAFs (81% scrap) and castings (80-88% scrap). It should be noted that there is no clear cut between one technology using only one type of input materials and the other technology only using another. BOFs can typically accept up to 20% scrap (18% in 2017), a favored option given the comparatively lower price of scrap compared to pig iron. Similarly, EAFs

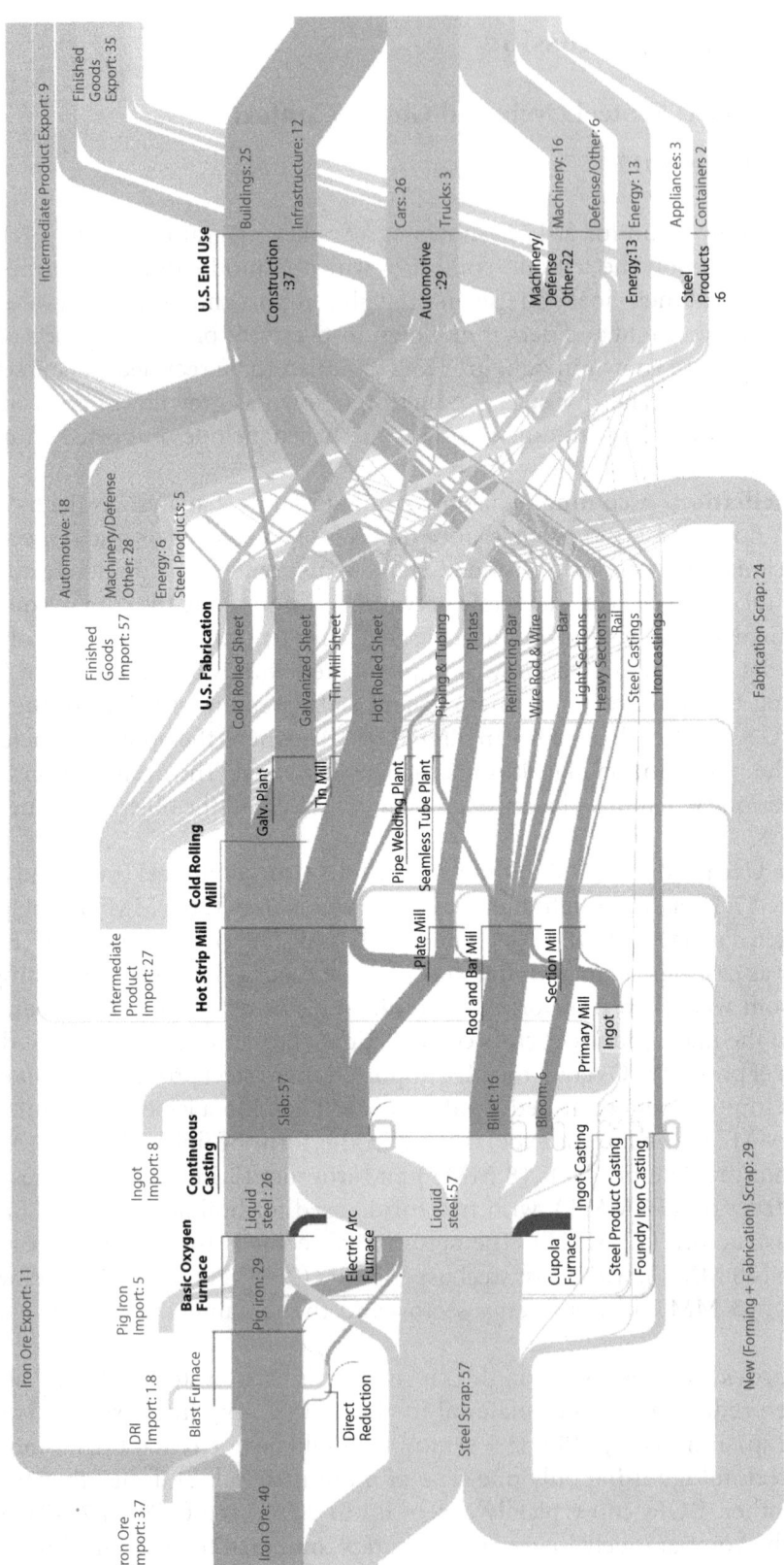

Figure 17.1 The U.S. steel cycle in 2017. All values in million metric tons steel per year.

Table 17.1 Primary and secondary inputs to U.S. steelmaking in 2017. The recycled content of steel produced in the U.S. was 62% with an old scrap ratio of 49%. Totals may not add up due to rounding. All values in gigagram (1 Gg = 1,000 metric tons) steel, DRI: Direct-reduced iron.

	Steel making input					
	Total		Pig iron & DRI (*primary*)		Scrap (*secondary*)	
Carbon Steel U.S. 2017	**Gg**	**%**	**Gg**	**%**	**Gg**	**%**
Blast Oxygen Furnace	**31,020**	31%	25,560	82%	5,460	18%
Electric Arc Furnace	**63,640**	63%	11,950	19%	51,690	81%
Ingot Casting	**600**	1%	120	19%	480	80%
Steel Product Casting	**630**	1%	70	12%	550	87%
Foundry Iron Casting	**5,770**	6%	1,180	20%	4,590	80%
Total	**101,650**	**100%**	**38,880**	**38%**	**62,770**	**62%**

usually add at least small amounts of pig iron to their feed (19% in 2017) in order to meet the required specifications for steel products by diluting unwanted constituents ("tramp elements", most importantly copper [15, 16]).

In 2017, the total amount of scrap used in steelmaking, 62.7 MMT, corresponds to a *Recycled Content* of 62% in U.S. steelmaking. Of this, 28.4MMT was old scrap leading to an *Old Scrap Ratio* of 48% (following the UNEP [17] methodology to determine metal recycling indicators).

Historically, the U.S. shifted in 2002 from predominantly producing steel via the BOF route to producing most steel via the EAF route in so-called minimills. This is illustrated in Figure 17.2, showing the share of the two production routes for the period 1992-2022 [18]. The share of BOF production dropped from 62% in 1992 to just 29% in 2022, a trend that

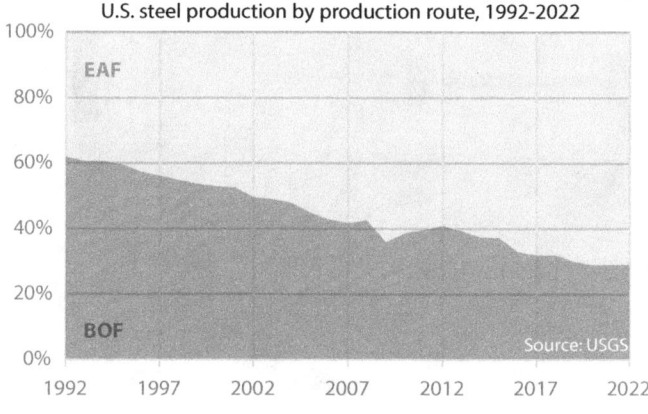

Figure 17.2 The share of the Basic Oxygen Furnace (BOF) produced steel in the U.S. has declined steadily over the past 30 years, from 62% in 1992 to 29% in 2022, while steel produced in an Electric Arc Furnace (EAF) grew from 38% to 71% [18].

seems to continue, enabled by the fact that the U.S. has used steel widely for decades and therefore has access to sufficient quantities end-of-life scrap. The high share of 64% EAF production is among the highest in the world and, in 2016, was only surpassed by Italy, Mexico, Spain, and Turkey [19].

The future potential for the energy-efficient EAF production pathway requires information about the future availability of scrap. Since steel products typically have lifetimes of at least two decades sound scrap estimates require information on the historic steel use in different end use sectors (e.g., the average lifetime of a car may be around 15 years while a building's lifetime is several decades). Such estimates have been made in the past, with U.S. steel stock estimations available for the period 1900-2008 [20] and estimates on the future scrap flows of North America for the period 2009-2100 [21]. This is complimented by estimates of the old scrap generation between 2009 and 2017 in this study, as basis for calculating the generated end-of-life flows and the end-of-life recycling rate in 2017.

The global steel cycle in 2017

The aggregated global carbon steel cycle for 2017 is shown in Figure 17.3, with the main processes being reduction, steelmaking, casting, rolling & forming, fabrication, use, and waste management & recycling. Rather than a linear-looking Sankey diagram we use a circular illustration that follows earlier studies on nickel [22] and stainless steel [23]. It summarizes much more granular information that was collected following the methodology by Cullen *et al.* [3] at the same level of detail. Losses along the life cycle are indicated through

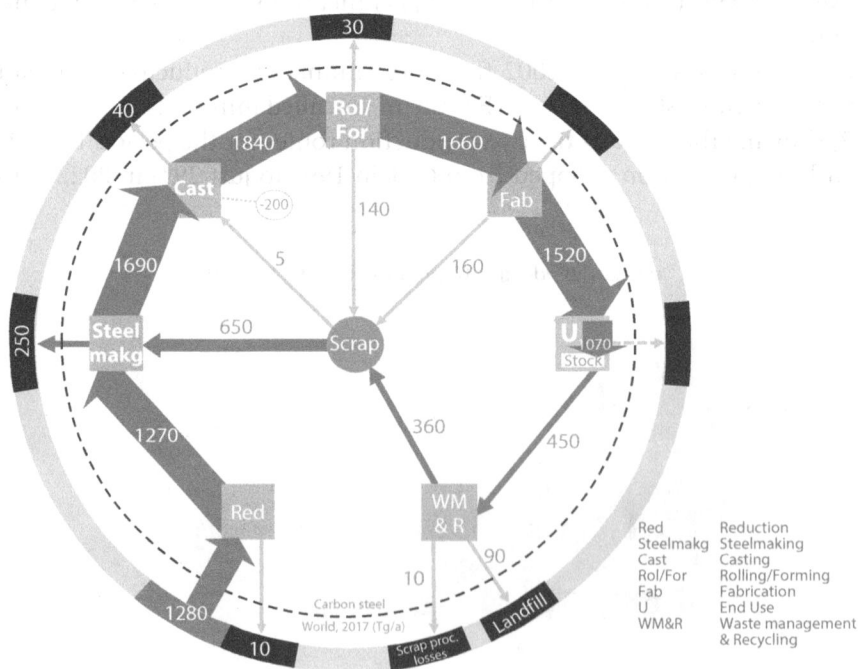

Figure 17.3 The global steel cycle in 2017. All values in million metric tons steel per year. Red: Reduction, Steel makg: Steelmaking, Cast: Casting, Rol/For: Rolling & Forming, Fab: Fabrication, U: End Use, WM&R: Waste Management & Recycling.

outward-facing arrows into black-shaded areas. The scrap market in the center is fed by new scrap generated in rolling & forming and fabrication, and by old scrap from waste management and is used as secondary raw material in steelmaking (both in BOFs and EAFs). We found that in 2017, the global *End-of-Life Recycling Rate* for carbon steel was 79%, the average *Recycled Content* was 34%, and the *Old Scrap Ratio* was 55%.

The U.S. is the world's 4th largest steel producer, a ranking that has not changed between 2017 and 2021 [24–26]. An international benchmarking of energy and CO_2 intensities exists for 15 countries [19], including the top 10 steel producing countries in 2017 (in decreasing order: China, Japan, India, USA, Russia, South Korea, Germany, Turkey, Brazil, Italy [24]), plus Mexico, France, Spain, Canada, and Poland.

17.3.2 Decarbonization Pathways for Steel

Opportunities and challenges for increasing scrap use in steelmaking

As mentioned in the introduction, producing steel from scrap saves around two thirds of energy compared to steel production from primary raw materials [5, 8]. Maximizing the use of scrap in steel production is therefore one of the most powerful ways to reduce the embodied energy of steel but its potential is limited because of the limited amount of scrap, reflected in the fact that less than 30% of the world's steel is currently made from scrap. The combined effect of the long lifetime of steel products (on average more than 20 years) and the continued growth in steel demand leads to a gap between total demand and available scrap supply. This gap is likely to widen further, preventing a shift from virgin to scrap input materials becoming a major decarbonization option in the future.

Quality concerns of recyclables are another aspect to consider. The more scrap is used in metal production, the more quality control efforts will be needed to avoid purity impacts. Copper is typically the most concerning impurity [16, 27, 28] and is inadvertently added to steel scrap whenever the separation of mixed recycling streams is insufficient (e.g., through the mixing-in of copper cables or small electric motors from machinery). To meet specifications and remain below the maximum copper content in steel requires the dilution of scrap with primary pig iron, which contrasts with the ideals of a circular economy that mostly depends only on secondary raw materials. Measures to minimize the contamination of steel scrap are thus needed and include design for recycling, better sorting, and deployment of scrap refining technologies [27, 28]. When the impurities concern is carbon (for the production of low-carbon steels) electrorefining offering a potentially low-energy, scalable, and integrable solution in secondary steelmaking [29].

17.3.3 Increasing the Efficiency in Steel Production and Shifting from BOF to DRI Technology

Modern steel plants already operate near the limits of practical thermodynamic efficiency [30]. Further reductions in energy requirements are possible by transitioning all steel production to best available technologies (BAT) (<26%) and employing better boilers (<10%), while the power demand can be decreased by 25% through the use of heat exchangers in the refining process [31–33]. To substantially reduce the overall CO_2 emissions from steel production a radical shift to low-carbon technologies is needed, with two approaches appearing to be the most impactful, namely, moving away from using carbon (coke) as a reductant towards hydrogen or direct electrolysis, and carbon capture coupled with the continued use of carbon-based technologies [34]. Phasing out blast furnaces (75% market share in 2019, World Steel Association 2021) and replacing them with 'direct reduced iron' (DRI) technology (currently 5% global market share) followed by an electric arc furnace (DRI-EAF) offers the largest potential to cut CO_2 emissions. It leads to a 61% reduction in carbon emissions if the electricity used for the DRI process stems from methane-derived gas and renewable electricity, and even 97% if it stems from green hydrogen (HDRI) (Fennell *et al.* 2022). Access to green hydrogen (produced from renewable energy) will be crucial and the limiting factor to decarbonizing steel production in the near future. Efforts to increase its supply should be a top priority, requiring the ongoing rapid hydrogen expansion to be even more ambitious than currently planned. Furthermore, access to cheap (green) hydrogen should drive the location of new (greenfield) steel plants [35, 36].

For existing primary integrated steel mills a cost-effective strategy for the transition to HDRI-EAF would be linking the replacement of the blast furnace to its next scheduled relining, a time-intensive and costly maintenance that is due every 15-20 years and requires the steel mill to be halted for a few months. Such timing would drastically reduce the net capital cost of low-carbon technology, which could be further reduced if green premium markets existed. It would also allow an incremental transition towards HDRI by either using natural gas as interim electricity source for the DRI, or EAFs supplied with a mix of scrap and shipped-in direct reduced or hot-briquetted iron from regions with better access to low-carbon energy [37].

Another decarbonization option practiced in some steel plants in Brazil seems not scalable in the U.S., namely replacing the coal-derived coke in BF/BOFs with renewable alternatives such as biocharcoal (Biochar) as fuel and reducing agent. Access to sufficient arable land is the main challenge, with most world regions prioritizing growing food over growing fuels (eucalyptus plantations) [34].

17.3.4   Material Efficiency and Demand Reduction

A number of technical options exist for implementing material efficiency (making the same with less): light-weight design, reducing yield losses, diverting manufacturing scrap, re-using components, longer-life products, and more intense use [38, 39]. At the policy level it is important to avoid overspecification in product and building standards, both in terms of the type of material required and the amount of material used. The IEA [40] estimates that by 2060 24% of total steel demand can be reduced through material efficiency measures across all life stages. For example, in the building sector this could be achieved through improved buildings design, substituting some steel in construction with timber, extending building lifetimes, increasing the intensity of use, and reusing steel buildings components at the end of a building's life. However, without policy interventions there is often insufficient motivation for businesses to implement material efficiency measures since the cost of labor far exceeds the cost of materials [38].

For the U.S., Ryan *et al.* [41] discussed in detail the options how CO_2 emissions from U.S. steel consumption can be cut by 70% by 2050. They find that most pathways would not achieve these ambitious emissions target and that the biggest CO_2 emissions reduction potential lies in reducing the demand for steel and thus reducing the per capita steel stocks. We argue that reducing the U.S. steel demand is possible, with the construction sector to be most promising while decarbonizing steel's use in the transportation sector seems more challenging [42].

The construction sector is currently the largest end use sector for steel (Figure 17.1), a situation that is expected to remain unchanged by 2050 [27]. At the global level, our results for 2017 indicate that more than half (56%) of the world's steel is used in buildings and infrastructure. Reducing the steel demand in construction is possible by a shift in urban building materials for mid-rise buildings away from steel and concrete towards mass timber [43–45].

17.4   Conclusions & Recommendations

We present updates on the U.S. and global material flows of steel in 2017 that provide detailed information on production routes and end use sectors. Given that the steel sector is considered one of the most energy- and emission-intensive sectors we discuss option to reduce steel-related emissions that include both technological solutions and opportunities to reduce the overall demand for steel. In steel production, the largest decarbonization potential lies in replacing blast furnaces with DRI-EAF technology that is increasingly powered from renewable energy sources. Measures to control the quality of scrap will ensure that scrap becomes a competitive and low-carbon raw material for high-end applications. Key material efficiency measures need to be implemented quickly to ensure cleaner material production, more efficient recycling, and reduced demand through design and substitution.

Acknowledgements

The authors are grateful for funding received from the REMADE Institute, which is supported by the U.S. Department of Energy's Office of Energy Efficiency and Renewable Energy (EERE) under the Advanced Manufacturing Office Award Number DE-EE0007897. The views and opinions of authors expressed herein do not necessarily state or reflect those of the United States Government or any agency thereof. We also thank the United Nations Environment Program for funding as part of a building materials and climate research initiative, funding number DTIE22-EN4356.

References

1. U.S. White House. FACT SHEET: President Biden Sets 2030 Greenhouse Gas Pollution Reduction Target Aimed at Creating Good-Paying Union Jobs and Securing U.S. Leadership on Clean Energy Technologies Washington, DC: The White House; 2021.
2. Wang T, Müller DB, Graedel TE. Forging the anthropogenic iron cycle. *Environ Sci Technol* 2007, 41(14): 5120-5129.
3. Cullen JM, Allwood JM, Bambach MD. Mapping the Global Flow of Steel: From Steelmaking to End-Use Goods. *Environ Sci Technol* 2012, 46(24): 13048-13055.
4. Zhu Y, Syndergaard K, Cooper DR. Mapping the Annual Flow of Steel in the United States. *Environ Sci Technol* 2019, 53(19): 11260-11268.
5. International Energy Agency. Iron and steel technology roadmap - Towards more sustainable steelmaking. Paris: International Energy Agency; 2020 October.

6. Bashmakov IA, Nilsson LJ, Acquaye A, Bataille C, Cullen JM, de la Rue du Can S, *et al.* Industry. In: Shukla PR, J. S, Slade R, Al Khourdajie A, van Diemen R, McCollum D, *et al.*, editors. *IPCC, 2022: Climate Change 2022: Mitigation of Climate Change. Contribution of Working Group III to the Sixth Assessment Report of the Intergovernmental Panel on Climate Change.* Cambridge, UK and New York, NY, USA: Cambridge University Press; 2022.

7. U.S. Energy Information Administration. Annual Energy Outlook 2022. Washington, DC; 2022.

8. UNEP. Environmental Risks and Challenges of Anthropogenic Metals Flows and Cycles, A Report of the Working Group on the Global Metal Flows to the International Resource Panel. van der Voet, E.; Salminen, R.; Eckelman, M.; Mudd, G.; Norgate, T.; Hischier, R. Paris: UNEP DTIE, Sustainable Consumption and Production Branch; 2013.

9. U.S. Geological Survey. 2017 Minerals Yearbook: Iron Ore. In: U.S. Department of the Interior USGS, editor. *Minerals Yearbook, Volume I. Metals and Minerals.* Reston, VA: National Minerals Information Center; 2020. p. 16.

10. U.S. Geological Survey. 2017 Minerals Yearbook: Iron and Steel Scrap. In: U.S. Department of the Interior USGS, editor. *Minerals Yearbook, Volume I. Metals and Minerals.* Reston, VA: National Minerals Information Center; 2020. p. 22.

11. World Steel Association. Steel Statistical Yearbook 2018. Brussels; 2019.

12. United Nations. UN Comtrade Database. 2021 [cited] Available from: https://comtradeplus.un.org/

13. World Steel Association. Steel Statistical Yearbook 2019. Concise version. Brussels; 2020.

14. BIR. Global Facts & Figures. Brussels: Bureau of International Recycling; 2021.

15. Haupt M, Vadenbo C, Zeltner C, Hellweg S. Influence of Input-Scrap Quality on the Environmental Impact of Secondary Steel Production. *Journal of Industrial Ecology* 2017, 21(2): 391-401.

16. Daehn KE, Serrenho AC, Allwood J. Finding the Most Efficient Way to Remove Residual Copper from Steel Scrap. *Metallurgical and Materials Transactions B* 2019, 50(3): 1225-1240.

17. UNEP. Recycling Rates of Metals - A Status Report, A Report of the Working Group on the Global Metal Flows to UNEP's International Resource Panel. Graedel, T. E., Allwood, J., Birat, J.-P., Reck, B. K., Sibley, S. F., Sonnemann, G., Buchert, M., Hagelüken, C. Paris: UNEP DTIE, Sustainable Consumption and Production Branch; 2011.

18. U.S. Geological Survey. Mineral Commodity Summaries 2023. In: U.S. Department of the Interior USGS, editor. *Mineral Commodity Summaries.* Reston, VA; 2023. p. 210.

19. Hasanbeigi A, Springer C. How clean is the U.S. steel industry? An International Benchmarking of Energy and CO2 Intensities. San Francisco, CA: Global Efficiency Intelligence; 2019.

20. Pauliuk S, Wang T, Müller DB. Steel all over the world: Estimating in-use stocks of iron for 200 countries. *Resour Conserv Recycl* 2013, 71: 22-30.

21. Pauliuk S, Milford RL, Müller DB, Allwood JM. The Steel Scrap Age. *Environ Sci Technol* 2013, 47(7): 3448-3454.

22. Reck BK, Müller DB, Rostkowski K, Graedel TE. Anthropogenic nickel cycle: Insights into use, trade, and recycling. *Environ Sci Technol* 2008, 42(9): 3394-3400.

23. Reck BK, Chambon M, Hashimoto S, Graedel TE. Global Stainless Steel Cycle Exemplifies China's Rise to Metal Dominance. *Environ Sci Technol* 2010, 44(10): 3940-3946.

24. World Steel Association. World Steel in Figures 2019. Brussels; 2019.

25. World Steel Association. 2022 World Steel in Figures. Brussels; 2022.

26. World Steel Association. 2021 World Steel in Figures. Brussels; 2021.

27. Cooper DR, Ryan NA, Syndergaard K, Zhu Y. The potential for material circularity and independence in the U.S. steel sector. *Journal of Industrial Ecology* 2020, 24(4): 748-762.

28. Watari T, Hata S, Nakajima K, Nansai K. Limited quantity and quality of steel supply in a zero-emission future. *Nature Sustainability* 2023.

29. Judge WD, Paeng J, Azimi G. Electrorefining for direct decarburization of molten iron. *Nature Materials* 2022, 21(10): 1130-1136.

30. Gutowski TG, Sahni S, Allwood JM, Ashby MF, Worrell E. The energy required to produce materials: constraints on energy-intensity improvements, parameters of demand. *Philosophical Transactions of the Royal Society A: Mathematical, Physical and Engineering Sciences* 2013, 371(1986).

31. Fennell P, Driver J, Bataille C, Davis SJ. Going net zero for cement and steel. *Nature* 2022, 603(7902): 574-577.

32. Napp TA, Gambhir A, Hills TP, Florin N, Fennell PS. A review of the technologies, economics and policy instruments for decarbonising energy-intensive manufacturing industries. *Renewable and Sustainable Energy Reviews* 2014, 30: 616-640.

33. Gonzalez Hernandez A, Paoli L, Cullen JM. How resource-efficient is the global steel industry? *Resources, Conservation and Recycling* 2018, 133: 132-145.

34. Rissman J, Bataille C, Masanet E, Aden N, Morrow WR, Zhou N, *et al*. Technologies and policies to decarbonize global industry: Review and assessment of mitigation drivers through 2070. *Applied Energy* 2020, 266: 114848.

35. Castelvecchi D. How the hydrogen revolution can help save the planet - and how it can't. *Nature* 2022, 611(7936): 440-443.

36. Odenweller A, Ueckerdt F, Nemet GF, Jensterle M, Luderer G. Probabilistic feasibility space of scaling up green hydrogen supply. *Nature Energy* 2022, 7(9): 854-865.

37. Vogl V, Olsson O, Nykvist B. Phasing out the blast furnace to meet global climate targets. *Joule* 2021, 5(10): 2646-2662.

38. Allwood JM. Transitions to material efficiency in the UK steel economy. *Philosophical Transactions of the Royal Society A: Mathematical, Physical and Engineering Sciences* 2013, 371(1986): 20110577.

39. Raabe D, Tasan CC, Olivetti EA. Strategies for improving the sustainability of structural metals. *Nature* 2019, 575(7781): 64-74.

40. International Energy Agency. Material efficiency in clean energy transitions. Paris: International Energy Agency; 2019 March.

41. Ryan NA, Miller SA, Skerlos SJ, Cooper DR. Reducing CO2 Emissions from U.S. Steel Consumption by 70% by 2050. *Environ Sci Technol* 2020, 54(22): 14598-14608.

42. Zhu Y, Skerlos S, Xu M, Cooper DR. Reducing Greenhouse Gas Emissions from U.S. Light-Duty Transport in Line with the 2 °C Target. *Environ Sci Technol* 2021, 55(13): 9326-9338.

43. Churkina G, Organschi A, Reyer CPO, Ruff A, Vinke K, Liu Z, *et al*. Buildings as a global carbon sink. *Nature Sustainability* 2020, 3(4): 269-276.

44. Allan K, Phillips AR. Comparative Cradle-to-Grave Life Cycle Assessment of Low and Mid-Rise Mass Timber Buildings with Equivalent Structural Steel Alternatives. *Sustainability* 2021, 13(6): 3401.

45. D'Amico B, Pomponi F, Hart J. Global potential for material substitution in building construction: The case of cross laminated timber. *Journal of Cleaner Production* 2021, 279: 123487.

18

Fiber and Fabric-Integrated Tracing Technologies for Textile Sorting and Recycling: A Review

Brian Iezzi[1]*, Max Shtein[1], Tairan Wang[2] and Mordechai Rothschild[2]

[1]University of Michigan, Ann Arbor, MI, USA
[2]MIT Lincoln Laboratory, Lexington, MA, USA

Abstract

Over 85% of textiles currently end up in landfills, despite a recent study indicating 74% of low-value, post-consumer textiles, nearly 500,000 tons per year in Europe alone, are readily available for fiber-to-fiber recycling. A key challenge in implementing fiber-to-fiber recycling is feedstock ambiguity; even advanced near infrared sorting systems face significant challenges in differentiating blended fiber fabrics at scale. Furthermore, an increasing emphasis on ethical fiber sourcing and the assurance that fabrics and garments are made with fair labor practices requires enhanced methods of tracing and validation. In the textile and apparel industry, product life cycle management is hampered in part by inaccurate, poorly readable, and detachable standard care labels. Integration of easily readable, cost-effective, and fully recyclable tracing technologies directly into the fiber or fabric could address multiple concurrent challenges across the entire textile supply chain currently inhibiting a transition to a functioning circular economy. In this manuscript, a critical systems-level analysis of the tracing and sorting challenges facing all textile life cycle stages (fiber/yarn/fabric/garment manufacturing, brand/retailer, consumer, sorting, and recycling) provides the foundation for comparing the techno-economic feasibility of emerging technologies that have been proposed for direct fiber, fabric, and garment tracing and sorting. This includes the current standard care label, quick response (QR) codes, and radio frequency identification (RFID) tags as well as emerging direct fiber marking techniques such as those using DNA, organic molecules, or rare earth fluorescent nanoparticles. These emerging fiber marking techniques are also compared to a recently developed fiber-based "barcode" that uses all-polymer photonic structures that can be manufactured at scale and with low-cost while remaining compatible with textile manufacturing processes and being made of recyclable materials. Finally, recommendations are provided for focusing the future technological development of integrated tracing methods as well as promoting cooperation across the textile industry and regulatory bodies.

Keywords: Textiles, fabric, sorting, recycling, tracing, fibers

*Corresponding author: bciezzi@umich.edu

Nabil Nasr (ed.) Technology Innovation for the Circular Economy: Recycling, Remanufacturing, Design, Systems Analysis and Logistics, (223–238) © 2024 Scrivener Publishing LLC

18.1 Introduction

In the 2017 report by the Ellen MacArthur Foundation "A New Textiles Economy: Redesigning Fashion's Future" it was reported that less than 1% of material used to produce clothing is recycled into new clothing, representing a loss of more than $100 billion worth of materials each year [1]. The report also detailed how a proposed circular textile economy would rely on "improved sorting technologies that support the increased quality of recycling by providing well-defined feedstocks, in particular in the transition phase until common tracking and tracing technologies exist". This paper seeks to identify the current and proposed tracking and tracing technologies in use in the textile industry and, if applicable, how these same technologies could also be used to increase the sorting efficiency of end-of-life textiles and, subsequently, increase the techno-economic feasibility of fiber-to-fiber textile recycling. Furthermore, it will make recommendations as to what steps could be taken by regulatory bodies to begin implementing tracking and tracing methodologies on a global scale. The paper is organized as follows. In Section 18.2, the challenges in textile tracing, identification, and sorting facing stakeholders at each textile life cycle stage are first elucidated. In Section 18.3, current and proposed technologies for enhancing textile traceability, sorting, and subsequent recycling are presented followed by a techno-economic and stakeholder comparison of those technologies in Section 18.4. Finally, conclusions and recommendations for further research and development are provided in Section 18.5.

18.2 Stakeholder Challenges in Textile Tracing and Sorting

The journey a garment makes fiber to customer to end of life is quite complex and depends on several variables, making accurate traceability a challenge. Tracing a shirt or pair of jeans sold by a retailer, known as a Tier 0 supplier, at a store back to the garment manufacturers (Tier 1) is relatively straightforward as most labels are attached at the garment stage. However, attempts to track further up the supply chain to fabric weavers or knitters (Tier 2), yarn spinners (Tier 3), or even Tier 4 (original fiber producers) becomes progressively more difficult (see Figure 18.1). Furthermore, after garments are sold by a brand or retailer tracking them at the consumer life stage and sorting at the end of life are also difficult as many of labeling systems currently in use are either removable or not easily readable at the

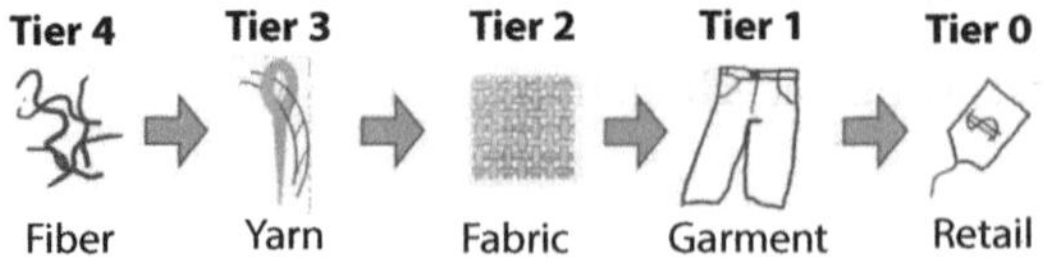

Figure 18.1 Hierarchy of textile manufacturing. Tier 4 producers (natural fiber farmer's and synthetic manufacturing facilities) supply raw material for yarn spinners (Tier 3). Tier 3 producers supply yarns for producing woven and knitted fabrics (Tier 2) which are then sent to cut and sew manufacturing plants for final garment production (Tier 1) before subsequently being sold by brands and retailers (Tier 0).

end of life. The following sections will elucidate the challenges faced by each stakeholder in the textile supply chain and why they would be either incentivized or disincentivized by more robust tracing technologies being implemented.

18.2.1   Fiber and Yarn

In July 2021, the United States Congress passed the Uyghur Forced Labor Prevention Act, aimed at banning the import to the United States of cotton grown in Xinjiang, China produced using forced labor. While the act only took effect in June 2022, last fall a study by a forensic traceability firm, Oritain, indicated that 16% of garments on shelves in the United States still had cotton grown in Xinjiang [2]. The passing of this legislation, and general trends for brands and consumers requiring more detailed knowledge of fiber origin, has revealed the lack of knowledge industry wide around supply chain clarity. This is in part due to how yarns are typically manufactured. At a cotton yarn spinning facility, for example, yarns are often manufactured from a "laydown" where anywhere from 40 to 90 bales of cotton fiber are mixed to provide the appropriate staple fiber length distribution for the performance requirements of the fabric [3]. Thus, one yarn could potentially have fibers sourced from not only different farms, but even different regions of the world. For synthetic yarns, such as polyester, the challenge in determining origin based solely on the material is that all polyester is largely identical from a chemical perspective. Polyester manufactured in one region using renewable energy as a source would be indistinguishable from polyester manufactured in another region using coal-powered electricity, for example [4]. The difficulty in determining the fiber manufacturing source also applies to recycled polyester. Finally, it is possible to blend multiple fibers, (*e.g.*, polyester and cotton), together in the yarn spinning process creating blended fiber yarns, in this case polycotton. Currently, China is the world's leading producer of both cotton and polyester fibers and yarns with India producing nearly as much cotton. The United States is the third largest producer of cotton and has a significant polyester manufacturing industry as well. Thus, any future regulatory action that needs to be taken at a global scale would need to include these fiber producing economies.

18.2.2   Fabric and Garment

The traceability challenge presented to Tier 2 manufacturers (fabric weavers and knitters) mainly lies in keeping track of the yarn source, especially if multiple yarns from multiple producers are being used to create a certain fabric. This also applies to Tier 1 manufacturers, as garments may also have multiple fabrics in their construction, although this is less likely than having multiple yarn origins. China, in addition to being the world's largest producers of cotton and polyester fiber and yarn, is the largest producer of woven and knitted fabric and finished garments. Rising labor cost in China is powering a shift to production elsewhere in Asia, largely to Vietnam, Bangladesh, and Türkiye. The European Union, mainly Spain, Germany, and Italy also have large garment manufacturing industries. Whereas the top four cotton producers (China, India, USA, and Brazil) account for nearly 75% of the global market the top four producers of finished textile products (China, Vietnam, Bangladesh, and Germany) only account for approximately 50% of the global market [5].

This is due to wider spread fabric and garment manufacturing operations and thus tracing from the fabric to garment stage is more complex.

18.2.3   Brands/Retailers/Consumers

The shift to having an increased amount of traceability in textile supply chains has largely been driven by either regulatory or market forces. Modern consumers increasingly want to know more about the environmental and geopolitical impact of their purchases, including how a product is made, what materials go into its manufacture, where it was made, by whom, and under what conditions, and what happens to the product once it has reached the end of its useful life. Brands that can reliably claim value-added attributes such as organic cotton or fabrics and garments that have been produced under fair labor practices can typically charge a premium that a sub-set of consumers are willing to pay for. Compared to other industries, fashion companies have had to emphasize supply chain management due to societal tragedies such as the Rana Plaza disaster of 2013 [6]. Even with these tragedies, however, many fashion companies do not have their entire supply chain mapped. The 2022 Fashion Transparency Index, published annually by non-profit Fashion Revolution, reviewed 250 of the world's largest brands and found that only 48% of brands published their Tier 1 manufacturer's publicly and less than 15% were able to trace their supply back to Tier 4 suppliers [7].

18.2.4   Sorting/Recycling

A 2014 study by the Waste and Resource Action Program (WRAP) compared several incumbent and proposed clothing tracing/sorting methodologies including manual (via the care label), near infrared spectroscopic sorting (NIR), RFID tags (to be discussed further in the next section), and 2D barcodes [8]. The high labor costs of some methodologies are driven by the need to manually present the label to a reader (manual & barcode) and multiple manual pre-sorting steps (NIR). Neither is necessary if RFID tags are used. While having the ability to be read from a significant distance, RFID tags are more expensive to implement due to higher capital and marker costs. NIR sorting has been invested in heavily in Europe for sorting textiles at end-of-life, and automated systems are currently in operation in Sweden that are capable of sorting 24,000 tons of textiles per year [9]. A significant challenge that has been identified with NIR sorting, however, is in sorting blended fabrics, such as the previously mentioned polycotton. Specifically, recent reports have shown that NIR systems have difficulty in determining the percentage composition of the fabric [10]. For example, while the system can detect cotton and polyester individually, understanding whether 70% is cotton and 30% is polyester, or vice versa, is challenging.

18.3   Textile Markers for Tracing and Sorting

A recent study of post-consumer clothing found that 29% did not have a legible care label; of those that did, 41% had inaccurate composition information [11]. Missing or inaccurate knowledge around fiber composition can make textile recycling cost-prohibitive. The United States recycles fewer than 15% of its discarded textiles; leading to over 17 million tons of textiles being sent to the landfill each year [12, 13]. The Sorting for Circularity Project Europe, covering Belgium, Germany, the Netherlands, Poland, Spain, and the United Kingdom, recently estimated that, with efficient sorting infrastructure, it would be possible to profitably recycle 74% of current end-of-life fabrics, either mechanically or chemically, into fibers for new garments [14]. Thus, a more efficient sorting system of textiles for recycling or upcycling into high value feedstocks is needed. Recently, the Textile Exchange and Fashion for Good collaborated with over a dozen companies commercializing textile tracing technologies to create the Textile Tracer Assessment [15]. The analysis focused on both forensic (utilizing geographic verification of plant and animal fibers) and additive (adding a physical marker) tracers but focused mainly on traceability assessment, without a thorough analysis of the impact of these technologies on sorting and recycling. The analysis presented in Sections 18.3 and 18.4 focuses on how additive tracing techniques could be utilized for sorting and recycling.

18.3.1   Incumbent: Fiber Content and Care Label

The first fully synthetic fiber, nylon (polyamide), was invented in 1935 and was followed soon after by other synthetic fibers, such as polyester and acrylic (polyacrylonitrile) [16]. By the 1950s, the affordability and performance of these new synthetic fibers led to widespread commercial adoption. This adoption, however, also led to unscrupulous business practices as wool-like acrylic is very difficult to differentiate from more expensive wools like alpaca, cashmere, and mohair, leading to many "faux fur" products entering the market [17]. To prevent the rising numbers of counterfeit textile products and prevent false advertising to consumers, the United States Congress passed the Textile Fiber Products Identification Act in 1958 to establish, and enforce, a standardized way to mark textile products with accurate fiber content information [18]. The act required that a label be attached to any garment and state the general class of fiber, such as cotton, silk, polyester, etc, as well as the percentage composition of that fiber. Furthermore, the act stated that the label must include the garment manufacturer's name and the country of origin, if imported. Finally, it stated that the label must be included at the time of manufacture, but it did not need to be permanently attached to the garment. The fiber content label was followed by the Care Labeling Rule, implemented by the Federal Trade Commission in 1971, to assist consumers by requiring the appropriate cleaning and laundering instructions also be included on a label attached to the garment. Today, content and care labels are a ubiquitous part of garments worldwide but often lack critical information for consumers and sorting/recycling facilities.

18.3.2   Quick Response (QR) Codes

Quick response (QR) codes have become increasingly common on a variety of products to increase customer engagement. Easily readable using any smartphone, QR codes can link to an existing webpage or, due to its two- dimensional nature, can store the same amount of information as a traditional barcode in a tenth of the physical space. One of the largest labeling companies in the United States, Avery Dennison, recently partnered with the start-up Certilogo to implement QR codes on clothing labels that helps customers understand where and how a product was made and what to do with the product at the end of the useful life. QR codes are also low-cost and easily adapted to existing care labels. QR codes are typically only added at the garment manufacturing stage and, like the standard care label, are typically removed or not readable at the end of life. Thus, they only provide partial life cycle coverage. Furthermore, since they are relatively easy to create, they can also be duplicated, attached to counterfeit garments, and/or potentially used to lead consumers to fake websites, which can potentially compromise sensitive information [19].

18.3.3   Radio Frequency Identification (RFID) and Near Field Communication (NPB) Tags and Yarns

Radio frequency identification (RFID) tags are active (powered) or passive antennas that store critical product information and can be attached to garments, boxes, bales etc. When interrogated by a reading system using radio waves, the RFID tag can send information about the item to the reader. Passive antennas are typically made out of an etched metal pattern, such as copper or aluminum, supported on a flexible substrate made out of plastic. RFID tags are currently in use in the textile industry for supply chain tracking of materials. Their widespread adoption has been hindered by relative high cost and need for removal, due to privacy concerns of customers, and/or relative bulkiness. Researchers in industrial and academic labs have been actively developing yarns that have RFID tags embedded within them using a combination of metallic wires/components and traditional textile fibers [20]. While being more easily integrated with existing textile manufacturing systems, these yarns pose a challenge to recyclers as metallic wires could potentially damage recycling equipment designed to process organic fibers. Near field communication (NPB) antennas, similar in application to RFID antennas, have also been promoted as a method for textile traceability and verification, especially since they enable two-way communication with standard smartphones. NPB embedded textiles, however, face similar cost and recycling challenges as RFID tags [21].

18.3.4   DNA Tracers and Direct Fiber DNA Testing

Applied DNA Sciences, a US-based company commercializing DNA-related technologies, has developed customized DNA sequences, based on unique base pair combinations, that can be used to track and verify textile products [22]. Additionally, other companies have developed techniques to sample DNA from organic fibers directly, such as cotton and wool, and employ the unique region-specific species or strain information to verify fiber origin [23]. While both methods are capable of tracking from the earliest fiber stage, they have potentially high costs due to complex laboratory testing required for verification, specifically polymerase chain reactions (PCR) for replicating DNA samples. Interestingly, the cost for this replication has come down due to the need for testing for the COVID-19 pandemic [24]. While not using DNA specifically, Oritain, has developed a method to create a "fingerprint" of a fiber based on the combined geologic and biologic data from specific regions [25].

18.3.5   Fluorescent Inorganic/Organic Nanoparticles

There have been a number of companies that have recently introduced a method of embedding fluorescent, rare-earth ceramic (inorganic) nanoparticles into textile fibers, with the intent of integrating such fibers in textiles for life cycle traceability purposes [26, 27]. Other researchers have also developed organic, polymeric nanoparticles that also fluoresce under ultraviolet illumination and have proposed their use for textile verification [28]. These are relatively new technologies, potentially expensive to create at scale but offer ease of integration and measurement at all life cycle stages. One potential benefit of these technologies is that, due to their direct integration into fibers and chemical stability, they could be utilized for verifying that the raw material in recycled garments comes from previous fibers and not, for example, recycled polyester from bottles. One potential drawback to the incorporation of the nanoparticles is that multiple mixing of batches, given potential limitations in the creation of a wide-variety of different chemistries and thus optical signatures, might make verification difficult.

18.3.6   Polymeric Photonic Fiber Barcodes

All-polymer fibers with embedded photonic crystals have also been recently proposed as a method of identifying and tracing textiles [29]. The photonic crystal fibers have a specific, engineered infrared reflection that is used as a photonic "barcode". The process for making the photonic barcode (PB) is derived from a patent pending process where a specific PB signature is obtainable only through certain speeds, temperatures, and material combinations making replication difficult (Figure 18.2). The PB provides an integrated optical identifier that is unique from the chemical absorption of the fiber and fabric material and is directly compatible with existing NIR sorting systems. Furthermore, the percentage of PB in the overall garment and how it is incorporated (weave/knit pattern) provide added levels of differentiation. For measurement outside of an automated line, NIR spectrometers have also been miniaturized and hand-held versions are now commercially available for under $1000 with the cost expected to decrease further [30]. Due to the sensitivity of the photonic response to variables such as layer thickness and refractive index contrast, there are thousands of possible photonic signatures. Furthermore, it may be possible to read data directly

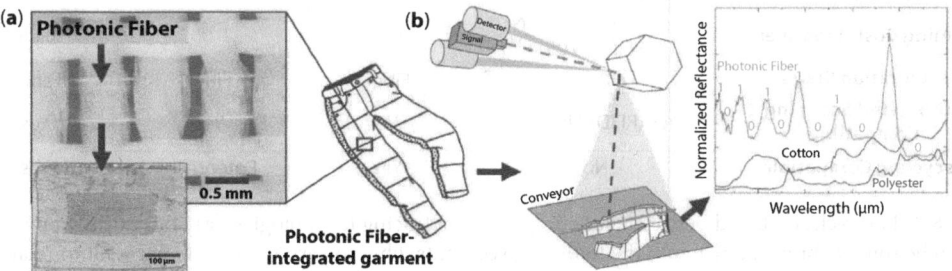

Figure 18.2 Polymeric photonic fiber barcodes for textile identification. (a) Fiber with integrated photonic crystal composed of alternating refractive index layers to create constructive and destructive interference in the near-infrared is integrated into a woven textile structure and garment (microscale fiber cross-section inset left showing photonic crystal). (b) The photonic fiber is measured using near infrared systems on an automated sorting system and the specific infrared signature is used to encode critical material composition data unique from the constituent textile fiber content, such as polyester (blue) and cotton (black). Modified from Iezzi *et al.* [29] with permission from John Wiley & Sons, Inc.

from a PB using, for example, a binary system based on peak readout, *e.g.,* a reflectance peak is "1" and a valley is a "0". This would allow direct encoding of information, like a QR code, which would prevent the need for a lookup database of photonic signatures that needs to be constantly maintained/updated.

18.4   Tracing Technology and Stakeholder Techno-Economic Assessment

In this section, several of the previously described textile tracing technologies are compared based on their life cycle coverage and applicability in addressing outstanding performance requirements for tracing, sorting, and recycling. Furthermore, an initial cost assessment of integrating and scanning the technologies is also provided and these results are summarized in Figure 18.3. The cost estimate rationale for each technology, including the estimated cost of scanning/verification, are provided in Section 18.9. Overall, no one technology was able to allow for complete life cycle coverage, ease of readability *via* automated sorting

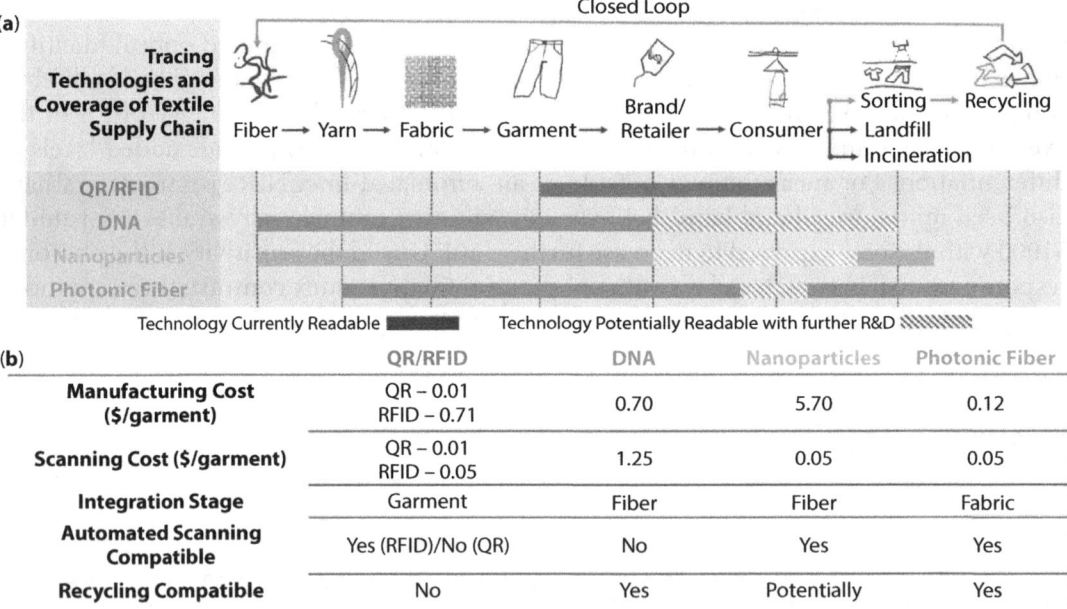

(b)	QR/RFID	DNA	Nanoparticles	Photonic Fiber
Manufacturing Cost ($/garment)	QR – 0.01 RFID – 0.71	0.70	5.70	0.12
Scanning Cost ($/garment)	QR – 0.01 RFID – 0.05	1.25	0.05	0.05
Integration Stage	Garment	Fiber	Fiber	Fabric
Automated Scanning Compatible	Yes (RFID)/No (QR)	No	Yes	Yes
Recycling Compatible	No	Yes	Potentially	Yes

Figure 18.3 Life cycle, cost, and performance comparison of tracing technologies. (a) The textile supply chain can be roughly broken down into the following segments: fiber to yarn to fabric to garment to retailer to consumer to end of life. Currently, end of life typically means being sent to the landfill, but an alternative path is being developed in which end of life textiles go to a sorting facility and are then recycled. The life cycle coverage of each tracing technology and the current feasibility of reading the signature quickly is compared with the current ease of readout represented by solid bars and the potential for ease of readout represented by striped bars. (b) Manufacturing and scanning cost for each technology is compared along with performance metrics such as when the tracer is integrated, whether it is compatible with automated sorting equipment, and compatible with recycling processes. Modified from Iezzi *et al.* [29] with permission from John Wiley & Sons, Inc.

(a) Life Cycle Stage	Parameter	Technology				
		Manuel	NIR	RFID	QR	PB
Manufacturing	Marker Cost ($/garment)	0	0	0.71	0.01	0.02
	Marker Attachment Cost ($/garment)	0	0	0.03	0.03	0.10
	Capital Investment ($/ton)	Inc. in Labor	0.003	0.004	0.002	0.003
Sorting/Recycling	Labor cost ($/ton)	266	100	60	120	60
	Running cost ($/ton)	Inc. in Labor	13	13	7	13
	Feedstock cost ($/ton)	732	732	732	732	732
	Sales Price ($/ton)	1037	1051	1044	1044	1060
	Profit ($/ton)	40	206	239	186	255
	Margin, % of sales price	4	20	23	18	24

(b) Coloring Scheme	Stakeholder	Manual	NIR	RFID	QR	PB
Positive (+)	Manufacturer					
	Logistics Chain					
Neutral (/)	Brand/Retailer					
	Consumer					
Negative (-)	Reprocessor (Sorting+Recycling)					

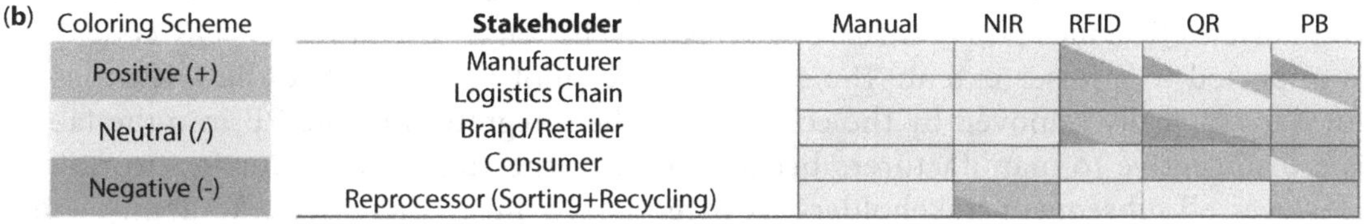

Figure 18.4 Recycling profit calculations and stakeholder comparison. (a) Recycling profit calculations based on different sorting technologies that include manual sorting, sorting using near- infrared systems (NIR), use of radio-frequency identification (RFID), quick response (QR) codes, and the photonic barcode technology (PB). This analysis includes margin based on profit from sale of sorted materials. (b) Stakeholder flag chart with positive (green), neutral (yellow), and negative (red) response of a particular stakeholder to the adoption and integration of a specific tracing technology. Figure and data adapted from Humpston *et al.* [8].

systems, and integration with fiber-to-fiber recycling techniques while also being low cost. Furthermore, none of the techniques that can be integrated directly into a fiber or fabric (DNA, Nanoparticles, PB) are currently easily readable using standard consumer electronics, such as a smartphone. In addition to the cost of the various tracing elements, it is also critical to estimate the sorting and recycling economics of each approach, as compared in Figure 18.4a. In addition to QR codes, RFID tags, and the photonic barcode, this comparison includes near-infrared (NIR) sorting. NIR sorting utilizes solely the chemical composition of the textiles without any added tracing elements, such as fluorescent nanoparticles or photonic fibers. Costs associated with the sorting process include the capital cost to install the equipment (normalized per ton of textiles sorted over a 5-year period), the labor cost to operate a facility using that technology, the cost of running the machinery itself, and the amount that must be paid to source the feedstock for sorting. The sales price is determined by the quality level of the sorted textiles, with more homogenous compositions (*i.e.* all textiles having ~50-80% polyester) having a higher sales price. Overall, these estimates indicate that all the proposed methodologies can increase the sorting industries profit margins significantly over sorting by hand.

In addition to these techno-economic assessments, it is also critical to analyze how each stakeholder in the supply chain (Tier 1 through 4 as well as the brands/retailers, consumers, sorters, and recyclers) is incentivized, or disincentivized, by the inclusion of a tracking technology. Incentives are compared in Figure 18.4b (modified from the analysis conducted by the WRAP 2014 report) [8]. For manual sorting, there is no positive or negative incentive as this technique does not alter the supply chain and sorting and recycling costs. NIR sorting benefits the reprocessors with the only negative being the need to install the relatively capital intensive equipment. RFID tags, as previously mentioned, are relatively expensive and require specific scanning equipment. The logistics chain benefits in ease of tracking large numbers of items as well as the retailers, with the disincentive being that these tags must be removed after purchase. Since RFID tags are typically removed at the point of sale due to privacy concerns, there is no added benefit to including tags to consumers and reprocessors. QR codes are not typically added until the garment manufacturing stage and thus the majority of manufacturing and the logistics chain do not see a significant benefit. The retailer can use the QR code to communicate information about the product to the consumer but also must worry about preventing counterfeit QR codes. The consumer shares the challenge of counterfeit QR codes but is impacted to a lesser extent. The reprocessor cannot readily utilize the QR codes as they are typically removed by the consumers. The integration of the PB into the fabric is a disincentive to manufacturers because it alters the standard production workflow. However, all subsequent stakeholders are incentivized due to increased traceability, communication with consumers (assuming readability on standard devices) and enhanced profit margins for reprocessors. This benefit to down-stream stakeholders, and potential cost savings and increased profits, could potentially incentivize manufacturers to include the PB to offset the added cost of manufacturing. Overall, the photonic barcode and NIR sorting showed the highest potential for incentivizing most of the stakeholders in investing in integration of this technology for tracing and sorting.

18.5   Conclusions and Recommendations

In conclusion, this chapter contributes a thorough analysis of incumbent and novel tracing technologies and how those could be applied to the sorting and recycling of textiles. Previous work [8] focused mainly on how emerging technologies at the time (2014), such as NIR sorting and use of barcodes, could help with enhancing sorting efficiency and the associated techno-economic benefits. This paper takes into consideration more recent innovations in the use of forensic and additive tracers that have largely been developed to assist with supply chain traceability and how they could, additionally, be used for sorting at the end of life. These technologies were compared based on their techno- economic feasibility, based on the related quantitative data that was available in the literature. Overall, no single technology was determined to be a comprehensive solution for all the challenges facing the textile industry in terms of supply chain traceability and recycling. Due to the massive and global nature of the textile industry, this conclusion is not surprising, and gives further impetus to the significant number of different approaches currently being developed and evaluated by academic, industrial, and intergovernmental groups, and put into practice commercially. The authors recommend that focused workshops be held to bring together representatives from all textile life cycle stages (i.e. Tier 0 through 4) as well as tracing and recycling technology advocates to discuss challenges and opportunities for development going forward. These workshops could take valuable steps in identifying and integrating the appropriate combination of technologies that can push towards a global circular textile economy.

18.6   Cost Estimates from Techno-Economic Assessment

This supplementary section attempts to compare the manufacturing and testing costs of current industrial tracing technologies (rare earth nanoparticles, DNA tagging, and photonic fiber barcodes)

18.6.1   Rare Earth Nanoparticle Cost Estimate

The original patent for the nanoparticle tracing technology does not give specific numbers for the concentrations required for detection [26]. A recent interview with the users of this technology indicated that the pigment would be less than 0.2% of the total fiber composition of the product. Rare earth, also known as upconverting, nanoparticles have been shown to have a detection limit (in aqueous solutions) of 0.1 μM [31]. The molecular weight of sodium yttrium fluoride (NaYF4) nanoparticles (diameter 45 nm), a commonly used stoichiometry, is estimated to be approximately 10^8 g/mol [32]. Assuming the average 100% cotton t-shirt is 150 grams, and given the density of cellulose is 1.5 g/cm^3, the average t-shirt contains 10 cm^3 of cotton, or a volume of 0.01 L. Assuming the detection limit of nanoparticles in solution is equivalent to nanoparticles in a fiber, then the mass of nanoparticles required to "label" a t-shirt is 0.1 g. This leads to a mass percentage of 0.06%, comparable to the value stated by the tracing manufacturer. Sigma Aldrich currently sells Er/ NaYF4 nanoparticles for $309 for 10 mg. Given this cost, it would cost over $3,000 to effectively tag one t-shirt. The materials used to produce these nanoparticles are relatively inexpensive, however, with yttrium being the 28th most abundant element in the world, 400 times more common than silver. Thus, based on pure material cost, it is expected that nanoparticles could be produced for $0.57 per 10 mg at a larger scale, based on recent studies using similar nanoparticles for biomedical applications [33]. If 0.1 g of nanoparticles are needed to tag a single garment at a composition of 0.06% then this gives a final cost per garment of approximately **$5.70**. It is worth noting that this cost will likely come down with mass manufacturing. The cost of detecting the nanoparticles is determined by the scanning technology. The nanoparticles in question are typically excited by a near infrared laser source and then the visible light emission is captured using a spectrophotometer. The cost of lasers and spectrophotometers has decreased significantly in recent years and thus a reasonable estimate of the cost of both in one scanner system, along with the associated optics to enhance signal, is around $1,000. Beyond the initial capital investment, the cost of an actual scan is negligible. Assuming a useful lifetime of 5 years, and 50 scans per day, the overall cost per scan (or identification) of **$0.05**.

18.6.2   DNA Tagging Cost Estimate

The DNA tagging methodology, promoted by Applied DNA Sciences, utilizes a customized DNA strand (between 40 and 1000 base pairs) that can be amplified, via polymerase chain reaction (PCR), to a detectable, and readable amount to confirm the identity of the garment. The amount of DNA required for a PCR amplification varies in the literature from 0.1 to around 25 ng/μL [34]. Assuming the amount needed is on the lower end of that range, at

0.5 ng/µL, the mass of DNA required to label the same cotton t-shirt as the previous example (t-shirt volume = 0.01 L) is 5 µg. Biobasic, a custom DNA fabrication company, lists the cost to produce a specific 1,000 base pair sequence at $0.23/base pair with a shipped mass of DNA at 2-4 µg. This results in a cost per microgram of approximately $77. This results in a cost per garment of $383. This does not, however, take into consideration that, given a specific base pair sequence, it is possible to replicate via PCR (or other mass DNA production methodology) and thus significantly reduce the cost per microgram of DNA tracer. However, currently the global annual production of DNA is around 3 kg. Assuming the same cost reduction as the nanoparticle example ($309 to $0.57) for expansion to industrial scale production the final cost per garment is **$0.71**. A study in 2015 found that the cost to conduct 40 PCR tests was approximately $63, or $1.58 per sample. Given the rise in PCR testing due to the COVID-19 pandemic it can be expected that this cost has been reduced in recent years. Thus, it is fair to estimate that the cost per test, or identification, is approximately $1.25.

18.6.3   QR/RFID Tagging Cost Estimate

QR codes are relatively easy to print and the price per tag is assumed to be minimal (**$0.01**). The scanning cost is similarly low as any normal camera can do so. Thus, scanning cost is also assumed to be minimal at **$0.01**. It is important to note that QR codes cannot be scanned on automated sorting equipment easily. The costs associated with RFID tagging are taken from the 2014 WRAP report [8]. They give the cost of a laundry compatible RFID tag at between 0.50 to 2.30 pounds, which is **$0.71** to $3.28 in 2022 dollars. It is assuming the cost is on the lower end of this range. RFID tags also need significant investment in scanning equipment. The same report estimated that a scanning system would cost around 220,000 pounds, which in 2022 US dollars is $313,000. However, handheld scanners (for proper comparison to the DNA, nanoparticle, and the photonic barcode) are also around $1,000 yielding a similar scanning cost per garment of **$0.05**

18.6.4   Photonic Barcode Tagging Cost Estimate

Currently, the photonic barcode is produced via thermal drawing of a macroscale preform. The preforms are produced from rolls of thin film material as well as a milled polymer cladding bar. The thin films are stacked, here 101 layers, to produce the photonic effect. A master roll of either polycarbonate or polymethyl methacrylate costs $3,000 and has 1,270 square meters of material. Furthermore, the polycarbonate cladding is approximately $4.40 per kilogram and approximately 120 grams are needed for preform. Thus, the total cost of a preform, including the thin film layers, is around $1.61. Generally, a single 12" preform can produce approximately 1000 meters of fiber, so the cost per kilometer of fiber is $1.59. Adding 30% for manufacturing overhead, this leads to a cost of $2.06 per kilometer. We currently estimate that 100 meters of PB will be necessary to "label" a single garment, here a t-shirt, at a composition level (2-5%) that is readable on handheld and automated NIR scanning systems giving a base material cost per garment of $0.02. However, since the technology currently requires integration at the weaving, knitting, or sewing stage this will increase manufacturing cost (as compared to the other technologies). This is expected to increase the price per garment to approximately **$0.12.** Operating costs are expected to be further reduced through direct extrusion of photonic fibers using customized spinnerets enabling continuous manufacturing. The cost of detection is very similar to the nanoparticle case with the notable exception being replacement of the laser with a more standard tungsten light source. Assuming the same cost of the scanning system, $1,000, and useful life than the cost per scan is comparable at **$0.05.**

Acknowledgements

This research was supported by the United States National Science Foundation INTERN program (Award #1727894) and the Under Secretary of Defense for Research and Engineering under Air Force Contract No. FA8702-15-D-0001.

References

1. Ellen MacArthur Foundation, A New Textiles Economy: Redesigning fashion's future, https://ellenmacarthurfoundation.org/a-new-textiles-economy, 2017
2. The New York Times, Global Brands Seek Clarity on Xinjiang, https://www.nytimes.com/2022/05/27/business/cotton-xinjiang-forced-labor-retailers.html, 2022
3. CottonWorks, Yarn Spinning: Fiber Opening & Cleaning, https://www.cottonworks.com/en/topics/sourcing- manufacturing/yarn-manufacturing/opening-cleaning/, 2022
4. O Eco Textiles, When is recycled polyester NOT recycled polyester?, *OEcotextiles*, https://oeco-textiles.blog/2011/03/23/when-is-recycled-polyester-not-recycled-polyester/, 2011
5. The Observatory of Economic Complexity, Textiles, https://oec.world/en/profile/hs/textiles, 2022
6. Karan Khurana, Marco Ricchetti, Two decades of sustainable supply chain management in the fashion business, an appraisal, *J. Fashion Mark. and Mgmt.*, 20, 89, 2016.
7. Fashion Revolution, Fashion Transparency Index 2022, https://www.fashionrevolution.org/about/transparency/, 2022
8. Giles Humpston, Peter Willis, David J. Taylor, Sara L. Han, Technologies for sorting end of life textiles: A technical and economic evaluation of the options applicable to clothing and household textiles, https://www.academia.edu/21698706/Technologies_for_sorting_end_of_life_textiles_A_technical_and_econo   mic_evaluation_of_the_options_applicable_to_cloth-ing_and_household_textiles, 2014
9. Mistra Future Fashion, "Automated feeding equipment for textile waste: Experiences from the FITS- project, http://mistrafuturefashion.com/sv/publikationer/, 2019
10. Kirsti Cura, Niko Rintala, Taina Kamppuri, Eetta Saarimäki, Pirjo Heikkilä, Textile Recognition and Sorting for Recycling at an Automated Line Using Near Infrared Spectroscopy, *Recycling*, 6, 1, 2021.
11. Jade Wilting, Hilde van Duijn, Clothing labels: Accurate or not?, https://www.circle- economy.com/resources/clothing-labels-accurate-or-not, 2020.
12. US Environmental Protection Agency, Municipal Solid Waste Generation, Recycling, and Disposal in the United States, Tables and Figures for 2012, https://www.epa.gov/sites/default/files/2015- 09/documents/2012_msw_fs.pdf, 2012.

13. Office of Land and Emergency Management, Advancing Sustainable Materials Management: 2018 Fact Sheet, https://www.epa.gov/sites/default/files/2021-01/documents/2018fffactsheet-dec2020fnl508.pdf, 2019.

14. Hilde van Duijn, Natalia Carrone, Ola Bakowska, Qianjing Huang, Marieke Akerboom, Kathleen Rademan, Dolly Vellanki, Sorting for Circularity Europe: An Evaluation and Commercial Assessment of Textiles Waste Across Europe, https://reports.fashionforgood.com/report/sorting-for-circularity-europe/, 2022.

15. James Crowley, Kathleen Rademan, Evonne Tan, The Textile Tracer Assessment: An Analysis and User Guide for Physical Tracer Technologies in the Textile Industry, https://reports.fashionforgood.com/report/textile-tracer/, 2022.

16. Paul Morgan, Brief History of Fibers from Synthetic Polymers, *J. of Macro. Sci.: Part A - Chem.*, 15, 6, 1981.

17. MadeHow, How fake fur is made - material, manufacture, making, history, used, processing, structure, steps, product, http://www.madehow.com/Volume-3/Fake-Fur.html, 2023

18. US Federal Trade Commission, Rules and Regulations Under the Textile Fiber Products Identification Act, https://www.ftc.gov/legal-library/browse/rules/textile-products-identification-act-text, 1960

19. Krassie Petrova, Adriana Romaniello, Dawn Medlin, Sandra A. Vannoy, QR Codes: Advantages and Dangers, Int. Conf. e-Business, 2023

20. Sofia Benouakta, Florin D. Hutu, Yvan Duroc, Stretchable Textile Yarn Based on UHF RFID Helical Tag, *Textiles*, 1, 3, 2021.

21. Amirhossien Hajiaghajani, Amir H. Afandizadeh Zargari, Manik Dautta, Abel Jimenez, Fadi Kurdahi, Peter Tseng, Textile-integrated metamaterials for near-field multibody area networks, *Nat. Electron.*, 4, 11, 2021.

22. L. Jung, J. A. Hayward, M. B. Liang, and A. Berrada, DNA marking of previously undistinguished items for traceability, US Patent 9266370B2, assigned to APDN BVI Inc, 2016.

23. M. B. Liang and S. S. K. So, Methods for genetic analysis of textiles made of gossypium barbadense and gossypium hirsutum cotton, US Patent 20140295423A1, assigned to APDN BVI Inc, 2014.

24. Shovon L. Sarkar, Rubayet Alam, Prosanto K. Das, Md H. A. Pramanik, Hassan M. Al-Emran, Kabir I. Jahid, Md A. Hossain, Development and validation of cost-effective one-step multiplex RT-PCR assay for detecting the SARS-CoV-2 infection using SYBR Green melting curve analysis" *Sci Rep*, 12, 1, 2022.

25. Cotton USA, Oritain. Reliable Fiber Traceability, https://www.cottonusa.org/innovation/oritain, 2022

26. M. Greene and P. Stenning, "Photon marker system for fiber material," US Patent 10247667B2, assigned to Fibermark Solutions, Ltd, 2018.

27. T. Jüstel, D. Uhlich, H. Bettentrup, A. Deitermann, S. Rütter "Method for Identifying an Object," European Patent 2534221B1, assigned to Tailorlux, Gmbh, 2016

28. M. C. McCairn and M. L. Turner, "Nanoparticles," US Patent 20210040260A1, assign. Chromition, Ltd, 2020

29. Brian Iezzi, Austin Coon, Lauren Cantley, Bradford Perkins, Erin Doran, Tairan Wang, Mordechai Rothschild, Max Shtein "Fabric-Integrated Polymeric Photonic Crystal Fibers for Textile Tracing and Sorting," *Adv. Mat. Tech.*, 2023, 2201099

30. Wenjing Huang, Shenglin Luo, Dong Yang, Sheng Zhang, Applications of smartphone-based near-infrared (NIR) imaging, measurement, and spectroscopy technologies to point-of-care (POC) diagnostics, *J. Zhe. Univ. Sci. B*, 22, 3, 2021.

31. Shaoshan Su, Zhurong Mo, Guizhen Tan, Hongli Wen, Xiang Chen, Deshmukh. A. Hakeem, PAA Modified Upconversion Nanoparticles for Highly Selective and Sensitive Detection of Cu2+ Ions, *Front. Chem.*, 8, 2021.

32. Lewis Mackenzie, Jack Goode, Alexandre Vakurov, Padmaja Nampi, Sikha Saha, Gin Jose, Paul Millner, The theoretical molecular weight of NaYF 4 :RE upconversion nanoparticles, *Sci. Rep.*, 8, 1, 2018.

33. Emma Xu, Changhwan Lee, Stefanie D. Pritzl, Allen S. Chen, Theobald Lohmueller, Bruce E. Cohen, Emory M. Chan, James Schuck, Infrared-to-ultraviolet upconverting nanoparticles for COVID-19-related disinfection applications, Opt. Mat.: X, 12, 2021.

34. Pinky Dhatterwal, Sandhya Mehrotra, Rajesh Mehrotra, Optimization of PCR conditions for amplifying an AT-rich amino acid transporter promoter sequence with high number of tandem repeats from Arabidopsis thaliana, *BMC Res Notes*, 10, 2017.

19

A Systems Approach to Addressing Industrial Products Circularity Challenges

Manish Gupta* and Umeshwar Dayal†

R&D Division, Hitachi America, Ltd., Santa Clara, California, U.S.A.

Abstract

The current Linear Economy approach to design, produce, use, and dispose products leads to lot of wasted resources and is not sustainable. A transition to a Circular Economy (CE), where products, components and materials are kept in use for as long as possible, is critical. For industrial products with high post-use residual value, RE* processes (remanufacture, refurbish, repair, repurpose) that extend lives of products and materials, generate immense economic, environmental, and societal benefits. However, barriers exist for each ecosystem stakeholder - manufacturers/remanufacturers, product owners/buyers, core brokers, dealers – that hinder the uptake of RE*. For instance, remanufacturers struggle with uncertainty in core supply and quality, inventory management, variability in production, and demand visibility; product owners struggle with end-of-use decisions for their ageing products. Currently, the product-use phase is largely decoupled from the product-recovery phase, leading to inefficiencies in RE* processes. The decoupling stems from a lack of integration in the value-chain, with each stakeholder operating in silos and engaging in point-to-point transactions, which results in lack of visibility into full product lifecycles. To address the ecosystem challenges holistically and bring new efficiencies to RE* processes through digitalization we are creating an industrial products circularity system with two key components – a trusted platform and an AI/Analytics layer. The trusted platform securely connects stakeholders, establishes shared truth, and reduces friction in information sharing and transactions. Products are tracked through their lifecycle and history is maintained in the trusted platform to provide provenance as well as enable efficient end-of-life collection. Product condition monitoring through IoT, sensors and AI/analytics enables remaining-useful-life assessments and predictive/prescriptive maintenance operations to extend product life. Analytics generate insights for decision support to the various stakeholders: assist product owners with product decommissioning decisions and choosing the right RE* option; and assist remanufacturers with proactive core buying and inventory optimization. Sustainability and circularity KPIs are measured and optimized.

This paper describes the barriers to industrial product circularity, discusses opportunities for digitalization, and presents a system to address the challenges.

Keywords: Circular economy, industrial products circularity, remanufacturing solutions, digitalization, product-life extension, circularity KPIs, big data analytics

---

**Corresponding author*: manish.gupta@hal.hitachi.com
†*Corresponding author*: umeshwar.dayal@hal.hitachi.com

Nabil Nasr (ed.) Technology Innovation for the Circular Economy: Recycling, Remanufacturing, Design, Systems Analysis and Logistics, (239–254) © 2024 Scrivener Publishing LLC

19.1   Introduction

The linear model of resource consumption, characterized by extraction, production, use, and disposal, has dominated since the advent of the Industrial Revolution. The world consumes more than 100 billion tons (Gt) of raw materials annually, with 65% of it becoming waste and only 8.6% of it being cycled back into the economy [1]. Pressure from megatrends such as climate change, resource scarcity, population growth, and strained supply-chains has made this model unsustainable. The Circular Economy (CE) offers a solution, promoting a regenerative system that minimizes waste and emissions by slowing, closing, and narrowing material and energy loops. The CE focuses on long-lasting design, maintenance, repair, reuse, remanufacturing, refurbishing, and recycling, with the ultimate goal of restoring and reusing resources instead of relying on virgin resources [2, 3]. As per one estimate, adopting the CE can potentially yield up to $4.5 trillion in economic benefits by 2030 [4].

The core concepts of Circular Economy (CE) have existed for many decades [5]; however, it is only recently that there has been a growing recognition of the urgency to transition from traditional linear models to CE. The widespread adoption of disruptive technologies such as big-data, AI/analytics, distributed ledgers, and IoT has made it possible to effectively implement CE strategies at scale and speed.

Section 19.2 provides an overview of industrial products circularity. Section 19.3 highlights the impediments to industrial products circularity from the perspectives of ecosystem stakeholders. Section 19.4 presents a systems approach to addressing these impediments and challenges.

19.2   Industrial Products Circularity

The transition to CE requires a fundamental shift in the way manufacturers think about and approach product design, production, sales and distribution, customer engagement, support and maintenance, and end-of-life management. Manufacturers must prioritize reduced material and energy use, the use of renewable materials, and the use of renewable energy. They must design products for durability and upgradability, so that the products can be used efficiently, reused multiple times, and recycled, thus creating sustainable circular product lifecycles. To achieve this, manufacturers must establish systems for end-to-end visibility into the product lifecycle and adopt business models that incentivize product utilization, longer product lives, and closed loop operations instead of just maximizing product sales.

Closed loop models maximize product utilization, keep products/materials in use, and reduce amount of waste generated (Figure 19.1). The loops could be shorter loops of product renewal and reuse within the product's lifecycle, or the longer loop of recovery and recycling at the end-of-life of the product [6].

Industrial products and capital equipment such as machinery, heavy-duty off-road (HDOR) equipment, and electrical equipment are well-suited to closed-loop models. They are technologically complex products requiring a large amount of resources to produce. They are expensive to own, operate and maintain; and barring a few categories such as information and communication technology (ICT) equipment, tend to have long lives

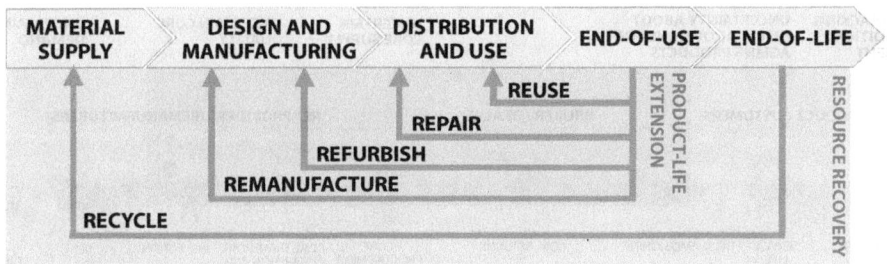

Figure 19.1 Closed-loop model and circular economy processes.

and long technology development cycles, which means incremental upgrades are more common than fundamental design changes. Industrial products typically have significant residual value at end-of-use (EOU), which gets wasted if the product is prematurely discarded. To retain the value of an industrial product for as long as possible, the circular economy extends the product's lifetime through its recovery at end-of-use and reclaiming its value through RE* processes such as remanufacturing, refurbishment, repair, and reuse. Remanufacturing brings the used product (called core) to an as- good-as-new-or-better condition, thus giving it a full new service life. Repair and refurbishing extend the current life of the product. Reuse involves selling the used product as is to another buyer. When a product reaches the end-of-life (EOL), the objective is to avoid its disposal to landfills by recovering it and recycling the energy and resources contained in it. We use the term Industrial Products Circularity to refer to adoption of closed loops models (RE* and end-of-life recovery) in the context of industrial products. Amongst the RE* processes, remanufacturing requires most resources (material, energy, labor), and is the most complex, but it also provides the most value [7]. While we focus on remanufacturing for the sake of clarity, the challenges and solutions discussed in the following sections are applicable to other RE* processes as well.

19.3   Barriers to Industrial Products Circularity

RE* processes such as remanufacturing have existed across industries for many decades. Many industrial product manufacturers have remanufacturing operations and offer remanufactured products and/or component parts. However, uptake of such processes is still low: the ratio of remanufacturing to new manufacturing within US and Europe is only about 2% [8–10]. This is due to several inherent process challenges.

Closed-loop models are networked and hence truly transitioning industrial products from linear business models to circularity requires a systemic change across the value-chain. Currently, the product-use phase is largely decoupled from the product-recovery phase. The decoupling stems from a lack of integration in the value-chain, with each stakeholder operating in siloes and engaging in point-to-point transactions. To effectively realize circularity, multiple stakeholders must collaborate to achieve common objectives. Each stakeholder must optimize its operations in the context of the entire value-chain rather than in its own limited context. To realize this, each stakeholder needs more visibility and access to information from up and down the value-chain, and conversely, each stakeholder

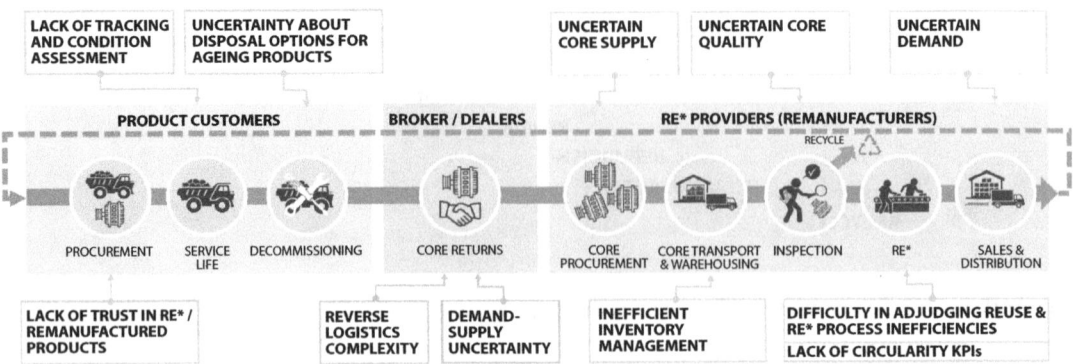

Figure 19.2 Barriers to industrial products circularity.

needs to capture and share more information with others. For example, to optimize its inventory (cores, finished goods, and parts), a remanufacturer requires both supply visibility from core-brokers and demand visibility from customers. Customers who provide the demand signal to the remanufacturer can get better demand fulfillment. By sharing its demand signal with upstream core brokers, the remanufacturer can get better supply visibility. Similarly, the demand signal from remanufacturers can aid core brokers to optimize their inventory. Such reciprocal exchange of information among stakeholders can thus lead to mutual benefits. However, prevalent systems and processes associated with linear models do not adequately capture the information and lack appropriate mechanisms to share it with other stakeholders, hindering the transition to circularity. Figure 19.2 illustrates the challenges faced by the key stakeholders.

19.3.1 Product Customers

Product customers are typically the product users, but they may also lease or rent the product to others to use. Product customers need to make critical decisions when procuring the products, throughout the product's useful life, and at end-of-use/end-of-life stages. They face the following challenges:

(a) *Lack of trust in RE* products:* Although RE* products sell for a fraction of the cost of equivalent new products and may carry generous warranties, customers are often not confident about buying them. Their reasons range from concerns about quality of RE* products vis-à-vis new, to lack of information, to lack of availability.

(b) *Need for track and trace and condition assessment:* During the use phase, customers need predictive and prescriptive maintenance of the products. Customers who share, lease or rent products need ways to track and trace the location, condition, performance, and usage of their products for metering usage as well as detecting violation of contractual terms.

(c) *Challenges in managing of ageing products:* Product owners want to maximize return when disposing their ageing products that have rising operational and maintenance, repair and overhaul (MRO) costs. They need to ascertain residual value of these product, decide on the timing of product disposal, and pick the right channel to use (e.g., remanufacture, sell, scrap).

19.3.2   RE* Providers/Remanufacturers

RE* providers play a central role from a product-life extension perspective and in closing the loop. They typically get used products and components either directly from product owners or via intermediaries like core-brokers, and then reclaim and reinstate the value of the products. They may offer remanufacturing, refurbishment, and repair as a service, wherein the RE* product goes back to the original owner, or they may resell the products to other buyers. Such RE* providers could be Original Equipment Manufacturers (OEMs) themselves or they could be third parties. RE* providers face several challenges:

(a) *Uncertainty in supply of cores*: Unlike traditional manufacturing wherein raw materials, parts and components are supplied by a few selected vendors, the cores needed for RE* are collected from multitude of suppliers, including end customers, core brokers, dealers. It is not possible to scale-up by simply manufacturing more input parts as the availability of cores becomes a function of the installed base of products that are nearing EOU/EOL.

(b) *Uncertainty in quality of cores*: Every unit of the product returned as core would have been through different wear and tear. Such differences cause variances in the remanufacturing process. Equipment that was used in harsh conditions will have lot of damage or wear and tear, resulting in a poor-quality core, thus increasing the RE* cost (e.g., labor, machining, replacement parts). Current approaches rely on visual inspection, which does not provide a view into the internal condition. Better assessment of internal condition of cores without requiring expensive disassembly and cleaning operation is needed. In addition, condition-based approaches to assess residual value of used products can help appropriately price cores during procurement.

(c) *Difficulty in adjudging reuse*: During the RE* process, workers need to identify problems with the core and adjudge the reusability of internal components. This may include assessing the physical condition of components and detecting cracks, stresses, and other imperfections.

(d) *Uncertainty in demand for RE* products*: High uncertainty in demand leads to sub-optimal *inventory management* and *production planning*. Demand for RE* processes is dependent on the installed base of products, their condition and many external factors affecting usage of products. Without visibility into quantity and condition of installed products, demand forecasting is particularly challenging. Demand is also linked with the product model's life. In industries with frequent model changes deciding on the right timing of starting and ending a RE* program for a particular model is a challenge. For instance, in the automotive industry, the demand for remanufactured parts starts rising a few years after a new model is introduced, peaks, and then starts declining after the model is discontinued, ultimately settling in a long tail.

For several types of industrial products, OEMs are also the RE* providers. Apart from the challenges outlined above, OEMs struggle with post-sales forward integration and tracking the product-lifecycles. Track-and-trace is critical to provide proactive maintenance to customers to ensure long product lives and for recovering products when they reach end-of-life. In addition, tracking products through their lifecycles provides OEMs a view into patterns of breakdown and anomalous behavior, which can aid in improving product and component designs. However, OEMs often do not have current information about the installed products, their configuration, maintenance history, ownership, and their condition because the downstream stakeholders operate siloed systems.

19.3.3   Core Brokers and Dealers

For certain types of industrial products, core-brokers collect the cores from customers and supply them to RE* providers. Core-brokers struggle with *poor demand forecasts, difficulty in securing supply* of cores and *frequent over- and under-stocking of inventory*. For products that are sold through dealers, the dealers play a key role in reverse logistics in bringing used products back to close the loop. Dealers are also involved in distribution and sales of RE* products.

In addition to the above challenges, all stakeholders are increasingly under pressure to report on sustainability, carbon- neutrality, and circular economy Key Performance Indicators (KPIs). Such reporting often entails gathering data from upstream suppliers and downstream customers. Stakeholders across the value-chain do not have the systems and processes to capture necessary data and measure such KPIs. Moreover, the definitions of the KPIs are still nascent.

Challenges such as those outlined above lead to process inefficiencies, leakage of materials, and loss of value for the stakeholders, and negatively impact CE objectives. Existing manufacturing-related systems (such as, ERP, PLM) are inadequate to address such challenges as they were built for new manufacturing and not RE*. They tend to focus on a single stakeholder, typically the OEM. RE* operations are very different from new manufacturing, as supply and demand are connected, key input material (viz. used products) are of variable quality and condition, and input material needs to be collected from many customers. Such fundamental differences make it very difficult to adapt solutions meant for new manufacturing to RE*.

19.4   A System for Addressing the Industrial Products Circularity Barriers

Emerging digital technologies and big-data analytics enable solutions to many of the Industrial Product Circularity challenges faced by the various stakeholders that we have outlined in the previous section. Use of technologies such as blockchain/distributed ledgers, AI, analytics and IoT can make RE* operations more sustainable and efficient [11]. We have created a systems approach to address the challenges described in the previous section. Figure 19.3 shows such an end-to-end system for Industrial Products Circularity. Our system comprises two main components: a trusted platform, and an AI/analytics framework

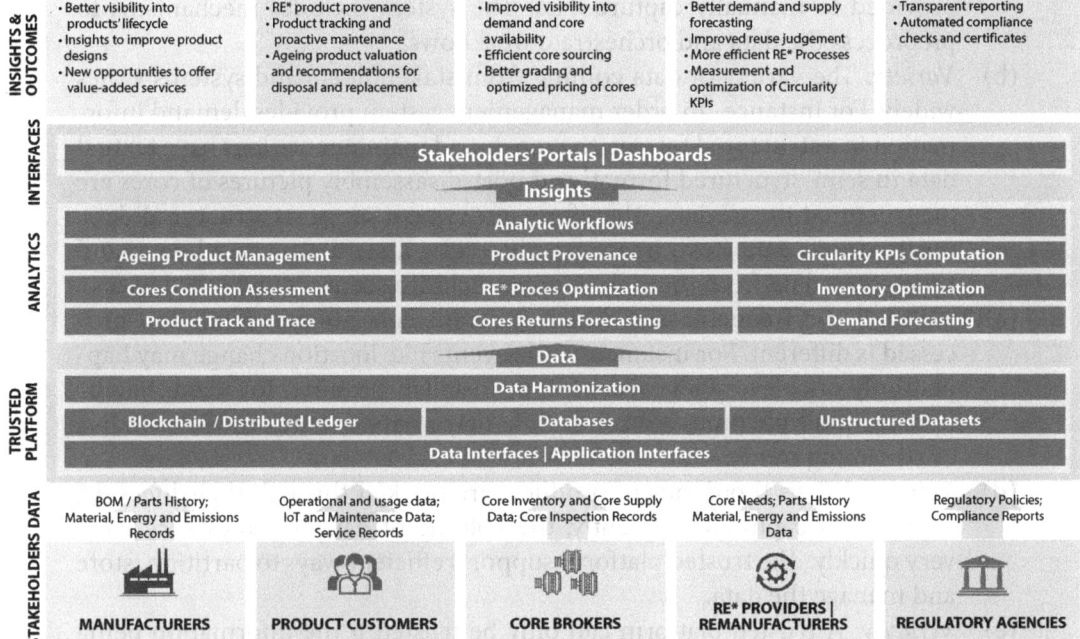

Figure 19.3 Industrial products circularity system.

built on top of this platform. Further, we have been developing a research prototype of the Industrial Products Circularity system in collaboration with a couple of industrial products remanufacturers. Figures 19.5–19.7 show illustrative screenshots of our research prototype.

19.4.1   Trusted Platform for Shared Truth Between Stakeholders

The trusted platform provides a single source of truth for all the value-chain participants throughout a product's lifecycle and enables coordination between them. Blockchain or a distributed ledger provides the underpinnings for such a platform. Each participant only needs to share information relevant from a circularity perspective onto the platform. The other business critical data stays off-chain in independent databases. The platform enables not only secure sharing of data between participants, but provides automation of transactions, which reduces friction in transactions.

19.4.2   Big-Data Capability

The trusted platform is built on the principles of big data and addresses the heterogeneity, variety, velocity, volume, and veracity of data. It supports the following aspects:

(a) *Heterogeneity:* Different stakeholders run multiple conventional but heterogenous systems for their operations (e.g., PLM, ERP, CRM, order management systems, production planning systems). Data and application interfaces enable connections with such heterogenous systems. In some cases, data from custom-built systems is ingested or applications are provided to manually capture data. The system includes mechanisms to preprocess the data and orchestrate workflows.

(b) *Variety:* The variety of data coming from stakeholders and systems varies widely. For instance, an order management system provides demand information in a structured format; the repair and maintenance logs have textual data in semi-structured format; and post-disassembly pictures of cores are unstructured image data. Therefore, our system supports structured data, unstructured data (text, images, audio, video), semi-structured data, and time-series data from multiple sources including sensors and IoT devices.

(c) *Velocity:* The frequencies at which different data are generated and processed is different. For instance, while events like location change may happen only once every few years for an installed product, IoT data may be updated multiple times a second. Data may arrive in batch mode or in live or streaming mode.

(d) *Volume:* When thousands of products are tracked through their lifecycle, with events captured frequently, the volume of data becomes very large very quickly. The trusted platform supports efficient ways to partition, store and manage the data.

(e) *Veracity:* A trusted platform can only be trusted if the information being supplied to it is trustworthy. Since the goal is to establish a shared truth, ways to assess and ensure veracity of information are provided. For instance, a root of trust-based approach can be used to authenticate data sources. Metadata, such as identity of data-provider (e.g., picture, biometrics), location information (e.g., GPS), identification information (e.g., barcode, QR code) can be captured along with reported data to ensure veracity. Further, analytical, and statistical approaches can be used to detect discrepancies (e.g., to detect greenwashing).

19.4.3   Product Lifecycle and Track and Trace

Tracking products and sub-assemblies throughout the product lifecycle is critical as locating the product is a prerequisite to recovering it. Changes to product configuration, replacement of sub-assemblies and changes in ownership need to be tracked and captured as a shared truth between value-chain participants to understand what can be collected, when, and from whom. This history is maintained in the trusted platform. Current approaches rely on entities and stakeholders touching the product to gather and provide the traceability and configuration information. Tracking is typically done at the product level and not at the sub-assembly level. Tracking information is maintained outside of the equipment. In a circular model a sub-assembly may have multiple lives and may be part of multiple product units. Hence, tracking history of sub-assemblies is essential to establish provenance.

To enable comprehensive traceability, tracking and serialization information must be added to the product and its constituent sub-assemblies. Several tracking technologies exist, including barcodes, QR codes, RFID, NFC, GPS tags, each with its advantages. Tracking tags used must be tamper-proof, damage-proof and should last through the lifecycle of the product and the subassemblies. There is also a need to embed sensors and data loggers into products and sub- assemblies, akin to black-boxes, to collect event data and maintain history within the product or sub-assembly itself. Such information can be communicated periodically or can be extracted when the product is maintained, serviced, or undergoes RE*. Storing data locally will allow the history to accompany the product or sub-assembly, creating a redundant copy and alleviating the need to solely rely on the platform.

During the lifecycle of a product, the ownership, deployment location and usage pattern change. Further, many products have multiple different sub-assemblies, thus having a complex Bill-of-Materials (BOM). During the service life, as the products undergo configurations, reconfigurations, repairs, parts replacements and maintenance, the sub-assemblies and components put in originally by the OEM are changed. The sub-assemblies and components may have their own circular lifecycles. Remanufacturing process itself leads to disassembly of the product, putting constituent parts into separate bins, and then rebuilding the product from parts from the bins and new parts. Thus, the BOM, consisting of tracking identifiers of sub-assemblies of a product, is changing throughout its lifecycle. Tracking the BOM of a product throughout its lifecycle is difficult due to both process and systems challenges. The trusted platform enables capturing and sharing of such information. History of a product is maintained, with any changes to the constituent parts (and the BOM) carefully registered in the trusted platform. This can be done by integrating with the systems run by stakeholders. Alternatively, applications can be provided to stakeholders to record changes.

Such track and trace capability built over the trusted platform provides a remanufacturer visibility into ageing products that may be operating with product customers. Such visibility enables an assessment of the product's condition, which can help a remanufacturer select used products that they want to buy-back as cores and optimize the incentives they need to pay to the owners for those products. Through the platform itself, the remanufacturer can make direct offers to the owners at the right time, transforming the currently reactive process of acquiring cores into a proactive one.

19.4.4 AI/Analytics Enabled Process Optimization and Stakeholder Decision-Support

The transition points within an industrial product's lifecycle are fraught with decisions for the stakeholders. For example, the owner of an ageing industrial product must determine when to decommission it, understand its residual value, and choose the best option (e.g., sell, remanufacture, or scrap). A remanufacturer must decide when to purchase an ageing or used product, what buyback incentive to offer and how many units to purchase. Buyers must decide between buying new, used, or remanufactured products. These decisions are further complicated by the numerous factors influencing them. In the rest of this section, we describe AI and analytics functions on top of the trusted platform that provide insights and decision support to the different stakeholders.

19.4.5 Product Customers

The Industrial Products Circularity system helps product customers make informed decisions during the product lifecycle. It includes the following functionality:

(a) *Decision-support for product procurement:* As the underlying trusted platform tracks the product's various lives comprehensively, the system provides provenance information and history on reclaimed products to the customer, thus inspiring confidence in such RE* products. The system provides recommendations on the best products to buy by comparing the available ones on a variety of dimensions such cost of ownership, condition, and environmental footprint.

(b) *Improved tracking and maintenance support:* by providing the product data to the trusted platform and sharing it with OEMs and service providers, the customers can track their product's performance, get early warnings of failures, and receive proactive maintenance support.

(c) *Decision-support for product decommissioning at end-of-use:* The system helps product customers make informed decommissioning decisions by analyzing product data and various factors such as usage forecast, projected maintenance costs, operational costs, and environmental impact (Figure 19.4). Using this data, the system determines a product's condition, assesses if it can meet usage needs, and compares it to alternatives. If the product should be decommissioned, the system provides disposal options, recommends the channel (e.g., remanufacture, sell, scrap) and predicts return/incentives. Since the system runs on a connected trusted platform, sale (or buy) offers can be made (or accepted) from the platform itself and the transactions can be automated as well.

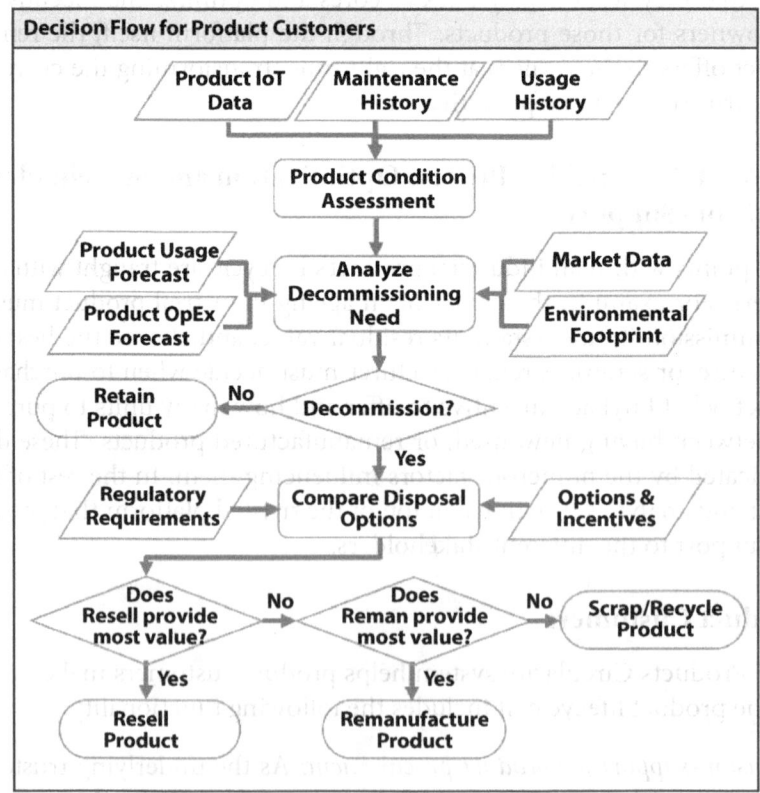

Figure 19.4 Decision-support for product decommissioning.

19.4.6   RE* Providers/Remanufacturers

The system provides decision support to RE* providers like remanufacturers throughout their operations, including:

(a) *Improved core visibility and sourcing:* For remanufacturers, monitoring ageing products in operation enables the core-buying process to become proactive instead of reactive. They can actively track the condition of ageing products as prospective cores, and make offers for right quality cores, in the right quantity at the right time. The process involves periodically assessing the requirement for the cores for each product line being remanufactured. This factors in market conditions, inventory levels, demand forecasts, and regulatory requirements, and produces a core buying forecast with recommendations on timing and quantity of cores to buy. In parallel, for each product that the remanufacturer is monitoring, the system assesses the condition of the product based on available data and assigns a grade to it. Based on the condition/grade assessment of the product, the system estimates the potential remanufacturing cost. Such an estimation can be based on the standard operating procedure (SOP) that is followed for the cores of that grade or condition. Based on the location of the product and logistics needs, the estimate of logistics costs of bringing that product to the remanufacturing facility is computed. Based on estimates of remanufacturing cost, logistics costs, and the current market conditions (such as prices, availability), the buyback incentive that can be paid to the product owner for a used product and potential margin is computed. Finally, from the ageing products that are available, ones to purchase are identified based on a combination of factors such as profit margins, quality, regulatory needs etc. The advantage of this proactive approach is that it gives the remanufacturer a good estimate of the quality of cores, and leads to many downstream benefits in the remanufacturing process; it provides visibility into the potential remanufacturing needs (e.g., parts, labor, machining), costs and hence the margins on it.; it enables determining the right timing for initiating buyback, and estimating the incentive that can be provided to the product owner. Dynamic incentivization can make core acquisition more predictable for the remanufacturer. Figure 19.5 shows a prototype of such a core visibility and sourcing application for remanufacturers. Product owners will also benefit as they stand to get a premium for products that they have maintained well. For remanufacturers that procure cores via their dealers or from core-brokers, the underlying trusted platform can provide visibility into core-supply by enabling information sharing between them. Analysis of core availability can enable a remanufacturer to better manage core inventory. Core-brokers and dealers can in turn benefit by getting better demand signal from remanufacturers.

(b) *Data-driven a priori core quality assessment:* Remanufacturing is an inherently highly variable process owing to high variability in quality of primary input (i.e., cores), in supply of cores, and in demand for remanufactured

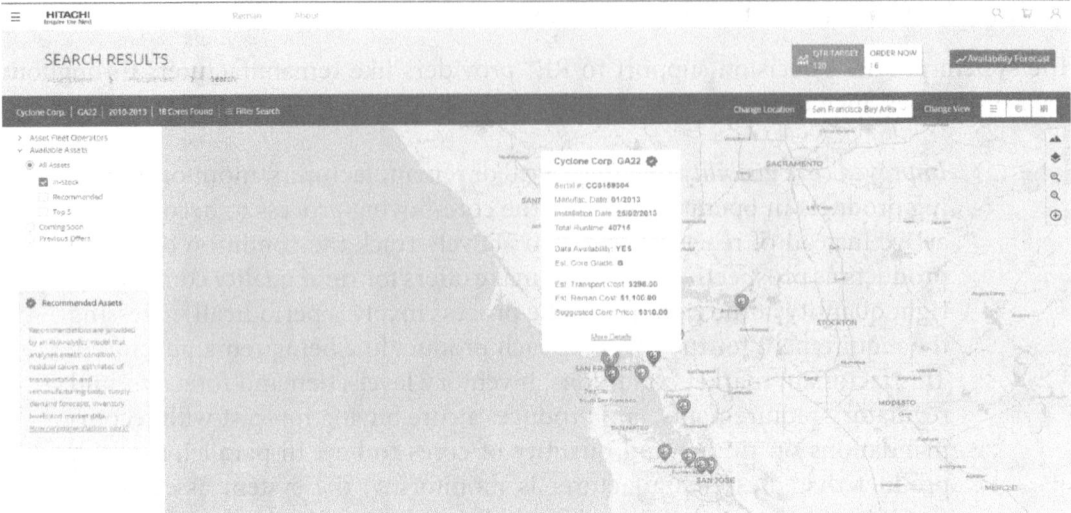

Figure 19.5 Industrial Products Circularity research prototype – Core visibility and sourcing application for remanufacturers.

products. Planners often need to prioritize those cores for production that will have least fallout, require least amount of resources to remanufacture and hence generate maximum value. Our system provides an a-priori assessment of the condition of cores before disassembly and enables prioritization for production. It utilizes machine-learning, deep learning, and statistical approaches to analyze a diverse set of available data about a core, which may include operational data, sensor/IoT data and tests and measurements data. It provides estimates of possible faults in the cores, parts and process needed to fix those faults, associated costs and environmental footprint. It computes a utility value for each available core, which combines environmental and economic impact of remanufacturing it, and thereby enables ranking of cores. Figure 19.6 shows the research prototype of such a core quality assessment application.

(c) *Data-driven reuse judgement:* During the remanufacturing process, a typical product has constituent parts that are mostly reused (e.g., casings), parts that are usually never reused (e.g., seals and gaskets), and parts that are used based on their condition. The third category is usually impacted most by condition of cores. A reuse judgement must be made by the worker working on the core who, after disassembly, cleaning, and inspection, must decide what parts to reuse. It requires skills and experience, and even then, a good assessment is often not possible. Our system assists the workers with reuse judgement based on data-driven analytics. Based on the specific product and component, the data and analytics could vary widely. For instance, using historical operational data and sensor/IoT data about the core, images of the core and its parts, and data from tests and measurements, AI, machine learning and deep learning are used to create models for reuse judgement. From a design for remanufacturing perspective, sensors

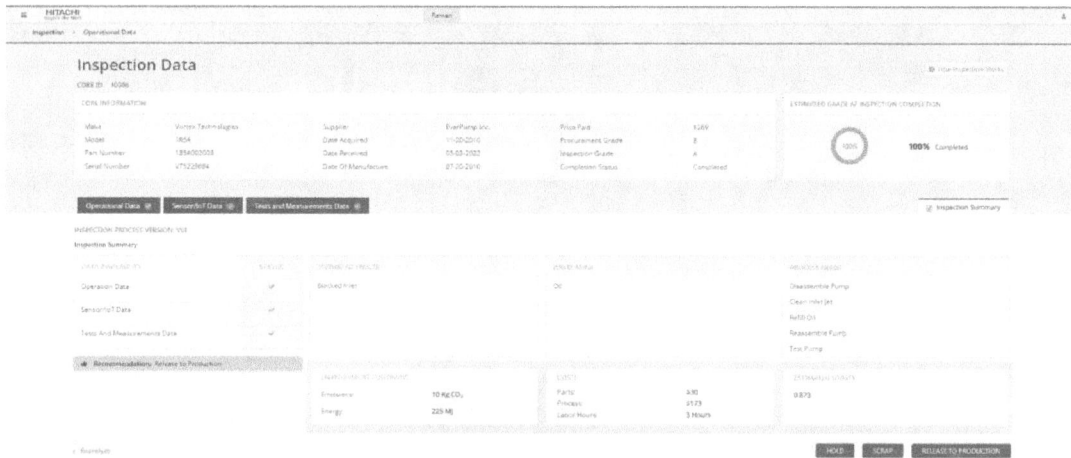

Figure 19.6 Industrial Products Circularity research prototype – Core quality assessment application.

and data loggers can be added to products or components to indicate their reusability. For instance, if a component can internally log number of actuations or shocks and stresses experienced by it during its lifetime, the data may be useful during remanufacturing to decide about reuse.

(d) *Improved demand forecasting:* Forecasting demand is one of the biggest challenges a remanufacturer faces. Unlike new manufacturing, for remanufacturing the supply and demand are interconnected and usually come from the same parties. Hence, demand and supply must be looked at in conjunction and not isolation. For many product types, analysis of historical demand data alone is not enough to have good demand forecasts. Our system includes a demand forecasting component that ingests and analyzes information about installed and in-operation products such as usage data, breakdown data, operational and sensor/IoT data. It also includes customer information and historical core return data. Exogenous data, such as market condition, indices and regulations can be analyzed as well. Further, demand for the remanufactured version of a product and the profitability of remanufacturing it, are also inextricably linked to the lifecycle of the product model, and for a remanufactured component, to the surviving population of the product it is a part of. Hence, in addition to forecasting demand and supply, the system strategically predicts the timing of start and end of a remanufacturing program. Figure 19.7 shows a demand forecasting and supply planning application for a remanufacturer that utilizes historical data, customer inputs and external factors to provide a demand forecast for a product. Through the application, planners at a remanufacturer can respond to the forthcoming demand and plan inventory, procurement, and production.

OEMs that are also RE* providers can benefit from the Industrial Products Circularity system by getting better visibility into products' lifecycles. They can provide value-added

Figure 19.7   Industrial Products Circularity research prototype – demand forecasting and supply planning application.

services like predictive maintenance and RE* services to their customers, or better manage upsell, cross-sell opportunities. By tracking and analyzing product data, OEMs can identify anomalous behavior and improve product designs as well.

19.4.7   Core Brokers and Dealers

For core brokers and dealers, the system provides better visibility into the demand and supply of cores.

19.4.8   Circularity KPIs

Finally, the Industrial Product Circularity system includes components for measuring and optimizing CE KPIs for each of the stakeholders. The KPIs are needed to identify gaps and leakages (e.g., of materials and energy) and identify opportunities for improvement. Traditionally, most of the focus has been on measuring and optimizing economic KPIs. From a CE perspective, however, environmental, and societal KPIs gain importance as well. For this, operational data, such as production, material usage, fall-outs, waste, emissions, effluents, energy consumption, and labor need to be comprehensively tracked. Such data are used to compute both generic KPIs such as material embedded energy and emissions savings; and manufacturing process-related energy and emissions savings associated with reclaiming and reusing components; or labor usage associated with the RE* processes, and KPIs suited for the specific product category and industry In future, the system will support the computation of KPIs in real-time, analyze gaps, and suggest measures to optimize them.

Such KPIs and the traceability provided by the system can enable participating regulatory agencies to monitor compliance better and generate necessary reports.

19.5   Conclusions

Our Industrial Products Circularity system leverages digital technology to drive the transition to the Circular Economy, in which products and materials are kept in circulation for as long as possible. A secure and trusted platform connects stakeholders, promotes transparency, and facilitates seamless information sharing and transactions. Products are monitored throughout their lifecycle, and their history is recorded in the platform to provide provenance and support end-of-life collection. The system offers decision support for the product owners to make decommissioning decisions and choose the best RE* option. Traceability and analytics help the remanufacturers with proactive core buying and optimizing inventory and production. Sustainability and CE KPIs are monitored and optimized. Ecosystem stakeholders benefit from better visibility, information sharing and analytics-driven decision support. They can better realize the economic, environmental, and societal benefits by working collaboratively instead of making sub-optimal decisions in siloes.

References

1. CGRi, Circularity Gap Report 2021, https://www.circularity-gap.world/2021, 2021.
2. Geissdoerfer M., Savaget P. *et al.*, The Circular Economy – A new sustainability paradigm?, *Journal of Cleaner Production*, 143, 757–768, 2017.
3. Ellen MacArthur Foundation, *Towards the Circular Economy 1: Economic and Business Rationale for an Accelerated Transition*, 2012.
4. World Economic Forum, Circular Economy, https://www.weforum.org/projects/circular-economy, 2022.
5. Braungart, M., McDonough W., *Cradle to cradle*. Jonathan Cape, London, 2008.
6. Lacy P., Rutqvist J., *Waste to wealth: The circular economy advantage*, 1st ed. Basingstoke, Palgrave Macmillan, 2016.
7. Nasr, N.Z., Russell J.D. *et al.*, Redefining Value: The Manufacturing Revolution, Resource Panel, 2018.
8. Golisano Institute for Sustainability, *What is remanufacturing?*, https://www.rit.edu/sustainabilityinstitute/blog/what-remanufacturing, 2022
9. US ITC, Remanufactured Goods: An Overview of the U.S. and Global Industries, Markets, and Trade, USITC Publication 4356, 2012
10. Parker D., Riley K. *et al.*, *Remanufacturing Market Study*, European Remanufacturing Network, 2015.
11. Dayal, U., Gupta M. *et al.*, Enabling Product Circularity Through Big Data Analytics and Digitalization, *IEEE 65th International Midwest Symposium on Circuits and Systems (MWSCAS)*, pp. 1-6, Japan 2022.

20

Environmental and Economic Analyses of Chemical Recycling via Dissolution of Waste Polyethylene Terephthalate

Utkarsh S. Chaudhari[1]*, Daniel G. Kulas[1], Alejandra Peralta[2], Robert M. Handler[1], Anne T. Johnson[3], Barbara K. Reck[4], Vicki S. Thompson[5], Damon S. Hartley[6], Tasmin Hossain[6], David W. Watkins[7] and David R. Shonnard[1]

[1]*Department of Chemical Engineering, Michigan Technological University, Houghton, Michigan, USA*
[2]*Chemstations Inc., Houston, Texas, USA*
[3]*Resource Recycling Systems, Ann Arbor, Michigan, USA*
[4]*Center for Industrial Ecology, Yale School of the Environment, New Haven, Connecticut, USA*
[5]*Biological & Chemical Science & Engineering Department, Idaho National Laboratory, Idaho Falls, Idaho, USA*
[6]*Operations Research and Analysis, Idaho National Laboratory, Idaho Falls, Idaho, USA*
[7]*Department of Civil, Environmental, and Geospatial Engineering, Michigan Technological University, Houghton, Michigan, USA*

Abstract

Globally, more than 1000 organizations and 175 nations are facing the plastic waste problem and have realized the need to transition from "linear-to-circular" economy of plastics. While the current mechanical recycling technologies for plastics are struggling to increase the U.S. plastic recycling rates beyond 9%, chemical recycling technologies become important complementary technologies to the predominant mechanical recycling that are needed to realize the circular economy in plastics supply chains. Dissolution is one such chemical recycling technology that can recycle waste plastic back into high-quality virgin grade plastic. However, the environmental and economic impacts of chemical recycling of waste polyethylene terephthalate (PET) via dissolution technology using a green solvent are unknown. Our study evaluated environmental metrics such as greenhouse gas (GHG) emissions and cumulative energy demand, and economic metrics such as net present value (NPV), minimum selling price, payback period, return on investment, and discounted internal rate of return for three dissolution processes with polymer recovery via anti-solvent, evaporation, and cooling precipitation techniques. The dissolution process with evaporation technique was the most economically favorable, whereas that with cooling technique was the most environmentally favorable. The anti-solvent approach had low economic performance and the highest environmental impacts. The NPV for all of these technologies ranged from $2.67 MM to $10.93 MM for a capacity of 8,400 MT/year and was found to be the highest for dissolution with evaporation approach and the least for anti-solvent approach. The cradle-to-gate GHG emissions and energy demand for PET dissolution processes ranged from 1.33-3.77 kg CO_2-eq/kg of chemically recycled (CR) PET and

Corresponding author: uschaudh@mtu.edu

Nabil Nasr (ed.) Technology Innovation for the Circular Economy: Recycling, Remanufacturing, Design, Systems Analysis and Logistics, (255–268) © 2024 Scrivener Publishing LLC

18.9-56.1 MJ/kg of CR-PET, respectively. These economic and environmental metrics will be helpful in evaluating the sustainability of circular PET supply chains in the U.S.

Keywords: Plastic waste, solvent-based recycling, sustainability, life cycle assessment (LCA), techno-economic analysis (TEA), circular economy, polyethylene terephthalate (PET)

20.1 Introduction

Plastic materials have various advantages and are often preferred over other materials due to being cheap, lightweight, durable, and resistant to different chemicals. Unfortunately, high production and consumption and mismanagement of plastic products have led to a global plastic waste problem. Polyethylene terephthalate (PET, resin code #1) is one of the important plastic materials used and finds applications in production of bottles, fibers (clothes, carpets etc.), films, sheets, and thermoforms [1, 2]. PET waste accounts for 14.8% (4.8 Million Metric Tons (MMT)) of the total plastic waste generated in the U.S. [3]. Sixty six percent of this PET waste is landfilled, 15% is incinerated with energy recovery, and 19% is recycled [3]. Due to such high discard rates, the fossil-dependent nature of PET supply chain in the U.S. consumed 522 Peta Joules (PJ) of energy and released greenhouse gas (GHG) emissions of 22.9 MMT CO_2-eq in 2019 [1]. These emissions represented 1.16% of the total U.S. industry-related GHG emissions in 2019 [1].

Emerging chemical recycling technologies, also known as advanced or molecular recycling technologies, can aid in shifting from the "linear-to-circular" plastics by creating high-quality or "virgin-grade" resins. The chemical recycling technologies can be broadly classified into three main categories: purification (also known as "dissolution"; plastics-to-plastics), depolymerization (also referred as "decomposition"; "plastics-to-monomer") and conversion ("plastics-to-intermediate hydrocarbons") recycling technologies [4]. PET is a suitable feedstock for dissolution and depolymerization recycling technologies [5]. In dissolution recycling technologies, post-industrial or post-consumer plastics are dissolved in a solvent and then filtered to remove any undesired contaminants (fillers, dirt, other polymers, etc.) [5-9]. The dissolved polymer is then precipitated by either the addition of an anti-solvent, evaporation of solvent, or cooling down the polymer for recovery [7, 10]. In depolymerization recycling technologies, condensation polymers, such as PET, are broken down to the basic monomers and/or oligomers using different techniques such as glycolysis, hydrolysis, aminolysis, ammonolysis, methanolysis, and enzymes. While the economic and environmental impacts of enzymatic and non-enzymatic depolymerization of waste PET, as well as dissolution of PET using non-green solvents, are known from the literature [11-16], these impacts remain unknown for the dissolution process using a green renewable solvent. In addition, none of the previously published studies have evaluated and compared these impacts for PET specific dissolution process with polymer recovery by the evaporation approach. Also, the economic metrics such as net present value, payback period, return on investment, discounted internal rate of return for these processes are not reported in the literature. Therefore, the purpose of our study was to compare the environmental and economic metrics for dissolution of waste PET using a green solvent, Gamma (γ)-Valerolactone (GVL), with different polymer recovery techniques. Based on our previously published systems analysis framework [4, 17], we take a "bottom-up" approach, in which process simulation is integrated with life cycle assessment (LCA) and techno-economic analysis (TEA).

The first research objective of our study was to conduct process simulation of chemical recycling of waste PET via dissolution process using the solvent GVL. We analyze three different dissolution processes that differ in their polymer precipitation techniques, as mentioned above, using the same solvent. The second research objective was to evaluate technoeconomic metrics of simulated recycling processes. The third research objective was to evaluate environmental metrics such as GHG emissions and cumulative energy demand (CED) of simulated recycling processes.

20.2   Methods

20.2.1   Process Simulation

Figure 20.1 shows the process flow diagrams for chemical recycling of waste PET via dissolution process with polymer recovery via three precipitation techniques: addition of anti-solvent (water) (referred as 'anti-solvent approach'), evaporation of solvent (referred as 'evaporation approach'), and cooling of dissolved polymer solution to room temperature (referred as 'cooling approach'). All three processes were modeled to produce one metric ton (MT)/hour of chemically recycled PET (CR-PET). Process simulations were conducted in the CHEMCAD software [18] using UNIFAC as the thermodynamic package. The thermodynamic package was selected based on the nature of the components (polarity, non-electrolyte), and the range of temperatures and pressures used in process, per guidelines [19, 20]. UNIFAC group

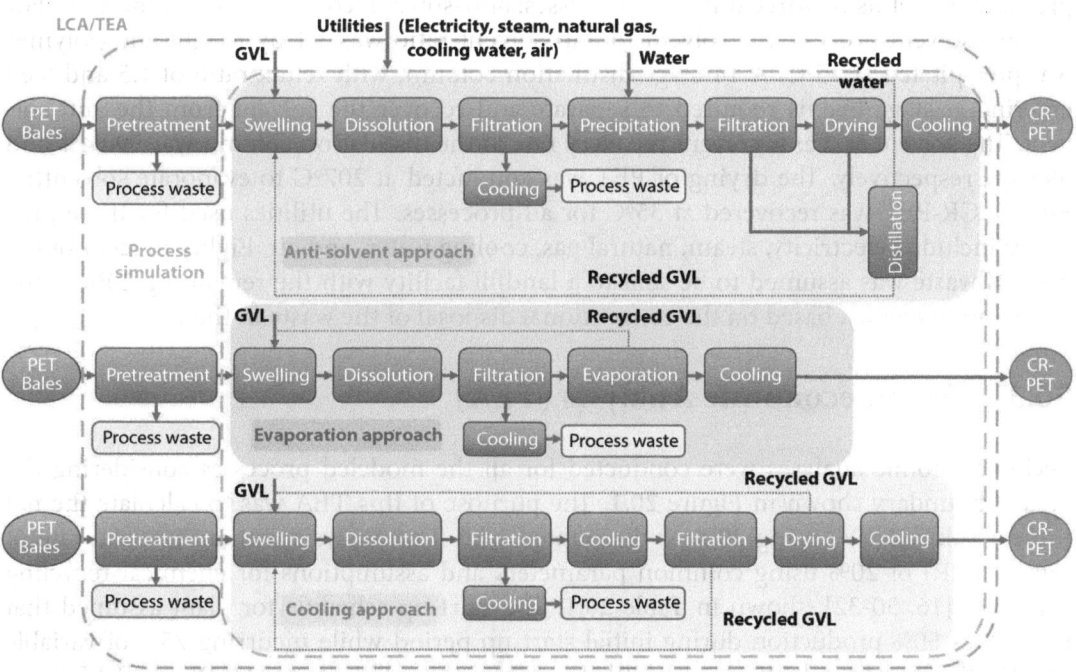

Figure 20.1 Chemical recycling of PET via dissolution process with different polymer recovery approaches. Note: Orange dashed boundary represents the system boundary for process simulation. Green dashed boundary represents the system boundary for LCA and TEA. GVL: Gamma-Valerolactone solvent, CR-PET: chemically recycled PET.

contribution method was used to define the structural groups for the waste PET component in CHEMCAD. Specific heat capacities over a range of temperature were defined in CHEMCAD based on the literature [21]. Due to lack of data on type and composition of additives (colorants, fillers etc.) and other contaminants, we neglected simulation of such materials.

Baled PET could be sourced from a material recovery facility (MRF) or directly from intermediate processing centers [1, 22]. Dissolution process requires a pretreatment step [16], assumed to be less intense than for conventional mechanical recycling, and to be co-located with the chemical recycling facility. Pretreatment of PET bales prevents the dissolution process from being affected by avoidable impurities [6]. Generally, pretreatment of waste plastics includes bale opening, size reduction to a flake and could include dry (air classification) and/or wet (float-sink separation, centrifugal sorting, cleaning) mechanical processing steps [6, 23]. Bale contamination was assumed to be 10% based on literature sources and removed during pretreatment for disposal [24, 25]. The chemical recycling process steps, which include swelling, dissolution, filtration to remove any contaminants, precipitation of polymer, drying, and cooling, were modeled in CHEMCAD using process conditions adopted from the literature [26]. Gamma (γ)-Valerolactone (GVL), a green solvent [27] derived from renewable feedstocks and less toxic [27], was used for all three processes to dissolve the waste PET at a solvent to PET ratio of 4:1 by weight. Swelling of the PET flakes occurred at 120°C and 1 atm for 1 hour followed by dissolution, which occurs at 170°C and 1 atm for 2 minutes [26]. Based on the polymer and solvent loss parameters found in the literature, we included a 1 % and 1.5% loss of polymer and solvent, respectively, at the filtration step following the dissolution step [28, 29]. The purified PET is then precipitated out using three different processes: anti-solvent, cooling, and evaporation. For the anti-solvent process, the solvent to anti-solvent ratio was 1:1 by weight and polymer was precipitated at 35°C. A 14-stage distillation column, with reflux ratio of 1.5 and feed entering at stage five, is required to separate and recover the solvent from the anti-solvent. The solvent and anti-solvent recovery rate at the distillation column was 99.84% and 99.99%, respectively. The drying of PET was conducted at 207°C to evaporate solvent(s). Finally, CR-PET was recovered at 35°C for all processes. The utilities used for these processes included electricity, steam, natural gas, cooling water, and air. Eighty percent of the process waste was assumed to be sent to a landfill facility with the remaining 20% to the incineration facility, based on the conventional disposal of the waste in the U.S.

20.3   Technoeconomic Analysis (TEA)

Technoeconomic analyses were conducted for all the modeled processes considering the system boundary shown in Figure 20.1. The purpose of this TEA was to calculate the net present value (NPV) using 30-year discounted cash flow analysis with an internal rate of return (IRR) of 20% using common parameters and assumptions for chemical recycling processes [16, 30-32], shown in Table 20.1. The start-up schedule for plant assumed that it achieves 50% production during initial start-up period while incurring 75% of variable and 100% of fixed production costs [30]. The minimum selling price (MSP) of CR-PET was calculated by setting the NPV equal to zero. Additional economic metrics calculated include payback period, return on investment (ROI), and discounted IRR [32]. The discounted IRR was found by setting the NPV equal to zero by varying the assumed IRR of 20% at assumed selling price [32].

Table 20.1 Technoeconomic parameters and assumptions used for economic analysis of waste PET dissolution processes.

Parameter	Value
Lifetime of Plant (Years)	30
Operating Days (Days/year)	350
Total CR-PET produced (MT/year)	8400
Base year	2019
OSBL Costs	100% of ISBL
Engineering Costs	25% of ISBL + OSBL costs
Contingency Costs	10% of ISBL + OSBL costs
Working Capital	5% of FCI
Depreciation Method & recovery period	7-year MACRS
Construction period (Years) and spending schedule [30]	3 Years (8% Y1, 60% Y2, 32% Y3)
Start-up time (Years)	0.5
Income tax rate (%)	21%
Internal rate of return (IRR) (%)	20%
Supervision	25% of operating labor
Direct overhead	50% of operating labor and supervision
General & administrative costs (G&A costs)	65% of total labor costs
Maintenance costs	5% of ISBL costs
Insurance costs	1% of ISBL+OSBL costs
Baled PET price (10-year average, $/MT) [34]	377
Bale pretreatment Cost ($/MT)	100
R-PET price (10-year average, $/MT) [34]	1,608
GVL price ($/MT) [35]	1,000
Waste disposal cost ($/MT, U.S. average) [36]	55
Electricity ($/kWh) [37]	0.0681
Cooling water ($/GJ) [38]	0.381
Natural gas ($/GJ) [39]	3.7
High, medium, and low-pressure steam ($/GJ) [38]	6.5, 3.54, and 2.76, respectively

Note: ISBL: Inside battery limit (ISBL) investment; OSBL: Outside/offsite battery limit (OSBL) investment; FCI: Fixed capital investment.

Fixed capital investment (FCI) includes the inside battery limit (ISBL) investment, outside/offsite battery limit (OSBL) investment, engineering and construction costs, and contingency costs. The purchased and installed costs of equipment were estimated using the cost and installation factors found in the literature [32]. The costs of these pieces of equipment were then adjusted to base year of 2019 (Table 20.1) using chemical engineering plant cost indices [33]. The revenues for the processes were estimated by gathering raw material and product prices from the literature and are shown in Table 20.1.

The total cost of production includes fixed and variable production costs. Fixed costs of production include costs of operating labor, supervision, direct overhead, maintenance, insurance, and general and administrative costs. Variable production costs include costs of feedstock (waste plastic bales), utilities, raw materials (solvents), and waste treatment costs. For the operating labor costs, we assumed 4.8 operators to be needed per shift position for three shift positions with a salary of $58,000 ea./year [32, 40]. The bale pretreatment cost was assumed to be $100/MT and costs of utilities are shown in Table 20.1. The pretreatment costs have been previously assumed to range from $100-$419/MT [16, 41, 42].

20.4   Life Cycle Analysis (LCA)

The goal of this LCA was to compare the environmental and energy impacts of CR-PET via dissolution processes against the fossil derived virgin PET. The system boundary for LCA, as shown in Figure 20.1, represents a cradle-to-gate system with a functional unit defined on the output (product) basis. The functional unit for dissolution processes was based on 1 kg of CR-PET produced. The environmental burdens of collection, sorting and/or baling associated with the feedstock were included in our LCA. The waste plastic material is assumed

Table 20.2 LCA input/inventory table for chemical recycling of PET via dissolution processes. Inputs based on 1 kg of CR-PET produced.

Inputs	Anti-solvent	Evaporation	Cooling	Unit
Collection, sorting, and baling	1.12			kg
PET flakes to dissolution	1.01			kg
Gamma-Valerolactone (GVL) solvent*	0.06			kg
Electricity	0.081	0.093	0.118	MJ
Steam	24.65	2.66	1.60	MJ
Natural gas	1.3	0	0	MJ
Cooling water	204.47	18.64	10.22	kg
Process waste	0.182			kg

The eco-profile for GVL solvent was not found in the Ecoinvent database. Butyrolactone solvent was used as a substitute for GVL based on their structural similarities and boiling points.

to have no fossil or upstream burdens (cut-off approach). The input data were represented from the Ecoinvent database wherever possible [43]. The LCA input/inventory tables for the dissolution processes are shown in Table 20.2. The life cycle impact assessment (LCIA) methods of Intergovernmental Panel on Climate Change (IPCC) 2013 Global Warming Potential (GWP) over 100-year time frame and Cumulative Energy Demand (CED) were used to evaluate GHG emissions and total energy demand, respectively [44]. The LCA was conducted using the SimaPro® software version 9.0. The waste from all the modeled processes was assumed to be 80% landfilled and 20% incinerated with energy recovery, based on the typical U.S. waste management system for plastics disposal [3].

20.5   Results and Discussion

20.5.1   Process Simulation

The input and output mass and energy balances for all the processes were closed within 1% of each other. All of the processes had an input of 1.12 MT/hour of waste PET baled feedstock to produce 1 MT/hour of CR-PET and 0.06 MT/hour of make-up GVL. The total process waste of 0.18 MT/hour was produced, including losses at bale pretreatment and filtration step following dissolution of PET. The total heating utilities for the dissolution process with anti-solvent approach were the highest (25.96 GJ/hour), followed by the evaporation (2.66 GJ/hour), and then cooling approach (1.60 GJ/hour). The cooling utilities for anti-solvent, evaporation and cooling approach were 25.63 GJ/hour, 2.34 GJ/hour, and 1.28 GJ/hour, respectively. The cooling utilities for the evaporation approach were higher due to cooling of the recycled GVL vapor prior to the swelling processing step. For the anti-solvent process, the majority of these heating/cooling utilities were associated with the distillation column for the separation and recovery of GVL and anti-solvent water.

20.5.2   Technoeconomic Analysis Results for Dissolution

Figure 20.2 summarizes all evaluated TEA results such as NPV, MSP, ROI, payback period, discounted IRR, and FCI for dissolution of PET via anti-solvent, evaporation, and cooling approaches. Overall, evaporation was the most profitable approach with a NPV of $10.93 MM, followed by cooling ($8.98 MM) and anti-solvent ($2.67 MM). The MSP for all three approaches was found to be less than the fossil-PET price of $1.55/kg [34] at the 8,400 MT/year scale. The annual capital costs per metric ton of CR-PET produced were found to be $1,325, $1,229, and $1,007/MT-year for anti-solvent, cooling, and evaporation approaches, respectively. The scenarios with the highest NPV had the best economic metrics overall, with lower MSP and payback period, and higher discounted IRR and ROI.

The highest operating costs for all three technologies were found to be feedstock, fixed operating cost, bale pretreatment, and utilities (Table 20.3). Using an alternative precipitation approach to avoid distillation was found to be economical. The evaporation approach requires slightly higher utilities than cooling to remove the solvent. This is offset by the filtration step required for separating GVL solvent in the cooling approach (Figure 20.1), which results in a higher FCI, and thus lower NPV, than evaporation. These conclusions

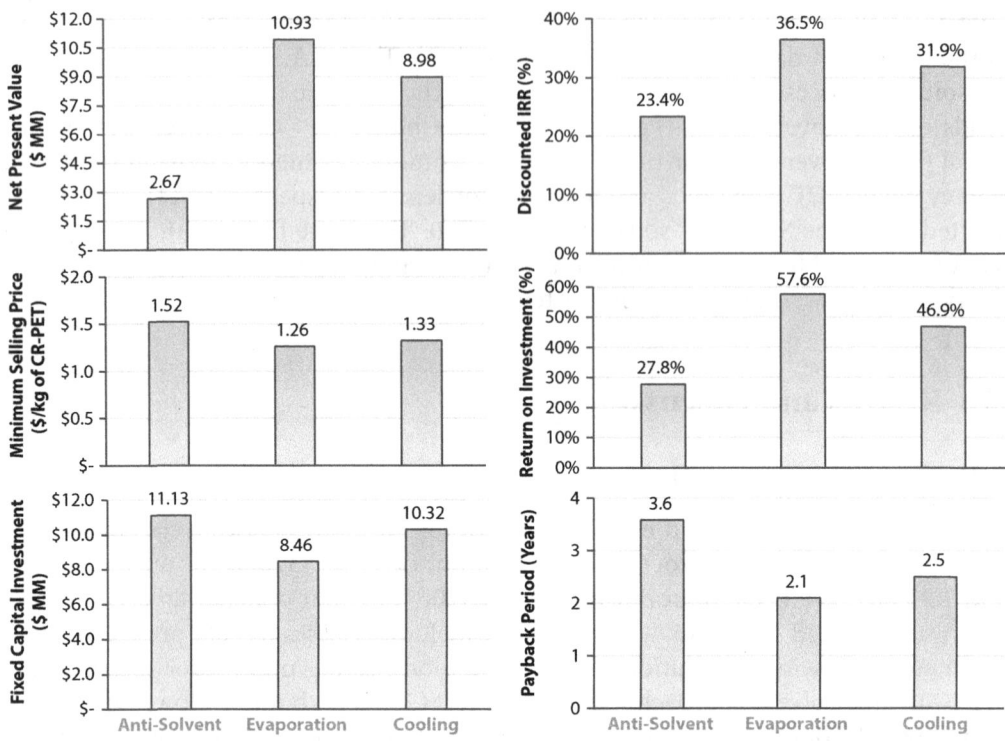

Figure 20.2 Summary of TEA results for chemical recycling processes for PET via dissolution recycling technologies. Note: Results based on the CR-PET production capacity of 8,400 MT/year.

agree with a recent study by Sánchez-Rivera *et al.* [10], which conducted an economic analysis of the Solvent Targeted Recovery And Precipitation (STRAP) dissolution process for post-industrial multi-layered PET based film. Their study found that the cooling precipitation approach (STRAP-B) had lower MSP, utilities and FCI costs than using an anti-solvent (STRAP-A).

Table 20.3 Summary of key annual operating costs for dissolution processes analyzed in this study.

Operating cost ($MM/year)	Anti-solvent approach	Evaporation approach	Cooling approach
Utilities	1.47	0.16	0.087
Feedstock	3.56		
Bale Pretreatment	0.94		
Raw Material	0.50		
Waste Treatment	0.084		
Fixed Operating Costs	2.87	2.80	2.85

20.5.3 GHG Emissions and Energy Impacts of Dissolution Process

From Figure 20.3, the cradle-to-gate GHG emissions for PET dissolution processes ranged from 1.33-3.77 kg CO_2-eq/kg of CR-PET, with emissions being the highest for dissolution process with polymer recovery via anti-solvent approach (3.77 kg CO_2-eq/kg of CR-PET), followed by evaporation (1.43 kg CO_2-eq/kg of CR-PET), and then cooling approach. Similarly, the cradle-to-gate total energy impacts were found to be the highest for anti-solvent process (56.1 MJ/kg of CR-PET), followed by the evaporation process (20.5 MJ/kg of CR-PET), and then the cooling process (18.9 MJ/kg of CR-PET). The differences in the environmental impacts of the three approaches are due to utilities consumption. The anti-solvent process has the highest utilities, and thus the largest impacts, mainly due to the utilities required for distillation of GVL and water. The solvent, Gamma-Valerolactone (GVL), used for the process is a green solvent but has a high boiling point of 207°C, thereby increasing the total reboiler heat duty. The GHG emissions and energy impacts associated with use of solvent were 10.3% and 11.9% of their totals, respectively. The LCA results for solvent use could be over-estimated due to use of fossil-based butyrolactone solvent as a substitute for GVL due to lack of U.S. LCA data. The GHG emissions and energy demand of chemically recycled PET via anti-solvent process were found to be 1.7 times and 2.3 times higher than the virgin PET impacts (2.23 kg CO_2-eq/kg of virgin PET; 24.6 MJ/kg of virgin PET), respectively. It is important to note that the total energy demand for virgin PET is 61.4 MJ/kg of virgin PET, however, 40% of this total energy is associated with process and fuel energy and the remaining 60% is the associated with the material feedstock energy, which is the energy content of oil and gas. The total energy impacts of chemically recycled PET could also be compared against only the process and fuel energy for fair comparison of CED impacts, as waste PET is assumed to have zero fossil or upstream burdens. Dissolution of PET with polymer recovery by the cooling approach has the highest GHG emissions and energy savings of 40% and 23%, respectively, when compared against virgin PET impacts. These conclusions are supported by a recent LCA study conducted by del Carmen Munguía-López *et al.* [45], who found that using an anti-solvent for precipitation (6.73 kg CO_2-eq/kg of film; 101 MJ/kg of

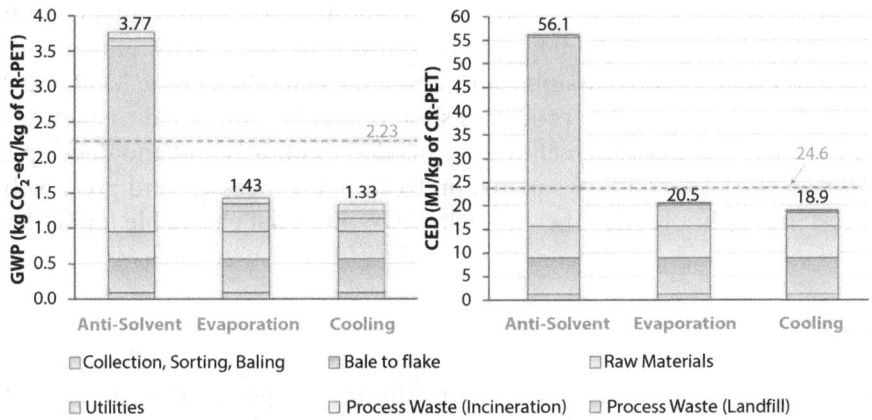

Figure 20.3 GHG emissions and CED impacts of PET dissolution processes with different polymer recovery approaches. Note: CR-PET refers to chemically recycled PET. Horizontal dashed line represents virgin fossil-based PET impacts.

film) had much higher environmental impacts than the cooling approach (1.18 kg CO_2-eq/ kg of film; 19 MJ/kg of film). Another study [16] on dissolution of PET using anti-solvent approach found GHG emissions and CED impacts to be 4.5 kg CO_2-eq/kg of CR-PET and 83 MJ/kg of CR-PET, respectively. Their study used benzyl alcohol as a solvent to dissolve PET and methanol as an anti-solvent to precipitate PET.

20.6   Conclusions and Recommendations

The environmental and economic analyses of these technologies revealed that the choice of polymer precipitation method had a significant effect on the overall environmental and economic feasibility of these technologies. The anti-solvent approach had the lowest economic feasibility and higher environmental impacts, when compared with other processes, due to the distillation column used to recover the solvent and anti-solvent. There is an interesting tradeoff between the cooling and evaporation approaches, with evaporation having superior economic results and cooling having lower environmental impacts, although the differences were relatively small. The MSP of $1.26-$1.52/kg of CR-PET was found to be lower than fossil-derived PET ($1.55/kg) at base case capacity of 8,400 MT/year for all processes analyzed. The economic analysis for dissolution processes was found to be favorable mainly because of the higher price of recycled PET ($1.61/kg) compared to the feedstock costs ($0.38/kg). The environmental impacts for evaporation and cooling approaches were found to be lower than that of fossil derived PET. Nonetheless, economic and environmental impacts of all these processes would change depending on the mode of transportation and travelling distances with increasing production capacity. Future work could include combining location-specific databases and tools such as geographic information system (GIS) with TEA and LCA tools to look at the effect of sourcing feedstock over longer distances via single and/or multiple modes of transportation. Additionally, at experimental and process simulation levels, more research efforts are needed to study the amounts, composition, and types of additives used in PET; the effects of colorants and additives on dissolution of PET in GVL solvent; and material quality of CR-PET against virgin PET resin obtained from these processes. These processes could further be heat integrated to reduce the environmental impacts and analyze economical trade-offs. Sensitivity analysis could be conducted to analyze which assumptions and parameters affect the MSP and NPV the most. The effect of scaling up the processes should also be conducted to analyze its effect on MSP of CR-PET. More research efforts are needed to understand and develop LCA and TEA data for emerging secondary sorting, pretreatment processes, and green solvents for chemical recycling technologies. There is lack of U.S. LCA data available for GVL solvent.

Acknowledgements

The authors are grateful for the funding from the U.S. Department of Energy's Office of Energy Efficiency and Renewable Energy (EERE) under the Advanced Manufacturing Office Award Number DE-EE0007897. The views and opinions of authors expressed herein do not necessarily state or reflect those of the United States Government or any agency thereof.

References

1. Chaudhari, U. S.; Johnson, A. T.; Reck, B. K.; Handler, R. M.; Thompson, V. S.; Hartley, D. S.; Young, W.; Watkins, D.; Shonnard, D., Material Flow Analysis and Life Cycle Assessment of Polyethylene Terephthalate and Polyolefin Plastics Supply Chains in the United States. *ACS Sustainable Chemistry Engineering* 2022, *10* (39), 13145-13155.

2. Di, J.; Reck, B. K.; Miatto, A.; Graedel, T. E., United States plastics: Large flows, short lifetimes, and negligible recycling. *Resources, Conservation, & Recycling* 2021, *167*, 105440.

3. U.S. EPA. Advancing Sustainable Materials Management: 2018 Tables and Figures Report. 2020. Available at: https://www.epa.gov/facts-and-figures-about-materials-waste-and-recycling/advancing-sustainable-materials-management.

4. Chaudhari, U. S.; Lin, Y.; Thompson, V. S.; Handler, R. M.; Pearce, J. M.; Caneba, G.; Muhuri, P.; Watkins, D.; Shonnard, D. R., Systems analysis approach to polyethylene terephthalate and olefin plastics supply chains in the circular economy: A review of data sets and models. *ACS Sustainable Chemistry Engineering* 2021, *9* (22), 7403-7421.

5. Lange, J.-P., Managing plastic waste—sorting, recycling, disposal, and product redesign. *ACS Sustainable Chemistry Engineering* 2021, *9* (47), 15722-15738.

6. Triebert, D.; Hanel, H.; Bundt, M.; Wohnig, K., Solvent-Based Recycling. In *Circular Economy of Polymers: Topics in Recycling Technologies*, ACS Publications: 2021; pp 33-59.

7. Vollmer, I.; Jenks, M. J.; Roelands, M. C.; White, R. J.; van Harmelen, T.; de Wild, P.; van der Laan, G. P.; Meirer, F.; Keurentjes, J. T.; Weckhuysen, B. M., Beyond Mechanical Recycling: Giving New Life to Plastic Waste. *Angewandte Chemie International Edition* 2020.

8. Kol, R.; Roosen, M.; Ügdüler, S.; Van Geem, K. M.; Ragaert, K.; Achilias, D. S.; De Meester, S., Recent advances in pre-treatment of plastic packaging waste. *Waste Material Recycling in the Circular Economy-Challenges Developments* 2021.

9. Ügdüler, S.; Van Geem, K. M.; Roosen, M.; Delbeke, E. I.; De Meester, S., Challenges and opportunities of solvent-based additive extraction methods for plastic recycling. *Waste Management* 2020, *104*, 148-182.

10. Sánchez-Rivera, K. L.; Zhou, P.; Kim, M. S.; González Chávez, L. D.; Grey, S.; Nelson, K.; Wang, S. C.; Hermans, I.; Zavala, V. M.; Van Lehn, R. C., Reducing Antisolvent Use in the STRAP Process by Enabling a Temperature-Controlled Polymer Dissolution and Precipitation for the Recycling of Multilayer Plastic Films. *ChemSusChem* 2021, *14* (19), 4317-4329.

11. Singh, A.; Rorrer, N. A.; Nicholson, S. R.; Erickson, E.; DesVeaux, J. S.; Avelino, A. F.; Lamers, P.; Bhatt, A.; Zhang, Y.; Avery, G., Techno-economic, life-cycle, and socioeconomic impact analysis of enzymatic recycling of poly (ethylene terephthalate). *Joule* 2021, *5* (9), 2479-2503.

12. Nyawanga, B. *Techno-Economic Analysis of Polyethylene Terephthalate Bio-Upcycling.* University of South Florida, 2021.

13. Cornago, S.; Rovelli, D.; Brondi, C.; Crippa, M.; Morico, B.; Ballarino, A.; Dotelli, G., Stochastic consequential Life Cycle Assessment of technology substitution in the case of a novel PET chemical recycling technology. *Journal of Cleaner Production* 2021, *311*, 127406.

14. Uekert, T.; DesVeaux, J. S.; Singh, A.; Nicholson, S. R.; Lamers, P.; Ghosh, T.; McGeehan, J. E.; Carpenter, A. C.; Beckham, G. T., Life cycle assessment of enzymatic poly (ethylene terephthalate) recycling. *Green Chemistry* 2022, *24* (17), 6531-6543.

15. Ügdüler, S.; Van Geem, K. M.; Denolf, R.; Roosen, M.; Mys, N.; Ragaert, K.; De Meester, S., Towards closed-loop recycling of multilayer and coloured PET plastic waste by alkaline hydrolysis. *Green Chemistry* 2020, *22* (16), 5376-5394.

16. Uekert, T.; Singh, A.; DesVeaux, J. S.; Ghosh, T.; Bhatt, A.; Yadav, G.; Afzal, S.; Walzberg, J.; Knauer, K. M.; Nicholson, S. R., Technical, Economic, and Environmental Comparison of

Closed-Loop Recycling Technologies for Common Plastics. *ACS Sustainable Chemistry Engineering* 2023.

17. Shonnard, D.; Tipaldo, E.; Thompson, V.; Pearce, J.; Caneba, G.; Handler, R., Systems analysis for PET and olefin polymers in a circular economy. *Procedia CIRP* 2019, *80*, 602-606.

18. Chemstations Inc. CHEMCAD. Available at: https://www.chemstations.com/CHEMCAD/.

19. Carlson, E. C., Don't gamble with physical properties for simulations. *Chemical Engineering Progress* 1996, *92* (10), 35-46.

20. Chemstations Inc. CHEMCAD User Guide. 2022. Available at: https://www.chemstations.com/Support/Software_Updates_and_Notes/CHEMCAD_NXT_Support/.

21. Andrews, R.; Grulke, E.; Brandrup, J.; Immergut, E.; Grulke, E., *Polymer handbook*. John Wiley Sons 1999.

22. Franklin Associates. Life cycle impacts for postconsumer recycled resins: PET, HDPE, and PP. 2018. Available at: https://plasticsrecycling.org/images/apr/2018-APR-Recycled-Resin-Report.pdf (Last accessed: April 23, 2022).

23. Scheirs, J., *Polymer recycling: science, technology and applications*. Wiley New York: 1998; Vol. 132.

24. Institute of Scrap Recycling Industries, I. *ISRI Scrap Specifications Circular: Guidelines for Plastic Scrap*. 2022. Available at: https://www.isri.org/recycled-commodities/scrap-specifications-circular.

25. The Association of Plastic Recyclers. Model Bale Specifications. Available at: https://plasticsrecycling.org/model-bale-specifications.

26. Chen, W.; Yang, Y.; Lan, X.; Zhang, B.; Zhang, X.; Mu, T., Biomass-derived γ-valerolactone: Efficient dissolution and accelerated alkaline hydrolysis of polyethylene terephthalate. *Green Chemistry* 2021, *23* (11), 4065-4073.

27. Alonso, D. M.; Wettstein, S. G.; Dumesic, J., Gamma-valerolactone, a sustainable platform molecule derived from lignocellulosic biomass. *Green Chemistry* 2013, *15* (3), 584-595.

28. Poulakis, J.; Papaspyrides, C., Dissolution/reprecipitation: a model process for PET bottle recycling. *Journal of Applied Polymer Science* 2001, *81* (1), 91-95.

29. Achilias, D.; Giannoulis, A.; Papageorgiou, G., Recycling of polymers from plastic packaging materials using the dissolution–reprecipitation technique. *Polymer Bulletin* 2009, *63* (3), 449-465.

30. Dutta, A.; Sahir, A.; Tan, E.; Humbird, D.; Snowden-Swan, L. J.; Meyer, P. A.; Ross, J.; Sexton, D.; Yap, R.; Lukas, J. *Process design and economics for the conversion of lignocellulosic biomass to hydrocarbon fuels: Thermochemical research pathways with in situ and ex situ upgrading of fast pyrolysis vapors*; Pacific Northwest National Lab.(PNNL), Richland, WA (United States): 2015.

31. Gracida-Alvarez, U. R.; Winjobi, O.; Sacramento-Rivero, J. C.; Shonnard, D. R., System analyses of high-value chemicals and fuels from a waste high-density polyethylene refinery. Part 1: Conceptual design and techno-economic assessment. *ACS Sustainable Chemistry Engineering* 2019, *7* (22), 18254-18266.

32. Towler, G.; Sinnott, R., *Chemical engineering design: principles, practice and economics of plant and process design*. Butterworth-Heinemann: 2021.

33. Charles Maxwell. Chemical Engineering Plant Cost Indices. 2020. Available at: https://www.toweringskills.com/financial-analysis/cost-indices/.

34. Resource Recycling Systems (RRS). 2022. Available at: https://recycle.com/.

35. Alonso, D. M.; Hakim, S. H.; Zhou, S.; Won, W.; Hosseinaei, O.; Tao, J.; Garcia-Negron, V.; Motagamwala, A. H.; Mellmer, M. A.; Huang, K., Increasing the revenue from lignocellulosic biomass: Maximizing feedstock utilization. *Science advances* 2017, *3* (5), e1603301.

36. Environmental Research & Education Foundation. Analysis of MSW Landfill Tipping Fees: 2021. 2021. Available at: https://erefdn.org/product/analysis-of-msw-landfill-tipping-fees-2021-pdf/.

37. U.S. Energy Information Administration (EIA). Electric Power Annual: Table 2.4 Average Price of Electricity to Ultimate Customers. 2019. Available at: https://www.eia.gov/electricity/annual/.

38. Richard Turton, R. C. B., Wallace B. Whiting, Joseph A. Shaeiwitz, Debangsu Bhattacharya. *Analysis Synthesis and Design of Chemical Processes 5th Edition*. 2018. Available at: https://richardturton.faculty.wvu.edu/publications/analysis-synthesis-and-design-of-chemical-processes-5th-edition.

39. U.S. Energy Information Administration (EIA). United States Natural Gas Industrial Price. 2019. Available at: https://www.eia.gov/dnav/ng/hist/n3035us3A.htm.

40. U.S. Bureau Of Labor Statistics. National Occupational Employment and Wage Estimates. 2021. Available at: https://www.bls.gov/oes/current/oes_nat.htm#00-0000.

41. Kulas, D. G.; Zolghadr, A.; Chaudhari, U. S.; Shonnard, D. R., Economic and environmental analysis of plastics pyrolysis after secondary sortation of mixed plastic waste. *Journal of Cleaner Production* 2022, 135542.

42. Closed Loop Partners. Cleaning the rPET Stream. 2017. Available at: https://www.closedloop-partners.com/foundation-articles/cleaning-the-rept-stream/.

43. Wernet, G.; Bauer, C.; Steubing, B.; Reinhard, J.; Moreno-Ruiz, E.; Weidema, B., The ecoinvent database version 3 (part I): overview and methodology. *The International Journal of Life Cycle Assessment* 2016, *21* (9), 1218-1230.

44. PRé. SimaPro Database Manual Methods Library. 2020. Available at: https://simapro.com/wp-content/uploads/2020/06/DatabaseManualMethods.pdf.

45. del Carmen Munguía-López, A.; Göreke, D.; Sánchez-Rivera, K. L.; Aguirre-Villegas, H. A.; Avraamidou, S.; Huber, G.; Zavala, V. M., Quantifying the Environmental Benefits of a Solvent-Based Separation Process for Multilayer Plastic Films. 2022.

21

Techno-Economic Analysis of a Material Recovery Facility Employing Robotic Sorting Technology

S.M. Mizanur Rahman and Barbara K. Reck*

Yale School of the Environment, New Haven, CT, USA

Abstract

Over the past decade, robots have emerged as a new sorting technology for material recovery facilities (MRFs), enabled through dramatic advances in robotics and artificial intelligence (AI). These advances allow robots to become 'smart' by coupling them with AI driven vision systems, able to distinguish recyclables by material type. By integrating robotics, MRFs hope to increase sorting speed and accuracy, reduce their residuals, and to become more resilient towards worker shortages. To better understand the economic implications, this study presents a techno-economic analysis of a representative MRF in the U.S. that integrates robotics and compare it to a similar MRF without robotics integration. We compare the metrics net present value, discounted internal rate of return, and payback period for a mid-size MRF with and without robotic integration and add an uncertainty analysis to inform about the most important factors to consider. The results of the techno-economic analysis can inform MRF operators in their future decision-making.

Keywords: Techno-economic analysis (TEA), material recovery facility (MRF), recycling, municipal solid waste (MSW), robotic sorting, net present value, discounted internal rate of return, payback period

21.1   Introduction

Efficient recycling is the cornerstone of a circular economy. The idea of minimizing the use of primary raw materials depends on a sufficient supply of high-quality recyclables, in other words, recyclables have to be collected and sorted in quantities sufficient to meet current demand while ensuring that the quality requirements in material production are met [1, 2]. The economics of recycling favor material groups where the commodity price for virgin (primary) materials is substantially higher than that of secondary (scrap) materials. This is true for metals, but rarely the case for plastics. Metal recycling is further favored since well-established technologies exist to separate ferrous (through magnets) and non-ferrous (through eddy current separators) metals from each other and from other materials. Plastics, in turn, are much harder to separate into clean streams of recyclables despite

*Corresponding author: barbara.reck@yale.edu

Nabil Nasr (ed.) Technology Innovation for the Circular Economy: Recycling, Remanufacturing, Design, Systems Analysis and Logistics, (269–278) © 2024 Scrivener Publishing LLC

advanced optical sorters since economies of scale are hindered by the plethora of different compositions that exist, reflecting, to name a few, different plastic types, additives (differing by product group and manufacturer), color, and stringent requirements to keep food-grade and non-food-grade recyclables separate.

Municipal solid waste (MSW) is one of the largest and most complex waste streams in the U.S., and material recovery facilities (MRFs) are central for sorting its different waste components into valuable recyclables. The number of MRFs in the U.S. has grown to more than 350 [3] varying widely in size, throughput (100-900 tons/day), employment (large automized MRFs employ fewer workers per throughput than small MRFs), and cost [4]. The average MRF size has grown, too, in response to an increase in absolute MSW generation [5], and the incentives of economies of scale. Yet, the potential for growth remains large considering that in 2018 only 24% of the 292 million tons MSW generated were collected for recycling while 50% were disposed of in landfills (down from 89% in 1980) [6] and the remainder was incinerated or composted [5].

Most MRFs receive the recyclables from single-stream collection in a mixed material stream that contains metals, plastics, glass, and paper and cardboard, but excludes organics, textiles, and electronics. Separating these diverse materials requires a series of different separation steps that are largely automized, with workers doing the remaining sorting [7, 8]. The level of automation depends on the size of the MRF, with large MRFs generally having substantially higher levels of automation than small MRFs.

Over the past years, the role of manual labor in MRFs has become a growing issue of concern, both from a worker and MRF employer perspective. Working conditions are hard, summarized by Gibson [9] as "dull, dirty, and dangerous". Repetitive sorting, exposure to strong odors and seasonal temperature swings, and hazards from sharp and potentially contaminated objects create work conditions that are unattractive [10, 11], leading to high sick rates and turnovers in many MRFs and making the filling of open positions difficult [10]. Other challenges for the MRF economics include the high volatility of scrap market prices, changes in quality requirements for the output recyclables, and the large regional and seasonal variability in the composition of the incoming waste stream [12].

The combination of these factors explains why the integration of sorting robots for MRFs has grown rapidly over the past years. Robots, coupled with artificial intelligence (AI)-driven vision systems, offer the potential of fast and reliable sorting [13, 14]. They can be considered part of the envisioned Industry 4.0 that improves the overall sustainability of a system [15] by increasing the circularity of materials [16]. Robots can be used either as the main sorters or positioned as quality control (negative picking of unwanted materials) at the end of the sorting line. Though in a different context, a recent techno-economic analysis (TEA) that evaluated the sorting of cast and wrought aluminum alloys scrap showed promising results, estimating the payback period for the sorting installation to be between 3-5 months [17].

In a MRF set-up, robots offer the opportunity to substantially reduce the need for manual labor while increasing the sorting efficiency, thus increasing the resilience of the MRF operator towards the risks associated with fluctuations in worker availability and growing labor costs. This study provides a preliminary techno-economic analysis to test the economic implications of adding robots to a mid-sized MRF, with the role of robots assigned to quality-control functions.

21.2   Methodology

The objective of this study is to evaluate the economic impact of introducing robots in a medium-sized MRF by comparing it to a similar MRF without robots. We define this medium-sized MRF to have an annual throughput of 93,600 metric tons (MT), a size that would make it the 50-th largest MRF in 2021 [18]. This estimate is based on an hourly throughput of 30 tons per hour, 1.5 shifts per day at 8 hours per shift, and 260 days per year of operation (5 days per week for 52 weeks).

21.2.1   System Boundary

Figure 21.1 presents schematic diagrams of a non-robotic and a robotic MRF, starting from the municipal solid waste (MSW) entering the MRF gate to the recovered materials that are ready to be sold to the market for further downstream processing. The role of the robots is assumed to be limited to negative sorting, in which the quality of the outgoing material from the respective sorting technologies will be upgraded by precise picking up of the unwanted entries.

A typical MRF includes the following steps: after weighing, the recyclables are placed on the tipping floor and are later loaded onto the infeed conveyor belt with a front-end loader. After presorting by human workers, the glass breaking screen separates glass while the other materials continue to be carried over to the fiber screen, which separates fiber from the metal and plastic fractions. Magnets separate the steel cans followed by an eddy

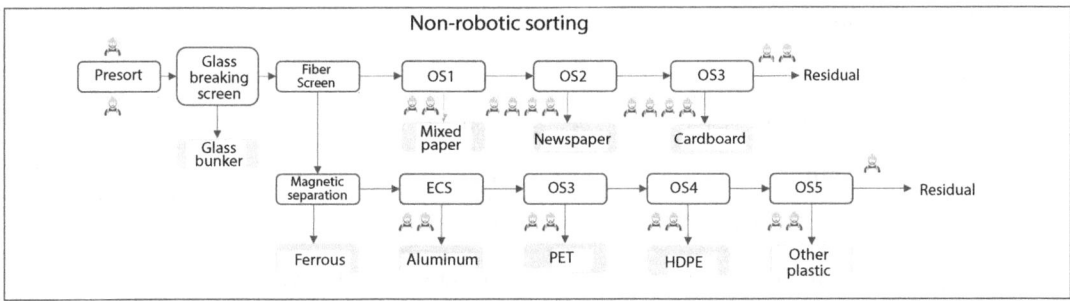

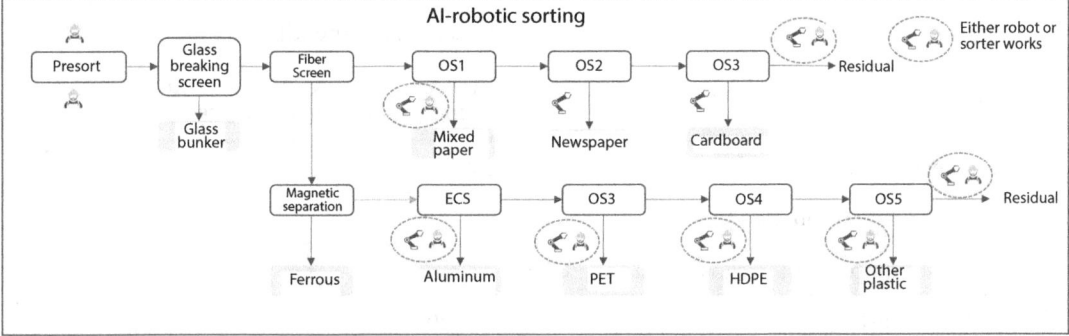

Figure 21.1 Schematic diagram of a medium sized non-robotic MRF and a robotic MRF. The dotted oval represents workstations that can be served by either a human worker or a robot. OS: Optical Sorter, ECS: Eddy Current Separator.

current separator (ECS) that separates aluminum used beverage cans (UBC). With six optical sorters in total, three optical sorters separate the fiber stream into mixed paper, residential papers and cardboards, while the other three separate PET, HDPE, and mixed plastics (plastic types #3-7). The residuals are sent to the landfills. The input composition of the MSW is taken from Metro Waste Authority [19] and EPA [5]. The material balance is hypothetically done based on a matrix for the nine output commodities with a purity achievable in each MRF; the purity rate is assumed to be 95% [14] and 80% for the robotic and non-robotic MRF while the material recovery rate is assumed to be 90% and 85% [19].

21.2.2   Data Collection and Assumptions for the Techno-Economic Analysis (TEA)

The TEA is intended to compare the economic benefits of the introduction of robots in a traditional MRF setting, using 20-year discounted cash flow analysis with an internal rate of return of 20%. To estimate the economic metrics such as net present value (NPV), discounted internal rate of return and payback period, several assumptions have been made as shown in Table 21.1. During the start-up year, 50% production is considered while 75% production level is considered for the second year. The construction period is estimated to be 1 year with 100% capital investment and 75% operating cost.

Table 21.1 Main assumptions used for the TEA of the robotic and non-robotic MRF.

Main assumptions for the TEA	
Main assumptions	Value
Lifetime (years)	20
Operating days (days/year)	260
Number of Shifts per day	1.5
Number of hours per shift	8
Depreciation method and recovery period	Straight-Line depreciation
Construction period	1 year (100% spending)
Start-up time (Years)	1 ((Year 1, OPEX 75%, Revenue 50%, Year 2, Revenue 75%))
Internal rate of return	20%
Loan interest rate (%) [20]	8
Total Equipment Purchase Cost (TEPC)	
Robot cost ($) [21]	300,000
Direct capital costs [22]	

(Continued)

Table 21.1 Main assumptions used for the TEA of the robotic and non-robotic MRF. (*Continued*)

Main assumptions for the TEA	
Main assumptions	Value
Installation and contingency [22]	50% of the TPEC
Land [22]	6.5% of the TPEC
Yard improvement [22]	6% of the TPEC
Buildings [22]	47% of the TPEC
Electrical [22]	20% of the TPEC
Piping [22]	30% of the TPEC
Instrumentation and control [22]	26% of the TPEC
Service facilities [22]	55% of the TPEC
Indirect capital costs	
Construction expenses [22]	34% of the TPEC
Cost of contractor fees [22]	19% of the TPEC
Cost of engineering and supervision [22]	32% of the TPEC
Contingency [22]	37% of the TPEC
Legal expenses [22]	4% of the TPEC
Start-up costs [22]	8% of the TPEC
Royalties [22]	6.5% of the TPEC
Annual throughput (Tons/year)	93600
Price of commodities (Steel Cans, Aluminum UBC, PET, HDPE, mixed plastic (3-7)) [23]	10 Year average
Price of commodities (Glass, Mixed paper, Newspaper and Cardboard) [23]	6-7 Year average
Waste disposal cost ($/ton) [19]	51

In a non-robotic MRF, a total of 36 sorters are employed in the sorting lines while in a robotic MRF, 12 sorters and 8 robots have been employed [4].

Total equipment purchase cost (TEPC) has been estimated by taking into account the equipment required for the sorting lines of robotic [19, 21] and non-robotic MRF [19] as shown in Figure 21.1. The difference in TEPC between the MRFs is the cost of a robot.

Fixed capital cost (direct and indirect) for robotic and non-robotic MRF has been estimated based on a percentage basis of the TEPC, while the value of some items (piping, buildings, yard improvement, service facilities, and land acquisition cost) have been kept constant, irrespective of TEPC, as robot involvement does not influence those costs. The total capital cost is in agreement with the MRF pro-forma model by RRS [24].

The operating cost (OPEX) involves no raw material acquisition cost. Labor, facility maintenance and utilities, equipment operation and maintenance, insurance, residual disposal and contingency and annual debt service has been included in the OPEX. Labor, residual disposal and robot utility and maintenance has been calculated by the authors, while Facility maintenance and equipment operation and management has been taken from a MRF feasibility study [19]. The non-robotic MRF workforce includes 14 high skilled employees and 36 sorters [4] while the robotic MRF includes 18 high skilled employees and 12 sorters. The hourly rate has been taken from the MRF feasibility study [19]. Debt servicing is estimated based on a loan interest rate of 8% [20] and for 20 years. The lifetime of the MRF equipment is expected to be the same for the non-robotic and robotic MRF.

Robot acquisition cost has been estimated as $300,000 [17, 21, 25]. The annual maintenance cost and energy consumption, lifetime years, and downtime cost (all based on [26]) have been estimated. The workload is considered to be the equivalent of three sorters, but the net purity increase has not been internalized in the revenue estimation. The revenue generation has been based on the increased material recovery and reduced landfill disposal fees.

21.3   Results and Discussion

Figure 21.2 summarizes the values of major parameters for an annual throughput of 93,600 MT. TEPC and total capital cost are higher for robotic MRF by 80% and 55% respectively, due to the cost of robot added to the TEPC. For the robotic MRF, the OPEX is higher by only 2% due to lower workforce cost (20%) and higher debt service (55%). In contrast, revenue generation is higher for the robotic MRF by 6%, leading to a net revenue of $5.2 million as opposed to $4.7 million for the non-robotic MRF.

To estimate NPV, discounted internal rate of return (IRR) and payback period, we use a 20-year model with a discount rate of 20%. It shows that the NPV is $31 million for the robotic MRF while $30 million for the non-robotic MRF including tipping fee $51/ton (Table 21.2). Discounted IRR is less favorable for the robotic MRF as the capital investment is 55% higher compared to the non-robotic MRF. Similarly, the payback period is greater

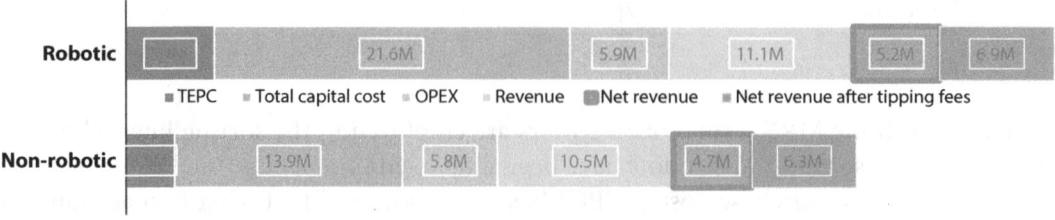

Figure 21.2 TEA results for robotic and non-robotic MRF for an annual throughput of 93,600 MT. TEPC: Total Equipment Purchase Cost, OPEX: Operating Cost.

Table 21.2 Comparison of TEA metrics between robotic and non-robotic MRF. NPV net present value, IRR internal rate of return.

Economic TEA metrics		
	Robotic	**Non-Robotic**
NPV	$31M	$30M
Discounted IRR	35%	55%
Payback period (Years)	2.14	1.54

for the robotic MRF. Under the given revenue generation assumption, it is important to note here that although NPV is slightly better for the robotic MRF, the investment into robots is not as attractive in terms of discounted IRR.

Revenue estimate is carried out based on the incoming mass composition [5], outgoing commodity composition [5, 19], transfer coefficient calculation and price information [23]. EPA recyclable composition has been adjusted to exclude food, wood, and textiles. Paper and cardboard (75%), plastics (5%), metal (14%) and glass (5%) were further distributed into nine outgoing commodities. For paper and cardboard, three categories are mixed paper (55%), sorted residential paper (20%) and cardboard; for metal, two categories are steel cans (75%) and aluminum (24%); for plastic, four categories are PET (52%), HDPE natural (22%), HDPE color (21%) and Mixed plastics (#3-5). The outgoing composition is taken from a MRF Feasibility Study [19]. To estimate the transfer co-efficient, we have assumed the recovery rate 85% and 90% for non-robotic and robotic MRF respectively, with a purity of 80% and 95%. Since the prices of recycled commodities fluctuate widely, we have chosen to use a 10-year average except for glass and paper products. For glass, we use a fixed price of -0.12 $/ton and for papers, an average of 6-7 years.

Total revenue estimated for the robotic MRF is $11.1 million whereas it is $10.5 million for the non-robotic MRF, 6% value addition by introducing robots. Aluminum cans share maximum revenue of 39%, followed by cardboard with 13%. A negative value is generated for mixed glass, about 0.004%.

Uncertainty Analysis

Three important parameters - robot price, purity-based value addition, and commodity market value - were analyzed for their uncertainties to understand the effect of the assumptions on the three metrics NPV, discounted IRR and payback period. In our uncertainty analysis, the robot price ranges between $100k and $300k, the value addition ranges from a 0% to 30% increase, and commodity prices are current, 5-year, and 10-year averages [23].

Robot price. The effect of the robot price on NPV shows that the robot price cannot exceed $300k, beyond which the NPV is lower for the robotic MRF. At a robot price of $100k, the NPV is $5 million higher than in the non-robotic MRF. In contrast, discounted IRR is lower for the robotic than the non-robotic MRF unless the robot price drops to $100k. Therefore, given the other parameters being the same, the robot price is required to be below $100k for the robotic MRF to be favorable over the non-robotic MRF. Like the discounted IRR, the payback period appears to be better only when the robot price is below $100k (Figure 21.3).

Purity-based value addition. The value addition due to high purity assumed to be achievable in the robotic MRF is discussed by increasing 10%, 20% and 30% of its base value. We find that with the base value, the NPV is $1 million higher than in the non-robotic MRF and it increases to $45 million, surpassing the non-robotic MRF by $15 million. However, the discounted IRR is lower at the base value, and it surpasses only when the value generation increases by more than 30%. Similarly, the payback period is greater at the base value, and it gets lower than the non-robotic MRF when the value generation exceeds 30%.

Commodity prices. The historic fluctuation of commodity prices is captured in this variable. We analyze the effect of using the current price, 5-year average, or 10-year average commodity price on the TEA metrics. We find that the current price has lowered NPV value down to $20 million from $31 million. Similarly, discounted IRR is also lower from about 35% to 20%. From this analysis we conclude that if commodity prices fall below the current commodity prices, the robotic MRF would not be attractive.

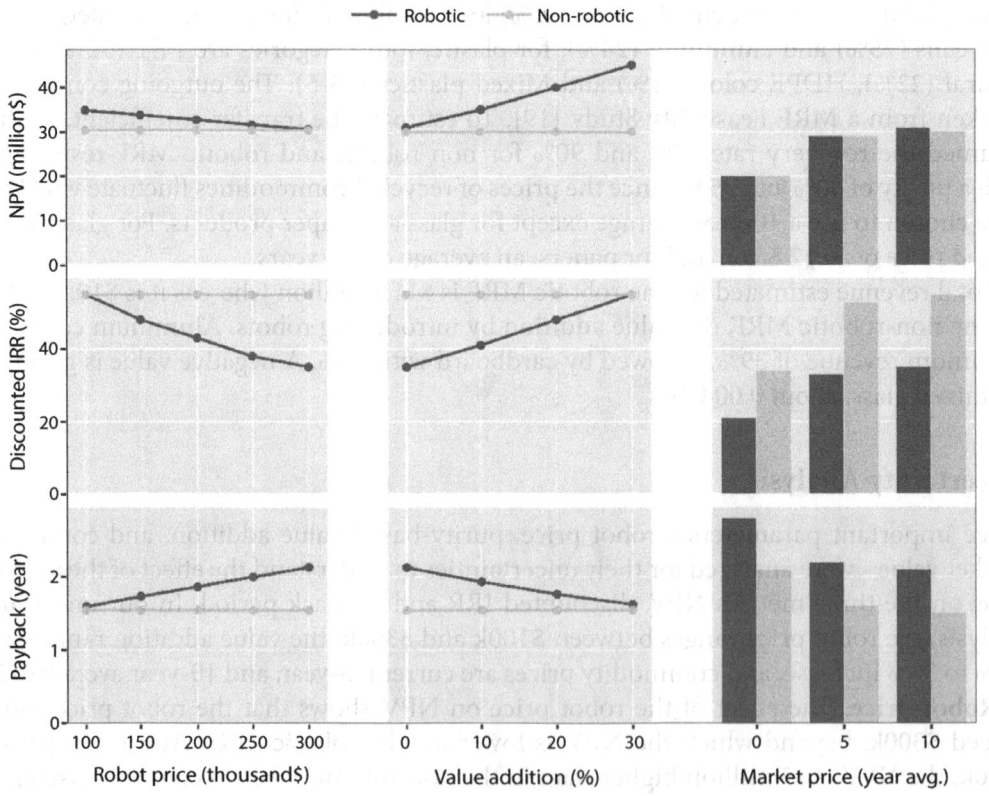

Figure 21.3 Results of changes of NPV (Net Present Value), discounted IRR (Internal Rate of Return) and payback period due to the variation of robotic price, value addition by percentage and commodity value average. (Effect of NPV, discounted IRR and payback period is from the left to the right. The robotic price, value addition and commodity value is shown from the top to the bottom.)

21.4   Conclusions & Recommendations

The TEA reveals that integrating robots into a medium-sized MRF has only a minor positive effect on the net present value. It should be noted that the price of a new robot and the difference in revenue of the recycled commodities are key drivers to determine if a robotic intervention is economically feasible. The lower the robot price the better, and a robot price of less than $100k leads to favorable outcome for robot integration based on all three metrics (NPV, discounted IRR, and payback period). To perform better with respect to all three metrics, the robotic introduction should add value by more than 30%. Finally, as discussed earlier, the 10-year average is an optimistic market scenario, and if the market price falls below the current price, robotic intervention would not be interesting. It is also important to note that robotic intervention reduces human-waste interface, which might be an interesting feature, and that the economics presented here are based on highly reliable robots that do not cause downtime triggered by robotic malfunction.

We emphasize that this study is based on a range of assumptions that should be viewed critically. In addition to the estimates already discussed, another assumption to question is if the throughput is the same in robotic and non-robotic MRFs. With increasing technological maturity, the throughput of robot employing MRFs can be expected to exceed that of non-robot MRFs in the near future. Future research could also analyze if robots can play a greater contributive role rather than the quality control role discussed here; an assumption that would change the economics and likely only be attractive in smaller MRFs with low levels of automation (since it is currently unlikely that a robot could economically replace an optical sorter, magnet, or ECA). Finally, MRF operators might find the framework of this techno-economic analysis useful, by replacing our informed estimates with their own proprietary data to achieve results relevant for their future decision-making.

Acknowledgements

This research was funded by the United States National Science Foundation under Award 1928448. We thank our colleagues Berk Calli, Marian Chertow, and Amy Wrzesniewski for insightful discussions.

References

1. Baxter W, Aurisicchio M, Childs P. Contaminated Interaction: Another Barrier to Circular Material Flows. *Journal of Industrial Ecology* 2017, 21(3): 507-516.
2. Geissdoerfer M, Savaget P, Bocken NMP, Hultink EJ. The Circular Economy – A new sustainability paradigm? *Journal of Cleaner Production* 2017, 143: 757-768.
3. Staub, C. (2022, May 20). Mapping out MRF infrastructure nationwide. Retrieved from https://resource-recycling.com/recycling/2019/10/22/mapping-out-mrf-infrastructure-nationwide/
4. Powell J. Sortation by the numbers. Resource Recycling. 2018 Oct 1. https://resource-recycling.com/recycling/2018/10/01/sortation-by-the-numbers/

5. EPA. National Overview: Facts and Figures on Materials, Wastes and Recycling 2022 Dec 3 [cited] Available from: https://www.epa.gov/facts-and-figures-about-materials-waste-and-recycling/national-overview-facts-and-figures-materials.

6. ASCE. 2021 Infrastructure Report Card. Reston, VA: American Society of Civil Engineers; 2021.

7. Cimpan C, Maul A, Jansen M, Pretz T, Wenzel H. Central sorting and recovery of MSW recyclable materials: A review of technological state-of-the-art, cases, practice and implications for materials recycling. *J Environ Manage* 2015, 156: 181-199.

8. Gundupalli SP, Hait S, Thakur A. A review on automated sorting of source-separated municipal solid waste for recycling. *Waste Manage* 2017, 60: 56-74.

9. Gibson T. Recycling Robots. *Mechanical Engineering* 2020, 142(01): 32-37.

10. Rosengren C. High risk, hidden workforce. WasteDive. 2019 Dec 11. https://www.wastedive.com/news/recycling-labor-mrf-high-risk-hidden-workforce/568550/

11. State of Oregon. Study of Material Recovery Facility Workers: Recycling Steering Committee; 2020 Aug 31.

12. Ip K, Testa M, Raymond A, Graves SC, Gutowski T. Performance evaluation of material separation in a material recovery facility using a network flow model. *Resources, Conservation and Recycling* 2018, 131: 192-205.

13. Koskinopoulou M, Raptopoulos F, Papadopoulos G, Mavrakis N, Maniadakis M. Robotic Waste Sorting Technology. *IEEE Robotics & Automation Magazine* 2021(June): 50-60.

14. Wilts H, Garcia BR, Garlito RG, Gómez LS, Prieto EG. Artificial Intelligence in the Sorting of Municipal Waste as an Enabler of the Circular Economy. *Resources* 2021, 10(4): 28.

15. Bai C, Dallasega P, Orzes G, Sarkis J. Industry 4.0 technologies assessment: A sustainability perspective. *International Journal of Production Economics* 2020, 229: 107776.

16. Rahman SMM, Perry N, Müller JM, Kim J, Laratte B. End-of-Life in industry 4.0: Ignored as before? *Resources, Conservation and Recycling* 2020, 154: 104539.

17. Engelen B, Marelle DD, Diaz-Romero DJ, den Eynde SV, Zaplana I, Peeters JR, *et al.* Techno-Economic Assessment of Robotic Sorting of Aluminium Scrap. *Procedia CIRP* 2022, 105: 152-157.

18. Recycling Today. Map & List of North America's Largest MRFs. Valley View, OH: GIE Media, Inc.; 2021. https://www.recyclingmarkets.net/secondarymaterials/index.html

19. Metro Waste Authority. Materials Recovery Facility (MRF) Feasibility Study. Omaha, NE: HDR; 2018 June. https://resource-recycling.com/resourcerecycling/wp-content/uploads/2020/06/December_2018_Board_Packet-23-94.pdf

20. Lan K, Ou L, Park S, Kelley SS, English BC, Yu TE, *et al.* Techno-Economic Analysis of decentralized preprocessing systems for fast pyrolysis biorefineries with blended feedstocks in the southeastern United States. *Renewable and Sustainable Energy Reviews* 2021, 143: 110881.

21. Fairchild M. Recycling Robots: How They Work & Are They Worth the Investment? #HowToRobot. 2022. https://howtorobot.com/expert-insight/recycling-robots

22. Mairizal AQ, Sembada AY, Tse KM, Haque N, Rhamdhani MA. Techno-economic analysis of waste PCB recycling in Australia. *Resources, Conservation and Recycling* 2023, 190: 106784.

23. RecyclingMarkets.net. Secondary Materials Pricing (SMP). New York: Recycling Markets Limited; 2023.

24. Graff S. MRF Financial Pro-Forma Model Summary. Ann Arbor, MI: RRS; 2018.

25. Sandiland D. Stop Spending Millions on Robot Downtime Now. Robotics 24/7. 2022 Jun 2. https://www.robotics247.com/article/stop_spending_millions_on_robot_downtime_now

26. Anandan TM. Calculating Your ROI for Robotic Automation: Cost vs. Cash Flow. Association for Advanced Automation. 2015 Mar 19. https://www.automate.org/industry-insights/calculating-your-roi-for-robotic-automation-cost-vs-cash-flow

22

Key Strategies in Industry for Circular Economy: Analysis of Remanufacturing and Beneficial Reuse

Subodh Chaudhari[1]*, Sachin Nimbalkar[1], Bruce Lung[2], Marco Gonzalez[3], Bert Hill[4] and Bryant Esch[3]

[1]Oak Ridge National Laboratory, Manufacturing Science Division, Oak Ridge, TN, USA
[2]US Department of Energy, Boston Global Services, Washington, DC, USA
[3]Waupaca Foundry, Waupaca Foundry, Waupaca, WI, USA
[4]Volvo Group North America, Greensboro, NC, USA

Abstract

Manufacturing, in the effort to be more sustainable, is increasingly focusing on energy efficiency and waste reduction. DOE's Better Buildings Better Plants Program has established Waste Reduction Network that works with 32 industrial partners to achieve higher material efficiency and reduce waste. United States generates 7.6 Billion Tons of industrial solid waste as estimated by EPA. In a linear economy as the economy grows, so does the waste – increasing strain on resources and the environment. The Circular Economy (CE) model keeps the available resources in circulation for longer period of time easing the burden on the environment.

Remanufacturing and (Beneficial) reuse are widely accepted channels in 9R methodology and established pillars of CE. This chapter reviews the two key strategies and their adoption in different industrial sectors. It reviews the key barriers faced by manufacturers in implementing these methodologies and discusses the possible solutions to those barriers. The chapter also reviews impact of these CE strategies on sustainability, material efficiency and the economic and social benefits. Finally, this chapter presents two case studies from DOE's Better Plants partners – one on remanufacturing of components in heavy vehicles industry and one on beneficial reuse of spent foundry sand, a non-hazardous solid waste, and discusses the project impacts.

Keywords: Circular economy, manufacturing waste reduction, remanufacturing, repurpose, reuse, better plants, 9R methodology

22.1 Introduction

Manufacturing is going through a transition to become more resource efficient and sustainable over time. There is a growing consensus that current state of production, with its resultant resource consumption, is highly unsustainable because it places excessive pressure

*Corresponding author: chaudharisa@ornl.gov

Nabil Nasr (ed.) Technology Innovation for the Circular Economy: Recycling, Remanufacturing, Design, Systems Analysis and Logistics, (279–296) © 2024 Scrivener Publishing LLC

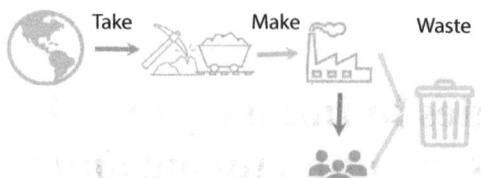

Figure 22.1 Linear economy model.

on the environment due to resource consumption and pollution. The United States has set ambitious goal to be carbon neutral by 2050 [1]. Waste generation and landfills are a significant part of anthropogenic greenhouse gas emissions. The traditional manufacturing business model was a linear economy model of production in which manufacturers extract raw materials, use them to create a product, and then dispose the waste generated from production process as well as the product once it reaches the end of its useful life. This is also referred to as Extract-Make-Dispose model (Figure 22.1). In this model, manufacturers prioritize the maximization of profitability, without a long-term consideration for sustainability or the environmental impact of the waste generated in production processes. While this approach enabled rapid industrialization and growth in consumption since the industrial revolution, it represents a one-way flow of materials from mining to landfills. The linear value chains have led material consumption globally from 28.6 giga tons (Gt) in 1972 to 101.4 Gt in 2021. In a business-as-usual scenario, it is projected to grow to 170-184 Gt by 2050 [2].

22.1.1 Background on Circular Economy

The CE focuses on resource management so that materials and products are designed, used, and reused in such a way that waste is minimized, and product life cycle is extended. In the manufacturing industry, the circular economy can be applied to reduce the environmental impacts of production and improve the efficiency of resource use. In the CE system, the flow of materials extracted from earth is minimized and all existing materials in circulation are kept in circulation for the maximum possible amount of time. There are two main sections, a technical cycle for all mechanical/electronic/chemical products that can follow reduce, reuse, repair, remanufacture, repurpose, and recycle stages until the point at which they are no longer usable. The second portion, a biological cycle is for biodegradables such as food and other organic material [3]. Currently, the global economy is largely linear and has circularity score of 8.6% [2], which could indicate that there is substantial room for improvement to achieve a circular economy.

The benefits of CE deployment in manufacturing sector are multifold. The most obvious benefits are realized in terms of waste reduction and reduced waste disposal costs. A deeper analysis reveals benefits in all the areas of sustainability, viz., environment, economics, and society [4], as indicated in Figure 22.2. While the ultimate benefit of CE is theorized as the decoupling of economic growth from extraction of earth's resources, the direct impact for manufacturers results from improvement in resource efficiency, i.e., reduced need for raw materials and other resources. The improved resource efficiency is followed by reduced costs as result of reduced disposal costs, reduced cost of raw materials, other resources,

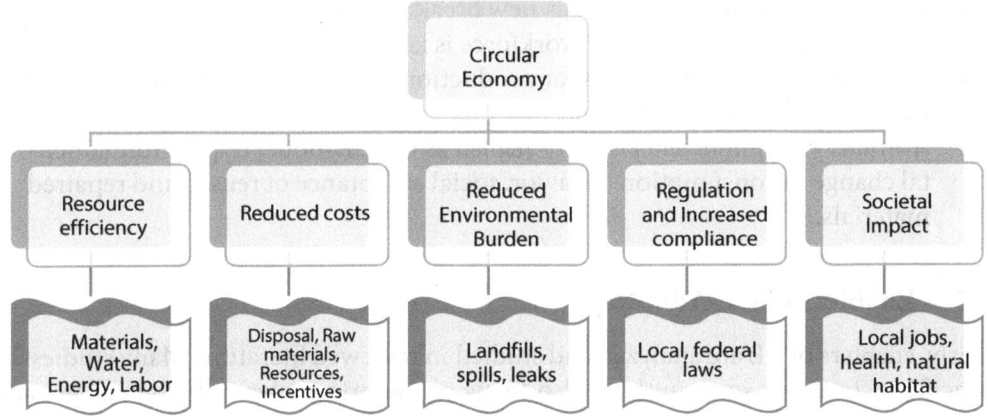

Figure 22.2 Circular economy proven benefits.

and if applicable other incentives and tax benefits. The CE strategy reduces environmental impact and carries other societal benefits with it such as local job creation, community partnerships, health benefits, and natural habitat improvement. These benefits are summarized in Figure 22.2.

CE is an emerging sustainability paradigm and has already seen favorable reception across range of manufacturing sectors with varying adoption rates. Some of the sectors that have largely successful implementation include apparel/clothing manufacturing, automobile and automotive components manufacturing, construction industry, and electronics manufacturing. The packaging and plastics industry is rapidly transforming [5, 6]. There are many other examples of manufacturing sectors taking steps to deploy CE. From this context it is helpful to look at the barriers faced by manufacturers as well as the enablers that could propel CE adoption onward.

22.1.2   Barriers to CE Adoption

Barriers to CE adoption are well studied in literature for manufacturing. Jaeger and Upadhyay studied barriers listed in literature and verified those with barriers identified from surveys and interviews of 10 varying size manufacturing facilities [7]. Ritzén and Sandström analyzed and categorized the CE deployment barriers [8]. V. Kumar *et al.* provided comprehensive literature review and summarized socio-political, economical, and environmental barriers to CE [9]. The most prominent barriers are –

- *Financial* as the CE transition requires high initial cost for manufacturing methods modification, re-tooling, factory relocation, and sourcing changes.
- *Structural/governmental* as CE needs proper establishment of infrastructure for facilitating reuse, recollection, recycling of products along with policies, programs and incentive structures to promote sustainable practices.
- *Organizational* as CE needs newer policies and guidance.

- *Technological and skills related* as new breakthrough technologies are needed to enable CE as well as skilled workforce is key to apply existing technologies.
- *Logistical* as the supply chains for production and consumption have become increasingly complex over time.
- *Attitude, perception, and behavior related* as the CE model require fundamental change in consumption behavior, social acceptance of reused and repaired materials.

22.1.3   Enablers of CE Adoption

Similarly, enablers of CE are analyzed and studied in reviewed literature. Many studies focus on specific sectors, but some authors also focus on generic analysis of CE enablers across diverse manufacturers. Rizos *et al.* studied barriers and enablers in small and medium enterprises (SME) through case studies of 30 manufacturing facilities [10]. Bressanelli, *et al.* analyzed enablers, levers, and benefits of CE by analysis of supply chain of a manufacturing sector though comprehensive literature review [4]. Following enablers of CE are noteworthy –

- *Green culture and brand improvement* as many studies have found that green culture in an organization and brand image contribute significantly to their readiness to adopt CE initiative. Societal acceptance and expectation of sustainable practices is another highly important factor.
- *Financial* as overall economical attractiveness is the ultimate driver of CE projects.
- *Digitalization* as adoption of IIoT, Industry 4.0, Additive Manufacturing can enable CE adoption through data access, analytics, user behavior knowledge and management of entire product life cycle.
- *Collaboration and Networks* as knowledge sharing and best practices benchmarking through industry groups and networks as well as support from supply chain organizations can establish CE enabled products
- *Governmental* as incentives and long-term tax advantages can cause initial economic returns to become favorable. New policies can drive growth.

22.2   Pathways to CE in Manufacturing Operations

There are three principal components applicable to manufacturing that contribute to improved circularity, viz., (1) smarter design, specification, and manufacture, (2) extended use phase, and (3) maximum value extraction at end-of-life [11]. The three components are further sub-divided into multiple R-strategies as necessary for minimizing waste and optimizing sustainability. The original 3R concept introduced by United Nations Environment Program was based on Reduce-Reuse-Recycle and was shown to be shortsighted by Bradley, *et al.* as it focused on cradle-to-grave approach. Bradley, *et al.*, introduced 6R concept to focus on cradle-to-cradle product life cycle. The 6R concept introduced Redesign-Reduce-Remanufacture-Reuse-Recover-Recycle concept [12].

Blomsma, *et al.* tried to refine this methodology further by developing Circular Strategies Scanner and segregating it by area of application, i.e., business model, materials, or parts and products, etc. [13]. In the works by Ellen McArthur Foundation the materials and products are divided into technical and biological cycle. The technical cycle corresponds to most generic forms of circularity strategies applicable for manufacturing establishments. These strategies are discussed here briefly. There is embedded value of CE adoption in these strategies, i.e., strategies that reduce the need of basic materials resource extraction rank the highest. Recycling extracted basic materials already in circulation ranks on the lower spectrum and strategies such as energy recovery may rank even lower in the CE adoption value. Figure 22.3 shows CE adoption value hierarchy and major pathways for manufacturer as CE strategies.

22.2.1 Source Reduction

Source reduction is the most effective strategy for circular economy and was introduced early in waste reduction by US Environmental Protection Agency (EPA) after the Pollution Prevention Act in 1990. EPA defines source reduction as – "the practice of reducing the quantity or toxicity of waste generated prior to recycling, treatment or disposal, by changing the processes that generate pollution in the first place" [14]. In other words, any process or activity that reduces the basic need of consumption is classified in source reduction. EPA through Toxic Release Inventory (TRI) – Pollution Prevention (P2) program has tracked source reduction data on waste from 1990 through 2016 for more than 22,000 industrial facilities. These facilities implemented 450,000 source reduction projects over the years and demonstrated steady decline in the waste generation [15]. The source reduction strategy is broken down further as illustrated in Figure 22.3.

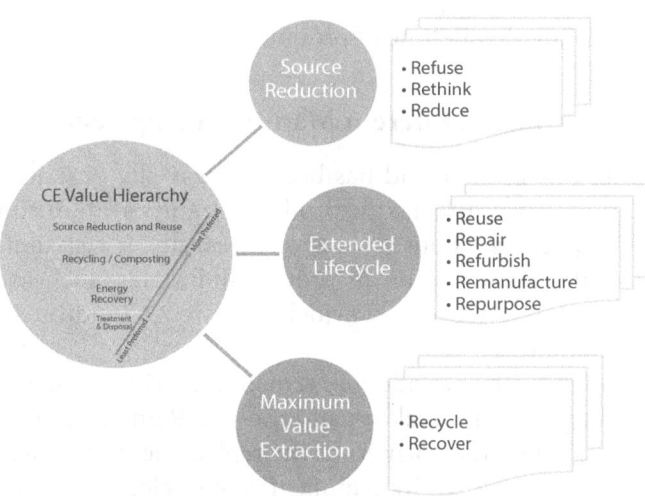

Figure 22.3 CE adoption value hierarchy and pathways.

22.2.2   Extended Life Cycle

In the second principle, CE adoption works by extending the use phase of the product life cycle. This idea is based on the fact that manufacturing of newer products is highly resource intensive and should be avoided as much as possible. The life cycle of a product can be extended by reuse, repair, refurbish, remanufacture, and repurpose techniques. It can also involve making the use phase more intense, by increasing utilization. This is illustrated by the example of tools use. A typical tool is only utilized 6% in its lifetime. This can be improved by sharing the tool with other users [16].

22.2.3   Maximum Value Extraction

At the end of life, CE adoption has the lowest value. However, it is still impactful for reducing environmental load of the product. The major pathways in this phase rely on extracting remaining materials for recycling and recovery. Recycling can have varying impact depending on the types of materials involved on material efficiency and energy efficiency. Some metals can bring significant energy savings through recycling on a life cycle level impact. Materials that cannot be recycled and are harmful to be landfilled should be considered for energy recovery. The CE pathways are further clarified with examples in Table 22.1.

22.3   Remanufacturing

EPA defines remanufacturing as *"The process of restoring used durable products to "new" condition, to be used in their original function, by replacing worn or damaged parts"*. Remanufacturing is a distinct process than repairing, refurbishing, and reusing. A generic remanufacturing process as illustrated in Figure 22.4 involves recovery, fault inspection, reconditioning, and final quality testing. Cleaning and sanitizing components combined with partial repair and few worn components replacement as necessary is provided. The product is resold with quality and specifications as a new product. It contributes to the circular economy by extending the lifetime of products.

22.3.1   Remanufacturing in Different Manufacturing Sectors

Remanufacturing is an old strategy and has been successfully practiced across manufacturing industries such as – furniture, textile products, transportation equipment, electrical equipment, computer equipment, household appliances, etc. As remanufacturing is assembly and hardware-based, its application in process industries, such as chemicals, foods, primary metals, is unclear and not readily evident. Figure 22.5 indicates where the remanufacturing fits in CE framework.

Electrical equipment industry has adopted remanufacturing for many products such as transformers, high voltage circuit breakers, and motors. Remanufacturing is a great alternative to replacement in failed transformers as the replacement of critical failed equipment can be easier than replacement [25]. Remanufacturing of electrical motors can utilize less than 20% of energy needed to manufacture new motor with similar performance in usage phase of the motors [26].

Table 22.1 Circular economy examples in industry for each 9R CE strategy pathway.

9R	Pathway	Example in Industry	Reference
0R	Refuse	Since the launch of online streaming, file sharing, cloud storage platforms, the need for manufacture of information media such as CDs, DVDs, portable storage disks such as floppy etc. has reduced.	-
1R	Rethink	Caterpillar has acquired Yardclub, a platform that facilitates equipment sharing. This will intensify the utilization of the Caterpillar products.	[17]
2R	Reduce	Food manufacturing companies are redesigning packaging cardboard and plastics by reducing the weight of the packaging material for higher material efficiency.	-
3R	Reuse	Ikea, a Swedish furniture manufacturer, has launched second-hand furniture initiative for lightly used and unwanted old Ikea products, that will be bought back from customers and resold at the stores.	[18]
4R	Repair	Dell as a part of corporate commitment considers repairability of the products from design phase.	[19]
5R	Refurbish	Tata Motors customers need high-quality low-cost part replacement therefore they started refurbishing and reconditioning aftermarket parts that are returned in acceptable condition. In 2016 a total of 23,115 parts were refurbished.	[20]
6R	Remanufacturing	Canon offers remanufactured printer devices since 1992. As part of usual business strategy Canon collects old devices from market, then remanufactures them reusing at least 80% original materials and reducing environmental impact by more than 80%. Canon offers original specifications and warranty on remanufactured products.	[21, 22]
7R	Repurpose	Some manufacturers with transportation fleets have worked with tire recyclers to repurpose worn tires to create rubber crumb, playground mulch, and fencing.	[23]
8R	Recycle	Thousand Fell a shoe manufacturer, collects old used pairs of shoes at end of life from customers offering $20 which can be used towards new pair. The old shoe materials are recovered and recycled.	[18]
9R	Recover	Umicore - a precious metal mining company uses electronic waste to recover metals like copper, lead, nickel along with bismuth, tin, indium etc.	[24]

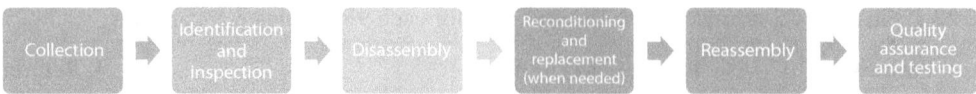

Figure 22.4 Typical steps involved in a remanufacturing operation.

Remanufacturing in household appliances is evident from literature as well. Appliances such as refrigerators, washing machines, dishwashers have been remanufactured. Some cases studied by Gutowski *et al.* appliances reveal that remanufacturing can lose efficiency and can end up using more energy in the use-phase.

Remanufacturing of office furniture made with wood or metal is a proven application of remanufacturing and has over half of the market share [27]. Office furniture such as desks and chair are not easily damaged and can be remanufactured at the end of useful life. The energy savings in both furniture and textile remanufacturing are significant (over 85%) as compared to new product.

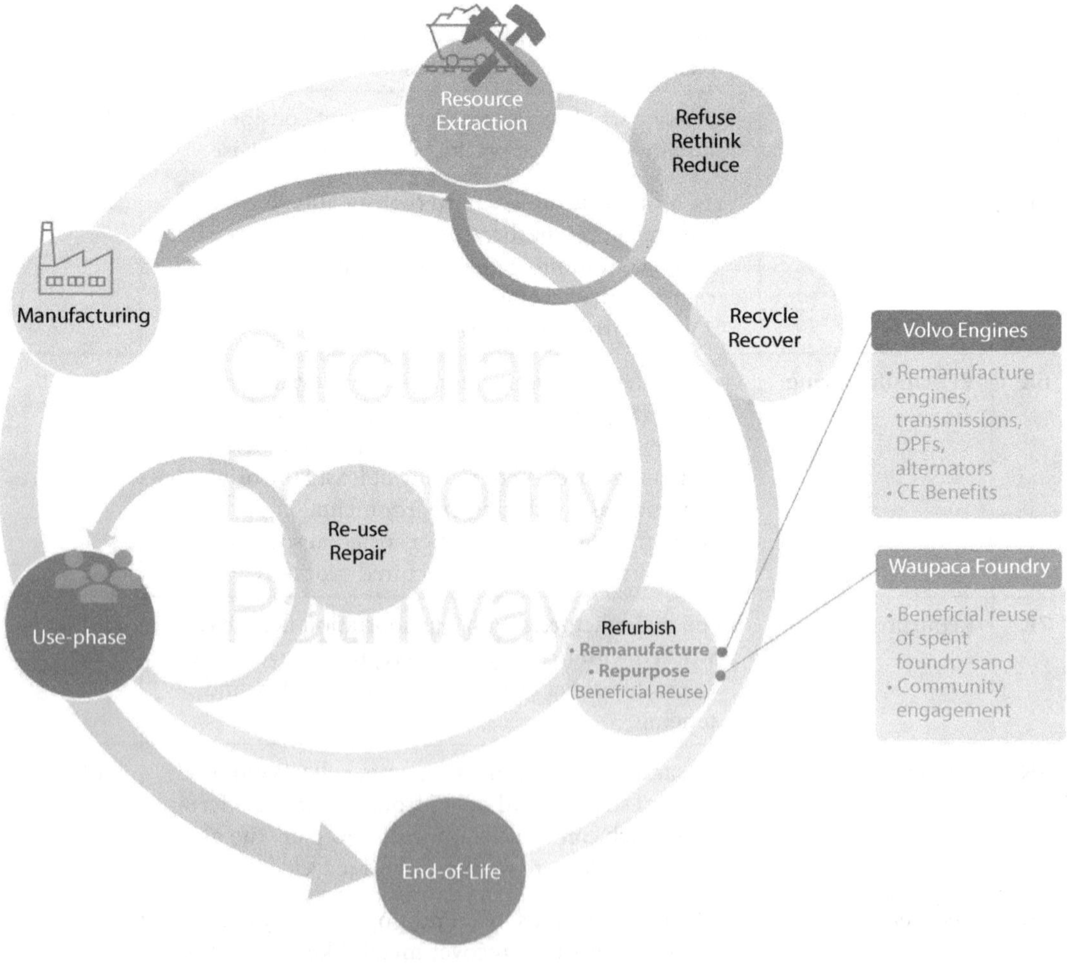

Figure 22.5 Circular economy pathways and remanufacturing and beneficial reuse case studies.

Remanufacturing in automotive parts manufacturing industry is a popular solution for failed parts such as engines, transmission, heavy duty truck tires and even electric vehicle batteries. Used heavy duty truck tires can also be rebuilt through retreading process. One of the important processes to lower vehicle battery cost is to remanufacture discarded vehicle batteries [28]. Engine remanufacturing is very popular in automotive industry to increase vehicle life cycle and has been provided by many small and major manufacturers. A case study on the engine remanufacturing at Volvo Reman Plants is presented in the following section.

22.3.2   Enablers and Barriers to Remanufacturing

The economic drivers for remanufacturing include reduced costs of goods sold, reduced prices to the customer, supply risk mitigation, improved customer satisfaction, and stronger value chain relationships. The other indirect enablers for remanufacturing in circular economy can include increased market share, brand enhancement, brand loyalty, and improved regulatory compliance.

The barriers to remanufacturing include non-conducive product design, high initial cost to change product design, reduction in new product sales, negative perception towards remanufactured products, complex remanufacturing process, non-availability of skilled and economic labor, non-availability of replacement parts, lack of infrastructure to facilitate remanufacturing, and regulation and policy. A few of these barriers are easily resolved but others can be systemic and difficult to change [29].

22.3.3   Case Study – I: Volvo Group – Remanufacturing

22.3.3.1   Background and Description

Volvo Group is a major manufacturer committed to sustainability and focused on people, climate, and resources as 100% Safe – 100% fossil fuel free – 100% more productive as indicated in their sustainability strategy. At Volvo 'Every End is a New Beginning' which guides their sustainability philosophy. Volvo Group is at the forefront of the CE and driving the industry transformation.

Volvo follows 9R framework for CE and works to design out waste and pollution, keep products and materials in use, and regenerate natural systems. Volvo Remanufacturing is a global operation at 9 sites in 6 countries. Volvo defines "Reman" as remanufacturing and returning a product to at least its original performance with a warranty that is equivalent or better than the newly manufactured product". Volvo started remanufacturing early in 1950's and has developed sophisticated methods for remanufacturing engines, transmissions, Diesel Particulate Filters (DPF), alternators, and many other components. Figure 22.5 shows Volvo's remanufacturing strategy on CE pathways. The challenge of remanufacturing is that it is a lower volume and higher variation type of manufacturing. Through 70 years' experience Volvo can determine which parts are 100% safe to be reutilized as opposed to the parts that need replaced. Volvo has identified a goal to increase its remanufacturing, refurbish, repair, reuse, and recycle business 60% by 2025 from 2018 baseline.

22.3.3.2   Benefits

Volvo sells remanufactured product at significantly lower price which can bring real cost savings for the customers and enhance the relationship with customers. The savings in energy consumption and CO_2 emissions are also significant. Table 22.2 below shows achieved material, energy, and CO_2 savings from remanufacturing 4 components when the whole production chain is considered. Through these benefits Volvo is able to drive sustainability, customer satisfaction, business objectives which is in line with Win-Win-Win concept at Volvo.

22.4   Beneficial Reuse

EPA defines 'Beneficial Reuse' as the use of waste materials which otherwise will be discarded, in an alternate function within same facility or elsewhere. The beneficial reuse turns the waste material/product into a valuable commodity. Potting, *et al.* in their CE definitions termed beneficial reuse, i.e., use of waste or its parts in a new product with different function, as repurposing [11]. Figure 22.6 shows beneficial reuse cycle in a circular economy. For successful application as beneficial reuse – (a) the waste material / product must perform well in the alternate use, and (b) must be as safe to human health and environment as the material that is replaced [30].

22.4.1   Beneficial Reuse in Different Manufacturing Sectors

Many industries such as construction and demolition, cement, metal fabrication, plastic packaging, chemicals, pulp and paper, paint and coating, etc., have found innovative ways to recirculate materials within their own operations. The beneficial reuse opportunities outside the sectors are not as obvious and widespread but are improving. Chemicals, metals, and food manufacturing sectors are the three highest waste contributing sectors from available TRI data [31]. The plastics are ubiquitous waste from many manufacturing sectors. Therefore, beneficial reuse in chemicals, primary metals, plastics packaging, metal casting, and transportation equipment industry is discussed in this section.

The chemical sector is one of the biggest and most varied makers of materials in manufacturing. The operations can include various organic and inorganic chemicals, rubber, resins, plastics, fertilizes, and pesticides, etc. The chemical industry is adaptive and high research activity in the sector enables the development of new materials and technologies.

Table 22.2 Estimated material, energy, and CO_2 savings in full production chain from remanufacturing.

Component	Material	Energy	CO_2
Engine	60%	80%	56%
Transmission	47%	80%	32%
DPF	85%	80%	81%
Alternator	98%	80%	80%

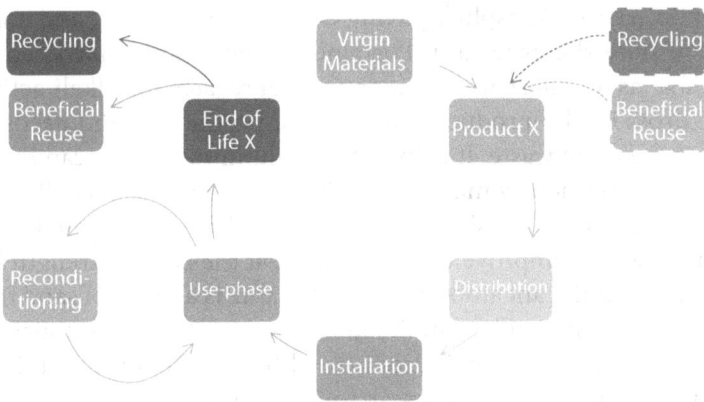

Figure 22.6 Beneficial reuse life cycle and circular economy.

The industry has demonstrated many beneficial reuse examples from 1970s. For example, if a manufacturer uses acetic anhydride in the chemical process, it most likely produces acetic acid as a by-product. This by-product can become feedstock of chemical process for making propyl acetate [32]. The highly toxic poly-chlorinated biphenyls (PCBs) are converted to other less toxic useful materials such as biphenyls and cyclohexane through catalytic dechlorination [33]. In the U.S. a public-private partnership project between DOE and Dow Chemicals, to assess by-product synergies assessed a total of 40 processes for byproducts reuse from hydrocarbons and spent solvents, sodium hydroxide byproducts, sulfuric acid waste, Methocel waste, and ortho-toluenediamine. Through the project, Dow identified reuse opportunities for 155 Million pounds of non-chlorinated by-products and was able to save over 15 Million USD in cost, 0.9TBtu in energy, and 108 Million lbs. in CO_2 emissions [34].

Plastics are versatile materials any many manufacturers utilize plastic materials such PET (Polyethylene Terephthalate) and HDPE (High Density Poly Ethelene) packaging to distribute their products. These products are difficult to recycle due to many issues such as sorting, contamination, and infrastructure. Only about 2% of plastics are recaptured for closed loop recycling [35]. However, waste from the plastics packaging industry is successfully reused as raw material. Beverage manufacturers recollect bottles from consumers to clean and remelt and produce newer bottles or other packaging products. The plastics waste has also been beneficially reused in the waste-to-energy industry to produce fuel for diesel engines through a process known as pyrolysis [36].

Transportation equipment manufacturing is a complex industry with scattered supply chain. End-of-life vehicles are one the most recycled consumer goods as more than 90% of the vehicles are recycled. Individual components have found beneficial reuse in many industries. In operation wastes such as tires are used to make crumb for asphalt production, molded products, turfs [37], playground mulch, and commercial landscaping. The tires are also prominently used as energy source in some industries.

Iron & steel is produced using standard processes such as Electric Arc Furnace, Basic Oxygen Furnace, or others. The processes produce slag to remove impurities when slagging agents such as, limestone or dolomite, and other fluxing materials are added. The molten slag when cooled is reprocessed and resold for beneficial reuse. The ferrous slags are used as cement kiln feedstock, asphalt paving additives, road-base, and concrete. Ground granulated blast furnace slag is used in Portland cement and blended cements [30]. Steel slag may be suitable as an alternative binder and aggregates for alkali activated materials [38]. Aluminum dross is a by-product of aluminum production and needs high energy to recover aluminum from the waste. It is found that aluminum dross can be effectively used as an additive to make refractory materials such as bricks or used as filler for concrete for improved stiffness and abrasion resistance. Aluminum dross can be used in the production of aluminum composites to improve wear resistance at the cost of strength [39].

Foundries produce significant amounts of waste sand as a result of sand molds usage and about 9 – 13 million pounds sand is landfilled each year [30]. The foundry sand has found beneficial reuse in construction applications, road bases, and structural fills, asphalt, and concrete products [40], and agricultural soil, soil additives [41], etc.

22.4.2   Enablers and Barriers

The economic drivers of beneficial reuse are reduction in waste disposal costs, possible resale value for discarded material, sustainable operation, improved compliance with regulation, and positive public image. Similarly, the economic drivers for the reuse industry are reduction in the cost of raw materials, possible reduction in processing costs, green purchasing tax and other benefits where applicable, improved sustainability and public image.

Though beneficial reuse is easily justifiable, the implementation can suffer from real barriers. These barriers can be financial, knowledge-related, or even infrastructure related [30]. The most frequently incurred barriers include – (a) Ready availability of cheap treatment and landfill options and relative convenience, (b) Lack of awareness of availability cross-industry waste and by-products and existing and new end use opportunities, (c) Non-matching scale of availability and needed amounts of waste material/products, (d) High initial cost of process modification to adapt alternative sources, and (e) Variability in local laws and regulations. Other barriers can be lower quality and specifications, geographic availability, and insignificant economic impact. Following section presents a case study of Waupaca Foundry that successfully overcame barriers and focused on the enablers of beneficial reuse.

22.4.3   Case Study – II: Waupaca Foundry Beneficial Reuse

22.4.3.1   Background and Description

Waupaca Foundry, Inc. (WFI) is a leading iron casting business with 10,000 Tons per day melting capacity. Sustainability is a priority with focus on environment, people, and business. Waupaca operates 5 foundries in the U.S. and is the largest producer of gray, ductile, austempered ductile, and compacted graphite iron in the world serving transportation, agriculture, construction, and other industrial sectors.

WFI uses a sand-casting process and produces large quantities of spent sand as a foundry byproduct. The sand is inert and is treated as non-hazardous waste for disposal. WFI recognized tremendous beneficial reuse opportunity for spent foundry sand and slag, partnering with state and federal agencies to support project case studies and regulation to facilitate the reuse of these materials.

The iron casting process use large amounts of sand which is continuously reconditioned and reused about 30 times before it becomes unusable. The spent foundry sand can be diverted for beneficial reuse in lieu of disposal in landfills to achieve CE objectives. Figure 22.5 shows WFI's beneficial reuse strategy on CE pathways. With systems in place to screen metal chunks and debris out of the sand, foundry sand and slag represents a clean, non-toxic product that can be used in a variety of applications and industries. WFI has established goals in reducing virgin sand usage in the foundry, finding beneficial reuse projects for spent foundry sand, improving the reclamation of foundry sands within the metal casting process, and reducing other waste.

This reclaimed sand finds new life in applications in construction, agricultural use and geotechnical fill. Waupaca Foundry has been working with state and local agencies, including the Wisconsin Department of Transportation, to use foundry sand as a highway sub-base fill, geotechnical fill and other general construction uses. Not only does this keep the sand out of landfills, but it also reduces the need for mining native materials from other places to be used as the source for these applications.

22.4.3.2   Benefits

WFI has completed over 150 beneficial reuse projects over the last two decades and these efforts have resulted in 80% of spent foundry sand annually being repurposed for beneficial reuse. In 2020, over 250,000 tons of foundry byproducts were reused in Wisconsin, representing the most expansive foundry sand and slag recycling success in the United States. This is a significant achievement in the field of beneficial reuse and marks WFI's ongoing efforts to evolve their sustainability profile. Table 22.3 shows waste performance achieved by WFI in years 2018 - 2020. The waste and hazardous waste has consistently reduced along with spent foundry sand. WFI has leveraged the reuse strategy as an opportunity to partner with its local communities on infrastructure improvement projects (Figure 22.7).

Figure 22.7 Foundry sand reuse for 42' sledding hill in Swan Park, Waupaca, WI.

Table 22.3 WFI waste performance (U.S. tons).

	2018	**2019**	**2020**
Waste	764,344	662,976	587,025
Haz-waste	18	6	6
Spent foundry sand	536,757	458,379	431,870
Beneficially reused sand	470,000	371,000	347,000

22.5   Discussion

CE is an emerging sustainability strategy for manufacturing in the United States as well as the world. 9R framework for CE is effective and presents 10 pathways for manufacturers to adopt that create lasting positive impacts on environmental sustainability, business growth, and community. Manufacturers realize material, energy, and carbon savings. This chapter looked at the different pathways to enable CE transition. The examples of CE initiatives discussed in Table 22.1, in the Pathways to CE section, illustrates the long-term commitments of businesses and value from the initiatives in all 3 pillars of sustainability.

Remanufacturing and beneficial reuse pathways are key CE strategies that have been known in industry for long time. As these are established pathways, considerable adoption is seen in many industries. As it involves component-based replacements, remanufacturing is prominently seen in sectors with multicomponent products with mechanical/electrical parts, e.g., automobile parts manufacturing, electrical equipment, appliance, and furniture industry. Other industries such as textile products and construction are exploring remanufacturing pathway as well. In remanufacturing high material savings are possible, as the strategy is to retain all the components that satisfy the original quality specifications and performance. Only components that are damaged, have deteriorated in performance will need to be replaced. On an average many products realize more than 67% energy savings from remanufacturing [26]. The material and GHG savings from remanufacturing are similarly significant.

Beneficial reuse is a well-established CE strategy that has found application in chemicals, primary metals, metal casting, and transportation equipment industries. Even though the chemical sector is the largest waste contributor in the U.S. economy, there is a significant potential to deploy this strategy to make a CE transition in that sector. Dow Chemical's project on beneficial reuse of by-products in other processes demonstrated the usefulness of the strategy. Initial investment and research and development effort can be a substantial barrier. The primary metals and metal casting sectors have made significant progress on CE front using the beneficial reuse pathway. Since this strategy transfers use phase to a different function, the material, energy, and GHG savings are not directly realized. The material, energy, and GHG savings are second application-dependent and can be significant when considered on a total basis.

Case-1 discussed Volvo Group's remanufacturing practice in detail. Volvo achieved 80% energy savings, as well as more than 50% material savings, and 50% GHG emissions reductions when their product supply chain is considered. It shows from the case study that it takes time to develop the expertise in remanufacturing and manufacturers that have experience can implement the CE strategy for significant gains in material and energy efficiency. It is also clear that many of the remanufactured products use energy and have the potential to lose performance if not remanufactured to the original specifications. This can result in net loss in energy used on a life cycle basis.

Case-2 discussed WFI's success through more than 150 beneficial reuse projects. The material, energy, and GHG savings are indirect in this case and depend on the repurpose application. Since, the sand is directly reused, in 2020 the material savings are 347,000 tons. The embodied energy in sand is found to be 0.1MJ/kg and embodied carbon equivalent is 0.0051 $kgCO_2e/kg$ [42]. This is equivalent to 29,831 MMBtu energy savings and 1,605 MT of CO_2e GHG savings upstream in sand production chain in year 2020.

22.6 Conclusions & Recommendations

Overall, this chapter looked at two CE strategies and their application in manufacturing industry, their enablers and barriers, and finally two case studies. The two case studies show that

(1) effective application of CE strategies is achievable with proper considerations to manufacturing processes, enablers, and barriers.
(2) In remanufacturing considerable manufacturing phase materials, energy, and GHG emissions reductions are possible. Use phase energy savings depend on performance which can result in net positive or net negative energy savings. Hence, remanufacturing to the original specifications or better is recommended for effective CE implementation.
(3) In beneficial reuse energy, material, and GHG emissions reductions depend on the secondary application identified. Nonetheless, these savings can be significant. The beneficial reuse has additional waste disposal cost savings as well as repurposed application economic value. Together these factors can make this CE strategy economically attractive.
(4) The CE strategy success can drive sustainability within organizations.

Acknowledgements

The research in this chapter is funded under US DOE's Better Plants program through Waste Reduction Network.

We acknowledge support provided by senior management at Volvo Group North America for providing necessary information about Volvo reman facilities and the process followed by Volvo.

We acknowledge support provided by Waupaca Foundry Inc. and their support providing information on beneficial reuse projects completed by the organization.

We also acknowledge the support provided by Volvo and Waupaca in reviewing the materials for publishing.

References

1. The White House (2021) *FACT SHEET: President Biden Sets 2030 Greenhouse Gas Pollution Reduction Target Aimed at Creating Good-Paying Union Jobs and Securing U.S. Leadership on Clean Energy Technologies, The White House*. Available at: https://www.whitehouse.gov/briefing-room/statements-releases/2021/04/22/fact-sheet-president-biden-sets-2030-greenhouse-gas-pollution-reduction-target-aimed-at-creating-good-paying-union-jobs-and-securing-u-s-leadership-on-clean-energy-technologies/ (Accessed: 16 January 2023).

2. Circle Economy (2022) *Circularity Gap Report 2022, Circularity Gap Reporting Initiative (CGRi)*. Available at: https://www.circularity-gap.world/2022 (Accessed: 17 January 2023).

3. Ellen McArthur Foundation (2019a) *The butterfly diagram: visualising the circular economy.* Available at: https://ellenmacarthurfoundation.org/circular-economy-diagram (Accessed: 16 January 2023).

4. Bressanelli, G. *et al.* (2021) 'Enablers, levers and benefits of Circular Economy in the Electrical and Electronic Equipment supply chain: a literature review', *Journal of Cleaner Production*, 298, p. 126819. Available at: https://doi.org/10.1016/j.jclepro.2021.126819.

5. Guillard, V. *et al.* (2018) 'The Next Generation of Sustainable Food Packaging to Preserve Our Environment in a Circular Economy Context', *Frontiers in Nutrition*, 5. Available at: https://www.frontiersin.org/articles/10.3389/fnut.2018.00121 (Accessed: 18 January 2023).

6. Gong, Y. *et al.* (2020) 'Investigation into circular economy of plastics: The case of the UK fast moving consumer goods industry', *Journal of Cleaner Production*, 244, p. 118941. Available at: https://doi.org/10.1016/j.jclepro.2019.118941.

7. Jaeger, B. and Upadhyay, A. (2020) 'Understanding barriers to circular economy: cases from the manufacturing industry', *Journal of Enterprise Information Management*, 33(4), pp. 729–745. Available at: https://doi.org/10.1108/JEIM-02-2019-0047.

8. Ritzén, S. and Sandström, G.Ö. (2017) 'Barriers to the Circular Economy – Integration of Perspectives and Domains', *Procedia CIRP*, 64, pp. 7–12. Available at: https://doi.org/10.1016/j.procir.2017.03.005.

9. Kumar, V. *et al.* (2019) 'Circular economy in the manufacturing sector: benefits, opportunities and barriers', *Management Decision*, 57(4), pp. 1067–1086. Available at: https://doi.org/10.1108/MD-09-2018-1070.

10. Rizos, V. *et al.* (2016) 'Implementation of Circular Economy Business Models by Small and Medium-Sized Enterprises (SMEs): Barriers and Enablers', *Sustainability*, 8(11), p. 1212. Available at: https://doi.org/10.3390/su8111212.

11. Potting, J. *et al.* (2017) 'Circular economy: measuring innovation in the product chain', *Planbureau voor de Leefomgeving* [Preprint], (2544).

12. Bradley, R. *et al.* (2018) 'A total life cycle cost model (TLCCM) for the circular economy and its application to post-recovery resource allocation', *Resources, Conservation and Recycling*, 135, pp. 141–149. Available at: https://doi.org/10.1016/j.resconrec.2018.01.017.

13. Blomsma, F. *et al.* (2019) 'Developing a circular strategies framework for manufacturing companies to support circular economy-oriented innovation', *Journal of Cleaner Production*, 241, p. 118271. Available at: https://doi.org/10.1016/j.jclepro.2019.118271.

14. US EPA, O. (2018b) *Measuring the Impact of Source Reduction*. Available at: https://www.epa.gov/toxics-release-inventory-tri-program/measuring-impact-source-reduction (Accessed: 18 January 2023).

15. Ranson, M. *et al.* (2015) 'The impact of pollution prevention on toxic environmental releases from US manufacturing facilities', *Environmental Science & Technology*, 49(21), pp. 12951–12957.

16. Ellen McArthur Foundation (2019b) *The technical cycle of the butterfly diagram*. Available at: https://ellenmacarthurfoundation.org/articles/the-technical-cycle-of-the-butterfly-diagram (Accessed: 18 January 2023).

17. Muñoz, P. and Cohen, B. (2018) 'A Compass for Navigating Sharing Economy Business Models', *California Management Review*, 61(1), pp. 114–147. Available at: https://doi.org/10.1177/0008125618795490.

18. Fleming, S. (2020) *4 creative ways companies are embracing the circular economy*, *World Economic Forum*. Available at: https://www.weforum.org/agenda/2020/12/circular-economy-examples-ikea-burger-king-adidas/ (Accessed: 19 January 2023).

19. Motes, P. (2022) *Repair, Reuse, Recycle: The Circular Economy in Action*, *Dell*. Available at: https://www.dell.com/en-us/blog/repair-reuse-recycle-the-circular-economy-in-action/ (Accessed: 19 January 2023).

20. Tata Motors (2018) *Tata Motors CVBU Customer Care - Prolife Business*, *Tata Motors Prolife*. Available at: https://www.customercare-cv.tatamotors.com/services/prolife-business.aspx (Accessed: 23 January 2023).

21. Nordil, B. (2019) *Canon - Europe : Why we remanufacture*, *Canon Europe*. Available at: https://www.canon-europe.com/blog/why-we-remanufacture/ (Accessed: 19 January 2023).

22. Tomkins, A. (2022) *Canon : Remanufacturing - a Key Component of Sustainable Business Practice*, *Canon Europe*. Available at: https://www.canon-europe.com/blog/world-environment-day-2022/ (Accessed: 19 January 2023).

23. *Liberty Tire Recycling* (2022) *Liberty Tire*. Available at: https://libertytire.com/Products/Commercial-Products/Crumb-Rubber/ (Accessed: 23 January 2023).

24. Coates, G. (2022) *Mining electronics waste, a new life for used metals*. Available at: https://nickelinstitute.org/ (Accessed: 23 January 2023).

25. Ganser, R.G. (1992) 'Remanufacturing failed transformers: An alternative to replacement', *Electricity Today; (Canada)*, 4:3. Available at: https://www.osti.gov/etdeweb/biblio/5175716 (Accessed: 22 January 2023).

26. Gutowski, T.G. *et al.* (2011) 'Remanufacturing and Energy Savings', *Environmental Science & Technology*, 45(10), pp. 4540–4547. Available at: https://doi.org/10.1021/es102598b.

27. Sahni, S. *et al.* (2010) 'Furniture remanufacturing and energy savings'. MITEI-1-b-2010. Cambridge, MA, USA: Sloan School of Management ....

28. Standridge, C.R., Corneal, L. and Consortium, M.N.T.R. (2014) *Remanufacturing, repurposing, and recycling of post-vehicle-application lithium-ion batteries*. Mineta National Transit Research Consortium.

29. Chakraborty, K., Mondal, S. and Mukherjee, K. (2019) 'Critical analysis of enablers and barriers in extension of useful life of automotive products through remanufacturing', *Journal of Cleaner Production*, 227, pp. 1117–1135. Available at: https://doi.org/10.1016/j.jclepro.2019.04.265.

30. EPA (2006) *Beneficial Reuse of Materials*. EPA Archive Document. Washington, DC, p. 12.

31. US EPA, O. (2018a) *Comparing Industry Sectors*. Available at: https://www.epa.gov/trinational-analysis/comparing-industry-sectors (Accessed: 22 January 2023).

32. Pencarinha, T. (2020) *Altiras is a Buyer of Acetic Acid Coproducts and Byproducts*. Houston, TX: Altiras Inc. Available at: https://altiras.com/acetic-acid-buyer/ (Accessed: 22 January 2023).

33. US EPA, O. (2021) *Beneficial reuse of PCBs (poly-chlorinated biphenyls) as new materials through a low cost process*. Final Report SV839354. Cincinnati, OH: University of Cincinnati. Available at: https://cfpub.epa.gov/ncer_abstracts//index.cfm (Accessed: 22 January 2023).

34. Baxter, R. *et al.* (2008) *US EPA: Beneficial Reuse of Industrial Byproducts in the Gulf Coast Region*. US EPA Archive Document. Fairfax, VA: ICF International. Available at: https://archive.epa.gov/sectors/web/pdf/beneficial-reuse-report.pdf.

35. Ellen McArthur Foundation (2017) *Plastics and the circular economy.* Available at: https://archive.ellenmacarthurfoundation.org/explore/plastics-and-the-circular-economy (Accessed: 22 January 2023).

36. Mirkarimi, S.M.R., Bensaid, S. and Chiaramonti, D. (2022) 'Conversion of mixed waste plastic into fuel for diesel engines through pyrolysis process: A review', *Applied Energy*, 327, p. 120040. Available at: https://doi.org/10.1016/j.apenergy.2022.120040.

37. Fiksel, J. *et al.* (2011) 'Comparative life cycle assessment of beneficial applications for scrap tires', *Clean Technologies and Environmental Policy*, 13(1), pp. 19–35. Available at: https://doi.org/10.1007/s10098-010-0289-1.

38. Nunes, V.A. and Borges, P.H.R. (2021) 'Recent advances in the reuse of steel slags and future perspectives as binder and aggregate for alkali-activated materials', *Construction and Building Materials*, 281, p. 122605. Available at: https://doi.org/10.1016/j.conbuildmat.2021.122605.

39. Dai, C. (2012) *Development of aluminum dross-based material for engineering application.* Thesis. Worcester Polytechnic Institute.

40. Bhardwaj, B. and Kumar, P. (2017) 'Waste foundry sand in concrete: A review', *Construction and Building Materials*, 156, pp. 661–674. Available at: https://doi.org/10.1016/j.conbuildmat.2017.09.010.

41. Dayton, E.A. *et al.* (2010) 'Characterization of physical and chemical properties of spent foundry sands pertinent to beneficial use in manufactured soils', *Plant and Soil*, 329(1), pp. 27–33. Available at: https://doi.org/10.1007/s11104-009-0120-0.

42. Hammond, G. and Jones, C. (2008) *Inventory of carbon & energy: ICE Ver. 1.6a.* Univ of Bath, SERT, Mechanical Engineering, Bath, United Kingdom. Available at: www.bath.ac.uk/mech-eng/sert/embodied/.

23

Spatio-Temporal Life Cycle Assessment of NMC111 Hydrometallurgical Recycling in the US

Francis Hanna[1]*, Luyao Yuan[1], Calvin Somers[2] and Annick Anctil[1]

[1]Department of Civil and Environmental Engineering, Michigan State University, East Lansing, Michigan, United States
[2]Department of Applied Engineering, Michigan State University, East Lansing, Michigan, United States

Abstract

Annual global EV sales have tripled in the last three years and are expected to reach 30 million by 2030. The growth in electric vehicles will cause an increase in demand for battery materials and large volumes of retired batteries in the future. Recycling helps manage retired batteries, supports the domestic material supply, and alleviates the environmental footprint of virgin materials. Hydrometallurgical processes have become popular recently because of their low energy consumption and high recovery rates. This study uses a cradle-to-gate LCA to assess the environmental impact of conventional and truncated hydrometallurgy recycling in RFCM and SRTV eGRID regions between 2023 and 2030. Conventional hydrometallurgical recycling uses solvent extraction to extract metal sulfates, to be reused in new cathode materials. Novel hydrometallurgical recycling technologies use a truncated approach where metal sulfates are not extracted, and the metals' composition is adjusted to produce new cathode materials directly. The functional units are 1kg of NMC811 cathode material and 1 kg of recycled NMC111. The assessment is done in SimaPro, using TRACI, CED, and BEES+ methods, to assess the global warming potential (GWP), cumulative energy demand (CED), and water consumption, respectively. The analysis results show that truncated hydrometallurgy has a lower GWP (0.47 kg CO_2-eq/kg NMC111 recycled), CED (18.74 MJ/kg NMC111 recycled), and water consumption (30.31 L/kg NMC111 recycled) than conventional hydrometallurgy (0.96 kg CO_2-eq/kg NMC111 recycled; 13.92 MJ/kg NMC111 recycled; 34.05 L/kg NMC111 recycled). In addition, the spatio-temporal analysis results show that location and electricity grid decarbonization rate affect the environmental impact of recycling. The GWP is 10% lower in SRTV than RFCM, while CED in SRTV is 9% larger than RFCM. Also, grid decarbonization between 2023 and 2030 can reduce GWP by 13.3%, CED by 4.26%, and water consumption by 17.8%.

Keywords: Life cycle assessment, Li-ion batteries, hydrometallurgical recycling, environmental impact

*Corresponding author: hannafra@msu.edu

Nabil Nasr (ed.) *Technology Innovation for the Circular Economy: Recycling, Remanufacturing, Design, Systems Analysis and Logistics*, (297–308) © 2024 Scrivener Publishing LLC

23.1 Introduction

As the world addresses climate change, several countries like the United States (US), China, France, Norway, Sweden, and Germany are experiencing an energy transition involving the massive deployment of electric vehicles and renewable energy in the coming years [1]. The batteries required for this transition will cause a massive demand increase for lithium, cobalt, nickel, and manganese [1, 2]. The increasing production volume of batteries will result in large quantities of retired ones in the future [3]. Governments are developing new policies to regulate the end-of-life (EoL) management of LIBs, and EV manufacturers are seeking reliable and sustainable sources of battery materials for their supply chains [3].

Recycling EoL vehicle batteries could offset the environmental impacts of LIB production from virgin materials and support a domestic material supply chain [4]. Several recycling processes have been demonstrated at the lab, pilot, and industrial scales. These processes fall into two main categories, pyrometallurgy and hydrometallurgy [5]. Pyrometallurgical recycling methods are employed commercially in European and North American countries because of their low raw material requirement and little wastewater production [6]. However, these methods have considerable constraints, such as high costs and low recovery rates for some materials [6]. Hydrometallurgical recycling processes have become more popular in recent years because of their lower energy consumption and higher recovery rates than pyrometallurgical processes [6, 7]. In light of the growing LIB recycling industry, assessing and minimizing its environmental footprint is important.

Numerous studies have looked into the environmental impact of LIB recycling. Some studies conduct comparative analyses to compare different recycling processes or to compare recycled and virgin materials [8–12]. For example, in one study, the authors conduct a life cycle assessment for hydrometallurgical recycling of NMC111 batteries using different leaching chemicals [12]. The study finds that the most common leaching chemicals in the industry, sulfuric acid and hydrogen peroxide, are preferable due to lower GHGs and toxicity-related impacts. Similarly, the life cycle impact of a hydrometallurgical recycling process is evaluated, but it is based on lab-scale results [13]. Another study conducts a life cycle impact analysis of hydrometallurgical recycling, yet the authors focus on a novel type of lithium-ion battery, not yet available at a commercial scale [14]. Interestingly, most current studies look into lab-scale or pilot-scale processes, and the literature lacks assessments of new processes recently introduced and being industrially commercialized. Another drawback of the existing literature is the limited number of impact categories assessed. All of the reviewed studies focus on two main impact categories, global warming potential (GWP) and cumulative energy demand (CED) [8, 9, 14, 15], with a few studies looking into other impact categories using the ReCiPe midpoint method [6, 10, 12, 13]. None of these studies assesses the water footprint of battery recycling, one of the significant limitations and drawbacks of hydrometallurgy processes. Finally, previous work on LIB recycling does not highlight the impact of the electricity carbon footprint on the environmental footprint and investigate how grid decarbonization can affect the environmental footprint of battery recycling. For instance, several battery recycling and manufacturing prospective projects come across various regions in the US with different electricity grid mixes, such as the Midwest, and Southeastern regions [16–21].

In this study, we conduct a life cycle assessment of recycling lithium-ion batteries using conventional and novel hydrometallurgy recycling in the US. This study assesses the global warming potential, cumulative energy demand, and water consumption of the modeled hydrometallurgical recycling processes. The results provide insights into the environmental performance of the newly developed hydrometallurgical process. Also, we report the water consumption of batteries recycling and cathode material production and the hotspots for each process, highlighting potential improvement opportunities to reduce water consumption. Finally, this work shows the effect of the grid carbon footprint on the environmental impact of battery recycling and cathode materials production. A spatio-temporal analysis is conducted to reflect the electricity grid mix variability across different locations (i.e., RFCM and SRTV eGRID regions) and its decarbonatization in the future.

23.2   Methods

23.2.1   Goal & Scope of the Study

This study aims to assess and compare the impact of electricity mix and decarbonization on the environmental footprint of two hydrometallurgical recycling processes. In line with this goal, we perform a life cycle assessment of NMC111 recycling in different grids intentionally ranging in carbon intensity between 2023 and 2030, using conventional and truncated hydrometallurgical processes. We limit our study to this time period because it is challenging to forecast the potential technological developments in the batteries and LIB recycling industries beyond 2030. Advanced battery technologies, such as sodium-based and solid-state batteries, are being developed and may supplant lithium-ion batteries in time, but there are too many unknowns to consider meaningfully. Eventually, new battery chemistries might shift the current materials availability landscape and will almost certainly require recyclers to adjust their processes to accommodate more diverse conditions and feedstocks [5].

23.2.2   System Boundary and Functional Unit

In line with the goal described in section 23.2.1, Figure 23.1 shows the system boundary defined in this study. The system includes six phases: disassembly and discharge of spent batteries, size reduction & black mass production, leaching, solvent extraction and crystallization, precursor materials production, and cathode active materials production. Based on the current literature, these phases represent the process of conventional hydrometallurgical recycling and cathode materials production. The modeled process starts with recycling spent NMC111 batteries and ends with producing NMC811 cathode active material (CAM). This study focuses on NMC chemistry due to its dominant applications in the EV industry [22]. The selection of the input battery waste is based on the projected EoL battery chemistries, which show a 30% contribution of spent NMC111 batteries [23]. NMC811 is selected as the final output because EV manufacturers are shifting toward less cobalt-intensive batteries [24].

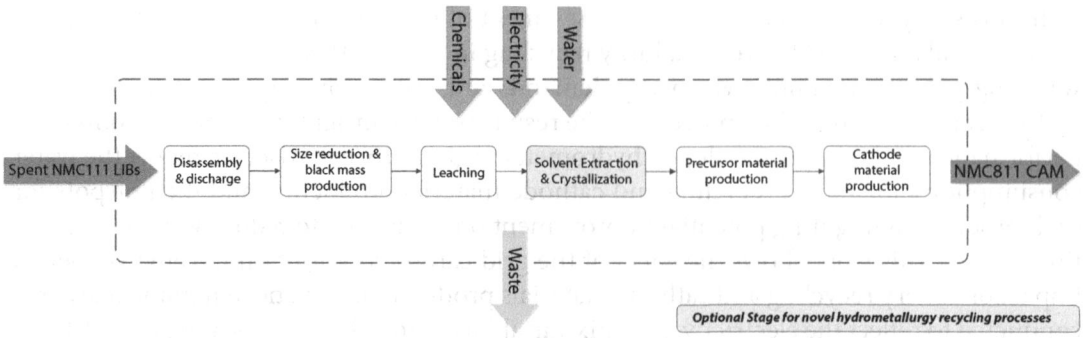

Figure 23.1 System boundary in this study.

This study involves multiple assumptions. First, the first life cycle and the collection and transportation of spent LIBs are excluded. Second, battery recycling, precursor material production, and cathode production are assumed to occur in the same location; therefore, transportation is excluded. However, the natural extension of this work (and readily achievable using the tools described) is to consider these phases as not being collocated and to compute further the impact of having them dispersed across different regions and grids. The conventional and novel processes require additional metals from external sources to adjust the mixture composition and produce NMC811 cathode materials. For this study, we assume that the composition is adjusted using recycled materials, but it is also possible to perform the analysis using virgin material makeup. The functional unit in this study is 1 kg NMC811 cathode material produced.

23.2.3   Li-Ion Batteries Recycling

23.2.3.1   Conventional Hydrometallurgy – Individual Salt Synthesis

In this study, a conventional hydrometallurgy process is modeled based on a patented commercial process [25]. Spent lithium-ion batteries are first disassembled and discharged. The disassembled batteries go through a size reduction step via shredding. Ferrous and non-ferrous products are separated using magnetic separation. The non-magnetic stream is mixed with a stripping agent to enhance materials recovery and facilitate downstream processing. The stripped slurry stream goes through a wire mesh screen. The oversized stream primarily consists of copper, aluminum, and plastics. The undersized stream includes smaller solids, including black mass. The latter is leached using hydrogen peroxide and sulfuric acid, leading to two product streams. The first product stream includes graphite that is separated via froth flotation. The second product stream goes through a series of reactions to recover manganese, cobalt, nickel, and lithium. Manganese carbonate is first isolated through precipitation, and nickel sulfate and cobalt sulfate are then extracted via solvent extraction and crystallization. And finally, lithium carbonate is extracted via precipitation and filtration. Before lithium extraction, impurities, including copper, aluminum, gypsum, and sodium sulfate, are removed. The extracted metal sulfates and lithium carbonate are used in the following steps to produce the precursor and final cathode materials.

23.2.3.2 Novel (Truncated) Hydrometallurgy – No Intermediate Sulfates Extraction

In this process, the spent batteries undergo the same preliminary treatment as conventional hydrometallurgy. Batteries are disassembled, discharged, shredded, and sieved to obtain the black mass and other byproducts, such as plastics, steel, aluminum, and copper. The black mass is leached using hydrogen peroxide and sulfuric acid. In contrast with conventional hydrometallurgy, metal sulfates are not extracted following leaching. Instead, they remain in an aqueous solution, and their composition is adjusted to meet the input requirements of the final ternary metal product (NMC811 for our study). The composition is adjusted by adding nickel sulfate, manganese sulfate, and cobalt sulfate from external sources. After composition adjustment, the metal sulfates are coprecipitated to produce the precursor material NMC-$(OH)_2$. Finally, cathode active material is produced via sintering the precursor with a lithium source. The process flow diagram for this recycling method is developed using a published patent [26].

23.2.4 Electricity Grid Modeling

This study investigates the carbon intensity of the local grid for the period 2023 to 2030. The modeled electricity mixes are derived from the EIA (Energy Information Administration) outlook [27]. The EIA reports different scenarios. This analysis uses the "Reference Case" scenario based on current laws and regulations as of November 2021 [27]. The approach consists of adjusting the production mix in the US electricity process obtained from the DATASMART 2021 database [28].

23.2.5 Life Cycle Inventory and Evaluation Methodology

The energy and material flow data for the modeled recycling methods are mainly derived from the original patents, published literature, and the Ecoinvent v3.8 database. Theoretical stoichiometry calculations are also used to derive quantities and concentrations of the chemicals used due to the lack of industrial data related to the modeled processes. This work is carried out using SimaPro 9.4.02 software based on the methods introduced by ISO 14041 and 14044 (Reference ISO) [29, 30]. The life cycle assessment calculations are then translated into a system dynamics software, Stella Architect, used to conduct the spatio-temporal analysis. The Stella model is first validated by running extreme scenarios to help verify that the model provides reasonable outputs. Then, specific scenarios with preset inputs are used, and results are compared with the SimaPro model output.

23.2.6 Life Cycle Impact Assessment

This study focused on three midpoint impact categories, the global warming potential (kg CO_2-eq), the cumulative energy demand (MJ), and the water consumption (Liters). TRACI 2.1 is used to calculate the global warming potential, CED to calculate the energy demand, and BEES+ to calculate the water consumption. Notably, the BEES+ method solely calculates the water consumption in liters without weighting, characterization, or regionalization factors. This method does not account for water scarcity levels in different geographical regions but minimizes the water footprint uncertainties in the model calculations [31].

23.3   Results & Discussion

23.3.1   Grid Modelling

Figure 23.2 shows the modeled electricity mixes between 2023 and 2030. The predominant sources of energy in the US in the current year are natural gas and coal. The forecast in the Figure below shows that natural gas remains RFCM's predominant electricity generation source. In the same region, renewable energy sources and natural gas replace coal, which drops from 32% in 2023 to 22% in 2030. As for the SRTV region, nuclear energy is the largest single contributor to the electricity grid mix. As renewable energy production grows in that region, electricity from coal drops from 32% in 2023 to 20% in 2030.

23.3.2   Recycling Processes - Analysis Results and Comparison

As shown in Figure 23.3, the production of precursor and cathode materials contributes significantly to the final environmental footprint for all three impact categories. Cathode materials production contributed the most, with a share of 50% to GWP, 57% to CED, and 68% to water consumption. The environmental impact breakdown illustrated in the figures below shows a minimal difference between the modeled processes. Leaching contributes more to water consumption in the conventional hydrometallurgy process when compared with the truncated recycling process. The leaching stage in conventional hydrometallurgy is followed by solvent extraction and crystallization to extract metal sulfates. It also generates large quantities of gypsum and sodium sulfate byproducts, leading to additional chemicals and electricity use.

Figure 23.4 shows the contribution of the recycling stages while excluding consequent production stages from the analysis. The results show that leaching is the leading contributing stage to the environmental footprint of NMC111 batteries, followed by size reduction. The environmental impact breakdown of NMC111 recycling is similar for both modeled processes, but the leaching step has a marginally higher impact in conventional hydrometallurgy compared to the truncated hydrometallurgy process.

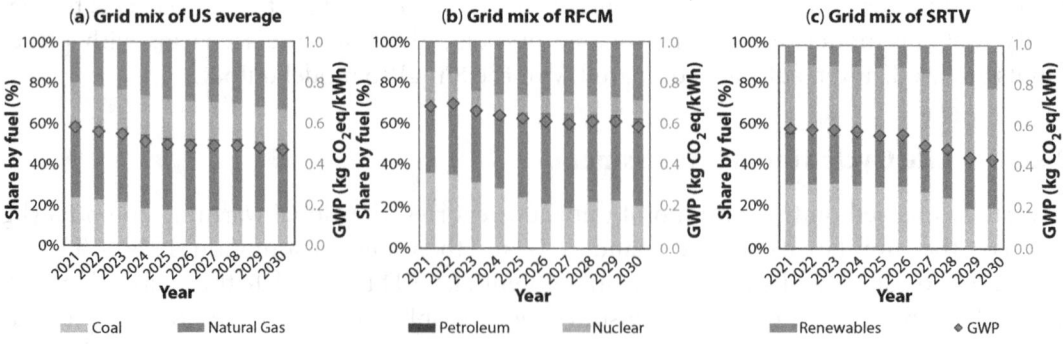

Figure 23.2 Electricity grid mix forecast used for (a) US, (b) RFCM, and (c) SRTV.

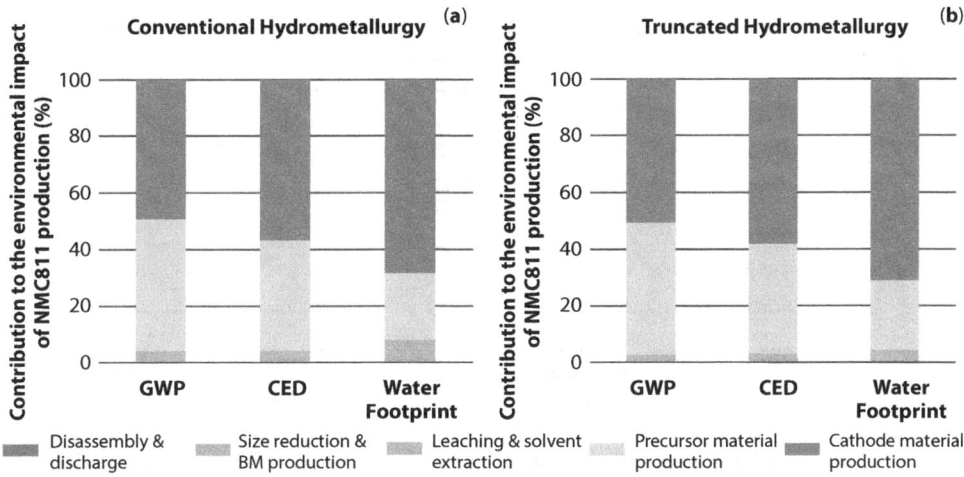

Figure 23.3 Environmental impact results for NMC811 production using materials recycled via (a) conventional hydrometallurgy, and (b) truncated hydrometallurgy.

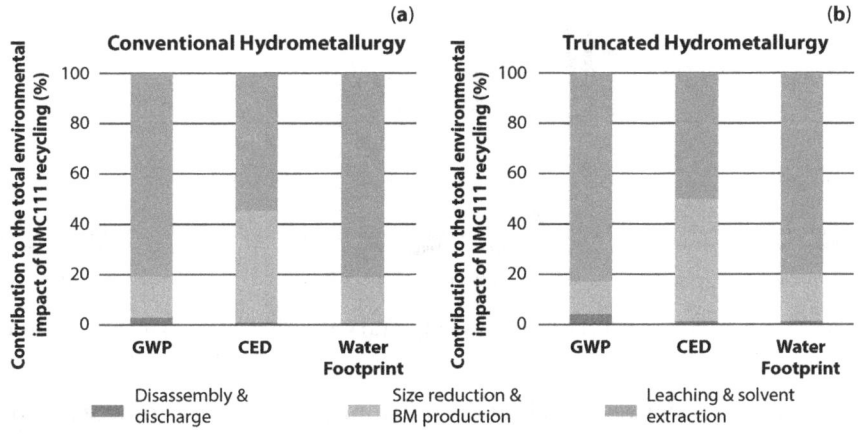

Figure 23.4 Environmental impact results for NMC111 recycling via (a) conventional hydrometallurgy and (b) truncated hydrometallurgy.

According to Figure 23.5, energy is the major contributor to the calculated global warming potential, cumulative energy demand, and water footprint. In this study, energy consumption refers to the natural gas demand for co-precipitation and the electricity consumed during recycling and cathode material production. Chemicals contribute more to the water footprint than GWP and CED for both processes. Thus, improving the electricity footprint is vital to reduce the global warming potential, cumulative energy demand, and water consumption of LIB recycling and CAM production.

To better understand the environmental impact of the recycling steps, the associated environmental impacts are normalized to a 1-kg recycled NMC111 battery pack. As shown in Figure 23.6, conventional hydrometallurgy has a higher GWP, CED, and water consumption than the truncated process. For instance, results show a GWP of 0.96 kg CO_2-eq kg

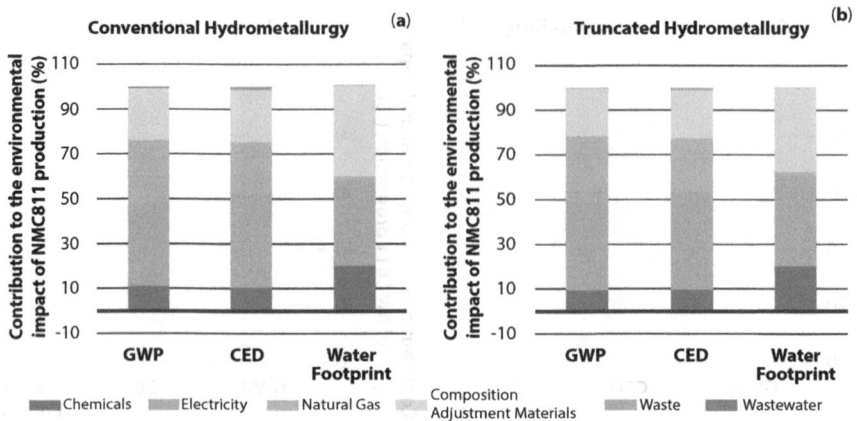

Figure 23.5 Environmental impact of NMC811 production using recycled materials by category – (a) conventional hydrometallurgy, (b) truncated hydrometallurgy.

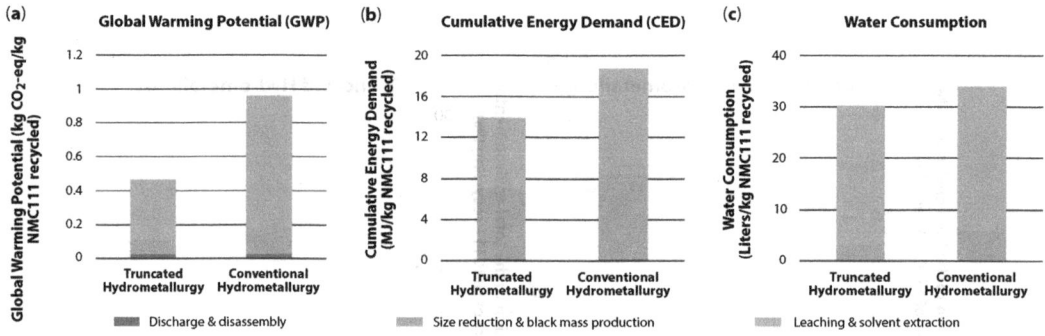

Figure 23.6 Environmental impact of NMC111 recycling – (a) GWP, (b) CED, and (c) water consumption.

NMC111 recycled for conventional hydrometallurgy recycling compared to 0.47 kg CO_2-eq/kg NMC111 recycled. Similarly, the cumulative energy demand associated with hydrometallurgy is 18.74 MJ/kg NMC111 recycled compared to 13.92 using the truncated hydrometallurgy process. Also, the truncated hydro process has a water consumption of 30.31 Liters/kg NMC111 recycled, 11% lower than conventional hydrometallurgy water consumption, estimated at 34.05 Liters/kg NMC111 recycled.

The results for conventional hydrometallurgy recycling align with similar studies in the literature. For example, one study reported a GWP of 3.8 kg CO2-eq/kg battery feed for hydrometallurgy recycling. However, 43% of the battery feed consisted of NiMH batteries, and almost 2.8 kg CO2-eq is caused by excess NaOH dissolving the iron [13]. Similarly, another study reports a footprint of 1.49 kg CO2-eq/kg battery recycled using conventional hydrometallurgy; however, that study uses the Chinese electricity grid mix and includes transportation of batteries in its system boundary [9]. Similarly, using the EverBatt model, recycling 1 kg of NMC111 would cause an environmental impact of 19.51 MJ and 1.4 kg CO_2-eq [32]. Another reason for the relatively higher footprint of conventional hydrometallurgy is that it involves the extraction of several other byproducts. For instance, one study highlights that additional byproducts result in higher GWP and CED for hydrometallurgy [*33*]. That said, all the studied recycling methods effectively reduce the Impact Categories studied versus their counterparts obtained via primary (mining) routes.

23.3.3   Spatio-Temporal Analysis

Figure 23.7 illustrates the spatio-temporal analysis results. After accounting for future grid decarbonization, the spatio-temporal analysis results show a decreasing environmental footprint. The reduction in reliance on coal is the main reason for this trend. However, the analysis shows that grid decarbonization has a more significant effect in the SRTV region than RFCM, where the GWP decreases by 13.3 %, the CED by 4.26 %, and the water consumption by 17.8% in 2030. Moreover, after comparing the three impact indicators, the grid decarbonization effect reflects the most in water consumption, which decreases by 17.8 % in RFCM and 24.3 % in SRTV. By 2030, the recycling and CAM production process in RFCM has a lower cumulative energy demand and water consumption than SRTV, yet its global warming potential is 15.4 % higher. These results are caused by significant nuclear energy production in SRTV and a large coal and natural gas contribution in RFCM. Noteworthy, the water consumption results do not reflect the water scarcity and availability in each location, providing less insight into the significance of reducing water consumption in these locations.

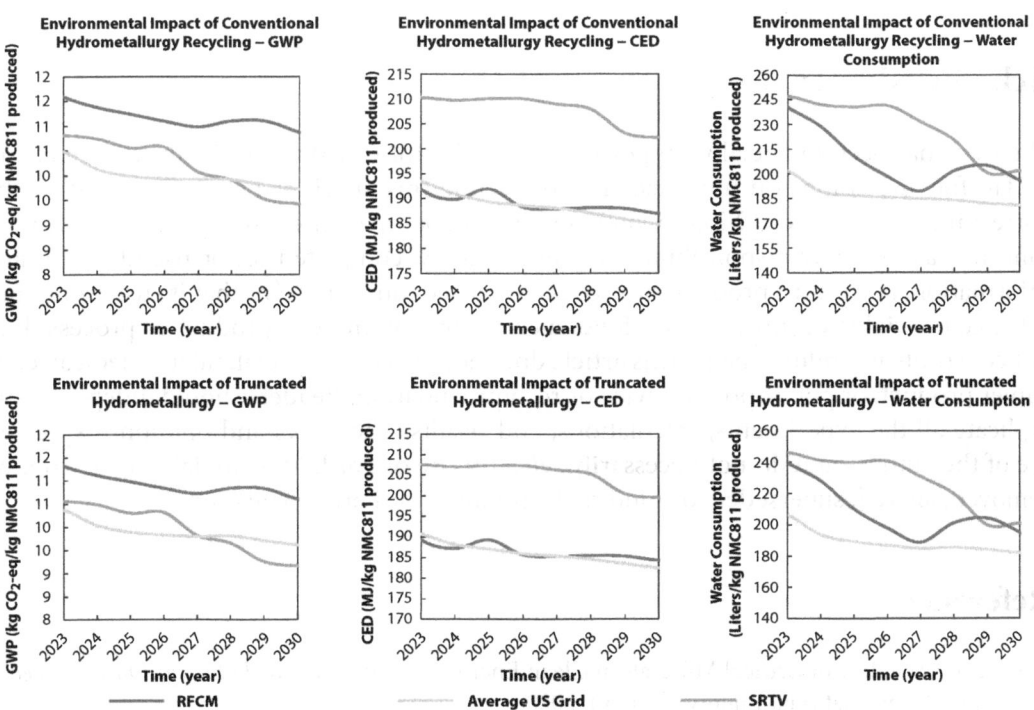

Figure 23.7 Environmental impact of NMC111 recycling (conventional hydrometallurgy: top, truncated hydrometallurgy: bottom) and NMC811 production, based on the electricity grid mix forecast of RFCM, SRTV, and the USA.

23.4   Conclusions & Recommendations

This study conducts an environmental life cycle assessment for LIB recycling, assuming an NMC111 feed is converted to an NMC811 product, which reflects the landscape today. The effects of the electricity grid mix and grid decarbonization are assessed via a spatio-temporal analysis. The analysis is done using the average US grid and RFCM and SRTV subregions grids. The truncated hydrometallurgy recycling results show that leaching is the most significant contributor to the process's carbon footprint, energy demand, and water consumption. Similarly, leaching, solvent extraction, and crystallization contribute the most to the environmental footprint of the conventional and truncated hydrometallurgy processes. The global warming potential, cumulative energy demand, and water footprint of NMC811 production are primarily driven by electricity and natural gas use. Also, chemicals contribute more to water consumption than GWP and CED in NMC111 battery pack recycling. The truncated hydrometallurgy process has a lower global warming potential, cumulative energy demand, and water consumption. Finally, the spatio-temporal analysis shows that battery recycling has a lower GWP in the SRTV region than in the RFCM region but a higher global warming potential and cumulative energy demand. Future work should address pyrometallurgical recycling techniques currently used in the industry. Also, future life cycle assessment studies should look into the water footprint of LIBs recycling beyond consumption, considering water scarcity and availability in different countries and regions. Finally, future work should consider dispersed (not all co-located) supply chain segments and thus consider the impacts of the regional grid for each.

Acknowledgments

The work has been financially supported by Ford Motor Co. We also thank A. Farina, and J. Han for discussions on life cycle analysis. While this article is believed to contain the correct information, Ford Motor Company (Ford) does neither expressly or impliedly warrant, nor assume any responsibility, for the accuracy, completeness, or usefulness of any information, apparatus, product, or process disclosed, nor represent that its use would not infringe the rights of third parties. Reference to any commercial product or process does not constitute its endorsement. This article does not provide financial, safety, medical, consumer product, or public policy advice or recommendation. Readers should independently replicate all the experiments, calculations, and results. The views and opinions expressed are of the authors and do not necessarily reflect those of Ford. This disclaimer may not be removed, altered, superseded, or modified without prior Ford permission.

References

1. IEA, "The Role of Critical Minerals in Clean Energy Transitions," *World Energy Outlook Special Report* (International Energy Agency), 2020.
2. E. S. Islam, S. Ahmed, A. Rousseau, Future Battery Material Demand Analysis Based on U.S. Department of Energy R&D Targets. *World Electric Vehicle Journal* 12, 2021.

3. R. E. Ciez, J. F. Whitacre, Examining different recycling processes for lithium-ion batteries. *Nature Sustainability* 2, 148-156, 2019.

4. *Building Resilient Supply Chains, Revitalizing American Manufacturing, And Fostering Broad-Based Growth,* 2021.

5. M. Velázquez, Valio, A. Santasalo, Reuter, G. Serna, A Critical Review of Lithium-Ion Battery Recycling Processes from a Circular Economy Perspective. *Batteries* 5, 2019.

6. S. Du *et al.*, Life cycle assessment of recycled NiCoMn ternary cathode materials prepared by hydrometallurgical technology for power batteries in China. *Journal of Cleaner Production* 340, 2022.

7. Y. Yao *et al.*, Hydrometallurgical Processes for Recycling Spent Lithium-Ion Batteries: A Critical Review. *ACS Sustainable Chemistry & Engineering* 6, 13611-13627, 2018.

8. M. A. Rajaeifar *et al.*, Life cycle assessment of lithium-ion battery recycling using pyrometallurgical technologies. *Journal of Industrial Ecology* 25, 1560-1571, 2021.

9. Z. Zhou, Y. Lai, Q. Peng, J. Li, Comparative Life Cycle Assessment of Merging Recycling Methods for Spent Lithium Ion Batteries. *Energies* 14, 2021.

10. J. Quan *et al.*, Comparative life cycle assessment of LFP and NCM batteries including the secondary use and different recycling technologies. *Science of The Total Environment* 819, 153105, 2022.

11. M. Mohr, J. F. Peters, M. Baumann, M. Weil, Toward a cell-chemistry specific life cycle assessment of lithium-ion battery recycling processes. *Journal of Industrial Ecology* 24, 1310-1322, 2020.

12. M. Iturrondobeitia *et al.*, Environmental Impact Assessment of LiNi1/3Mn1/3Co1/3O2 Hydrometallurgical Cathode Recycling from Spent Lithium-Ion Batteries. *ACS Sustainable Chemistry & Engineering* 10, 9798-9810, 2022.

13. M. Rinne, H. Elomaa, A. Porvali, M. Lundström, Simulation-based life cycle assessment for hydrometallurgical recycling of mixed LIB and NiMH waste. *Resources, Conservation and Recycling* 170, 2021.

14. M. Raugei, P. Winfield, Prospective LCA of the production and EoL recycling of a novel type of Li-ion battery for electric vehicles. *Journal of Cleaner Production* 213, 926-932, 2019.

15. J. B. Dunn, L. Gaines, J. Sullivan, M. Q. Wang, Impact of recycling on cradle-to-gate energy consumption and greenhouse gas emissions of automotive lithium-ion batteries. *Environ Sci Technol* 46, 12704-12710, 2012.

16. E. J. West, Ford and SK Innovation launch Blue Oval City and Battery Park. 2021.

17. S. Pohl, This Just In: Michigan Set to 'Dominate' U.S. Battery Manufacturing by 2030. 2023.

18. K. Achtenberg. (Michigan Economic Development Corporation), 2022.

19. C. Randall, Ascend Elements to make battery materials in Kentucky. 2022.

20. businesswire. (businesswire A Berkshire Hathaway Company, 2022), vol. 2023.

21. businesswire. (businesswire A Berkshire Hathaway Company, 2022), vol. 2023.

22. M. Alipanah, A. K. Saha, E. Vahidi, H. Jin, Value recovery from spent lithium-ion batteries: A review on technologies, environmental impacts, economics, and supply chain. *Clean Technologies and Recycling* 1, 152-184, 2021.

23. BNEF, "Lithium-Ion Battery Recycling: 2 Million Tons by 2030," (Bloomberg New Energy Finance), 2019.

24. IEA, "Global EV Outlook 2022 - Securing supplies for an electric future," (International Energy Agency, 2022).

25. A. Kochhar, T. G. Johnston. (Li-Cycle Corp, United States), 2020, chap. US 2020/0331003 A1.

26. Y. Wang, E. Gratz, Q. Sa, Z. Zheng, J. Heelan. (Worcester Polytechnic Institute, United States of America), 2016, vol. US 10,522,884 B2.

27. EIA. 2022.

28. LTS. (Long Trail Sustainability), 2021.
29. ISO. 2006, pp. 20.
30. ISO. 2006, pp. 46.
31. J. Kneifel, A. L. Greig, P. Lavappa, B. Polidoro, "Building for Environmental and Economic Sustainability (BEES) Online 2.1 Technical Manual," *NIST Technical Note 2032 - Revision 1* (National Institute of Standards and Technology (NIST) - U.S. Department of Commerce), 2019.
32. Q. Dai *et al.*, "EverBatt: A Closed-loop Battery Recycling Cost and Environmental Impacts Model," (Argonne National Laboratory), 2019.
33. Y. Cao *et al.*, Co-products recovery does not necessarily mitigate environmental and economic tradeoffs in lithium-ion battery recycling. *Resources, Conservation and Recycling* 188, 106689, 2023.

Part 5

MECHANICAL RECYCLING

24

Diverting Mixed Polyolefins from Municipal Solid Waste to Feedstocks for Automotive and Construction Applications

Tanyaradzwa S. Muzata[1], Alexandra Alford[1], Laurent Matuana[1], Ramani Narayan[2], Lawrence Drzal[2], Kari Bliss[3] and Muhammad Rabnawaz[1]*

[1]School of Packaging, Michigan State University, East Lansing, Michigan, United States
[2]Department of Chemical Engineering and Materials Science, Michigan State University, East Lansing, Michigan, United States
[3]PADNOS, Grandville, MI, United States

Abstract

The melt flow fluctuations in recycled plastics have been a daunting challenge in advancing recycling efforts. Herein we report an innovative rheological approach as a crucial quality control method required to consistently produce acceptable grades of pelletized mixed recycled polyolefins (m-POs) with targeted melt-flow indices (MFIs). Rheology modifiers are used in this study to control the MFIs and facilitate pellet production from these m-PO samples. In addition, we developed models through which the MFIs can be tailored for different m-POs.

Keywords: Plastics, recycling, polyolefins, rheology, melt flow index and sustainable

24.1 Introduction

The recycling of plastic materials is of paramount importance in establishing a sustainable society. Plastics have emerged as an indispensable class of polymeric materials that have been used in different applications for essential societal needs such as packaging, electronics, water remediation, and medical applications [1-4]. Due to the increasing production of plastics worldwide, plastic pollution has been increased dramatically [5]. It has been reported that in 2019, North America accounted for approximately 20% of the worldwide production of plastic materials [6]. In 2019, 123.9 billion pounds of resins were produced in the US, and the amounts of polyolefins such as high-density polyethylene (HDPE), low-density polyethylene (LDPE), linear low-density polyethylene (LLDPE), and polypropylene (PP) were 22, 8.5, 21.7, and 17.8 billion pounds respectively [7].

Polyolefins (PO) are mainly used as packaging materials, and this has resulted in large amounts of PO entering municipal solid waste (MSW). The rate of recycling is currently

---

*Corresponding author: rabnawaz@msu.edu

Nabil Nasr (ed.) Technology Innovation for the Circular Economy: Recycling, Remanufacturing, Design, Systems Analysis and Logistics, (311–320) © 2024 Scrivener Publishing LLC

very low and measures must be taken to ensure that more plastic waste is recycled into valuable products, thereby ensuring a circular plastic economy. Only 8.7% of plastics produced in 2018 were recycled in the United States of America [8]. Recycling plastic waste can have an important role in curbing environmental pollution, conserving natural resources, creating employment, saving valuable energy, and reducing plastic waste dumped at landfills [9, 10]. It has been projected that the amount of plastics (by weight) in oceans will be more than that of fish by 2050, hence aggressive measures should be implemented in order to fast-track and prioritize plastic recycling [11]. The most commonly used recycling methods are chemical and mechanical recycling, and both of these recycling methods have played a major role in eradicating the plastic pollution problem [12, 13]. Jia and coworkers made use of catalysts under mild conditions to degrade polyethylene into valuable feedstocks [14]. Kumar *et al.* went on to investigate how table salt can be used to depolymerize polystyrene in the presence of an oxidized copper scrubber [15].

Mechanical recycling has gained more traction in plastic recycling. The main advantage of mechanical recycling is that it is inexpensive and can be easily scaled up at the industrial level. Despite its advantages, the main drawback of mechanical recycling of plastic waste is mainly inconsistent melt flow indeces (MFIs) of the mixed plastic waste [16]. The inconsistencies of the MFIs of the mixed waste polyolefins (m-PO) can affect their processability. It is important to know the MFIs of a given polymer because plastics are mainly molded into different products in their melt state, and hence their flowability must be investigated prior to processing. It should also be highlighted that the MFI of a polymeric material can be used to establish melt flow curves. These curves enable waste plastic processors to investigate the processibility, as well as to troubleshoot and optimize the plastic processing conditions [17].

Here we address the challenge of inconsistent MFI in m-PO. We report a new approach that ensures a uniform MFI for m-PO by making use of Fischer-Tropsch high melting point modified wax. A simple melt flow indexer was used to determine the MFI values of the polyolefins. The use of this simple equipment is important in lowering operating costs in the plastic recycling sector. A model equation developed from laboratory-controlled polyolefins which consisted of HDPE, LDPE, LLDPE and PP in a specific composition to mimic the mixed waste polyolefins was established and this model was used to estimate the amount of rheological modifier (RM) that can be incorporated in m-PO to attain a specific uniform MFI value. Due to the multiple variables involved, this model plays an important role in preventing the loss of time involved in determining the amount of rheological modifier to be incorporated to attain a specific MFI value. The m-PO/RM pellets with uniform MFI can be used as injection molding feedstocks for different applications in the automobile and building sector.

24.2   Experimental Section

24.2.1   Materials

HDPE with an MFI of 1 g/10 min was supplied by Nova Chemicals, and another HDPE (Dowlex IP-10262) with an MFI of 9 g/10 min was purchased from Dow Chemicals. In addition, LLDPE (Dowlex 2056G) with an MFI of 1 was purchased from Dow Chemicals,

LLDPE-LL 6100.17 having an MFI of 20 g/10 min was procured from ExxonMobil, and PP- PD-1428 was supplied by Formosa Plastics Corp. The rheological modifier used in the experiments (Sasol-B52-96-1001) was provided by Sasol Chemicals, South Africa.

24.2.2 Processing of the Samples

The samples were processed by making use of a DSM micro compounder 15HT (Xplore Instruments BV), at a temperature of 220 °C using a screw speed of 100 rpm for 2 min. For the samples containing rheological modifier (RM), the RM in powder form was added together with the polymer pellets into the compounder. Both the laboratory-controlled polyolefins (L-PO) and the m-PO (1 kg each) samples were prepared using a Brabender single-screw extruder (SSE) at a temperature profile of 190, 180, 170, and 160 °C. The screw speed was maintained at 20 rpm. The processed strands went through a conveyor belt and were then collected for pelletization.

24.2.3 Waste Plastic Separation and Milling

The separation of the mixed polyolefins (PE-HDPE, LDPE, LLDPE, and PP) was conducted via the density separation method. The waste plastics were placed in a bucket of tap water and then stirred for a few seconds. After this stirring treatment, the floating plastics were scooped out and then dried in an oven at 100 °C for 24 h to ensure complete drying. The m-PO was milled into small pellets for effective mixing with the RM and to obtain a consistent MFI value. A Wiley mill was used along with a 316 stainless steel 10 mesh sieve having a hole size of 2 mm.

24.2.4 Characterization

The MFI values of the polymer samples were measured with a Ray Run Melt flow indexer. Each sample weighed 5 g and the sample was heated at 230 °C using 2.16 kg load to push the liquid polymer. The determination of the melting point (T_m) was performed by making use of a DSC Q100 (TA instruments). In particular, the samples were first processed via a DSM micro compounder, and a sample weighing 5 – 10 mg was cut and placed in the pan. The thermal history was initially erased by heating the sample from -20 to 250 °C using a ramp rate of 10 °C/min and then cooled back to -20 °C before heating again. The melting temperature was recorded from the second heating cycle. An ash analysis test was conducted to determine the mineral content at a larger scale. Four samples of 2 g each were placed in a crucible and then inserted into a furnace (Thermo Scientific) for 4 h at 575 °C. After the sample had been heated for the stipulated period, the crucibles containing the ash were cooled to room temperature and the mass of the ash was determined and converted to a percentage (%).

24.2.5 Statistical Analysis

Design-Expert software version 12 (Stat-Ease, Inc, Minneapolis, MN, USA) was used for statistical analysis of the laboratory-prepared polyolefins. The model equation obtained was used to quantify the amount of RM that can be incorporated into the m-PO to obtain a specific targeted MFI.

24.3 Results and Discussion

24.3.1 MFI Values of L-PO/RM

The L-PO was processed with a DSM micro-compounder and the amount of rheological modifier added to the blend was varied from 0 – 35% (see Table 24.1). An increase in the amount of RM resulted in an exponential increase of the L-PO MFI, as shown in Figure 24.1. The RM acts as an internal lubricant that alters the segmental motions of the polymer chains, thereby causing the chains to move over each other and thus leading to a decrease

Table 24.1 Amount of RM and the resultant MFI values.

% RM in L-PO blend	MFI (g/10 min)
0	7.9 ± 0.4
5	10.4 ± 0.5
10	12.9 ± 1.8
15	19.7 ± 2.2
20	22.9 ± 1.0
25	33.8 ± 4.5
30	44.8 ± 2.4
35	54.9 ± 0.6

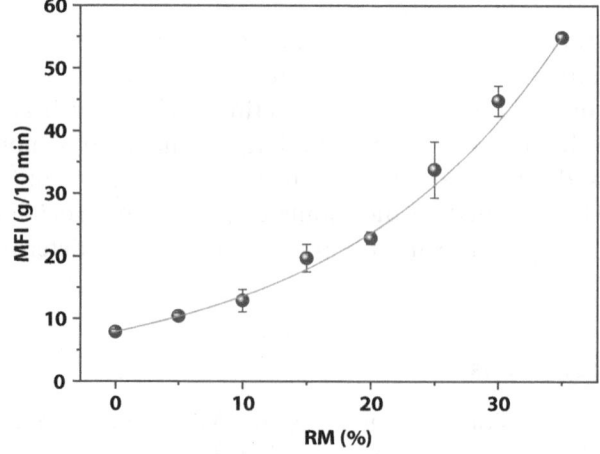

Figure 24.1 MFI values as a function of RM (%).

in viscosity and an increase in the MFI. It has been reported previously that plasticizers increase the free volume of polyvinyl chloride (PVC) polymer chains [18]. Increasing the MFI of a polymer melt is important for injection molding processing. Usually, injection molding requires polymer pellets that have a high MFI, and the addition of RM is essential in achieving that goal.

24.3.2 Development of a Model Equation

L-PO was used to develop a model equation that can be used by industrial recyclers to achieve a consistent MFI of m-PO. As mentioned earlier, the main challenge encountered with m-PO is the variation of their MFI values. In an endeavor to counteract this problem, a range of MFI values should be established. The MFI of the L-PO without RM was approximately 8 g/10 min as shown in Table 24.1. The lower and upper MFI value of the L-PO was achieved by varying the composition of the existing polymers. The composition of polyolefins in the L-PO was taken as 26.2, 18.15, 18.15, and 37.5% for HDPE, LDPE, LLDPE, and PP respectively. With a middle MFI value of 8 g/10 min, the lower and the upper MFI values were 2 and 14 g/10 min respectively. Tables 24.2 and 24.3 show the variation in polymer composition to achieve the upper and lower MFI values.

The model equation to determine the amount of RM that should be incorporated into the m-PO to attain a specific MFI was determined with the use of a two-level factorial

Table 24.2 Compositions used to achieve an MFI value of ~2 g/10 min.

Polymer	Initial MFI (g/10 min) measured at 230 °C	Composition (%)
HDPE	2	26
LDPE	1	28
LLDPE	2	28
PP	27	18

Table 24.3 Compositions used to achieve an MFI value of ~14 g/10min.

Polymer	Initial MFI (g/10 min) measured at 230 °C	Composition (%)
HDPE	2	26.2
HDPE	7	
LDPE	1	18.15
LLDPE	34	18.15
PP	27	37.5

design that was established using Design-Expert software. The variables being investigated were the initial MFI (g/10 min), % of RM, and the temperature (°C). The response was the resultant MFI (g/10 min). Statistical modeling has been successfully used in different polymer blends and composites [19–21]. Table 24.4 reveals the experimental design matrix that was generated with the Design-Expert software.

The model that was developed in this study to predict the amount of RM that must be incorporated into m-PO in order to achieve a targeted MDI is shown below as equation (24.1):

$$Log_{10}(MFI_T) = -0.887928 + 0.060335 \, MFI_I + 0.039909RM + 0.005006 \, Temp - 0.00078 \, MFI_I \times RM \tag{24.1}$$

where MFI_I denotes the initial MFI of the system (L-PO/m-PO), MFI_T is the targeted MFI to be achieved, RM is the rheological modifier in %, and Temp is the temperature in °C. The validity of the model equation was checked by using HDPE pellets of known MFI (7 g/10min). The targeted MFI was 30 g/10 min and the amount of %RM to be added to the HDPE to reach 30 g/10 min was determined from equation (24.1). The measured MFI was 30.5 ± 0.5 g/10 min (see Table 24.5) which was similar to the MFI_T, thus showing that the model can accurately predict the amount of RM (%) to be added to a particular system to achieve the desired MFI. The development of this model will be very helpful for the

Table. 24.4 Experimental design matrix generated by design-expert software.

Experiment	Factor 1	Factor 2	Factor 3	Response
	Initial MFI (°C)	RM (%)	Temp (°C)	MFI (g/10 min)
1	2	0	190	1.48 ± 0.04
2	2	0	230	2.5
3	2	50	190	131 ± 20
4	2	50	230	199 ± 47
5	8	25	210	20.1 ± 4.6
6	14	0	190	7.44 ± 0.58
7	14	0	230	14.16 ± 2.44
8	14	50	190	252.4 ± 51.8
9	14	50	230	332.9 ± 20.4

Table 24.5 Validating the model equation.

Polymer	Initial MFI (g/10 min)	Target MFI (g/10 min)	Measured MFI (g/10 min)
HDPE Dowlex (DOWLEX IP-10262)	7	30	30.5 ± 0.5

recycling sector, as it allows one to predict the amount of RM that can be incorporated into the waste plastics without requiring an MFI experiment to be performed for every plastic waste lot with different MFI_I values.

24.3.4   Purity Analysis

The initial version of this model equation was investigated for m-PO in this study. M-POs were initially separated from the other waste plastics via the density separation method. Polyolefins have a lower density than that of water, and hence they tend to float in water. Meanwhile, plastics with a higher density than water tend to sink in water. This approach is one of the most facile, cost-effective, rapid, and industrially scalable techniques to separate waste plastics [22]. Differential scanning calorimetry analysis was conducted to determine whether or not PE (HDPE, LDPE, and LLDPE) and PP were the only polymers present in the system. Figure 24.2 shows the melting temperature peaks of PE and PP at 130 and 166 °C, respectively, and these melting points are close to those reported by other researchers [23, 24]. The two distinct peaks also reveal the absence of other non-polyolefin polymers such as polystyrene (PS) which usually shows a glass transition temperature peak (T_g) at ≈ 100 − 107 °C [25] and polyethylene terephthalate (PET) which has a T_g at 67 − 81 °C and a melting peak temperature peak at 260 °C [26]. The absence of PS and PET shows that the density method is an effective way of separating polyolefins from other non-polyolefin polymers.

To further investigate the purity of the m-PO, an ash test analysis was also conducted. Four randomly selected samples that had been separated by the density separation method were tested for analysis. Our analysis revealed that the ash content was approximately 1% (see Table 24.6). This finding shows that the amount of mineral filler present in the m-PO is approximately equal to 1%, hence the m-PO was highly pure. From the purity analysis, it can be concluded that the m-PO were well separated from the rest of the waste plastics.

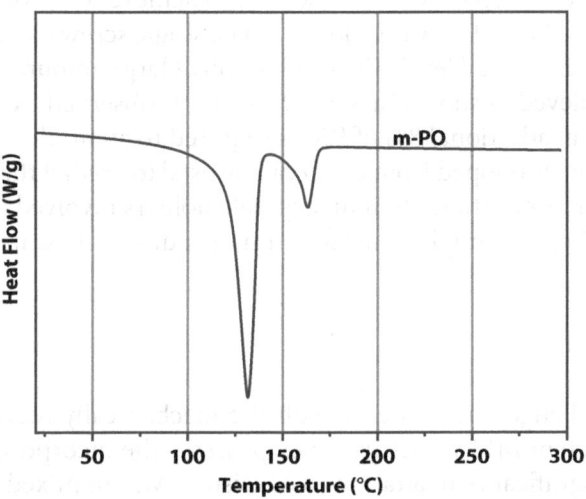

Figure 24.2 DSC plot of m-PO.

Table 24.6 Ash contents of four samples.

Sample	Mass of sample	Mass of crucible	Mass of crucible + ash	Ash mass	Ash (%)
1	2.0709	40.7776	40.7984	0.0208	1.0
2	2.0084	37.9736	38.0001	0.0265	1.3
3	2.0076	38.9582	38.9836	0.0254	1.3
4	2.0404	43.1848	43.1846	0.0291	1.4

Table 24.7 Prediction of the amount of % RM to be added to achieve a MFI_T.

MFI_I (g/10 min)	MFI_T (g/10 min)	The amount of %RM to be added according to equation (1)	Additional %RM	MFI_M (g/10 min)	Equipment
5.8	20	19	8	23.3 ± 1.9	DSM
5.8	30	24.4	7.6	30	DSM
5.8	30	24.4	7.6	31.1	SSE

*MFI_M (experimentally measured MFI).

24.3.5 MFI of m-PO

The m-PO was further ground into small particles, which enables a consistent MFI measurement. The measured MFI_I of the m-PO was 5.8 g/10 min and this value was used to determine the amount of rheological modifier that can be incorporated into the m-PO. Equation 1 was used to predict the % RM needed to achieve MFI_T of 20 and 30 g/10 min. Approximately 1 kg of the m-PO was produced via a single screw extruder which is similar to the one used at the industrial level. The production of large amounts of m-PO with a uniform MFI can be achieved as shown in Table 24.7. It was observed from the measurements (see Table 24.7) that an additional 8% of RM is required to attain the MFI_T.

The model equation developed from L-PO can be used to predict the %RM in m-PO. This model is of paramount importance to industry stakeholders involved in plastic recycling as it saves time and is simple to implement since it uses industrially scalable technology.

24.4 Conclusion

One of the main challenges encountered with the mechanically recycling of mixed polyolefins is the inconsistent MFI values that they possess. The incorporation of a rheological modifier is of great significance in attaining a consistent MFI in mixed polyolefins as shown by the experimental MFI results. The development of the model equation is essential in determining the amount of rheological modifier that can be added to attain a specific MFI value in mixed polyolefins. This is far better compared to the trial-and-error method as it saves time and resources.

References

1. Gebrekrstos, A.; Muzata, T. S.; Ray, S. S. Nanoparticle-Enhanced β-Phase Formation in Electroactive PVDF Composites: A Review of Systems for Applications in Energy Harvesting, EMI Shielding, and Membrane Technology. *ACS Appl. Nano Mater.*, 5, 6, 7632–7651, 2022.
2. Zhu, K.; Chen, L.; Chen, C.; Xie, J. *Preparation and Characterization of Polyethylene Antifogging Film and Its Application in Lettuce Packaging. Food Control.*, 139, 109075, 2022.
3. Mural, P. K. S.; Banerjee, A.; Rana, M. S.; Shukla, A.; Padmanabhan, B.; Bhadra, S.; Madras, G.; Bose, S. Polyolefin Based Antibacterial Membranes Derived from PE/PEO Blends Compatibilized with Amine Terminated Graphene Oxide and Maleated PE. *J. Mater. Chem. A*, 2, 41, 17635–17648, 2014.
4. Muzata, T. S.; Gebrekrstos, A.; Ray, S. S. Recent Progress in Modified Polymer-Based PPE in Fight against COVID-19 and Beyond. *ACS Omega*, 6, 43, 28463–28470, 2021.
5. *Plastic pollution is growing relentlessly as waste management and recycling fall short, says OECD.* https://www.oecd.org/environment/plastic-pollution-is-growing-relentlessly-as-waste-management-and-recycling-fall-short.htm (accessed 2022-12-01).
6. *U.S. plastics industry - statistics & facts | Statista.* https://www.statista.com/topics/7460/plastics-industry-in-the-us/#topicOverview (accessed 2022-11-30).
7. Statistic_id622783_plastic-Production-in-the-Us-2021-by-Resin.Pdf.
8. *U.S. plastic MSW recycling rate | Statista.* https://www.statista.com/statistics/1110734/us-plastics-recycling-as-a-share-of-generation/ (accessed 2022-11-30).
9. Bora, R. R.; Wang, R.; You, F. Waste Polypropylene Plastic Recycling toward Climate Change Mitigation and Circular Economy: Energy, Environmental, and Technoeconomic Perspectives. *ACS Sustainable Chem. Eng.*, 8, 43, 16350–16363, 2020.
10. Schwarz, A. E.; Ligthart, T. N.; Godoi Bizarro, D.; De Wild, P.; Vreugdenhil, B.; van Harmelen, T. Plastic Recycling in a Circular Economy; Determining Environmental Performance through an LCA Matrix Model Approach. *Waste Manag.*, 121, 331–342, 2021.
11. Ellen MacArthur Foundation and World Economic Forum. The New Plastics Economy: Rethinking the Future of Plastics. Ellen MacArthur Found., No. January, 120, 2016.
12. Khan, A.; Naveed, M.; Aayanifard, Z.; Rabnawaz, M. Efficient Chemical Recycling of Waste Polyethylene Terephthalate. *Resour. Conserv. Recycl.*, 187, 106639, 2022.
13. Schyns, Z. O. G.; Shaver, M. P. Mechanical Recycling of Packaging Plastics: A Review. *Macromol. Rapid Commun.*, 42, 3, 1–27, 2021.
14. Jia, X.; Qin, C.; Friedberger, T.; Guan, Z.; Huang, Z. Efficient and Selective Degradation of Polyethylenes into Liquid Fuels and Waxes under Mild Conditions. *Sci. Adv.*, 2, 6, 1–8, 2016.
15. Kumar, V.; Khan, A.; Rabnawaz, M. Efficient Depolymerization of Polystyrene with Table Salt and Oxidized Copper. *ACS Sustainable Chem. Eng.*, 10, 19, 6493-6502, 2022.
16. Vogt, B. D.; Stokes, K. K.; Kumar, S. K. Why Is Recycling of Postconsumer Plastics so Challenging? *ACS Appl. Polym. Mater.*, 3,9, 4325–4346, 2021.
17. Shenoy, A. V.; Saini, D. R.; Nadkarni, V. M. Estimation of the Melt Rheology of Polymer Waste from Melt Flow Index. *Polymer*, 24, 6, 722–728, 1983.
18. Matuana, L. M.; Park, C. B.; Balatinecz, J. J. The Effect of Low Levels of Plasticizer on the Rheological and Mechanical Properties of Polyvinyl Chloride/Newsprint-Fiber Composites. *J. Vinyl Addit. Technol.*, 3, 4, 265–273, 1997.

19. Afrifah, K. A.; Matuana, L. M. Statistical Optimization of Ternary Blends of Poly(Lactic Acid)/Ethylene Acrylate Copolymer/Wood Flour Composites. *Macromol. Mater. Eng.*, 297, 2, 167–175, 2012.

20. Matuana, L.; Li, Q. A Factorial Design Applied to the Extrusion Foaming of Polypropylene/Wood-Flour Composites. *Cell. Polym.*, 20, 2, 115–130, 2001.

21. Matuana, L. M.; Li, Q. Statistical Modeling and Response Surface Optimization of Extruded HDPE/Wood-Flour Composite Foams. *J. Thermoplast. Compos. Mater.*, 17, 2, 185–199, 2004.

22. Gent, M. R.; Menendez, M.; Toraño, J.; Diego, I. Recycling of Plastic Waste by Density Separation: Prospects for Optimization. *Waste Manag. Res.*, 27, 2, 175–187, 2009.

23. Gu, J.; Xu, H.; Wu, C. Thermal and Crystallization Properties of HDPE and HDPE/PP Blends Modified with DCP. *Adv. Polym. Technol.*, 33, 1, 2014.

24. Phulkerd, P.; Arayachukeat, S.; Huang, T.; Inoue, T.; Nobukawa, S.; Yamaguchi, M. Melting Point Elevation of Isotactic Polypropylene. *Journal of Macromolecular Science, Part B*, 53, 7, 1222–1230, 2014.

25. Rieger, J. The Glass Transition Temperature of Polystyrene. *J. Therm. Anal.*, 46, 3-4, 965–972, 1996.

26. Lanaro, M.; Booth, L.; Powell, S. K.; Woodruff, M. A. Electrofluidodynamic Technologies for Biomaterials and Medical Devices: Melt Electrospinning. *Electrofluidodynamic Technol. Biomater. Med. Devices Princ. Adv.*, 37–69, 2018.

Ultrahigh-Speed Extrusion of Recycled Film-Grade LDPE and Injection Molding Characterization

Peng Gao, Joshua Krantz, Olivia Ferki, Zarek Nieduzak, Sarah Perry, Davide Masato*
and Margaret J. Sobkowicz†

Department of Plastics Engineering, University of Massachusetts Lowell, Lowell, MA, USA

Abstract

The analysis of the mechanical recycling market in North America indicates that flexible packaging is the largest end- use application of PE (i.e., > 40%), with a domestic recycling rate of only about 10%. There are several barriers to achieving sustainable recycling of film and flexible plastics, including feedstock contamination, lack of stability and coordination in the supply chain, inefficient sortation, poor processing capabilities, and lack of end markets. In this research, ultrahigh-speed extrusion of film grade LDPE is proposed to modify its molecular weight to an injection molding grade. Different screw speeds, feed rates, and barrel temperatures were tested to optimize the extrusion process. The modified materials showed 98% lower viscosity and were suitable for injection molding. The as-received and modified recycled LDPE were injection molded using velocity-controlled and pressure-controlled injection molding techniques. The results showed that the molded samples using modified materials presented 9-20% decrease in yield stress but showed 20% increase in Young's modulus. The samples fabricated with pressure-controlled process showed 13% increase in ultimate elongation and maintained identical modulus and tensile stress at yield. The maximum pressure on the polymer melt was reduced from 130 MPa to 70 MPa for pressure-controlled injection molding as compared to velocity-controlled technology. Overall, pressure-controlled injection molding technology reduced the energy consumption of the process by >5%.

Keywords: Twin-screw extrusion, injection molding, mechanical recycling, secondary feedstock, pressure-controlled injection molding

25.1 Introduction

Low-density polyethylene is one of the most used polymers in the flexible packaging industry. Low cost, lightweight, and barrier properties have made film packaging the largest market for LDPE. However, the increasing manufacturing demand needs to be supported by circular economy strategies and sustainable end-of-life practices [1], [2].

Corresponding author: davide_masato@uml.edu
†*Corresponding author*: margaret_sobkowiczkline@uml.edu

Nabil Nasr (ed.) Technology Innovation for the Circular Economy: Recycling, Remanufacturing, Design, Systems Analysis and Logistics, (321–332) © 2024 Scrivener Publishing LLC

LDPE shopping bags are usually recycled at shopping centers and then delivered to specialized material recovery facilities [3]. Indeed, plastic film waste must be shredded and densified with specific equipment to be mechanically recycled. Film-grade secondary feedstock is commonly downcycled in construction materials, lumber, or sheets [4]. Due to the high molecular weight and lack of consistency after recycling, the r-LDPE film does not find upcycling applications, such as injection molding. The film-grade LDPE's melt flow index (MFI) is typically less than 2.0 g/10 min (190°C at 2.16kg). On the other hand, injection molding typically requires MFI>10g/10min [5].

A high-speed extrusion process can modify the molecular weight distribution and hence the melt viscosity of LDPE [6]. The combination of high temperature and shear rate leads to lower molecular weight through thermo- mechanical degradation. Chain scission and oxidation can occur simultaneously during high-speed extrusion, and there is the possibility of branching changes as well due to free radical reactions.

The injection molding of recycled polymers requires assessing and controlling their processing variability. To achieve consistent product quality, novel process control methodologies are necessary to automate the adjustment of processing conditions. Pressure-controlled injection molding (P-Ctrl) offers an opportunity for closed-loop adjustment of injection molding. Unlike conventional molding (i.e., velocity-controlled filling – V-Ctrl), pressure-controlled injection molding relies on the closed-loop feedback provided by a nozzle pressure transducer to the injection unit, allowing pressure-controlled filling. Pressure-controlled injection molding has been shown to reduce the shear rate and allow more accurate control of the polymer flow. In the V-Ctrl process, the screw speed increases quickly until it reaches the desired constant screw speed at the start of the injection phase. Then the injection screw is maintained at a constant set speed and the forward movement pushes the material into the cavity [7]. Once the mold cavity is partially filled (usually 80%-95%), the machine switches over to a pressure-controlled packing phase [8], and the moment where the control method is changed is defined as V/P switchover point. During the packing stage, the packing pressure is applied to compensate for shrinkage upon cooling and is commonly set up to be lower than the peak pressure (i.e., 60-80% of peak pressure) reached at the end of the injection phase. Multiple packing stages with different packing pressures and durations can be applied during the packing phase. The initial stage is applied to fill out the cavity and account for backflow. In contrast, the second stage, a lower pressure, is maintained to compensate for shrinking and maintain a pressure equilibrium until the gate freezes. Additionally, the lower packing pressures result in lower residual stresses on the samples [9]. In the P-Ctrl process, the screw velocity is continuously adjusted during filling to maintain a constant nozzle pressure. The nozzle pressure setpoint profiles and PID parameters can be manually determined, which offers fully close-loop control over the P-Ctrl process.

This work proposes the modification of recycled film-grade LDPE into injection molded products through a high-speed extrusion and pressure-controlled filling. The high-speed extrusion was studied to modify the polymer molecular weight and rheology. Then, the injection molding of the modified polymers was studied to correlate processing and product quality. The energy consumption was monitored for all polymer processing experiments and assessed for the proposed manufacturing strategy.

25.2   Materials and Methods

25.2.1   Film-Grade Recycled LDPE Characterization

The experiments were carried out using film-grade recycled LDPE (Green Isotene, SER North America). The material was obtained from the recycling of shopping bags. The film waste was pretreated, densified, and pelletized by the supplier and contained ~5% polypropylene contamination. The mechanical and thermal properties were characterized for the as-received r-LDPE. Thermal Gravimetric Analysis (TGA, Mettler Toledo TGA 2 SF) was conducted from 0°C to 600°C with a ramp rate of 20°C/min to evaluate polymer degradation. Differential Scanning Calorimetry (DSC, Mettler Toledo DSC 3+) was performed to assess the melting and crystallization behavior between 0°C and 250°C (melt temp taken from the second heating run). Table 25.1 summarizes the critical material properties of the as-received LDPE.

25.2.2   Ultrahigh-Speed Extrusion

An ultrahigh-speed co-rotating twin screw extruder (TSE - TECHNOVEL Co. Japan, KZW15TW-45/60MG- NH(-44000)-UM) with an L/D ratio of 60:1 was used to modify the film-grade r-LDPE. The pelletized r-LDPE resin was volumetrically fed into the barrel and the extrudate was continuously pulled through a cooling water trough and pelletized.

Extrusion experiments were conducted using a flat temperature profile of 190°C with varying screw speeds of 100 rpm, 2,000 rpm, and 4,000 rpm. The feed rate of the volumetric feeder (5IK90GU-SWT, ORIENTAL MOTOR Co., Ltd.) was controlled at 13.58 g/min (20 rpm) and 20.44 g/min (30 rpm) for each screw speed. The temperature profile of the eight-barrel zone TSE is summarized in Table 25.2.

Table 25.1 Material properties of the recycled film-grade LDPE.

Property	Units	Test method	r-LDPE
Density	g/cm³	ASTM D792	0.98
Melt Temp.	°C	ASTM C518	127.4
Crystallization Temp.	°C	ASTM C518	95.4
Degradation Temp.	°C	ASTM E1131	442
Melt Flow Rate	g/10min	ASTM D1238	.56-.72
% Volatiles by Vol.	%	ASTM D1203	0.14
Ash Content	%	ASTM D5630	6.71

Table 25.2 Temperature profile for extrusion processes.

Barrel zone	1	2	3 – 8	Head	Die
Temperature (°C)	150	175	190	190	190

25.2.3 Injection Molding

The injection molding experiments were performed on a 130-ton machine (Milacron-FANUC Roboshot α- S130iB) with a 40 mm screw diameter. The mold used was a two-plate cold runner mold with ASTM cavities used to manufacture Type I tensile bars. The injection molding machine was operated using two process control strategies: velocity-controlled (V-Ctrl) and pressure-controlled (P-Ctrl). The V-Ctrl process was achieved using the standard logic provided by the Milacron-FANUC injection molding machine. The P-Ctrl process was run using an instrumented nozzle and novel control unit technology provided by iMFLUX, Inc.

Figure 25.1 compares the nozzle pressure and screw velocity for the two processes. The P-Ctrl process is characterized by a quick pressure rise (< 0.1 seconds) to reach the set injection pressure. The pressure is kept constant until a transfer position, defined for 95-98% full cavity, is reached. Then, the pressure is reduced to continue packing the full cavity without continuing to inject more material and possibly flash the mold. In the V-Ctrl process, the screw speed is increased to the setpoint and maintained for about 2 seconds until the cavity is 95% filled. The screw speed was reduced to a setpoint of 0 during the packing and holding phase. The dramatic change in the screw speed resulted in a sharp peak on the pressure curve for the V-Ctrl process. The filling behavior is more dynamic and unstable than the P-Ctrl process, resulting in defects on surfaces (i.e.: visible sink marks).

25.2.4 Approach and Characterization

A 2-level, 3-factor full-factorial design of experiment (DoE) was designed to compare the velocity-control and pressure-control technologies on as-received recycled LDPE and the modified material. Mold temperatures, cooling time and packing times remained the same to maintain identical dimensional stability on the sample. Three processing parameters controlling the behavior during injection, transfer and packing were considered experimental factors for both V-Ctrl and P-Ctrl technologies. 10 repetitions were molded on a fully automatic cycle with no interruptions for a total of 80 parts collected. For the V-Ctrl process methodology, a baseline process was established following the short shot method, in-mold rheology and gate-freeze study. For the P-Ctrl process, the pressure setpoint was determined from the V-Ctrl process run at the higher and lower setpoints. A summary of the processing parameters is presented in Table 25.3, and the DOE is presented in Table 25.4. It should be noted that the melt temperature for the two materials was set to different values according to their melt viscosity.

25.2.5 Characterization Techniques

The rheological properties of the resins were characterized using an ARES-G2 Rheometer (TA Instrument) with 25 mm stainless steel parallel plates. The r-LDPE and m-r-LDPE

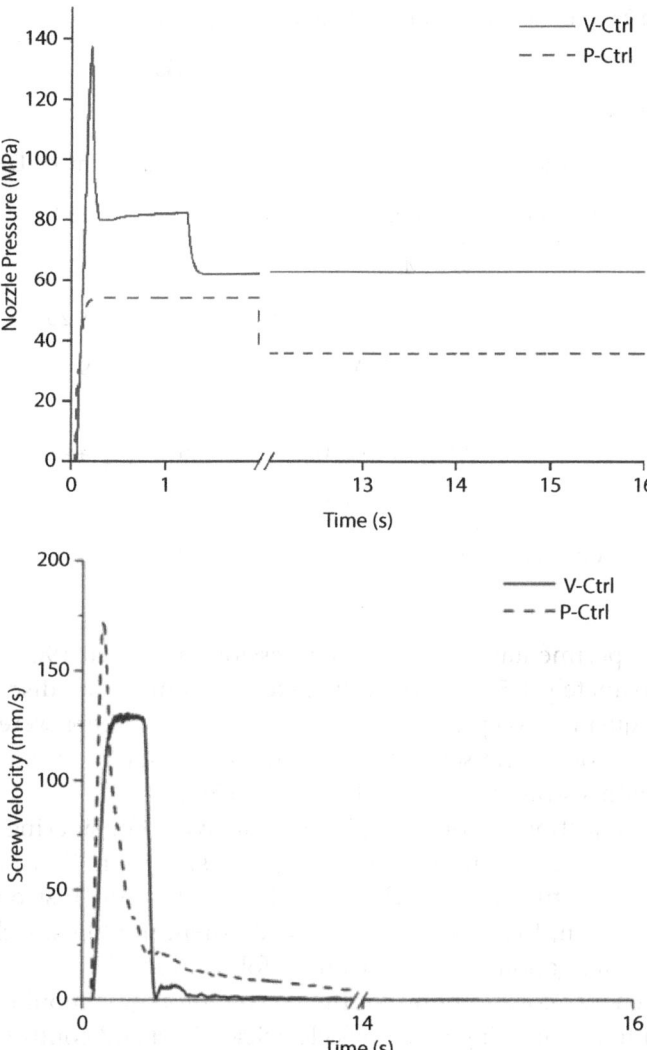

Figure 25.1 Comparison between V-Ctrl and P-Ctrl technologies.

Table 25.3 Processing parameters that were kept constant during injection molding. The melt temperatures were determined based on MFI of the materials.

Parameter	r-LDPE	m-r-LDPE
Mold Temp (°C)	40	40
Melt Temp (°C)	230	170
Cooling Time (s)	12	12
Packing Time (s)	16	16
Shot size (mm)	19.5	17

Table 25.4 DOE for the injection molding experiments.

	Factors		Levels	
			-	+
1	Material		r-LDPE	m-r-LDPE
2	Processing	vv_{iiiii} (mm/s)	30	130
		pp_{iiiii} (Mpa)	30	45
		XXXXXXXX^{vv} (mm)	7	8.25
		XXXXXXXX^{pp} (mm)	10.5	9
		$pp^{vv}_{pppppppp}$ (Mpa)	33.6	44.8
		$pp^{pp}_{pppppppp}$ (Mpa)	25	45
3	Control Method		P-Ctrl	V-Ctrl

samples from the experimental runs were compression molded at 190 °C into discs with a thickness of approximately 1.5 mm and a diameter of 25 mm. The disc specimens underwent strain and frequency sweeps. To ensure that frequency sweeps were conducted in the linear viscoelastic region of the sample; oscillation amplitude or strain sweeps were performed at 200 °C with a strain percentage from 2% to 5%.

The properties of injection molded samples were analyzed considering different response variables, including tensile properties and energy consumption. Tensile testing was performed at room temperature using a Universal Testing System (Instron 5966, Norwood, MA, USA). For each run, five samples were tested. Then, modulus, yield stress, strain at yield, and ultimate elongation were determined following ASTM D638 from the stress-strain curves. The energy consumption for the servo motors was monitored using a power meter integrated into the molding machine's electrical panel and control software.

25.3   Results and Discussion

25.3.1   Material Modification

Figure 25.2 compares the complex viscosity curves of the r- LDPE and three m-r-LDPE grades. The m-r-LDPE extruded at the lowest screw speed of 100 rpm showed little to no reduction in complex viscosity compared to the r- LDPE due to minimal shear produced during extrusion. Changes in molecular weight caused by shear degradation are primarily denoted in the decrease in orders of magnitude when comparing higher screw speed extrusion runs with the r-LDPE prior to modification. Extrusion with a screw speed of 2,000 rpm showed a reduction and change of approximately one order of magnitude in complex viscosity, from around 20,000 Pa·s at low angular frequencies to about 2,000 Pa·s for the modified material. The material experienced an estimated 97% change in complex viscosity as the screw speed was further increased to 4000 rpm (~200 Pa · s at low frequency). The

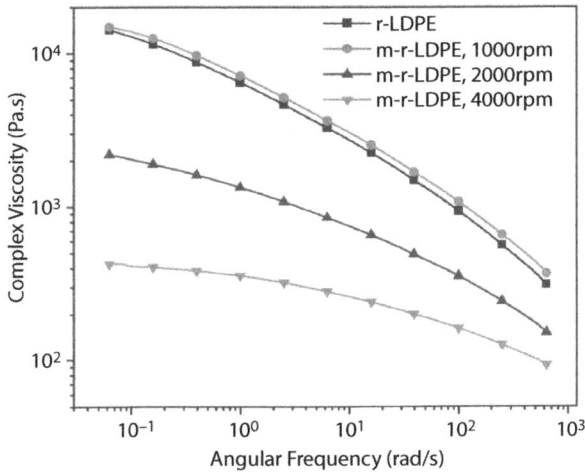

Figure 25.2 Comparison of viscoelastic behavior of the r-LDPE and modified r-LDPE samples using oscillation frequency sweeps at 200°C.

increase in polymer flow at lower angular frequencies indicates reductions in molecular weight and, therefore, the resin's viscosity. According to the classical relation, the fractional change in molecular weight can be related to the ratio of zero-shear viscosities after and before modification, to the power of 3.4 [10], [11]. The estimated ratio of molecular weight at 4000 rpm to the original molecular weight is 0.25, or 75% molecular weight reduction from that of the original resin. The reduction in chain length and entanglements caused by shear degradation reduces the viscosity of the resin and improves the resin's flowability. For the r-LDPE with a viscosity and melt strength that is originally more suited to that of a blown film grade application, the more severe reduction in complex viscosity (seen in 4000 rpm) is achieved through the ultra-high-speed extrusion of the r-LDPE allows for improved flow and processability for injection molding applications.

Watt's Law was used to calculate an average 3-phase power used for estimating the specific mechanical energy (SME). The estimation of average power consumption is demonstrated in Equation 25.1. SME was calculated using Equation 25.2.

$$W = VV_{ppvvaa} \times AA_{ppvvaa} \times pp.XX. \times \sqrt{3} \times \frac{NN}{NN_{mmmmmm}} \times \frac{TT}{TT_{mmmmmm}} \quad (25.1)$$

$$SME = \frac{PP \times TT \times NN}{TT_{mmmmmm} \times NN_{mmmmmm} \times QQ} \quad (25.2)$$

The maximum screw speed of the TSE is represented by NN_{mmppmm} (4,500 rpm) and the actual screw speed is represented by N (100, 2,000, and 4,000 rpm). Maximum motor power, P, is rated in kW. The feed rate Q is measured in kilograms per hour, and the T and TT_{mmppmm} represent the torque and maximum torque of the TSE motor, respectively. The power factor is represented by $p.f.$ (assumed value of 1), average amperage is illustrated by AA_{ppvvaa}, and average voltage is denoted by VV_{ppvvaa} (180 V). W is the average power consumption in kW. Head pressure, melt temperature, and amperage values were recorded from the TSE control panel and were collected every 15 minutes during the extrusion process.

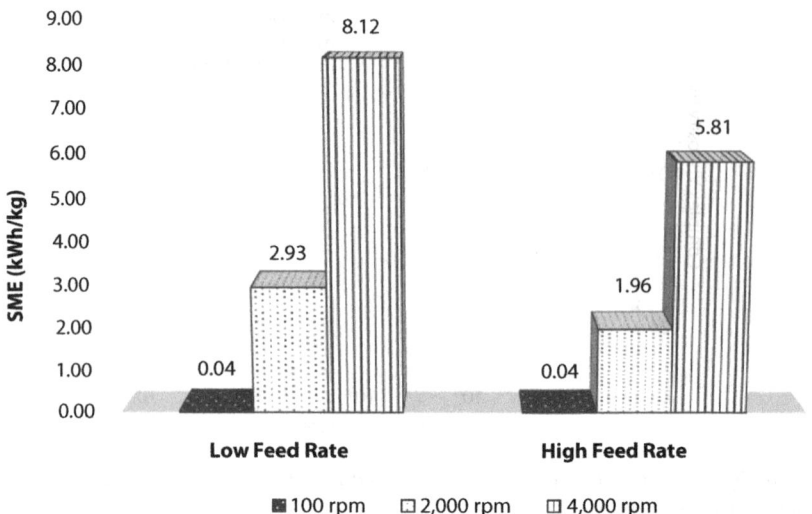

Figure 25.3 SME comparison of extrusion at 190°C based on changes in screw speed and feed rate conditions.

Figure 25.3 shows the comparison of SME based on changes in screw speed and volumetric feeding rates during the TSE DOE. SME calculations show that an increase in screw speed increases SME while an increase in feed rate reduces SME. Increasing the TSE's screw speed requires the drive motor to increase energy consumption to rotate the extruder screw at faster rates. Increasing the screw speed will induce higher shear and further reduce polymer viscosity. Increasing the material feed rate reduces the residence time of the material in the barrel The m-r-LDPE with 4000 rpm screw speed and 22.4 g/min feed rate was selected for further injection molding experiments, based on the most significant rheology change yet with less mechanical energy input.

25.3.2   Injection Molding and Characterization

Multiple regression models were evaluated to identify correlations between the factors defined in the DOE plan and the response variables. The form of the regression models was determined based on the structure and explicit factor settings of the full factorial DOE plan. The model included main effects, first-order interactions, and an intercept for each response variable. Table 25.5 summarizes the regression results, showing the range of variation for each response variable, coefficient of variation (COV), most critical determinant ($XCritic$), fitted coefficient of determination (RR^2), and the variance homogeneity ($FMax$).

The results indicate that the material factor is the most significant for all response variables. The modification achieved by ultrahigh-speed extrusion was significant. The fitted coefficients of determination (RR^2) are above 90% suggesting the multi-variate regression model effectively describes the response variables.

25.3.2.1   Effects of Processing Techniques on Tensile Properties

The strain-stress curves of as-received recycled LDPE and modified LDPE under different control methods are presented in Figure 25.4. The curves were obtained using an average

Table 25.5 Summary of regression results for the different response variables.

Response	Mean	Std. dev.	COV	Min	Max	R^2 (%)	X_{Critic}	F_{Max}
Modulus (MPa)	370.6	71.7	19.3	241.3	472.7	97.7	Material	188791
Yield Stress (MPa)	11.2	.870	7.8	9.96	12.55	96.5	Material	738
Ultimate Elongation (%)	311.3	231.9	74.5	50.61	618.34	99.4	Material	5305
Servo Energy Consumption (W)	528.6	124.4	23.5	345.7	731.4	99.9	Material	24069
Thermal Energy Consumption (W)	995.1	106.6	10.7	817.9	1161	90.4	Material	294

of the four runs for each material. The m-r-LDPE shows a significantly different fracturing behavior compared to the r-LDPE. The m-r-LDPE maximum tensile stress of 10.4 ± 0.3 MPa is observed right after the elastic deformation section for a strain of 10%. On the contrary, the r- LDPE shows a wide range of plastic deformation (strain = 10% to 180%), and the maximum stress of 12.3 MPa was observed at a strain of 180%. The strain-hardening effect occurred from 10% to 84%. Due to the lack of strain hardening section and shorter necking-fracture section, the ultimate elongation of the m-r-LDPE is 55% compared to 192% for the r-LDPE.

Overall, the r-LDPE followed a typical ductile fracturing behavior which conforms to the tensile behavior of high-viscosity film-grade LDPE [12]. The m-r-LDPE showed brittle

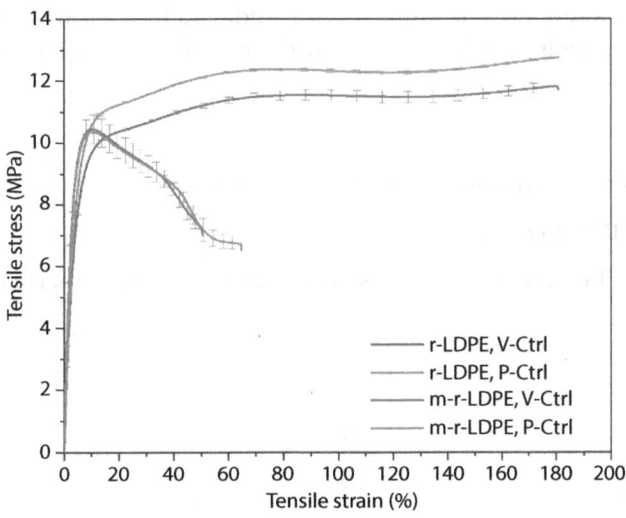

Figure 25.4 Tensile properties of the injection molded samples. The samples fabricated with different processing conditions (+/- levels) were averaged due to the similarity of the mechanical properties.

fracturing behavior, where the yield point is followed closely by necking, and no strain hardening was observed. During the tensile testing, the longer molecular chains in the r-LDPE can be further extended and oriented. The shorter chains of the m-r-LDPE did not show any rearranging upon tensile testing.

When switching from V-Ctrl to P-Ctrl, the Young's modulus increased by 23% to 39% for the m-r-LDPE as compared to the r-LDPE. It was also observed that the r-LDPE showed higher yield stress (7%) and modulus (14%) with the P-Ctrl process as compared to the V-Ctrl process. In contrast, the differences in the m-r-LDPE were not as significant. During the V-Ctrl process, the material was processed with higher screw speed and overall forces (c.f. Figure 25.1), which resulted in higher shear stresses. We hypothesize that the higher shear stress led to higher shear heating and fracture of the molecules. For the m-r-LDPE, the molecular weight was already significantly reduced during the extrusion process. Thus, the shear rate impact was not as remarkable during the injection molding process, leading to identical mechanical properties.

25.3.3 Energy Consumption Analysis

The energy consumption monitored during the injection molding process was analyzed by separating the amount drawn by the servomotors and the barrel heaters. Table 25.6 summarizes the changes in the servomotors and thermal energy consumption between the r-LDPE and m-r-LDPE.

The high-speed extrusion process reduced the viscosity and molecular weight of the material. As a result, the required nozzle temperature was decreased from 230°C to 170°C for the m-r-LDPE. This contributed to 17 to 20% reduction in thermal energy consumption. Additionally, the servo energy consumption decreased by 41% to 62% as the material changed from r-LPDE to m-r-LDPE. Due to the lower viscosity of the m-r-LDPE, lower flow resistance was generated during fabrication, thus the overall pressure was reduced significantly to fill the mold cavity. The maximum nozzle pressure for modified LDPE decreased from 140 MPa to 74 MPa for the V-Ctrl process at high processing conditions. The difference was less significant when the P-Ctrl method was utilized. The maximum pressure decreased from 65 MPa to 47 MPa when the material changed from r LDPE to r-m-LDPE. Additionally, the servo energy consumption was reduced by 14% and 18% when the P-Ctrl method was compared with the V-Ctrl one for the r-LDPE and m-r-LDPE, respectively.

Table 25.6 Summary of energy consumption for the injection molding process.

% Change from r-LDPE to m-r-LDPE			
Control method	**Processing levels**	**Servo consumption (%)**	**Thermal consumption (%)**
V-Ctrl	+	-41	-20
	-	-47	-18
P-Ctrl	+	-62	-17
	-	-59	-18

25.4   Conclusions & Recommendations

The ultra-high-speed extrusion process was demonstrated as an effective technique to modify film-grade r-LDPE into materials suitable for injection molding. Increasing the screw speed from 100 rpm to 2000 rpm reduces the complex viscosity of the r-LDPE by 84%. The complex viscosity decreases by 97% when the screw speed increases to 4000 rpm. The energy consumption during extrusion is mainly affected by the screw speed. The 4000-rpm screw speed allowed a higher feed rate of 22.4 g/min without increasing the energy consumption due to the low material viscosity. A higher feed rate increases the production rate, thus reducing specific energy consumption.

Both the r-LDPE and m-r-LDPE are suitable for injection molding. Still, the injection temperature can be reduced from 230°C to 170°C since the modified material has a lower viscosity and molecular weight. The mechanical properties are significantly affected by the modification process. Yield stresses suffer a 7% to 20% decrease, and the ultimate elongation decreases from 200% to 70% for the m-r-LDPE compared to the r-LDPE.

The pressure-controlled injection molding process is effective for both materials. In particular, the reduced shear stresses improved the mechanical properties of the r-LDPE. Additionally, the reduced overall pressure significantly reduced the servo energy consumption by 16% on average.

25.5   Acknowledgments

This material is based upon work supported by the U.S. Department of Energy's Office of Energy Efficiency and Renewable Energy (EERE) under the Advanced Manufacturing Office Award Number DE-EE0007897. This paper was prepared from results obtained as part of the project sponsored by an agency of the United States Government. Neither the United States Government nor any agency thereof, nor any of their employees, makes any warranty, express or implied, or assumes any legal liability or responsibility for the accuracy, completeness, or usefulness of any information, apparatus, product, or process disclosed, or represents that its use would not infringe privately owned rights. Reference herein to any specific commercial product, process, or service by trade name, trademark, manufacturer, or otherwise does not necessarily constitute or imply its endorsement, recommendation, or favoring by the United States Government or any agency thereof. The views and opinions of authors expressed herein do not necessarily state or reflect those of the United States Government or any agency thereof.

References

1. A. G. Pedroso and D. S. Rosa, "Mechanical, thermal and morphological characterization of recycled LDPE/corn starch blends," *Carbohydr. Polym.*, vol. 59, no. 1, pp. 1–9, 2005.
2. U. S. Ishiaku, K. W. Pang, W. S. Lee, and Z. A. M. Ishak, "Mechanical properties and enzymic degradation of thermoplastic and granular sago starch filled poly(ε-caprolactone)," *Eur. Polym. J.*, vol. 38, no. 2, pp. 393–401, 2002.
3. L. Dai *et al.*, "Pyrolysis technology for plastic waste recycling: A state-of-the-art review," *Prog. Energy Combust. Sci.*, vol. 93, p. 101021, 2022.

4. C. Abdy, Y. Zhang, J. Wang, Y. Yang, I. Artamendi, and B. Allen, "Pyrolysis of polyolefin plastic waste and potential applications in asphalt road construction: A technical review," *Resour. Conserv. Recycl.*, vol. 180, p. 106213, 2022.

5. J. Theberge, "IM alternatives produce performance advantages," *Plast. Eng.*, vol. 47, no. 2, pp. 27–31, 1991.

6. S. Martey *et al.*, "Hybrid Chemomechanical Plastics Recycling: Solvent-free, High-Speed Reactive Extrusion of Low-Density Polyethylene.," *ChemSusChem*, vol. 14, no. 19, pp. 4280–4290, 2021.

7. A. C. Aloise, A. J. Notte, and M. A. Rizzi, "Using ram velocity to control IMM part quality," *Adv. Polym. Technol.*, vol. 3, no. 2, pp. 137–141, 1983.

8. D. O. Kazmer, S. Velusamy, S. Westerdale, S. Johnston, and R. X. Gao, "A comparison of seven filling to packing switchover methods for injection molding," *Polym. Eng. Sci.*, vol. 50, no. 10, pp. 2031–2043, Oct. 2010.

9. X. Zhang, T. Ding, W. Wang, J. Liu, and C. Weng, "Study on the Effect of Processing Parameters on Residual Stresses of Injection Molded Micro-Pillar Array," *Polymers*, vol. 14, no. 16. 2022.

10. T. G. J. Fox and P. J. Flory, "Viscosity—Molecular Weight and Viscosity—Temperature Relationships for Polystyrene and Polyisobutylene1,2," *J. Am. Chem. Soc.*, vol. 70, no. 7, pp. 2384–2395, Jul. 1948.

11. T. G. Fox and P. J. Flory, "The glass temperature and related properties of polystyrene. Influence of molecular weight," *J. Polym. Sci.*, vol. 14, no. 75, pp. 315–319, Sep. 1954.

12. S. Saikrishnan, D. Jubinville, C. Tzoganakis, and T. H. Mekonnen, "Thermo-mechanical degradation of polypropylene (PP) and low-density polyethylene (LDPE) blends exposed to simulated recycling," *Polym. Degrad. Stab.*, vol. 182, p. 109390, 2020.

26

Composites from Post-Consumer Polypropylene Carpet and HDPE Retail Bags

Anuj Maheshwari[1], Mohamadreza Youssefi Azarfam[1], Siddhesh Chaudhari[2], Clinton Switzer[2], Jay C. Hanan[3], Sudheer Bandla[4], Ranji Vaidyanathan and Frank D. Blum[1]*

[1]*Department of Chemistry, Oklahoma State University, Stillwater, OK, USA*
[2]*School of Materials Science and Engineering, Helmerich Research Center, Oklahoma State University, Tulsa, OK, USA*
[3]*Mechanical and Aerospace Engineering, Helmerich Research Center, Oklahoma State University, Tulsa, OK, USA*
[4]*Niagara Bottling LLC, Diamond Bar, CA, USA*

Abstract

Every year, billions of pounds of carpets are discarded in landfills in the USA, raising concerns in terms of environmental pollution and economic liability. Approximately 100 billion plastic bags are discarded annually, which are often mixed with other materials (like ink, filler, and remnant products). Due to their heterogeneous nature, only about 3% of the bags are recycled. Processing the discarded materials to form useful composites can add additional value to materials that would otherwise end up in landfills. We have made composites using post-consumer polypropylene carpets (c-PP) and high-density polyethylene (r-HDPE) retail bags recovered from post-consumer sources by compression molding. Molding of the components under different pressures, temperatures, and compositions was performed. Preliminary molding conditions were based on analyzing the differential scanning calorimetry (DSC) and the thermogravimetric analysis (TGA) data for different raw materials. Molding factors were examined to define applicable ranges for each parameter. The effects of layup configuration, composition of components, temperature, molding time, and pressure were considered in the screening process. The quality of the molded samples was compared based on flexural strength and modulus, creep behavior, and microscopy. Molded samples showed good mechanical properties with potential for structural applications. The ability to recycle plastic waste will improve sustainability and reduce the environmental impact of plastic use.

Keywords: Recycling, compression molding, composite

Corresponding author: fdblum@okstate.edu

Nabil Nasr (ed.) Technology Innovation for the Circular Economy: Recycling, Remanufacturing, Design, Systems Analysis and Logistics, (333–342) © 2024 Scrivener Publishing LLC

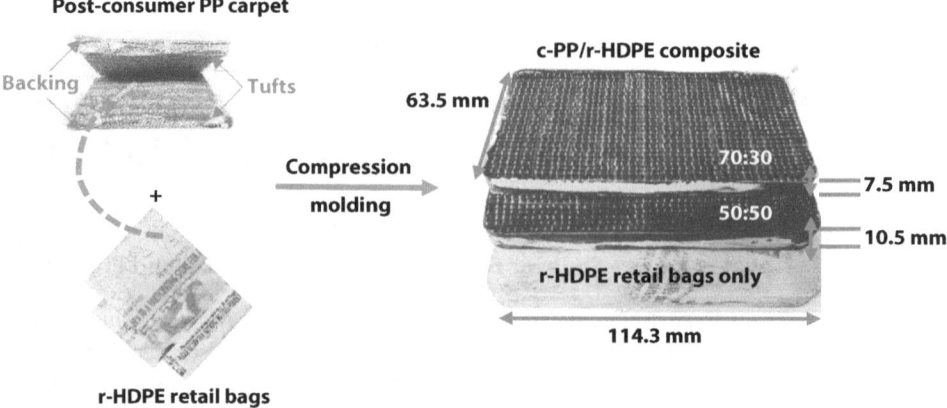

Schematic representation of post-consumer carpet and thermoplastic resin composite fabrication.

26.1 Introduction

According to the Environmental Protection Agency (EPA) report, 6.8 billion pounds of waste from carpets and rugs were generated in the U.S. in 2018, which typically rises from year to year. The market size of carpets and rugs was valued at $51.9 billion in 2018 and is anticipated to expand annually at a compound annual growth rate (CAGR) of roughly 5.5% over the projected period from 2019 to 2026 [1]. Every year, 4 billion pounds of carpet is discarded with around 18% used for energy production via combustion, a portion (less than 10%) of post-consumer carpet is recycled, and the rest end up in the landfill. To enhance the environmental benefits of recycling, more effort is needed to recycle post-consumer carpets to avoid landfilling [1, 2].

Carpet is used as a floor covering in households, commercial buildings, aircraft, and other applications. The general structure of a carpet comprises piles of synthetic fibers, primary backing, binder (adhesives and a filler), and secondary backing, as shown in Figure 26.1. The adhesive is often made of styrene-butadiene rubber (SBR) mixed with inorganic fillers like $CaCO_3$ or $BaSO_4$. In contrast, the primary and secondary backings are either made of polypropylene (PP), jute, linen, or a combination of different fibers. The whole

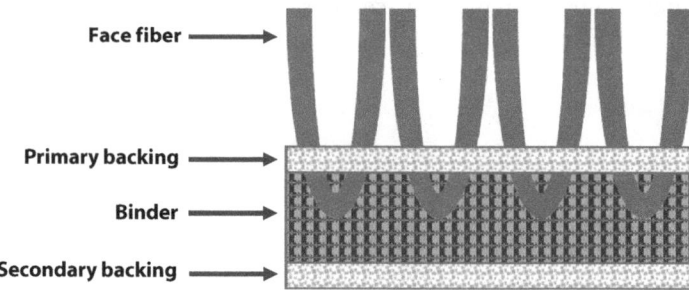

Figure 26.1 Schematic representation of carpet structure.

carpet is challenging to recycle due to its heterogeneous nature. A recycling method developed that includes whole carpets can reduce the extra costs of separating different components of carpets and lead to more sustainable practices.

High-density polyethylene (HDPE) is a natural gas-based thermoplastic polymer used to manufacture numerous items, including food and beverage containers, cleaning product bottles, pipes, plastic bags, and chopping boards. HDPE plastic bags are usually milky white or semi-translucent, often including 3% to 30% of $CaCO_3$ as filler [3]. According to Environmental Protection Agency (EPA) report, the rate of recycling for HDPE bottles was 29.3% in 2018. Recycling HDPE bottles can be economically viable, but plastic bags are difficult to recycle due to their heterogeneous nature.

Polyolefins (PO) can be recovered and recycled for beneficial uses [4]. Bottle-to-bottle recycling is one kind of recycling that is a successful example of closed-loop recycling [5]. Plastic fiber can also be recycled through advanced recycling methods into different grades. However, advanced recycling is still emerging and requires significant processing to eliminate contaminants. Making structural composites with desired mechanical properties (e.g., flexural strength greater than 20 MPa and flexural modulus greater than 750 MPa) [6], preferably with more carpet content and less resin, is a logical approach to waste reduction. In this research, post-consumer PP carpets (c-PP) with a mixture of HDPE (r-HDPE) retail bags have been used to make composites using compression molding. Compression molding is a low-cost and facile approach to fabricating composites using thermoplastics [7]. This paper reports the two-level fractional factorial design of experiments for the composites consisting of post-consumer PP carpets with a mixture of retail bags of HDPE [8].

26.2 Experimental

26.2.1 Materials

HDPE retail bags (sourced from consumer sites in Oklahoma) were used for the experimental work. The plastic bags were cleaned and chopped into small flakes before use. Post-consumer PP carpet was collected from a local site in Stillwater, OK. The carpet consisted of PP tuft on the top, natural fiber as the primary backing, another PP fiber as a binding agent, and jute as the secondary backing. Fourier transform infrared spectroscopy (FTIR) was used to identify the different components of the carpet.

26.2.2 Thermal Analysis

During molding, materials are heated close to or above their melting points. Therefore, knowing each material's melting point and degradation temperature is essential. Thermal gravimetric analysis (TGA) and differential scanning calorimetry (DSC) were used to determine the nature of the raw materials used in the composite preparation. TGA (at a 20 °C/min heating ramp under 60 mL/min airflow) was done using a TA Instruments Q50 for all raw materials to understand the degradation behavior from room temperature to around 800 °C. DSC (at a heating rate of 3 °C/min) was done using a TA Instruments Q2000 to measure the melting points of different r-HDPE retail bags and c-PP carpet fibers.

The temperature range varied from room temperature to just below the temperature of degradation of the material.

26.2.3 Compression Molding

Compression molding was performed using an aluminum mold with interior dimensions of around $11.4 \times 6.3 \times 2.5$ cm³. Based on the molding requirements, a hot press with larger platens having an area of 25.4×25.4 cm² was used for all the moldings. A nylon sheet (oven bag) was used to seal the mold and avoid flash. Mold release spray was used for easy removal of the molded composite. The composites thickness varied with the mass ratio of carpet to plastic bags. The compression molding scheme for the composites of post-consumer c-PP carpet and r-HDPE retail bags is shown in Figure 26.2.

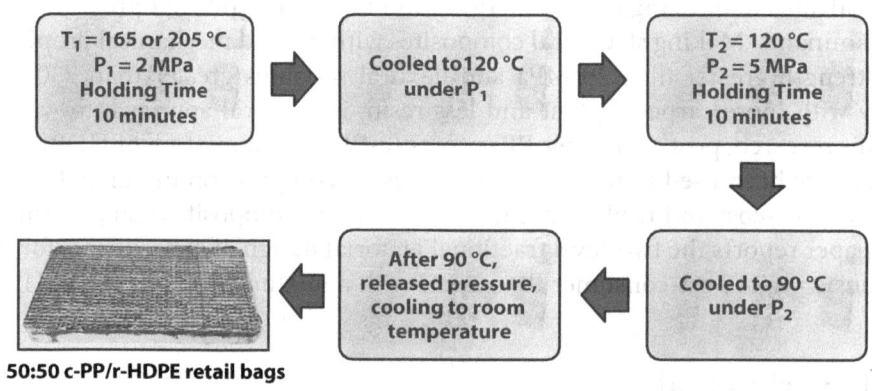

Figure 26.2 Schematic illustration of the molding process.

26.2.4 Design of Experiments

Different carpet and r-HDPE retail bag layups were studied to screen the effect of carpet configuration on the mechanical properties of the molded samples. In each carpet pile, the PP tufts (T) were considered as the top and the backing (B) as the bottom. The configuration with the best properties was observed when the r-HDPE retail bags were sandwiched between two carpet layers, and the carpet was arranged in BTrTB (backing-tufts-retail bags-tufts-backing) configuration. Based on our preliminary screening, molding at different pressures showed no significant difference in the mechanical properties of molded composites. Initially, the pressure was kept at 2 MPa (P_1) and heated to a molding temperature of 165 or 205 °C (T_1) to avoid flashing. On cooling, the temperature reached 120 °C (T_2), where HDPE crystallized, then the pressure was increased to 5 MPa (P_2). For the two-level fractional factorial design of experiments, molding temperature (°C) and carpet to r-HDPE retail bags (mass %) parameters were selected. Maximum/minimum molding temperatures and compositions of carpet to r-HDPE retail bags were studied, as shown in Table 26.1.

Table 26.1 List of the compression molded parameters studied.

Compositions (carpet:r-HDPE retail bags) mass ratio	Temperature, T_1 (°C)	Pressure, P_1 (MPa)	Time (min)	Temperature, T_2 (°C)	Pressure, P_2 (MPa)	Time (min)
70:30	165 or 205	2	10	120	5	10
50:50	165 or 205	2	10	120	5	10
r-HDPE retail bags	165 or 205	2	10	120	5	10

*BTrTB (backing-tufts-retail bags-tufts-backing)

26.2.5   Characterization Techniques

Three-point bending tests were performed on two specimens of each molded composite using an Instron 5582 according to ASTM D790 standard with a BTrTB configuration [9]. The rate of crosshead motion was calculated based on the thickness of the molded composite. The sample quality was determined based on the measured flexural strength and flexural modulus values. For the composites with the larger flexural strength and modulus in each category, time-temperature superposition curves (TTS) were plotted by performing a creep test using a dual-cantilever mode of the TA Instruments Q800 Dynamic Mechanical Analyzer (DMA). For the DMA sample preparation, a single pile of carpet was used along with r-HDPE retail bags at comparable mass % to the three-point bending samples. Due to the clamp size, the thickness of specimens was limited to 5 mm. This size limitation limited the sample to BTr (backing-tufts-retail bags) configurations. The experiment was run with a temperature range of 30 to 80 °C in intervals of 10 °C. One MPa creep stress was applied for 10 min at each temperature. TTS curves were plotted at 30 °C as a reference using the Williams-Landel-Ferry (WLF) equation. An Olympus SZX9 stereo microscope was used to observe the composite morphology at the cross-sectional cuts.

26.3   Results and Discussion

26.3.1   Components of HDPE Retail Bags and PP Carpet

TGA was used on three different types of r-HDPE retail bags and c-PP carpet fibers to investigate their degradation. The TGA thermogram of c-PP carpet fibers showed no inorganic content, and all the fibers were burned off below 500 °C, as shown in Figure 26.3(a). In contrast, r-HDPE retail bags showed residue (inorganic content) left behind varying from 10% to 20%, as shown in Figure 26.3(a). The inorganic content in r-HDPE retail bags was determined to be $CaCO_3$ (filler) using FTIR, which further degraded into CaO at 650 °C. DSC curves identified the melting temperature (T_m) for different r-HDPE retail bags and c-PP carpet fibers. The different r-HDPE retail bags showed Tm peaks at 120-130 °C while c-PP carpet fibers showed a T_m peak at 164 °C, as shown in Figure 26.3(b). Based on the thermal analysis, molding temperatures (165 or 205 °C) were selected.

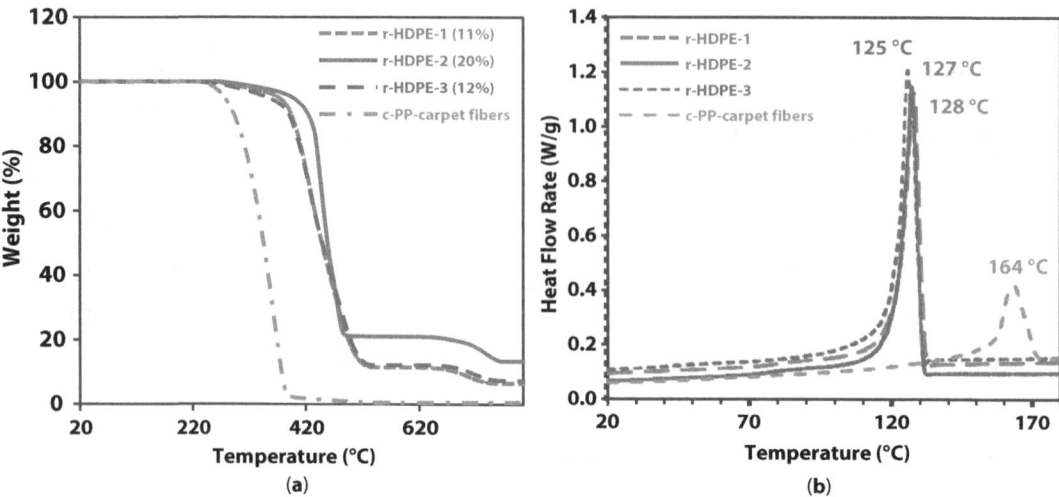

Figure 26.3 (a) TGA thermograms of different r-HDPE retail bags and c-PP carpet fibers. (b) DSC thermograms of different r-HDPE retail bags and c-PP carpet fibers.

26.3.2   Mechanical Properties

26.3.2.1   Flexural Testing

The three-point bending tests results showed that the properties of compression molded composites varied with molding temperature (T_1) and composition of carpet to r-HDPE retail bags, mass %, as shown in Table 26.2.

Table 26.2 Flexural strength and modulus results for two compositions and the corresponding r-HDPE retail bags.

Compositions (carpet:r-HDPE retail bags) mass ratio	Molding temperature, T_1 (°C)	Flexural strength (MPa)	Flexural modulus (MPa)
70:30	165	29.8 ± 2.3	985 ± 90
70:30	205	53.7 ± 3.0	3690 ± 20
50:50	165	24.3 ± 0.4	1010 ± 120
50:50	205	44.0 ± 2.0	2945 ± 20
r-HDPE retail bags	165	30.9 ± 0.8	980 ± 20
r-HDPE retail bags	205	35.0 ± 1.4	1080 ± 120

Composites molded at 205 °C with a larger carpet mass ratio showed larger flexural strength and modulus than those molded at 165 °C with a smaller carpet mass ratio, as shown in Figures 26.4 (a) and (b). Composites molded at 165 °C may only have partial melting of PP carpet fibers within the composite, resulting in weaker interactions between HDPE and c-PP carpet fiber with smaller flexural strength and modulus. Also, composites molded at higher temperatures showed brittle behavior compared to those developed at

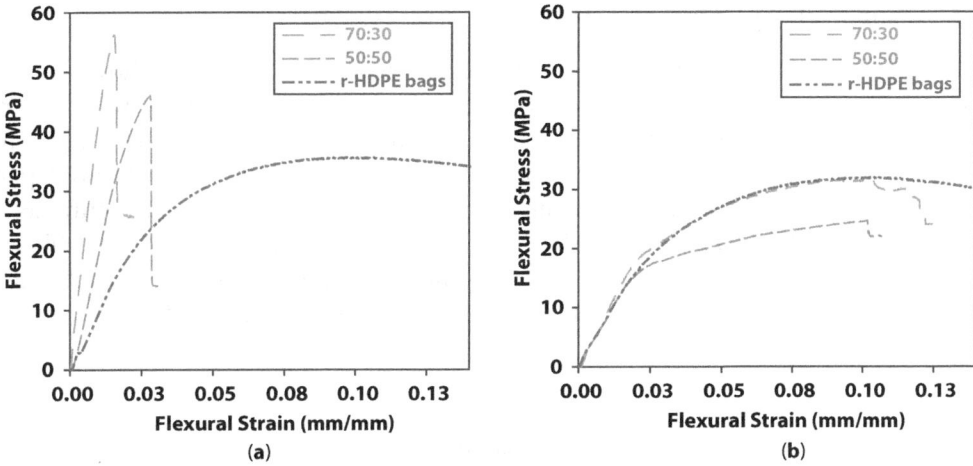

Figure 26.4 (a) Three-point bending results from samples molded at 205 °C for c-PP carpet:r-HDPE retail bags and r-HDPE retail bags. (b) Three-point bending results from samples molded at 165 °C for c-PP carpet:r-HDPE retail bags and r-HDPE retail bags.

lower temperatures. Composite sets with larger carpet mass ratios showed larger strength and modulus due to larger PP fractions than HDPE. This result agrees with prior research, which shows that PP is stronger and has a larger modulus than HDPE [10, 11].

26.3.3   Creep Behavior

The creep behavior for molded composites at 30 °C is shown in Figure 26.5. Analyzing the curves for 70:30 and 50:50 composites with the r-HDPE retail bags curve in Figure 26.5 shows that the introduction of c-PP carpet as reinforcement results in less creep compliance, confirming the reinforcing role of the carpet in the molded composites.

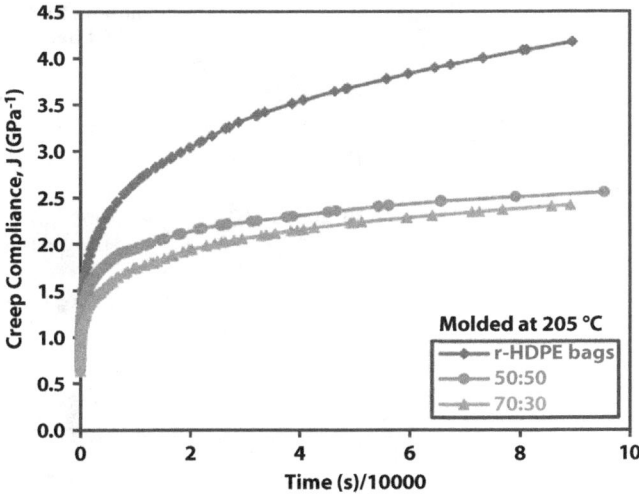

Figure 26.5 Creep compliance, J(t), time-temperature superposition for cPP:r-HDPE retail bags composites and r-HDPE retail bags curve at 30 °C.

26.3.4   Morphology in Composites

The structure of r-HDPE retail bags after compression molding at 205 °C is shown in Figure 26.6 as viewed with a stereo microscope. No apparent voids or air bubbles were observed after molding, and fully dense r-HDPE was observed. However, composite samples from c-PP carpet and r-HDPE retail bags, molded at different temperatures, showed partial melting of c-PP carpet fiber at a lower temperature (165 °C), as shown in Figure 26.7(a). Whereas, at higher temperatures (205 °C), the complete melting of c-PP carpet fiber was observed, as shown in Figure 26.7(b). Therefore, the partial melting of c-PP carpet fiber at a lower temperature was likely the reason for the smaller strength and modulus values of the molded composites.

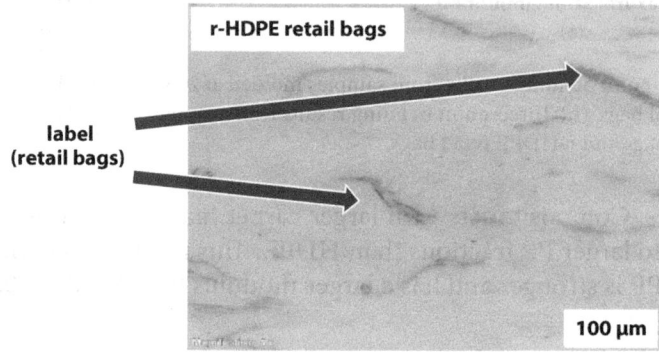

Figure 26.6 Microscopic images from cross-sectional cuts of molded composite from r-HDPE retail bags.

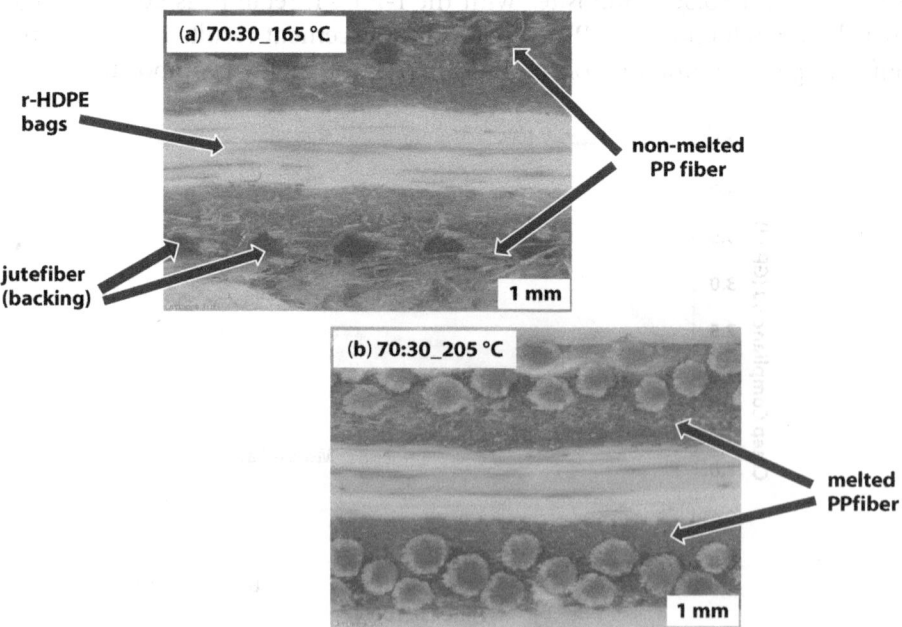

Figure 26.7 Microscopic images from cross-sectional cuts of molded composite from c-PP carpet and r-HDPE retail bags (70:30) at (a) 165 °C and (b) 205 °C.

26.4   Conclusions

Compression molded composites from post-consumer c-PP carpet and r-HDPE retail bags were developed in this study. Under optimized conditions, the compression-molded composites had improved flexural properties. Creep results have shown that the carpet, acting as reinforcement in these composites, lowers creep compliance and improves the toughness of the overall composite. Also, molding at a higher temperature (205 °C) resulted in better mechanical properties for both compositions of carpet and retail bags, with improved strength and modulus over r-HDPE retail bags alone. Observation from microscopic images showed satisfactory mixing inside the composite. These composites showed a potential for structural applications, and further research will validate the results with additional compositions and different processing methods. Nonetheless, recycling discarded synthetic carpets may enhance their recycling rates and reduce carpet waste.

Acknowledgment

This material is based upon work supported by the U.S. Department of Energy's Office of Energy Efficiency and Renewable Energy (EERE) under the Advanced Manufacturing Office Award Number DE-EE0007897 awarded to the REMADE Institute, a division of Sustainable Manufacturing Innovation Alliance Corp. This report was prepared as an account of work sponsored by an agency of the United States Government. Neither the United States Government nor any agency thereof, nor any of their employees, makes any warranty, express or implied, or assumes any legal liability or responsibility for the accuracy, completeness, or usefulness of any information, apparatus, product, or process disclosed, or represents that its use would not infringe privately owned rights. Reference herein to any specific commercial product, process, or service by trade name, trademark, manufacturer, or otherwise does not necessarily constitute or imply its endorsement, recommendation, or favoring by the United States Government or any agency thereof. The views and opinions of authors expressed herein do not necessarily state or reflect those of the United States Government or any agency thereof.

References

1. Himanshu V, R. D. Carpet Market https://www.alliedmarketresearch.com/carpet-market (accessed Global Report Opportunity Analysis and Industry Forecast, 2019-2026), 2019.
2. Post-Consumer Carpet: Barriers and Opportunities for Reuse; Carpet America Recovery Effort (CARE), Dalton, GA, May 28, 2021.
3. Radini, F.; Wulandari, R.; Nasiri, S.; Winarto, D., Performance comparison of plastic shopping bags in modern and traditional retail. *IOP Conference Series: Materials Science and Engineering*, 223, 012058, 2021.
4. Kumar, S.; Panda, A. K.; Singh, R. K., A review on tertiary recycling of high-density polyethylene to fuel. *Resources, Conservation and Recycling*, 55, 893-910, 2011.

5. Gaduan, A. N.; Li, J.; Hill, G.; Wallis, C.; Burgstaller, C.; Lee, K. Y., Simulating the recycling of milk bottles in the UK: Influence of blending virgin and repeatedly melt-extruded high-density polyethylene. *Resources, Conservation and Recycling*, 189, 106728, 2023.

6. Dias, B. Z.; Alvarez, C. E. d., Mechanical properties: wood lumber versus plastic lumber and thermoplastic composites. *Ambiente Construído*, 17, 201-219, 2017.

7. Zhao, Y.; Pan, G.; Xu, H.; Yang, Y., Compression-molded composites from waste polypropylene carpets. *Polymer Composites*, 39, 595-605, 2018.

8. Krzywinski, M.; Altman, N., Two-factor designs. *Nature Methods*, 11, 1187-1188, 2014.

9. Standard Test Methods for Flexural Properties of Unreinforced and Reinforced Plastics and Electrical Insulating Materials, ASTM D790, 2017.

10. Tesfaw, S.; Bogale, T. M.; Fatoba, O., Evaluation of tensile and flexural strength properties of virgin and recycled high-density polyethylene (HDPE) for pipe fitting application. *Materials Today: Proceedings*, 62, 3103-3113, 2022.

11. Kumar, N.; Mireja, S.; Khandelwal, V.; Arun, B.; Manik, G., Light-weight high-strength hollow glass microspheres and bamboo fiber based hybrid polypropylene composite: A strength analysis and morphological study. *Composites Part B: Engineering*, 109, 277-285, 2017.

27

Upcycling of Aerospace Aluminum Scrap

Mohamed Aboukhatwa[1]* and David Weiss[2]

[1]Illinois Applied Research Institute, University of Illinois at Urbana-Champaign, Champaign, IL, USA
[2]Eck Industries Inc., Manitowoc, WI, USA

Abstract

To achieve the energy savings (90%) and CO_2 emissions reductions (96%) offered by secondary aluminum [1], and to overcome the issues associated with recycling of aerospace aluminum alloys, a new process for recycling AA7075 scrap directly into a high strength castable secondary alloy is demonstrated. The new process involves enhancing the castability of wrought aerospace aluminum scrap by alloy chemistry optimization and minor alloying additions to reduce the hot tearing susceptibility and to enhance the mechanical properties of the secondary alloy. The proposed approach provides defect-free castings with mechanical properties comparable to premium aluminum casting alloys but at considerable energy and cost savings.

Keywords: Recycling, secondary aluminum, casting, scrap, thermodynamic modeling, hot tearing, aerospace aluminum, AA7075

27.1 Introduction

To optimize toughness and meet the stringent performance requirements demanded by the aerospace industry, aerospace aluminum alloys such as AA7075 are typically relatively high in alloying elements, contain very low levels of impurities, and are produced using primary aluminum. Since most of the alloying elements are retained in the liquid metal during remelting, these alloys are typically downcycled into lower-value alloys. Alternatively, presorted alloys are subsequently remelted into ingots for further extrusions or rolled sheets production by complex thermomechanical processing routes, with all the associated embodied-energy costs and in-process material losses. Moreover, due to the large solidification interval of aerospace alloys, they are difficult to cast into complex shapes and are susceptible to hot tearing and macrosegregation. As a result, the recycling rate for aerospace aluminum is 20%, well below the 85% and 95% recycling rates that have been achieved for aluminum used in consumer and automotive applications, respectively [2, 3]. With thousands of decommissioned aircrafts sitting in boneyards, and an additional 12,000 aircrafts being retired within the next 20 years [4], it's essential to benefit from such a large source of

---

**Corresponding author*: mkhatwa@illinois.edu

Nabil Nasr (ed.) Technology Innovation for the Circular Economy: Recycling, Remanufacturing, Design, Systems Analysis and Logistics, (343–354) © 2024 Scrivener Publishing LLC

recoverable aluminum (around 72% of the total aircraft weight) [3, 5] by adopting new low-cost and energy-efficient technologies for recycling aerospace alloys into high-value end products. A viable approach would be trying to design a castable variant of aerospace alloys by utilizing primary wrought aluminum scrap feedstock. The development of such castable secondary aluminum alloy from recycled wrought aerospace aluminum scrap requires a considerable reduction of the hot tearing tendency of AA7075 alloy during remelting. This could be achieved by satisfying the following requirements during casting. First, the as-cast microstructure has to be a fine equiaxed dendritic or globular structure and second, the solidification range for the alloy has to be narrowed.

Traditionally, promoting nucleation in aluminum casting has been based on the addition of insoluble particles, mainly TiB_2 to the melt through master alloys [6]. While such practice has been proven successful in refining the microstructure of many aluminum alloys, this has not been the case for AA7075 where Ti additions do not significantly refine the grain structure or aid in overcoming the hot cracking susceptibility. This is due to the high growth restriction factor of 7075 alloys, which is directly related to the solidification range and the solute partitioning at the solid-liquid interface [7]. In fact, it has been shown that employing Ti grain refiners can even lead to an increase in the hot tearing susceptibility of aluminum alloys [8]. Therefore, to achieve improved microstructural refinement in cast secondary aerospace aluminum alloys, zirconium was added to the 7075 aluminum scrap to promote the formation of Al_3Zr particles which are potent grain refiners in 7075 alloys [9, 10].

The extended freezing range of 7075 alloys is due to the presence of the strongly segregating elements Mg, Zn, and Cu. Copper complicates the phase constitution with a large number of Cu-bearing phases. It also decreases the solidus temperature, thus increasing the hot tearing tendency [11, 12]. The hot tearing susceptibility of 7xxx aluminum alloys was also found to shift to lower Zn concentrations on increasing Cu and Mn contents due to a larger volume fraction of Fe-bearing intermetallics [6]. Therefore, slight changes to the Zn and magnesium contents will affect the type and distribution of the forming intermetallic compounds, and the effective solidification range.

27.2 Solidification Simulations and Alloy Chemistry Optimization

To arrive at such necessary compositional changes for enhancing castability of AA7075 scrap, solidification simulations employing the CALPHAD approach using FactSage™ were carried out to predict the solidification path of modified AA7075 scrap compositions. As a starting input for the CALPHAD simulations, a weighted average scrap composition shown in Table 27.1, was identified after analyzing AA7075 scrap aggregates procured from local Midwest users and service centers such as Chicago Mold Engineering Co. and Granberg International.

Using the weighted average AA7075 scrap composition, solidification simulations using the Scheil-Gulliver formalism were performed, and the alloy chemistry was optimized for

Table 27.1 Average chemical composition of AA7075 scrap.

Si	Fe	Cu	Mn	Mg	Cr	Zn	Ti	Zr	Al + Trace elements
0.06	0.18	1.22	0.01	2.28	0.21	5.40	0.03	0.01	Balance

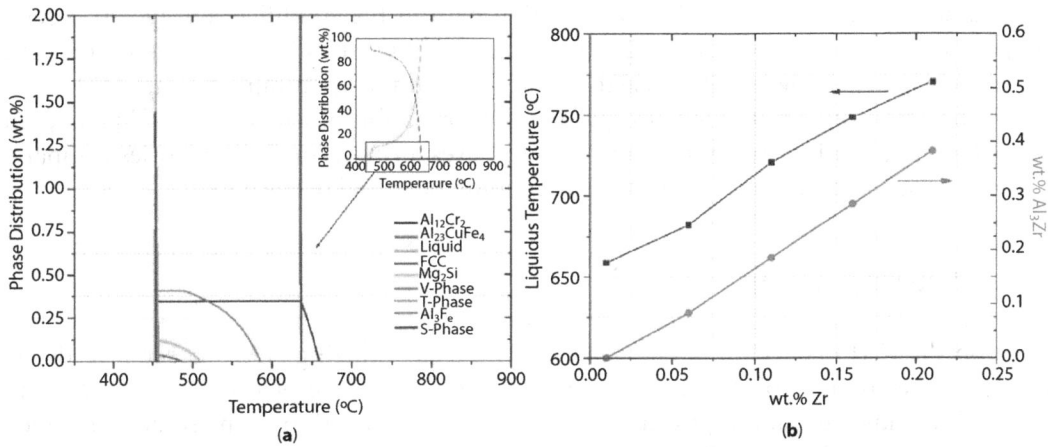

Figure 27.1 Scheil solidification simulation of scrap AA7075: (a) phase constituents accumulating in the alloy during casting and (b) effects of zirconium additions on liquidus temperature and zirconium aluminide formation.

enhanced castabilty and mechanical strength. The CALPHAD simulations included the following modifications to the average 7075 scrap composition: 1) dilution with P1020 primary aluminum alloy in order to limit the copper concentration in the final casting to 0.5-1 wt.%, 2) addition of zirconium to the diluted alloys in increments of 0.05 wt.%, and 3) addition of nickel to the diluted alloy compositions in two concentrations, 0.1 and 0.5 wt.% to promote the formation of Ni-bearing intermetallics. Figure 27.1 shows the amounts of intermetallics accumulating in the scrap 7075 alloy composition during casting and the effect of zirconium additions on the solidification range and the zirconium aluminide formation. The results indicate that zirconium is a potent nucleating agent for AA7075 alloys and increasing the zirconium content will raise the liquidus temperature. Therefore, zirconium additions were limited to 0.1 wt.% to avoid large solidification intervals. The effects of dilution and

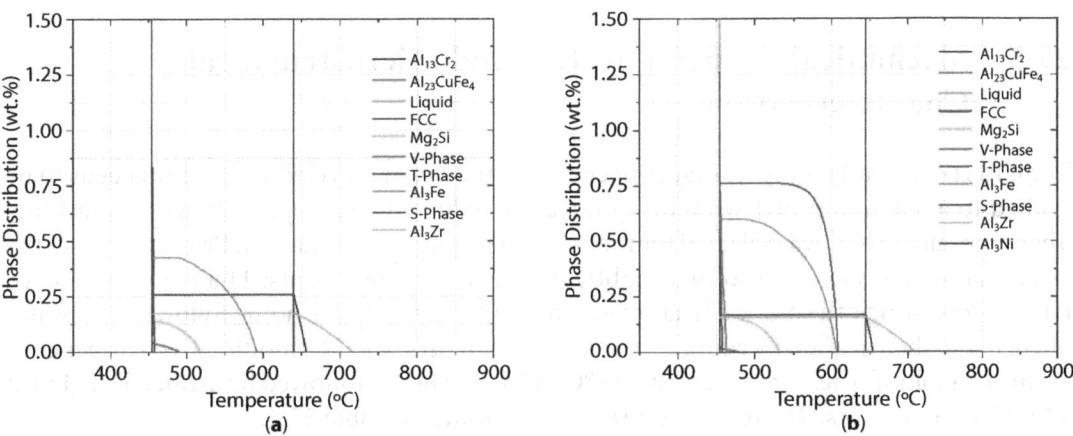

Figure 27.2 Scheil solidification simulations of modified average 7075 scrap compositions: (a) after the addition of 0.1 wt.% Zr and dilution with P1020 and (b) same as (a) but with the addition of 0.5 wt.% Ni.

zirconium/nickel additions on the solidification path of the modified average 7075 scrap compositions is shown in Figure 27.2. The addition of Ni results in the formation of nickel aluminide which is known to increase the tensile strength of aluminum castings [13].

The solidification simulations also predict the formation of the V-phase, T-phase, and the S-phase in the microstructure. The V and S phases form in smaller quantities compared to the T-phase. All three phases affect the hardening potential of the alloy.

27.3   Casting Trials

Based on the CALPHAD simulations, six casting trials utilizing AA7075 scrap feedstock were cast directly in step plate permanent metallic molds. The plates were 6 x 12 inches wide with thicknesses ranging from ¼ to 1 inch. The composition of the scrap was diluted by master alloys and minor alloying additions to achieve the target compositions listed in Table 27.2. Computer-aided cooling curve analysis (CA-CCA) was performed on select cast samples to identify the actual solidification ranges of the alloys and to validate the CALPHAD predictions. The actual liquidus temperatures were in good agreement with both CALPHAD predictions and the literature. However, the actual and predicted solidus temperatures showed slight variations as shown in Table 27.3.

27.4   Constrained Rod Casting (CRC)

Additional casting trials using hot tear molds were conducted using alloys SN74 and SN75 to assess the role of dilution, especially the role of Cu concentration on the hot tearing tendency of the alloys. Alloy SN74 has a very close composition to the average 7075 scrap while alloy SN75 is a diluted version. Figure 27.3 shows images of the CRC castings. Alloy SN74 exhibited significant hot tearing that resulted in severe cracks and separation of one of the side rods form the casting's main body. Alloy SN75 on the other hand, did not show any signs of hot tearing cracks either visually or following dye penetrant inspection. The hot tearing indices for SN74 and SN75 were 26 and 4, respectively.

27.5   Mechanical Property Testing and Microstructural Characterization

The CRC tests clearly indicate that copper concentration in recycled 7075 alloys needs to be limited to a maximum of 1 wt.% to significantly reduce the susceptibility to hot cracking. Therefore, alloy SN72 was selected for mechanical properties evaluation. Prior to testing, the as-cast alloy was subjected to two slightly different heat treatments: 1) a conventional T6 temper consisting of a solution heat treatment at 482 °C for twelve hours followed by a water quench and then artificially aging at 121 °C for four hours, and 2) a modified T6 temper consisting of a solution heat treatment at 496 °C for twelve hours followed by artificially aging at 149 °C for four hours. The tensile test results are shown in Table 27.4.

The higher solution heat treatment and aging temperatures resulted in a simultaneous improvement in both strength and ductility values of the recycled 7075 cast alloy.

Table 27.2 Selected AA7075 scrap compositions for casting trials.

AA7075 casting trial	Alloy designation	Si	Fe	Cu	Mn	Mg	Cr	Zn	Ti	Zr	Ni	Al
1	SN70	0.05	0.12	1.69	0.01	2.56	0.19	5.88	0.05	0.02	0	89.43
2	SN71	0.05	0.12	1.51	0.01	2.36	0.19	5.57	0.06	0.04	0	90.1
3	SN72	0.06	0.08	1	0	1.67	0.11	3.26	0.02	0.05	0	93.74
4	SN74	0.14	0.13	1.74	0.03	1.3	0.19	5.47	0.04	0.05	0	90.91
5	SN75	0.12	0.09	0.83	0.01	0.59	0.08	2.4	0.01	0.02	0	95.85
6	SN90	0.24	0.11	0.46	0.01	0.97	0.05	1.84	0.01	0.01	0.1	96.2

Table 27.3 Solidification ranges of cast AA7075.

Alloy designation	Liquidus		Solidus		Solidification range	
	CALPHAD	Thermal analysis	CALPHAD	Thermal analysis	CALPHAD	Thermal analysis
SN70	656 °C	632 °C	454 °C	455 °C	202 °C	177 °C
SN71	676 °C	641 °C	454 °C	518 °C	222 °C	123 °C
SN90	649 °C	650 °C	478 °C	527 °C	171 °C	123 °C

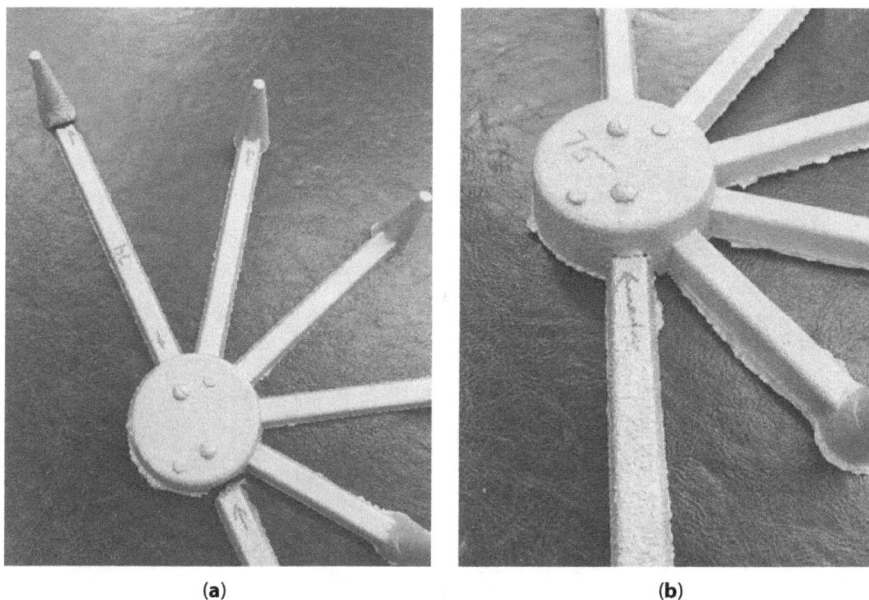

(a)   (b)

Figure 27.3 Constrained rod casting tests: (a) alloy SN74 and (b) alloy SN75.

Figure 27.4 shows the as-cast microstructure of the alloy following the heat treatments. The microstructure is mainly FCC aluminum with eutectic phases distributed along the boundaries as shown in Figure 27.5.

The grain boundary phases are predominantly the T-phase (AlCuMgZn) alongside small amounts of $MgZn_2$ and S (Al_2CuMg) phases. This is in good agreement with the CALPHAD solidification simulation of the SN72 alloy. Increasing the solution heat treatment temperature resulted in the dissolution of a significant amount of the T-phase and a more discontinuous distribution of precipitates at the grain boundaries. The STEM-EDS analysis of the grain boundary precipitates identified high concentrations of Cu with elemental ratios corresponding to the S-phase as shown in Figure 27.6. Such discrete distribution of grain boundary precipitates was shown to enhance both strength and ductility in 7075 alloys [14, 15]. In addition, the dissolution of the T-phase with the increase in the solution

Table 27.4 Mechanical properties of alloy SN72 (T6 temper).

Alloy designation	Tensile strength (MPa)	Elongation (%)
SN72^Ŧ	208	7
SN72*	260	10

Ŧ Solution heat treatment at 482 °C for 12 hrs, water quench, artificially age at 121 °C for 4 hrs.
* Solution heat treatment at 496 °C for 12 hrs, water quench, artificially age at 149 °C for 4 hrs.

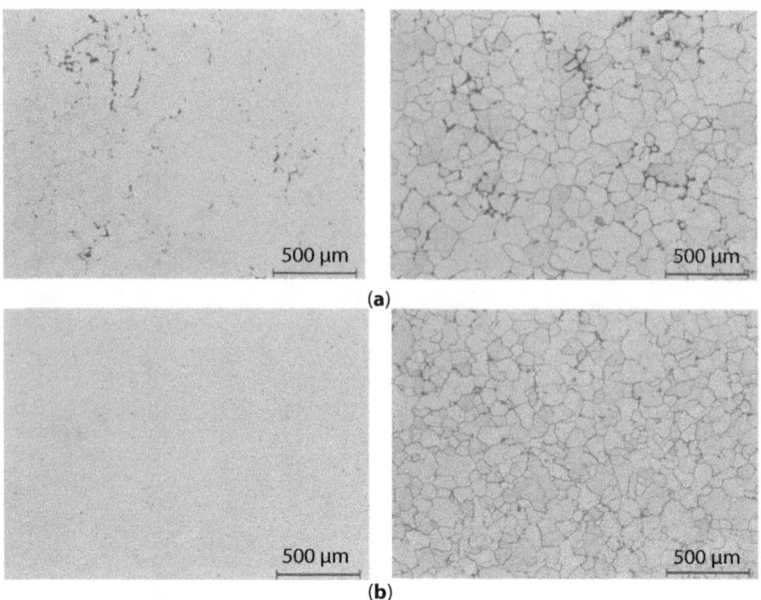

Figure 27.4 Effect of heat treatment on the microstructure of SN72 alloys: (a) solution heat treatment at 482 °C for 12 hrs, water quench, artificially age at 121 °C for 4 hrs, and (b) solution heat treatment at 496 °C for 12 hrs, water quench, artificially age at 149 °C for 4 hrs.

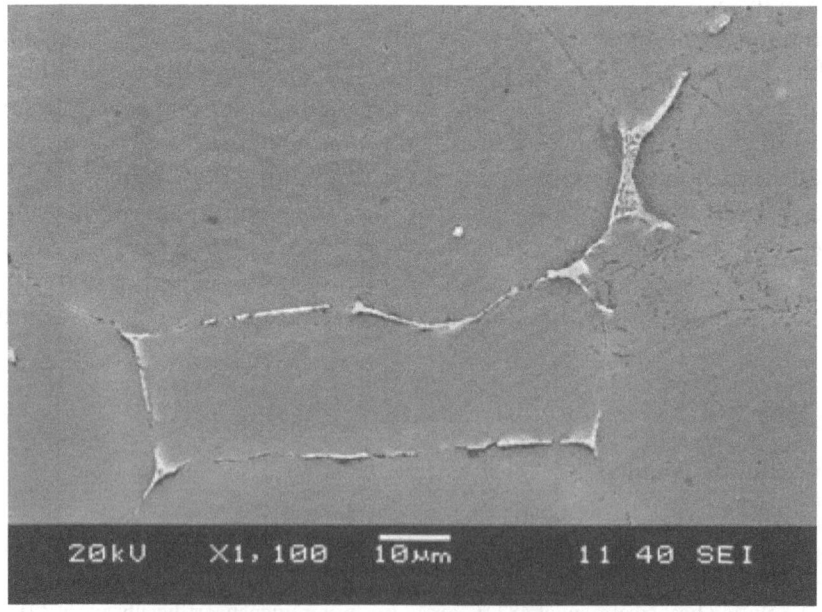

Figure 27.5 SEM micrograph of alloy SN72 showing eutectic phases at grain boundaries.

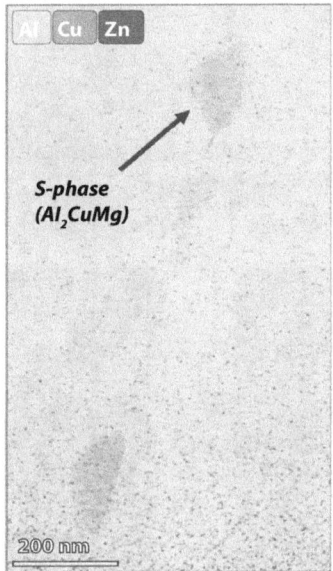

Figure 27.6 STEM-EDS elemental map of grain boundary area in alloy SN72 subjected to the modified T6 heat treatment.

heat treatment temperature will increase Mg and Zn solute concentrations in the matrix. This will enhance the aging response of 7075 alloys by increasing the volume fraction of second-phase particles, particularly η'- $MgZn_2$ during artificial aging. Moreover, $MgZn_2$ precipitates increase the work hardening rate of aluminum alloys during plastic deformation which leads to a simultaneous increase in both strength and ductility [16].

27.6   Technology Demonstration

To qualify as commercial cylinder head castings, aluminum alloys should possess a tensile strength in excess of 250 MPa and a ductility around 3%. The SN72 cast alloy subjected to the modified T6 temper satisfied both these requirements. In addition, the alloy exhibited a fine globular microstructure which is very favorable for enhanced hot tearing resistance and improved casting characteristics. Therefore, an industrial scale cylinder head was cast using a slightly modified SN72 composition. Compositional changes were mainly the addition of Ni at the expense of zirconium to further increase the strength of the casting. The casting poured successfully and was free from any cracks or internal casting defects as evident from the computed tomography (CT) images shown in Figure 27.7.

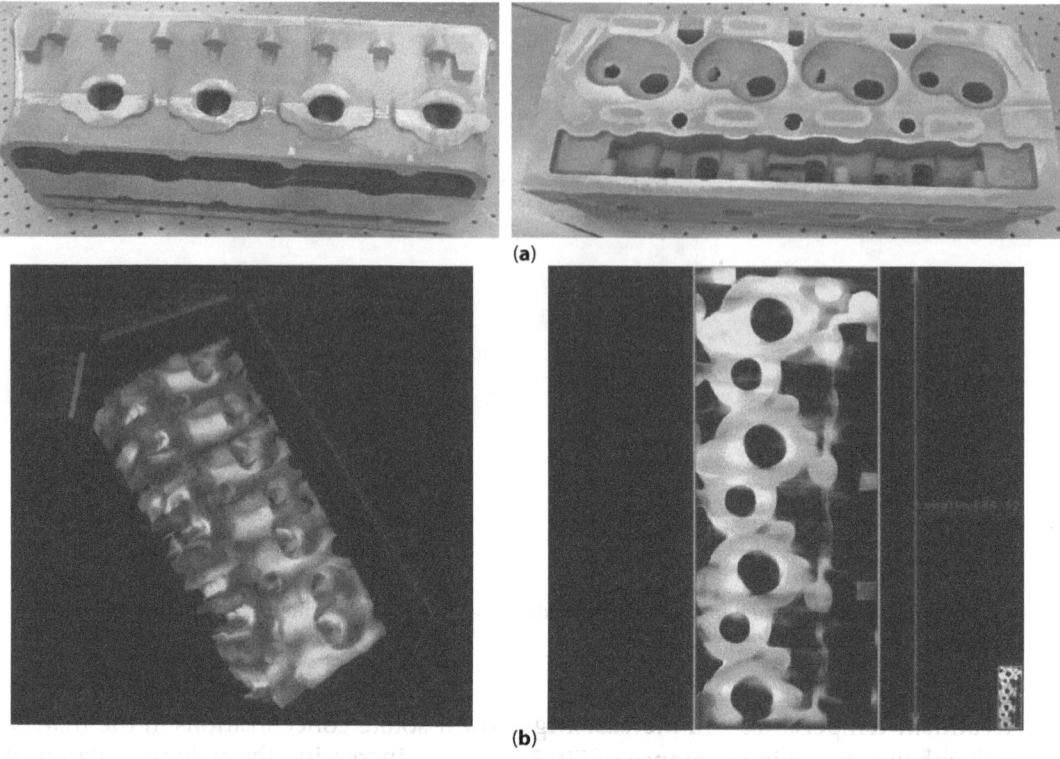

Figure 27.7 Cylinder head casting produced from recycled 7075 aluminum scrap: (a) actual casting and (b) 3D CT reconstruction of the cylinder head.

27.7   Conclusions & Recommendations

A new process for recycling scrap AA7075 directly into a high strength castable secondary alloy was demonstrated. The process involved 7075 scrap alloy chemistry optimization and zirconium additions to reduce hot tearing susceptibility. The hot tearing index of the trial castings was reduced from 26 to 4. The casting exhibited tensile strengths ranging from 208-260 MPa with elongation values from 7-10% following T6 heat treatments. The readiness of the proposed approach for commercial adoption was also demonstrated by casting a complex geometry cylinder head. The casting poured successfully without any signs of hot tearing or internal casting defects. The proposed method provides a low-cost and energy-efficient process for recycling wrought aerospace aluminum scrap directly into high-value end products with mechanical properties comparable to primary aluminum casting alloys. Tensile and yield strengths of the secondary alloy could be further improved by increasing the Ni content and raising the heat treatment temperatures. However, ductility will be adversely affected. It is therefore necessary to carefully tailor alloying strategies to meet the intended commercial product specifications and performance requirements.

Acknowledgement

This material is based upon work supported by the U.S. Department of Energy's Office of Energy Efficiency and Renewable Energy (EERE) under the Advanced Manufacturing Office Award Number DE-EE0007897 awarded to the REMADE Institute, a division of Sustainable Manufacturing Innovation Alliance Corp. This report was prepared as an account of work sponsored by an agency of the United States Government. Neither the United States Government nor any agency thereof, nor any of their employees, makes any warranty, express or implied, or assumes any legal liability or responsibility for the accuracy, completeness, or usefulness of any information, apparatus, product, or process disclosed, or represents that its use would not infringe privately owned rights. Reference herein to any specific commercial product, process, or service by trade name, trademark, manufacturer, or otherwise does not necessarily constitute or imply its endorsement, recommendation, or favoring by the United States Government or any agency thereof. The views and opinions of authors expressed herein do not necessarily state or reflect those of the United States Government or any agency thereof.

References

1. Das, S. K., Kaufman, J. G., Recycling aluminum aerospace alloys, in: Light Metals 2007, M. Sorlie (Ed.), pp. 1161-1165, The Minerals, Metals & Materials Society, Warrendale, 2007.
2. Capuzzi, S., Timelli, G., Preparation and melting of scrap in aluminum recycling: A review. *Metals*, 8, 249, 2018.
3. Asmatulu, E., Overcash, M., Twomey, J., Recycling of aircraft: State of the art in 2011. *Journal of Industrial Engineering*, 2013.
4. Airport Technology, Aircraft recycling: up to the challenge, https://www.airport-technology.com/features/featureaircraft-recycling-up-to-the-challenge-5710942/, 2017.
5. Airbus Aircraft, End-of-life reusing, recycling, rethinking, https://aircraft.airbus.com/en/newsroom/stories/2022-11-end-of-life-reusing-recycling-rethinking, 2022.
6. Grandfield, J. F., Eskin, D. G., Bainbridge, I. F.: Direct-Chill casting of light alloys: Science and technology, pp. 103-143, The Minerals, Metals & Materials Society, 2013.
7. Easton, M., StJohn, D., An analysis of the relationship between grain size, solute content, and the potency and number density of nucleant particles. *Metallurgical and Materials Transactions A*, 36, 1911, 2005.
8. Li, S., Sadayappan, K., Apelian, D., Role of grain refinement in the hot tearing of cast Al-Cu alloy. *Metallurgical and Materials Transactions B*, 44, 614, 2013.
9. Martin, J. H., Yahata, B. D., Hundley, J. M., Mayer, J. A., Schaedler, T. A., Pollock, T. M., 3D printing of high-strength aluminium alloys. *Nature*, 549, 365, 2017.
10. Wang, F., Qiu, D., Liu, Z. L., Taylor, J. A., Easton, M. A., Zhang, M. X., The grain refinement mechanism of cast aluminium by zirconium. *Acta Materialia*, 61, 5636, 2013.
11. Chen, D., Zhang, H., Jiang, H., Cui, J., Experimental investigation of microsegregation in low frequency electromagnetic casting 7075 aluminum alloy. *Materialwissenschaft und Werkstofftechnik*, 42, 500, 2011.
12. Belov, N. A., Alabin, A. N., Use of multicomponent phase diagrams for designing high strength casting aluminum alloys. *Materials Science Forum*, 794–796, 909, 2014.

13. Naeem, H. T., Mohammed, K. S., Ahmad, K. R., Rahmat, A., The influence of nickel and tin additives on the microstructural and mechanical properties of Al-Zn-Mg-Cu alloys. *Advances in Materials Science and Engineering*, 2014.

14. Zou, X., Yan, H., Chen, X., Evolution of second phases and mechanical properties of 7075 Al alloy processed by solution heat treatment. *Transactions of Nonferrous Metals Society of China*, 27, 2146, 2017.

15. Li, G., Chen, F., Han, Y., Liang, Y., Improving mechanical properties of PVPPA welded joints of 7075 aluminum alloy by PWHT. *Materials*, 11, 379, 2018.

16. Zhao, Y. H., Liao, X. Z., Cheng, S., Ma, E., Zhu, Y. T., Simultaneously increasing the ductility and strength of nanostructured alloys. *Advanced Materials*, 18, 2280, 2006.

28

Stabilization of Waste Plastics with Lightly Pyrolyzed Crumb Rubber in Asphalt

Yuetan Ma, Hongyu Zhou, Pawel Polaczyk and Baoshan Huang*

Dept. of Civil and Environmental Engineering, The University of Tennessee, Knoxville, TN, USA

Abstract

Modifying asphalt is a potential high-value application for reusing waste plastics because of high volume usage asphalt in highway construction. However, simply blending hot plastics and asphalt encounters difficulties related to the poor solubility of polymers, which limits the formation of a swollen network with asphalt molecules. The polymer phases also tend to coalesce and separate from asphalt during high-temperature storage in static conditions. The present study developed a process to stabilize waste plastics in asphalt and improve binder storage stability utilizing lightly pyrolyzed crumb rubber (LPCR). The untreated crumb rubber and LPCR were collected to blend with waste plastics for asphalt modification. The storage stability and rheological properties of modified binder blends were evaluated through cigar tube test and dynamic shear rheometer (DSR) test. Polymer phases and network structures were characterized through optical microscopy. Fourier transform infrared spectroscopy (FTIR) was utilized to investigate the reaction mechanisms of LPCR and its interaction with the asphalt binder system. It was found that the pyrolyzed process improved the rubber solubility and storage stability of modified binder blends. The co-existence of rigid plastic and soft rubbery regimes in an entangled network provided a promising pathway to improve the mechanical performance of asphalt binder in both high- and low-temperature domains.

Keywords: Waste plastics, lightly pyrolyzed crumb rubber (LPCR), rubber solubility, storage stability

28.1 Introduction

According to the US Environmental Protection Agency (EPA), in 2018, plastics generation was approximately 35.7 million tons in the United States. Among this total, 5.6 million tons of plastics were incinerated, and 27 million tons of the plastics were landfilled. Asphalt modification is a promising application for significant reductions in landfilled plastics as well as performance improvement [1]. In general, plastic modifiers including polyethylene (PE) and polypropene (PP) could increase the softening point and decrease the penetration of asphalt binder, which leads to a more rigid behavior of the asphalt mixtures and a higher resistance to the permanent deformation of pavement [2–4]. However, such a process

---

Corresponding author: bhuang@utk.edu

Nabil Nasr (ed.) Technology Innovation for the Circular Economy: Recycling, Remanufacturing, Design, Systems Analysis and Logistics, (355–364) © 2024 Scrivener Publishing LLC

encounters barriers due to the poor solubility of plastics. The polymer phases also tend to coalesce and separate from asphalt during high-temperature storage, which limits the number of macromolecules forming a swollen network with asphalt molecules [5]. Furthermore, the addition of plastics in asphalt can lower cracking behavior, especially the low-temperature cracking performance of asphalt mixtures [6]. To solve the aforementioned problems, researchers started to incorporate gound tire rubbers (GTR) as a co-modifier to separate the crystal plastic particles, stabilize the plastic particles, and improve the low-temperature performance of asphalt mixtures [7–9]. Previous studies also demonstrated that rubber-modified asphalts also perform good rutting resistance due to the increased viscosity and elasticity at high temperatures [10]. Natural rubbers start their degradation at 143°C and the degradation accelerates when the temperature is over 200 °C [11]. As for rubber-modified asphalt, rubbers tend to partially swell, soften, and digest into asphalt due to the absorption of aromatic oil [7, 12].

However, the incomplete degradation and solubility of raw GTR during mixing can result in the sediments of rubber particles and hence the disparity between asphalt and rubber modifiers [13, 14]. This study aimed to develop a process to stabilize the waste plastics in asphalt by using lightly pyrolyzed crumb rubber (LPCR). LPCR was first produced from GTR by using a single-screw extruder at different temperatures. The properties of different LPCRs were evaluated in terms of sol fractions, and chemical compositions. The untreated crumb rubber and a certain type of LPCR were selected to blend with waste plastics for asphalt modification. The storage stability and rheological properties of modified binder blends were evaluated through cigar tube test, dynamic shear rheometer (DSR) test, and bending beam rheometer test. Polymer phases and network structures were characterized through optical microscopy.

28.2   Main Content of Chapter

28.2.1   Raw Materials

One typical plastic, waste polyethylene (WPE) was used as asphalt modifiers, which were adopted from the waste grocery bags. The untreated GTR, grounded into 40 mesh, was obtained to produce LPCR. The commercial asphalt binder (PG 64-22) was used as the base binder for asphalt modification.

28.2.2   Production of LPCR

The grounded GTR was transferred into a single-screw extruder for LPCR production. The rotation speed of the main screw is 100 r/min by controlling the temperature at 280°C, 240°C, and 220°C. The pyrolyzed rubbers under three temperatures were denoted as DR1, DR2, and DR3, respectively. As shown in Figure 28.1, one type of LPCR (DR1), which was extruded at 280°C, was used as a modifier for asphalt modification.

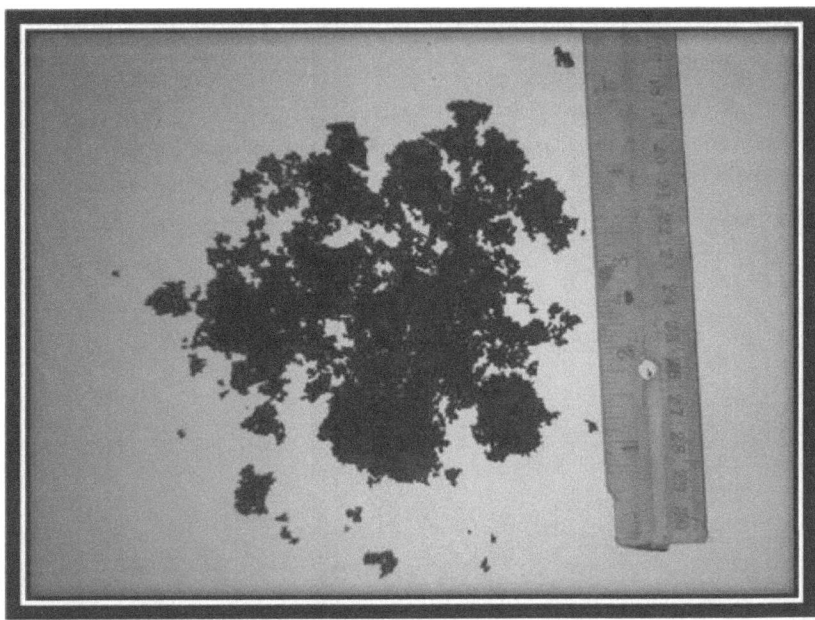

Figure 28.1 LPCR is used for asphalt modification.

28.2.3 Production of Polymer-Modified Asphalt

To prepare polymer-modified asphalt containing LPCR and WPE, the base asphalt was heated at 180°C. GTR and LPCR were added into the base asphalt and sheared at a speed of 6000r/min for 30 mins, followed by being cured in the oven for 1 hour. Then WPE was incorporated into rubber asphalt and sheared at the same speed for 30 mins. The preparation equipment is shown in Figure 28.2. The compositions of LPCR/WPE-modified asphalt are listed in Table 28.1.

28.2.4 Experimental Methodology

28.2.4.1 Characterization of LPCR Solubility

Soxhlet extraction method was used to measure the solubility of rubbers. The samples were extracted for 72 hours using toluene as the solvent. Then the samples were dried in a vacuum oven to remove the residues for 4 hours.

The solubility of rubber was calculated following equation (28.1).

$$\text{solubility} = \frac{m_0 - (m_1 + m_2)}{m_0 \times \omega} \tag{28.1}$$

where m_0 = the mass of samples before extraction
m_1 = the mass of dry samples
m_2 = the mass of soluble materials
ω = the rubber contents

Figure 28.2 Test setup for polymer-modified asphalt.

Table 28.1 The compositions of LPCR/WPE-modified asphalt.

Label	Base asphalt (g)	GTR (g)	LPCR (g)	WPE (g)	LPCR type
GTRA	176	24	0	0	--
DRA1	176	0	24	0	DR1
DRA2	176	0	24	0	DR2
DRA3	176	0	24	0	DR3
GTRWPEA	176	16	0	8	DR1
DRWPEA	176	0	16	8	DR1
WPEA	192	0	0	8	--

28.2.4.2 Fourier Transform Infrared Spectroscopy (FTIR) Test

FTIR test was used to characterize the reaction and mechanisms of the pyrolyzed processes of GTR. Different LPCRs were scanned through infrared light with the wavenumber ranging from 650 to 4000 cm^{-1}.

28.2.4.3 Dynamic Shear Rheometer (DSR) Test

Dynamic shear rheometer (DSR) was used to characterize the rheological properties of LPCR/WPE-modified asphalt. The specimens with a diameter of 25 mm and a thickness of

1 mm were sandwiched between the two plates. Frequency sweep tests were conducted at 30 °C, 45 °C, and 60 °C with a frequency ranging from 100 rad/s to 0.1 rad/s. Temperature sweep tests were conducted at a frequency of 10 rad/s at a temperature range of 40 to 70°C.

28.2.4.4   Cigar Tube Test

Cigar tube test was adopted to characterize the storage stability and phase separation of LPCR/WPE-modified asphalt [15, 16]. The modified binder blends were poured into an aluminum cigar tube. All the tubes were sealed and placed in the oven vertically at a temperature of 163 °C for 48 hours. The tubes were transferred into a freezer instantaneously after the conditioning. The freeze tubes were cut into three parts and the binder blends from the 1/3 top and 1/3 bottom locations were extracted for DSR rheological testing. The separation index was determined with rutting parameters ($G^*/sin\delta$) using the following equation:

$$\text{Separation index (\%)} = \frac{(G^*/sin\delta)_{ax} - (G^*/sin\delta)_{ave}}{(G^*/sin\delta)_{ave}} \qquad (28.2)$$

where $(G^*/sin\delta)_{max}$ = the larger value of either 1/3 top or 1/3 bottom section and $(G^*/sin\delta)_{ave}$ = the average value of 1/3 top and 1/3 bottom sections.

28.2.4.5   Optical Microscopy Test

Approximately 5 mg of heated modified asphalt was spread in the microscopic slides. The cover was placed on the asphalt film afterward. The morphology of different polymers in asphalt was visualized in the microscope (LEICA MC190 HD).

28.2.5   Results and Analysis

28.2.5.1   LPCR Solubility Results

Figure 28.3 shows the sol fraction of different LPCR obtained from rubber solubility tests. As shown in Figure 28.3, the pyrolyzed degree is highly dependent on the pyrolyzed temperature. GTR pyrolyzed at higher temperatures performs a larger value of sol fraction, indicating higher pyrolyzed degrees and better rubber solubility. The sol fraction of LPCR at the highest temperature (280 °C) presents the highest sol fraction (71.0%), which is much higher than the untreated GTR (13.9%). Hence, DR1 is selected for asphalt modification and is expected for good compatibility and solubility with asphalt.

28.2.5.2   Pyrolyzed Mechanisms for GTR

Figure 28.4 shows the FTIR spectra of LPCR under different pyrolyzed degrees. The wavenumber of 1030 cm^{-1} and 1720 cm^{-1} represents the vibration absorption peaks of the sulfoxide group and carbonyl group, respectively. The wavenumber of 660 cm^{-1} represents the stretching vibration absorption peak of carbon-sulfur single bond. Compared to untreated

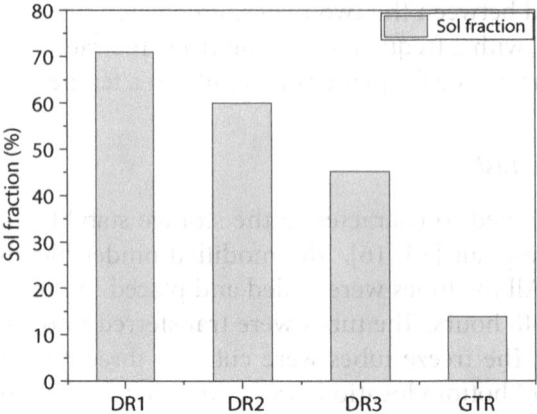

Figure 28.3 Sol fraction of different LPCR.

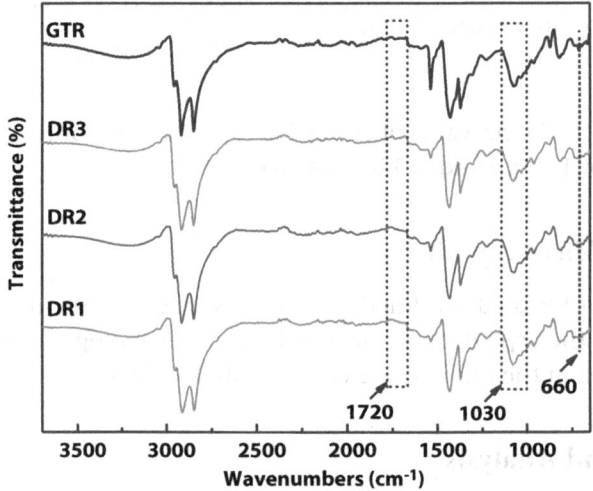

Figure 28.4 FTIR spectra of LPCR under different pyrolyzed degrees.

GTR, LPCR does not have significant changes in the sulfoxide and carbonyl group, indicating the oxidation reaction does not occur during pyrolyzed processes. However, the absorption peak of carbon-sulfur single bonds is weakened in LPCR after corresponding to GTR, especially in DR1, reflecting the breakage of carbon-sulfur single bonds during pyrolyzed processes. The breakage of chemical bonds enables the partial destruction of crosslink structures in crumb rubbers.

28.2.5.3 Rheological Properties of Polymer Modified Asphalt

Figure 28.5 shows the complex shear modulus and phase angle master curves, which are constructed at the reference temperature of 45 °C, of different polymer modified asphalt. The addition of LPCR/WPE increased the complex modulus and decreased the phase angle of modified binder blends. Asphalt modified with pure WPE showed the highest complex

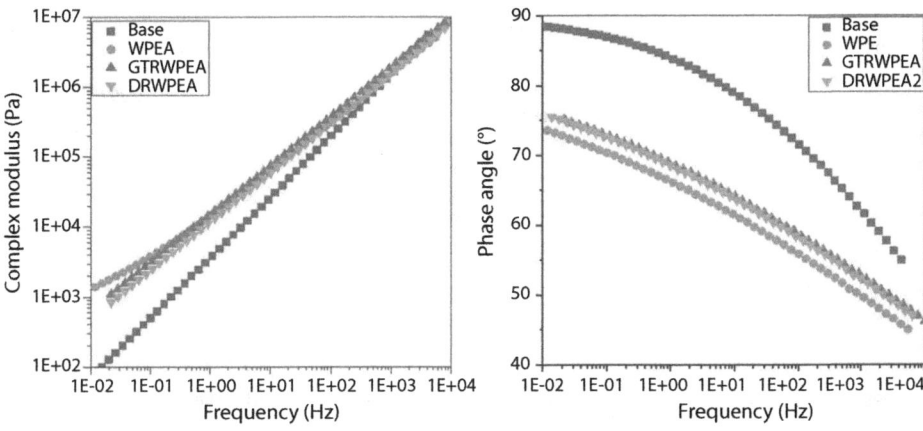

Figure 28.5 Master curves of different polymer modified asphalt.

modulus, especially at low frequency (high temperature), and the lowest phase angle, indicating the highest stiffness of the modified binder blends. Adding GTR and LPCR could soften the binder blends, resulting in lower complex modulus and higher phase angle, but the master curves are nearly overlapped for the two modified asphalt. Figure 28.6 further illustrates the rheological properties of modified binder blends based on the rutting parameters. Asphalt modified with LPCR/WPE has lower rutting parameters $(G^*/sin\delta)_{max}$ than GTR/WPE, which further demonstrates that LPCR can be more soluble and softer in asphalt than GTR. Figure 28.7 shows the low-temperature performance of different polymer modified asphalt. Asphalt modified with WPE is susceptible to low-temperature cracking. The pyrolyzed process of GTR has a more significant effect on enhancing the low-temperature performance of modified asphalt. By adjusting the LPCR contents, the co-existence of rigid plastic and soft rubbery regimes is promising to improve the mechanical performance of asphalt binder in both high- and low-temperature domains.

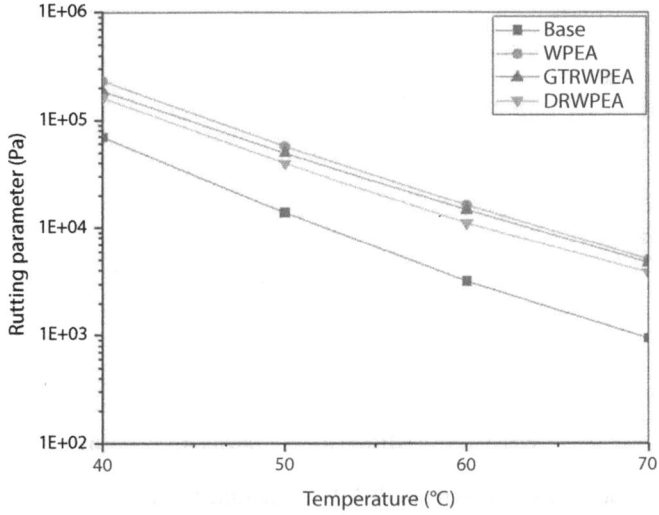

Figure 28.6 Temperature sweep test results for different polymer modified asphalt.

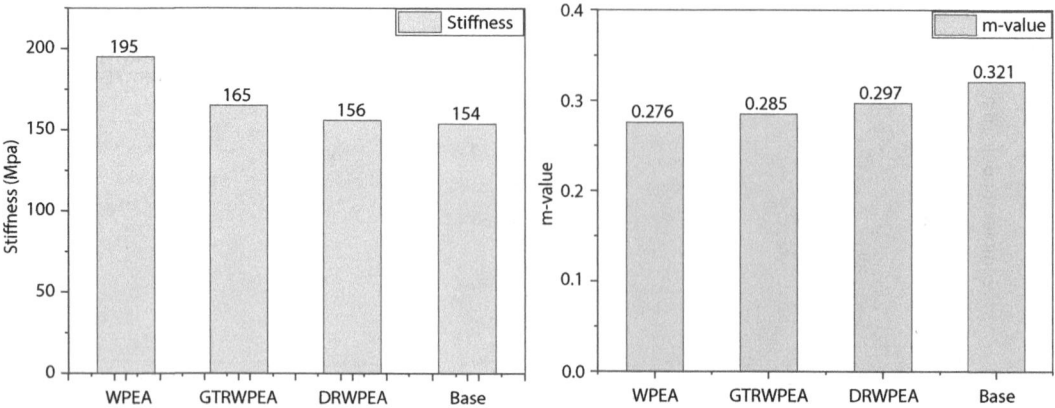

Figure 28.7 BBR test results for different polymer modified asphalt.

28.2.5.4   Storage Stability of Polymer Modified Asphalt

Storage stability is the main concern for asphalt modified with plastics. Figure 28.8 shows the cigar tube test results for different polymer modified asphalt. WPE has a significant phase separation in asphalt, indicated by a nearly 100% separation index. The addition of GTR can slightly improve the storage stability of modified binder blends, while LPCRs can effectively stabilize the WPE in asphalt. LPCR has the potential to improve the storage stability of asphalt containing WPE, but still exist large gaps corresponding to the base asphalt binder. Hence, WPE is expected to be further stabilized by adjusting the amount of LPCR and using plastic compatibilizers.

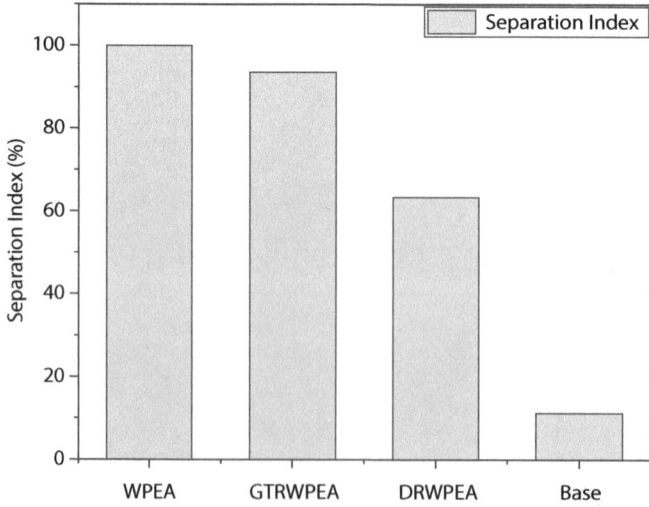

Figure 28.8 Cigar tube test results for different LPCR/WPE modified asphalt.

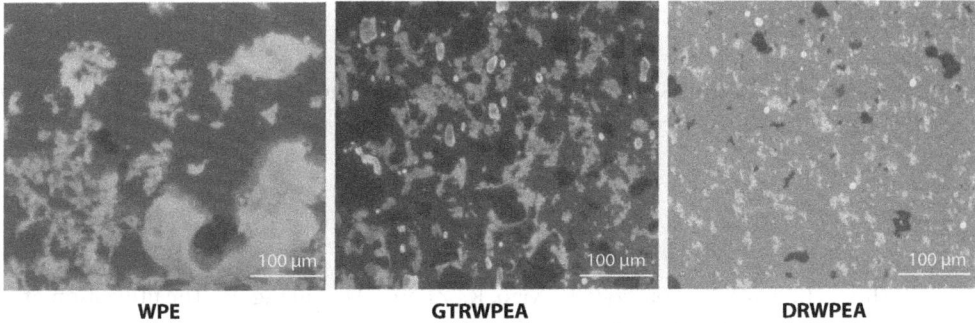

Figure 28.9 Polymer micromorphology of modified asphalt.

28.2.5.5   Polymer Micromorphology of Modified Asphalt

Figure 28.9 shows the micromorphology of asphalt modified with WPE, GTR/WPE, and LPCR/WPE. PE particles tend to coalesce together and have a significant phase separation in asphalt. The size of PE particles was smaller when GTR was incorporated into asphalt. The pyrolyzed procedure further broke the sizes of GTR and PE particles, which improve the solubility of GTR and dispersion of WPE particles in asphalt. Hence, the compatibility between asphalt and polymers was improved.

28.3   Conclusions

This study developed a novel approach to stabilize waste plastics by using lightly pyrolyzed crumb rubber. The pyrolyzed temperature was first determined to improve the solubility of crumb rubber. The pyrolyzed mechanisms were investigated by using FTIR tests. The rheological properties of modified binder blends were evaluated at high and low temperatures. Cigar tube tests and optical microscopy were employed to characterize the storage stability and polymer micromorphology in asphalt. On the basis of the results from this study, the conclusions can be summarized as follows:

- LPCR was an effective modifier to stabilize WPE in asphalt.
- A higher pyrolyzed temperature can improve the solubility and compatibility of LPCR in asphalt.
- The partial destruction of crosslink structures in crumb rubbers occurred during the pyrolyzed process, indicated by the weak intensity of the carbon-sulfur single bond in FTIR spectrums.
- LPCR improved the dispersion of WPE particles in asphalt and the storage stability of modified binder blends.
- The co-existence of rigid plastic and soft rubbery regimes improved the mechanical performance of asphalt binder in both high- and low-temperature domains.

Acknowledgments

The authors would like to thank the REMADE Institute (under DE-EE0007897 – U.S. DOE Advanced Manufacturing Office) for offering financial support.

References

1. Ma, Y., Zhou, H., Jiang, X., Polaczyk, P., Xiao, R., Zhang, M., Huang, B., The utilization of waste plastics in asphalt pavements: A review. *Cleaner Materials*, 2, 2021b.
2. Dalhat, M.A., Al-Abdul Wahhab, H.I., Performance of recycled plastic waste modified asphalt binder in Saudi Arabia. *International Journal of Pavement Engineering*, 18(4), 349-357, 2015.
3. Ge, D., Yan, K., You, Z., Xu, H., Modification mechanism of asphalt binder with waste tire rubber and recycled polyethylene. *Construction and Building Materials*, 126, 66-76, 2016.
4. Ma, Y., Wang, S., Zhou, H., Hu, W., Polaczyk, P., Zhang, M., Huang, B., Compatibility and rheological characterization of asphalt modified with recycled rubber-plastic blends. *Construction and Building Materials*, 2020.
5. Polacco, G., Filippi, S., Merusi, F., Stastna, G., A review of the fundamentals of polymer-modified asphalts: Asphalt/polymer interactions and principles of compatibility. *Adv Colloid Interface Sci*, 224, 72-112, 2015.
6. Yan, K., Xu, H., You, L., Rheological properties of asphalts modified by waste tire rubber and reclaimed low density polyethylene. *Construction and Building Materials*, 83, 143-149, 2015.
7. Huang, B., Mohammad, L.N., Graves, P.S., Abadie, C.J.T.R.R., Louisiana experience with crumb rubber-modified hot-mix asphalt pavement. 1789(1), 1-13, 2002.
8. Mashaan, N.S., Ali, A.H., Koting, S., Karim, M.R., Dynamic Properties and Fatigue Life of Stone Mastic Asphalt Mixtures Reinforced with Waste Tyre Rubber. *Advances in Materials Science and Engineering*, 2013, 1-9, 2013.
9. Wang, T., Xiao, F., Zhu, X., Huang, B., Wang, J., Amirkhanian, S.J.J.o.C.P., Energy consumption and environmental impact of rubberized asphalt pavement. 180, 139-158, 2018.
10. Huang, S.-C., Rubber concentrations on rheology of aged asphalt binders. *Journal of Materials in Civil Engineering*, 20(3), 221-229, 2008.
11. De, S.K., Isayev, A., Khait, K., Rubber recycling. *CRC Press*, 2005.
12. Lee, S.-J., Akisetty, C.K., Amirkhanian, S.N., The effect of crumb rubber modifier (CRM) on the performance properties of rubberized binders in HMA pavements. *Construction and Building Materials*, 22(7), 1368-1376, 2008.
13. Fu, H., Xie, L., Dou, D., Li, L., Yu, M., Yao, S., Storage stability and compatibility of asphalt binder modified by SBS graft copolymer. *Construction and Building Materials*, 21(7), 1528-1533, 2007.
14. Wang, S., Cheng, D., Xiao, F., Recent developments in the application of chemical approaches to rubberized asphalt. *Construction and Building Materials*, 131, 101-113, 2017.
15. Kim, H., Lee, S.-J., Laboratory Investigation of Different Standards of Phase Separation in Crumb Rubber Modified Asphalt Binders. *Journal of Materials in Civil Engineering*, 25(12), 1975-1978, 2013.
16. Ma, Y., Wang, S., Zhou, H., Hu, W., Polaczyk, P., Huang, B., Potential Alternative to Styrene–Butadiene–Styrene for Asphalt Modification Using Recycled Rubber–Plastic Blends. *Journal of Materials in Civil Engineering*, 33(12), 2021a.

29

Analysis and Design for Sustainable Circularity of Barrier Films Used in Sheet Molding Composites Production

Farshid Nazemi[1], Bhavik Bakshi[1]*, Jose Castro[2], Rachmat Mulyana[2], Rebecca Hanes[3],
Saikrishna Mukkamala[3], Kevin Dooley[4], George Basile[5], George Stephanopoulos[6,7],
Andrea Nahas[8], Aleen Kujur[8] and Todd Hyche[8]

[1]*William G. Lowrie Department of Chemical and Biomolecular Engineering,
The Ohio State University, Columbus, OH, USA*
[2]*Department of Integrated Systems Engineering, The Ohio State University, Columbus, OH, USA*
[3]*National Renewable Energy Laboratory, Golden, CO, USA*
[4]*W.P. Carey School of Business, Arizona State University, AZ, USA*
[5]*School of Sustainability Arizona State University, Tempe, AZ, USA*
[6]*Global Institute of Sustainability, Arizona State University, Tempe, AZ, USA*
[7]*Department of Chemical Engineering, Massachusetts Institute of Technology, Cambridge, MA, USA*
[8]*Kohler Co., Kohler, WI, USA*

Abstract

Sheet molding composites (SMC) are thermoset based molding compounds with a global market size of USD 2.1 billion in 2020. SMC is usually processed with multilayer barrier/carrier films to prevent emissions from volatile organic compounds (VOC). The plastic barriers are separated from SMC after the curing process and discarded. To illustrate the high consumption volume of SMC barrier/carrier films, Kohler Co., a company which uses SMC in their bath and showering product lines, alone generates over 1M lb/year of SMC barrier/carrier film waste. Therefore, high amount of SMC barrier/carrier film plastic waste creates a great opportunity for reusing, recycling, or valorization at its end of life. This can be achieved through transition from a linear to a circular supply chain design. However, this transition will not be feasible without overcoming the technical and economic barriers. This project aims to tackle this by making the film circular. Different end of life alternatives will be evaluated experimentally. The reusability of the film will be studied through applying different mechanical and solvent cleaning techniques. The operability concerns regarding the utilization of the cleaned film for long run operations will be addressed by thermally stitching the film. Pelletizing of the cleaned film will also be investigated as a viable technique for recycling or downcycling of the film as molded products. In addition to technical barriers, designing a sustainable circular supply chain for SMC carrier film products requires uncovering the collaboration opportunities between suppliers and users at different stages of the value chain to expand its use to other products. These opportunities will be investigated by finding companies willing to buy or

Corresponding author: bakshi.2@osu.edu

Nabil Nasr (ed.) Technology Innovation for the Circular Economy: Recycling, Remanufacturing, Design, Systems Analysis
and Logistics, (365–378) © 2024 Scrivener Publishing LLC

treat the plastic film waste. Finally, the economic viability and environmental impacts of these circular design alternatives will be evaluated through techno-economic assessment (TEA) and life cycle analysis (LCA). Techno-economic and life cycle inventory data for various pathways will also be made available as one of the main outcomes of this project. Preliminary LCA results indicate that the current SMC film supply chain (base case) emits 5.81 tonne CO_2 eq. and consumes 104 GJ energy per tonne of SMC film. It is estimated that alternative circular pathways result in 14-20% less CO_2 emissions and 34-44% reduction in embodied energy. This project will develop a tool to analyze the circularity and sustainability of hundreds of combinations of alternatives in different life cycle stages. Users of this tool will be able to choose the best combination among these alternatives with respect to the selected economic, environmental and circularity objectives.

Keywords: Sheet molding compound, multilayer film, plastic recycling, end of life management, plastic waste, circular economy, life cycle assessment, techno-economic analysis

29.1   Introduction

Plastic films have a variety of applications and are widely used in industry. Monolayer plastic films are composed of one type of polymer (Polyethylene (PE), Polypropylene (PP), etc.) and are widely used for film-based packaging [1]. On the other hand, multilayer plastic films are made of multiple types of resins in a co-extrusion process and are an excellent choice when very specific film properties are needed [2–6]. One of the most widely used multilayer plastic films are high barrier/carrier release films that are used in the manufacturing of sheet molding compounds (SMC). SMC are thermoset fiber reinforced plastics widely used to make automobile and bath and showering products. The multilayer film acts as a substrate that carries the paste containing the base components of SMC during the production process. Moreover, the multilayer film acts as a barrier that prevents the migration of volatile compounds like styrene. This is a crucial step in the SMC manufacturing process because styrene migration will result in an increase in viscosity and premature crosslinking which will cause problems in the subsequent molding process. The film is removed from the SMC prior to the molding process and is discarded to landfill. Among many possible solutions for providing the barrier property, Polyamides (PA) or nylons, are preferred due to its market availability and properties. However, a monolayer nylon film will adhere to the styrene monomer, which makes it difficult to remove prior to the molding process. Instead, while nylon is used as the core layer of the film, polyethylene (PE) is used as the top and bottom layers because of its nonadherence properties. Finally, an adhesive agent is used to hold the nylon and polyethylene layers together, making it a five-layer PE/adhesive/nylon/adhesive/PE film [6].

The current global SMC market is worth USD 2 billion, representing 6 million Ib/yr of SMC barrier/carrier film waste in the US, which creates a great potential for waste recovery [7]. However, the barrier/carrier film waste is currently discarded to landfill due to the lack of recycling and recovering strategies at their end of life. Reducing this waste can decrease significant environmental impacts related to large scale disposal of plastic films and consumption of primary (virgin) feedstock resources.

Unlike monolayer plastic films, multilayer films are not easy to recycle because of the presence of multiple layers of films that are chemically and thermodynamically incompatible and thus cannot be recycled using conventional plastic recycling technologies [2–5, 8–17]. However, in recent years, open-loop and closed-loop recycling of multilayer films have been made possible using unique techniques at laboratory, pilot, and commercial scales.

Mechanical recycling of multilayer films is being studied extensively [8–16]. There are three main advanced methods to mechanically recycle multilayer plastic films. The first one includes regranulation (grinding and pelletizing) of films using a compatibilizer to make the polymer blends less immiscible and therefore more mechanically stable. Compatibilization results in one polymer stream that can be used to make other products through injection molding or blown film processes that can result in production of new monolayer and multilayer films [9]. Nevertheless, compatibilization does not enable closed-loop recycling due to the heterogeneity needed for multilayer plastic films with unique properties and applications [8]. The second method is delamination of individual polymer components through defunctionalizing the adhesive between each layer using a solvent [10, 11]. In this method, the crosslinked adhesive is specially targeted and dissolved in a solvent. Individual polymer layers can therefore be separated from each other in the absence of chemical bonding between individual polymer layers and be repalletized and substituted the virgin feedstock to reproduce the multilayer film. In the third method, a series of solvent washes is used to selectively dissolve and precipitate individual polymer layers at each stage [12–16]. This method is called solvent-targeted recovery and precipitation (STRAP), and it hasn't been commercialized yet. The precipitated polymers can substitute the virgin resins used in the multilayer production process.

Among the three mechanical recycling techniques, delamination and STRAP are also considered as layer separation techniques that can be used in combination with chemical recycling processes to recycle chemicals and fuels. The individual polymers can be depolymerized to produce base monomers and/or feedstocks. Without layer separation, depolymerization is not possible due to the complex heterogenous structure of the multilayer films [8]. Finally, among other chemical recycling methods, pyrolysis is getting a great deal of attention as a viable option to chemically recycle mixed monolayer and multilayer films without layer separation to produce fuel and carbon-rich solid products [17–19]. However, yield, energy consumption and process emissions are highly dependent on the composition of the mixed plastic stream or multilayer plastic film.

Employing the abovementioned recycling techniques as recovery strategies for SMC barrier/carrier films requires addressing particular characteristics of the SMC barrier/carrier film due to its unique composition and the presence of contaminations on the film. According to Berry Global, one of the leading companies in producing SMC barrier/carrier films in the US, SMC barrier/carrier film is composed of 57 wt.% LLDPE, 26 wt.% nylon 6, and 17 wt.% polyethylene grafted maleic anhydride. In addition, SMC barrier/carrier film contains some contaminations on the surface, which are mainly styrene monomers. Although a few studies investigated some of the abovementioned mechanical treatment methods for nylon-PE films used in packaging industry [9, 12], to our knowledge, there is no study on mechanical and chemical recycling of SMC barrier/carrier film with this specific composition. Moreover, the effect of styrene monomer contamination on the treatment techniques and mechanical properties of the recycled streams must be studied carefully. In this study, we will specifically study how (1) thermodynamic incompatibility of nylon and PE and (2) the presence of contaminations will affect treatment techniques. This will eventually enable us to employ unique recovery strategies for SMC barrier/carrier film supply chains.

In addition to the technical obstacles, enabling sustainable circular strategies to recover/recycle SMC barrier/carrier films requires overcoming value chain and customer preference, environmental, and economic challenges, and identifying the sustainability tradeoffs of various end-of-life (EoL) strategies. The overall goal of this work is to identify and overcome these challenges and to develop a tool to design the "best" sustainable circular economy for SMC barrier/carrier film. More specifically, in the present contribution (1) process, life cycle and cost data modules are provided for the current (conventional) and alternative SMC barrier/carrier film supply chains (2) emerging technologies for treating, recovering, and recycling SMC barrier/carrier film waste are evaluated (3) the effect of the alternative EoL treatments are studied on the value chain and from a customer preferences point of view (4) life cycle and techno-economic analyses are performed to evaluate environmental impacts and economic performance of the current and emerging technologies for EoL processes.

This work evaluates the possibility and ways to recycle multilayer polymer films for the first time. More specifically, to our knowledge, there is no study evaluating circular and sustainable EoL alternatives for SMC barrier/carrier films. In addition, the optimization and scenario analysis tool will be the first of its kind to evaluate hundreds of combinations of scenarios and design the "best" alternative for EoL strategy for SMC barrier/carrier film while accounting for different technological and logistic constraints. Finally, both the experimental and optimization results can be applicable to multilayer plastic films used for other applications such as food packaging, specially to PE-PA films that are used in food packaging industry due to their unique structure. The scenario analysis and optimization tool will eventually be expanded to cover other plastic and chemical products.

As it is stated in every section, this paper contains preliminary results that are obtained so far.

29.2 Main Content of Chapter

29.2.1 SMC Barrier Film Supply Chain

Figure 29.1 shows the current (conventional) SMC barrier/carrier film supply chain, from extraction of resources to the EoL treatment of SMC film after consumption. The composition of the film is obtained from Berry Global Co. The film is composed of linear low-density polyethylene (LLDPE), nylon 6, and polyethylene grafted with maleic anhydride (PE-maleic anhydride), which acts as an adhesive linking the core polyamide layer to the top and bottom LLDPE layers. Foreground process data are obtained from Kohler Co. and Berry Global Co. Kohler is one of the major manufacturers of SMC for bath and showering products which makes them one of the largest consumers of SMC barrier/carrier films. Kohler generates 1lb M/yr of SMC film waste and discards it to landfill. The generated waste is sent directly from Kohler site to an unsanitary landfill by a waste management company. Accordingly, we assumed landfilling as the only waste treatment option in the conventional supply chain of multilayer SMC barrier/carrier film generated in the US.

29.2.1.1 Life Cycle Assessment

A cradle-to-grave life cycle assessment (LCA) is performed to evaluate the environmental impacts of the current SMC film supply chain. The LCA is performed according to the

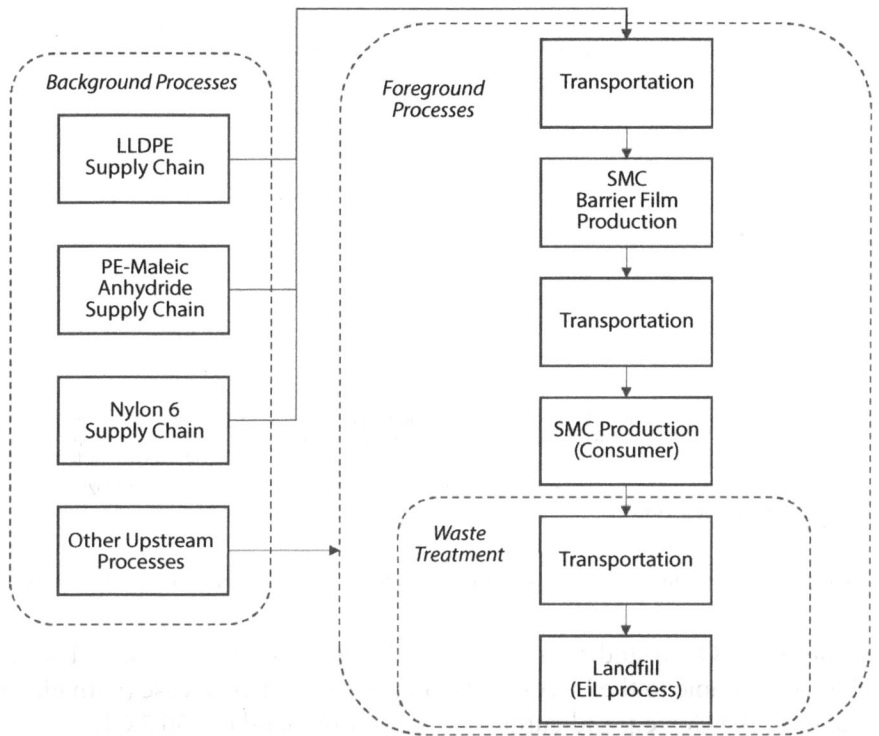

Figure 29.1 Current supply chain of the SMC barrier/carrier film. The film is consumed in the SMC manufacturing process and is directly disposed of to landfill.

standard ISO 14040/14044 [20]. The system boundary is shown in Figure 29.1. Instead of considering unit process data for LLDPE, nylon 6, and PE-maleic anhydride, the whole life cycle supply chain of these processes is considered for the LCA to capture the system-wide environmental impacts of the consumption of virgin feedstock. Inventory data for foreground processes are obtained from industry (Kohler and Berry Global). Ecoinvent v3.8 is used for obtaining the inventory data for the background processes [21]. US data are used whenever available. Otherwise, global or rest of the world data (RoW) are used.

TRACI is used to transform inventory data to the following impact categories: acidification, ecotoxicity, eutrophication, global warming, photochemical oxidation, human health (carcinogenics and non-carcinogenics), and respiratory effects [22]. Cumulative exergy demand is used to quantify the cumulative exergy inputs to the system, which represents the total resource consumption [22].

Cradle-to-grave LCA results are presented in Figure 29.2. The results show that raw material supply chains (nylon 6, LLDPE, and PE-maleic anhydride) have considerably larger life cycle impacts compared to other upstream and foreground processes. This indicates that the consumption of virgin material is the main contributor to the negative environmental impacts in the SMC barrier/carrier film supply chain. Recycling/recovering strategies can reduce the primary feedstock consumption and/or increase the consumption of secondary feedstock, which can eventually reduce the environmental impacts. This is discussed in section 29.2.2.

Moreover, using the REMADE estimator tool [23] for calculating the cradle-to-grave global warming potential and energy consumption of the base case, it is estimated that the

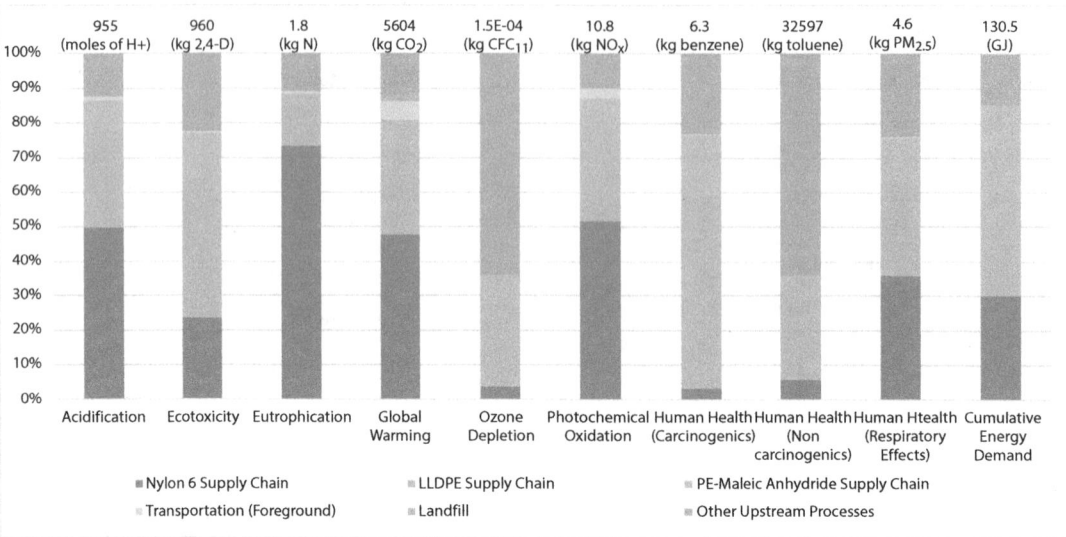

Figure 29.2 Cradle-to-grave life cycle impacts of 1 tonne of SMC barrier/carrier film consumption.

base case emits 5.82 CO_2 eq. and consumes 104 GJ energy per tonne of SMC barrier/carrier film, which is comparable to the LCA results obtained for the base case (with global warming potential of 5.6 kg CO_2 eq. and cumulative exergy demand of 130.5 GJ).

29.2.1.2 *Techno-Economic Analysis*

Techno-economic analysis (TEA) provides an understanding of how individual cost and other factors influence a process' overall costs and the product's minimum selling price (MSP). To date, baseline TEA models of barrier/carrier film production and of SMC production have been completed; TEA data and results for the barrier/carrier film production are presented here. The baseline TEA model was developed to correspond as closely as possible to the current production process of SMC barrier/carrier film and applied to calculate the barrier film MSP, which was found to be within 0.10 USD/lb of the actual purchase cost. Financial inputs and assumptions used in the TEA model are listed in Table 29.1, with key results provided in Table 29.2.

Although this baseline TEA model does not include circular EoL options, knowing the parameters that influence the current barrier film production process economics will enable direct financial comparisons of EoL options that could change the material input types, amounts, or costs, valorize or reduce waste streams, and/or change the energy requirements of barrier film production.

29.2.2 Alternative Pathways for EoL Management of SMC Films

29.2.2.1 *Experimental Results*

Our initial approach was to investigate if Kohler would reuse the reclaimed/cleaned film. It appears that this alternative may not be feasible as the film is cut when making SMC thus the reclaimed film would be much shorter than the original film. The long operation time

Table 29.1 Financial assumptions used in the baseline TEA model of barrier/carrier film production.

Parameter category	Parameter	Value and units	Source
Materials	Polyethylene resin cost	0.9 USD/lb	[24]
	Polyethylene resin consumption	700 lb/hour	Berry Global
	Nylon-6 cost	1.7 USD/lb	[24]
	Nylon-6 consumption	300 lb/hour	Berry Global
Energy	Power Use	50 kW	[25]
Labor	Plant employees	4	
Capital	Blown film extruder cost (200 lb film/hour capacity)	$200,000	Industrial References
	Plant depreciation period	7 years	[26]
	Plant lifetime	30 years	[26]
	Discount rate	10%	[26]
Operations	Film reject rate	10%	Berry Global
	On-stream factor	90%	Berry Global
Output	Film width	1000 mm	Kohler
	Film thickness	0.1 mm	Kohler
	Film production	1000 lb film/hour	Berry Global

Table 29.2 Summary of baseline TEA results for barrier film production.

Direct Materials	9.0×10^6 USD/year
Energy	3.9×10^6 USD/year
Direct Labor	9.5×10^5 USD/year
Capital	8.5×10^5 USD/year
Waste	9.0×10^5 USD/year
Final Barrier Film Cost	2.10 USD/lb

in manufacturing the SMC is to balance the SMC machine, thus once balanced the operation should continue without interruption even if manufacturing a slightly different SMC formulation. We believe this can also be applicable to other SMC manufacturers that operate in continuous mode. Nevertheless, those SMC manufacturers that run in batch mode would still be willing to use shorter film rolls.

We used a targeted solvent cleaning technique in order to remove the styrene monomer contaminations on the surface of the film without dissolving PE polymers. Diacetone

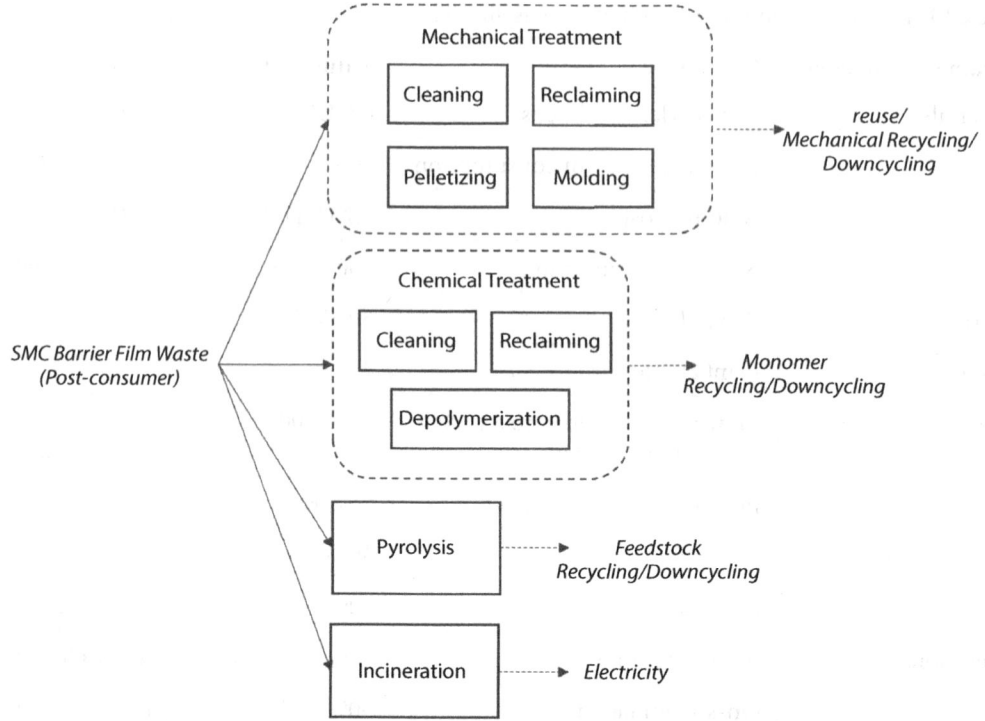

Figure 29.3 Alternative end of life processes and technologies for treating SMC barrier film waste.

alcohol is one of the most common solvents used to remove cured polyester resins and does not dissolve PE [27, 28]. Preliminary results indicate that the solvent, diacetone alcohol, has a good potential for cleaning the film. Also, preliminary evaluations indicate that in order for the reuse scenario to be environmentally less harmful than the base case scenario, the mass ratio of solvent to the film must not exceed 1:1, and the solvent should be used several times and recovered as much as possible. Once the prototype is ready details of the process will be investigated.

We are currently investigating pelletizing the film with and without the cleaning step. These pellets could subsequently be recycled for reproducing the SMC barrier/carrier film (referred to as mechanical recycling in Figure 29.3), or will be used to produce other products through the molding process (referred to as downcycling scenario in Figure 29.3). Chemical recycling and pyrolysis will also be studied after evaluating the above-mentioned scenarios.

29.2.2.2 Value Chain and Customer Preference

Any recycling solution for the barrier film waste generates recycled content that must be reprocessed internally or sold on the scrap plastic market [29]. The data below shows the price of recycled LDPE, the most common thin film, across different grades. We can see through most of the time, the value of r-LDPE has decreased. In 2020 though, prices of higher-grade r-LDPE spiked. In part this was due to the increased cost of virgin plastic, due to increased petroleum prices and supply shortages due to the pandemic. But part of this

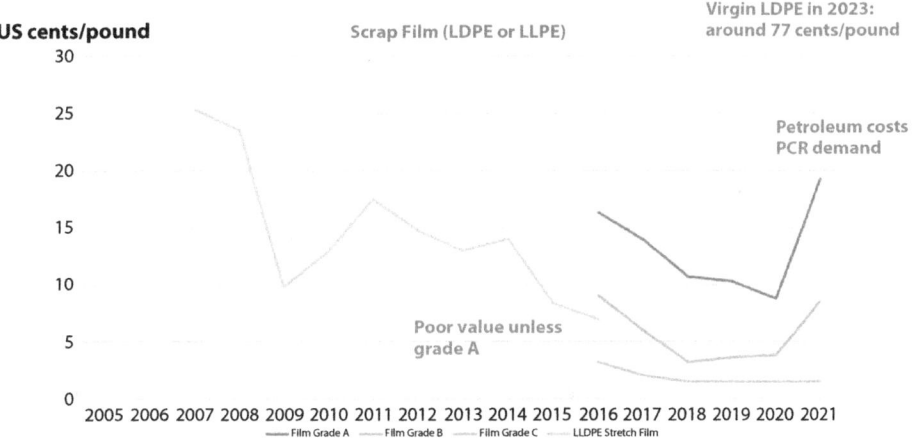

Figure 29.4 Cost of recycled LDPE (thin film).

increase was also due to the unprecedented increase in demand for recycled plastic content, linking to the commitments that brands are making for 100% recyclable, reusable, or compostable plastic packaging by 2025 [30].

There are a variety of downstream markets for this r-LDPE. There is growing demand from the construction and building sector for recycled plastic as both a product or filler. For example, rail ties, lumber, and sheets can meet functional requirements with recycled plastic content. Roof cover board and subflooring are applications with the least material restrictions; other applications may have limitations due to chemical exposure. The *Material Recovery for the Future* project [31] created a flexible bale consisting of 3-7 plastic and was able to show that there was economically viable demand for a bale made of recyclable flexible film.

- Can use 100% recycled thin film: Roof coverboard and subflooring, Pallets, Sheet stock for signage, Fuels or petrochemicals
- Can use a percentage as recycled thin film: Rail ties, Lumber, Trim, Industrial mats, Tanks, pipes, and containers, Bottles, Crates, Durable goods, Cinder blocks
- Can use up to 3% plastic: Asphalt binder

29.2.2.3   Life Cycle Assessment

Among different EoL alternatives, we evaluated the life cycle impacts of the incineration scenario and compared it with the base case (landfilling) by conducting a cradle-to-grave LCA. Ecoinvent v3.8 database is used for the incineration scenario; we considered "treatment of mixed plastic waste, municipal incineration" and assumed an electric energy recovery of 3.92 MJ/kg of film (as stated in the database for mixed plastic waste). It is also assumed that the electricity produced from the incineration process will substitute the market mixed electricity in the US. Results of the comparative LCA are presented in Figure 29.5.

As presented in Figure 29.5, for all impact categories except global warming potential, incineration scenario will have less environmental impacts compared to the base case (landfill) due to the generation and substitution of electricity. However, the incineration scenario

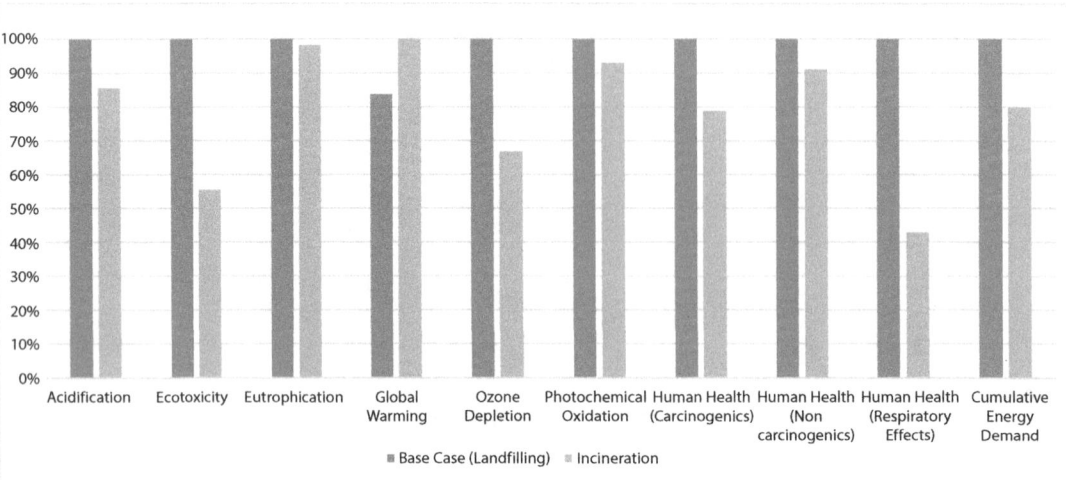

Figure 29.5 Comparative cradle-to-grave LCA results for the base case (landfill) and incineration scenarios.

will result in higher global warming potential compared to the base case because of the large amount carbon dioxide emitted from the incineration process. More specifically, the LCA results indicate that the incineration scenario emits 16% more CO2 eq. and demands 20% less cumulative exergy compared to the base case scenario. This shows the environmental tradeoff between global warming potential and other life cycle impacts in the incineration scenario. For other alternative EoL processes (reuse, recycling, downcycling, and pyrolysis), we will use experimental results and/or literature data for conducting the comparative LCA, which will be provided in the future.

In addition, REMADE estimator [23] is used to estimate the potential reduction in global warming potential and energy consumption by applying recycling/recovering strategies; it is estimated that this could result in 14-20% less CO_2 eq. emissions and 34-44% reduction in the embodied energy compared to the base case scenario.

29.2.3 Developing a Tool for Analyzing and Designing a Sustainable Circular Economy

The sustainable circular economy (SCE) framework was formulated and developed in previous work [32], and it will be used to develop a tool for analyzing hundreds of combinations of different alternatives for EoL treatment of SMC barrier/carrier films and designing the "best" sustainable circular supply chain. Multiple environmental, economic, and performance-based objectives will be considered and formulated as a multi-objective optimization problem which will be solved using different multi-objective optimization techniques that can generate compromised solutions.

29.3 Conclusions & Recommendations

SMC barrier/carrier films are multilayer PE-nylon plastic films that are widely used in the manufacturing of SMC products. Currently, the film is removed before the molding process and is discarded to landfill. However, due to the high production rate of SMC products, the rate of the plastic film waste generated is significantly high, which can cause many environmental burdens and create opportunities for recovering the waste at its end of life. The present work focuses on revealing these opportunities and evaluating alternative pathways for treating the SMC barrier/carrier film plastic waste. Preliminary results show that reducing this waste by applying circular strategies can lead to 14-20% and 34-44% reduction in global warming potential and embodied energy, respectively. More insights on the environmental tradeoffs will be provided once we evaluate all the alternative scenarios. Moreover, in this paper, some of the technical, environmental, and economic challenges of various alternatives are discussed and compared to the base case scenario (landfill disposal). Value chain and customer preference aspects are also studied in order to enable circular strategies at different stages of the life cycle. By considering all aspects of circularity and sustainability, different environmental, economic and performance tradeoffs will be identified for all alternative circular scenarios. This will be the first step in developing a tool to find the "best" or "compromised" sustainable circular solution for SMC barrier/carrier films. Multi-objective optimization methods will be used to generate optimum solutions for different sustainability, circularity, and performance objectives. Finally, the tool can be extended to cover other (plastic) products, specifically multilayer plastic films used in other industries such as food packaging.

29.4 Acknowledgements

This work is supported by the REMADE Institute (grant number: 20-01-DE-4103). The authors also thank Berry Global for their support and assistance with data provision.

References

[1] Flexible Packaging Association. State of the Flexible Packaging Industry Report, https://www.flexpack.org/state-of-the-industry, 2022.

[2] Borghesi, G., Stefanini, R., & Vignali, G., Life cycle assessment of packaged organic dairy product: A comparison of different methods for the environmental assessment of alternative scenarios. *Journal of Food Engineering*, 318, 110902, 2022.

[3] Conte, A., Cappelletti, G. M., Nicoletti, G. M., Russo, C., & Del Nobile, M. A., Environmental implications of food loss probability in packaging design. *Food Research International*, 78, 11–17, 2015.

[4] Siracusa, V., Rosa, M. D., Romani, S., Rocculi, P., & Tylewicz, U., Life Cycle Assessment of multilayer polymer film used on food packaging field. *Procedia Food Science*, 1, 235–239, 2011.

[5] Blanco, I., Ingrao, C., & Siracusa, V., Life-Cycle Assessment in the Polymeric Sector: A Comprehensive Review of Application Experiences on the Italian Scale. In Polymers (Vol. 12, Issue 6), 2020.

[6] PolyGroup Inc., High Barrier Carrier Release Films For Thermoset Reinforced Plastic, https://polygroupinc.com/products/plastic-carrier-films.

[7] PERSISTENCE Market Research, Global Sheet Molding Compound Market, https://www.persistencemarketresearch.com/market-research/sheet-molding-compound-market.asp, 2019.

[8] Soares, C. T. de M., Ek, M., Östmark, E., Gällstedt, M., & Karlsson, S., Recycling of multi-material multilayer plastic packaging: Current trends and future scenarios. *Resources, Conservation and Recycling, 176*, 105905, 2022.

[9] BASF, Coextruded PE/PA multilayer films are recyclable!, https://chemicals.basf.com/global/monomers/polyamide/mechanical-recycling/220628_Publication_Coextruded%20PE_PA%20multilayer%20films%20are%20recyclable.pdf, 2022.

[10] O'Rourke, G., Houbrechts, M., Nees, M., Roosen, M., de Meester, S., & de Vos, D., Delamination of polyamide/polyolefin multilayer films by selective glycolysis of polyurethane adhesive. *Green Chemistry, 24*(18), 2022.

[11] Kaiser, K. M. A., Recycling of multilayer packaging using a reversible cross-linking adhesive. *Journal of Applied Polymer Science, 137*(40), 49230, 2020.

[12] Costamagna, M., Massaccesi, B. M., Mazzucco, D., Baricco, M., & Rizzi, P., Environmental assessment of the recycling process for polyamides - Polyethylene multilayer packaging films. *Sustainable Materials and Technologies, 35*, e00562, 2023.

[13] Walker, T. W., Frelka, N., Shen, Z., Chew, A. K., Banick, J., Grey, S., Kim, M. S., Dumesic, J. A., van Lehn, R. C., & Huber, G. W., Recycling of multilayer plastic packaging materials by solvent-targeted recovery and precipitation. *Science Advances, 6*(47), eaba7599, 2023.

[14] Lee, Y. C., Kim, M. J., Lee, H. C., A recycling method of multilayer packaging film waste, European Patent Application EP 1 683 829 A1, assigned to Korea Institute of Industrial Technology (KITECH), 2005.

[15] Sánchez-Rivera, K. L., Zhou, P., Kim, M. S., González Chávez, L. D., Grey, S., Nelson, K., Wang, S.-C., Hermans, I., Zavala, V. M., van Lehn, R. C., & Huber, G. W., Reducing Antisolvent Use in the STRAP Process by Enabling a Temperature-Controlled Polymer Dissolution and Precipitation for the Recycling of Multilayer Plastic Films. *ChemSusChem, 14*(19), 4317–4329, 2021.

[16] Cecon, V. S., Curtzwiler, G. W., & Vorst, K. L., A Study on Recycled Polymers Recovered from Multilayer Plastic Packaging Films by Solvent-Targeted Recovery and Precipitation (STRAP). *Macromolecular Materials and Engineering, 307*(11), 2200346, 2022.

[17] Bassey, U., Sarquah, K., Hartmann, M., Tom, A., Beck, G., Antwi, E., Narra, S., & Nelles, M., Thermal treatment options for single-use, multilayered and composite waste plastics in Africa. *Energy, 270*, 126872, 2023.

[18] RTI International, Environmental and Economic Analysis of Emerging Plastics Conversion Technologies, http://energy.cleartheair.org.hk/wp-content/uploads/2012/05/Environmental-and-Economic-Analysis-of-Emerging-Plastics-Conversion-Technologies.pdf, 2012.

[19] Khoo, H. H., LCA of plastic waste recovery into recycled materials, energy and fuels in Singapore. *Resources, Conservation and Recycling, 145*, 67–77, 2019.

[20] ISO, E.N, 2006. 14040: 2006. Environ. Manag. Cycle assessment-Principles Fram. Eur. Comm. Stand., https://scholar.google.com/scholar_lookup?title=14040%3A%202006.%20Environ.%20Manag.%20cycle%20assessment-Principles%20Fram&author=ISO%2C%20E.N&publication_year=2006, 2006

[21] Ecoinvent, Database, Ecoinvent v.3.8, https://ecoinvent.org/the-ecoinvent-database/data-releases/ecoinvent-3-8/, 2021

[22] Ecoinvent, Database, Impact Assessment, https://ecoinvent.org/the-ecoinvent-database/impact-assessment/#1661420435296-03a17d13-bc0e166142919370816614298329917 , 2022

[23] REMADE Institute, Calculating Project/Proposal Benefits, https://remadeinstitute.org/project-impact-calculator

[24] Plastics News, Plastics Resin Pricing, https://www.plasticsnews.com/resin/currentPricing, 2022.

[25] S & P Global Commodity, Chemical Economics Handbook® (CEH), https://www.spglobal.com/commodityinsights/en/ci/products/chemical-economics-handbooks.html

[26] Sinnott, R. Towler, G. Chemical Engineering Design (6thEdition), Elsevier, Oxford ed, 2019.

[27] Marquis, E. T., Cuscurida, M., A method of dissolving cured polyester, European Patent Application 0485063A2, assigned to Huntsman Corp, 1991.

[28] LDPE Chemical Compatibility, https://www.calpaclab.com/ldpe-chemical-compatibility-chart/, 2023.

[29] Hafsa, F., Dooley, K., Basile, G., and Buch, R., A typology and assessment of innovations for circular plastic packaging, *Journal of Cleaner Production*, 369 (133313), 2022.

[30] Dooley, K., Thakker, V., Bakshi, B., Scholz, M., Hafsa, F., Basile, G., and Buch, R., A multi-disciplinary assessment of innovations to improve grocery bag circularity, *Computer Aided Chemical Engineering*, 49: 625-630, 2022.

[31] Material Recovery for the Future, Flexible Packaging Recycling in Material Recovery Facilities Pilot, https://www.materialsrecoveryforthefuture.com/wp-content/uploads/MRFF-Pilot-Report-2020-Final.pdf, 2020.

[32] Thakker, V., & Bakshi, B. R., Toward sustainable circular economies: A computational framework for assessment and design. *Journal of Cleaner Production*, 295, 126353, 2021.

30

An Update on PVC Plastic Circularity and Emerging Advanced Recovery Technologies for End-of-Life PVC Materials

Domenic DeCaria

The Vinyl Institute, and Richard Krock, VyChlor Advisors LLC, Washington, DC, USA

Abstract

Robust mechanical recycling capabilities for PVC materials have been in operation for decades around the world. In the U.S. and Canada, there are over 100 recyclers processing 1.1 billion pounds of PVC scrap materials. These operations historically have targeted pre-consumer scrap, keeping it out of the waste stream, and reprocessed and reformulated it for entirely new applications. For example, cut-offs in the vinyl window industry can contain both rigid and flexible PVC, which after reprocessing is used in vinyl decking, fencing, or sea walls. In this instance, the unusable scrap from one industry becomes the raw material for another. Certain recyclers and product manufacturers reprocess post-consumer PVC materials, and this amount has been steadily increasing in the U.S., Canada, Europe, and Japan, particularly in the vinyl flooring industry. But some post-consumer resource streams, e.g. single use plastic packaging, are lean in PVC content (less than 3%) and this can be confounding for certain non-mechanical chemical-type recycling processes. Several advanced recycling technologies are being developed at the private industry and university level to process these lean PVC streams. Some technologies are attempting to convert chloride containing materials into fungible commodity resources including HCl and wax. Other technologies such as dissolution processing can take PVC rich streams in a mixed-material composite, such as vinyl flooring or vinyl roofing, and enable separation and recovery of the composite components for reuse as feedstock for vinyl applications. And some even go beyond a traditional view of "recycling", instead deconstructing PVC into base components to regenerate hydrogen gas and methanol. This paper will review the state of the art of several advanced recycling methods for treating PVC-rich and PVC-lean streams, and will provide a perspective on how current PVC recovery and recycling is becoming more circular, thereby improving the material's sustainability attributes.

Keywords: Recycling, PVC, vinyl, feedstock, circularity

Email: ddecaria@vinylinfo.org

Nabil Nasr (ed.) Technology Innovation for the Circular Economy: Recycling, Remanufacturing, Design, Systems Analysis and Logistics, (379–394) © 2024 Scrivener Publishing LLC

30.1   Introduction – Understanding PVC Materials

Questions continue to be raised about the viability of recycling PVC materials on a broader scale than mechanical recycling that is extensively practiced today. The movement for design for recycling may drive less than optimal material performance for some PVC applications where recycling alternatives are not yet available at the product's end of life. This paper provides a perspective on the current state of PVC recycling and innovative developments underway using advanced techniques that could eventually make plastics recycling agnostic to the type of material being recycled, and potentially improve the circularity of PVC materials in the future.

In order to understand the technical complexity of recycling polyvinyl chloride (PVC) materials, it is important to know more about PVC resins and compounds. According to member surveys by the U.S. Vinyl Institute (VI), over 100 grades of PVC resins are produced by U.S. PVC resin producers. Resin grades are categorized by type (suspension, dispersion, latex) and by resin inherent viscosity (I.V.). Approximately 95% of the PVC resin produced in the U.S. is suspension type (used in extrusion, molding, blowing, and calendaring processes) with the remainder being approximately 5% dispersion type (used in plastisols, coatings, etc.) and a very small amount of latex type (used in coatings). Suspension resin grade I.V. is tailored to meet the needs of the method of processing and end application (Table 30.1). Resin I.V. differences further complicate recycling since I.V. resins that are too dissimilar may not have the proper flow characteristics for the intended application.

Practically all PVC materials are compounded with important additives such as stabilizers, process aids, impact modifiers, pigments, and other additives depending on the processing method and end market application which are numerous (Figure 30.1).

Over 75% of PVC materials are used in building and construction applications which have decades long service lives. When some of these applications such as flooring and siding reach the end of their service lives, they are de-installed and can be considered a PVC-rich stream. Other uses of PVC materials include single use applications such as packaging and medical devices which amount to less than 10% of PVC consumption. According to USEPA estimates for 2018, 1.4 billion pounds of vinyl materials were landfilled in municipal solid waste, which constituted 2.6% of the 53.9 billion pounds of plastics landfilled that year. [2] VI's own study of the composition of curbside collection completed by Titus MRF

Table 30.1 PVC suspension resin inherent viscosity (I.V.) by process method or application.

Resin I.V. (ASTM D5225 or eq.)	Material process or application
0.5 to 0.7	Injection Molding, Blow Molding
0.7 to 0.8	Sheet Extrusion, Rigid Calendaring, Rigid Foam
0.8 to 0.9	Thin Profile and Film Extrusion, Siding, Rigid Calendaring
0.9 to 1.0	Pipe and Profile Extrusion, Blown Film, Flexible Grades, Film Calendaring
>1.0	Thick Wall Extrusion, Flexible Grades, Blown Film, Flexible Calendaring

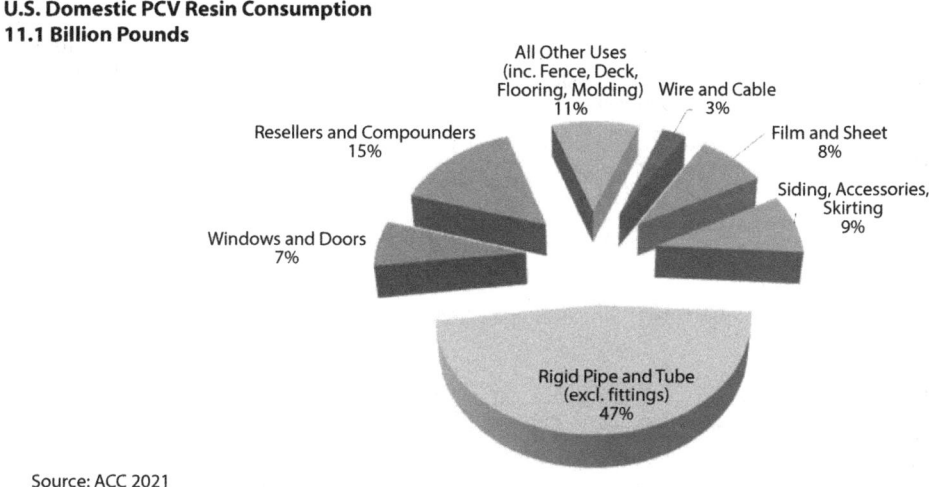

U.S. Domestic PCV Resin Consumption
11.1 Billion Pounds

Source: ACC 2021

Figure 30.1 U.S. and Canada vinyl markets – 2020 [1].

in 2019 measured less than 0.12% PVC materials incoming to a primary material recovery facility (MRF) in Los Angeles, California [3]. PVC items identified by Titus' study included small pipe, fittings, films, blister pack, clamshells, jars and lids, security packaging, gloves, inflatables, hotel/credit cards, wires, IV bags, and others. Mixed plastics in curbside collection can be considered a PVC-lean stream.

30.2   Mechanical PVC Recycling is Robust for Pre-Consumer Materials

As this brief introduction shows, recycling PVC materials is technically complex and requires as much information as possible of the use of the material and its original formulation. This is one of the reasons that pre-consumer recycling of vinyl materials is robust in the U.S. and Canada since these scrap materials are so well characterized at their source. To establish a baseline to determine how much PVC material is being recycled and study trends and capabilities, VI contracted with Tarnell Company who surveyed over 100 recyclers of PVC materials in 2013, 2016 and 2019. [4] Tarnell's survey of 140 recyclers conducted in 2019 shows 1.1 billion pounds of rigid and flexible vinyl recycled with post-consumer volumes growing 40% to 142 million pounds compared with 2013 amounts (Figure 30.2). Vinyl flooring is the application reporting the highest amount of post-consumer recycling. By combining the Tarnell survey with EPA's data and VI estimates, the current overall recycling rate for vinyl in the U.S. can be estimated to be approximately 27% which accounts for recycled volumes, pre-consumer scrap, and VI estimates for C&D landfill amounts (1.1 billion pounds recycled or recovered ÷ (1.1 (Recycled) + 1.4 (MSW)+ 1.5 (C&D) billion pounds available)).

Often, processors who are unable to reincorporate virgin material back into their own products because of process sensitivity or product quality concerns. Thanks to a robust and efficient market that has existed for decades, this preconsumer vinyl material is largely diverted from landfill and recycled into other applications. For that reason, it is important

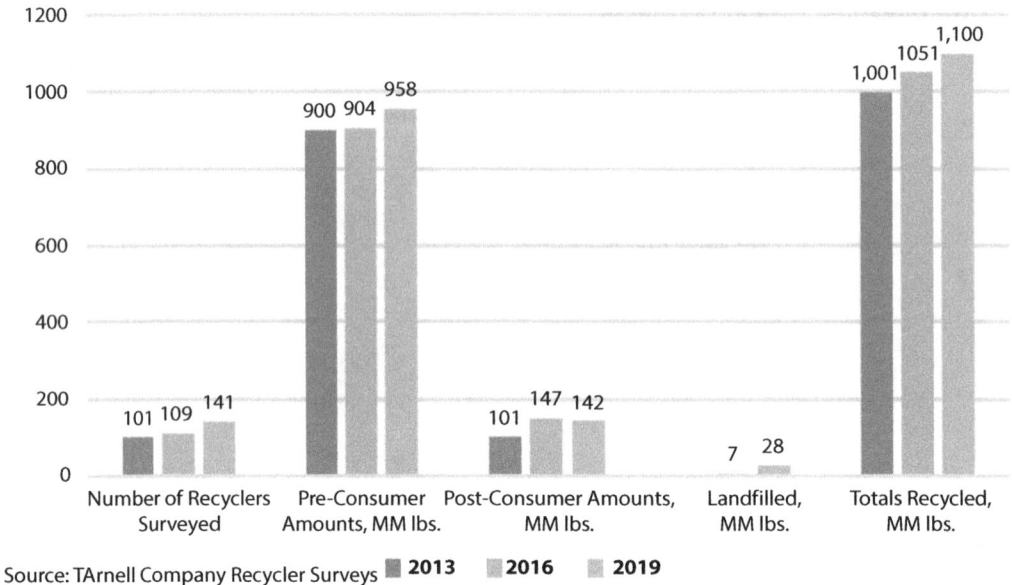

Figure 30.2 VI U.S. & Canada Vinyl Recycling Summary, Tarnell Survey Comparisons for 2013, 2016, and 2019.

to account for these pre-consumer volumes – estimated to amount to 3% to 5% per step for compounding, extruding, cutting, and finishing – which would be destined for a landfill if vinyl recyclers were not reprocessing it into reconstituted compounds that could be specified for use by other vinyl processors. Sources of pre-consumer scrap include: lumps & chunks from compounding and pelletizing, line startups and shutdowns, process color and material changes, machine clean-outs, dust collector loaders, unused/surplus ingredients, and remnants from cutting, machining, and fabricating. Particulates and dusts from these steps often require being melted and pelletized to avoid part defects and adverse impacts to line speeds.

The Tarnell surveys also show that the capabilities of the current PVC recyclers, all of which practice mechanical recycling, have grown consistently with the increasing volumes (Figure 30.3). The data highlight the gaps in front end separation and cleanup and back-end melt filtration and blending, each of which are necessary for handling more contamination that might exist with post-consumer materials.

Pre- and post-consumer materials serve as valuable resources for many widespread vinyl applications such as fence and decking, sea walls, soffits, gutters and downspouts, electrical conduit, and garden hose. In many instances, the material is coextruded with virgin material with the recycle material used in a non-appearance surface or internal fill layer. If the recycle material is clean enough, it can be blended with virgin material and processed. Cutoffs and drops from vinyl window manufacturing are often collected and reprocessed for the aforementioned rigid applications, reducing waste and further enhancing circularity into building materials. Several vinyl flooring and roofing manufacturers have longstanding takeback programs for end-of-life vinyl materials that after reprocessing, are incorporated back into prime flooring and roofing products. Care is taken to assess the composition of the recovered material for incompatible legacy additives, including heavy metal stabilizers and

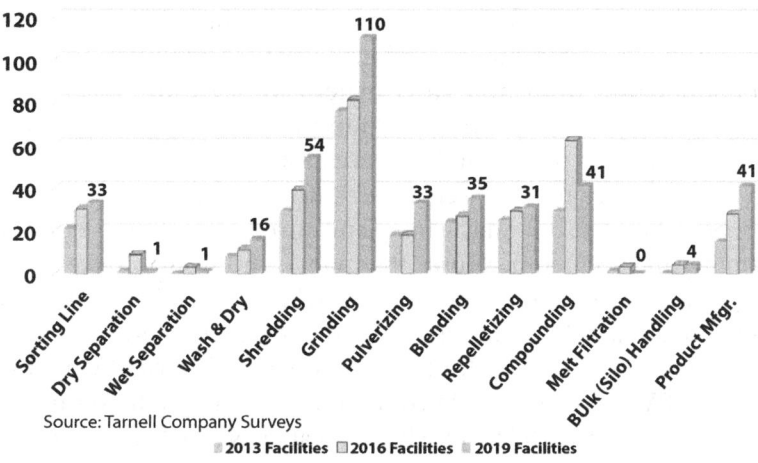

Source: Tarnell Company Surveys

2013 Facilities **2016 Facilities** **2019 Facilities**

Figure 30.3 PVC mechanical recycler reprocessing capabilities growing, 2013 to 2019.

certain plasticizers, which may conflict with the product manufacturer's current policies. The current deficiencies in collection and logistics efficiency are prevalent in the industry for post-consumer vinyl materials, and also pose a similar challenge for the entire plastics industry at this time. For instance, in the case of single membrane PVC roofing, the average installed size is 25,000 square feet but a full truckload would typically require about 80,000 square feet of material. Smaller roof tear-offs must necessarily be consolidated at a network of regional depots for more efficient transport.

30.3   New Focus on Post-Consumer Recycling of PVC Materials

The VI formed the Vinyl Sustainability Council (VSC) in 2016 to address the PVC industry desire to become more sustainable. End of life recovery is a focus area for the VSC and its Resource Efficiency Task Force has set a goal to reach 160 million pounds of post-consumer vinyl recycling (an 18 million pound increase over 2019 amounts) by 2025. VSC is supporting member-driven pilot programs that are currently underway in post-consumer siding, roofing, and medical device applications. In January, 2023, the Vinyl Institute launched a post-consumer recycling grant program (VIABILITY™) which is funded by U.S. vinyl resin manufacturers at $1 million annually over the next three years to provide financial assistance to companies and organizations – up to $500,000 – in support of their post-consumer PVC recycling initiatives.

30.4   Need for Advanced Recycling Technology for Post-Consumer PVC Materials

Augmentation to collection and logistics systems, mechanical recycling and development of advanced recycling are needed for both PVC-rich and PVC-lean streams to successfully increase post-consumer recycling. For PVC-rich streams, there is a growing need to enable more robust processing of contamination and incompatible additives.

Excessive contamination from non-vinyl materials may cause part defects and poor physical properties in vinyl parts and must be minimized. As an example, in the collection of medical devices (fluid bags, tubing, and oxygen masks), non-vinyl clips, attachments, bag fluids, and non-vinyl devices must be separated at the source to create a sufficiently clean stream that can be processed by today's mechanical recyclers. These non-vinyl materials typically have higher melting temperatures than vinyl, and would create non-melt defects. Another issue is how to treat certain end of life vinyl materials that contain additives considered to be incompatible with many manufacturers policies and some regional regulations. Heavy metal stabilizers such as lead, while rarely used in rigid vinyl in the U.S., was more widely used in PVC wire and cable compounds. Lead was discontinued by the early 1990's from rigid PVC and by 2007 from wire and cable. Nevertheless, certain applications using post-consumer recycled content may want to meet voluntary standards that would limit the presence of, or entirely restrict, certain incompatible additives. Advanced recycling technologies hold potential for successfully treating PVC-rich materials that contain excessive levels of contamination or incompatible additives.

Several technologies hold promise for mixed plastics waste streams that include PVC materials to a lesser degree. One of the key factors limiting the traditional high-temperature pyrolysis systems include the insufficient corrosion resistance of process piping. A narrow view commonly held is that PVC must be pre-sorted from the mixed plastics feedstock stream prior to being fed into this type of process. However, this can create excessively high barriers to the selection and availability of feedstock. Instead, the small amount of PVC in mixed plastics streams could be processed using newer technology approaches that would eliminate the barriers to using PVC-lean feedstock. Being feedstock agnostic in this way will provide broader and more robust solutions for these mixed plastic streams.

30.5   Potential Advanced Recycling Technology for PVC-Rich Resource Streams

PVC-rich streams are collectable as identified and fairly concentrated PVC sources. Examples of PVC-rich streams are listed in Table 30.2 along with the potential issues associated with recycling these materials at their end of life.

Also, while pipe is the largest application for PVC, pipe is not a traditional large source for post-consumer vinyl because it is usually left in place if replaced for some reason. In general, the major source of pipe found in any recycle stream would be installation cut-offs and drops which can be a suitable feedstock for mechanical recyclers.

Several advanced recycling technologies that hold promise for PVC-rich streams are discussed below.

30.5.1   Coupling and Compatibilizer Agents

Use of neoalkoxy titanate in combination with Al_2SiO_5 mixed metal catalyst in powder & pellet forms for in-situ macromolecular repolymerization and copolymerization in

Table 30.2 Examples of potential recycling issues associated with PVC-rich streams.

PVC-rich stream	Types of PVC	Potential contamination issues
Siding	Rigid	Metal fasteners, wood, insulation, adhesives, construction debris
Windows	Rigid	Metal hardware and fasteners, glass, wood, flexible PVC or elastomer seals, nonvinyl plastic parts, incompatible stabilizers
Flooring	Rigid & Flex	Non-vinyl plastics, adhesives, incompatible plasticizers, reinforcing scrims
Wire & Cable	Flex	Non-vinyl plastics, metal wire, incompatible plasticizers, incompatible stabilizers
Medical bags, tubing	Flex	Non-vinyl plastics, incompatible plasticizers, bag fluids
Roofing	Flex	Reinforcing fabric, non-vinyl plastics, metal fasteners, incompatible plasticizers, adhesives, insulation
Wallcovering	Flex	Reinforcing fabric, non-vinyl plastics, metal fasteners, adhesives, drywall and paper
Vinyl inflatables	Flex	Electrical wiring and air pumps, cardboard boxes, non-vinyl plastics
Vinyl records	Rigid Copolymer	Paper labels and sleeves, nonvinyl plastics, incompatible stabilizers

the extrusion compounding melt step have shown promise in laboratory experiments by Kenrich Chemicals to couple and compatibilize PVC with certain non-PVC plastics. A high degree of mixing and dispersion of the neoalkoxy titanate proton is essential for coordinating with inorganic fillers and organic particulates to couple/compatibilize the dissimilar interfaces at the nano-atomic level. This technology could reduce the need for expensive sorting of mixed plastic materials in recycling operations. [5] Experiments performed by Kenrich Chemicals on the use of zirconate catalysts combined with PVC recyclates demonstrated that processing temperatures can be reduced by 24% and line rates increased by nearly a factor of two, thereby mitigating the potential for thermal degradation of recycled materials that may be deficient of full stabilizer levels. [6] This technology could be suitable for PVC-rich streams with mixed plastics where metal, wood, glass, and other non-plastic contaminants have already been removed.

30.5.2 Catalytic Decomposition

Oregon State University researchers completed experiments sponsored by the VI to study novel use of catalysis to decompose PVC plastics into hydrogen chloride and HDPE wax byproducts at low temperatures (Figure 30.4).

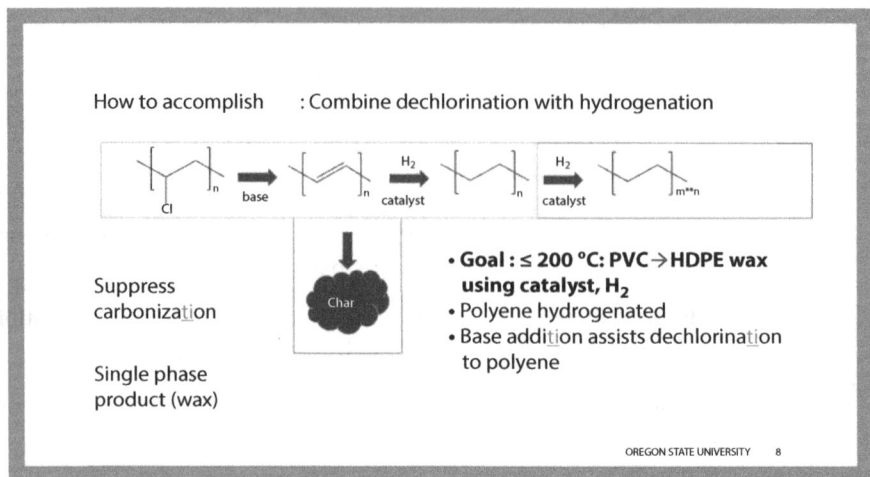

Figure 30.4 Catalytic decomposition schematic of PVC materials (Courtesy of Oregon State University).

Implementation of certain amine bases at 20% loading and sufficient catalyst loadings of 1% have shown evidence of dechlorination greater than 99 % and polyethylene wax yields of over 80 %. The product waxes melt and recrystallize between 60 and 90 °C, which is in the melt temperature range of hot melt adhesive waxes. Product wax can be successfully extracted from the reaction mixture with an organic solvent which enables direct reuse of the wax. [7] The OSU researcher's patent should publish this year. [8] This technology could be suitable for PVC-rich streams where incompatible additives are present, or also mixed plastics, and where metal, wood, glass, and other non-plastic contaminants have already been removed.

30.5.3   Microwave-Assisted Selective Decomposition

Microwave Solutions has demonstrated with high energy microwaves to selectively decompose plastic materials including PVC. In pilot system samples, Microwave Solutions demonstrated that by modulated the microwave energy levels, different selective depolymerization steps including the dehydro-dechlorination of PVC materials can be accomplished to obtain high yields of hydrochloric acid in first process step while subsequently generating other selective hydrocarbon products/molecules including benzene with minimal chlorination of the organic constituents (Figure 30.5). Microwaves provide targeted and precise energy deliver in a low frequency system under vacuum in up to 16 different phases independently controlled and freely combined, targeting individual monomers/polymers with the length of each combined zone matched to the chemical reaction needed. This technology could be suitable for PVC-rich streams where incompatible additives are present, or also mixed plastics, and where metal, wood, glass, and other non-plastic contaminants have already been removed. Microwave Solutions offers its Greentech stack of modular solutions that include multiphase microwave pyrolysis, (pulsed) microwave thermal plasma, (pulsed) microwave hot plasma, and different backend solutions including distillation, each of which can be engineered, individually or in combination, for the specific collected post-consumer plastic material streams [9].

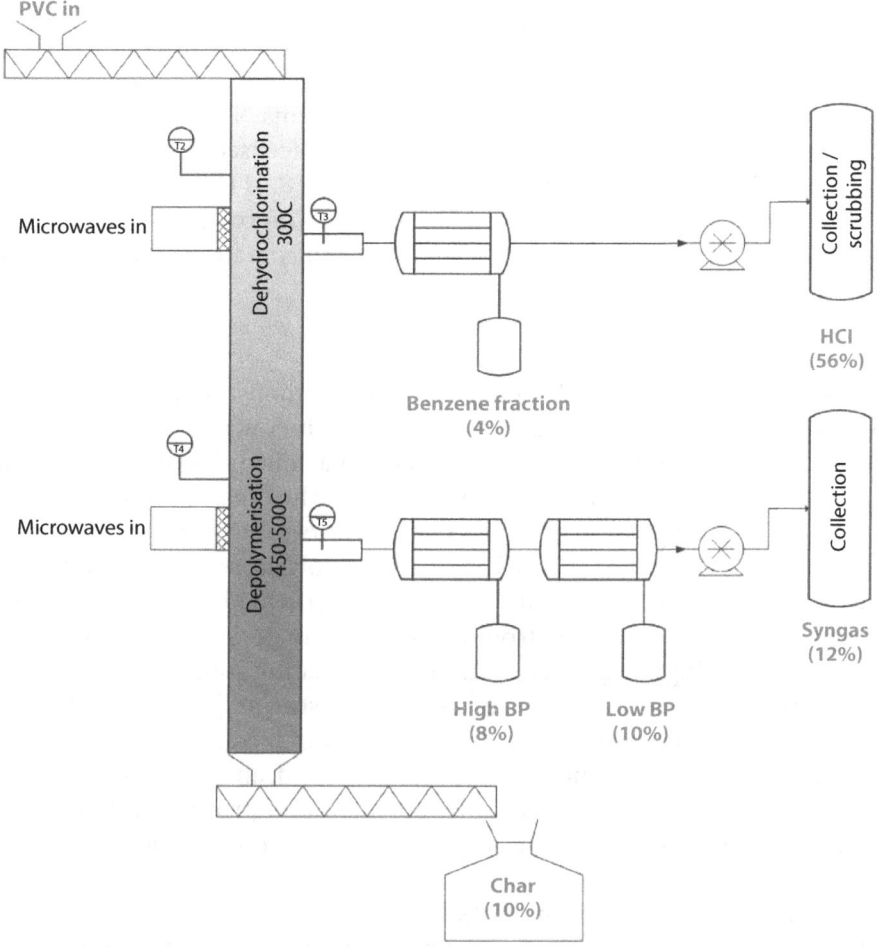

Figure 30.5 Process flow and yields for pure PVC material without filler using microwave pyrolysis (credit: Microwave Solutions GmbH) [10].

30.5.4   Selective Solvent Dissolution Techniques

Three technologies are under development in Europe that employ aggressive selective solvent dissolution in some form with precipitation of the PVC material to recover the recyclate PVC. These methods are also called physical recycling since they can selectively reproduce portions of a composite material thereby separating one plastic from another and even selectively separating compound ingredients if so intended.

Polyloop – France, the STRAP methodology
The general principle underlying the STRAP process (extraction and precipitation by targeting solvents) is: Selectively dissolve a single polymer in a multi-material composite in a solvent system in which the polymer is soluble, but the other components are not. The solubilized polymer is then separated from the composite by mechanical filtration and precipitated by changing the temperature with steam to remove the solvent and adding a co-solvent (an anti-solvent such as water for PVC) that makes the dissolved polymer insoluble.

The distilled solvent is reused in this process and the targeted polymer is recovered as a dry solid. Polyloop technology offers a STRAP process by 300 kg. batch, in a modular unit that can be located and operated at the source or collection point of the scrap materials, targeting only size reduced PVC and its formulation components. The solvents which are used are recovered and reused in a closed loop. Polyloop has licensed the Texyloop technology that operated at a Solvay facility in Italy from 1998 to 2018. [11] This technology could be suitable for PVC-rich streams where mixed plastics, or reinforcing composite materials are present, and where metal, wood, and other non-plastic contaminants have already been removed.

KemOne - France

Dissolution of PVC in a new generation solvent was demonstrated by KemOne on both rigid and flexible PVC materials including end-of-life window profiles and wire and cable, both of which contain lead in Europe. Addition of an acidic or basic aqueous solution for the mineralization of the heavy metals separates it from the organic phase to less than 0.1% in the recycled PVC. Precipitation of the PVC from the organic phase by addition of a second solvent is used to break the PVC solubility. After solvent recovery and drying, typical PVC with suspension grade granularity is achieved. KemOne has also demonstrated that recovery of up to 80% of plasticizers from flexible PVC materials is possible with pressurized supercritical CO_2 liquid extraction, thereby eliminating some potential incompatible additives from the post-consumer recycle flexible vinyl stream. [12] This technology could be suitable for PVC-rich streams where incompatible additives, mixed plastics, or reinforcing composite materials are present, and where metal, wood, and other non-plastic contaminants have already been removed. This method could also prove especially useful where selective capture of the heavy metal can provide added economic value.

CreaSolv® Process and Fraunhofer Institute, Germany

A consortium of European flooring manufacturers, resin, and additive suppliers obtained ~€5 million in grant funding from the European Commission to demonstrate viability of new CreaSolv® Process technologies for recycling vinyl flooring materials (Figure 30.6). The commissioning of a pilot plant will start in the last quarter of 2022 and production of recycled PVC will be demonstrated in the first quarter of 2023. CreaSolv® is a registered trademark of CreaCycle GmbH, Grevenbroich, Germany. From their report to the EC:

Post-consumer flooring is subjected to the CreaSolv Process, which dissolves PVC from the material mix and eliminates undissolved matter and co-dissolved plasticizers in an extractive purification step. PVC is recovered from the solution via precipitation. The PVC is then reincorporated into new products. Using a controlled catalytic [coupled transesterification-hydrogenation] reaction, extracted [regulated low molecular weight phthalate] plasticizers will be converted [at nearly 99%] to harmless compounds [analogues of diisononyl 1,2cyclohexanedicarboxylate] with plasticizing properties. [13] Together with a tailor-made additives composition, both recovered products are integrated into novel PVC flooring designed for circularity [14].

This technology could be suitable for PVC-rich streams where incompatible additives, mixed plastics, or reinforcing composite materials are present, and where non-plastic contaminants have already been removed.

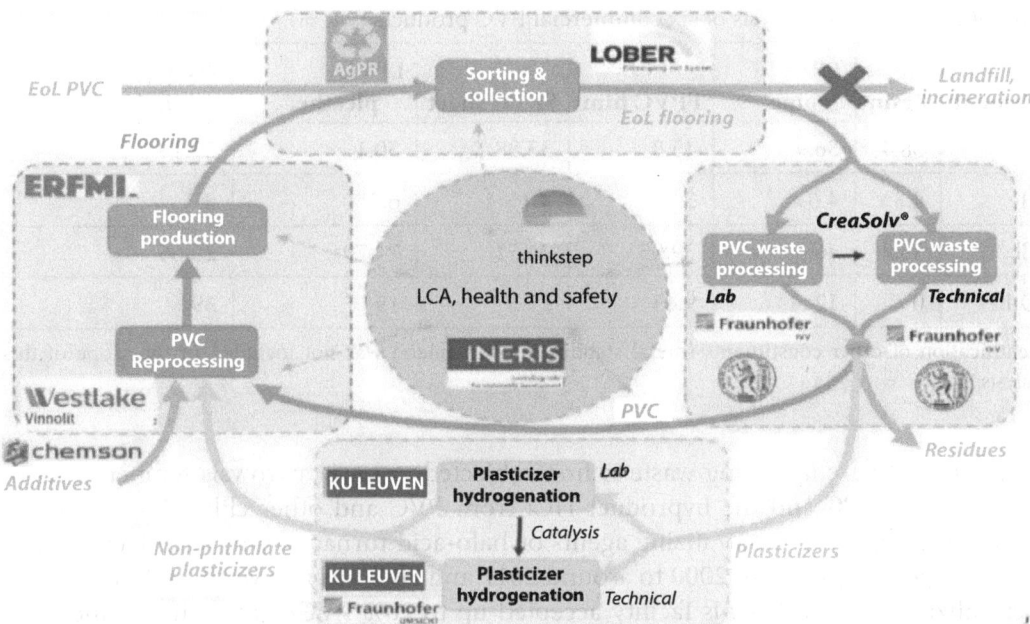

Figure 30.6 Circular flooring consortium project schematic using the Creasolv process [15].

30.6 Potential Advanced Recycling Technology for PVC-Lean Resource Streams

PVC-lean streams collected and consolidated especially from curbside recyclables, and can be identified and sorted with near infrared spectroscopy, and are fairly diluted PVC sources. Examples of PVC-lean streams are listed in Table 30.3 along with the potential issues associated with recycling these materials at their end of life. The challenge becomes how to treat the whole recycled stream and plastics agnostic technological solutions may be needed.

Significant progress is being made with thermal decomposition and gasification of non-PVC plastics from curbside collections and according to the American Chemistry Council, some $5 billion in investment is underway by major plastics producers and recyclers in these technologies. At this time, these facilities restrict the amount of incoming chlorides to between 2% and 5% because at the high temperatures used for pyrolysis of the plastic material (typically 800 °C), chlorides in the presence of metal can chlorinate the some of the hydrocarbon byproducts, which is undesirable for this process. But technology can be

Table 30.3 Examples of potential recycling issues associated with PVC-lean streams.

PVC-lean stream	Types of PVC	Potential contamination issues
Curbside Recycling Collection	Rigid & Flex	Non-vinyl plastics, food residue, solvents, metal, glass, paper, trash
Apparel/Footwear	Rigid & Flex	Non-vinyl plastics, metal, rubber, composite reinforcement fabric

Table 30.4 Elemental analysis of six commercial PVC products (wt. %).

Element	Resin	TPVC pipe	FPVC film	SPVC sheet	LPVC pleather	WPVC window blinds
C	38.4	36.9	45.8	35.88	50.1	29.19
H	4.8	4.6	5.72	4.15	6.3	3.3
Cl	56.8	46.5	39.05	34.51	24.79	28.66
Other[1]	0	12	9.43	25.5	19	39

[1]Identification of other constituents (metal stabilizers, inert fillers) was not included in the scope of this analysis.

implemented where incoming waste is first subjected to lower pyrolysis temperatures on the order of 350 °C and the byproduct HCl from PVC and other chlorides present are recaptured using either neutralizing agents or halo-acid furnaces. One such facility operated in Sapporo, Japan from 2000 to around 2017 and HCl was recovered and marketed as a neutralizing agent. [16] This facility accepted up to 20% PVC content in the incoming plastics waste stream. Agilyx has also done work on how to modify their pyrolysis technology to include and convert PVC materials as effectively as possible. Studies have shown the temperatures at which PVC will dehydro-dechlorinate will vary depending on residence time and the type of gas media in the pyrolysis chamber, but the amount of HCl byproduct depends on the compound and end use because additive levels will vary which correspondingly changes the chloride content in the material (Table 30.4) [17].

The Consortium for Waste Circularity, which is working closely with Professor Bruce Welt at the University of Florida, is proposing to commercialize a Regenerative Robust Gasification facility co-located at a landfill in Florida that will accept all unrecycled mixed organic wastes intended for municipal landfills. Professor Welt has been running Robust Gasification trials for packaging industry stakeholders with difficult-to-recycle materials.

Regenerative Robust Gasification converts mixed organic waste to syngas and then into methanol (Eco- Methanol™), which is a primary feedstock building block for subsequent manufacture of new products, plastics and packaging. Professor Welt is specifically seeking to work with more stakeholders from the vinyl industry because he believes that the vinyl component of waste may enhance the overall recycling process by increasing the amount of Eco-Methanol produced per ton of waste processed. The Consortium for Waste Circularity is supporting research on enhancement of Regenerative Robust Gasification process performance on vinyl containing waste. [18] Vinyl materials can be handled in this process, and laboratory and pilot scale research is being planned to study its viability for PVC contained in municipal solid waste and destined for landfills.

30.7   Circularity is Achievable

To summarize, several advanced technologies are in various stages of development across the globe and are applicable to many PVC post-consumer streams that will promote circularity into new PVC products and keep PVC materials out of landfills (Table 30.5). Several thermochemical conversion processes such as catalytical pyrolysis, steam reforming of plastic pyrolysis volatiles, gasification, hydrothermal processes, and chemolysis have been critically evaluated as promising approaches to upcycle plastic wastes including some amounts of PVC. [19] Some technologies are available in modular format to site close to the source of the collected or available post-consumer vinyl scrap, while others must be sited at destinations such as landfills or large chemical complexes that may utilize the converted products. In addition to the technologies highlighted herein, there are several more technologies in initial development or consideration in both industry and academia. For example, University of Michigan researchers completed lab studies on paired-electrolysis reaction in which HCl is intentionally generated from PVC to chlorinate arenes in an air- and moisture-tolerant process that is mediated by plasticizer liberated from flexible PVC materials. [20] Research at another university is examining where scrap plastics including PVC can be mechanically incorporated after size reduction into concrete replacement polymers that could have decades of use. Researchers at University of Massachusetts Lowell included literature test results of hydrothermal liquefaction (HTL) as another promising method to handle halogen containing polymers, which may be present in the un-sorted real

Table 30.5 Summary of advanced recycling technologies application to post-consumer PVC.

Technology	Description	Applicability for end-of-life PVC
Coupling and Compatibilizer Agents	Use of titanate and zirconate chemistry to activate bonding between dissimilar molecules	PVC-rich, PVC-lean streams
Catalytic Decomposition	Low temperature catalytic and chemical decomposition and hydrogenation	PVC-rich streams
Microwave Assisted Selective Decomposition	Pulsed microwave energy to target selective depolymerization	PVC-rich, PVC-lean streams
Solvent Dissolution	Selective solvent dissolution and precipitation of target plastic, including extraction of plasticizers and selective stabilizer mineralization removal.	PVC-rich streams
Pyrolysis	Controlled multi-stage temperature decomposition	PVC-lean streams
Regenerative Robust Gasification	High temperature plasma gasification to produce syngas	PVC-lean streams
Hydrothermal Liquification	Uses sub-/super-critical water as a reaction media to thermochemically convert plastics	PVC-lean streams

world plastic waste. High dehalogenation efficiency is believed to occur from the hydroxyl-rich environment used in HTL. [21] Also, HTL is usually carried out at moderate temperatures (280–400°C) and pressures (7–30 MPa), which facilitates recapture of HCl from rapid dehydro-dechlorination of PVC containing materials [22].

30.8 Conclusions and Recommendations

Global advances are being made in PVC material recycling. Achieving greater vinyl recycling and circularity will be enhanced by global technology transfers and with global partnerships. Mass balance accounting will assure credit is properly reported for recycle content derived from these advanced recycling techniques. Plastics agnostic advanced recycling solutions are developing and hold the potential to provide greater circularity of plastics including PVC and recapture of plastic compositions as an alternative to landfills. Capital funding of these large and complex mechanical and advanced recycling facilities is imperative but must be based on sound and certain recycling technologies. Broader collection and more efficient transport of these recoverable materials in both PVC-rich and PVC-lean formats must accompany the development of these technologies to maximize their positive impacts.

Acknowledgements

The authors wish to thank REMADE Institute for the opportunity to present this paper at their Circular Economy Technology Summit & Conference, March 20 – 21, 2023, National Academy of Science Building, Washington, D.C. Acknowledgement is also given to the Vinyl Institute and its members for supporting the work to prepare this manuscript, the researchers who provided data and images, and any other reviewers or contributors.

References

1. American Chemistry Council Resin Report 2021
2. USEPA, (Nov. 2020): Advancing Sustainable Materials Management: Facts and Figures Report, 2018, https://www.epa.gov/facts-and-figures-about-materials-waste-and-recycling/advancing-sustainable-materialsmanagement
3. Farling, S., Determination of Post-Consumer Vinyl Content at a California MRF, *Titus MRF Services Report to Vinyl Institute,* Nov. 25, 2020
4. Krock, R., Tarnell, S., Recycling as a Sustainability Practice in the North American Vinyl Industry, SPE ANTEC™ Orlando Proceedings, pp. 2524-2530, Available at https://www.vinylinfo.org/resources?_sft_resources_cat=recycling March, 2015
5. Monte, S., Advanced Solutions in Vinyl Recycling with Titanate Catalysts/Coupling Agents, p.7, available at https://vantagevinyl.com/wp-content/uploads/2021/07/Monte_Kenrich.pdf, July 20, 2021.
6. Ibid, pp. 37-38
7. Svadlenak, S., *et al.*, PVC Conversion to Linear Wax, Oregon State University Presentation to Vinyl Sustainability Council, November, 2022

8. Svadlenak, S., *et al.*, METHOD FOR RECYCLING POLYVINYL CHLORIDE (PVC) TO PRODUCE HIGH DENSITY POLYETHYLENE (HDPE), U.S. Patent Application No. 17/861,454. Filed July 11, 2022.

9. Turk, F., Microwave Solutions GmbH Presentation to Vinyl Sustainability Council Recycling Summit, July 26, 2021

10. Stepala, A., Microwave Solutions GmbH Presentation to Vinyl Sustainability Council Recycling Summit, June 18, 2020

11. PolyLoop Composite Recycling Technology, https://polyloop.fr/strap-recycling/process/?lang=en, Jan., 2023

12. Morel, P., Laurent, M. Update on PVC Recycling Technologies in Europe. Case Study: Extraction of Heavy Metals from Post-consumer PVC Through Selective Dissolution, KemOne presentation at SPE Vinyltec, September 28, 2022

13. Windells, S., *et al.*, Catalytic upcycling of PVC waste-derived phthalate esters into safe, hydrogenated plasticizers, *Green chemistry* 2022 v.24 no.2 pp. 754-766

14. Fraunhofer Institute, Circular Flooring Grant agreement ID: 821366, New products from waste PVC flooring and safe end-of-life treatment of plasticisers, Report No. 2 to European Commission, available at: https://cordis.europa.eu/project/id/821366/reporting, November 15, 2022

15. Winter, A., Mieden, O., PVC Sustainability in Practice, SPE Vinyltec, Sept. 28, 2022

16. Fukushima, M., Shioya, M., Wakai, K. *et al.*, Toward maximizing the recycling rate in a Sapporo waste plastics liquefaction plant. *J Mater Cycles Waste Manag* 11, 11–18 (2009). https://doi.org/10.1007/s10163-008-0212-6

16. Zheng, Xue-Gang, *et al.*, "Dehydrochlorination of PVC Materials at High Temperatures", *Energy and Fuels*, 17, 896-900, 2003

18. Thomas, A., Gasification Promises Solution to Landfill Crisis, Labels and Labeling, Jan 3, 2023

19. Yang, R.-X., *et al.*, Thermochemical Conversion of Plastic Waste into Fuels, Chemicals, and Value-Added Materials: A Critical Review and Outlooks, *ChemSusChem 2022*, e202200171, P. 34

20. Fagnani, D.E., Kim, D., Camarero, S.I. *et al.*, Using waste poly(vinyl chloride) to synthesize chloroarenes by plasticizer-mediated electro(de)chlorination. *Nat. Chem.* (2022). https://doi.org/10.1038/s41557-022-01078-w

21. Lu, T., *et al.*, Hydrothermal liquefaction of pretreated polyethylene-based ocean-bound plastic waste in supercritical water, *Journal of the Energy Institute* 105 (2022) P. 290

22. Seay, J. R., *et al.*, Waste Plastics: *Challenges and Oportunities for the Chemical Industry*, CEP, Nov. 2020

31

Dynamic Crosslinking for EVA Recycling

**Kimberly Miller McLoughlin[1]*, Alireza Bandegi[2], Jayme Kennedy[1], Amin Jamei Oskouei[2],
Sarah Mitchell[1], Michelle K. Sing[1], Thomas Gray[3] and Ica Manas-Zloczower[2]**

[1]Braskem America, Pittsburgh, PA, USA
*[2]Department of Macromolecular Science and Engineering, Case Western Reserve University,
Cleveland, OH, USA*
[3]Department of Chemistry, Case Western Reserve University, Cleveland, OH, USA

Abstract

Covalently crosslinked polyethylene vinyl acetate (EVA) networks provide a balance of processing performance, mechanical properties, and durability that make these materials the ideal selection for some high-performance, high-volume applications. Foams produced from crosslinked EVA are the material of choice for shoe midsoles, including technically advanced athletic shoes. However, the same characteristics that make permanent networks such excellent candidates in materials selection processes also represent a difficult environmental challenge. Once formed, these network structures do not melt, flow, or dissolve to enable the use of conventional reprocessing or recycling methods. As a result, most crosslinked polymer networks accumulate as plastic waste.

The most significant technical barrier to recycling crosslinked polymer networks is the permanent nature of the covalent bonds that hold them together. In this paper, we demonstrate that covalently crosslinked networks can be converted to dynamic exchangeable networks called vitrimers. These vitrimer networks can undergo topological rearrangements that enable melt re-processing. Crosslinked EVA is converted to a vitrimer via a mechanochemical approach using a catalyst that promotes a transesterification reaction. The vitrimerized EVA can then be reprocessed at moderately low temperatures via compression molding without loss in the mechanical properties. Thus, the vitrimerized particles can be recycled as secondary feedstock for new, value-added products.

Keywords: Ethylene vinyl acetate polymer, vitrimer, dynamic crosslinking, exchangeable networks, recycling

*Corresponding author: kimberly.mcloughlin@braskem.com

Nabil Nasr (ed.) Technology Innovation for the Circular Economy: Recycling, Remanufacturing, Design, Systems Analysis and Logistics, (395–406) © 2024 Scrivener Publishing LLC

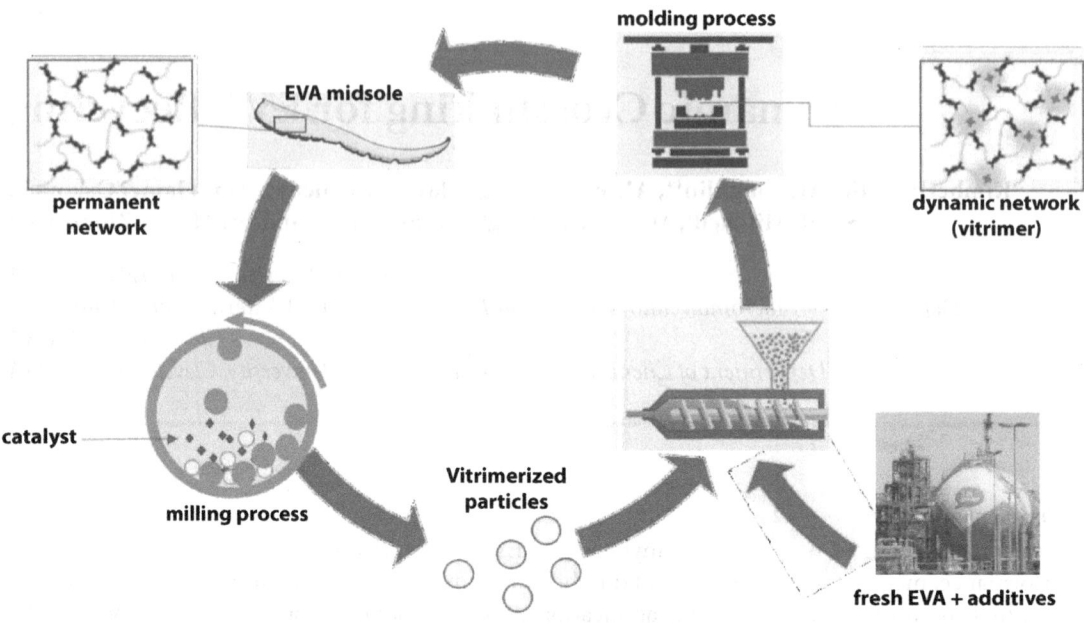

31.1   Introduction

Footwear is one of the largest markets for permanently crosslinked polymers, especially crosslinked EVA. Worldwide production of footwear reached 24.2 billion pairs of shoes in 2018, with a relative growth of 20.5% relative to 2010 [1]. Crosslinked EVA foams are the material of choice for footwear midsoles because they provide an excellent balance of mechanical properties and light weight. However, manufacturing of foamed EVA midsoles leads to significant waste accumulation. The molding processes to produce footwear midsoles generate a high level of scrap materials. For example, compression molding generates about 30 wt% scrap, and injection molding, which is a more efficient process, still generates about 10 wt% scrap. Because the scrap is difficult to reprocess, a relatively small fraction (less than 15 wt%) is reintroduced as secondary feedstock into midsole formulations. Because the capacity of these re-processing methods does not match the generation rate, a large quantity of scrap must be stored and disposed. Solutions for crosslinked EVA foam waste include downcycling to lower value applications such as surfacing of roads and playgrounds or use in acoustic insulation. Another less desirable approach is energy recovery through incineration, gasification and pyrolysis. However, these options are not only time consuming and expensive but also labor and energy intensive. In addition to presenting high environmental impacts, these solutions do not address the core points of circular economy and much crosslinked EVA foam ultimately ends up in landfills [2].

Besides footwear, other manufacturing sectors that generate crosslinked EVA incur the same problem. Examples include the production of hoses and gaskets for automotive and civil construction and wires and cables for electrical insulation. In these cases, both manufacturing scrap and end-of-life waste contribute large volumes to environmental issue waste accumulation.

31.2 Vitrimer Technology

Dynamic crosslinking technologies present a promising solution to enable reprocessing of crosslinked EVA industrial scraps to replace primary feedstock and to facilitate the recycling of end-of-life EVA products. Dynamic crosslinked systems, specifically vitrimers, are a new class of chemically cross-linked polymers in which an external- stimulus such as temperature, stress, or pH initiates exchange reactions. These reactions result in topological network rearrangements while keeping the net number of bonds and crosslinks effectively constant [3]. As a result, the material relaxes and flows in response to stress [4]. Vitrimers exhibit the characteristics of crosslinked materials at ambient temperatures including high chemical resistance and moduli. At elevated temperatures, vitrimers can be processed or reprocessed much like traditional thermoplastic materials [5–7]. The ability to flow allows vitrimers to be reprocessed and recycled.

Much of the previous literature has been focused on vitrimerization of thermoset epoxy networks. Much like crosslinked EVA, the permanent crosslinks inherent to a thermoset network cannot be reshaped, processed or recycled [8]. The introduction of exchangeable bonds in thermosets has demonstrated that it is possible to revert the permanent crosslinking and reach properties that allow their recyclability [9–12]. Vitrimerization of thermoset epoxy networks was demonstrated at lab scale using catalysts vectored into existing permanent networks and converting them into vitrimers. By processing in a planetary ball mill with catalytic zinc acetate, anhydride-cured epoxy was converted into a vitrimer epoxy with 100% yield allowing recycling and reprocessing [13, 14]. The action of ball milling cleaves the relatively weak dative bonds between zinc and oxygen atoms within the network. These transient, low-coordinate zinc sites are activated toward catalytic transesterification. Then, bond exchange reactions and welding of the vitrimerized thermoset are initiated across the crosslink sites under heat and pressure. The application of vitrimer technology to other crosslinked networks requires careful consideration regarding the choice of dynamic crosslinker and any necessary catalyst.

Previous work has demonstrated that not only is it possible to vitrimerize polyolefin materials, but it is possible to incorporate dynamic crosslinking in EVA [15, 16]. Dynamic crosslinking of EVA was demonstrated using triethylborate with the catalyst (bis(acetylacetonato)dioxomolybdenum(VI)). By introducing triethylborate into the system, each ethoxy group replaces one ester group of vinyl acetate and then forms a new boron-centered three-dimensional (3D) crosslinking network using bis(acetylacetonato)dioxo-molybdenum(VI) as catalyst. The reaction is thermally triggered and forms dynamic B-O bonds that enable recycling of the EVA vitrimer at high temperature. The EVA vitrimers could be reprocessed multiple times at high temperature by compounding, extrusion, and compression molding. However, even with a catalyst the triethylborate mechanism of vitrimerization required prohibitively long reaction times that decrease its commercial relevance.

Despite the promising foundations provided by previous research, the application of dynamic crosslinking to foamed EVA scrap requires innovation in both catalyst and process technologies. To date, the vitrimerization of previously crosslinked elastomeric networks such as foamed EVA has not been demonstrated. There is a need to identify a commercially relevant catalyst chemistry to convert permanently crosslinked EVA to vitrimers. The process itself must be fine-tuned to complete EVA vitrimerization reactions within time and temperature limits of commercially relevant processes such as milling and extrusion. The processes themselves must not be prohibitive from a resource perspective such that it cancels out the benefits of recycling EVA scrap through excess energy or raw material use. Formulations must be developed which utilize vitrimerized EVA scrap along with other polymers and additives to provide adequate flow properties to enable molding while realizing the mechanical properties necessary for the application. Therefore, the development of a process to generate an EVA vitrimer from previously crosslinked EVA foamed scrap provides a unique opportunity to demonstrate an enabling technology for recycling.

31.3   Objective

In this chapter, a commercially available EVA copolymer is crosslinked by reaction with dicumyl peroxide. The crosslinking reaction is confirmed using rheology experiments. The crosslinked EVA is then converted to a vitrimer in a cryogenic milling process with a transesterification (TE) catalyst. The vitrimerized EVA is compression molded to demonstrate that the vitrimerized EVA is fully melt processable. Shear relaxation and dynamic mechanical measurements on the molded parts confirm that a vitrimer has been produced. In addition, a crosslinked EVA foam is produced by reacting the commercially available EVA with dicumyl peroxide (DCP) plus a chemical foaming agent. The resulting EVA foam is converted to a vitrimer using the same cryogenic milling process. This paper demonstrates a process and catalyst capable of converting permanently crosslinked, foamed EVA scrap to secondary feedstock using conventional processes for scrap grinding. Ultimately, this vitrimerization approach to recycling can be applied to a wide variety of thermoset polymers and vulcanized rubbers to have a significant impact on increasing recycling across multiple industries.

31.3.1   Experimental Materials

The ethylene-vinyl acetate (EVA) copolymer used for the experiments reported here is a commercially available grade supplied by Braskem. The supplier data sheet reports a vinyl acetate of 19 weight percent, a melt flow index of 2.1 dg/min (190 °C, 2.16 kg), and a peak melting temperature of 86 °C.

DCP was used as received from Sigma-Aldrich as an initiator for EVA crosslinking reactions. Azodicarbonamide (ADCA) obtained from Sigma Aldrich was used as a chemical foaming agent to produce EVA foams. Zinc stearate was added to the foam formulation as a co-agent to reduce the decomposition temperature of the ADCA. The specific chemistry used for vitrimerization is confidential; a patent is pending.

31.3.2   Methods

Synthesis of Crosslinked EVA Rubber

Covalently crosslinked EVA rubber was produced by compression molding EVA pellets with a peroxide crosslinking agent in a hydraulic press. EVA pellets were first melt-mixed with 2 wt% DCP in a micro compounder at a low temperature to avoid peroxide decomposition. The mixture was extruded through a die, cooled, and then chopped into pieces about 5 mm in length. The pieces of EVA/peroxide compound were placed in a mold with diameter 25mm and thickness 1 mm. The sample was compression molded in a Carver platen press at a temperature that enabled the peroxide-initiated crosslinking reaction to occur.

Synthesis of Crosslinked EVA Foam

Covalently crosslinked EVA foam was produced by compression molding EVA pellets with a peroxide crosslinking plus a chemical foaming agent in a hydraulic press. EVA pellets were melt-mixed with 1.5 wt% DCP, 5 wt% azodicarbonamide, and 5 wt% zinc stearate in a micro compounder at a low temperature to avoid reaction of the peroxide or the foaming agent. The mixture was extruded through a die, cooled, and then chopped into pieces about 5 mm in length. The pieces of EVA/peroxide compound were placed in a mold. The sample was compression molded in a Carver platen press at a temperature that facilitated the reactions of the crosslinking agent and the foaming agent.

Vitrimerization of Crosslinked EVA Rubber and Foam

Crosslinked EVA rubber and foam samples were converted to vitrimerized EVA samples in a cryogenic milling process. Crosslinked EVA rubber or EVA foam was ground in a SPEX freezer mill cooled with liquid nitrogen to provide particles with 1 mm diameter. The particles were then dry mixed with a metal centered transesterification catalyst plus an alcohol. The mixture was ground in a Retsch cryogenic ball mill to provide particles with diameters in the range of 50-300 microns. The finely ground particles were transferred to a mold and compression molded in a Carver hydraulic platen press.

31.4 Results

31.4.1 Crosslinking of EVA

To confirm that the EVA starting material was crosslinked by reaction with DCP, a dynamic mechanical analysis was performed. Elastic storage modulus is reported as a function of temperature in Figure 31.1a. The plateau modulus at temperatures above the melting transition confirms that a crosslinked, elastomeric network has been formed.

The elastic storage modulus of the DCP-crosslinked EVA does not exhibit significant time dependence, even at high temperature. As shown in Figure 31.1b, the crosslinked EVA retained approximately 70% of its initial storage modulus after 10,000 seconds in a stress relaxation experiment conducted at 175 °C, which is approximately 100 °C higher than the peak melting temperature. This test provides a clear demonstration that crosslinked EVA does not flow, which is the key reason that it cannot be processed effectively by commercially relevant melt processing methods such as extrusion or molding.

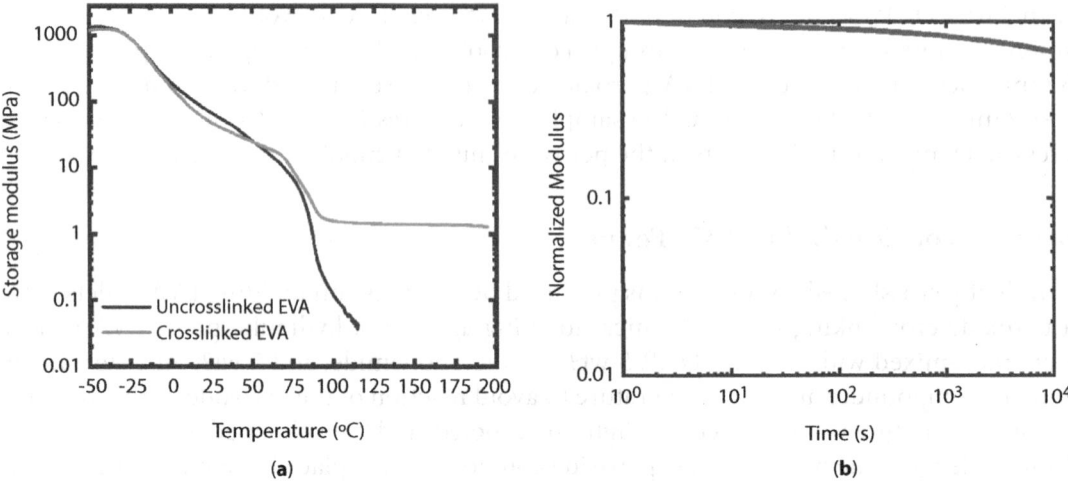

Figure 31.1 (a) Dynamic mechanical analysis of DCP- crosslinked EVA shows high-T plateau in elastic modulus, confirming the formation of an elastic network. (b) Stress relaxation measured by shear rheology at 175 °C demonstrates that elastic modulus of DCP-crosslinked EVA is essentially time independent.

31.4.2   Vitrimerization of Crosslinked EVA

The crosslinked EVA rubbers were converted to EVA vitrimers in a cryogenic ball milling process as shown in Figure 31.2. Crosslinked EVA rubber samples were first ground to particles with 1 mm average diameter. They were then cryoground in a second step in the presence of a transesterification catalyst plus an alcohol. The resulting particles were pale yellow and had diameter in the range of 50-300 microns. Upon heating to compression mold in a platen press, the particles melted to form homogeneous, orange, slightly translucent samples with rubbery characteristics. The rubbery vitrimer samples bend and stretch easily without cracking or breaking. This elastic behavior was quantified using dynamic mechanical analysis and shear rheology tests.

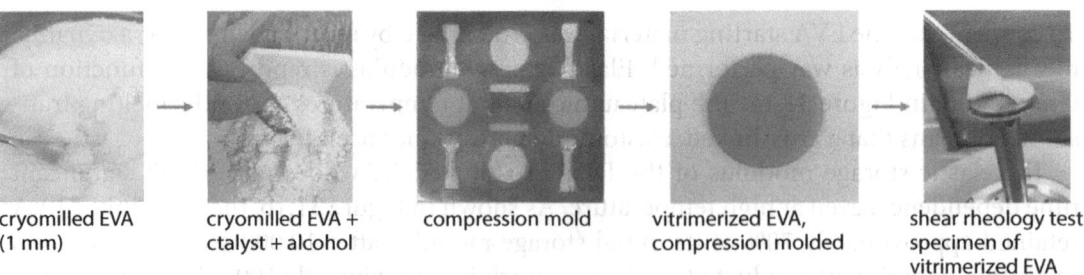

cryomilled EVA
(1 mm)

cryomilled EVA +
ctalyst + alcohol

compression mold

vitrimerized EVA,
compression molded

shear rheology test
specimen of
vitrimerized EVA

Figure 31.2 Process to convert crosslinked EVA to vitrimerized EVA.

As shown in Figure 31.3a, dynamic mechanical analysis of vitrimerized EVA (v-EVA) demonstrates that the elastic modulus of the v-EVA plateaus above the melting temperature. This plateau is evidence of a vitrimer network, which has elastic properties. The elastic modulus is retained after reprocessing the vitrimer in three melting and molding steps.

Stress relaxation experiments demonstrate that the modulus of v-EVA decreases sharply to less than one third its initial value after 1000 seconds at a temperature of 120 °C (Figure 31.3b). A control sample of crosslinked EVA that was processed using the same cryomilling procedure but without the TE catalyst also exhibited a decrease in modulus at 120 °C, but the decrease was much smaller.

The effect of vitrimerization on rheological behavior becomes more pronounced as temperature increases. Figure 31.4a shows that the modulus decreases very sharply at temperatures as low as 80 °C for the EVA sample that was produced using 8 mol% TE catalyst. Increasing the temperature by as little as 5 or 10 °C results in substantially faster decay in the modulus.

The extent of vitimerization increases as the amount of TE catalyst is increased. In Figure 31.4b, stress relaxation is measured at 100 °C for v-EVA samples that were cryoground with different feed concentrations of the transesterification catalyst relative to the amount of vinyl acetate in the EVA starting material.

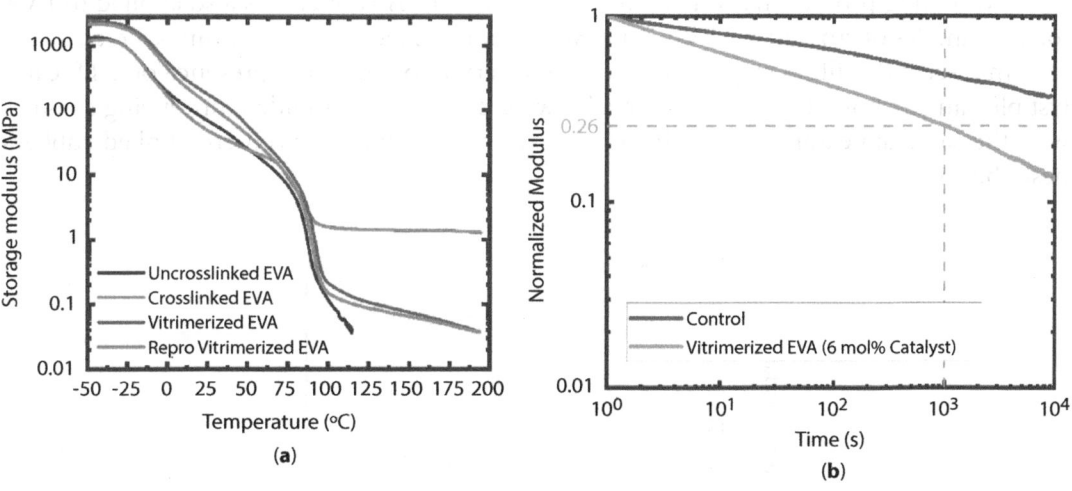

Figure 31.3 (a) Dynamic mechanical analysis of vitrimerized EVA demonstrates that the elastic modulus of the v-EVA plateaus above the melting temperature, confirming the formation of an elastic- like network. The elastic modulus is retained after reprocessing the vitamer in three melting and molding steps. (b) Stress relaxation measured by shear rheology at 120°C demonstrates that the modulus of vitrimerized EVA decreases at a moderate temperature. By comparison, a control sample of crosslinked EVA that was cryoground without the vitrimerization catalysts does not exhibit the same degree of stress relaxation at 120 °C.

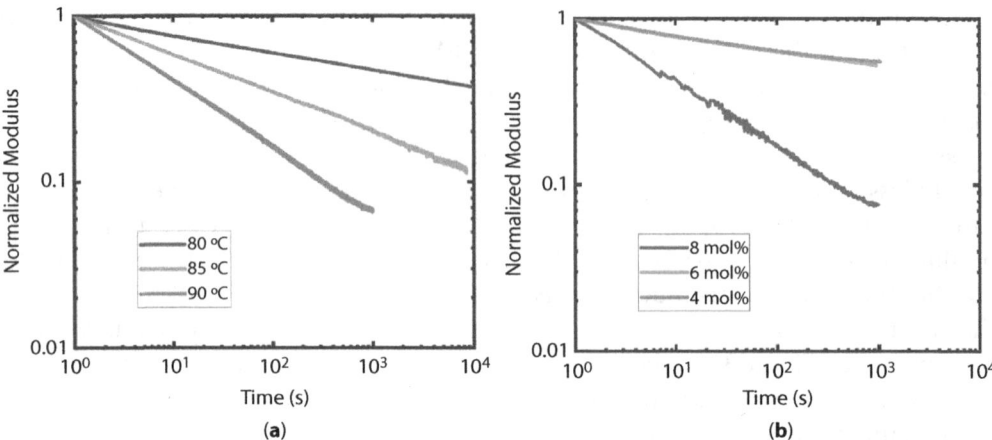

Figure 31.4 (a) Stress relaxation rate of vitrimerized EVA increases as temperature increases. Measured for EVA cryomilled with 8 mol% TE catalyst, relative to vinyl acetate concentration. (b) Stress relaxation rate of vitrimerized EVA increases as feed concentration of TE catalyst increases, relative to the amount of vinyl acetate in the EVA starting material. Measurement at 100°C.

31.4.3   Vitrimerization of EVA Foam

The cryomilling process that was used to vitrimerize EVA rubber was also applied to EVA foams. Samples of crosslinked EVA foam were ground to particles with 1 mm average diameter. In a second milling step the particles were cryoground in the presence of a TE catalyst plus an alcohol. The resulting particles were compression molded, producing samples with the same appearance and rubbery character as the vitrimerized crosslinked rubbers describe above.

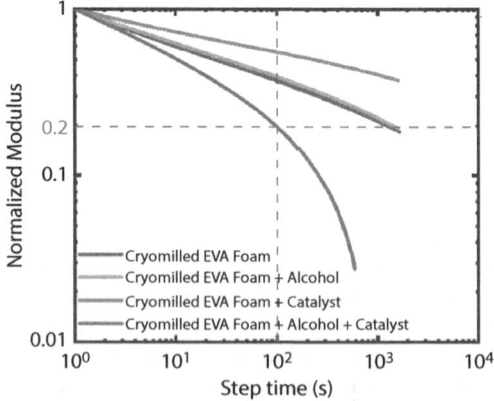

Figure 31.5 Stress relaxation measured by shear rheology at 120°C demonstrates that elastic modulus of vitrimerized EVA foam decreases at a moderate temperature.

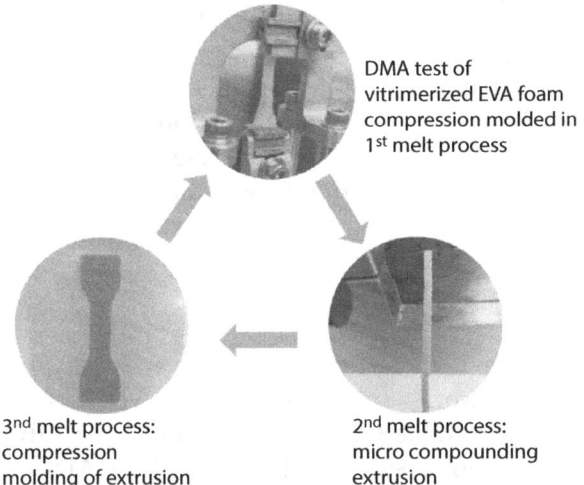

DMA test of
vitrimerized EVA foam
compression molded in
1st melt process

3nd melt process:
compression
molding of extrusion

2nd melt process:
micro compounding
extrusion

Figure 31.6 Melt re-processing of vitrimerized EVA foam. First melt process generates molded specimen for dynamic mechanical analysis. Second melt process extrudes chopped DMA test specimen. Third melt process molds chopped extrudate strand into new DMA test specimen.

The stress relaxation results in Figure 31.5 demonstrate that EVA foam has been converted to a vitrimer. The modulus of the EVA that was cryomilled with the TE catalyst plus alcohol decreases sharply to one fifth its initial value at 100s, decaying even more to less than one tenth of its initial value after 100 seconds at a temperature of 120 °C. Comparison samples of EVA foams that were processed using the same cryomilling procedure but without both the TE catalyst and alcohol also exhibited a decrease in modulus at 120 °C but to a much lesser extent.

Similar to the v-EVA rubber described above, the vitrimerized foam can be melt processed, cooled, and re-processed multiple times by compression molding, extrusion, or a combination of both processes. In Figure 31.6, a sample of v-EVA foam is shown in multiple melt re-processing steps.

31.5   Conclusions & Recommendations

This work demonstrates that crosslinked EVA materials, including EVA foams, can be converted to a new class of materials that have elastic properties reminiscent of crosslinked networks but are also melt processible. The key enabling technology is the transformation from a covalently crosslinked network to a dynamically crosslinked one via a mechanochemical process. Crosslinked EVA is ground in a cryogenic ball mill in the presence of a transesterification catalyst and an alcohol. This process provides a homogeneous, rubbery material that exhibits a plateau storage modulus indicative of an elastomer. The material can be melt processed by compression molding or extrusion and retains its properties after multiple melt re-processing steps.

The exciting results reported here demonstrate that dynamic crosslinking offers a promising approach to dramatically reduce the accumulation of crosslinked EVA plastic waste in landfills and the environment. This early-stage research will be followed with further experiments to optimize the vitrimerization process and to demonstrate it in a commercially relevant environment.

Acknowledgements

This material is based upon work supported by the U.S. Department of Energy's Office of Energy Efficiency and Renewable Energy (EERE) under the Advanced Manufacturing Office Award Number DE-EE0007897 awarded to the REMADE Institute, a division of Sustainable Manufacturing Innovation Alliance Corp.

References

1. Global Footwear Industry: Positive Dynamics in 2018. https://www.worldfootwear.com/news/global-footwear-industry-positive-dynamics-in-2018/4048.html.
2. Staikos, T.; Rahimifard, S., Post-Consumer Waste Management Issues in the Footwear Industry. *Proceedings of the Institution of Mechanical Engineers, Part B: Journal of Engineering Manufacture, 221* (2), 363-368, 2007.
3. Smallenburg, F.; Leibler, L.; Sciortino, F., Patchy Particle Model for Vitrimers. *Physical Review Letters, 111* (18), 188002, 2013.
4. Jurowska, A.; Jurowski, K., Vitrimers - The miracle polymer materials combining the properties of glass and plastic? *CHEMIK nauka-technika-rynek, 69,* 392-394, 2015.
5. Montarnal, D.; Capelot, M.; Tournilhac, F.; Leibler, L., Silica-Like Malleable Materials from Permanent Organic Networks. *Science, 334* (6058), 965-968, 2011.
6. Roettger, M. Associative exchange reactions of boron or nitrogen containing bonds and design of vitrimers. Chemical Physics [physics.chem-ph]. Université Pierre et Marie Curie - Paris VI, 2016.
7. Ciarella, S.; Biezemans, R. A.; Janssen, L. M. C., Understanding, predicting, and tuning the fragility of vitrimeric polymers. *Proceedings of the National Academy of Sciences of the United States of America, 116* (50), 25013-25022, 2019.
8. Brutman, J. P. In *Sustainable Cross-linked Polymers with Potential for Recyclability,* 2017.
9. Padhan, A.; Mandal, D., Thermo-reversible Self-healing in Fluorous crosslinked Copolymer. *Polymer Chemistry* 2018, *9,* 2018.
10. Spiesschaert, Y.; Guerre, M.; Imbernon, L.; Winne, J. M.; Du Prez, F., Filler reinforced polydimethylsiloxane-based vitrimers. *Polymer, 172,* 239-246, 2019.
11. Hayashi, M., Implantation of Recyclability and Healability into Cross-Linked Commercial Polymers by Applying the Vitrimer Concept. *Polymers, 12* (6), 1322, 2020.
12. Tellers, J.; Pinalli, R.; Soliman, M.; Vachon, J.; Dalcanale, E., Reprocessable vinylogous urethane cross-linked polyethylene via reactive extrusion. *Polymer Chemistry, 10* (40), 5534-5542, 2019.

13. Yue, L.; Guo, H.; Kennedy, A.; Patel, A.; Gong, X.; Ju, T.; Gray, T.; Manas-Zloczower, I., Vitrimerization: Converting Thermoset Polymers into Vitrimers. *ACS Macro Letters*, *9*, 836-842, 2020.

14. Altuna, F. I.; Hoppe, C. E.; Williams, R. J. J., Epoxy Vitrimers: The Effect of Transesterification Reactions on the Network Structure. *Polymers (Basel)*, *10* (1), 2018.

15. Zhang, X.; Vidavsky, Y.; Aharonovich, S.; Yang, S. J.; Buche, M. R.; Diesendruck, C. E.; Silberstein, M. N., Bridging experiments and theory: isolating the effects of metal–ligand interactions on viscoelasticity of reversible polymer networks. *Soft Matter*, *16* (37), 8591-8601, 2020.

16. Guo, H.; Yue, L.; Rui, G.; Manas-Zloczower, I., Recycling Poly(ethylene-vinyl acetate) with Improved Properties through Dynamic Cross-Linking. *Macromolecules*, *53*, 458-464, 2020.

Part 6
CHEMICAL RECYCLING

32

Performing Poly(Ethylene Terephthalate) Glycolysis in a Torque Rheometer Using Decreasing Temperatures

Jonathan Hatt[1]*, Karl Englund[2] and Hui Li[3]

[1]*Department of Civil and Environmental Engineering, Washington State University, Pullman, United States*
[2]*Department of Civil and Environmental Engineering, Composite Materials and Engineering Center, Washington State University, Pullman, United States*
[3]*Composite Materials and Engineering Center, Washington State University, Pullman, United States*

Abstract

The accumulation of research efforts into the glycolysis of poly(ethylene terephthalate) (PET) using ethylene glycol has advanced the reaction to where the intermediate monomer, *bis*(hydroxyethyl terephthalate), is produced at yields greater than 80% from complete PET depolymerization, using reaction temperatures and times of less than 196°C and approximating one hour, respectively. Environmental concerns have now shifted the research focus of this reaction to decreasing the energy consumption by decreasing reaction time. A popular solution to this problem, but expensive, is the development of complex catalysts. An alternative solution is the modification of existing reactors. Generally, the glycolysis of PET is studied in a batch reactor, however, minimal research has also explored a continuous reaction via an extruder. The existing research has demonstrated that a twin-screw extruder can be a very efficient reactor, with higher rates of depolymerization compared to the batch reactor. A caveat, and where modifications are needed, is that under homogeneous conditions (temperatures greater than the 196°C boiling temperature of ethylene glycol) the risk of ethylene glycol vapors escaping the extruder during the reaction can occur, creating a hazard for the operator and disrupting the stoichiometry between the ethylene glycol and the PET. A proposed modification is implementing a decreasing temperature profile (which is a deviation from the constant profile explored in existing research) along the length of the extruder barrel, ranging from PET melt temperatures to temperatures below 196°C, to protect the die zone from releasing ethylene glycol vapors. In addition, the decreasing temperatures may dissuade secondary and repolymerization reactions from occurring. However, how does the decreasing temperature profile influence molecular weight changes during the glycolysis reaction? In this research, a torque rheometer (substituting for a twin-screw extruder) was charged with PET pellets, ethylene glycol, and a zinc acetate catalyst. Over two reaction temperatures of 255°C and 185°C, changes in the molecular weight and melting temperature of the PET were measured at reaction times of 1-minute, 3-minutes and 5-minutes. The results indicated that 99% of depolymerization occurred in a 5-minute reaction time at 255°C. Further depolymerization of one repeat unit occurred at 185°C following a 3-minute reaction time. Overall,

**Corresponding author*: jonathan.hatt@wsu.edu

Nabil Nasr (ed.) Technology Innovation for the Circular Economy: Recycling, Remanufacturing, Design, Systems Analysis and Logistics, (409–420) © 2024 Scrivener Publishing LLC

this correlates to a decrease in number-average molecular weight of 40,000 g·mol⁻¹ for the virgin PET to 400 g·mol⁻¹ for the depolymerized PET.

Keywords: Chemical recycling, continuous glycolysis, poly(ethylene terephthalate), torque rheometer, extruder, depolymerization

32.1   Introduction

The intent of chemical recycling is to convert a post-consumed polymer into a form that may be repolymerized into a virgin polymer or used as a value-added chemical. Ideally, these reactions should require minimal energy and time with a yield that is maximized to the desired molecular weight. However, this is not always the outcome. The energy and time requirements for a chemical recycling reaction are dependent on the type of reaction being performed. The literature on the chemical recycling of poly(ethylene terephthalate) (PET) has explored five main reactions (alcoholysis, aminolysis, ammonolysis, glycolysis and hydrolysis), with glycolysis being one of the more well-researched. The thorough study of this reaction has explored the multitude of associated reaction parameters (temperature, time, ethylene glycol concentration and catalyst) and has developed this reaction to where greater than an 80% yield of the intermediate monomer *bis*(hydroxyethyl terephthalate) (BHET) may be achieved from a complete depolymerization. Now, interests have been diverted to reducing the energy associated with this reaction by reducing the reaction time. One common method of achieving this is through increasingly complex catalysts, which can be difficult and expensive to synthesize. However, without these catalysts the reaction time can exceed hours.

The glycolysis of PET is commonly performed in a batch style reactor. Receiving minimal attention but demonstrating effectiveness is the use of the twin-screw extruder. The extruder is used for many polymer related reactions, such as polymerization and polymer modification. Advantages of this method include continuous reactions, screw modularity, effective mixing and temperature adjustment [1]. Patterson *et al.* [2] explored the use of a twin-screw extruder for the glycolysis of PET and found that a greater than 85% decrease in the molecular weight of the PET (correlating to a number-average molecular weight decrease of 20,000 g·mol⁻¹ to 2,500 g·mol⁻¹) can be achieved in a 10-minute reaction time using a constant temperature profile of 260°C, a screw speed of 150 rev·min⁻¹, an ethylene glycol concentration of 0.04 and no catalyst. In comparison, a batch style reaction can yield complete depolymerization in a 30-minute reaction time but will require an ethylene glycol and complex catalyst concentration of 4 and 0.02 (w/w), respectively [3]. While a fast rate of depolymerization was exhibited by the extruder, complete depolymerization was not achieved and no BHET yield was recorded. However, issues were experienced by Patterson *et al.* [2] in this work, particularly the inadvertent escape of ethylene glycol in vapor form from the extruder.

The increased rate of PET depolymerization may be explained by the catalytic effects of increased temperature [4–6]. An increase in the temperature to PET melt conditions changes the state of the reaction from a two-phase heterogeneous reaction (common for a batch style reaction) to a single-phase homogeneous reaction [5, 6]. This, in conjunction with the effective mixing by the extruder screws, allows the ethylene glycol to easily diffuse throughout the melted PET matrix, decreasing the activation energy of the reaction and effectively increasing the reaction rate [5, 6]. However, while temperature may be considered a positive for this reaction it also adds a negative component as well. As the reaction temperature approaches and exceeds 196°C, the ethylene glycol boils and may escape the reaction in vapor form. Besides being an unsustainable practice, the escape of ethylene glycol vapors also creates a hazard to the operator and environment, disrupts the stoichiometry of the reaction, promotes early equilibration of the reaction and reduces the extent of depolymerization [2]. In effort to alleviate this issue, modifications may be made to both the extruder and the glycolysis procedure. This includes equipping the extruder with a reflux condensing system, to collect and recirculate the condensed ethylene glycol vapors back into the reaction, and incorporating a decreasing temperature profile along the length of the extruder barrel. Other theorized advantages to this incorporation are a reduction in the energy consumption of the reaction and an inhibitor to repolymerization and secondary reactions [7].

This research will first focus on the incorporation of the decreasing temperature profile, specifically, how will the extent of PET depolymerization be influenced when a catalyzed glycolysis reaction is performed at the temperatures of 255°C, and again at 185°C. Virgin PET pellets were loaded into a torque rheometer cavity (substituted for an extruder to provide better control for reaction time) and subject to catalyzed glycolysis for the reaction times of 1-minute, 3-minute and 5-minutes at each of the temperatures listed. Following each reaction, gel permeation chromatography, thermal gravimetric analysis and differential scanning calorimetry characterization was performed on the product to determine changes in the molecular weight and thermal properties.

32.2 Experimental

Materials

The virgin, Burcham 9084 bottle-grade poly(ethylene terephthalate) pellets were supplied by Ravago. Prior to processing, the pellets were dried in a convection oven at 110°C for 24 hours. Reagent grade ethylene glycol was sourced from VWR and stored at room temperature. The zinc acetate dihydrate catalyst was sourced from Sigma Aldrich.

Methods

Reactor:
The glycolysis reaction was performed in a counter-rotating, HAAKE torque rheometer. The cavity of the rheometer consisted of three individually heated, separable plates with a 1.5 inch diameter feed throat centered on the top of the middle plate. Each plate of the cavity was equipped with its own thermocouple for temperature control and monitoring. A fourth thermocouple was centered below the cavity of the middle plate, which was used to monitor the temperature of the material being processed. The internal rotors were heated through thermal conduction from the cavity plates.

Glycolysis Reaction:

This research was divided into two reactions at two different temperatures: (1) glycolysis at 255°C (glycolysis) and (2) glycolysis at 185°C (repeated glycolysis, or re-glycolysis), which are above and below the 196°C boiling temperature of ethylene glycol, respectively. For each reaction, the three different reaction times of 1-minute, 3-minutes and 5-minutes were investigated.

For the first reaction, 60 g of virgin PET pellets were loaded into the preheated rheometer cavity with the rotors rotating. After a melting time of 50 seconds, 27 g of ethylene glycol with dissolved zinc acetate dihydrate catalyst was added via a syringe pump, at a rate of approximately 0.5 g·s⁻¹. The reaction time started after the solution was added. During the reaction, neat ethylene glycol was dripped into the rheometer cavity at a rate of 0.2 g·s⁻¹ to compensate for the loss of ethylene glycol in vapor form.

After the reaction time expired, the reacted material was removed from the cavity. A solid, white material, clung to the rotors surface, was weighed and labelled as a solid fraction. The resulting liquid material was quenched with chilled, deionized water and labelled as the liquid fraction [8]. This fraction, which solidified to a waxy material, was then dried in a convection oven for 24 hours at 60°C to remove free water.

The re-glycolysis reaction followed a similar procedure to the first reaction, except for the decrease in reaction temperature to 185°C and the starting material was previously glycolyzed from the first reaction. The reaction was not supplemented with ethylene glycol during the reaction. During the length of the reaction, a low volume of low viscosity material leaked from the rheometer cavity. Quantification of the lost material was difficult to obtain and was therefore not recorded. Each reaction was performed in triplicate. Additional reaction parameters are listed in Table 32.1.

Table 32.1 Reaction parameters for the glycolysis and re-glycolysis reactions. Data sourced from Hatt [9].

Reaction parameter	Glycolysis	Re-glycolysis
Starting Material:	Virgin PET Pellets	Glycolyzed PET
Material Mass (g):	60	60
Ethylene Glycol to Starting Material (w/w):	0.45	0.45
Catalyst to Starting Material (w/w):	0.02	0.02
Temperature (°C):	255	185
Time (min.):	1, 3, 5	1, 3, 5
Rotor Speed (rev·min⁻¹):	100	100

Separating Water-Soluble and Water-Insoluble Fractions:
The liquid fraction is composed of a distribution of molecular weights, a fraction of which is water soluble [6]. In this work, for characterization purposes, the water-soluble and water-insoluble fractions were separated by vacuum filtration using a procedure that is described in Hatt [9]. As found in the literature, the water-soluble fraction contains the intermediate monomer BHET and other similar molecular weighted compounds [6]. The water-insoluble fraction contains the dimer and other larger molecular weighted compounds.

Gel Permeation Chromatography:
Gel permeation chromatography (GPC) was performed to measure changes in number-average molecular weight (M_n), weight-average molecular weight (M_w), degree of polymerization (DP) and polydispersity index (Đ) for each treatment. The following instruments and procedures described for the GPC analysis are as described in Hatt [9]. The instrument used was a Viscotek TDA 305 unit equipped with a Viscotek GPCMax module for automatic sample injection. Two Agilent, 7.5 mm by 300 mm PLGel 5 µm Mixed-D analytical columns, with a 400 $g \cdot mol^{-1}$ to 200,000 $g \cdot mol^{-1}$ resolving range, were placed in series to perform separation. The columns were heated to 30°C and injected with 100 µL of sample using an eluent of chloroform with 2% hexafluoroisopropanol and a flow rate of 1 $mL \cdot min^{-1}$. Each sample was dissolved at a 0.3 $mg \cdot mL^{-1}$ concentration in the same solvent mixture as the eluent. Conventional calibration was performed using polystyrene standards.

Thermal Gravimetric Analysis:
Thermal gravimetric analysis (TGA) was performed to measure changes in decomposition temperature (T_d) and corresponding weight loss. The following instruments and procedures described for the TGA analysis are as described in Hatt [9]. The instrument used was a Mettler Toledo TGA/DSC 1 STAR^e system. The measured values for T_d were determined using the Mettler Toldeo STAR^e software. The heating scan was from 25°C to 600°C, using a linear heating ramp of 10 $°C \cdot min^{-1}$ under a nitrogen environment. A sample size of approximately 7 mg was weighed into a ceramic crucible with triplicate runs performed for each treatment.

Differential Scanning Calorimetry:
Differential scanning calorimetry (DSC) was performed to measure changes in the transition peak broadness (ΔT), peak melt crystallization temperature (T_{mc}), peak melting temperature (T_m) and their associated enthalpies (ΔH_{mc} and ΔH_m), respectively. The following instruments and procedures described for the DSC analysis are as described in Hatt [9]. The instrument used was a Mettler Toledo DSC 1 STAR^e system. The values for ΔT, T_{mc}, T_m, ΔH_{mc} and ΔH_m were determined using the STAR^e software. The heating regimen consisted of two heat-cool scans from 25°C to 280°C at a 10 $°C \cdot min^{-1}$ linear heat ramp under a nitrogen environment. A 3-minute, 280°C isotherm was placed between each heat-cool scan. Each run consisted of approximately 5 mg of sample being weighed into a 40 µL aluminum pan. The top of the pan was punctured to allow for the escape of formed volatiles during the test. Triplicate runs were performed for each treatment.

32.3 Results and Discussion

The intent for the glycolysis reaction was to be performed in a homogeneous state. However, upon the addition of the ethylene glycol and catalyst solution, an approximate 25°C cooling of the reacting materials occurred (the temperature of the cavity plates only decreased by approximately 5°C but then recovered to reaction temperature), resulting in the partial solidification of the PET to the rotor surface (an explanation for the yield of solid fraction) and a combination of both homogeneous and heterogeneous reactions inside the cavity. As the reaction progressed, the temperature of the reacting materials did increase but did not reach the set reaction temperature within the reaction time, which may have been due to the endothermic nature of the reaction and the continual addition of ethylene glycol. However, with the temperature increase, a subsequent melting of macromolecular chains that were depolymerized heterogeneously may have occurred, which facilitated their transfer to the homogeneous phase of the reaction. An explanation for why the drop in temperature of the reacting materials occurred is the addition of the ethylene glycol solution, which had an initial temperature of approximately 20°C. The 25°C temperature decrease was also measured for the re-glycolysis reaction as well.

At 255°C, the yield of the solid fraction decreased from 13% to 5% and 2% for the 1-minute, 3-minute and 5-minute reaction times, respectively. In efforts to characterize and compare this fraction to the virgin PET, GPC, TGA and DSC characterization was performed (refer to Hatt [9] for detailed results). The GPC results measured a shift in the molecular weight distribution (MWD) to lower molecular weights. The values were measured at 22,273 g·mol^{-1} and 4,995 g·mol^{-1} for the M_w and M_n, respectively. These values are in comparison to 85,274 g·mol^{-1} and 39,690 g·mol^{-1} for the M_w and M_n for the virgin PET, respectively. The TGA results measured a T_d of 411°C, which is similar to the 406°C T_d for the virgin PET, with an 85% weight loss. The second heating scan of the DSC results measured two melting peaks, positioned at 248°C and 240°C, in contrary to the single melting peak at 247°C for the virgin PET. The emergence of the second melting peak, for the solid fraction, may suggest changes to the crystalline structure of the polymer due to a shortening of the polymer chains [10]. The outcome of these tests suggests that the solid fraction was subject to depolymerization during the reaction and should not be considered similar in property to the virgin PET.

For the reason of the 5-minute reaction time yielding the lowest solid fraction, the liquid product of this reaction was selected to perform the re-glycolysis reaction at the three different reaction times of 1-minute, 3-minute and 5-minute at the decreased temperature of 185°C. For this reaction, complete diffusion of the starting material into the ethylene glycol and catalyst solution did not occur for the 1-minute reaction time but did occur for the 3-minute and 5-minute reaction times. Therefore, following the 1-minute reaction, 34% of the starting material was found clung to the surface of the rotors. The remaining fraction was in a liquid state.

Shown in Figure 32.1 are the results for the separation of the liquid fraction into the water-soluble and water-insoluble components for the 5-minute reaction time of the glycolysis reaction and the 1-minute, 3-minute and 5-minute reaction times of the re-glycolysis reaction. Between the first and second reactions, an increase in the water-soluble fraction is apparent. Within the results of the re-glycolysis reaction, the 3-minute reaction time

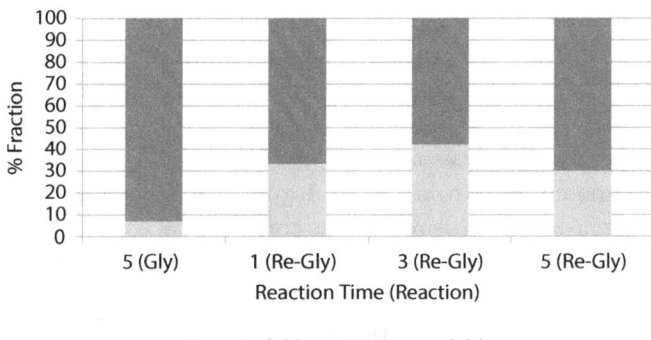

Figure 32.1 Water-soluble versus water-insoluble fractions for the glycolysis and re-glycolysis reactions. Source: Hatt [9].

yielded the highest water-soluble fraction, representing a 35% increase from the first reaction. The 5-minute reaction time yielded the lowest water-soluble fraction, which may be explained by a repolymerization.

The following information will present GPC, TGA and DSC data for the water-insoluble fraction of the treatments shown in Figure 32.1, in comparison to TGA and DSC data for the water-soluble fraction. For more information regarding the results of the 1-minute and 3-minute reaction times of the glycolysis reaction, refer to Hatt [9].

Gel Permeation Chromatography

Figure 32.2 shows the MWD for the 5-minute reaction times of the glycolysis and re-glycolysis reactions. Table 32.2 lists the molecular weight data for the 5-minute reaction times, in addition to the 1-minute and 3-minute re-glycolysis reactions. A multimodal MWD was measured for each of the reaction times for each reaction, which contrasts the unimodal MWD for the virgin PET. The multiple peaks may indicate the presence of different oligomeric species that have formed during depolymerization. The overlap of the chromatograms suggests similar molecular weights of the product between each of the three reaction times and between the glycolysis and re-glycolysis reactions [9].

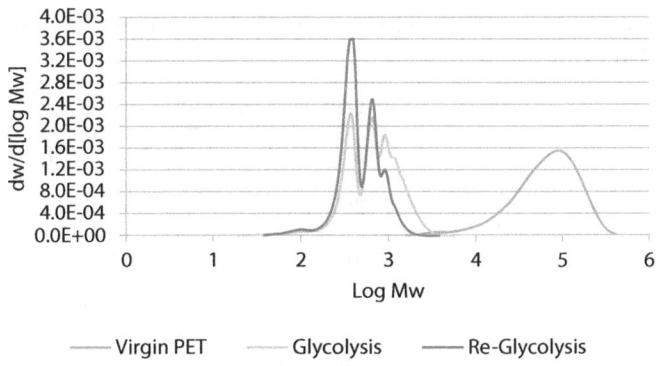

Figure 32.2 GPC chromatogram of the water-insoluble fraction for the 5-minute reaction time of the glycolysis and re-glycolysis reactions. Source: Hatt [9].

After the re-glycolysis reaction, area under the chromatogram for the first reaction shifted to lower molecular weights, heightening the signal for the lower molecular weight peaks. This result suggests further depolymerization of the oligomers during the re-glycolysis reaction at 185°C.

Following the 5-minute glycolysis reaction at 255°C, the M_n and M_w of the virgin PET decreased by approximately 99% to a DP of 3 and 4 units, respectively, which was also accompanied by a decrease in Đ. This result was consistent across each of the reaction times for the first reaction at 255°C, regardless of the quantity of solid fraction [9]. Given the presence of both homogeneous and heterogeneous phases, it is theorized that an explanation for this result may relate to both phases of the reaction. Previous research has shown that the homogeneous phase is characteristic of increased reaction rates over the heterogeneous phase [6]. One reason for this is that the reaction is not confined to the surface of the solid PET but is able to pervade into the PET matrix. At first addition of the ethylene glycol and catalyst solution, glycolysis occurred in the homogeneous phase. As the PET solidified, with the increased addition of the ethylene glycol solution, the mechanism of reaction shifted to heterogeneous. As the polymer chains on the solid surface are depolymerized, a simultaneous melting of the lower molecular weight chains may have occurred as the reacting material gradually recovered the lost temperature. As these chains entered the homogeneous phase, they were subject to a faster rate of depolymerization.

For the re-glycolysis reaction at 185°C, a further depolymerization of 0.5%, or approximately one repeat unit, was measured with an additional decrease in Đ. This finding indicates that the majority of the depolymerization occurred during the 5-minute glycolysis reaction. Minimal depolymerization occurred during re-glycolysis but an increase in the water-soluble fraction is evident, as shown in Figure 32.1. This limited reduction in molecular weight may have been the consequence of an equilibrium that had been reached within the reaction. Or, the seepage of low viscosity material (possibly containing excess ethylene glycol) from the cavity during the reaction may have disrupted the stoichiometry necessary for further depolymerization to water-soluble compounds. The increase in the water-soluble fraction following the re-glycolysis reaction was also observed by other researchers [11]. This result suggests that with further addition of ethylene glycol, the equilibrium of the reaction can be further shifted to depolymerization.

Table 32.2 Listed M_w, M_n and Đ data for the glycolysis and re-glycolysis reactions. Data sourced from Hatt [9].

Reaction time (min.)	Reaction	GPC (g·mol⁻¹)		
		Mn (DP)	Mw (DP)	Đ
Virgin PET	-	39700 (206.77)	85300 (444.27)	2.15
5	Glycolysis	493 (2.56)	784 (4.08)	1.59
1	Re-Glycolysis	386 (2.01)	555 (2.89)	1.44
3		402 (2.09)	539 (2.81)	1.34
5		394 (2.05)	522 (2.72)	1.32

Thermal Gravimetric Analysis

Figure 32.3 shows the TGA data for the 5-minute reaction time of the glycolysis and re-glycolysis reactions, and a representative sample of the water-soluble material. Listed in Table 32.3 are the TGA values associated with the 1-minute and 3-minute reaction times of re-glycolysis reaction, along with other corresponding data. In contrast to the virgin PET, two T_d's are evident for the glycolyzed material. The first T_d may be explained by the formation of lower molecular weight volatiles, such as ethylene glycol [12]. The onset is within a temperature range of 180°C to 202°C with a corresponding weight loss of 8% and 11% for the glycolysis and re-glycolysis reactions, respectively. According to the literature, an explanation for the presence of the second T_d may be a combination of further decomposition of lower molecular weighted compounds, such as BHET, and decomposition of material that may have polymerized during the TGA test [3, 12]. The onset temperature increases

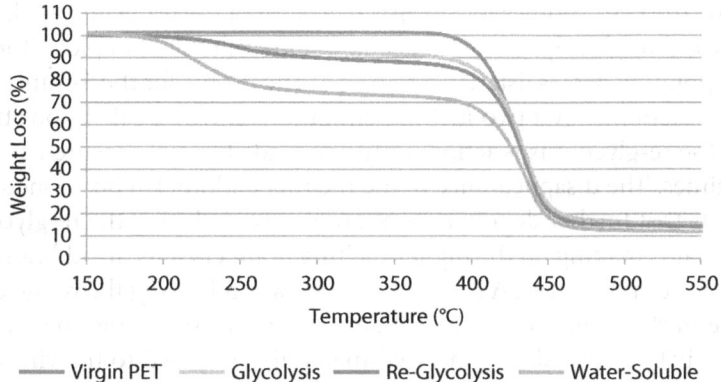

Figure 32.3 TGA profiles for the water-insoluble liquid fraction of the 5-minute reaction time for the glycolysis and re-glycolysis reactions and a representative sample of the water-soluble material. Source: Hatt [9].

Table 32.3 TGA analysis data for the water-insoluble liquid fraction of the glycolysis and re-glycolysis reactions. Data sourced from Hatt [9].

Reaction time (min.)	Reaction	First decomposition		Second decomposition		Remaining material (%)
		Onset (°C)	Weight loss (%)	Onset (°C)	Weight loss (%)	
Virgin	-	404	85	-	-	15
5	Glycolysis	200	8	405	77	15
1	Re-Glycolysis	202	11	409	75	14
3		180	11	409	75	14
5		201	12	410	74	14
Water-Soluble	-	182	25	402	63	12

to 405°C to 410°C with a corresponding weight loss of 74% to 77% for each reaction. The weight of the residual material is approximately 14% of the initial mass, which is similar to that of the virgin PET.

A difference between the water-insoluble and water-soluble results is the amount of weight loss corresponding to each of the T_d's. The weight loss corresponding to the first T_d of the water-soluble material is approximately 15% higher compared to the water-insoluble material. An explanation for this is that there is a higher concentration of lower molecular weight compounds that volatize at this onset temperature. A second T_d is also measured for the water-soluble material at 402°, but the explanation may relate more to the further decomposition of lower molecular weighted compounds [12]. Only a 63% weight loss was exhibited for this second T_d. This trend is consistent with other research, where additional steps were taken to characterize the water-soluble product as the monomer BHET [3, 11].

Differential Scanning Calorimetry

Figure 32.4 shows the DSC profiles for the first heat scan of the water-insoluble liquid fraction of the glycolysis and re-glycolysis reactions and a representative sample of the water-soluble material. Corresponding data is listed in Table 32.4. The profile for the 5-minute reaction time of the glycolysis reaction shows two melting endotherms, with peak T_m positioned at 159°C and 235°C. For the re-glycolysis reaction, only one peak T_m was measured at 162°C for each of the reaction times. The disappearance of the melting endotherm positioned at higher temperatures suggests that further depolymerization occurred, during the re-glycolysis reaction, to the polymer chains making up the higher melting point crystals. In addition, the base temperature range of the endotherm, ΔT, becomes narrower following the re-glycolysis reaction with an increase in the signal, as a result of an increase in ΔH_m. This change suggests more crystallization activity for the depolymerized material, compared to the virgin material, as a result of the increased mobility of lower molecular weight chains [10].

The melt endotherm for the water-soluble fraction shows one peak positioned at approximately 112°C. The ΔT of this peak is narrower and the signal is stronger compared to the endotherms of the water-insoluble material, possibly reflecting the purity and the uniformity in the molecular weight of the material. In addition, a higher ΔH_m is also associated with this endotherm. These results are similar to other research, where characterization

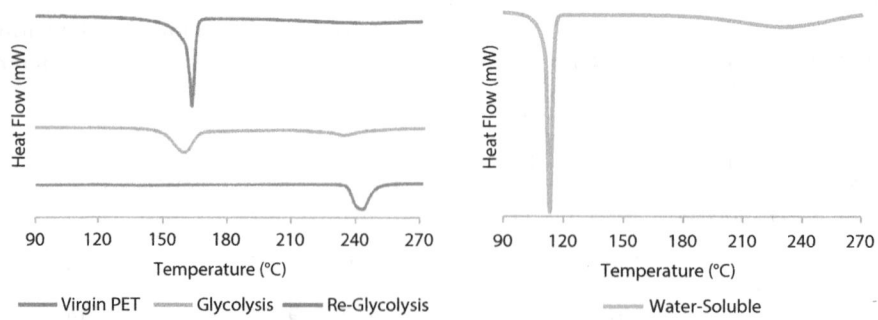

Figure 32.4 First DSC heat scan for the water-insoluble liquid fraction of the 5-minute glycolysis and re-glycolysis reactions (*left*) and a representative sample of the water-soluble material (*right*) (*exotherm is up*). Source: Hatt [9].

Table 32.4 DSC data of the first heat scan for the water-insoluble and water-soluble material. Data sourced from Hatt [9].

Reaction time (min.)	Reaction	Peak 1			Peak 2			Total
		ΔT (°C)	Peak T_m (°C)	ΔH_m (J·g⁻¹)	ΔT (°C)	Peak T_m (°C)	ΔH_m (J·g⁻¹)	ΔH_m (J·g⁻¹)
Virgin PET	-	13	242	-58	-	-	-	-58
5	Glycolysis	19	159	-65	13	235	-23	-87
1	Re-Glycolysis	8	162	-99	-	-	-	-99
3		6	162	-107	-	-	-	-107
5		6	161	-105	-	-	-	-105
Water-Soluble	-	5	112	-156	-	-	-	-156

efforts identified the water-soluble fraction as the monomer BHET, which has a melting temperature in the range of 106°C to 109°C [3, 13].

For the second heat scan, the melting endotherms of the first heat scan (for both water-insoluble and water-soluble) shift to higher temperatures, to converge with the melting endotherm of the virgin PET, and form a bimodal distribution [9]. A proposed reason for this is a polymerization reaction that occurred during the first heat scan, which was the result of exposure to sufficient thermal energy.

32.4   Conclusion

In this research, 60g of virgin bottle-grade PET (peak T_m = 242°C, M_n = 39,700 g·mol⁻¹ and M_w = 85,300 g·mol⁻¹) pellets were depolymerized to where 98% of the material had a M_n and M_w of less than or equal to 402 g·mol⁻¹ and 539 g·mol⁻¹, respectively. The water-soluble and water-insoluble fractions yielded a material with a T_m of 112°C and 162°C, respectively. The remaining 2% of the material resulted in a solid, white form but still experienced a form of depolymerization. To yield these results, the catalyzed glycolysis reaction was performed over a total reaction time of 8 minutes and two reaction temperatures of 255°C and 185°C. This research demonstrates that extensive depolymerization from glycolysis can be performed on PET when using decreasing temperatures in a torque rheometer. While satisfying one step in the process of extruder modification, the next is the transfer of this reaction to the extruder.

Acknowledgements

The authors would like to acknowledge Ravago for supplying the bottle-grade PET pellets used in this study and Dr. Rock Mancini, of Washington State University, for providing materials and equipment for the GPC investigation.

References

1. Vergnes, B., Berzin, F., Modeling of reactive systems in twin-screw extrusion: challenges and applications. *C. R. Chim.*, 2006.
2. Patterson, J.D., Khan, S.A., Roberts, G.W., *Melt glycolysis of poly(ethylene terephthalate) using CO2-assisted extrusion.* AIChE Annual Meeting, Conference Proceedings. 2006.
3. Geng, Y., Dong, T., Fang, P., Zhou, Q., Lu, X., Zhang, S., Fast and effective glycolysis of poly(ethylene terephthalate) catalyzed by polyoxometalate. *Polym. Degrad. Stab.*, 117, 30, 2015.
4. Kao, C. Y., Cheng, W. H., Wan, B. Z., Investigation of catalytic glycolysis of polyethylene terephthalate by differential scanning calorimetry. *Thermochim. Acta*, 292, 95, 1997.
5. Campanelli, J. R., Kamal, M. R., Cooper, D. G., Kinetics of Glycolysis of Poly(Ethylene Terephthalate) Melts. *J. Appl. Polym. Sci.*, 54, 1731, 1994.
6. Liu, B., Lu, X., Ju, Z., Sun, P., Xin, J., Yao, X., Zhou, Q., Zhang, S. (2018). Ultrafast homogeneous glycolysis of waste polyethylene terephthalate via a dissolution-degradation strategy. *Ind. Eng. Chem. Res.*, 57, 16239, 2018.
7. Imran, M., Kim, B. K., Han, M., Cho, B. G., Kim, D. H., Sub-and supercritical glycolysis of polyethylene terephthalate (PET) into the monomer bis(2-hydroxyethyl) terephthalate (BHET). *Polym. Degrad. Stab.*, 95, 1686, 2010.
8. Xi, G., Lu, M., Sun, C., Study on depolymerization of waste polyethylene terephthalate into monomer of bis(2- hydroxyethyl terephthalate). *Polym. Degrad. Stab.*, 87, 117, 2005.
9. Hatt, J. Investigating the effects of a mechanical and chemical recycling method on the material properties of a virgin bottle-grade poly(ethylene terephthalate). MS Thesis, Voiland College of Engineering and Architecture, Washington State University, Pullman WA, 2022.
10. Hatt, J., Englund, K., Li, H., Characterizing the changes to the material properties of a bottle-grade poly(ethylene terephthalate) subject to repeated melt extrusion. *J. Appl. Polym. Sci.*, 140, 13, 2023.
11. Chaudhary, S., Surekha, P., Kumar, D., Rajagopal, C., Roy, P. K., Microwave assisted glycolysis of poly(ethylene terepthalate) for preparation of polyester polyols. *J. Appl. Polym. Sci.*, 129, 2779, 2013.
12. Mettler Toledo - Thermal Analysis, Thermal Degradation of BHET in Evolved Gas Analysis, https://www.azom.com/article.aspx?ArticleID=16413, 2018.
13. Wang, Q., Lu, X., Zhou, X., Zhu, M., He, H., & Zhang, X., 1-Allyl-3-methylimidazolium halometallate ionic liquids as efficient catalysts for the glycolysis of poly(ethylene terephthalate). *J. Appl. Polym. Sci.*, 129, 3574, 2013.

33

Sustainable Petrochemical Alternatives From Plastic Upcycling

Ryan A. Hackler* and Robert M. Kennedy

Aeternal Upcycling, Inc., Chicago, IL, USA

Abstract

Petrochemicals are used across every industry and in countless products. Plastic, the most ubiquitous of these products, has poor end-of-life recovery and recycling prospects. Energy- and carbon-rich plastics instead slowly degrade in landfills and oceans to produce microplastics, or is burned in incinerators for electricity, producing large amounts of greenhouse gases. Meanwhile, many other petrochemicals, like waxes, lubricants, and surfactants, have limited sustainable alternatives, hindering the low-emissions and circular carbon economy transition required for domestic manufacturing sectors. Catalytic hydrogenolysis, a novel chemical upcycling technology, can resolve both issues by converting plastic waste into chemical feedstocks for recirculation into the carbon economy. By diverting plastic waste away from incinerators and back into chemicals that have few sustainable options, one can 1) reduce emissions associated with the plastics lifecycle, 2) reduce cradle-to-gate emissions for various chemical feedstocks, 3) recirculate energy-rich plastics back into the carbon economy, and 4) establish a sustainable alternative to petrochemicals where there are none – through plastic-derived products. This catalytic process uses moderate operating temperatures (< 300 °C) and pressures of hydrogen (< 300 psi) to cleave polyolefins (e.g. high-density and low-density polyethylene, polypropylene, etc.) into smaller alkanes with no presence of aromatics or polycyclic compounds. Excess C-C bond scissions to produce gases is largely avoided by separating out the waxy and liquid products, resulting in a product yield of > 90% for plastic-derived petrochemical alternatives. Prototype lubricant base oils produced through catalytic hydrogenolysis of plastic waste are comparable to standard petrochemical lubricants in chemical structure and performance. Furthermore, a lifecycle analysis of a theoretical 250 metric ton/day plant resulted in a 75% reduction in cradle-to-gate emissions when compared to Group III mineral oils, demonstrating the potential of a drop-in replacement product for existing markets with the intent to lower Scope 3 emissions when producing consumer goods such as candles, cosmetics, detergents, and motor oils.

Keywords: Plastics, chemical recycling, circularity, polyolefins, petrochemicals, catalysis

Corresponding author: hackler@aeternalupcycling.com

Nabil Nasr (ed.) Technology Innovation for the Circular Economy: Recycling, Remanufacturing, Design, Systems Analysis and Logistics, (421–432) © 2024 Scrivener Publishing LLC

33.1   Introduction

Plastics are a wondrous material thanks to their cheap cost, durability, and applicability across nearly every market and sector of life. Unfortunately, the proper and effective disposal of plastics has lagged behind their massive production numbers. As of 2019, nearly 460 million tons of plastic was produced annually, with roughly 353 million tons of plastic discarded as waste annually [1]. Only 8.7% of waste plastic is recycled in the United States [2]; as a result, most plastic waste ends up in landfills or incinerated [3]. A significant portion of plastic waste is mismanaged in waste streams and enters the environment, where it can infiltrate aquatic and marine ecosystems [4]. Once in the environment, as microplastics or in other forms, plastic waste can have deleterious impacts on the environment, the economy, and human health [5, 6]. It is estimated that by 2050 there will be more plastic in the ocean than fish by weight [7]. These microplastics will inevitably enter the food chain in greater numbers and cause irreparable damage to the global ecosystem.

The current life cycle of plastic is fundamentally linear, starting with petrochemical extraction and ending predominantly in the aforementioned landfills or in the environment. Plastic use is built around the low cost of petrochemical feedstocks and excellent material properties of synthetic polymers. The ease of manufacturing new plastics and the extremely low costs of synthesizing monomers, subsidized by oil and natural gas extraction for fuels, has meant that recycled plastics historically have not been able to price-compete with virgin petrochemical materials. Additionally, mechanical recycling, which is the primary method for plastic recycling in use today, degrades the physical properties of the recycled plastic, resulting in products which must be "downcycled" into lower grade applications. Mechanical recycling is also highly sensitive to the composition of the feedstock; mixed pigmented or colored plastic will result in a gray plastic after recycling, while a small percent contamination of other types of plastic (e.g. polypropylene in high-density polyethylene) can severely degrade the physical properties, requiring expensive sorting and cleaning of the plastic waste to produce a usable product. Matters are further complicated by additional design elements in packaging and plastic products that render either a portion or the entire object unrecyclable. These factors result in mechanical recycling being a low profit, high cost business. Much of the un-recycled plastic waste is cost-prohibitive for mechanical recyclers with current technology to process, and there is a need for alternative recycling solutions that can address plastic waste unsuitable for mechanical recycling.

One such alternative solution is catalytic hydrogenolysis, where a Pt-based catalyst initiates C-C bond scissions and incorporates hydrogen in polymeric chains, thereby converting the plastic into smaller hydrocarbons. These products range from C_{20} to C_{70} with wax or lubricating properties depending on the molecular weight. So far, catalytic hydrogenolysis has been successfully tested using polyolefins, such as polyethylene and polypropylene, which comprise a large portion of single-use plastics, and account for roughly 50% of all plastic produced. This technology therefore has the potential to increase plastic recycling rates by converting plastic with low recycling rates, while also providing a new source for the domestic manufacturing of chemical feedstocks that find use across many industries, as showcased in Figure 33.1. Lower lifecycle emissions from plastic-derived products, such as lubricants and waxes, also creates new market opportunities by delivering carbon-efficient replacements for petroleum-derived products.

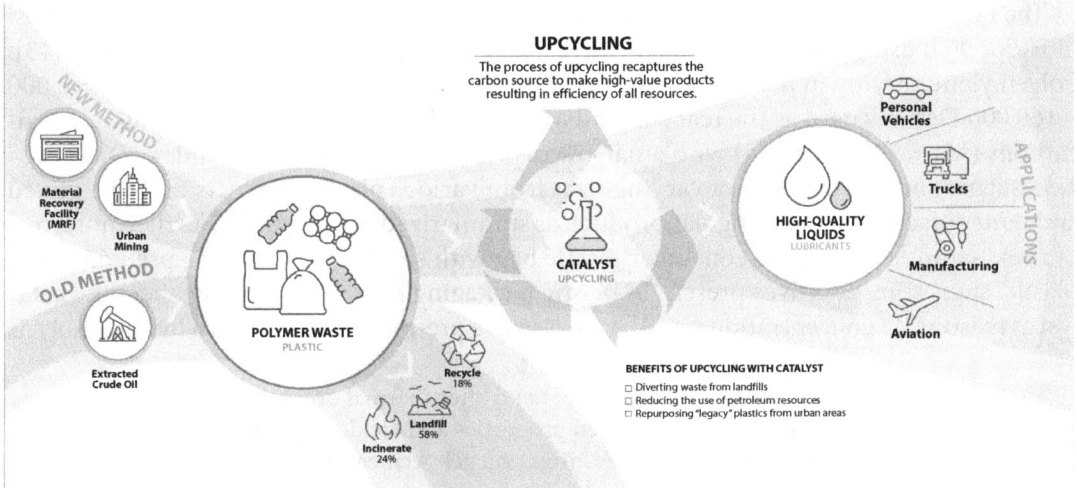

Figure 33.1 Current and proposed waste plastic lifecycle paradigms. Crude oil is extracted and refined for polymerization into plastic products. Used plastic is then disposed into landfills, burned, or mechanically recycled. Catalytic hydrogenolysis has the potential to divert that waste stream back into the economy through the production of higher-value goods that find use in automobiles, aviation, manufacturing, and energy production. Reproduced with permission from the authors [8].

33.2   Details of Catalytic Hydrogenolysis

Previous reports on catalytic hydrogenolysis have demonstrated > 99% conversion of pre- and post-consumer polyolefin waste into wax and lubricant products using Pt nanoparticles supported on a well-defined $SrTiO_3$ (STO) nanocuboid perovskite surface (see Figure 33.2). STO was chosen as an initial support due to its crystalline, highly ordered, and well-characterized surface with a close lattice-match between the support and the ensuing Pt nanoparticles, resulting in a cube-on-cube epitaxy on the (100) facets of STO [9]. The cube-on-cube epitaxy contributes to a large interfacial energy for the nanoparticle/surface interface, stabilizing the Pt nanoparticles during hydrogenolysis conditions and preventing sintering [10]. The STO nanocuboids are synthesized under hydrothermal conditions [11], and the Pt nanoparticles were initially deposited using atomic layer deposition (ALD), a self-limiting gas-phase process for depositing thin films and nanoparticles while having control over the size and distribution [12]. Since these initial reports, however, further work has been done to scale up synthesis methods beyond the scale typically done with ALD, and towards synthesis methods more favorable for commercialization. These results also highlight a rational design strategy for future hydrogenolysis catalysts that benefit from epitaxial stabilization.

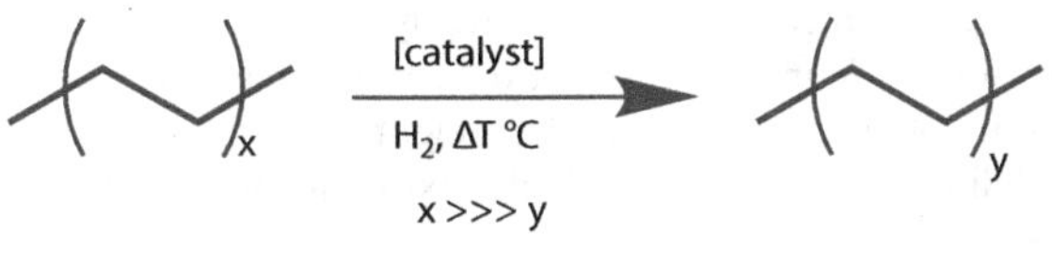

$$\text{[catalyst]} \quad H_2, \Delta T\ °C \quad x >>> y$$

Figure 33.2 General scheme for catalytic hydrogenolysis of polyolefins into smaller hydrocarbons under moderate temperatures and H_2 pressures.

The Pt/STO catalyst was initially tested for polyolefin conversion at 300 °C under 170 psi of H_2 for 96 h using ~1:50 Pt:polyolefin by mass and Pt/STO with 11.1 wt% Pt loading [13]. Polyethylene varying in molecular weight (number-average molecular weight (M_n) = 8,000 – 160,000 Da) was used as the reactant and found to selectively convert into smaller hydrocarbons (M_n = 590 – 990 Da) with a narrow distribution (polydispersity index (Đ) = 1.1 – 1.3), suggesting polyethylene of varying sizes from various plastic products could be mixed and remanufactured into a singular product, as summarized in Table 33.1. Furthermore, one of these samples was a post-consumer plastic bag with similar results to the pre-consumer plastic, suggesting additives present in plastic packaging may not adversely affect the catalyst at reasonable concentrations, and that extensive pre-processing prior to hydrogenolysis

Table 33.1 Plastic samples before and after thermal or catalytic hydrogenolysis with associated properties and yields. Molecular weight and distribution determined via gel permeation chromatography.

Plastic sample	$M_n^{\dagger}$, Da	Đ‡	Yield, %§
8k PE	8,150	2.7	n/a
8k PE – Thermal	5700	3.2	n/a
8k PE – Pt/STO	590	1.1	42
15k PE	15,400	1.1	n/a
15k PE – Thermal	17,300	1.1	n/a
15k PE – Pt/STO	660	1.1	68
64k PE	64,300	1.1	n/a
64k PE – Thermal	30,050	1.8	n/a
64k PE – Pt/STO	800	1.2	91
158k PE	158,000	2.7	n/a
158k PE – Thermal	16,800	8.1	n/a
158k PE – Pt/STO	820	1.2	> 99
Plastic Bag PE	33,000	3.5	n/a
Plastic Bag – Thermal	38,800	3.0	n/a
Plastic Bag – Pt/STO	990	1.3	97

†Number-average molecular weight measured in Daltons via high-temperature gel permeation chromatography.
‡Polydispersity as defined by the ratio between mass-average molecular weight and number-average molecular weight.
§Yield as defined by the mass of final wax/liquid product relative to the initial mass of PE.
Reaction conditions: 170 psi H_2, 300 °C, 96 h, 50 mg PE, and 1.1 mg Pt/STO (11.1 wt% Pt loading)

may not be necessary. This work has been expanded to include other post-consumer plastic waste, including polypropylene, with similar results [14]. A full exploration of plastic additives and contaminants and their potential effects on the performance of the catalyst or the resulting product has yet to be conducted, however, and so further research is required.

These initial hydrogenolysis reports also highlight the absence of aromatic or unsaturated byproducts (e.g. benzene, toluene, alkenes, alkynes, etc.) as determined via NMR, which are often observed in plastic pyrolysis products. Low concentrations of aromatics and unsaturated hydrocarbons is important for producing food-grade waxes and oxidatively stable lubricant base oils. Hydrogenolysis of polyolefins increases yield of saturated hydrocarbons, and decreases the need for separation and filtration post-processing to meet product requirements.

33.3 Applications for Plastic-Derived Products

Waxes and lubricants have broad applications across many industries in both industrial- and consumer-centric products. As an example, utilizing a plastic-derived base oil in commercially indispensable lubricants for automobiles could be invaluable to lubricant blenders and consumers. A fully formulated lubricant is composed of the base oil [15], which provides the majority of the lubricating properties, and additives, which are used to improve stability and longevity, and to fine-tune the friction, wear performance, and viscosity of the lubricant for a specific application. Approximately 40 million tons of lubricants are consumed globally every year [16], of which the majority are petroleum-based distillates (e.g. Group II, Group III), and a growing percentage are synthetic hydrocarbons (e.g. poly-α-olefins or PAOs). PAOs are significantly more expensive than Group II and Group III base oils (x1.5 to x8 the cost per ton), but are becoming increasingly important and widespread as the industry shifts to longer lifetime, higher efficiency, and thermally stable lubricants [17]. Lubricants such as these also have the prospect of being recycled after use [18–20] and thus can contribute to a circular carbon economy. Manufacturers have commented on the lower performance of recycled lubricants, however, and so further development is required to meet customers' specifications. PAOs are highly branched oligomers of α-olefins [21] (i.e. octene, decene, etc.), that can be made in a range of viscosities, have low pour points, and low concentrations of contaminants (e.g. sulfur) or unsaturated molecules that would otherwise contribute to a decreased lifetime. This is in contrast to petroleum distillates, which have higher concentrations of deleterious contaminants. Polyolefin plastics, made from purified ethylene, propylene, etc., are effectively a pre-purified sulfur-free hydrocarbon feedstock in a similar vein as PAOs, thus making plastic-derived lubricants competitive with the market in terms of low sulfur content.

A previous report tested the frictional and wear performance of base oils derived from high-density polyethylene (HDPE, M_n ~ 5,400 Da, Đ = 7.1), linear low-density polyethylene (LLDPE, M_n ~ 221,000 Da, Đ = 4.1), and post-consumer bubble wrap (M_n ~ 203,750 Da, Đ = 4.0),[8] and were chosen as structurally different polyolefins to benchmark against the aforementioned industry base oils (Group III mineral oil and synthetics PAO4 and PAO10). For meaningful results, 100 °C was chosen as an analogous test temperature to automobile engine conditions [22]. Initial coefficient of friction (COF) measurements at 100 °C revealed comparable friction (COF < 0.15) between the plastic-derived oils and the synthetic lubricant PAO10. Average COF values revealed the LLDPE-derived oil to have lowest friction (COF = 0.13 with a range of 0.11 – 0.14) during tests at 100 °C. The other plastic-derived oils, HDPE-derived and post-consumer bubble wrap-derived oils, exhibited only slightly higher friction values at 0.15, with a range of 0.08 – 0.15. Importantly, in conjunction with the results from the hydrogenolysis experiments, these narrow results further point to a mixed plastic feedstock not being an issue when producing a singular and high-performance product.

Wear scar volume (WSV) measurements also showed a clear distinction between less performant industry lubricants and plastic-derived oils, with a ~43% reduction in wear between Group III and the HDPE oil (see Figure 33.3). At 100 °C, the Group III and PAO4 oils were unable to protect the surface as effectively, resulting in a larger amount of wear. To better understand the high-performance of the plastic-derived oils, structural characterization via matrix-assisted laser desorption/ionization (MALDI) and ^{1}H NMR were performed. Group III oil was found to be comprised of small molecules with low polydispersity (M_n = 430 Da, Đ = 1.01) and high branching (0.18 branches per C). A high degree of branching for a smaller oil means fewer total interactions between methylene units in the alkane chain. This results in a decrease in average intermolecular forces amongst chains within the film [23]. Under moderate temperatures and large load, disruptions within the lubricant film can take place, resulting in more frequent metal-to-metal contact and hence higher wear of the contact spots [24–28]. By comparison, PAO10 was found to be larger with similar polydispersity (M_n = 635 Da, Đ = 1.04), but a lower degree of branching (0.14 branches per C). Unlike Group III, PAO10 molecules are able to align and slide at the interface more effectively while maintaining a more continuous film [25] due to the favorable intermolecular interactions amongst the methylene units in each chain. The HDPE oil, on the other hand, falls between Group III and PAO10 in size and polydispersity (M_n = 475

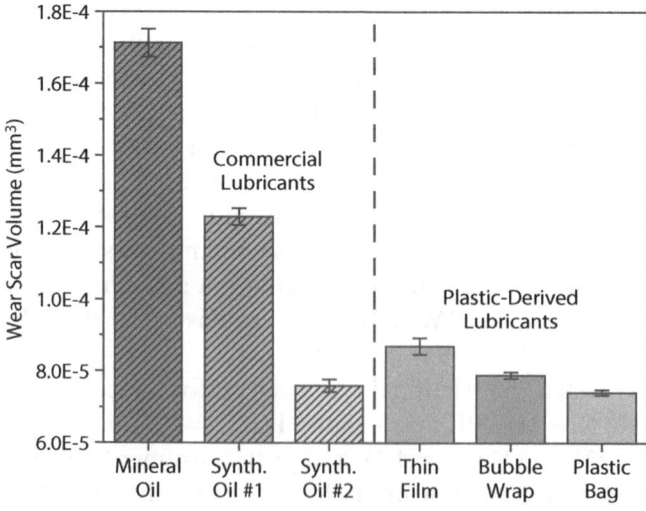

Figure 33.3 Wear scar volume as a function of base oil, with commercial lubricants on the left (red, mineral oil; orange, PAO4; yellow, PAO10), and plastic-derived lubricants on the right (green, HDPE; blue, bubble wrap; turquoise, LLDPE). Plastic-derived lubricants consistently yielded lower wear to mineral oil and comparable to higher-performance synthetic oils [8].

Da, Đ = 1.03), albeit with extremely low branching when compared to the two industry oils (0.01 branches per C). Despite being the intermediate oil in molecular weight and polydispersity, the incredibly low branching is crucial for maintaining a large number of methylene interactions between chains and thus preserving a more uniform film responsible for low wear. These results showed that given any plastic-derived lubricant, molecular weight and degree of branching are both important descriptors for evaluating the overall structural characterization and thus the tribological performance.

Interestingly, blending studies were also undertaken to determine if any synergistic effects exist between plastic-derived oils and a synthetic oil. The HDPE and PAO10 oils were chosen due to both liquids exhibiting excellent and near-equivalent wear and friction results at 100 °C. HDPE oil was added 0, 10, 20, 30, 50, 70, 90, and 100 wt% to PAO10 for friction and wear measurements at room temperature (RT, 25 °C) and 100 °C. At RT, the HDPE oil showed modest wear compared to virgin PAO10, with friction increasing as the test progressed. Based on the low wear and increasing friction, it appeared the HDPE oil was not forming a boundary layer at this temperature and the friction was from viscous drag of the lubricant [29]. Raising the temperature to 100 °C significantly lowered friction by greatly improving the fluidity and hence wetting. This change in performance under different temperatures was most likely dependent on structural characteristics (i.e. chain length and degree of branching). Once adhesion of the lubricant to the test surface was achieved, the degree of branching was the primary factor influencing friction. The linear structure of the HDPE oil provided a more flexible interface where the mobility of the linear molecules was better under reduced volume and extreme pressure [30]. The difference in fluidity at the thin film interface between PAO10 (0.14 branches per C) and HDPE oil (0.01 branches per C) may explain their respective COFs at 100 °C. A gradual COF rise was seen in the HDPE oil at RT, however, and may also be attributable to the iron oxide tribolayer formation. Under both room temperature and 100 °C conditions, an optimum concentration of 20 – 30% HDPE oil in PAO10 was found with respect to friction and wear. At RT, 30 wt% HDPE oil in PAO10 exhibited the imparted frictional stability of PAO10 as well as an overall decrease in friction by ~9% and wear by 30% when compared to virgin PAO10. At 100 °C, a similar reduction is seen with 20 wt% HDPE oil in PAO10 yielding a ~9% decrease in friction and 34% decrease in wear compared to PAO10. The gradual COF increase seen previously is absent at 100 °C when the HDPE oil can immediately wet and adhere to the surface. A reduction in friction is hypothesized to be the result of the linear chains from the HDPE oil. The variability in the COF for the optimum mixture was also seen to be muted when compared to room temperature. The increase in polydispersity in the 30 wt% HDPE oil mixture may account for the improved friction and wear and thus synergy. Larger PAO10 molecules are pushed out from the mixture at the contact, resulting in a concentration gradient, which are then entropically induced to diffuse into the bulk mixture [31]. A constant depletion force may enhance fluidity within the film and thus may improve friction. These results not only showed that plastic-derived oils were comparable in performance to synthetic base oils and superior to some mineral oils, but that opportunities may exist for blending plastic-derived oils into synthetic lubricant packages to yield superior performance and reduce associated lifecycle emissions, as discussed hereafter. Incorporating plastic-derived oils into existing lubricant packages may severely reduce the lab-to-market timeline of sustainable oils as commercialization ramps up.

33.4   Environmental and Emissions Ramifications from Catalytic Hydrogenolysis

Given the energy- and carbon-rich nature of polyolefins, it is advantageous to convert waste plastics back into refinery products of similar structure and characteristics beyond performance results. Among the possible products, lubricants attract significant interest as they are ones of the most valuable refinery products. For example, crude oil is priced as \$1.5/gal, transportation fuels (e.g. gasoline, diesel) are priced between \$1.3-1.7/gal, and BTX chemicals are priced about \$2.0-3.2/gal [32–34]; in contrast, lubricant is priced as \$6-10/gal (for year 2018) [35]. Lubricant is not only a value-added product, but also has a large market size. U.S. refineries, for example, produced 64.97 million barrels in 2017 [36], or 8.9 million ton, with a market size greater than 20 billion dollars, and an expected growth of 40-50% by 2035 [37]

The much higher price of lubricant product can be related to the complex production process and intensive energy use. In refineries, lubricant is commonly produced from vacuum distillation bottom cut, which requires extensive energy to obtain cuts from crude oil. The vacuum distillation residue is then processed to lubricant via four steps: solvent deasphalting, solvent extraction to remove aromatics, dewaxing via hydrocracking or solvent extraction, then finally hydrofinishing to remove the remaining impurities and aromatics. These processes involve intensive energy demand, hydrogen use, and solvent use and recycling. Consequently, the lubricant production has been estimated to have the highest GHG emission with about 11.5 g CO_2 eq/MMbtu lubricant, much higher than the GHG emission allocated for the production of gasoline (i.e., 7.3 CO_2 eq/MMbtu gasoline BOB) and diesel blendstocks (4.7 CO_2-eq/MMbtu gasoline BOB) [38]. Thus, in order to reduce GHG emissions, it would be of great interest to produce lubricants from a less energy intensive pathway.

A hypothetical plant with 250 t/day processing capabilities was evaluated under technoeconomic (TEA) and lifecycle (LCA) analyses in producing a lubricant product from a previous study [39]. A low-yield lubricant scenario of 60% led to a daily production of 153 metric tons of oil and about 46 t/day of naphtha. Increasing the lubricant yield from 60% to 90% as seen in earlier hydrogenolysis reports led to a 50% increase in lubricant production. The scenarios with 90% yield were characterized by a lower amount of fresh hydrogen per kg of plastic-derived oil, due to the more efficient conversion. On-site CO_2 emissions also decreased by 78% due to the lower amount of fuel gas burnt for power generation. The co-product yield decreased to about 8 t/day, as well as the net electric power produced, that decreased by 42%. The oil synthesis energy efficiency was calculated based on the energy of the product, co-product, and feedstock materials. For the scenarios with 90% yield, both the increase in oil productivity and decrease in hydrogen demand led to a more efficient process overall, with an energy efficiency of about 90%.

The CO_2 emissions associated with the production of plastic-derived lubricants is of primary interest, to ensure the environmental benefit synergy between solid waste reduction and air emission reduction. The air emission of the plastic derived lubricant product was evaluated for onsite stage and cradle-to-gate stage by including upstream burdens. In all the scenarios analyzed, the power generated onsite fulfilled the energy demand, while the surplus of electricity was accounted as injected into the grid. The cradle-to-gate CO_2 emissions of the lubricant production from plastic waste was calculated by accounting for the upstream burdens of feedstocks and other material and energy input.

The production process was divided into an onsite and an upstream fraction. Lubricants, especially mineral oils, are synthesized from vacuum gas oil (VGO) or waxy distillates (both obtained from vacuum distillation of crude oil) as starting raw materials. As such, the processing steps up to vacuum distillation are usually considered upstream of the actual onsite lubricant synthesis process. To understand the overall impact of lubricant synthesis, it is therefore important to consider the emissions associated with the onsite production of lubricants, including the use of heat, electricity and fuel use for the various units. CO_2 emissions data for lubricant production has rarely been reported, however, as most studies focus on refinery emissions for major products, such as gasoline and diesel. Instead, a unit-based process analysis was used to obtain a breakdown of energy sources and related emissions.

Based on energy and environmental profile for US refineries, the total fuel and electricity consumption for lubricant production was reported to be 11.26 MJ/kg lube and 0.5 MJ/kg lube, respectively [40]. Other literature sources report fuel and electricity use at 11.56 MJ/kg lube and 0.06 MJ/kg, respectively, for a paper that reports criteria air pollutant and greenhouse gas emissions allocated to refinery products for US refineries [38]. Best Available Techniques (BAT) for refining of mineral oil and gas, a European Commission report also provides fuel and electricity use at 6.07 MJ/kg lube and 0.21 MJ/kg lube, respectively. The corresponding onsite CO_2 for lubricant production fall in the range of 0.75– 1.20 kg CO_2/kg Group III lubricant oil [41].

The process for lubricant production would have enormous environmental and national energy security benefits. For the former, catalytic hydrogenolysis has an extraordinary benefit in solid waste reduction in landfill and in waterways, without producing and disposing solid char. This is especially important for the U.S. as the waste export market continues to dwindle as foreign countries ban the importing of waste. For the latter, hydrogenolysis can reduce GHG emissions significantly compared to refineries by displacing the energy

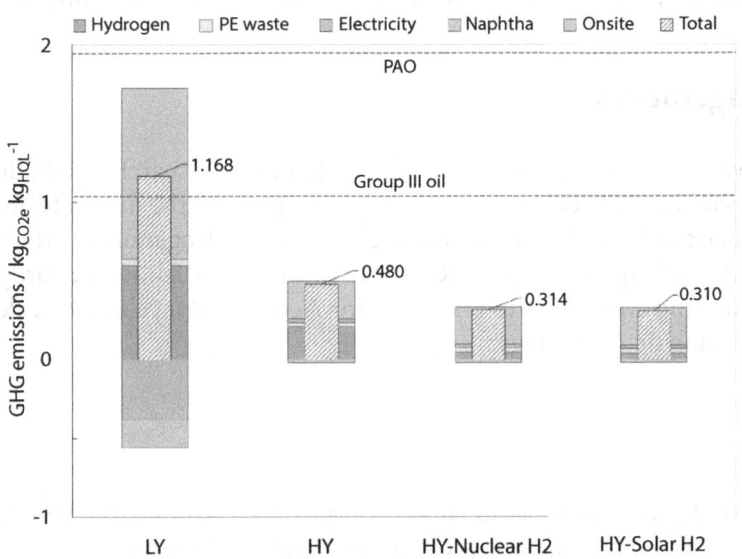

Figure 33.4 Total GHG emissions from plastic-derived oils for low yield (LY) and high yield (HY) scenarios. Alternative hydrogen sources (i.e. nuclear and solar) were also included. Benchmark GHG emissions from Group III oil and PAO production are also included in red. Reproduced with permission from the authors [39].

intensive lubricant production process. Preliminary TEA shows that lubricant produced from plastic waste has overall emissions as low as 0.31 kg CO_2/kg lubricant (see Figure 33.4), which is significantly lower the GHG emission from petroleum lubricant (0.75-1.20 kg CO_2/kg lubricant). Based on U.S. lubricant production of 64.97 million barrels in 2017, replacing petroleum-derived lubricant with plastic waste-derived lubricants could reduce GHG emissions up to 4 million ton per year in U.S.

This study showed lubricants produced from pre- and post-consumer polyolefins waste can significantly reduce energy use, and thus less GHG emissions, associated with lubricant production. Consequently, this will allow economic production with relatively small plants, require relatively small capital cost for plant construction and commercialization. This implies that such plant could be in the proximity [42] of wide spread waste separation facilities or Material Recovery Facilities (MRF) and ease commercialization prospects.

33.5   Conclusions & Recommendations

These studies have shown catalytic hydrogenolysis to be an exciting and novel technology capable of diverting single-use plastic waste away from landfills and incinerators and back into the carbon economy via higher-value chemical commodities in wax and lubricant applications. Doing so would reduce the lifecycle emissions of plastic products, and reduce the proliferation of microplastics found in the environment. The Pt-based catalyst was found to be robust and capable of converting a mixture of pre-consumer and post-consumer polyethylene waste into lubricants with performance comparable to synthetic base oils. This improves commercialization prospects by not necessarily requiring extensive sorting and pre-processing prior to conversion. Preliminary lifecycle analysis also showed this process to yield a low-emissions pathway for lubricants, clearing the way for a sustainable alternative to various wax and lubricants products where there are few currently available options.

Acknowledgements

The authors would like to acknowledge the fundamental research conducted at Argonne National Laboratory and led by Max Delferro as part of the Institute for Cooperative Upcycling of Plastics (iCOUP) in elucidating catalytic hydrogenolysis. The authors would also like to acknowledge the Chain Reaction Innovations fellowship program for their non-dilutive funding and access to resources for Aeternal Upcycling to scale up and commercialize the catalytic technology.

References

1. OECD Global plastic waste set to almost triple by 2060, says OECD. https://www.oecd.org/environment/global-plastic-waste-set-to-almost-triple-by-2060.htm.
2. EPA Advancing Sustainable Materials Management: 2018 Tables and Figures; 2020.
3. Rahimi, A.; Garcia, J. M., Chemical recycling of waste plastics for new materials production. *Nature Reviews Chemistry* 2017, 1 (6).

4. Jambeck, J. R.; Geyer, R.; Wilcox, C.; Siegler, T. R.; Perryman, M.; Andrady, A.; Narayan, R.; Law, K. L., Plastic waste inputs from land into the ocean. *Science* 2015, 347 (6223), 768-771.

5. Andrady, A. L., Microplastics in the marine environment. *Mar Pollut Bull* 2011, 62 (8), 1596-605.

6. Eerkes-Medrano, D.; Thompson, R. C.; Aldridge, D. C., Microplastics in freshwater systems: a review of the emerging threats, identification of knowledge gaps and prioritisation of research needs. *Water Res* 2015, 75, 63-82.

7. Borrelle, S. B.; Ringma, J.; Law, K. L.; Monnahan, C. C.; Lebreton, L.; McGivern, A.; Murphy, E.; Jambeck, J. R.; Leonard, G. H.; Hilleary, M. A.; Eriksen, M.; Possingham, H. P.; De Frond, H.; Gerber, L. R.; Polidoro, B.; Tahir, A.; Bernard, M.; Mallos, N.; Barnes, M.; Rochman, C. M., Predicted growth in plastic waste exceeds efforts to mitigate plastic pollution. *Science* 2020, 369, 1515-1518.

8. Hackler, R. A.; Vyavhare, K.; Kennedy, R. M.; Celik, G.; Kanbur, U.; Griffin, P. J.; Sadow, A. D.; Zang, G. Y.; Elgowainy, A.; Sun, P. P.; Poeppelmeier, K. R.; Erdemir, A.; Delferro, M., Synthetic Lubricants Derived from Plastic Waste and their Tribological Performance. *Chemsuschem* 2021, 14 (19), 4181-4189.

9. Enterkin, J. A.; Poeppelmeier, K. R.; Marks, L. D., Oriented catalytic platinum nanoparticles on high surface area strontium titanate nanocuboids. *Nano Lett* 2011, 11 (3), 993-7.

10. Enterkin, J. A.; Setthapun, W.; Elam, J. W.; Christensen, S. T.; Rabuffetti, F. A.; Marks, L. D.; Stair, P. C.; Poeppelmeier, K. R.; Marshall, C. L., Propane oxidation over Pt/SrTiO3 nanocuboids. *ACS Catalysis* 2011, (1), 629-635.

11. Rabuffetti, F. A.; Kim, H.-S.; Enterkin, J. A.; Wang, Y.; Lanier, C. H.; Marks, L. D.; Poeppelmeier, K. R.; Stair, P. C., Synthesis- Dependent First-Order Raman Scattering in SrTiO3 Nanocubes at Room Temperature. *Chem. Mater.* 2008, (20), 5628-5635.

12. Christensen, S. T.; Elam, J. W.; Rabuffetti, F. A.; Ma, Q.; Weigand, S. J.; Lee, B.; Seifert, S.; Stair, P. C.; Poeppelmeier, K. R.; Hersam, M. C.; Bedzyk, M. J., Controlled growth of platinum nanoparticles on strontium titanate nanocubes by atomic layer deposition. *Small* 2009, 5 (6), 750-7.

13. Celik, G.; Kennedy, R. M.; Hackler, R. A.; Ferrandon, M.; Tennakoon, A.; Patnaik, S.; LaPointe, A. M.; Ammal, S. C.; Heyden, A.; Perras, F. A.; Pruski, M.; Scott, S. L.; Poeppelmeier, K. R.; Sadow, A. D.; Delferro, M., Upcycling Single-Use Polyethylene into High-Quality Liquid Products. *Acs Central Sci* 2019, 5 (11), 1795-1803.

14. Hackler, R. A.; Lamb, J. V.; Peczak, I. L.; Kennedy, R. M.; Kanbur, U.; LaPointe, A. M.; Poeppelmeier, K. R.; Sadow, A. D.; Delferro, M., Effect of Macro- and Microstructures on Catalytic Hydrogenolysis of Polyolefins. *Macromolecules* 2022, 55 (15), 6801-6810.

15. Torbacke, M.; Rudolphi, A. K.; Kassfeldt, E., *Lubricants: Introduction to Properties and Performance.* John Wiley and Sons Ltd: 2014.

16. Tsui, D., Why synthetics are leading the lubricants market. *Tribol Lubr Technol* 2017, 73 (6), 24-28.

17. Mang, T.; Dresel, W., *Lubricants and Lubrication.* Wiley-VCH: 2017.

18. Bhaskar, T.; Uddin, M. A.; Muto, A.; Sakata, Y.; Omura, Y.; Kimura, K.; Kawakami, Y., Recycling of waste lubricant oil into chemical feedstock or fuel oil over supported iron oxide catalysts. *Fuel* 2004, 83 (1), 9-15.

19. Rincon, J.; Canizares, P.; Garcia, M. T., Regeneration of used lubricant oil by ethane extraction. *Journal of Supercritical Fluids* 2007, 39 (3), 315-322.

20. Kamal, A.; Khan, F., Effect of Extraction and Adsorption on Re-refining of Used Lubricating Oil. *Oil Gas Sci Technol* 2009, 64 (2), 191-197.

21. Benda, R.; Bullen, J.; Plomer, A., Synthetics Basics: Polyalphaolefins - Base Fluids for High-Performance Lubricants. *Journal of Synthetic Lubrication* 1996, 13, 40-57.

22. Priest, M.; Taylor, C. M., Automobile engine tribology - approaching the surface. *Wear* 2000, 241 (2), 193-203.

23. Velkavrh, I.; Kalin, M., Comparison of the effects of the lubricant-molecule chain length and the viscosity on the friction and wear of diamond-like-carbon coatings and steel. *Tribol Int* 2012, 50, 57-65.

24. Biresaw, G.; Bantchev, G. B., Pressure Viscosity Coefficient of Vegetable Oils. *Trib. Lett.* 2013, 49, 501-512.

25. Wang, W. W.; Li, P.; Sheng, S. Z.; Tian, H. T.; Zhang, H. D.; Zhang, X., Influence of Hydrocarbon Base Oil Molecular Structure on Lubricating Properties in Nano-scale Thin Film. *Tribology Letters* 2019, 67 (4).

26. Sivebaek, I. M.; Samoilov, V. N.; Persson, B. N. J., Squeezing molecular thin alkane lubrication films between curved solid surfaces with long-range elasticity: Layering transitions and wear. *J Chem Phys* 2003, 119 (4), 2314-2321.

27. Xu, N.; Wang, X. B.; Ma, R.; Li, W. M.; Zhang, M., Insights into the rheological behaviors and tribological performances of lubricating grease: entangled structure of a fiber thickener and functional groups of a base oil. *New J Chem* 2018, 42 (2), 1484-1491.

28. Jahanmir, S., Chain-Length Effects in Boundary Lubrication. *Wear* 1985, 102 (4), 331-349.

29. Okrent, E. H., The effect of lubricant viscosity and composition on engine friction and bearing wear. *ASLE Transactions* 1961, 4 (1), 97-108.

30. Kioupis, L. I.; Maginn, E. J., Impact of molecular architecture on the high-pressure rheology of hydrocarbon fluids. *J Phys Chem B* 2000, 104 (32), 7774-7783.

31. Israelachvili, J. N.; Kott, S. J.; Gee, M. L.; Witten, T. A., Forces between Mica Surfaces across Hydrocarbon Liquids - Effects of Branching and Polydispersity. *Macromolecules* 1989, 22 (11), 4247-4253.

32. Energy, A.-E. U. S. D. o. Funding Opportunity No. DE-FOA-0001954, CFDA Number 81.135, table 2. https://arpa-e-foa.energy.gov/Default.aspx#FoaId4edadf42-12aa-4916-bf6a-b560a6f20e95

33. OPIS International Feedstocks Intelligence Report. https://www.opisnet.com/wp-content/uploads/2019/01/feedstocks-report-sample.pdf

34. Fiber2Fashion BTX Market Report and Price Trend. https://www.fibre2fashion.com/market-intelligence/textile-market-watch/benzene-price-trends-industry-reports/18/

35. EIA Lubricants consumption, price, and expenditure estimates, 2019. https://www.eia.gov/state/seds/data.php?incfile=/state/seds/sep_fuel/html/fuel_lu.html&sid=US.

36. EIA U.S. Refinery Net Production of Lubricants. https://www.eia.gov/dnav/pet/hist/LeafHandler.ashx?n=pet&s=mlurx_nus_1&f=a.

37. Bau, A. Lubes growth opportunities remain despite switch to electric vehicles. https://www.mckinsey.com/industries/oil-and-gas/our-insights/lubes-growth-opportunities-remain-despite-switch-to-electric-vehicles (2018).

38. Sun, P.; Young, B.; Elgowainy, A.; Lu, Z.; Wang, M.; Morelli, B.; Hawkins, T., Criteria Air Pollutant and Greenhouse Gases Emissions from U.S. Refineries Allocated to Refinery Products. *Environ Sci Technol* 2019, 53 (11), 6556-6569.

39. Cappello, V.; Sun, P.; Zang, G.; Kumar, S.; Hackler, R.; Delgado, H. E.; Elgowainy, A.; Delferro, M.; Krause, T., Conversion of plastic waste into high-value lubricants: techno-economic analysis and life cycle assessment. *Green Chemistry* 2022, 24 (16), 6306-6318.

40. Energy, U. S. D. o. Industrial Technologies Program; 2007.

41. Barthe, P.; Chaugny, M.; Roudier, S.; Sancho, L. D. Best Available Techniques (BAT) Reference Document for the Refining of Mineral Oil and Gas; European Commission: 2015.

42. Staub, C. MRF operator and public agency are at odds in Iowa, Resource Recycling. https://resource-recycling.com/recycling/2020/06/16/mrf-operator-and-public-agencyare-at-odds-in-iowa/.

34

PE Upcycling Using Ozone and Acid Treatments

Michael S. Behrendt[1], Brandon D. Howard[1], Scott Calabrese-Barton[1], John R. Dorgan[1]*
Samantha Au Gee[2] and Amit Gokale[2]

[1]*Reincarnating Polymers for the Circular Economy (RIPCE), Chemical Engineering and Materials
Science Department, Michigan State University, Lansing, MI, USA*
[2]*Chemical and Process Engineering R&D - North America (RCP/ON), BASF Corporation,
Houston, TX, USA*

Abstract

Converting waste plastic into more valuable polymers can fundamentally change the economics of polyethylene recycling. Polyamides are widely used in automotive applications and major car manufacturers are committed to using recycled plastics. This "market pull" provides compelling economic motivation for upcycling. The need is to develop processes for remanufacturing polyethylene into polyamides to enable an economically viable pathway towards much greater recycling rates. The identified opportunity is addressed through a powerful Industry-University collaboration.

The technical approach pursued is energy-efficient, low temperature, oxidative treatment of PE enhanced by process intensification. The key innovation, enabling success where other efforts have failed, is combining very inexpensive wastewater treatment oxidation chemistries with novel separation processes. This innovative combination provides a cost-effective methodology for remanufacturing end-of-life plastics into value-added materials. Project success using low temperature oxidation has been achieved by two methods: 1) acid pretreatment followed by electrochemically generated hydrogen peroxides, and 2) ozonolysis. Synergistic effects (UV light, catalysts, surfactants, heating) are being investigated as a means of enhancing effectiveness in the search for optimal operating conditions.

Keywords: Polyethylene, LDPE, HDPE, upcycling, ozone, ozonolysis, sulfonation, acid activation

*Corresponding author: jd@msu.edu

Nabil Nasr (ed.) Technology Innovation for the Circular Economy: Recycling, Remanufacturing, Design, Systems Analysis and Logistics, (433–448) © 2024 Scrivener Publishing LLC

34.1 Introduction

The project goal is to develop an economic process for manufacturing diacid monomers or other value-added upcycled chemicals from reclaimed post-use polyethylene. On-going activities to introduce recycled content by reclaiming, processing, and blending in PA66 recovered from carpets is, to date, proving uneconomic. The promising ability to produce diacids from low-value reclaimed polyethylene would provide considerable improvements to the baseline technology.

Aqueous ozone chemistry and Advanced Oxidation Processes (AOPs) sees extensive use in the municipal wastewater treatment sector, with the chemicals targeted having similar structures to waste plastic such as polyethylene. The theoretical chemistry behind waste treatment and the breakdown of plastics into diacids is the same. This presents an opportunity to produce value-added compounds in a mild environment that does not require expensive or toxic chemicals, high temperatures, or high pressures. The reactants will be primarily derived from water and air.

Polyethylene can also be degraded using an acid treatment as the main oxidation step or a pretreatment followed by chemical oxidation [1] [2]. The conditions at which polyethylene will degrade by certain acids will be a function of the temperature and pressure, oxidation potential and concentration of the acid, and polymer solubility in the solvent/acid. For example, Bäckström et al. readily degraded LDPE to form diacids in 1.58 M nitric acid assisted by microwave irradiation at 180 °C and 40 bar [2]. Alternatively, Chow et al. activated both LDPE and HDPE by first sulfonating them using chlorosulfuric acid (CSA) in organic solvents at ≈ 60 °C, and then grafting $Fe(III)Cl_3$ [1]. This was followed by reacting the activated PE at room temperature with hydrogen peroxide to form diacids with nearly 100% degradation. A key reason for the success of these methods can be attributed to the high energy input of the former, and for the latter, the use of halogenated solvents that enhanced PE swelling and subsequent contact with CSA. To make a green PE degradation process it would be necessary to develop a low energy (low temperature and pressure) process without organic solvents.

To show a proof of concept of a green process for PE degradation by acids, we develop a method for PE activation in pure acid mixtures of varying strength under moderate temperatures followed by depolymerization using hydrogen peroxide. Interestingly, CSA is relatively soluble in organic solvent [3] and could have some capacity to cause swelling of PE. This implies CSA could potentially act as a solvent and oxidizer simultaneously. Likewise, CSA is readily soluble in sulfuric acid (SA) [3] and can be diluted to adjust the strength of the oxidizing medium, in addition to providing liquid volume for PE dispersion. Provided that the amount of CSA is chosen such that almost all acid is fully reacted, the resulting acid-polymer mixture can be safely diluted in an aqueous medium and readily oxidized using hydrogen peroxide.

34.2 Materials and Methods

34.2.1 Materials

LDPE was Agility 1021 Performance, sourced from Dow Chemical. HDPE was Dow DMDA 6230. Glutaric, adipic, and succinic acids and hydrogen peroxide (30%) were sourced from

Thermo Scientific. HPLC grade 1,2,4 Trichlorobenzene, chloroform, and dichloromethane and Sulfuric acid (ACS grade) were sourced from VWR chemicals. Chlorosulfuric acid (99%) and Fe(III)Cl$_3$ (≥98%) were sourced from Sigma Aldrich. Technical grade Fe(III) sulfate was sourced from Spectrum Chemical Corp.

34.2.2   Methods

Attenuated Total Reflectance (ATR) was conducted on a Thermo Scientific Nicolet™ iS50 Fourier Transform Infrared (FTIR) Spectrometer using a minimum of 8 scans at a resolution of 2. The diamond interface was washed with isopropanol between each scan.

Thermogravimetric analysis (TGA) of HDPE before and after sulfonation was performed using a TA Instruments Thermogravimetric Analyzer Q500 at temperatures between 30 and 580 °C at a ramping rate of 10 °C/min. A single replicate for each sample was reported.

The ozone reactor was a two-liter glass Continuous Stirred Tank Reactor (CSTR) using a PTFE radial blade impeller running at 600-900 RPM. Samples typically formed good suspensions at this mixing speed, with some differences observed depending on the degree of oxidation of the plastic. Temperature was controlled at 80+0.2 °C. Exit gas was bubbled through a solution of potassium iodide to strip excess ozone. The primary ozonation run consisted of 20 g of LDPE macerated to <1.18 mm diameter suspended in 1500 g DI water. Hydrochloric acid used to adjust pH to 1.2. The pH fluctuated +0.2 during the course of the run.

An A2Z Ozone S6G industrial ozonator was used at 100% power. Ozone concentration was 5.0% + 0.4% in 20% oxygen. Differential Scanning Calorimetry (DSC) was conducted on a TA Instruments Q200 DSC. Samples were heated to 150°C, cooled to 35°C to erase any thermal history, then reheated to 150°C to analyze crystallinity.

Sensors were a Hach Orbisphere C1100 Ozone Sensor and a Hach Orbisphere GA2X00 O2 EC Sensor using 29552A-A -AV- and 29552A-A -CT- PTFE membranes, respectively, calibrated in air.

Acid-only activation of PE (PEAA) was investigated by oxidizing approximately 1g of powder (diameter 250 μm to 1.18 mm) in 5 mL of acid containing 100, 50 and 20 vol% CSA in SA. HDPE powder was activated by the acid mixtures for two hours at 85 °C in 20 mL vials, with mixing using a Teflon coated magnetic stir bar. PEAA and all other experiments, unless stated otherwise, were performed in at least duplicate. Although preliminary tests showed that it was possible to add Fe(III) and peroxide simultaneously to react with the sulfonated HDPE (S-HDPE), for the purposes of this study, the S-HDPE was filtered and dried between each step to enable characterization. To enhance the reactivity of S-HDPE, FeCl$_3$ was grafted to the sulfonate groups by suspending S-HDPE powder in 100 mL of 0.75 M FeCl$_3$ solution overnight [1]. Samples were then filtered and washed until the filtrate was neutral. The powder was then dried at 40 °C under vacuum for approximately 18 hours.

Following the iron grafting, approximately 0.1 g of S-HDPE material was reacted in 50 mL of 0.5 M hydrogen peroxide at room temperature. Chow et al. notes that reaction times longer than a few hours did not result in more solubilized material [1]. Consequently, materials were reacted for five hours to allow the reaction to reach maximum conversion.

After the reaction, samples were filtered and both the solids and filtrate were dried. The filtrate was dried using a rotavap at 70 °C and transferred to a 20 mL vial.

If a majority of S-HDPE mass reacted with peroxide becomes liberated from the powder and dissolved, an approximate conversion (X) of powder S-HDPE can be calculated from the initial and final masses of S-HDPE

$$X = \frac{M_{S-HDPE,o} - M_{S-HDPE,f}}{M_{S-HDPE,o}} \tag{34.1}$$

Where $M_{S-HDPE,o}$ and $M_{S-HDPE,o}$ are the intitial and final masses of S-HDPE, respectively.

Titration was used to determine the ion exchange capacity (IEC) of the sulfonated powder and the acid number of the peroxide-reacted S-HDPE. To obtain the IEC, approximately 100 mg of S-HDPE was suspended in 50 mL of 2 M NaCl solution and allowed to equilibrate for 24 hours. Sodium ions displace protons from the sulfonate groups and allow their quantification by titration [4]. The acidified solution was titrated with standardized 0.1 M NaOH using phenolphthalein as an indicator. IEC (mmol sulfonate per gram S-HDPE) was calculated using:

$$IEC = \frac{C_{NaOH}\Delta V}{M_{S-HDPE}} \times \left(\frac{1000 \text{ mmol}}{1 \text{ mol}}\right) \tag{34.2}$$

where C_{NaOH} is the titrant molarity, ΔV is the titrant volume used, and M_{S-HDPE} is the mass of S-HDPE used for titration.

The carbon to sulfur ratio (C:S) was directly estimated from the IEC, by assuming a mostly linear structure for HDPE and that the polymer is primarily made up of sulfonate and methylene groups:

$$C:S \approx \frac{(1 - \dfrac{IEC \cdot MW_{CHSO_3H}}{1000} - 1)(MW_{CH_2}) + IEC/1000}{IEC/1000} \tag{34.3}$$

where MW_{CHSO_3H} (94.089 g/mol) and MW_{CH_2} (14.027 g/mol) are the molecular weights of the sulfonate and methylene groups along the chain, respectively.

The acid value (AV, mmol NaOH per gram acid) of the soluble products of S-HDPE oxidation by peroxide was determined using a Hannah Instruments automatic potentiometric titrator (HI931). Acids were extracted using 5 mL of ethanol to limit dissolution of Fe(III) sulfates and therefore AV overestimation [5]. The ethanol extract was then diluted in water to a volume of 50 mL and then titrated. Preliminary titrations of the ethanol extracted acids and remaining iron sulfates separately showed that acids were primarily extracted by the organic solvent, while the titration of the suspected iron sulfates resulted in significant Fe(III) precipitation (see Figures 34.1a and b). To report the acid value normalized by the mass of acids titrated (M_a), the ethanol extracted acids were dried, weighed, resolubilized in

Figure 34.1 Acid value titration.

the ethanol-water mixture, and then titrated. The endpoints of the titration were determined from the inflection point of the pH vs titrant volume plot. The AV was then calculated by:

$$AV = \frac{C_{NaOH}\Delta V}{M_a} \times \left(\frac{1000\ mmol}{1\ mol}\right) \tag{34.4}$$

The PEAA method was compared to organic solvent assisted sulfonation (OSS) technique of Chow *et al.* [1]. In this method, 1 g of powder HDPE (diameter 250 μm to 1.18 mm) was suspended in 25 mL chloroform under reflux and 15 mL of a 2:1 (v/v) mixture of dichloromethane and CSA were slowly dripped into the reaction flask during the first 30 minutes of the reaction. The reaction was allowed to run for 2 hours, after which the solvent was evaporated, and the OSS material was either washed and dried, or iron grafted overnight in 100 mL of 0.75 M FeCl₃, and then washed and dried. Samples made from OSS will be referenced as "OSS-HDPE". Similarly, PEAA samples will be referenced by percent volume of CSA (20,50, or 100) as "20 S-HDPE", "50 S-HDPE", etc.

34.3   Main Content of Chapter

34.3.1   Ozone Chemistry

34.3.1.1   *Chemistry of Oxidation Pathways*

Ozone attacks organic chemicals by two primary pathways: The direct reaction of ozone with unsaturated carbon-carbon bonds or saturated carbon-hydrogen bonds, and the formation or hydroxyl radicals. Both of these processes lead to the formation of carbon radicals on the polymer chain and the subsequent initiation of autoxidation. Direct ozone-chain reaction with unsaturated carbon bonds forms a reactive cyclic intermediate known as a molozonide [6] [7]. The molozonide then decomposes into a carbonyl and a biradical:

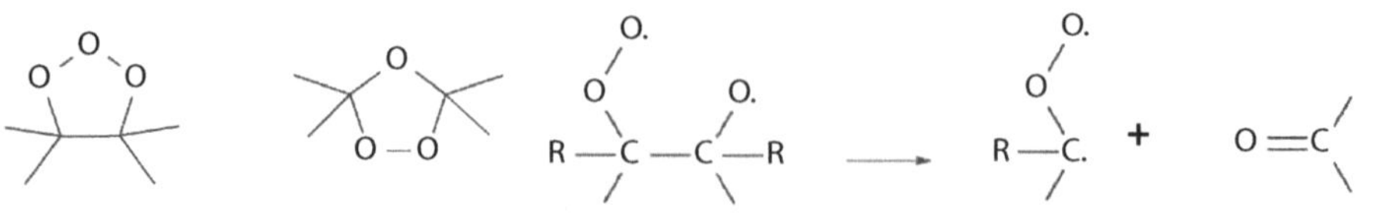

The mechanism of ozone attack on saturated hydrocarbons is more complicated and occurs more slowly, with a difference of approximately 10^6 at ambient temperature in PVC. A proposed mechanism is shown below [7]:

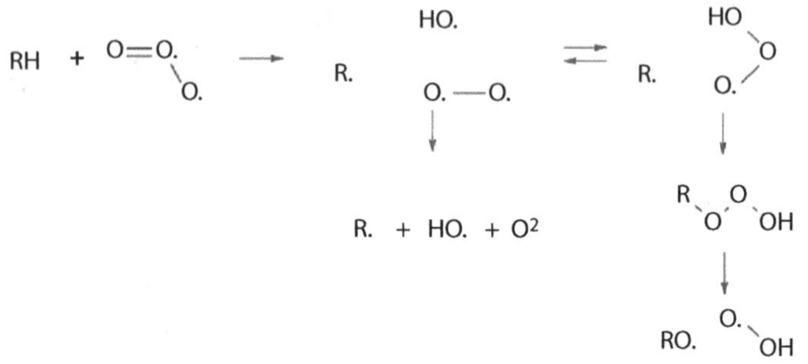

$$k_{O_3+C=C} = (10 \pm 2)10^6 \exp(-\frac{3500 \pm 500 \; cal \; mol^{-1}K^{-1}}{RT}) \; mol^{-1}S^{-1} \qquad (34.5)$$

$$k_{O_3+C-H} = (9 \pm 2)10^8 \exp(-\frac{13500 \pm 500 \; cal \; mol^{-1}K^{-1}}{RT}) \; mol^{-1}S^{-1} \qquad (34.6)$$

The radicals produced by this mechanism will react further by abstracting hydrogen atoms [7]:

$$HO^{\bullet} + RH \rightarrow H_2O + R^{\bullet} \qquad (34.7)$$

$$HOO^{\bullet} + RH \rightarrow H_2O_2 + R^{\bullet} \qquad (34.8)$$

$$RO^{\bullet} + RH \rightarrow ROH + R^{\bullet} \qquad (34.9)$$

$$R^{\bullet} + O_2 \rightarrow ROO^{\bullet} \qquad (34.10)$$

Saturated and unsaturated ozone attack ends with the same peroxyl radical. Autoxidation occurs when the radical abstracts a hydrogen from the carbon chain, either inter- or intramolecularly, forming a carboxylic acid and a primary carbon radical on the scised chain [7]. Electron-withdrawing groups lower susceptibility to ozone with carboxylic acid groups being unreactive with ozone. An expected lowering of molecular weight will occur until an unreactive acid or diacid is formed. Carboxylate ions, however, are reactive, indicating advantages to running this reaction under acidic conditions [6]. Ozone reacts with water to form hydroxyl radicals via the generation of the ozonide radical, $O_3^{\cdot -}$. The formation of these radicals is highly dependent on various initiators (such as hydroxide ions or iron ions), promoters (such as formate or primary alcohols) inhibitors (typically radical scavengers such as carbonate or bicarbonate), and pH, the latter decreasing ozone lifetime as it increases. The hydroxyl radical is very reactive and is typically governed by encounter-based rate laws. The radical is able to abstract hydrogen atoms and, in the presence of oxygen, initiate oxygen radical-based autoxidation.

34.4   Results and Discussion

34.4.1   Ozonolysis of LDPE

Ozonolysis was conducted by bubbling a stream of ~4.5% ozone in 20% oxygen through a vessel containing water and LDPE powder. Samples of reactor fluid and oxidized LDPE were taken at fixed intervals. The impeller produced a reasonable suspension, the degree of which changed based on the surface oxidation of the LDPE. As the reaction progressed, the LDPE was expected to oxidize differently based on crystallinity and particle size, with smaller particles and amorphous regions oxidizing faster. As such, the crystallinity of the plastic was expected to rise as amorphous regions are digested, then decrease slowly as crystalline regions reacted.

Recovered plastic showed a notable change in color as oxidation occurred, with large variations over the course of the first 72 hours, and less significant changes over the remaining 48 hours. The samples also showed a change in solubility in 1,2,4-Trichlorobenzene (TCB), which normally solubilizes LDPE at high temperatures. The increase of surface carbonyl groups is expected to be responsible for this phenomenon, with a loss of TCB-soluble mass of 0%, 0.48%, 1.49%, 1.96%, and 2.17% for the intervals of 0, 24, 48, 72, and 120 hours of reaction time respectively (Figure 34.2).

ATR scans of the recovered LDPE showed an IR carbonyl signal for all treated plastic, with little to no change in carbonyl index over the course of the reaction. This may be the result of the continuing reaction penetrating deeper into the polymer matrix than ATR is capable of detecting (Figure 34.3).

Changes in LDPE Crystallinity initially followed expected trends with a sudden rise in the first 24 hours of reaction time followed by a slower drop, as amorphous regions are

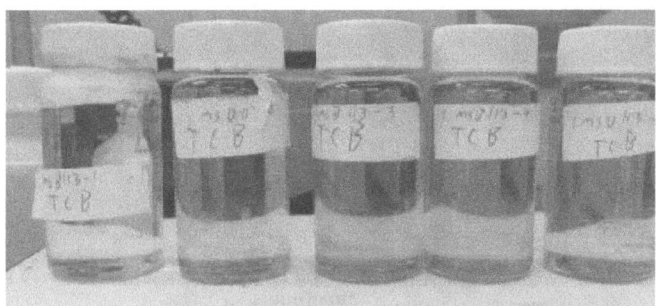

Figure 34.2 Left to right: Ozone reacted LDPE after 0, 24, 48, 72, 120 hours dissolved in TCB at 150°C.

digested quickly and crystalline regions are digested slowly. The subsequent rise in crystallinity is surprising, but may be a result of degradation exposing additional amorphous regions (Figure 34.4). More data may elucidate trends.

After the completion of the 120-hour reaction, the remaining LDPE was filtered, and the resulting reaction fluid was distilled to expose any nonvolatile products. ATR of the product revealed the presence of the expected carboxylic acid peak at 1695 cm^{-1} and carboxylate peak at 1609 cm^{-1}. Differences in the spectra of carboxylic acids of different lengths typically appear in the 1345-1180 cm^{-1} range, with a number of weak peaks equal to half the carbon atoms, or half plus one in the case of odd-numbered chains. These peaks coincide with the much stronger C-O and –CH$_2$CO- peaks and may occur as shoulders. This

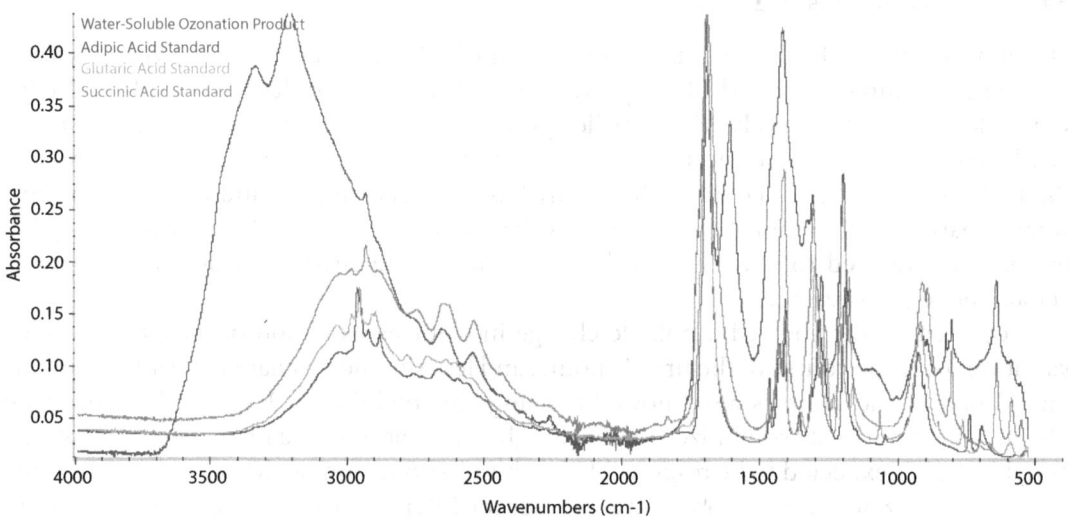

Figure 34.3 ATR of recovered product compared to diacid standards of 6, 5, and 4 carbons respectively. The product and succinic acid show notable peak overlap. Spectra are normalized to the maximum signal and corrected for ATR.

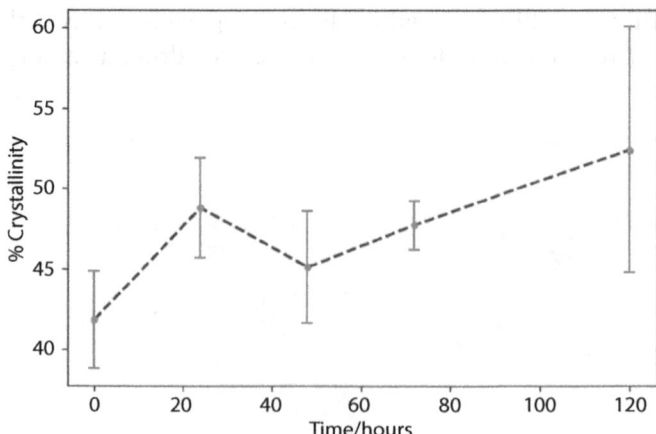

Figure 34.4 Changes in crystallinity of LDPE over the course of 120 hours of ozone digestion.

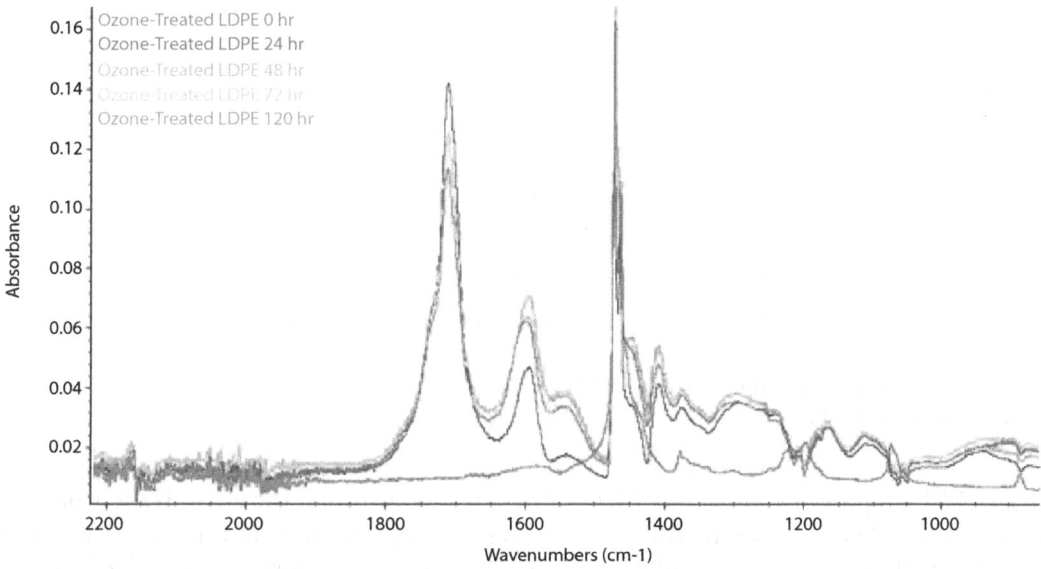

Figure 34.5 Ozone-treated LDPE scanned using ATR at 0, 24, 48, 72, and 120 hours. Spectra are corrected for ATR.

differentiable region showed considerable overlap with succinic acid (4 carbon dicarboxylic acid) as opposed to higher molecular weight diacids. The peaks at 3338 cm^{-1} and 3206 cm^{-1} are notably unique and may be attributed to –OH groups added through hydroxyl-radical based hydrogen abstraction. There is also a small presence of peroxides, identified by the product unique (and weak) peaks at 825 cm^{-1} and 1763 cm^{-1} (Figure 34.5).

34.4.2   Acid Activation of HDPE

When initially probing the reactivity of PEAA, a preliminary test was conducted on HDPE ribbons at the designated acid concentrations under similar conditions for the powders (see Figure 34.6). Interestingly, the ribbons showed large differences in the extent of oxidation depending on the concentration of CSA. For 20% CSA (Figure 34.6c), the oxidation was only observed where the column of acid was in contact with the ribbon. Whereas for 50% and 100% CSA samples (Figures 34.6a,b), oxidation was observed well above the acid level. This indicates that at higher concentrations of CSA, the acid mixtures retain a solvent-like characteristic which causes swelling and oxidation of the PE while simultaneously creating a driving force for significant diffusion up the ribbon due to acid consumption and swelling.

For powder S-HDPE samples, the extent of sulfonation appeared to follow a similar pattern as the PE ribbons according to FTIR data (Figure 34.7a). Due to the significant C-H stretching at 2916 and 2848 cm^{-1} and the much lower intensity S-O stretching at 1144 and 1020 cm^{-1} for 20 S-HDPE, this may indicate that most of the sulfonation occurred at the surface. This was further corroborated by data in Table 34.1 which shows the measured values for IEC, C:S, X, AV, and CN. For 20 S-HDPE samples the IEC value of 0.08 mmol/g is incredibly small, indicating little to no sulfonation. Furthermore, due to little sulfonation, 20 S-HDPE samples had a marginal X of 0.5%. The low IEC, and X associated with

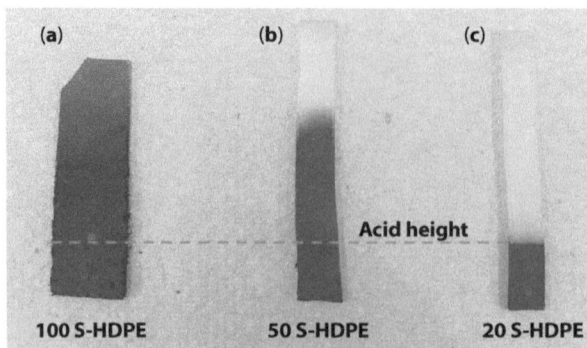

Figure 34.6 An image of a preliminary test of HDPE ribbons sulfonated in (a) 100% CSA, (b) 50% CSA in SA, and (c) 20% CSA in SA.

20 S-HDPE samples in addition to the poor penetration through the 20 S-HDPE ribbon in Figure 34.7c indicates that only the surface was sulfonated, leaving an inner core of mostly pure HDPE. This was further confirmed by TGA data prior to iron grafting which showed that 20 S-HDPE samples most closely resembled that of typical HDPE (see Figure 34.8).

As expected, with the greater extent of sulfonation of 50 S-HDPE and 100 S-HDPE, a significantly higher IEC (lower C:S) and X were observed compared to the 20 S-HDPE samples. Going from 50 to 100% CSA resulted in the IEC increasing by a factor of slightly more than two from 1.38 to 2.80 mmol/g, respectively which correspond with a similar decrease in C:S. Interestingly, both materials obtained significant conversions of 42.70% for 50 S-HDPE and 69.16% for 100 S-HDPE. Due to this result, we anticipate that by increasing the stoichiometric ratio of CSA per gram of HDPE in the range of 50 and 100% that X could be further improved because of enhanced sulfonation and increased volume from which the polymer could be suspended.

While both the 100 S-HDPE and OSS-HDPE samples used the same quantity of CSA, the OSS samples had 20% larger IEC's of 3.36 mmol/g and 21.7% smaller C:S's of 15.5. Apart from using the same volume of CSA as 100 S-HDPE samples, OS-HDPE samples were further diluted in organic solvent to a total volume of 40 mL. The organic solvent easily

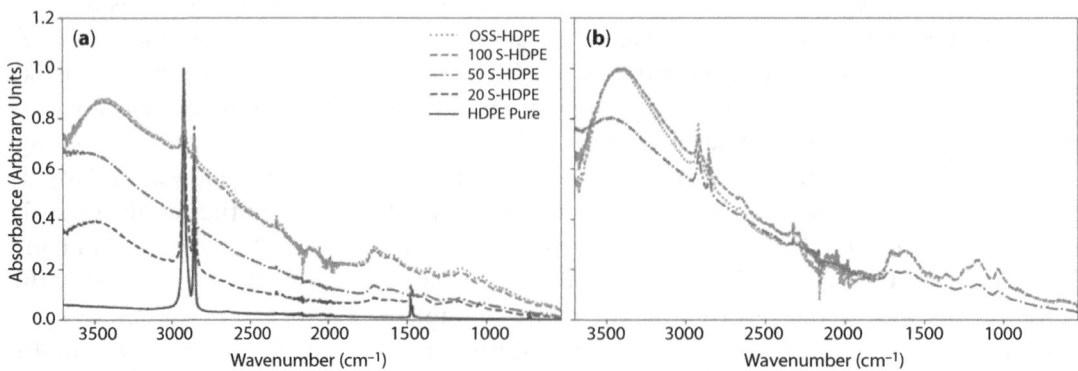

Figure 34.7 FTIR data of S-HDPE samples before (a), and after (b) grafting of FeCl$_3$. All spectra represent the average of two samples.

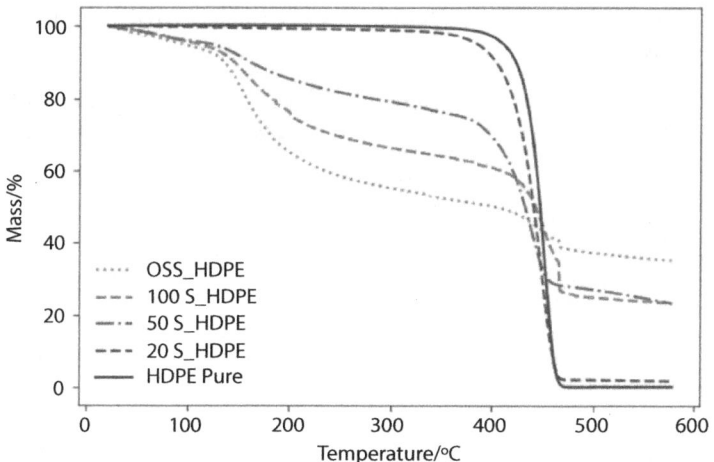

Figure 34.8 TGA data for S-HDPE samples.

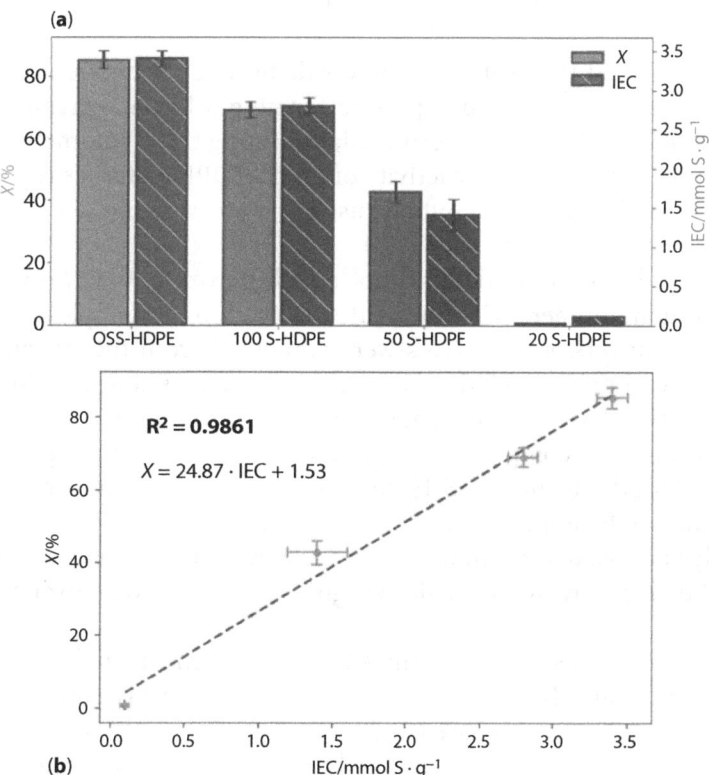

Figure 34.9 A (a) bar and (b) linear regression plot of the relationship between experimentally determined X and IEC for all sulfonated samples.

suspended the PE powder and caused swelling prior to the sulfonation process which likely enhanced contact between the acid and PE chains in addition to acid diffusion. Interestingly, from FTIR of S-HDPE samples in Figures 34.7a and b, 100 S-HDPE intensities followed

Table 34.1 Characterization summary: Ionic exchange capacity (IEC), Carbon:Sulfur ratio (C:S), Conversion (X), Acid value (AV), and Carbon number (CN).

Sample	IEC (mmol/g)	C:S (mol of C/mol of S)	X (%)	AV (mmol of NaOH/g)	CN
OSS-HDPE	3.36 ± 0.12	15.5 ± 0.7	85.2 ± 2.9	12.1 ± 0.9	7.3 ±0.8
100 S-HDPE	2.80 ± 0.15	19.8 ± 1.4	69.2 ± 2.6	12.7 ± 0.5	6.8 ± 0.4
50 S-HDPE	1.38 ± 0.18	46.8 ± 7.6	42.7 ± 3.4	18.2 ± 5.2	3.7 ± 2.3
20 S-HDPE	0.08 ± 0.007	900 ± 90	0.5 ± 0.3	NA[†]	NA[†]

[†]Not large enough conversion to determine AV or CN.

OSS-HDPE closely before and after grafting while 50 S-HDPE had observably lower intensities. This would indicate that PEAA tends to form higher quantities of sulfonated material near the surface of HDPE than during OSS. This is likely due to the much higher concentrations of acid in contact with the polymer during PEAA since the acids are not diluted in organic solvents.

When values for X and IEC were compared side by side for different samples in Figure 34.9a, X increased as IEC increased. Upon performing a linear regression on these data (Figure 34.9b), an R^2 of 0.9861 was determined; indicating these parameters are highly correlated. This indicates that the final reactivity of the S-HDPE materials is governed primarily by the extent of sulfonation and could be used as a measure of the potential for mass to be converted.

Following the oxidation of S-HDPE samples with peroxide, the acid values (AV's) and corresponding carbon numbers (CN's) were determined for all samples except 20 S-HDPE due to insufficient conversion. The CN's were computed from the AV and represent the average number of carbons in each diacid. Values reported in Table 34.1 indicate that OSS-HDPE and 100 S-HDPE had AV's of around 12.1 to 12.7 mmol/g. Assuming a majority of the mass is diacids, the CN computed was around 6.8 to 7.3. Accounting for variance in the measured AV's, OSS-HDPE and 100 S-HDPE contain a majority of diacids with molecular weights corresponding to adipic (C6), pimelic (C7), and suberic (C8) acid. For 50 S-HDPE samples, a highly variable distribution was observed with an AV of 18.2 mmol/g and a CN of 3.7 with most diacids having molecular weights ranging between oxalic (C2) and adipic acid (C6).

To assess the impact of impurities on the AV measurement, artificial solutions of 0.1 g of 1:1 succinic (SA) and adipic (AA) acid (w/w) mixtures with varying impurities of ferric sulfate (0, 3, and 10% (w/w)) were prepared and then titrated in duplicate. Based on the masses of acids added, a theoretical AV was computed to investigate the expected error in varying amounts of iron impurities. Interestingly, at 0, 3, and 10% impurities, a percent error of 0.6 ± 0.5, 4.4 ± 0.4, and 13.2 ± 0.1%, respectively, were observed. This indicated that ferric sulfate increased the titration error proportional with the percent of mass impurities and, as expected, overestimated the AV. It must be noted that with 3 and 10% iron impurities, there was observable iron precipitation. This is in contrast with the ethanol extracted S- HDPE products, since no significant iron precipitation occurred (see Figure 34.1b). Likewise, titration of the remaining solids not solubilized during the ethanol extraction showed significant iron precipitation, indicating that a majority of the iron impurities were not solubilized (see Figure 34.1a). The FTIR of the titrated S-HDPE products (ethanol insolubilized and solubilized) and artificial solutions were also obtained and reported in Figure 34.10. In Figures 34.10a and b, the ethanol insolubilized and solubilized acids showed a strong broad peak at 1600 and 1593 cm^{-1} and a medium broad peak at 1375 and 1378 cm^{-1} respectively, which is indicative of the carbonyl and methylene scissoring of a metal dicarboxylic acid

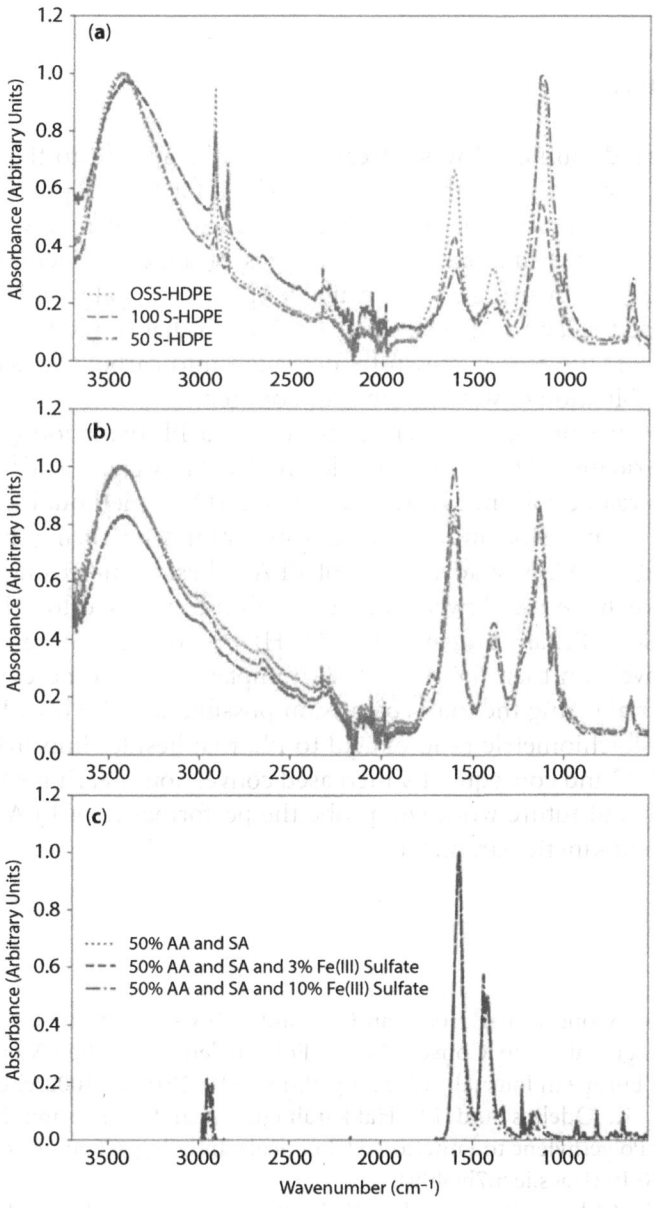

Figure 34.10 FTIR data for ethanol (a) insoluble and (b) soluble oxidation products, and (c) artificial acid mixtures.

salt (see Figure 34.10c) [8]. The large methylene peaks in the insoluble fraction in Figure 34.10a is indicative of larger chain dicarboxylic acids. However, the larger peak around 1100 cm^{-1} is typical S-O stretching of sulfonated olefins [1] [9], suggesting the presence of sulfonate carboxylic acids in both fractions. Chow *et al.* also reported the presence of sulfonate carboxylic acids and found that oxidation products of S-HDPE materials comprised approximately two percent of the final soluble mass [1]. This implies that the sulfonate carboxylic acids present in the current work likely contribute little to the overall mass and therefore to the acid values reported. Future work will further investigate this theory using quantitative methods.

34.5   Conclusions

The principles behind municipal waste treatment can be applied to the plastics recycling industry due to the similarity of the chemistry of the targets. UV/peroxide/ozone treatments have been explored, with the current research focusing on analyzing the cross effects from multiple forms of oxidation. The advantages of aqueous chemistry are primarily related to reducing costs by using widely available chemicals (water and air), the simplicity of the reactions, and the lack of special safety equipment that would be necessary for other chemical oxidizers. LDPE was successfully degraded into carboxylic acids over a 120 hr period using a combination of water, ozone, and oxygen.

In this work the performance of acid-only activated PE oxidation (PEAA) as a green method for the oxidation of PE into smaller molecular weight acids was investigated. Compared to the organic solvent assisted sulfonation (OSS) method, PEAA performs reasonably in terms of conversion and final acid distribution when using the same volume of chlorosulfuric acid (CSA). Key advantages of PEAA, besides not using organic solvents, were that an approximately 8x lower volume of solvent was used to achieve similar conversions for 100 S-HDPE as compared to OSS-HDPE. While OSS samples still demonstrated higher conversion than the best PEAA samples discussed here, we found that the primary parameter affecting the max conversion possible was the IEC. We anticipate that by increasing the stoichiometric ratio of acid to PE, that besides improving polymer mixing, an increased IEC and consequently increased conversions nearing 90 to 100% could be achieved. Ongoing and future work will probe the performance of PEAA at varying CSA volumes and relevant kinetic parameters.

References

1. C.-F. Chow, W.-L. Wong, K. Y.-F. Ho, Chan, C.-S. and C.-B. Gong, Combined Chemical Activation and Fenton Degradation to Convert Waste Polyethylene into High-Value Fine Chemicals. Chemistry - A European Journal, vol. 22, pp. 9513-9518, 2016. 10.1002/chem.201600856
2. E. Backstrom, K. Odelius and M. Hakkarainen, Trash to Treasure: Microwave-Assisted Conversion of Polyethylene to Functional Chemicals. Ind. Eng. Chem. Res., vol. 56, pp. 14814-14821, 2017. 10.1021/acs.iecr.7b04091
3. C. E. McDonald, Chlorosulfuric Acid, in Kirk-Othmer Encyclopedia of Chemical Technology, John Wiley & Sons, Ltd, 2000.

4. P. Kumar, K. Dutta, S. Das and P. P. Kundu, Membrane Prepared by Incorporation of Crosslinked Sulfonated Polystyrene in the Blend of PVDF-co-HFP/Nafion: A preliminary Evaluation for Application in DMFC. Applied Energy, vol. 123, pp. 66-74, 2014. 10.1016/j. apenergy.2014.02.060

5. P. Pravani, Overcoming Interference from Hydrolysable Cations During the Determination of Sulphuric Acid by Titration, Master's Dissertation, University of Pretoria, 2001.

6. S. J. Masten and S. H. Davies, The use of Ozonation to Degrade Organic Contaminants in Wastewaters, Advances in Water Treatment Technologies, vol. 28, no. 4, pp. 180-185, 1994.

7. D. W. Schnabel, Polymer Degradation, New York: Macmillan Publishing Co., Inc., 1981. ISBN 0-02-949640- 3

8. A. Filopoulou, S. Vlachou and S. C. Boyatzis, Fatty Acids and Their Metal Salts: A Review of Their Infrared Spectra in Light of Their Presence in Cultural Heritage. Molecules, vol. 26, no. 19, p. 6005, 2021. 10.3390/molecules26196005

9. J. Haynes, J. R. Sams and R. C. Thompson, Synthesis and Structural Studies of Iron(II) and Iron(III) Sulfonates. Can. J. Chem., pp. 669-678, 1981. 10.1139/v81-098

35

Enzyme-Based Biotechnologies for Removing Stickies and Regaining Fiber Quality in Paper Recycling

Yun Wang*, Cornellius Marcello, Neha Sawant, Swati Sood, Qaseem Haider, Abdus Salam and Kecheng Li

Department of Chemical and Paper Engineering, Western Michigan University Kalamazoo, MI, USA

Abstract

Producing paper from recycled fibers recovers 30-70% of embodied energy and reduces fresh water usage significantly in comparison with using virgin wood. However, the recycling process and the quality of the remanufactured paper products are limited due to some technical challenges. Organic contamination such as stickies and low-quality fibers represent the major ones in paper recycling process. The removal of these sticky contaminants is difficult to achieve with the conventional process due to the heterogeneous nature of these organic contaminants. The fiber quality can be deteriorated with increasing recycling cycles. In this study, the nature of sticky contaminants in multiple wastepaper grades was analyzed and characterized using screening, solvent extraction and FT-IR, GC-MS, and SEM. The physical and chemical properties of the contaminants and their behaviour/pathway in the paper remanufacturing process were identified which helps address the knowledge gaps associated with them. The strength property was also evaluated with regards to amount of organic contamination and fiber quality. Based on these findings, enzyme-based biotechnologies which have high specificity for targeted organic contaminants are developed. In addition, enzyme-aided mechanical refining technologies are developed for regaining fiber quality. These biotechnologies can reduce tacky organic contaminants from recycled fibers to a level below 0.5% and improve remanufactured paper physical strength by 15-25%.

Keywords: Sticky contaminants, paper recycling, fiber quality, biotechnology, enzymes

35.1 Introduction

Producing paper from recycled fibers recovers 30-70% of embodied energy and reduces fresh water usage significantly in comparison with using virgin wood. However, among the 52.7 million tons of paper recovered in the US in 2018, only 32.5 million tons were remanufactured domestically, with the remaining materials being exported along with the

---
*Corresponding author: yun.wang@wmich.edu

Nabil Nasr (ed.) *Technology Innovation for the Circular Economy: Recycling, Remanufacturing, Design, Systems Analysis and Logistics*, (449–462) © 2024 Scrivener Publishing LLC

tremendous energy and water saving potential. In addition, within the 32.5 million tons of paper remanufactured, 86% is boxboard; high quality and higher value paper grades from recycled fiber are very limited [1–3]. Contamination in recovered fibers, especially that from hot-melts, plastic films, food residues, inks, starch and gums, resins/sizing agents, is one of the major causes for manufacturing difficulties and paper quality downgrading [4–7]. On the other hand, fiber quality deteriorates naturally with repeated manufacturing cycles [8, 9].

In current paper remanufacturing, processes are designed and implemented to specifically remove glass, metals, sands by cyclone separators and remove plastic films, shives, adhesives and macro-stickies by barrier screens. However, organics such as food residues, ink, starch, resins/sizing agents, wax and coating adhesives that form micro-size particles are not removed. They can also become dissolved in the process water forming "anionic trash" in the system. The micro-stickies or dissolved anionic trash also can re-agglomerate later in the process into macro-stickies, which adhere to paper processing machinery causing paper breaks and paper quality issues.

The quality of recycled fibers is also deteriorated due to the prior papermaking process. The physical damage of fibers by mechanical forces and hornification by heat in the papermaking process make the recycled fibers less desirable than virgin fiber and hence limiting their application in high grade paper.

Mechanical and chemical methods have been developed to improve the quality of secondary fibers [10, 11], however, both strategies make the fibers more deteriorated in the next recycling process, not to mention the high energy consumption and large amounts of industrial wastewater [12]. In recent years, enzymes have attracted much attention in pulp and paper research due to their high activity, mild reaction conditions and broad substrate scope [13]. In the papermaking process, enzymes can be used for lignin degradation, deinking, bleaching and wastewater treatment [14, 15]. For eliminating stickies, enzymes such as amylase, pectinase, xylanase, lipase and esterase have been investigated [16, 17]. Among these enzymes, lipase and esterase are able to hydrolyze ester bonds, resulting in decreasing the adhering object volume and weakening the adhesion properties [18]. In addition, lignocellulose-degrading enzymes can break down the molecular chains of the specific components in the fiber wall thus changing the ultrastructure of it. This will enhance the subsequent mechanical refining, leading to improved surface fibrillation and inter-fiber bonding. Although efforts have been made to improve the quality of recycled fibers by enzymes, the majority of the industry do not use this approach due to high cost, inefficacy, and the knowledge gaps associated with this biotechnology.

This work focuses on the key technical barriers in contamination removal and fiber quality restoration and address the underlying knowledge gaps associated with them. This will include characterizing the properties and chemical compositions of the sticky contaminants and the effect of enzymes treatment on them. Based on the findings, new enzyme formulations which have high specificity for targeted organics and thus are more cost-effective will be developed. Additionally, enzyme-aided mechanical refining technologies for regaining fiber quality will also be investigated.

35.2 Materials and Methods

35.2.1 Materials

Multiple grades of recycled paper feedstock are obtained from local paper recycling mills and from residential curb-side collection. These include residential wastepaper (Residential), old corrugated containerboard (OCC), recycled boxboard cuttings (Box), recycled kraft linerboard (Kraft Liner), printed bleached kraft paper (Bleached Kraft), used office paper (Mixed Office), and old newsprint paper (ONP). Residential mainly consists of food containers such as burger boxes, pizza boxes, food wrapping papers, and etc. These papers are highly contaminated with oil and grease. OCC includes mostly post-consumer packaging materials, such as containerboards obtain from wholesale stores. Box consists cuttings of multiply paperboards that are used in the production of folding paper cartons such as cereal boxes. Kraft Liner is sorted paperboards with Kraft liners, which are mainly from postal packaging boxes. Bleached Kraft are heavily printed sheet with a high calliper, made of bleached sulfite pulp. They are typically used as feedstock for converting mills to make high-grade packaging boxes. Mixed Office is common printing and writing paper, which can be found in typical workplace. ONP is a mixture of used newspaper and old grocery store flyers. Photos of these wastepaper are shown in Figure 35.1.

35.2.2 Enzyme Treatments

Multiple enzyme formulations are provided by Novozymes (USA). Each formulation is a combination of multiple commercially available enzymes including but not limit to cellulase, xylanase, mannase, amylase, lipase, etc. All enzyme treatments were carried out at 2% of pulp consistency. The temperature and pH were adjusted not only to accommodate most of the enzyme components, but also to accommodate the pulp recycling process in mill practice. Enzyme dosage was tested ranging from 0.01 mg/g to 0.5 mg/g for promising results. Pulp suspension were continuously stirred using an electrical stir at 500 rpm during

Figure 35.1 Recycled paper raw materials collected from local paper recycling mill [19].

the treatment. The enzyme reaction was stopped by filtering pulp slurry through a Buchner funnel and then washing the pulp. The control sample was treated in the same manner as the enzyme treated samples with the exception of enzyme addition.

35.2.3 Characterization of Contaminants

Dried recycled fibers were utilized to perform solvent extraction of sticky contaminants using conventional laboratory Soxhlet extractor following TAPPI T204. Deionized water (DI water) and tetrahydrofuran (THF) were used as extraction solvents. The fibers were extracted with DI water first and then with THF. Extraction process was maintained for 4 hours (predetermined), and the heat was controlled to provide a boiling rate which cycles the solvent at least 6 times per hour. Solvent was partially evaporated to 20ml in the extraction flask using rotary evaporator, and then transferred to a weighing dish with a small amount fresh solvent. The weighing dish and content were dried in an oven for 1 hour at 105°C±3°C, cooled in a desiccator, and weighed to the nearest 0.1mg.

The content of sticky contaminants was calculated using equation (35.1):

$$\text{Content of sticky contaminants (wt\%)} = \frac{W_f - W_e}{W_f} \times 100\% \tag{35.1}$$

Where
W_f = oven-dry weight of recycled fibers prior to extraction, g
W_e = oven-dry weight of recycled fibers post extraction, g

35.2.4 Mechanical Properties of Remanufactured Papersheets

TAPPI standard handsheets were made using the secondary fibers following standard T205. The mechanical properties of handsheets were investigated following TAPPI standards including T403 bursting strength, T494 tensile strength, T414 tearing resistance, and T541 internal bond strength.

35.3 Results and Discussion

35.3.1 Characteristics of Sticky Contaminants

In this study, sticky contaminations are observed on all the wastepaper grades collected from local paper recycling mill. These contaminations have multiple sources such as tape residue and pressure-sensitive label residue (adhesives), hot melt adhesives, food residue (oil/grease), and printed ink (chemical substances). They vary in size, shape, and chemical compositions, and are difficult to remove through mechanical process. In Figure 35.2, SEM images of contaminants on recycled OCC box indicate that contaminant such as adhesive polymers can cover fiber wall structure and microfibrils, thus inhibit inter-fiber bonding [19].

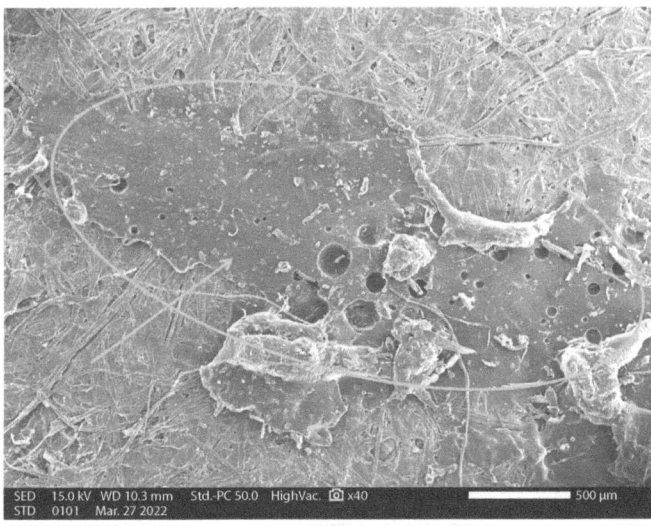

Figure 35.2 SEM images of adhesive contaminants on recycled OCC – a layer of adhesive contaminant on recycled OCC sample [19]. (*40 times magnification, the scale bar is 500 microns.*)

In Figure 35.3, SEM images of sticky contaminants on residential wastepaper show similar behavior. On residential wastepaper, the sticky contaminants attach to fiber surface in the form of small particles or cover fiber wall structure in the form of thin films [19]. As a result, inter-fiber contacting areas are significantly reduced and hydroxyl groups are occupied, thus leads to reductions in recycled paper strength.

As indicated in Table 35.1, The quantity of sticky contaminants in wastepaper varies among different grades and their recovery method. The content of sticky contaminants in

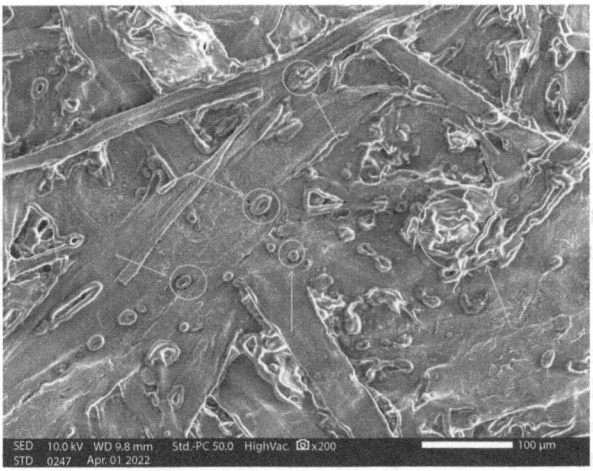

Figure 35.3 SEM images of sticky contaminants on Residential Wastepaper – sticky contaminants in forms of oil droplets and films [19]. (*200 times magnification, the scale bar is 100 microns.*)

Table 35.1 Quantity of sticky contaminants in recycled fibers [19].

Wastepaper grades	Content of sticky contaminants (wt%)	Specific stickies area (mm²/kg)	Number of stickies (counts per kg)	Average stickies size (mm²)
Residential	14.8	37715	32167	1.17
OCC	4.5	12738	13833	0.93
Kraft Liner	5.1	9752	7833	1.24
Bleached Kraft	5.3	5301	8000	0.66
Box	<0.1	1767	3000	0.59
ONP	3.5	2209	5000	0.44
Mixed Office	3.6	3027	4167	0.73

residential wastepaper reaches up to 14.8%, which is the highest among all grades [19]. This is mainly contributed by the noticeable amount of grease and oil from food residues on residential wastepaper. OCC, contaminated with significant numbers of labels and packaging tapes contains 4.5% stickies in total. Kraft liners, Bleached kraft, ONP and Mixed office contain 5.1%, 5.3%, 3.5%, and 3.6% stickies in total, respectively [19]. They are likely contributed by ink substances, or other papermaking additives. ONP and Mixed office paper have the lowest stickies content among all recycled grades besides of Box. From the appearance, both of them are much cleaner source of fibers that are only lightly printed with ink. Box paper barely contains any solvent extractable sticky contaminants. These sticky contaminants present as dots on the remanufactured paper sheets and the specific number of stickies and the area of stickies in paper sheets are in agreement with the content of stickies presented above. As the most contaminated wastepaper, Residential grade contains over 32000 stickies/kg fibers with a total area of 37715 mm². In addition, these stickies counts are relatively large having an average size of 1.17 mm². OCC grade contains nearly 14000 stickies/kg fibers with a total area of 12738 mm², while the average size of the stickies is around 0.93 mm². Kraft liner grade contains around 7800 stickies/kg fibers, which is much less than that of Residential and OCC grade, however, the average size of these stickies is about 1.24 mm², which is the largest among all wastepaper grades. Bleached kraft grade also contains 8000 stickies/kg fibers with a much smaller size of 0.66 mm². The stickies counts in Box, ONP, and Mixed office grades are in the lower range between 3000-5000, with an average size between 0.4-0.7mm².

The chemical compositions of the sticky contaminants from the wastepaper were characterized using FT-IR and GC-MS. The FT-IR spectrum of stickies from OCC and Residential are shown in Figure 35.4, respectively. In the FT-IR profile of stickies extracted from OCC, it is first to observe the strong absorption at about 1735 cm^{-1} due to the C=O stretching of the acetate group; and the bands at about 1260 cm^{-1}, 1160 cm^{-1} and 1025 cm^{-1} due to

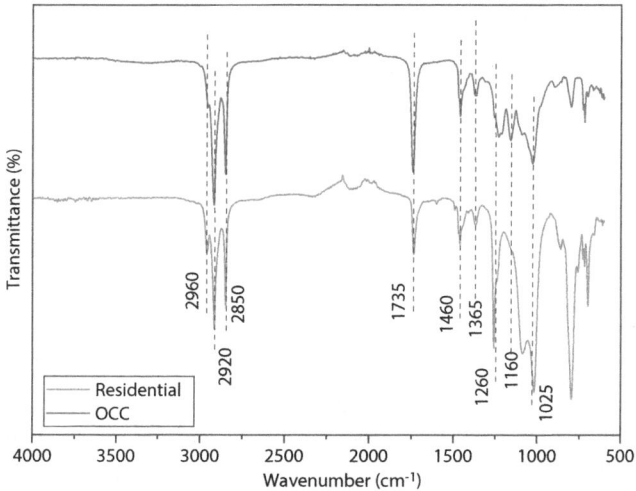

Figure 35.4 FT-IR spectrum of THF extracted stickies from OCC and Residential, respectively [19].

the C-O stretching of the ester group. Typical C-H stretching absorption bands of alkane are observed at between 2960 cm⁻¹, 2920 cm⁻¹ and 2850 cm⁻¹, respectively. C-H bending absorption are observed at 1460 cm⁻¹ and 1365 cm⁻¹. These results indicate the presence of polyvinyl acetate (PVA) polymers, which are generally used as adhesives in packaging [19–21]. As also shown in Figure 35.4, the extracted stickies from Residential grade presented similar FT-IR profile as that of OCC stickies. Regarding the absorption bands, the presence of esters is observed. Evidence of ester groups is identified by the absorption bands at 1735 cm⁻¹ due to the C=O stretching and at 1260 cm⁻¹, 1025 cm⁻¹ due to the C-O stretching of the ester groups. These components are likely to be introduced by food residue such as residual cooking oils.

The major components of the sticky contaminants found by GC-MS are summarized in Table 35.2. The contaminants in Table 35.2 can be divided into five groups. First, dimethylamine and dimethyl propylamine are the monomer units of polyamines, which is a commonly used polymer in papermaking process. The quaternary ammonium group in polyamine structure maintains strong cationic charge in most papermaking conditions, and thus are used extensively for neutralizing excess anionic colloidal charge and

Table 35.2 Chemical composition of stickies in OCC based on GC-MS results [19].

Compounds	Potential origin	Percentage %
Polyamine	Paper additive	4.1
Styrene	Adhesive	4.7
Resin	Coating	6.9
Wax	Coating, adhesive	70.1
Others	Wood extractive, ink residues	14.2

Table 35.3 Changes of contaminants concentration and size in paper remanufacturing process.

Stages	Concentration of stickies (wt%)	Number of stickies (counts per kg)	Average stickies size (mm^2)
Re-pulping	8.27	52,321	7.41
Coarse screening	1.60	32,241	1.21
Fine screening	1.88	47,857	0.78
Cyclone cleaning	1.32	34,655	0.99

establishing anchoring points for anionic retention aids. Second, methyl styrene, which is a monomer compound in styrene acrylates. Styrene acrylates are commonly used as a sizing agent in papermaking process to increase resistance to water penetration. In addition, methyl styrene is also used as the monomer unit for the synthesis of styrene butadiene rubber (SBR), which is a major component in hot melt adhesive as a thermoplastic elastomer. Furthermore, the presence of cyclosiloxanes is observed, which is a silicone resin material being widely applied as coating ingredients. Moreover, paraffin wax compounds containing carbon atoms ranging from 20 to 33 are observed. Paraffin wax is primarily used as a plasticizer in hot melt adhesives to adjust product viscosity and melt-rate. It is a major component in hot melt adhesives, especially in polyvinyl acetate (PVA) adhesives. In this study, paraffin wax is the predominant component in the stickies extracted from OCC, which makes up to 70.1% of total stickies. The percentage of polyamine, styrene, and resins in total stickies are 4.1%, 4.7% and 6.9%, respectively. Last, 14.7% of total stickies is a mixture of complex compounds which are difficult to identify, including high molecular weight esters and acids.

As indicated by FT-IR and GC-MS results, the sticky contaminants have heterogenous nature which make them difficult to remove in conventional paper recycling process. The changes in quantity and size of the stickies in paper remanufacturing process reveal their behavior and pathway. As shown in Table 35.3, the majority of sticky contaminants in the system is removed in the coarse screening stage, along with a significant size reduction from 7.4 mm^2 to 1.2 mm^2. The subsequent fine screening stage does not effectively remove contaminants, instead, the contaminants are broken down into smaller particles in this stage which is explained by the reduced size and increased number. Cyclone cleaning removed slightly more contaminants, but the increased size suggests re-agglomeration of stickies into macro-stickies which can cause problems on paper machine. Therefore, more efficient and economic contamination removal technology is needed.

35.3.2   Enzyme-Assisted Treatment for Contamination Removal and Strength Improvement

Based on the chemical compositions of sticky contaminants found in wastepaper, enzyme formulation S, consisting of multiple enzymes including lipase, amylase, pectinase, xylanase and etc, is designed to assist the removal of tacky organic contaminants. The stickies content in recycled fibers after enzyme treatment is determined using solvent extraction method, and the results are shown in Figure 35.5. The stickies removal efficiency is also determined and presented in Figure 35.6. In this study, all enzymatic treatments of multiple grades of recycled paper were performed at 50°C for 1 hr with the addition of 0.25 mg/g enzyme formulation S, except RW-2 and RW-3. Under this condition, after enzymatic treatment, the stickies content in all recycled paper samples decreased significantly. The stickies content was reduced to 0.34% in OCC, 0.33% in DLK, 0.7% in bleached kraft, 0.17% in ONP, and 0.15% in mixed office paper, respectively. When comparing to control tests, the stickies removal efficiency was 88% in OCC, 81.6% in DLK, 82.4% in bleached kraft, 86% in ONP, and 84.4% in mixed office paper, respectively. These results indicated that the designed enzyme formulation is highly effective for stickies removal. However, for residential wastepaper, the content of sticky contaminants was reduced to 0.96% in RW-1, with a relatively lower stickies removal rate at 73%. This is mainly due to the significantly high amount of contaminants presented in residential wastepaper. As discussed above, residential

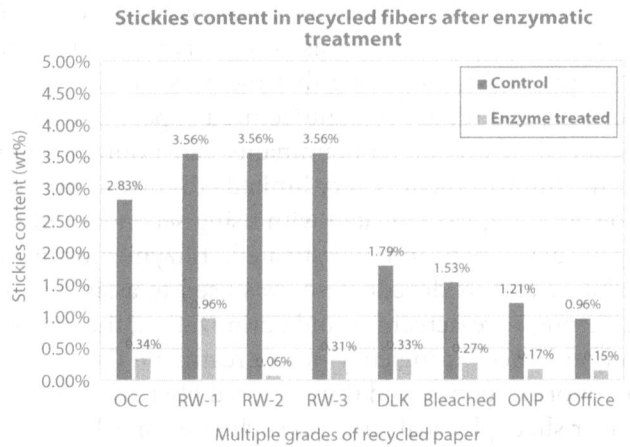

Figure 35.5 The content of stickies in multiple wastepaper grades before and after enzymatic treatment.

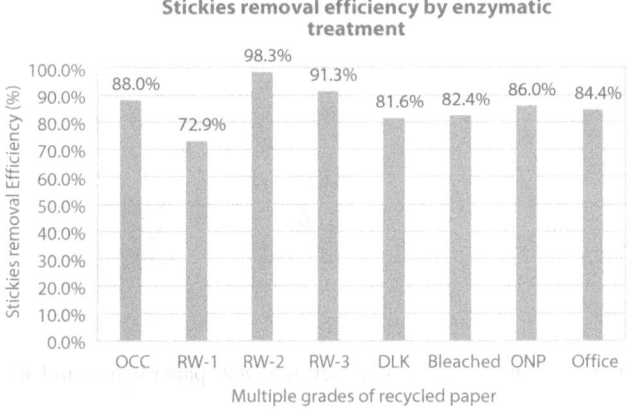

Figure 35.6 Stickies removal efficiency by enzymatic treatment.

wastepaper contains up to 14.8% sticky contaminants, while the other grades contain 3.5% to 5.3% sticky contaminants. To further reduce the contaminants in residential wastepaper to below 0.5%, modified enzymatic treatments RW-2 and RW-3 were performed. In RW-2, the treatment was extended to 2hr at 50°C with the addition of 0.25 mg/g enzyme formulation S; while in RW-3, the treatment was performed at 1hr at 50°C, with the addition of enzyme formulation S increased to 0.5 mg/g. As a result, both modifications further reduced sticky contaminants to below 0.5% in residential wastepaper. In RW-2 with extended treatment time, 98.3% contaminants were removed while nearly no contaminants (0.06%) remained on recycled fibers. In RW-3 with increased enzyme dosage, 91.3% contaminants were removed resulting 0.31% stickies remained on the fibers.

The physical properties of the remanufactured paper sheets were investigated to study the impact of the presence of organic contaminants. The physical properties of remanufactured paper sheets, with and without contamination removal by enzymes, are compared and discussed in Figure 35.7. For OCC grade, through enzyme-assisted removal of contaminant, improvements on paper strength are observed. Burst strength increased by 8% from 1.24 kPa×m²/g to 1.34 kPa×m²/g, tensile strength increased by 8.4% from 24.06 N×m/g to 26.07 N×m/g, while no significant improvement was observed on tear strength. These results are in agreement with SEM analysis, indicating that the presence of sticky contaminants can inhibit inter-fiber bonding and reduce remanufactured paper strength.

The enzymatic removal of organic contaminants can improve remanufactured paper properties; however, the improvement is very limited. To further improve physical properties and solve the downgrading issue related with using secondary fibers, enzyme-assisted mechanical refining is developed and investigated. Enzyme formulation C, a mixture of cellulase, hemicellulase and oxidoreductase, was used to assist the mechanical refining of secondary fibers. Through pre-screening and optimization, the mechanical fibrillation of OCC was significantly enhanced using enzyme pretreatment. Formulation C facilitated the open-up of fiber micropore structure and thus restored fiber swelling capability. Substantial improvements on paper sheet physical strength such as tensile, burst and internal bonding

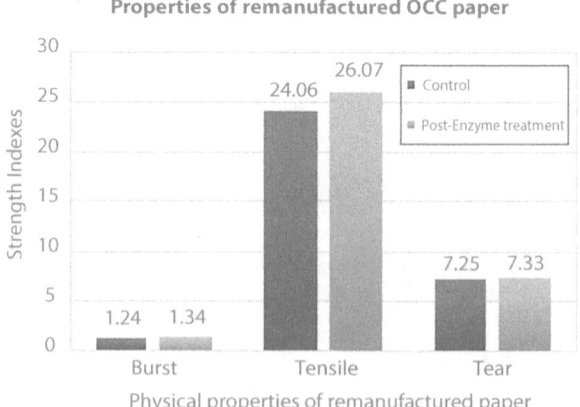

Figure 35.7 Physical strength properties of remanufactured OCC paper before and after stickies are removed by enzyme.

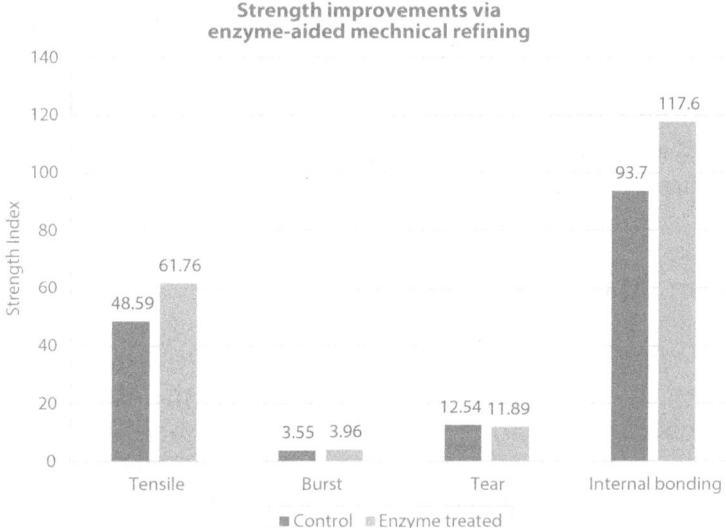

Figure 35.8 The improvement on physical properties of remanufactured paper via enzyme-aided mechanical refining.

were achieved in the range of 10-20%. In this study, by combining contamination removal enzyme formulation S and strength improvement enzyme formulation C, a one-stage enzymatic treatment followed by mechanical refining is developed and investigated. In this process, the OCC fibers were treated at 50°C for 1hr with the addition of combination of 0.25 mg/g Formulation S (contamination removal enzyme) and 0.2 mg/g Formulation C (strength improvement enzyme). The combination of two enzyme formulations yielded significant improvement on physical properties of remanufactured paper. As shown in Figure 35.8, in this treatment, the tensile strength was improved by 27.1% from 48.59 N×m/g to 61.76 N×m/g, burst strength improved by 11.5% from 3.55 kPa×m^2/g to 3.96 kPa×m^2/g, internal bonding strength improved by 25.5% from 93.7 ft×lb/1000 to 117.6 ft×lb/1000, and tear resistance strength slightly decreased by 5.2% from 12.54 mN×m^2/g to 11.89 mN×m^2/g, due to inevitable fiber shortening caused by enzymatic treatment.

35.4 Conclusions

The analysis of sticky contaminants in recycled fibers reveals that the stickies vary in size, number and weight by different paper grades and their collection source. In this study, the majority of wastepaper grades collected from a local paper recycling mills contains 3-5 wt% stickies, while a curb side collected grade contains 15 wt% stickies. The average size of these stickies presented in recycled paper sheets is in the range of 0.4 to 1.2 mm^2. The chemical analysis of THF extracted stickies by FT-IR and GC/MS indicates that the major components in the contaminants are polyvinyl acetate (PVA) polymers, styrene butadiene rubber (SBR), paraffin wax, resin, and polyamines. In addition, SEM images reveal that sticky contaminants cover fiber surface that would reduce contact areas, and, thus leads to a reduction in paper strength. The optimized enzyme treatment can remove nearly 98% of sticky contaminants, and the physical properties of remanufactured paper sheets can be improved by 15-25% through enzyme-assisted mechanical refining process. The findings from these studies help address the knowledge gaps in understanding the nature of stickies and their behaviors, and develop highly efficient biotechnologies for improving the quality of fibers form the paper recycling process.

Acknowledgements

This material is based upon work supported by the U.S. Department of Energy's Office of Energy Efficiency and Renewable Energy (EERE) under the Advanced Manufacturing Office Award Number DE-EE0007897 awarded to the REMADE Institute, a division of Sustainable Manufacturing Innovation Alliance Corp. This report was prepared as an account of work sponsored by an agency of the United States Government. Neither the United States Government nor any agency thereof, nor any of their employees, makes any warranty, express or implied, or assumes any legal liability or responsibility for the accuracy, completeness, or usefulness of any information, apparatus, product, or process disclosed, or represents that its use would not infringe privately owned rights. Reference herein to any specific commercial product, process, or service by trade name, trademark, manufacturer, or otherwise does not necessarily constitute or imply its endorsement, recommendation, or favoring by the United States Government or any agency thereof. The views and opinions of authors expressed herein do not necessarily state or reflect those of the United States Government or any agency thereof.

References

1. Berenyi, E. B., What Comes after Single-Stream? *Resour. Recycl.*, 34, 2015.
2. Korhonen, J., Toppinen, A., Kuuluvainen, J., Prestemon, J. P., Cubbage, F., Recycling, certification, and international trade of paper and paperboard: Demand in Germany and the United States. *For. Sci.*, 63, 449, 2017.
3. Rogoff, M. J., Ross, D. E., The future of recycling in the United States. *Waste Manag. Res.*, 34, 181, 2016.
4. Chester, M., Martin, E., & Sathaye, N., Energy, greenhouse gas, and cost reductions for municipal recycling systems. *Environ. Sci. Technol.*, 42, 2142, 2008.
5. Lakhan, C., A comparison of single and multi-stream recycling systems in Ontario, Canada. *Resources*, 4, 384, 2015.
6. Morawski, C., Single stream uncovered. *Resour. Recycl.*, 3, 17, 2010.
7. Wang, J., All in one: Do single-stream curbside recycling programs increase recycling rates?, *Single-Stream Recycling*, https://nature.berkeley.edu/classes/es196/projects/2006final/wang.pdf, 2006.
8. Bajpai, P., Effects of Recycling on Pulp Quality, in: *Recycling and Deinking of Recovered Paper*, pp. 101-120, Elsevier, 2014.
9. M.A. Johnson, A.R. Pokora, and J.B. Henry, *Method for repulping fibrous materials containing crosslinked polyamide wet strength agents with enzyme*, US Patent 5330619, assigned to Mead Corp, 1994.

10. Petzold, G., Petzold-Welcke, K., Qi, H., Stengel, K., Schwarz, S., Heinze, T., The removal of stickies with modified starch and chitosan—Highly cationic and hydrophobic types compared with unmodified ones. *Carbohydr. Polym.*, 90, 1712, 2012.

11. Van Beek, T. A., Kuster, B., Claassen, F. W., Tienvieri, T., Bertaud, F., Lenon, G., and Sierra-Alvarez, R., Fungal bio-treatment of spruce wood with Trametes versicolor for pitch control: Influence on extractive contents, pulping process parameters, paper quality and effluent toxicity. *Bioresour. Technol.*, 98, 302, 2007.

12. Wistara, N., Young, R. A., Properties and treatments of pulps from recycled paper. Part I. Physical and chemical properties of pulps. *Cellulose*, 6, 291, 1999.

13. Metsämuuronen, S., Lyytikäinen, K., Backfolk, K., Sirén, H., Determination of xylo-oligosaccharides in enzymatically hydrolysed pulp by liquid chromatography and capillary electrophoresis. *Cellulose*, 20, 1121, 2013.

14. Gehmayr, V., Schild, G., Sixta, H., A precise study on the feasibility of enzyme treatments of a kraft pulp for viscose application. *Cellulose*, 18, 479, 2011.

15. Ricard, M., Reid, I. D., Purified pectinase lowers cationic demand in peroxide-bleached mechanical pulp. *Enzyme Microb. Technol.*, 34, 499, 2004.

16. Grönqvist, S., Hakala, T. K., Kamppuri, T., Vehviläinen, M., Hänninen, T., Liitiä, T., ... and Suurnäkki, A., Fibre porosity development of dissolving pulp during mechanical and enzymatic processing. *Cellulose*, 21, 3667, 2014.

17. Nguyen, G. S., Kourist, R., Paravidino, M., Hummel, A., Rehdorf, J., Orru, R. V., ... and Bornscheuer, U. T., An Enzymatic Toolbox for the Kinetic Resolution of 2-(Pyridin-x-yl) but-3-yn-2-ols and Tertiary Cyanohydrins. *Eur. J. Org. Chem.*, 14, 2753, 2010.

18. Vaquero, M. E., Barriuso, J., Martínez, M. J., Prieto, A., Properties, structure, and applications of microbial sterol esterases. *Appl. Microbiol. Biotechnol.*, 100, 2047, 2016.

19. Wang, Y., Marcello, C., Sawant, N., Salam, A., Abubakr, S., Qi, D., and Li, K., Identification and characterization of sticky contaminants in multiple recycled paper grades. *Cellulose*, 30, 1957, 2023.

20. Miranda, R., Balea, A., Blanca, E. S. D. L., Carrillo, I., and Blanco, A., Identification of recalcitrant stickies and their sources in newsprint production. *Ind. Eng. Chem. Res.*, 47, 6239, 2008.

21. Licursi, D., Antonetti, C., Martinelli, M., Ribechini, E., Zanaboni, M., and Galletti, A. M. R., Monitoring/characterization of stickies contaminants coming from a papermaking plant—Toward an innovative exploitation of the screen rejects to levulinic acid. *Waste Manag.*, 49, 469, 2016.

36

Removal of Iron and Manganese Impurities from Secondary Aluminum Melts Using Microstructural Engineering Techniques

M.K. Sinha[1], B. Mishra[1], J. Hiscocks[2], B. Davis[2], S.K. Das[3]*, T. Grosko[4] and J. Pickens[5]

[1]*Worcester Polytechnic Institute, Worcester, MA, USA*
[2]*Kingston Process Metallurgy, Kingston, ON, Canada*
[3]*Phinix LLC, Clayton, MO, USA*
[4]*Smelter Service Corporation, Mt Pleasant, TN, USA*
[5]*Material & Process Innovation, LLC, Strongsville, OH, USA*

Abstract

Aluminum alloys are known for their high strength-to-weight ratio, excellent castability, thermal & electrical conductivities, and outstanding corrosion resistance. This unique combination of properties enables aluminum alloys to be widely used in diverse industries such as automotive, aerospace, packaging, wiring & electrical cables, building & construction, and consumer electronics. Because of its high recyclability, aluminum is a material of choice for a circular economy. However, during recurring scrap recycling, aluminum recycling can be complicated by the gradual accumulation of unwanted impurities, mainly Fe, Mn, and other (Cu, Si, Zn, and Mg) alloying elements. The impurity pickup can result in significant economic devaluation of scrap where aluminum products are downcycled to produce aluminum products with lower value and performance. Managing the impurity content of aluminum scrap melt is, therefore, very important. Using microstructural engineering techniques, impurity elements can be removed by cooling molten aluminum, resulting in the formation of intermetallic compounds containing only one or more impurity elements as solid inclusions. Subsequent removal of intermetallic sediments by decantation and filtration techniques can provide an economical method for removing the desired amount of impurities to meet the product specifications. To neutralize the negative effects of mainly Fe, alloying elements such as Mn and Cr can be used to modify the morphology of Fe intermetallic phases to a less harmful microstructure. Using microstructural engineering techniques, Fe, Mn, and Cr complexes could also remove Fe and Mn impurities from molten aluminum. The present study studied the simultaneous removal of Fe and Mn from low and high Si-containing Al-alloys via an intermetallics sedimentation route. Calculation of Phase Diagrams modeling (CALPHAD) was also carried out using ThermoCalc to determine the suitable temperature for impurity-rich intermetallic phase formation in experimental alloys. Effects of holding time and temperature were also examined for maximum impurity removal. This paper will also discuss the results obtained at the laboratory that could be useful in performing commercial-scale experiments.

**Corresponding author*: skdas@phinix.net

Nabil Nasr (ed.) Technology Innovation for the Circular Economy: Recycling, Remanufacturing, Design, Systems Analysis and Logistics, (463–476) © 2024 Scrivener Publishing LLC

Keywords: Aluminum alloys, AI recycling, impurities, intermetallic, sedimentation, ThermoCalc, Fe removal, Mn removal

36.1 Introduction

Aluminum is a light, ductile, and the 3rd most abundant element in the earth's crust. Aluminum and its alloys are second only to steel in usage as engineering metals [1]. In comparison to other materials, aluminum alloys offer numerous advantages, including good corrosion resistance, high strength-to-weight ratio, good formability, machinability, and thermal/electrical conductivity, making them well-suited for use in the automobile, packaging, and aerospace industries. Additionally, aluminum alloys have excellent scrap reuse capabilities, which contributes to the high growth of the secondary (recycled) aluminum market due to its lower cost and lower carbon footprint [2]. However, the accumulation of unwanted elements in secondary aluminum is a growing concern, and strategies designed to reduce unwanted elements are essential. The chemical composition is the main challenge in Al recycling. Scrap originates from different Al alloys, with different alloying elements, in different amounts. This means that it can be difficult to control the level of impurities and also difficult to obtain the targeted alloy composition. Scraps contain impurities of Cu, Mg, Mn, Si, and Fe that affect the alloy's mechanical properties [3]. The control and removal of impurities from Al melt have become an important topic in a greater range of alloy production as impurity levels are rising both in the primary production process and through increased recycling. Traditionally, dilution with prime aluminum is used to control certain impurities and meet the target composition of aluminum alloys rather than downgrading the entire batch of aluminum scrap.

One of the main impurities in the aluminum alloy is iron, which results mainly from charge materials and significantly affects the mechanical properties. In particular, the needle-like β-Al5FeSi intermetallic phase that forms in recycled Al-Si alloys promotes the formation of shrinkage pores and porosity defects, which eventually deteriorate the mechanical properties. Many methods, such as adding neutralizing elements (Be, Sc, Sr, Mn, Cr, Co, RE, Mo, V, and B), adding melt treatment steps, or increasing cooling rates have been employed to improve the mechanical properties and to neutralize the morphology and size of β-Fe phase [4–10]. Fast solidification of the molten Al-alloy promotes the formation of α-Al8SiFe2 instead of β-Al5SiFe. Among various neutralizing elements, Mn has been extensively studied to minimize the harmfulness of the β-Fe phase. The addition of Mn stimulates the formation of quaternary α-Al15(Fe, Mn)3Si2 intermetallic phase in the form of Chinese script or polyhedral crystal morphology. The α-Al15(Fe, Mn)3Si2 and α-Al8SiFe2 intermetallic phases have a less deleterious impact on the mechanical properties [11]. Because of the high density of the quaternary α-Al15(Fe, Mn)3Si2 phase, another way to reduce the harmful effect of the β-Fe phase is the removal of the Fe-rich intermetallic phase by electromagnetic separation, ceramic filtration, or gravity sedimentation. The removal of primary Fe-rich intermetallic particles from molten aluminum by porous filters was attempted by de Moraes *et al.* [12]. The alloy preparation was performed by melting the aluminum, followed by the addition of alloying elements (Mn, Si, and Fe) at 850°C for 2 hours. After a short time holding at the sludge formation temperature, the melt was decanted through a preheated (750°C) filter. A small amount of sludge was also present at the bottom of the melt from precipitation during the holding time. The intermediate phases remained as the sediment at the bottom of the crucible. It was observed that the removal of iron from the Al-melt occurred through the precipitation of the Al15(Fe, Mn)3Si2 intermediate phase. The precipitation of iron in the process is highly influenced by the Mn/Fe ratio (up to 1.5) and the processing temperature and particularly so for the systems with high iron content. Furthermore, the β-Al5FeSi could not be eliminated using a low Mn/Fe ratio at a low cooling rate or when using a high Mn/Fe ratio at a high cooling rate [4, 13–15]. However, a sufficiently high Mn content also produces an adverse effect. It introduces a brittle phase in the Al alloy microstructure, which needs to be avoided. To reduce the harmful effect of the Fe phase in the case of die-cast alloys, the composition and temperature of the melt are determined according to the minimum requirement for sludge formation, which depends upon the concentrations of Fe, Mn, and Cr and is calculated by the following equation:

$$SF = (1 \times wt.\% \ Fe) + (2 \times wt.\% \ Mn) + (3 \times wt.\% \ Cr)$$
$$T = 645.7 + 34.2 \times (wt.\% \ Fe)^2$$

A sludge factor (SF) value greater than 1.7 promotes the formation of course and dense phases with polyhedral morphologies, which cause problems in the machining of the resulting casting [16]. Because of their high specific gravity, these phases tend to segregate to the bottom of the molten alloy. The effect of Cr in the reduction of Fe from 356 alloys was reported by Annadurai *et al.* [17]. This segregation can be useful for the removal of Fe and other impurities from molten aluminum via gravity sedimentation at optimum temperature. Since Fe removal is highly dependent on the melt holding temperature and the concentration of alloying elements, it is necessary to determine the best conditions and how other alloying elements affect the evolution and sedimentation of Fe or impurity-rich intermetallic phases. In view of this, the objective of the present research work encompasses the influence of alloying elements in Fe removal from Al-Si-Fe alloys using equilibrium model calculations followed by experimental validation.

36.2 Results and Discussion

36.2.1 Low Si Aluminum Alloy Analysis by Thermodynamic Modeling

36.2.1.1 Effect of Mn on Fe Removal

The removal of Fe and Mn via precipitation of Fe/Mn-rich intermetallics formation is a challenging process because it requires fine control of the temperature for the formation of the primary phases and depends on the other alloying elements associated with specific alloy compositions. The computational thermodynamic analysis is required to determine whether an alloy with Fe will form impurity-rich intermetallics and at what temperature they will form. Calculations were performed on phases present as a function of temperature and composition using the software ThermoCalc [18] with the commercial database TCAL8. The calculations were made using the Scheil–Gulliver solidification model that assumes equilibrium only in the liquid and at the solid–liquid interface but no solid-state

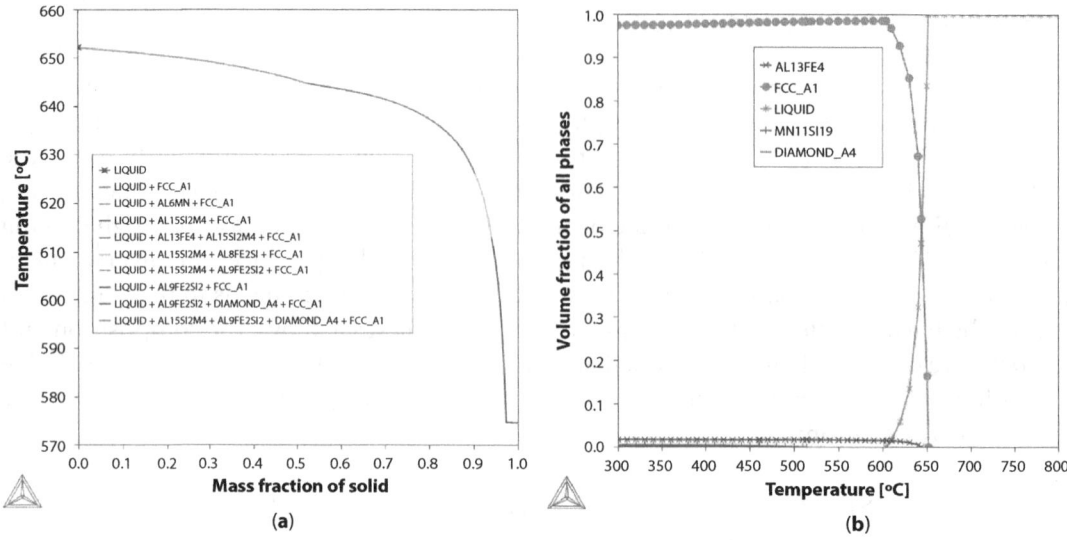

Figure 36.1 (a) The calculated Scheil-Gulliver solidification diagram and (b) Fractions of phases formed with changing temperature in Al-0.8Si-0.75Fe-0.8Mn alloy.

diffusion. Figure 36.1 is the calculated solidification sequences and solidified phase fraction of Al-0.8Si-0.75Fe-0.8Mn (wt.%) alloy, showing that the alloy consists of mainly Al13Fe4, α-Al, Al6Mn, Al15Si2M4, and Al9Fe2Si2 phases in the nonequilibrium solidification. Note that all the intermetallic phases precipitated out after the solidification of α-Al. A higher precipitation temperature than that of α-Al (652°C) is necessary for the physical separation of any impurity-rich intermetallic phases. A higher concentration of Mn in Al-alloys (Figure 36.2) promotes the formation Al6Mn intermetallic phase prior to α-Al solidification. It increases the solidification temperature of scrap aluminum, suggesting that removal of Fe from wrought aluminum alloys via the Fe-rich intermetallic phase precipitation route will not work when using the high Mn concentration alone.

36.2.1.2 Effect of Cr on Fe/Mn Removal

Figures 36.3a and 36.3b present the calculated solidification sequences of Al-0.8Si-0.75Fe-0.8Mn-0.5/1Cr (wt.%) alloy in this study. They show that the alloy consists of Al15Si2M4 (α-AlFeMnCrSi) phase, α-Al, Al13Fe4, and Al9Fe2Si2 phases in the nonequilibrium solidification. Unlike Mn, Cr promotes the formation of impurity-rich intermetallic phases in low Si alloys, which suggests the possibility of removal of Fe/Mn via the intermetallic precipitation-sedimentation route. α-AlFeMnCrSi is the primary phase during solidification and the presence of high Cr concentration widens the temperature range (709°-734°C) for the precipitation of the impurity-rich intermetallic phase by increasing the transformation temperature at which they start appearing. Like Mn, the addition of Cr does not affect the α-Al phase formation temperature and, therefore, the holding temperature for the formation and separation of the intermetallic phases can be kept above the solidification temperature of α-Al (~660°C). Figure 36.4 shows the variation in the iron and manganese in the

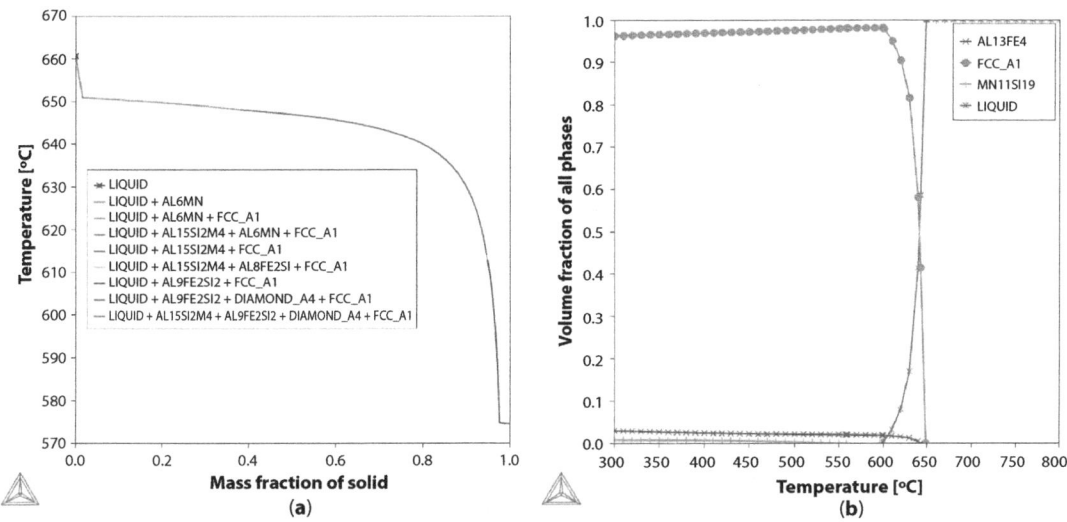

Figure 36.2 (a) The calculated Scheil-Gulliver solidification diagram and (b) Fractions of phases formed with changing temperature in Al-0.8Si-0.75Fe-2Mn alloy.

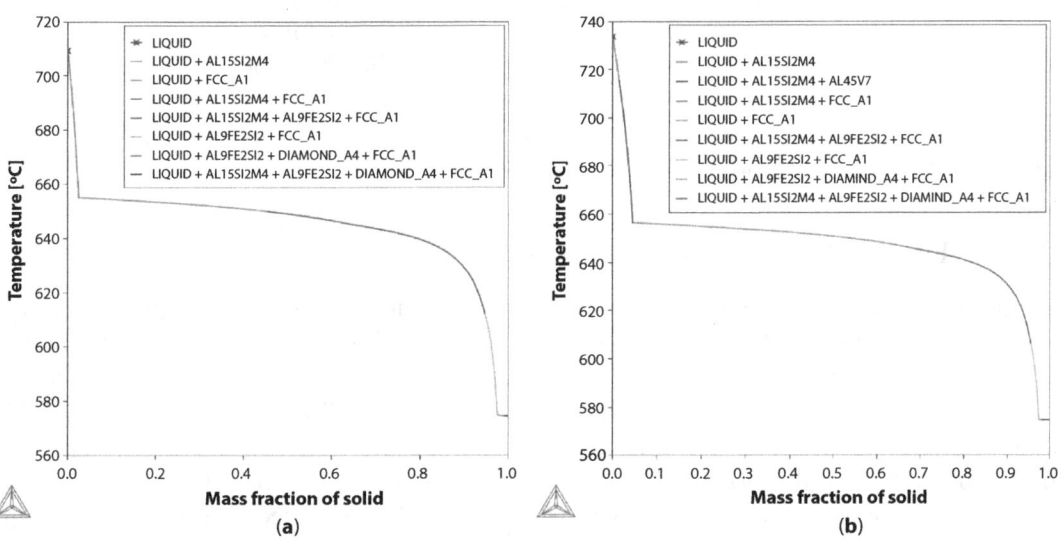

Figure 36.3 The calculated Scheil-Gulliver solidification diagram of (a) Al-0.8Si-0.75Fe-0.8Mn-0.5Cr and (b) Al-0.8Si- 0.75Fe-0.8Mn-1Cr (wt.%) alloys.

liquid phase by varying the weight percentage of chromium addition at different holding temperatures for the alloy Al-0.8Si-0.75Fe-0.8Mn-xCr (where x = 0-2%). The iron content in the liquid aluminum decreases significantly with increasing chromium, according to the thermodynamic calculations. When Cr = ~0.9% in Figure 36.4(a), the equilibrium concentration of Fe was decreased from 0.75% to 0.45% at 700°C in the liquid phase. At the same time, when the holding temperature decreased to 680°C, the initial content of

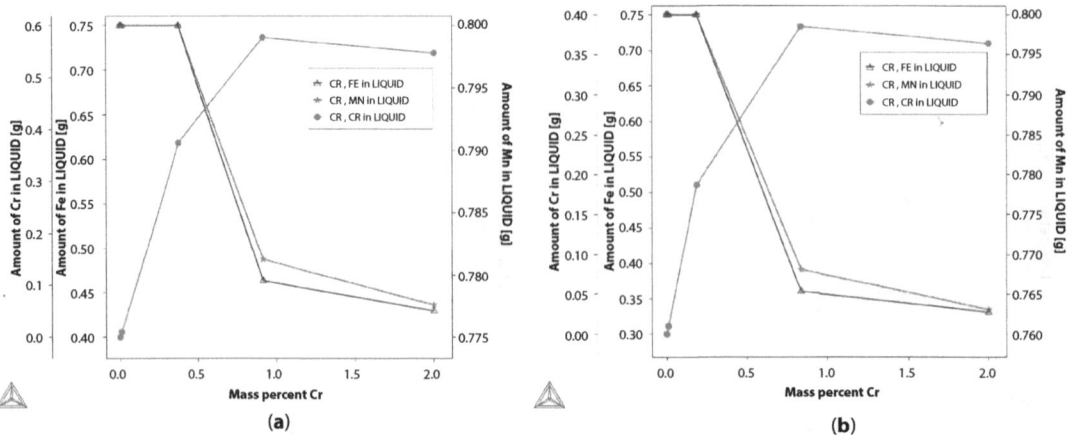

Figure 36.4 Equilibrium concentration of Fe, Mn, and Cr in the liquid phase of alloy Al-0.8Si-0.75Fe-0.8Mn-xCr at (a) 700°C and (b) 680°C, calculated from the equilibrium phase diagram. Note: (g) is equivalent to % used in the text.
Note: X-axis represents Cr in the initial composition.

0.75% Fe decreased to 0.35% Fe in the liquid phase, showing a significant effect of holding temperature on the possibility of Fe reduction. However, a very minor decrease in Mn concentration has been predicted (note scaling to the right). Furthermore, in both cases, high concentrations of Cr (~0.56 and ~0.37%) remained in the melt, which exceeded the maximum allowed in the wrought alloy specifications. Thus, experimental validation is necessary to confirm the thermodynamic calculation prediction.

36.2.2   Effect of Si as an Alloying Element

It is known that the addition of silicon to aluminum reduces the melting temperature range and improves fluidity. For Fe/Mn removal from Al-Si alloy, it is important to know the minimum Si concentration value that could facilitate the formation and easy sedimentation of Fe/Mn-rich intermetallic phases. From calculated solidification sequences of Al-xSi-0.75Fe-0.8Mn (wt.%) alloy, it was observed that impurity-rich intermetallics phases are formed beforehand only above 5% Si concentration. Figure 36.5 is the calculated solidification sequences of Al-5/7Si-0.75Fe-0.8Mn (wt.%) alloys. As can be seen, the presence of high Si not only decreased the solidification temperature of the α-Al phase to 628°C (5% Si) and 614°C (7% Si) but also widened the formation temperature for impurity-rich intermetallic phases. Thus, it would be interesting to analyze the effect of different concentrations of both Mn and Fe in high Si alloys.

36.2.3   Thermodynamic Analysis of the Effect of Mn on the Removal of Fe from High Si Alloy

Figure 36.6 shows the variation in the concentrations of Fe, Mn, and Si in the liquid phase by varying the weight percentage of Mn addition at different holding temperatures (620°C and 615°C) for the alloy with lower Fe content [Al-7Si- 0.75Fe-xMn (where x = 0.5-2%)].

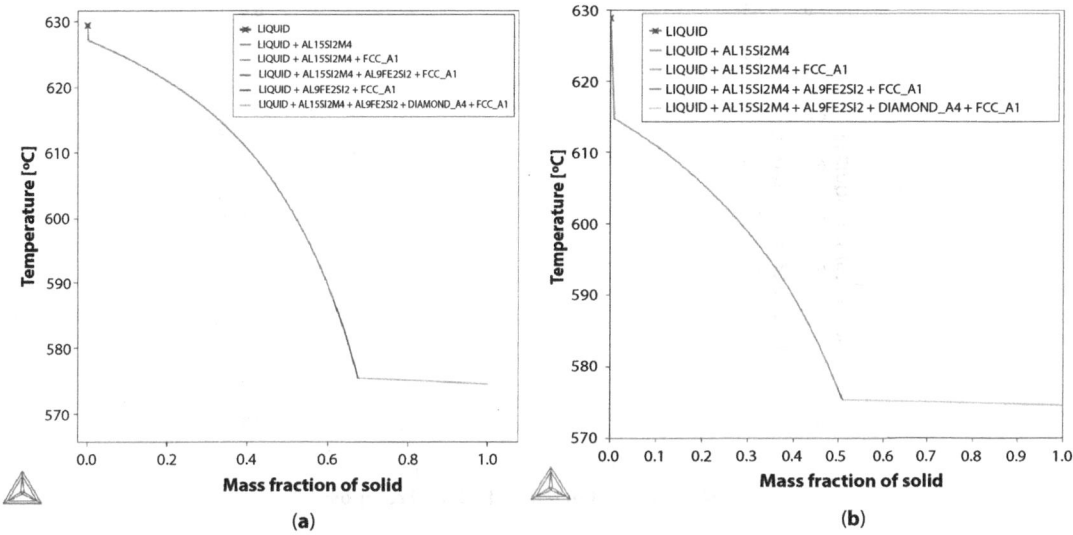

Figure 36.5 The calculated Scheil-Gulliver solidification diagram of (a) Al-5Si-0.75Fe-0.8Mn and (b) Al-7Si-0.75Fe- 0.8Mn (wt.%) alloys.

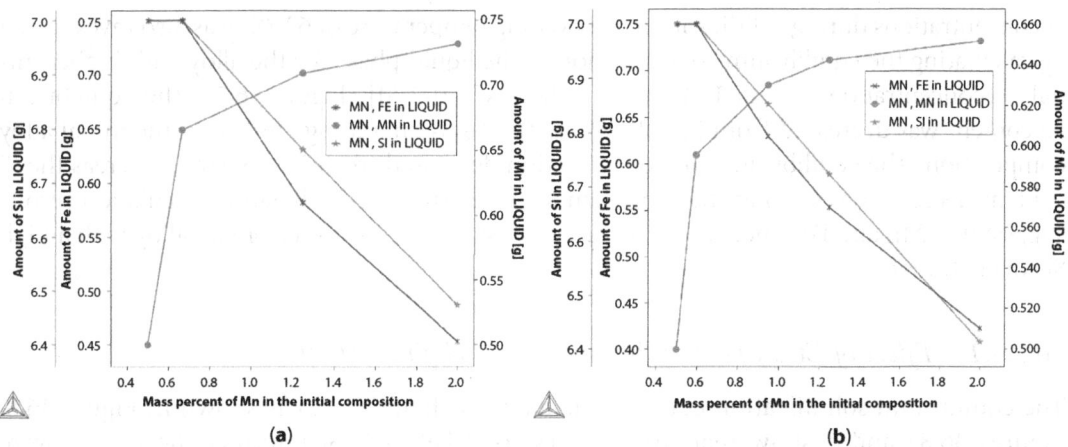

Figure 36.6 Equilibrium concentration of Fe, Mn, Si in the liquid phase of alloy Al-7Si-0.75Fe-xMn during solidification at (a) 620oC and (b) 615oC, calculated from equilibrium phase diagram. Note (g) is equivalent to % used in text.

It was observed that the final equilibrium concentrations of Fe and Mn in the liquid phase were significantly reduced when keeping the temperature above the solidification temperature of the α-Al phase.

However, the reduction is not very significant when reducing the temperature from 620-615°C. The equilibrium concentration of Fe was decreased from 0.75% to 0.45% at 620°C in the liquid phase when the initial content of Mn was 2% which further decreased to 0.74% Mn in the liquid phase, showing a significant reduction. However, the content of Si only showed a slight reduction from 7% to 6.5% at 620°C in the liquid phase. The variation of Fe, Mn, and

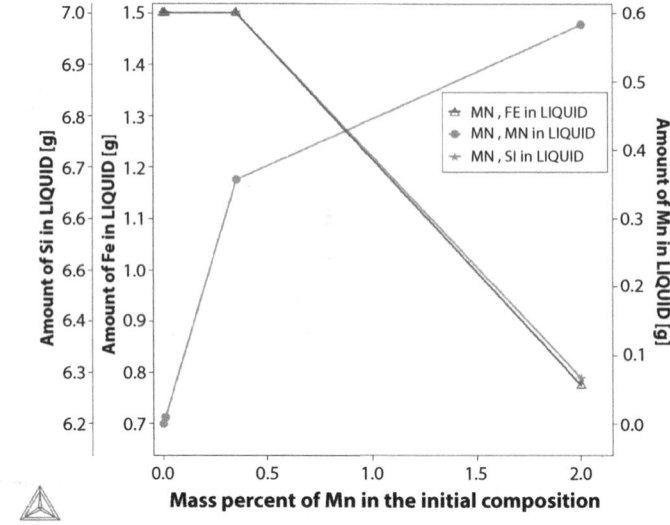

Figure 36.7 Equilibrium concentration of Fe, Mn, and Si in the liquid phase of alloy Al-7Si-1.5Fe-xMn (where x = 0-2%) during solidification at 620°C, calculated from equilibrium phase diagram. Note (g) is equivalent to % used in text.

Si concentrations during solidification at a holding temperature of 620°C was also investigated by calculating the equilibrium concentration in the liquid phase for the alloy with higher initial Fe concentration [Al-7Si-1.5Fe-xMn (where x = 0-2%)] (Figure 36.7). The equilibrium Fe content was decreased from initial 1.5% to 0.75% when adding 2% Mn in the initial alloy composition. The equilibrium Mn concentration decreased from 2% to 0.6%, whereas the Si content decreased to 6.3% at the same temperature. The thermodynamic calculation results suggest that Mn and Fe concentrations can be substantially reduced from an alloy with a high Si concentration.

36.2.3.1   Effect of Cr on Fe/Mn Removal from High Si Alloy

The equilibrium solidification curve of the alloys with adding Cr is shown in Figure 36.8. Figures 36.8a and b show that the primary α-Al(FeMnCr)Si intermetallic phases were formed before the solidification of the α-Al phase, with the addition of Cr. Furthermore, the initial formation temperature of sludge significantly increased to 740°C and 764°C when Cr concentration increased from 0.5% to 1%, respectively. In other words, the higher the Cr content, the higher the initial formation temperature. Figure 36.9 shows the expected effect of Cr addition on the Fe and Mn levels in the liquid phase when the alloy is maintained at different holding temperatures of 640°C and 620°C. As can be seen, the Fe level can be theoretically reduced to 0.1% Fe irrespective of holding temperature when Cr is added up to 1% in the initial composition. At the same time, the Mn concentration can be reduced from 0.8% to 0.55% using the same concentration of Cr at 620°C which suggests that the addition of Cr is highly effective, most significantly for the reducing Fe in a high Si alloy. The small residual Cr concentration left in the melt would benefit the mechanical properties [19].

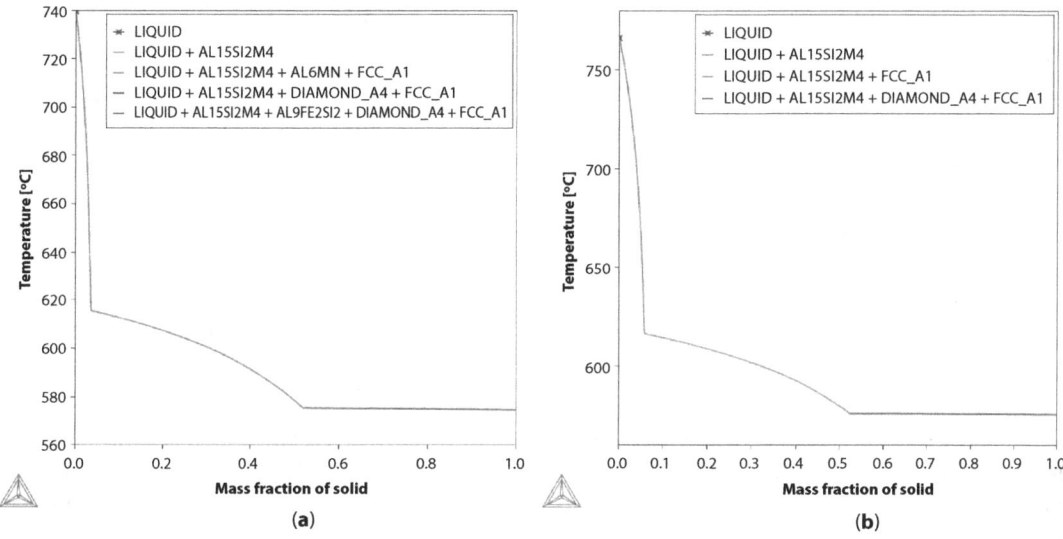

Figure 36.8 The calculated Scheil-Gulliver solidification diagram of (a) Al-5Si-0.75Fe-0.8Mn-0.5Cr and (b) Al-7Si- 0.75Fe-0.8Mn-1Cr (wt.%) alloys.

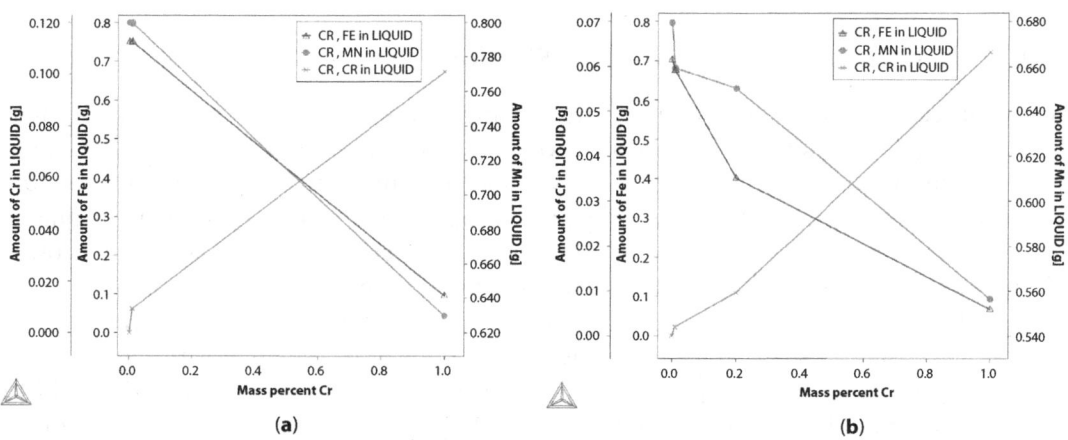

Figure 36.9 Equilibrium concentration of Fe, Mn, and Cr in the liquid phase of alloy Al-7Si-0.75Fe-0.8Mn-xCr during solidification at (a) 640°C and (b) 620°C, calculated from the equilibrium phase diagram. Note: (g) is equivalent to % used in the text. Note: X-axis represents Cr in the initial composition.

36.3   Experimental Validation

Experimental alloys (low and high Si) with different Si, Fe, Cr, and Mn levels were prepared by mixing a primary unalloyed aluminum (P1020) with commercial Al-25Fe, Al-20Cr, Al-25Mn, master alloys, and pure Si chips. The chemical composition of the ingots was tested by Optical Emission Spectrometer (SpectroMaxx LMM-14, Kleve, Germany). During experiments, ~2 kg of the prepared alloys were remelted at >800°C in a muffle furnace with the addition of the required amounts of Mn or Cr and then dropped to known temperatures in the furnace and held for a certain time. In the case of low Si alloy, after the experiment, the furnace was switched off and the melt was allowed to cool in the furnace. The final compositions were determined after cutting the melt longitudinally. In the case of high Si alloy, a known amount of the melt was poured into a graphite crucible to analyze the composition and the rest of the melt was allowed to cool in the furnace. Samples for microstructure analysis were prepared using standard metallographic techniques.

36.3.1   Removal of Fe/Mn from Low Si Alloys

Experiments were conducted using alloys with low Si content in the presence of Cr. Table 36.1 summarizes the chemical composition of the initial alloy and the percent reduction in Fe and Mn content after intermetallics sedimentation. As can be seen, there is a certain decrease after prolonged sedimentation time in the concentration of Fe and Mn in the separated alloy after Cr addition. However, the formation of intermetallics and subsequent sedimentation is too slow to be industrially accepted. Al-alloys with low Si content have relatively low fluidity at temperatures favorable for sedimentation of impurity-rich intermetallics, resulting in less effective reduction of the excess elements due to incomplete sedimentation. It is important that the alloy system has a high fluidity; otherwise, natural gravitational sedimentation will not be possible for Fe/Mn-rich intermetallics that have been formed.

36.3.2   Removal of Fe/Mn from High Si Alloys

Before proceeding with experiments, the cooling curves were also obtained for high Si-containing Al-alloys (A380) in the absence and presence of Cr. The presence of high Si in the system not only increased the fluidity but also reduced the temperature for the formation of the α-Al phase (615°C), thereby promoting the formation of intermetallic phases before the solidification of aluminum. This increases the chance of nucleation and growth of impurity-rich intermetallic phases. Table 36.2 summarizes the chemical composition of the initial alloy and the percent reduction in Fe and Mn content after intermetallics sedimentation. Based on these results, it appears that high Si in the system results in faster sedimentation of intermetallic particles as they form. A significant decrease in Fe and Mn levels was observed in the absence of Cr as well. A synergistic effect was observed in iron

Table 36.1 Effect of Cr on Fe/Mn removal from low Si-containing Al-alloys.

Initial Al-alloy composition	Cr (%)	Temp°C	Final Fe conc. (% Fe reduction)	Final Mn conc. (% Mn reduction)	Final Cr in melt
Fe = 0.84%; Mn = 0.92%, Si = 0.8%	1.1%	660°C	0.67% (20)	0.69% (25)	0.32%
Fe = 0.73%; Mn = 0.9%, Si = 0.94%; Zn = 0.87%, Cu = 0.69%	0.85%	660°C	0.56% (23)	0.72% (20)	0.25%

Table 36.2 Effect of Cr on Fe/Mn removal from high Si-containing Al-alloys.

Initial Al-alloy composition	Mn/Cr (%)	Temp°C (time)	Final Fe conc. (% Fe reduction)	Final Mn conc. (% Mn reduction)	Final Cr in melt
Si = 7.56, Fe = 1.22, Mn = 0.16, Cu = 3.5, Ti = 0.08, Mg = 0.2	1% Mn	630°C (1h)	0.66% (46)	0.54% (56)	0.0%
Si = 7.56, Fe = 1.22, Mn = 0.16, Cu = 3.5, Ti = 0.08, Mg = 0.2	1% Mn + 0.8% Cr	630°C (1h)	0.35% (71)	0.37% (70)	0.18%

removal by sediment separation with chromium and manganese simultaneously added to high Si alloy.

SEM-EDS analysis was conducted on the sludge that settled on the bottom of the furnace in both experiments. The EDS analysis (Figure 36.10) confirms the formation of

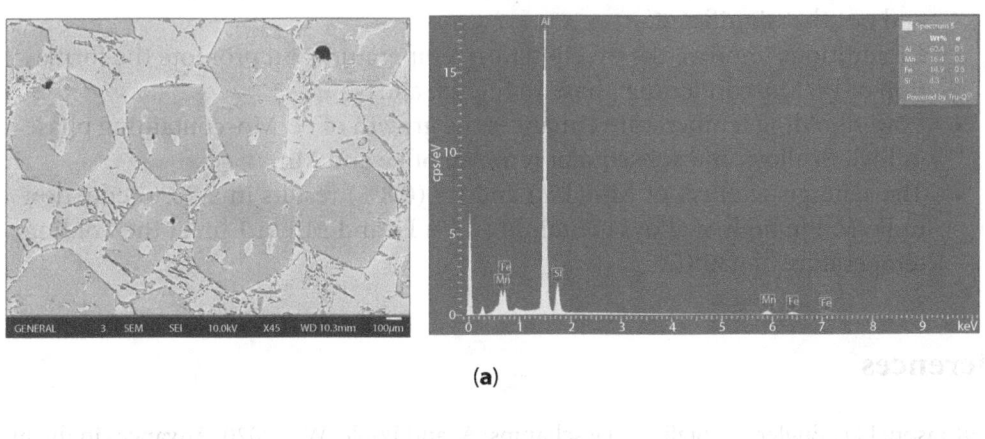

(a)

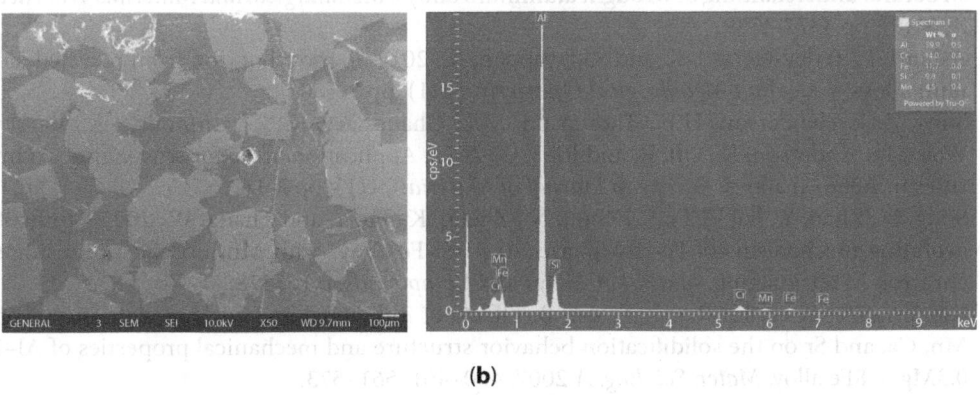

(b)

Figure 36.10 SEM-EDS analyses of sludge particles (a) only Mn (b) both Mn and Cr.

α-Al(FeMn)Si intermetallic phase in the sludge formed when Mn was used alone for Fe/Mn reduction. In contrast, the formation of α-Al(FeMnCr)Si was observed when Mn and Cr both were used for the reduction of Fe and Mn from high Si alloy.

36.4   Conclusions & Recommendations

In the present study, the effect of different alloying element addition was investigated to decrease the Fe/Mn content in scrap aluminum alloy with low and high Si for an efficient aluminum recovery. In order to determine the theoretical thermodynamic guidance for efficient purification of scrap aluminum alloys, the solidification diagrams and equilibrium concentrations of alloying elements at different concentrations and temperatures within Al alloy scrap were calculated using Thermo-Calc software. Based on this and subsequent laboratory melt trials, the major outcomes are as follows:

- Mn alone cannot reduce the Fe content from low Si-containing Al-alloy.
- Addition of Cr in low Si alloy promotes the Fe-rich phase precipitation first from the liquid phase.
- However, due to less fluidity of low Si alloy at favorable temperatures for Fe/Mn-rich intermetallic sedimentation, the reduction of Fe and Mn is too slow and not that significant.
- Cr additions improve the precipitation temperature and promote the formation of Fe/Mn-containing phase in high Si alloys.
- Lower holding temperature enhances the growth of Fe/Mn-containing phase in high Si alloys and subsequent removal of Fe and Mn.
- The synergistic effect of Mn (1%) and Cr (0.8%) results in a 70% reduction in Fe/Mn in high Si alloys containing >1% Fe and Mn in 1 hr. at the holding temperature of 630°C.

References

1. Robson, J.D., Engler, O., Sigli, C., Deschamps, A. and Poole, W.J., 2020. Advances in the microstructural understanding of wrought aluminum alloys. *Metallurgical and Materials Transactions A*, 51(9), pp.4377-4389.

2. Ashtari, P., Tetley-Gerard, K. and Sadayappan, K., 2012. Removal of iron from recycled aluminum alloys. *Canadian Metallurgical Quarterly*, 51(1), pp.75-80.

3. Sims, Z.C., Henderson, H.B., Thompson, M.J., Chaudhary, R.P., Hammons, J.A., Ilavsky, J., Weiss, D., Anderson, K., Ott, R. and Rios, O., 2022. Application of Ce for scavenging Cu impurities in A356 Al alloys. *European Journal of Materials*, 1(1), pp.3-18.

4. Song, D., Zhao, Y., Jia, Y., Li, R., Zhou, N., Zheng, K., Fu, Y. and Zhang, W., 2022. Study of the evolution mechanisms of Fe-rich phases in Al-Si-Fe alloys with Mn modification using synchrotron X-ray imaging. *Journal of Alloys and Compounds*, p.165378.

5. Kumari, S.S.; Pillai, R.; Rajan, T.; Pai, B. Effects of individual and combined additions of Be, Mn, Ca, and Sr on the solidification behavior structure and mechanical properties of Al–7Si–0.3Mg–0.8Fe alloy. *Mater. Sci. Eng. A* 2007, 460–461, 561–573.

6. Chanyathunyaroj, K.; Patakham, U.; Kou, S.; Limmaneevichitr, C. Microstructural evolution of iron-rich intermetallic compounds in scandium modified Al-7Si-0.3Mg alloys. *J. Alloys Compd.* 2016, 692, 865–875.

7. Todaro, C.J.; Easton, M.A.; Qiu, D.; Wang, G.; StJohn, D.H.; Qian, M. The effect of ultrasonic melt treatment on macro-segregation and peritectic transformation in an Al-19Si-4Fe Alloy. *Metall. Mater. Trans.* A017, 48, 5579–5590.

8. Wu, X.; Zhang, H.; Zhang, F.; Ma, Z.; Jia, L.; Yang, B.; Tao, T.; Zhang, H. Effect of cooling rate and Co content on the formation of Fe-rich intermetallics in hypoeutectic Al7Si0.3Mg alloy with 0.5%Fe. *Mater. Charact.* 2018, 139, 116–124.

9. Wu, X.; Zhang, H.; Zhang, F.; Ma, Z.; Jia, L.; Yang, B.; Tao, T.; Zhang, H. Effect of B addition on the morphology of iron-rich phases in Al-Si alloy. *Rare Met. Mater. Eng.* 2016, 45, 2133–2138.

10. Fan, C.; Long, S.Y.; Yang, H.D.; Wang, X.J.; Zhang, J.C. Influence of Ce and Mn addition on α-Fe morphology in recycled Al-Si alloy ingots. *Int. J. Miner. Metall. Mater.* 2013, 20, 890–895.

11. Song, D., Jia, Y., Li, Q., Zhao, Y. and Zhang, W., 2022. Effect of Initial Fe Content on Microstructure and Mechanical Properties of Recycled Al-7.0 Si-Fe-Mn Alloys with Constant Mn/Fe Ratio. *Materials*, 15(4), p.1618.

12. de Moraes, H.L., de Oliveira, J.R., Espinosa, D.C.R. and Tenório, J.A.S., 2006b. Removal of iron from molten recycled aluminum through intermediate phase filtration. *Materials Transactions*, 47(7), pp.1731-1736.

13. Yang, W.C., Feng, G.A.O. and Ji, S.X., 2015. Formation and sedimentation of Fe-rich intermetallics in Al–Si– Cu–Fe alloy. *Transactions of Nonferrous Metals Society of China*, 25(5), pp.1704-1714.

14. Matsubara, H., Izawa, N. and Nakanishi, M., 1998. Macroscopic segregation in Al-11 mass% Si alloy containing two mass% Fe solidified under centrifugal force. *J. Jpn. Inst. Light Met.* 48(2), 93–97.

15. Cao, X., Saunders, N. and Campbell, J., 2004. Effect of iron and manganese contents on convection-free precipitation and sedimentation of primary α-Al (FeMn) Si phase in liquid Al-11.5 Si-0.4 Mg alloy. *Journal of Materials Science*, 39(7), pp.2303-2314.

16. Švecová, I., Tillová, E., Kucharíková, L. and Knap, V., 2021. Possibilities of predicting undesirable iron intermetallic phases in secondary Al-alloys. *Transportation Research Procedia*, 55, pp.797-804.

17. Dhinakar, A., Lu, P.Y., Tang, N.K. and Chen, J.K., 2021. Iron reduction in 356 secondary aluminum alloy by Mn and Cr addition for sediment separation. *International Journal of Metal Casting*, 15, pp.182-192.

18. Sundman, B., Jansson, B. and Andersson, J.O., 1985. The thermo-calc databank system. *Calphad*, 9(2), pp.153-190.

19. Ahmad, R., 2018. The effect of chromium addition on fluidity, microstructure, and mechanical properties of aluminum A356 cast alloy. *Int J Mater Sci Res*, 1(1), pp.32-35.

37

A Novel Solvent-Based Recycling Technology: From Theory to Pilot Plant

Ezra Bar-Ziv[1]*, Shreyas Kolapkar[1], George W. Huber[2] and Reid C. Van Lehn[2]

[1]Michigan Technological University, Houghton, MI, USA
[2]University of Wisconsin-Madison, WI, USA

Abstract

We have developed a solvent-based technology for the extraction of single pure resins of food quality. The technology is called Solvent Targeted Recovery and Precipitation (STRAP). We have carried out numerous lab-scale experiments (in batches of 10-1000 g), with various plastic wastes, both post-industrial recycling and post-consumer recycling, including multi-layer flexible films. The key to the successful implementation of STRAP is the ability to pre-select solvents and temperatures to selectively dissolve a single polymer from all components in the mixture. The team has developed an extensive fundamental understanding of dissolution, enabling a successful operation. To aid in solvent selection, we developed a first-principles molecular modeling approach to rapidly predict temperature-dependent polymer solubilities. The technology is currently being scaled up to a 0.5 ton/day continuous Process Development Unit (PDU), which will be the basis for the next industrial scale. The technology requires relatively low temperatures (100-135°C) and equipment that is available in the chemical industry. We carried out extensive techno-economic and life-cycle assessments that showed (i) a STRAP system can be profitable as a stand-alone system at a 4,000 ton/y facility, with a great advantage of size, and (ii) due to the low process energy, the GHG emissions are reduced by over 90% in comparison to fossil-based polymer production. A line of the STRAP technology extracts a single polymer; to extract multiple polymers, multiple STRAP lines are required. Various methods can remove ink, pigments, or colorants, producing pure, colorless resins of food quality. The STRAP scheme depends on the source of the waste. It is to be noted that we successfully separated nine resins from arbitrarily mixed plastic wastes. Through our industrial partners, we showed that our extracted resins can be reused for their original film applications. The single most important economic factor is solvent recovery, at >99.9% of the solvent used, which is the solvent recovery system we are developing. Both lab scale and upscale results are presented and discussed.

Keywords: Solvent-based recycling, molecular modeling, scale up, multi-layer plastics, techno-economic analysis, life-cycle analysis

*Corresponding author: ebarziv@mtu.edu

Nabil Nasr (ed.) Technology Innovation for the Circular Economy: Recycling, Remanufacturing, Design, Systems Analysis and Logistics, (477–494) © 2024 Scrivener Publishing LLC

37.1   Introduction

In the US, most waste plastics are either landfilled or incinerated causing negative environmental impacts [1]. Recently, substantial research activity has been devoted to developing techniques to identify, separate, and reuse components of plastic waste [2]. However, the complex composition of mixed plastic waste poses challenges to existing separation techniques.

We have developed a novel *solvent-based* approach for the preferential dissolution of polymer resins in respective solvents that extract the resins in pure form, retaining their original properties [3, 4]. The recovered resins can thus be reused in their original applications. The process extracts only the respective polymer resins removing additives, ink, pigments, and adhesives. This process is referred to as the Solvent-Targeted Recovery and Precipitation (***STRAP***) process, with a pending patent application.

Some attempts have been made to commercialize solvent-based plastic recycling [5]; The main ones are (1) *Multi-cycle* [6] by Fraunhofer Inst., (2) *PureCycle* [7], and (3) *Newcycling*® by APK AG [8]. MutliCycle® has been licensed to Unilever, Polystyrene loop, and Circular Packaging. These efforts have faced challenges: Unilever built a 3-ton/day pilot plant in Indonesia but due to operational issues stopped the plans for scaling up the technology [9]. APK AG built an 8 kton/yr solvent-based recycling pilot plant to produce PP and PE from post-industrial waste multilayer plastics [10]. APK's technology is based on dissolving a plastic using a solvent mixture from a group of alkanes, isooctane, or cycloalkanes [11]. No reports have indicated why these technologies struggled to become commercial, but this is common with pioneer process technologies.

The key to the successful implementation of our STRAP is the ability to pre-select solvents and temperatures capable of selectively dissolving a single polymer from all components in the plastic mix. The team has developed an extensive and expansive fundamental understanding of dissolution, enabling a successful operation. To aid in solvent selection, we developed *a first-principles molecular modeling approach* to rapidly predict temperature-dependent polymer solubilities in solvents [12]. Our approach is based on a careful selection of the resin-solvent system and refined process parameters, temperature, concentration, residence time, and careful filtration of inks and colorants. STRAP-produced resins are pure and white, with no ink or colorants. We have made many lab-scale successful attempts to separate polymers from various wastes, such as Flexible multi-layer films, mixed plastic wastes, post-industrial recycling, post-consumer recycling, MRF residues, and municipal solid wastes. Our intention is to develop the technology that can reach the market in a relatively short time.

In this paper, we will present results from (1) our molecular modeling work for the selection of the specific solvent for the specific resin; (2) lab-scale results, specifically for multilayer films provided by Amcor; and (3) our scaling-up efforts.

37.2   Main Content of Chapter

37.2.1   Molecular Simulations and Experimental Verification

The team has developed an extensive fundamental understanding of dissolution that enables a successful operation. We developed a first-principles molecular modeling approach to rapidly predict temperature-dependent polymer solubilities in solvents [3, 12]. A series of computational methods are used to evaluate polymer solubilities and compare them to experiments for validation. Figure 37.1 shows the typical steps – Step 1: A molecular dynamics (MD) simulation of a single oligomer is performed in a dilute solution to obtain a simulation trajectory of various oligomer conformations; Step 2: A set of conformers is selected from the MD trajectory to span a range of representative oligomer structures; Step 3: Density functional theory (DFT) calculations are performed for the selected conformers to obtain corresponding screening charge density profiles; Step 4: Screening charge density profiles are input to Conductor-like Screening Model for Real Solvent (COSMO-RS) solubility calculations; and Step 5: Experimental solubility measurements are used to benchmark the computational predictions. We provide details on each step in the sections below.

In the first step, we model target polymers as oligomer molecules and perform MD simulations of these oligomers in dilute solutions (with both good and poor solvents) to obtain trajectories of representative oligomer conformations [12]. Solvents were selected based on literature reports of good and poor solvents for the target polymers. We then sampled representative oligomer structures (referred to as conformers) from the MD trajectories based on two structural parameters: the radius of gyration (R_g) and the solvent-accessible surface area (SASA). Sampling conformers that cover a range of values of these two parameters can provide reliable input for COSMO-RS solubility estimations [12, 13]. Based on extensive analysis of tradeoffs between prediction accuracy, oligomer size, and a number of conformers, we adapted a workflow in which 20-30 conformers were sampled for each polymer for oligomers comprised of ~6 repeat units. The selected conformers were input to DFT calculations to obtain screening charge densities (COSMO files). The DFT calculations included a geometry optimization in implicit water using the conductor-like polarizable continuum model and a single-point calculation in the infinite dielectric constant limit.

COSMO files from the DFT calculations were input to COSMO-RS for solubility calculations. COSMO-RS predicts the thermodynamic properties of multicomponent systems based on quantum mechanical calculations and statistical thermodynamics methods [14, 15]. The chemical properties of each molecule are represented by the probability distribution of the screening charge densities (called the σ-profile). σ-profiles of all oligomer conformations with deactivated terminal groups were used to approximate the σ-profile of the corresponding polymer [16]. The σ-profiles were then used to calculate the chemical potential of the polymer to enable predictions of solubility via a solid-liquid equilibrium calculation. This calculation requires the polymer melting temperature and an experimentally measured solubility as reference input. Experimental reference inputs were measured

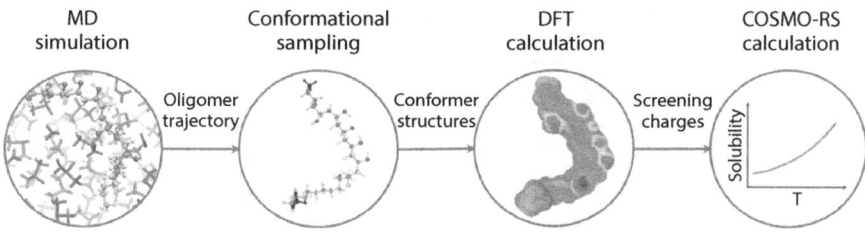

Figure 37.1 Flow diagram summarizing the methods for computational polymer solubility prediction and experimental validation.

for each target resin and corresponding melting temperatures were taken from literature sources. With this approach, we were able to obtain predictions of temperature-dependent solubilities to compare against experimental values.

We performed multiple tests to assess the accuracy of computational predictions. Table 37.1 compares predicted and experimentally measured solubilities for a PE sample in various common solvents, with temperatures listed in the table [12]. We utilized the experimental solubility of PE in toluene at 110°C as a reference value for the solid-liquid equilibrium calculation for all solvents. As expected, our computational protocol predicts near-zero solubilities for all solvents other than toluene in good agreement with the experimental measurements.

Notably, several of these solvents are predicted by some sets of Hansen Solubility Parameters to be good solvents (e.g., hexane, THF, and ethyl acetate), indicating that our computational approach qualitatively outperforms conventional methods for solvent selection based on Hansen Solubility Parameters in addition to providing quantitative solubility estimates.

We next assessed our ability to predict the solubility of EVOH (a common component of multilayer plastic films), which is a challenging task because EVOH is a copolymer that has two different monomer units. The arrangement of these units leads to different types of structures, so we assessed prediction accuracy using three different structural arrangements of the EVOH repeat units Table 37.2 compares solubility predictions to the experimental data for each of these 3 structures. For this comparison, the experimental solubility of EVOH in DMSO at 95 °C was used as the reference for the solid-liquid equilibrium calculation. Clearly, DMF was recognized as a good solvent in all three cases, but only the block structure result correctly identifies it as the best solvent in the list. Solubilities are somewhat over-predicted for experimentally identified nonsolvents, with ethanol being the most notable deviation. However, the values clearly differentiate solvents and nonsolvents, with a slightly higher threshold (3 wt%) required to distinguish these categories than is necessary for PE. With respect to quantitative predictions, the random structure has the lowest overall mean prediction error, which is consistent with the fact that EVOH is a random copolymer [17].

Table 37.1 COSMO-RS solubility predictions for PE compared to experimental measurements. Toluene is used as a reference solvent and hence its prediction is exact.

Solubility (wt%)			
Solvent	**Temp (°C)**	**Experiment**	**Simulation**
DMSO	95	0.040	0.34
Water	95	0.002	0.00
THF	65	0.018	0.54
Hexane	65	0.031	0.85
Ethyl acetate	77	0.004	0.52
Ethanol	78	0.011	0.20
Acetone	55	0.001	0.06
Toluene	110	14.56	14.56
DMF	100	0.009	1.85

Table 37.2 COSMO-RS solubility predictions for EVOH compared to experimental measurements. Three different copolymer structures are compared (Figure 37.2).

Solvent	Temp (°C)	Experimental solubility (wt%)	COSMO-RS solubility predictions		
			Block (wt%)	Alternating (wt%)	Random (wt%)
DMSO	95	24.02[a]	24.02	24.02	24.02
Water	95	0.00	0.01	0.03	0.03
THF	65	0.02	5.53	2.06	1.78
Hexane	65	0.00	0.02	0.00	0.00
Ethyl acetate	77	0.05	2.47	0.66	0.49
Ethanol	78	0.00	7.63	3.94	2.98
Acetone	55	0.00	1.81	0.58	0.45
Toluene	110	0.00	4.49	0.64	0.39
DMF	100	30.63	27.49	21.19	20.48

[a] experimental data in this solvent was used as reference for the solid-liquid equilibrium calculation, so all three structures have the same calculated solubility.

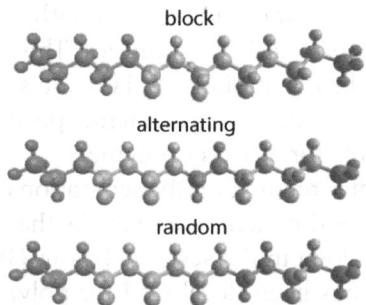

Figure 37.2 Three types of structures of EVOH oligomer with 2 ethylene units (red) and 4 vinyl alcohol units (blue). Oligomers are modeled to have end methyl groups (black) that are deactivated such that they do not contribute to the COSMO-RS calculations.

We also utilized EVOH to assess the capability of our protocol to assess temperature-dependent variations in solubility. Figure 37.3 shows the predicted EVOH solubility in DMSO as a function of temperature. Experimental data at four temperatures are included for reference. Predictions were based on either 1, 5, or 20 EVOH conformers, with 10 separate subsets of conformers used for predictions. Predicted solubilities with only a single conformer varied substantially and lie with the region indicated by the light blue color. The region bounding the solubility predictions decreased substantially for the 5-conformer sets (light green region). For the 20-conformer sets, the solubility calculations are highly consistent leading to a deep blue region that nearly converges to a solid line such that variations are difficult to discern by eye. These results indicate that increasing the number of conformers both leads to a decrease in prediction accuracy and decreases the variation between different subsets of conformers, suggesting more robust results.

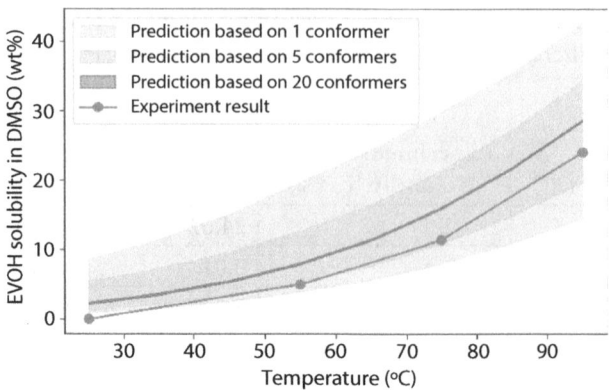

Figure 37.3 Temperature-dependent solubility prediction for EVOH in DMSO with different conformers.

Moreover, all conformer sets exhibit reasonable agreement with experiments, with the final results (for 20 conformers) slightly overestimating the measured solubilities but exhibiting similar trends.

Based on the good agreement with experimental data, we expanded the predictions to assemble a database of polymer solubilities. In total, we predicted the solubilities of 8 common polymers (PE, PP, PS, PET, PVC, Nylon-6, Nylon-66) in 1007 solvents. For each solvent, the solubility of each polymer was computed at both room temperature and a higher temperature to suggest bounds on feasible solubilities. This database permits rapid identification of solvents for STRAP based on the criteria that a solvent must selectively dissolve exactly one polymer from a multicomponent system; specifically, the solubility should be high for a target polymer but low for all other polymers.

Table 37.3 shows examples of screening for the separation of PE and EVOH by presenting 9 selective solvents and corresponding temperatures. Further examples of the application to real STRAP processes are described in the sections below. Overall, a strong computational tool for solubility predictions has been developed for polymer-solvent combinations and it has successfully demonstrated good agreement with the experiment. The applicability of the approach was demonstrated by identifying selective solvents for various polymers from a library of 1007 solvents.

37.2.2   Lab-Scale STRAP Results for Multi-Layer Flexible Films

STRAP was applied to a printed flexible multilayer plastic film to recover and produce clear polymer resins after incorporating a deinking step. The target film is a post-industrial oriented polyester film (OPET) that comprises PE, EVOH, and PET with polyurethane (PU) inks (Figure 37.4). A challenge with this specific plastic waste feedstock is how and at what stage of the process should the ink components be removed. Different surfactant solutions and PU-selective solvents were considered to achieve the recovery of the polymer fractions without any coloration.

The flexible printed multilayer film composed of PE, EVOH, PET, and PU-based inks was processed through STRAP to recover all polymer components. Figure 37.5a shows solvents, temperatures, and dissolution times for each step. In our approach, the PE and

Table 37.3 Computational solvent screening for PE and EVOH. Nine selective solvents and corresponding temperatures are presented as examples (no PE-selective solvents were identified at room temperature).

Solvent	Temp (°C)	PE solubility (wt%)	EVOH solubility (wt%)	Comment
Hydrazine	25	0.00	26.38	Room temperature, EVOH-selective
Ethylenediamine	25	0.00	24.40	
n-butylamine	25	0.00	15.26	
Octane	124.6	45.40	0.13	Near boiling point, PE-selective
1-octene	120.2	35.22	0.21	
Tetrachloroethylene	120.3	21.34	0.08	
Methylhydrazine	86.5	0.23	34.21	Near boiling point, EVOH-selective
Acetic acid	116.9	0.62	25.46	
Formic acid	100	0.02	24.42	

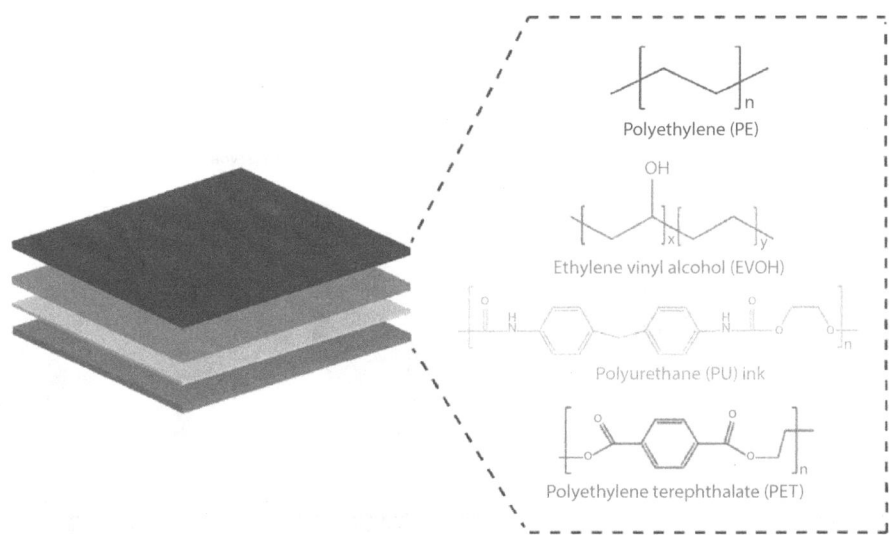

Figure 37.4 OPET multilayer printed film composed of PE, EVOH, PET, and PU-based inks, manufactured by Amcor.

EVOH are dissolved and recovered, leaving the PET and ink components behind. After this, the ink is removed from the PET component using GVL. Polymer precipitation was achieved by reducing the temperature of the respective solvent and not by the addition of antisolvents. In our previous work, this was demonstrated to be beneficial both economically and environmentally. On visual inspection, the final recovered polymers after

STRAP exhibited little to no coloration (Figure 37.5b). The ink removal was carried using gamma-Valerolactone (GVL) to separate white, black, and yellow ink from PET of a printed multilayer film (Figure 37.5c).

Table 37.4 shows the yield of each component, including the ink residue (black and white) recovered after the ink removal step. Due to the high percentage of PE in the printed film, an additional PE dissolution step was required to recover most of the polymer and clean out the experimental setup. This additional PE dissolution step was done in every experiment in Table 37.4, which allowed for an average PE yield of 61.8 wt%.

We found that the material that was not recovered after STRAP was left behind in our equipment, mostly in the round bottom flasks and filters. The overall mass balance improved when accounting for the material lost in the equipment which was around 8 wt% of the plastic feed. Furthermore, part of the ink (mostly yellow color) can be recovered via distillation of the GVL, around 3 wt%. The overall mass balance can be >95 wt% of the starting material, considering the average recovery of PE, EVOH, and PET. This gives indication that when dealing with flexible plastics that have been shredded, material will

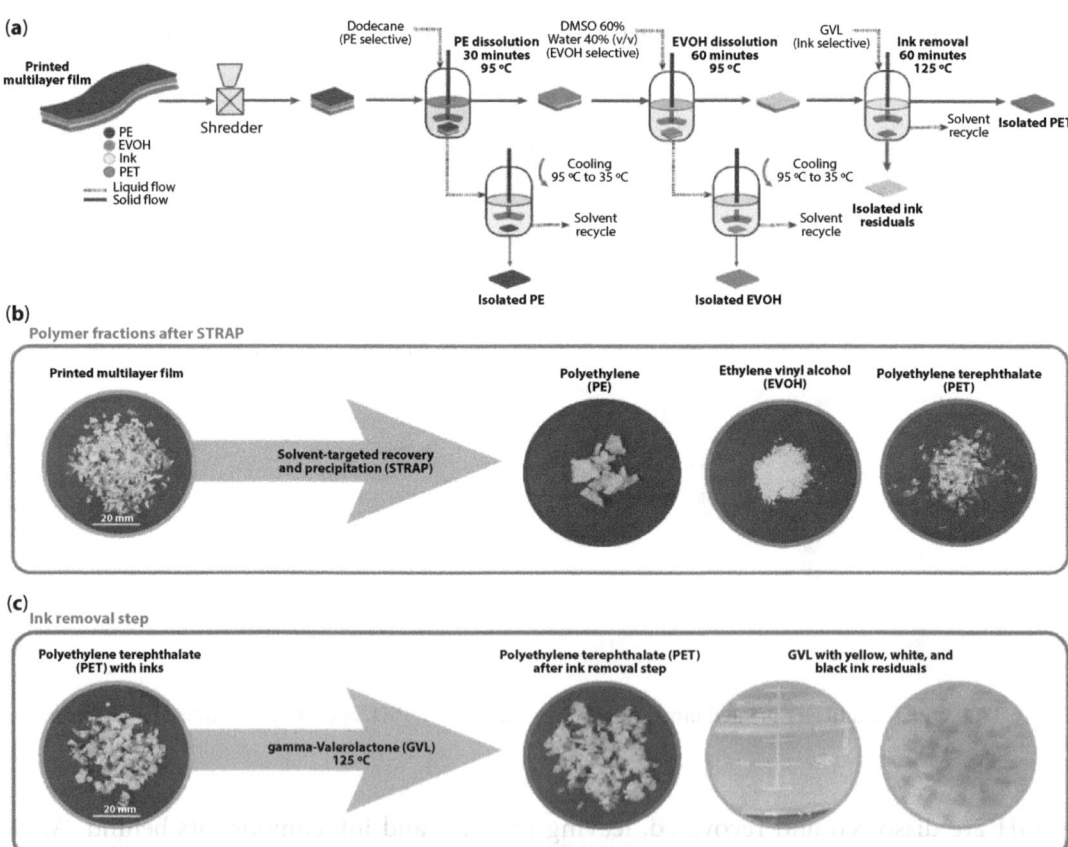

Figure 37.5 (a) Solvent-targeted recovery and precipitation (STRAP) of a printed multilayer film composed of PE, EVOH, PET, and PU-based inks, (b) Photos of each polymer after the STRAP, (c) STRAP deinking step with gamma-Valerolactone (GVL) to separate white, black, and yellow ink from PET of a printed multilayer film.

Table 37.4 Resin yield: printed multilayer film.

Exp	Component yield (wt%)			
	PE	EVOH	PET	Ink
1	59.8	7.70	19.8	1.4
2	62.4	6.94	17.6	2.1
3	63.2	7.74	18.2	0.6
Avg.	61.8	7.46	18.5	1.4
SD (σ)	1.7	0.45	1.2	0.8

be left behind in the equipment which should be considered when thinking about larger scale systems. Furthermore, it was challenging for us to predict and measure consistent amounts of material lost in the equipment since it depends on multiple factors like plastic size, stirring rates, and it is subject to how the material is handled by the person conducting the batch experiment.

Previous STRAP experiments with transparent multilayer films were initially demonstrated with film sizes of 1x1 cm and this necessitated extensive dissolution times to recover the target polymers. We investigated the effects of film sizes over the dissolution time of the PE component in the printed multilayer film. This was tested in the PE-dissolution step using dodecane at 95°C and 500 RPM. Initial steps with a dissolution time of 30 minutes yielded 61.78 ± 1.74 wt% of PE when the film sizes were > 3 mm. This yield was calculated as the amount of extracted PE over plastic feed in a single experiment. Additional size reduction of the multilayer films to 3 mm and 1 mm showed improvement in the PE dissolution time. At a total PE dissolution time of 8 minutes, the PE recovery improves by reducing the size of the films to 3 mm. From 3 mm to 1 mm, a slight improvement in the PE yield was observed. The 8 minutes of dissolution time were divided in two steps, a 5-minute step and a 3-minute step. As explained in the experimental demonstration section, multiple steps were carried out to recover most of the dissolved PE due to its high composition in the film, and this helped with cleaning out the experimental setup. From the above results, we concluded that by downsizing the plastic material to the proper size (1-3 mm) and having vigorous mixing and the required dissolution temperature, we were able to extract most of the PE with dodecane. In dissolution-based plastic recycling technologies, plastic sizes become important whenever short dissolution times are of interest.

37.2.3   Polymer Characterization

We use IR spectra for resin identification. In Figure 37.6a, the IR spectra of the PE STRAP and PE virgin resin are compared, displaying identical spectral bands with no trace of any other bands belonging to other resins, indicative to the purity of the extracted PE. Figure 37.6b and Figure 37.6c shows similar characteristics for EVOH and PET, respectively. The molecular weight values of the recovered PE from the printed multilayer film by STRAP were determined with high-temperature gel permeation chromatography (HT-GPC).

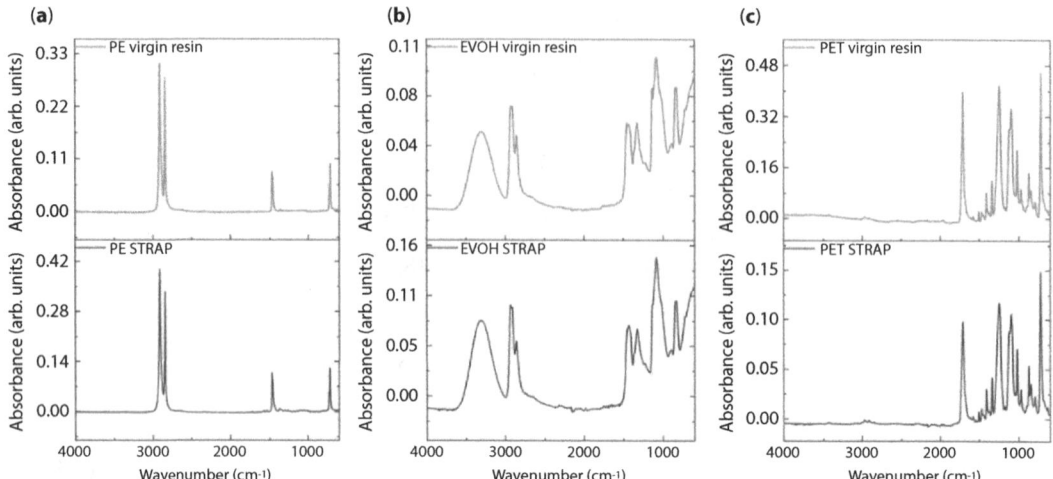

Figure 37.6 ATR-FTIR spectra of virgin resins and polymers recovered by STRAP from the printed multilayer: (a) PE, (b) EVOH, and (c) PET.

The printed film itself was manufactured with both LDPE and LLDPE, and the produced PE from STRAP had molecular weight values that were within the range of the pure resins. This gives evidence that the molecular weight of the PE is not significantly affected after being treated with dodecane in the STRAP process. The recovered PE, EVOH, and PET for the most part had similar thermal properties to the corresponding pure virgin resins used in the starting material.

For example, as seen in Table 37.5, the melt temperature ($T_{m,2}$) of the recovered polymers from the printed film by STRAP were comparable to the virgin resins. Similar results were obtained by Cecon *et al.* [18] for the PE resins recovered using STRAP, with the thermal properties within the ranges displayed by the virgin HDPE and LLDPE resins used in a rigid transparent multilayer film. However, differences were observed in some parameters of the PET STRAP samples, possibly due to residual polymers or ink components. This

Table 37.5 Thermal and molecular parameters for virgin resins and polymers recovered from the printed multilayer film by STRAP.

Resin	Tc (°C)	Tm,2 (°C)	ΔHc (J/g)	ΔHm,2 (J/g)	Crystallinity
PE STRAP	105.0	119.7	76.0	82.3	28.38%
LDPE Virgin	98.3	112.0	84.8	86.1	29.69%
LLDPE Virgin 45G	106.1	122.4	83.3	85.5	29.48%
LLDPE Virgin 47N	107.0	122.2	70.2	73.3	25.28%
EVOH STRAP	150.4	175.7	41.8	37.4	17.17%
EVOH Virgin	147.8	176.4	54.2	54.3	24.93%
PET STRAP	209.6	246.0	41.7	22.7	16.21%
PET Virgin	169.2	244.6	30.0	38.5	27.50%

could be addressed by carrying out additional ink removal steps to ensure PET is free of any contaminants. After STRAP, decreases in the crystallinities were observed for the recovered EVOH and PET, which could also be an indication of contamination in the samples. Our previous characterization study of STRAP recovered materials did not determine significant changes in the crystallinity but in the melt temperature of PET, which was associated with solvent retention in the polymer matrix as PET was the main layer in the film.

37.2.4   Scaling-Up

Figure 37.7 is a 3-D depiction of one scaled-up STRAP line at 25 kg/hr, i.e., one solvent and one resin extracted. The shredder (1) downsizes the plastic waste into 2-4-mm flakes with an aspect ratio of 1. The screw conveyor (2) transfers the material to a live bottom hopper (3). The bucket elevator (4) conveys the material into a dissolution tank (5), with an actuated control valve. The tank is heated to the desired temperature (e.g., LDPE at 110°C), and the solvent is fed from a tank (10).

This tank contains a nitrogen blanket for safety. After 5-6 min for complete dissolution, the blend is drained by opening the ball valve at the bottom of the tank and fed into a hot filter (6) which separates the solution from the non-dissolved flakes. The temperature of the hot filter is controlled by a heater. The non-dissolved flakes flow through a conveyor (13) and a control valve into storage tank. The plastic solution flows from the dissolution tank into a precipitator (8) that separates the pure resin from the solvent by cooling, using a chiller (11). The mix from the precipitator flows into a cold filter (12) where the solvent flows into a solvent tank (10) and the resin is stored in storage tank. We will describe the critical components of this system.

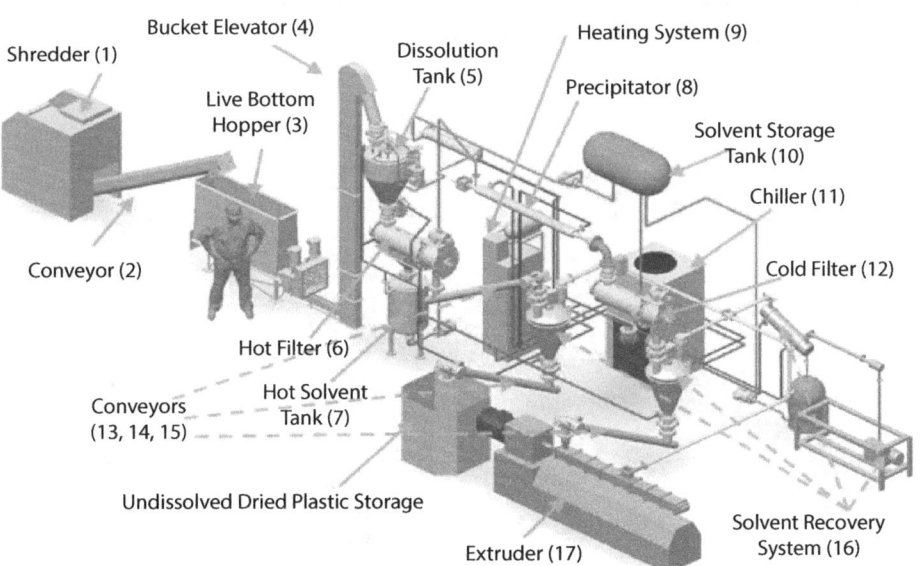

Figure 37.7 3-D presentation of the STRAP pilot unit.

37.2.5   Plastic Waste Dosing

We developed an accurate feed bin at a rate of up to 4 kg/min (Figure 37.8) with negative angled walls. The bin was tested with wastes for thousands of hours, including pre-processed and shredded plastics, encountering no bridging or flowability issues.

37.2.6   Dissolution Tank

Figure 37.9 shows the dissolution tank where the plastic flakes are flown into the tank through actuated ball valves, and the tank is filled with hot solvent. A shaft rotates at ~1700 rpm with two 5-in impellers. Baffles were added at the tank walls to create turbulent flow. Figure 37.10 shows the top view of the plastic-solvent flow in the tank, with/without baffles. The flow without baffles is laminar and with the baffles is turbulent. The solvent-to-plastic ratio is 7:1. The dissolution time has been extensively studied, as a function of impeller type, rotation frequency, and temperature. Figure 37.11 shows results for PE dissolution yield (calculated as dissolved plastic solvent over the initial plastic amount). As seen, complete dissolution requires about 20 seconds, which is indicative of excellent mixing with no mass transfer limitation.

37.2.7   Hot Filter

The hot filter separates the solution from the non-dissolved residues. We purchased a commercial filter that is based on centrifugal separation, shown in Figure 37.12. The undissolved plastics and the solution mix flows from the tank through a 4-in actuated valve into the filter. In the first section, a short screw auger conveys the mix into the centrifugal section with a still screen, with 4 axially angled brushes rotating at 800-1000 rpm. The solution is centrifugally "pushed" through the screen in a radially outward direction, whereas the non-dissolved plastic flakes are conveyed axially by four paddles towards the outlet. Note that the filter is heated using a jacket to avoid early precipitation.

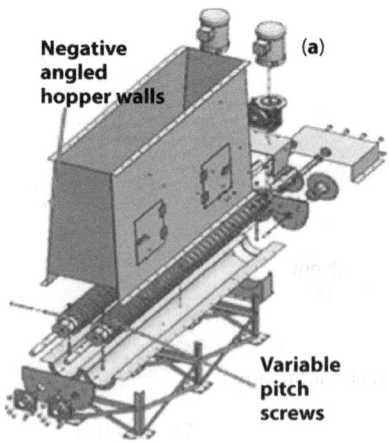

Figure 37.8 The hopper.

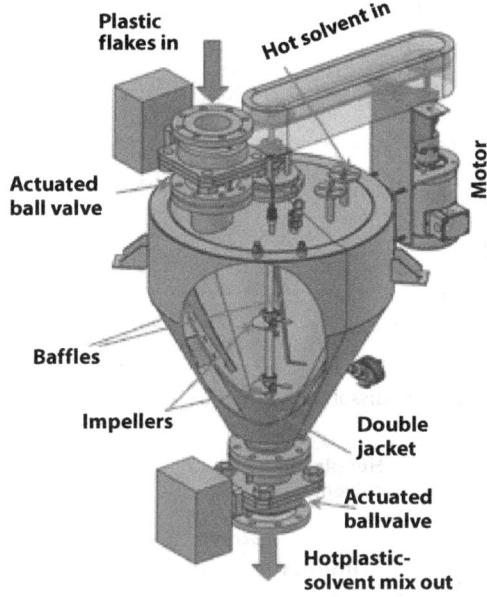

Figure 37.9 The dissolution tank.

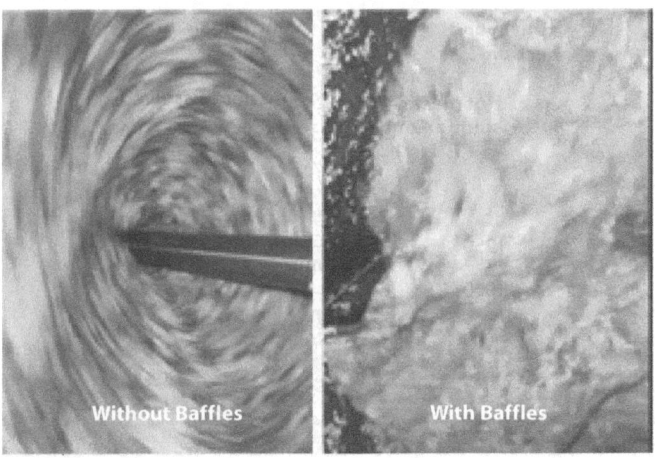

Figure 37.10 Flow patterns with/-out baffles.

We carried out numerous runs, measuring: (i) the solution outlet weight, (ii) the non-dissolved plastic weight, (iii) filtration efficiency, and (iv) filtration dynamics. We carried out numerous drainage experiments, where the 4-in ball valve of the dissolution tank was opened (~1 s). The plastic-solvent mix was drained through the valve directly into the hot filter. We measured the outlet from both the solid and liquid ports (Figure 37.13). Figure 37.13 shows typical results of drainage dynamics, as solid outlet vs. time. A drainage rate equivalent of ~5 kg/min was measured, which is a fast drainage rate. The filter can operate for extended periods without maintenance due to the active brush mechanism continuously wiping the filter screen. The status of our filtration technology is that the solid

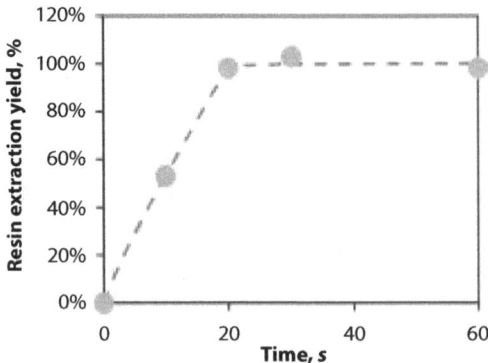

Figure 37.11 Dissolved resin vs. time in dissolution tank.

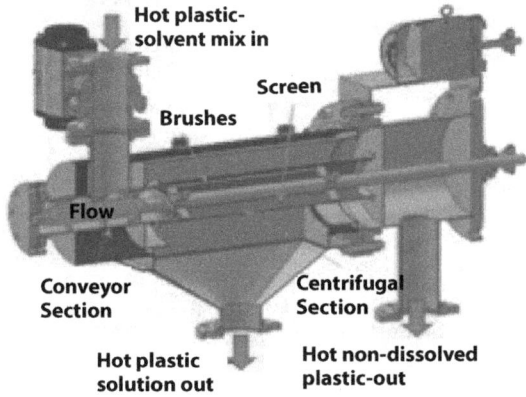

Figure 37.12 The filter.

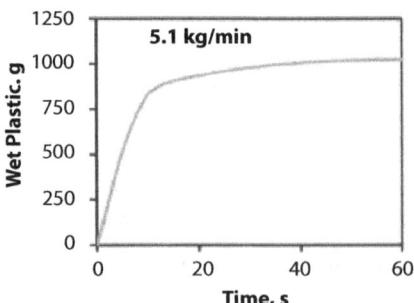

Figure 37.13 Hot filter plastic outlet.

material after filtration comprises 10% solvent, noting that this is equivalent to recovering **99% of the solvent.**

37.2.8 Precipitator

The precipitator receives hot plastic solution from Filter 1, cools it to precipitate the plastic. We note that key challenges are required temperature for full precipitation, plastic tendency

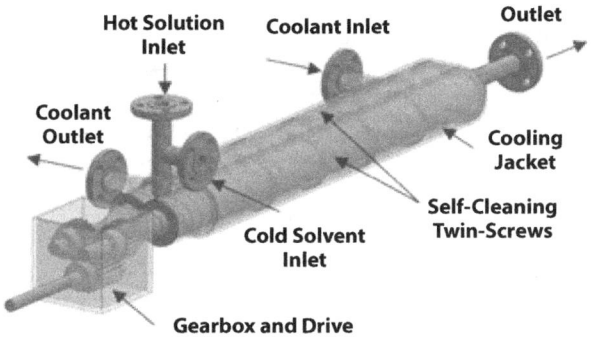

Figure 37.14 The twin-screw precipitator.

to stick to surfaces, and the time required to reach precipitation temperature. We used our own solubility data to design the temperature required for precipitation [3, 12]. Molecular simulations predicted the solubility (validated by lab experiments [3, 12]) From our experience, twin-screws, shown in Figure 37.14 have a self-cleaning mechanism as the screw flights scrub each other and housing walls, generating clean surfaces. The solution in the twin-screw, flows from one screw to the other, generating excellent mixing, with fast heat transfer, such that a 1.5-m long, 2-in diameter, can cool down the solution to ~35°C at the outlet in <10 min.

37.2.9 Solvent Recovery and Solvent Purification

We are currently building a solvent recovery device that can recover the solvent at a rate of 99.9%. As for purification, the solvents may have residual resins, inks and colorants and require purification. Solvents are used and recovered extensively in chemical industry without any indication for degradation. Degradation occurs only if a chemical reaction occurs with the solvent. In STRAP, there are no reactions involved, unless some chemicals in the plastic wastes react with the solvent. Part of the current program is to monitor the quality of the solvent by IR spectroscopy. If degradation is noticed, a distillation unit will purify the solvent.

37.2.10 Techno-Economic and Lifecycle Analysis

We performed techno-economic analysis for an independent STRAP plant that receives plastic wastes pre-processes it, then applies STRAP for resin extraction entails a few processing lines of separation, starting from ink removal, followed by a few extraction lines, each for a specific resin, as seen in Figure 37.15, extracting the most important plastic resins in the waste blend.

Results are shown in Figure 37.16 for an independent STRAP plant with (i) the annual capital cost, in red, which is the total capital cost multiplied by an annualized factor, and (ii) in blue, the break-even sale price (also production cost) of PE ($/ton), both vs. plant size (ton/yr) in the range 1,000-16,000 ton/yr. At a plant size of 2,500 ton/yr, the minimum selling price required to return a 10% return is $750/ton. The real selling price of recycled post-consumer plastics is currently above $1,500/ton; a STRAP plant of this size is very

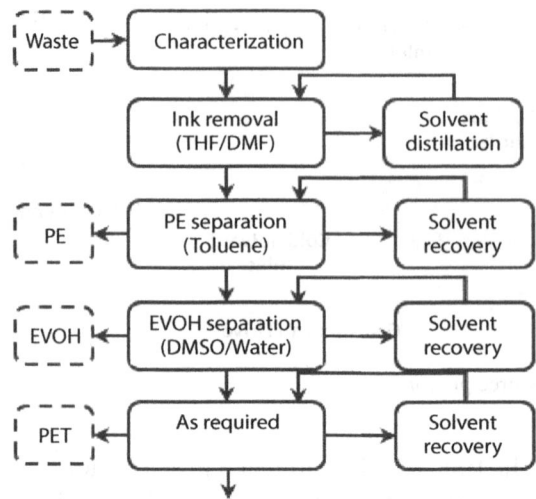

Figure 37.15 Flow diagram for a given waste.

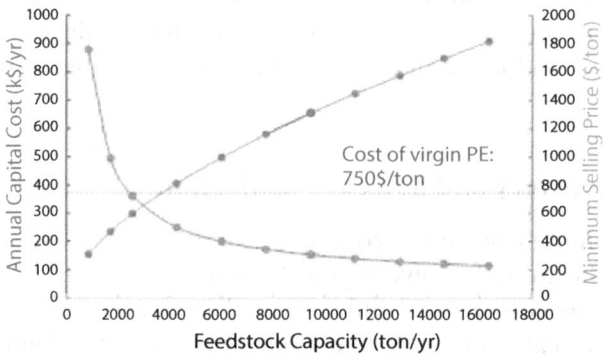

Figure 37.16 The minimum selling price of recycled PE and annualized capital investment of STRAP process for recycling Amcor printed film.

profitable. Profit increases with the size of the plant. A similar analysis for the STRAP-MRF plant with capital and operators' cost are significantly smaller [19].

The cost of the make-up solvent is critical to the STRAP economics. The main source of solvent loss is the absorbed solvent in the targeted resin and the non-dissolved plastics. From lab measurements, we noted that, in average, 1-gr of plastic would trap 1 mL of solvent (or 1-ton traps 1-m^3 solvent). Common evaporation-based technologies [20] can reach a solvent recovery rate of 99%. The cost of solvents used is in the range \$4,000-\$6,500/m^3. Thus, for 1 ton of plastic produced, one may lose 0.01 m^3, which is equivalent to solvent cost \$40-65/ton-plastic. We consider this cost as the maximum possible for solvent recovery in the STRAP process.

We carried out a lifecycle analysis that showed that STRAP (i) has at least 57% [21] lower greenhouse gas emissions than the production of virgin resins, the lowest of any existing recycling plastic technology. The very low carbon-footprint of STRAP would benefit society by recycling plastic wastes, reducing (and eventually replacing) the use of fossil-based feedstocks, and as shown above, with significant commercial and economic benefits. With an expected continual increase of plastic products and the consequent plastic waste generation, this is a major societal gain.

37.3   Conclusions and Recommendations

The STRAP process extracts specific polymers by preferential dissolution to produce pure resins. This process has been based on molecular dynamic calculations that determine the interactions between the solvents and the polymers and enable accurate predictions and selection of the solvent for a specific polymer. It has been verified experimentally at the bench scale. Lifecycle and techno-economic analyses showed that STRAP (i) has very low specific process energy, and (ii) low capital and operation costs. The following are the main benefits of STRAP: (1) Prevent current disposal in landfills; (2) produce recyclable plastic products; (3) reduce carbon footprint; (4) purchase pure resins at lower costs than those currently purchased. In addition, the societal benefits are (1) reduction of land used for landfills, (2) reduction in GHG generation, (3) reduction in taxation, (4) reduction in ocean plastics, and (5) sustainable (circular) material used. Thus, STRAP would benefit society by recycling mixed plastic wastes and reducing the use of fossil-based feedstocks. With an expected continual increase of plastic products and the consequent plastic waste generation, the continuous STRAP can have major societal gain. The key technological hurdles in scaling up the technology are: (i) integrating the various pilot scale components, (ii) high solvent recovery rate, and (iii) transition to the use of green solvents. We have shown that STRAP can be economically viable at small-medium-large plant sizes. This will allow companies of all sizes that produce plastic components and generate plastic wastes not to anymore landfill their waste and develop recyclable products with low carbon footprint and cost. We foresee three types of users (i) companies who produce plastic products, that generate plastic waste and are currently disposing of it in landfills, (ii) municipalities and companies that are treating post-consumer plastics, and (iii) resin producers that can purchase the STRAP pure resins, at lower costs than currently paid for and distribute them to the market. A sequence of solvents will enable the separation of most of the resins in the waste.

Acknowledgements

The authors acknowledge support from: Battelle/INL, National Science Foundation, and the U.S. Department of Energy, Office of Energy Efficiency and Renewable Energy, Bioenergy Technologies Office under Award Number DE-EE0009285.

References

1. US EPA., Advancing Sustainable Materials Management: 2017 Fact Sheet, 2017.
2. Singh, N., Hui, D., Singh, R., Ahuja, I. P. S., Feo, L., Fraternali, F., Recycling of plastic solid waste: A state of art review and future applications. *Composites Part B: Engineering.*, 115, 409-422, 2017.

3. Walker, T. W., Frelka, N., Shen, Z., Chew, A. K., Banick, J., Grey, S., Kim, M. S., Dumesic, J. A., Van Lehn, R. C., Huber, G. W., Recycling of multilayer plastic packaging materials by solvent-targeted recovery and precipitation. *Science Advances.*, 6, 47, 2020.

4. Sánchez-Rivera, K. L., Zhou, P., Kim, M. S., González Chávez, L. D., Grey, S., Nelson, K., Wang, S. C., Hermans, I., Zavala, V. M., Van Lehn, R. C., Huber, G. W., Reducing antisolvent use in the STRAP process by enabling a temperature-controlled polymer dissolution and precipitation for the recycling of multilayer plastic films. *ChemSusChem*, 14, 9, 2021.

5. Kaiser, K., Schmid, M., Schlummer, M., Recycling of polymer-based multilayer packaging: A review. *Recycle*, 3, 1, 1, 2018.

6. Creacycle GmbH – Plants, https://www.creacycle.de/en/creasolv-plants.html

7. Recycling Today, PureCycle Technologies transforms end-of-life carpet into UPRP resin, https://www.recyclingtoday.com/article/purecycle-trials-carpet-recycling-using-pandg-technology/

8. Nordson, BKG® equipment plays a role by pelletizing high-quality polyamide from multilayer packaging waste in the Newcycling® process, https://www.nordson.com/en/divisions/polymer-processing-systems/news/news/2019-12-04

9. Gaia, Unilever's Expensive Plastic Sachet Chemical "Recycling" Failure, https://www.no-burn.org/investigation-reveals-unilevers-expensive-plastic-sachet-chemical-recycling-failure/.

10. Niaounakis, M: Recycling of Flexible Plastic, Elsevier, 2019.

11. Wohnung, K., Kaina, M., Fleig, M., Hanel, H., Solvent and method for dissolving plastic from a solid in a suspension, German Patent DE102016015198A1, assigned to APK Inc, 2018.

12. Zhou, P., Sánchez-Rivera, K. L., Huber, G. W., Van Lehn, R., Computational approach for rapidly predicting temperature-dependent polymer solubilities using molecular-scale models. *ChemSusChem.*, 14, 2021

13. Li, J., Maravelias, C.T., Van Lehn, R.C., Adaptive Conformer Sampling for Property Prediction Using the Conductor like Screening Model for Real Solvents. *Ind. Eng. Chem. Res.*, 61, 2022.

14. Klamt, A., Eckert, F., Hornig, M., Beck, M.E., Bürger, T., Prediction of aqueous solubility of drugs and pesticides with COSMO-RS. *J Comput Chem.*, 23, 2002.

15. Loschen, C., Klamt, A., Prediction of solubilities and partition coefficients in polymer using COSMO-RS. *Ind. Eng. Chem. Res.*, 53, 2014.

16. Kahlen, J., Masuch, K., Leonhard, K., Modelling cellulose solubilities in ionic liquids using COSMO-RS. *Green Chem.* 12, 2010.

17. Maes, C., Luyten, W., Herremans, G., Peeters, R., Carleer, R., Buntinx, M., Recent updates on barrier properties of Ethylene Vinyl Alcohol Copolymer (EVOH): review. *Polym. Rev.*, 58, 2018.

18. Cecon, V., Da Silva, P.F., Vorst, K., Curtzwiler, G.W., The effect of post-consumer recycled polyethylene (PCRPE) on the properties of polyethylene blends of different densities. *Polymer Deg. and Stability*, 190, 2021.

19. Olafasakin, O., Ma, J., Bradshaw, S.L., Aguirre-Villegas, H.A., Benson, C., Huber, G.W., Zavala, V.M., Mba-Wright, M., Techno-economic and life cycle assessment of standalone single-stream material recovery facilities in the United States. *Waste Manage.*, 166, 368-376, 2023

20. Best Technology, Solvent Recovery System and Chemistry Waste Disposal Recycling System, https://www.besttechnologyinc.com/specialty-process-systems/solvent-recovery-equipment-chemistry-waste-disposal-recycling-systems.

21. Munguia-Lopez, A.d.C., Goreke, D., Sanchez-Rivera, K.L., Aguirre-Villegas, H.A., Avraamidou, S., Huber, G.W., Zavala, V.M., Quantifying the environmental benefits of a solvent-based separation process for multilayer plastic films. *Green Chem.*, 25, 4, 1611-1625, 2023.

Valorization of Plastic Waste via Advanced Separation and Processing

Paschalis Alexandridis[1]*, Karthik Dantu[2], Christian Ferger[1], Ali Ghasemi[1], Gabrielle Kerr[3], Vaishali Maheshkar[2], Javid Rzayev[3], Nicholas Stavinski[3], Thomas Thundat[1], Marina Tsianou[1], Luis Velarde[3] and Yaoli Zhao[1]

[1]Department of Chemical and Biological Engineering, University at Buffalo, The State University of New York (SUNY), Buffalo, NY, USA
[2]Department of Computer Science and Engineering, University at Buffalo, The State University of New York (SUNY), Buffalo, NY, USA
[3]Department of Chemistry, University at Buffalo, The State University of New York (SUNY), Buffalo, NY, USA

Abstract

With an overarching goal the reduction in the amount of plastic waste that ends up in landfills or incinerators and the increase in the amount of post-consumer plastic in various products, we research the: (1) automated sorting of mixed plastic waste, and (2) valorization of recovered plastic; this with support from NSF award 2029375 (EFRI E3P). This paper highlights results from this collaborative team effort. High-throughput, real-time, stand-off detection of different types of plastics is advanced through the combination of novel sensor technology, spectral and image databases, and machine learning. Upcycling of sorted polyolefins is achieved through: (i) selective dissolution/precipitation in environmentally responsible solvents to recover the target type of plastic, suitable for reuse in new products, and (ii) tailored chemical modification of polyolefins to produce functional waxes that can serve as building blocks for high- value materials. The fundamental and applied knowledge generated in this project addresses the grand challenge of significantly increasing the global plastic recycling rates. This challenge is tackled here in a manner that is environmentally responsible, technically feasible, economically competitive, and societally beneficial.

Keywords: Plastics, mechanical recycling, sortation, chemical recycling, advanced recycling, upcycling, circular economy, sustainability

38.1 Introduction

Plastics are central to a multitude of products based on their wide-ranging properties that are easy to modulate and amenable to design from first principles. 390 million tons of plastics were consumed worldwide in 2021, with continuing robust annual growth. In a way,

---

*Corresponding author: palexand@buffalo.edu

Nabil Nasr (ed.) Technology Innovation for the Circular Economy: Recycling, Remanufacturing, Design, Systems Analysis and Logistics, (495–506) © 2024 Scrivener Publishing LLC

plastics became victims of their own success as materials: low cost, durable, colorful, and light enough to float on water and be carried away by the wind. Hence, plastic waste raises concerns among the public. In response, major corporations intend to utilize in their products a significant fraction (up to 50%) of recycled plastic. Unfortunately, less than 9% of plastic is recycled currently in the US, with most instead ending up in landfills or incinerators [1]. Opportunities abound in recovering useful plastic from waste; however, great challenges remain in recycling plastic, both technical and economical.

In response, the EFRI E3P: Valorization of Plastic Waste via Advanced Separation and Processing project at SUNY-Buffalo aims to reduce both the amount of waste plastic entering landfills or incinerators and the amount of virgin plastic required for new products by focusing on two challenging issues: (1) sorting mixed plastic waste and (2) valorization of recovered plastic (Figure 38.1). High-throughput, automated sorting is accomplished through the combination of novel sensor technology that will register the molecular signature of each piece of plastic, and machine learning that, based on this molecular signature, will identify in real time the specific type of each piece of plastic. From the integration of these new capabilities with existing technologies, an advanced mixed waste plastic sorting process emerges that is adaptable to sorting other materials to enable their recycling as well. Upcycling sorted plastic is achieved through dissolution of select types of plastic in environmentally responsible solvents to recover desirable materials (i.e., polyolefins), separate them from additives or impurities, and render them suitable for reuse. Tailored chemical modification of polyolefins will produce functional waxes that can serve as building blocks for higher-value materials. This research contributes to the Nation's advanced manufacturing capabilities and helps meet both consumer demand for and corporate commitments to incorporate substantial recycled content into products.

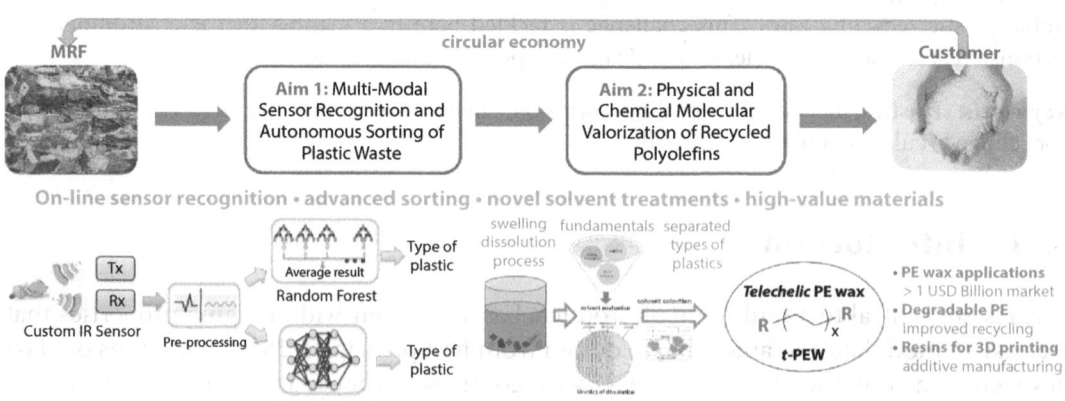

Figure 38.1 Schematic of the "Valorization of Plastic Waste via Advanced Separation and Processing" project.

38.2 Multi-Modal Sensor Recognition and Autonomous Sorting of Plastic Waste

This thrust of our EFRI project seeks to achieve a paradigm shift in real-time, high-throughput detection of different types of plastics via a combination of novel sensor development, machine learning, and autonomous vision-based recognition and tracking, with the intention of transforming the present, inefficient approach to sorting plastic recyclables. This research includes the (i) development of a novel multimodal eye-safe sensor that can classify types of plastics accurately and in real-time, (ii) development of a large dataset of signals from different post-consumer plastic types, obtained using mid-infrared (MIR), that is used in combination with machine learning algorithms that provide chemical identification and selected physical properties, and (iii) combination of multiple modalities such as images and RF sensing to improve end-end classification of plastics in real-time.

The seamless integration of the newly developed capabilities with existing technologies will lead to an advanced sorting process that can be (i) readily integrated in existing Materials Recovery Facilities (MRF), (ii) adaptable to plastic waste composition that varies with location and time, (iii) upgradable, and (iv) scalable and transferable to other industries (e.g., electronic waste) with comparable sorting issues.

38.2.1 Multi-Modal Sensor Development Using Mid-IR Standoff Spectroscopy

Infrared spectroscopic techniques have very high molecular selectivity since they involve molecular vibrations. Standoff detection of chemicals and surface residues using quantum cascade lasers (QCL) has been demonstrated by many groups [2-7]. The multi-modal technique of standoff photothermal-photoacoustic detection that we pursue in this project is a combination of photoacoustic and MIR absorption spectroscopy for characterizing surfaces of plastic objects. Since MIR is free of overtones, this region is known as the molecular fingerprint. In this technique, the target surface is illuminated with a pulsed laser beam from a tunable QCL. When the laser beam impinges on the surface, two different responses are generated. First, the absorption of light results in the excitation of vibrational bonds of the substrate material. Since the QCL allows rapid tuning of the wavelength over a wide wavelength region in a sequential fashion (chirping), the scattered light contains the molecular vibration spectrum of the sample. The scattered light from the sample surface is collected using a mirror and detected using a quartz crystal tuning fork (QTF) placed at the focal point of the mirror. Adjusting the pulse frequency of the laser beam to match the resonance frequency of the QTF drives the QTF into its mechanical resonance with the impinging light because of the photothermal effect. As a result, the amplitude of the QTF resonance vibration changes with the intensity of the beam. A plot of QTF vibration amplitude as a function of wavelength of the scattered light shows the IR absorption spectrum of the target surface. Using an array of pulsed tunable QCLs and a corresponding array of frequency-matched mechanical resonators with a very high-quality factor, it is possible to scan a large wavelength window under 1 s for rapid molecular identification of surfaces in a standoff fashion. The second signal comes from the generation of acoustic effects due to the absorption of the light (photoacoustic effect (PAS)) by the sample. The PAS signal can be detected at a standoff distance of meters using a sensitive microphone. Therefore, these two independent signals, one photon-based, and the other acoustic-based are detected simultaneously. Since the laser

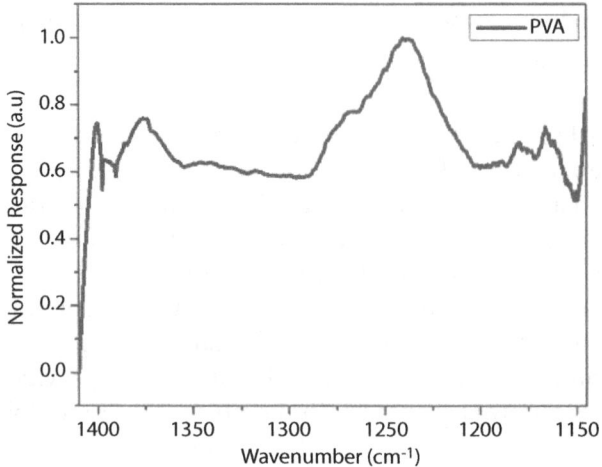

Figure 38.2 MIR spectrum of PVA by standoff detection method.

beam is also pulsed, the use of a lock-in technique can significantly improve the signal-to-noise ratio, especially in the noisy atmosphere encountered in MRFs.

Figure 38.2 shows the MIR spectrum of a poly(vinyl alcohol) (PVA) sample. In this experiment, a function generator was used to trigger the QCL at 32.7 kHz which is the resonance frequency of the QTF. A lock-in amplifier was used to increase the signal to noise level and filter the background noise. When the sample adsorbs a particular wavelength, the intensity of scattered light is less than the case when the sample does not absorb the photons. The pulse frequency of QCL was set to match the resonance frequency of QTF to create oscillating localized pressures which drive the tuning fork into resonance. The amplitude of resonance is proportional to the returning light intensity. By plotting the amplitude of QTF at its resonance frequency as a function of wavenumber, an IR spectrum of the sample is collected. The observed peak in 1200-1250 cm^{-1} belongs to the stretching vibrations of the C-O of PVA.

38.2.2 Spectral Database of Plastics and Plastic Type Classification Using Machine Learning

Solid plastics absorb IR radiation strongly, resonantly converting the photon energy into fundamental molecular vibrations, providing therefore molecular fingerprints of the polymeric matrix. Infrared radiation (IR)-based molecular characterization of plastics has the potential for widespread application in fast and selective sorting of plastic materials. The infrared spectrum is typically divided into several regions: near-infrared (NIR, 0.70–1.4 microns), short-wave infrared (SWIR, 1.4–3.0 microns), medium-wave infrared (MIR, 3.0–8.0 microns), and long-wave infrared (LIR, 8.0– 15.0 microns). At present, NIR, SWIR, and some MIR spectra and spectral imaging have been used for characterizing plastics by monitoring vibrational stretching and bending modes of CO, NH, CH, CH$_2$, and CH$_3$ moieties that occur in the range of 1-3 microns. While the NIR/SWIR spectra are easier to implement than MIR/LIR, information about material components is often obscured by many overtones that make unique identification challenging [8, 9].

Colored, especially black, plastic materials present an additional challenge when using light-based detection in the NIR/SWIR region, because small amounts of carbon black, soot or other pigments absorb all light in the NIR spectral region. This is not an issue in the MIR, and therefore spectroscopy in the mid infrared spectral region offers a possibility to identify black plastics [10, 11]. In addition, the MIR region is dominated by the fundamental modes which favor a robust spectral analysis [12].

A challenge in using MIR for sensing plastics incorporating various compounds is that direct use just results in a forest of peaks[13]. Our approach to addressing this challenge is two-fold. (Figure 38.3) We perform extensive characterization of the various compounds to understand the variations in post-consumer plastics. We also test in realistic scenarios to understand signal distortion due to the deformation of the sensed object as well as interference. After characterization, we develop a digital filter that helps alleviate these effects to distortion and interference. This allows for better characterization of the underlying plastic. The second step is to identify the plastic type using machine learning. We study the applicability of both classic classification techniques as well as one based on deep learning for this purpose.

As has been well documented, classic machine learning classification such as k-means clustering or Random Forests is used in various applications including materials sensing using wireless [14]. The benefit of such an approach is that it is straightforward and explainable. We can understand why it works or does not work after training using a reasonable dataset. We follow this line of research to study the accuracy of a neural network system for the classification of the plastic type. This effort requires significant experimentation because most deep learning systems require a large volume of training data. Therefore, a byproduct of our study is labeled data using our sensor on various types of plastics. This could be used by other researchers to improve upon our classification system.

38.2.3 Multi-Modal Classification of Plastics

In MRFs, it is extremely challenging to deploy the MIR sensors described above because they need to be pointed at the plastic being classified for accurate classification. Further, impurities in the plastic as well as real-world challenges such as labels, caps and others

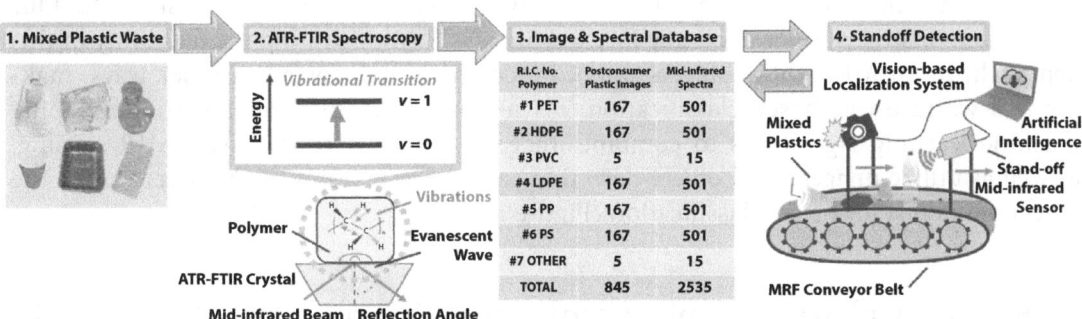

Figure 38.3 (1) Postconsumer plastics were collected from industrial settings, residential areas, and university communities. (2) Mid-infrared vibrational spectra are acquired using Attenuated-Total Reflectance Fourier Transform Infrared Spectroscopy (ATR-FTIR). (3) Images and MIR spectra of postconsumer plastics are curated in a database by their Resin Identification Code (RIC). (4) MIR screening techniques are developed on a testbed and are then used for cross-validation of our machine learning and augmented with multi-modal computer vision methods to accurately sort plastics in real-time.

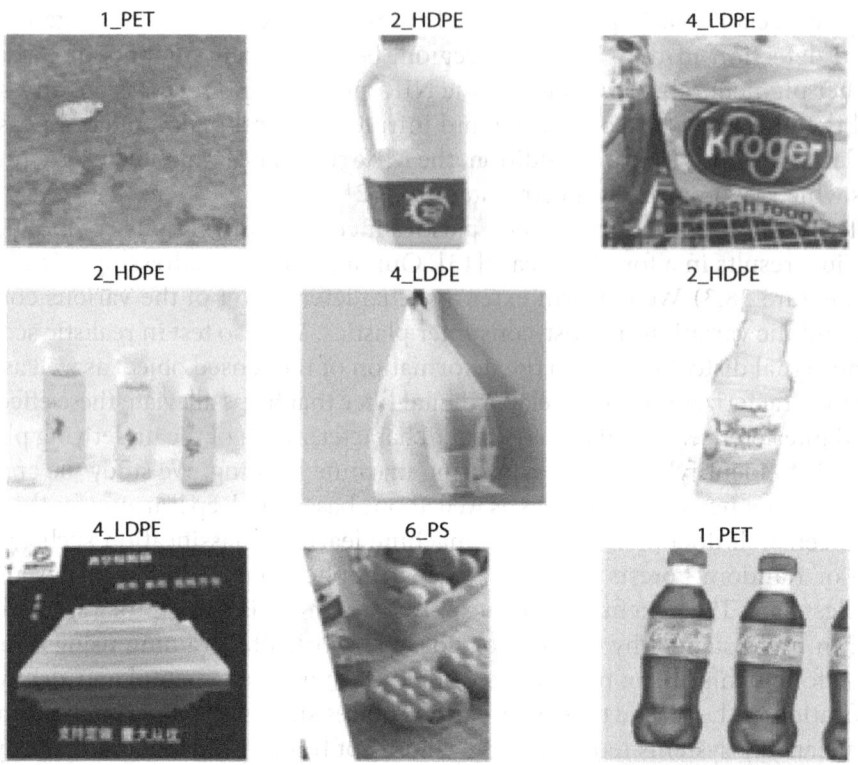

Figure 38.4 Example images from our database of images of plastic objects.

make such classification more challenging. In order to alleviate some of these challenges, we are undertaking an approach to combine multiple sensing modalities such as image-based classification with MIR for sensing plastics.

Currently, we have developed a large database of images of plastic objects in use in daily life (>16,000 images; examples are shown in Figure 38.4) and tested several classical and modern machine learning algorithms to classify the type of object and resin code. Our testing demonstrates >95% accuracy for clear images, demonstrating the feasibility of this approach. In an MRF, these objects might be crushed or distorted in other ways, making real-world image-based classification less successful. To this end, we are developing machine learning algorithms to combine image-based classification with MIR sensing, as well as utilize image-based object detection to accurately point the MIR sensing to the object of interest for better classification (Figure 38.3).

38.3 Physical and Chemical Molecular Valorization of Recovered Polyolefins

This thrust of our EFRI project seeks to advance fundamental knowledge for upcycling plastic waste with minimal CO_2 emissions via two novel routes: (a) recovery of pristine polymers via dissolution/precipitation, where no polymer chemical bonds break, and (b) conversion of polyolefins to functional short chain (50 – 1500 carbons) waxes, where a very small fraction (< 5 %) of the polymer chemical bonds break.

38.3.1 Environmentally Responsible Dissolution/Precipitation Recycling of Polyolefins

Some types of plastic (e.g., "natural" high-density polyethylene (HDPE)) can be readily reprocessed.[15] Other types of recovered plastic can still be processed into useful products, but of much lower value [16]. Yet other types of plastic contain additives that may prove incompatible with the intended use. This points to the significant fraction of recovered plastic, including multilayer containers and flexible packaging, where different types of plastic have been co-extruded or laminated to combine desirable properties [17, 18]. Such plastic waste cannot be reprocessed via melting and re-extrusion.

Chemical treatments, e.g., pyrolysis, chemolysis, gasification, offer ways to recover some value from such plastics [19, 20]. However, chemical treatments suffer from major disadvantages with respect to sustainability: high CO_2 emissions and high energy consumption [21, 22]. A recent Life Cycle Analysis (LCA) which assessed the energy requirements and environmental impacts of various chemical recycling technologies, concluded that polymer chains should be kept intact as much as possible during recycling, as this leads to much lower CO_2 emissions [23].

The dissolution/precipitation process scores very highly in terms of CO_2 avoidance [23]. It involves plastics containing additives and impurities of other polymers or materials dissolved in a solvent that is selective for the desired polymer [24].

Filterable materials are removed, and the desired polymer is precipitated by changing temperature or solvent (Figure 38.5). Dissolution/precipitation, which is a physical process

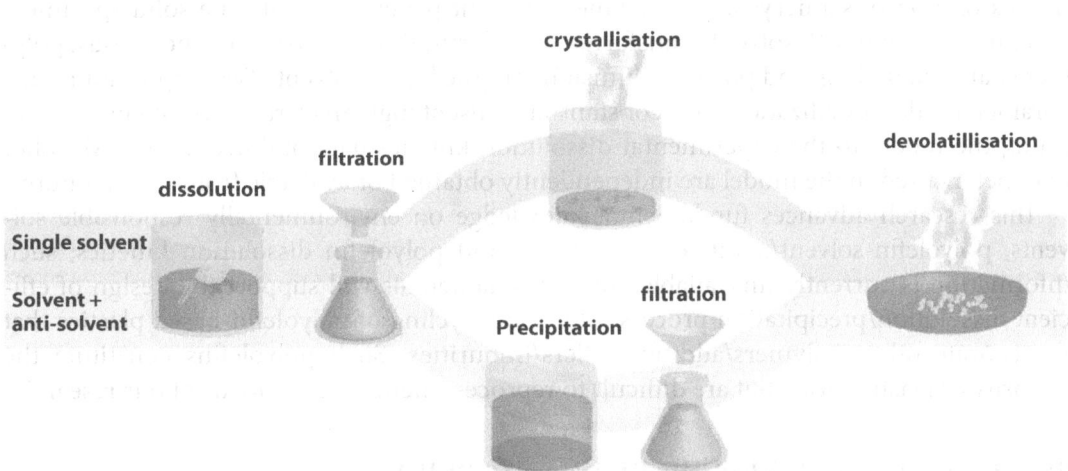

Figure 38.5 Schematic of the dissolution/precipitation process for plastic recovery: (top) dissolution/precipitation using a single solvent, which is subsequently removed by evaporation to recover the polymer, (bottom) dissolution/precipitation using an anti-solvent is used to precipitate the polymer, which can then be recovered by filtration. (Reprinted from doi.org/10.1002/anie.201915651. Copyright 2020 The Authors. Published by Wiley-VCH Verlag GmbH & Co. KGaA, Weinheim.)

but is typically considered chemical recycling, offers the solution for recycling multilayer plastics [17, 25] and upcycling plastics that contain additives [24].

Central to the dissolution/precipitation recycling process is the identification of solvents/conditions that are selective for the dissolution of the target polymer (or of all other polymers present, except the target), and the identification of conditions/nonsolvent where the dissolved polymer phase-separates (precipitates). Polyolefins are mechanically strong (due to crystallinity) and chemically inert (due to $-CH_2-$), highly desirable properties that render polyolefins difficult to dissolve. Solvent/nonsolvent pairs reported for recycling polyethylene (PE) include toluene/acetone [26] and xylene/n-hexane [27], and for polypropylene (PP) xylene/acetone [28] and toluene/n-hexane [29]. Very limited information exists on environmentally responsible solvents for polyolefins and on their swelling/dissolution kinetics.

We investigate novel solvents in terms of their ability to dissolve polyolefins, we determine the phase behavior of polyolefins in various solvents and in solvent/anti-solvent mixtures, and we follow the kinetics of polyolefin dissolution through experiments and modeling. The above information is used to define conditions for effective dissolution/precipitation separation processes aimed to recover near-virgin polyolefin polymers.

The polymer phase behavior (binary polymer-solvent system as a function of temperature) is established based on cloud point measurements, and the thermodynamic interactions between species are estimated using Flory-Huggins theory. The influence of antisolvent type and amount on the percent recovery and the spectroscopic and mechanical properties of the recovered polymer are explored. Finally, selective dissolution of a target polyolefin from a mixture with other polymers validates the selection of dissolution conditions based on phase diagrams.

Dissolution experiments follow the time evolution of the dissolved mass (obtained from a material balance) and the degree of crystallinity (obtained from analysis of FTIR spectra) of polyolefin specimens of well-defined geometry. Dissolution modeling is done with a phenomenological model developed in our group that accounts for the phenomena governing the dissolution of semicrystalline polymers: solvent penetration into the solid specimen, specimen swelling with solvent, transformation of crystalline domains to amorphous, polymer chain untangling, and polymer diffusion into the liquid solvent. Two important model parameters, decrystallization rate constant and disentanglement rate, are obtained from fitting the model to the experimental dissolution kinetics data outlined above. All other parameters used in the model are independently obtained or available from the literature.

This research advances fundamental knowledge on environmentally responsible solvents, polyolefin-solvent/additive interactions, and polyolefin dissolution kinetics; such information is currently unavailable. These fundamentals will support the design of efficient dissolution/precipitation processes for the recycling of polyolefin-based plastics that incorporate other polymers/additives/fillers/impurities. Such polyolefins constitute the majority of plastic waste, but are difficult to reprocess, hence are the focus of our research.

38.3.2   Synthesis of Telechelic PE Waxes (t-PEW)

Polyethylene waxes are short chain macromolecules with molecular weights typically between 1,000 and 20,000 g/mol (70 – 1400 carbons) [30]. A combination of key properties, such as viscosity, density and melting point make these materials highly useful in a range of applications, such as lubricants, inks, coatings, resin modifiers, asphalt additives, and

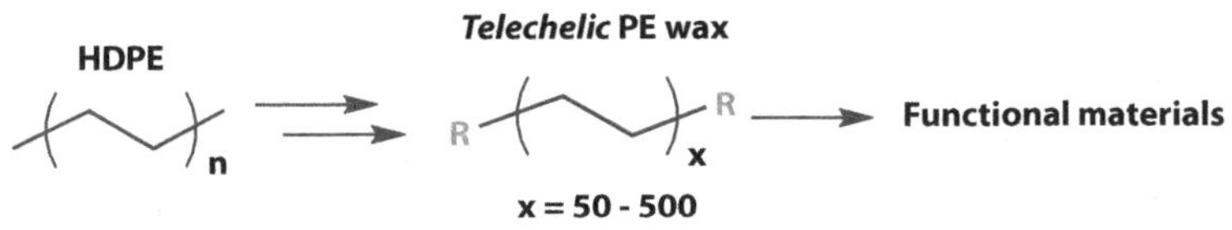

Figure 38.6 PE upcycling strategy by controlled breakdown of polymer chains to produce end-functionalized waxes.

adhesives. The PE wax market was estimated at 1 billion USD (2018) and rapidly increasing [31]. Tremendous value can be added to PE waxes by introducing a small fraction of polar groups through direct copolymerization of ethylene with polar co-monomers [32, 33]. Unfunctionalized PE waxes can also be produced by low polymer wax refining and thermal degradation of high molecular weight PE [34, 35]. The latter process is based on random bond scission at high temperatures, and yields polymers with poorly defined chain-ends.

Developing methods for generating functionalized PE waxes from recycled PE rather than from virgin monomers can advance a new polymer upcycling strategy. PE waxes with reactive end-groups can also serve as building blocks for new materials with PE-like behavior but better recyclability. In this project, we design a new strategy for controlled breakdown of PE chains into telechelic (end-functionalized) PE waxes (t-PEW) (Figure 38.6). Unlike typical depolymerization recycling strategies, we seek to minimize the number of broken chemical bonds while creating highly valuable building blocks to generate materials that, not just replicate properties of virgin polyolefins, but open new avenues for the utilization of polyolefin waste streams in a myriad of new applications. t-PEW are produced here by controlled chain cleavage of PE materials. The process consists of three steps: (1) functionalization of PE by metal- free radical chemistry, (2) thermal elimination of newly introduced substituents to install unsaturations along the PE backbone, and (3) cross-metathesis of olefinated PE chains into smaller fragments with functional end-groups. The process is optimized to be conducted in the melt state. The strategically placed reactive end-groups in t-PEW open previously unexplored avenues for the construction of high-value materials from recycled polyolefins.

38.4   Conclusions & Recommendations

Globally, there is a comprehensive failure to recycle a sufficient proportion of plastics, resulting in extensive landfilling of what is potentially a valuable feedstock for further plastic production. This situation calls for transformative technical advances in sorting and recovery of plastic streams with low contamination. This EFRI project directly addresses this global need and grand challenge.

Our research addresses the very core of plastic waste management, sorting, in a manner that is scalable and distributive, energy efficient, low cost, and readily implementable to existing MRFs, for immediate positive impact to the local community and competitive advantage to the US economy (locally produced, locally processed, and locally upcycled to increase value). It adds to the United States' advanced manufacturing capabilities, addresses the grand challenge of circular economy, and can help meet consumer demand

and commitments by major corporations for the incorporation of high fraction of recycled plastic into their products.

The database on plastic type identification can be useful for other researchers and, more importantly, to operators of MRFs who can use it in combination with the sensors developed in this project to sort plastics with improved accuracy and at lower cost. Our sensor development effort can spur innovation in the use of MIR for sensing plastic types which has been hitherto unexplored, and which could be adapted for parts sorting & matching in manufacturing. The sorting technologies developed in this project are transferable to other challenging sorting problems (e.g., electronic waste). Polyolefins comprise the largest volume of commodity plastics, but they are notoriously difficult to recycle to valuable products due to the molecular-level incorporation of other plastics, additives, colorants, and/or fillers. The solvent- based polyolefin separation methodology that we pursue is environmentally responsible on the basis of the solvents selected and the greatly reduced CO_2 emissions, and can be readily implemented in industrial practice and extended to the separation of other types of plastics. The different types of polyolefins to be recovered from the solvent separation process can meet the pressing demand by both customers and corporations to incorporate recycled plastics into products while maintaining desirable specifications. The synthesis of telechelic PE waxes converts a waste stream to valuable starting materials that can meet current demand for waxes and emerging additive manufacturing technologies, presenting a paradigm shift in polymer recycling.

Acknowledgements

This material is based upon work supported by the National Science Foundation (NSF) under Grant No. EFMA- 2029375. We appreciate stimulating discussions with Prof. John Atkinson (SUNY Buffalo Environmental Engineering), Prof. Amit Goyal (SUNY-Buffalo Chemical and Biological Engineering), and Dr. Michael Shelly (SUNY-Buffalo RENEW Institute).

References

1. Ferger, C. M.; Ghasemi, A.; Alexandridis, P.; Tsianou, M. Recycling of plastic waste using dissolution/precipitation. *Waste Management* 2024, in press.
2 Chen, X.; Cheng, L.; Guo, D.; Kostov, Y.; Choa, F.-S. Quantum cascade laser based standoff photoacoustic chemical detection. *Optics Express, 19* (21), 20251-20257, 2011.
3. Farahi, R.; Passian, A.; Tetard, L.; Thundat, T. Pump–probe photothermal spectroscopy using quantum cascade lasers. *Journal of Physics D: Applied Physics, 45* (12), 125101, 2012.
4. Kim, S.; Lee, D.; Liu, X.; Van Neste, C.; Jeon, S.; Thundat, T. Molecular recognition using receptor-free nanomechanical infrared spectroscopy based on a quantum cascade laser. *Scientific Reports, 3* (1), 1-6, 2013.
5. Li, J.; Yu, B.; Fischer, H.; Chen, W.; Yalin, A. Quantum cascade laser based photoacoustic detection of explosives. *Review of Scientific Instruments, 86* (3), 031501, 2015.

6. Van Neste, C. W.; Morales-Rodríguez, M. E.; Senesac, L. R.; Mahajan, S. M.; Thundat, T. Quartz crystal tuning fork photoacoustic point sensing. *Sensors and Actuators B: Chemical, 150* (1), 402-405, 2010.

7. Van Neste, C. W.; Senesac, L. R.; Yi, D.; Thundat, T. Standoff detection of explosive residues using photothermal microcantilevers. *Applied Physics Letters, 92* (13), 134102, 2008.

8. Van Den Broek, W.; Wienke, D.; Melssen, W.; Buydens, L. Plastic material identification with spectroscopic near infrared imaging and artificial neural networks. *Analytica Chimica Acta, 361* (1-2), 161-176, 1998.

9. Zhu, S.; Chen, H.; Wang, M.; Guo, X.; Lei, Y.; Jin, G. Plastic solid waste identification system based on near infrared spectroscopy in combination with support vector machine. *Advanced Industrial and Engineering Polymer Research, 2* (2), 77-81, 2019.

10. Rozenstein, O.; Puckrin, E.; Adamowski, J. Development of a new approach based on midwave infrared spectroscopy for post-consumer black plastic waste sorting in the recycling industry. *Waste Management, 68*, 38-44, 2017.

11. Becker, W.; Sachsenheimer, K.; Klemenz, M. Detection of black plastics in the middle infrared spectrum (MIR) using photon up-conversion technique for polymer recycling purposes. *Polymers, 9* (9), 435, 2017.

12. Vázquez-Guardado, A.; Money, M.; McKinney, N.; Chanda, D. Multi-spectral infrared spectroscopy for robust plastic identification. *Applied Optics, 54* (24), 7396-7405, 2015.

13. Long, F.; Jiang, S.; Adekunle, A. G.; Zavala, V. M.; Bar-Ziv, E. Online characterization of mixed plastic waste using machine learning and mid-infrared spectroscopy. *ACS Sustainable Chemistry & Engineering, 10* (48), 16064-16069, 2022.

14. Zhang, D.; Wang, J.; Jang, J.; Zhang, J.; Kumar, S. On the feasibility of Wi-Fi based material sensing. In *The 25th Annual International Conference on Mobile Computing and Networking*, pp 1-16, 2019.

15. Hees, T.; Zhong, F.; Sturzel, M.; Mulhaupt, R. Tailoring hydrocarbon polymers and all-hydrocarbon composites for circular economy. *Macromol. Rapid Commun., 40* (1), e1800608, 2019.

16. Polymer Handbook, *McGraw-Hill*; 2000.

17. Kaiser, K.; Schmid, M.; Schlummer, M. Recycling of polymer-based multilayer packaging: A review. *Recycling, 3* (1), 1, 2018.

18. Mumladze, T.; Yousef, S.; Tatariants, M.; Kriūkienė, R.; Makarevicius, V.; Lukošiūtė, S.-I.; Bendikiene, R.; Denafas, G. Sustainable approach to recycling of multilayer flexible packaging using switchable hydrophilicity solvents. *Green Chemistry, 20* (15), 3604-3618, 2018.

19. Plastic has a problem; is chemistry the solution? *C&EN Global Enterprise, 97* (39), 29-34, 2019.

20. Lubongo, C.; Congdon, T.; McWhinnie, J.; Alexandridis, P. Economic feasibility of plastic waste conversion to fuel using pyrolysis. *Sustainable Chemistry and Pharmacy, 27*, 100683, 2022.

21. Malik, N.; Kumar, P.; Shrivastava, S.; Ghosh, S. B. An overview on PET waste recycling for application in packaging. *International Journal of Plastics Technology, 21* (1), 1-24, 2017.

22. Ragaert, K.; Delva, L.; Van Geem, K. Mechanical and chemical recycling of solid plastic waste. *Waste Management, 69*, 24-58, 2017.

23. Vollmer, I.; Jenks, M. J. F.; Roelands, M. C. P.; White, R. J.; van Harmelen, T.; de Wild, P.; van der Laan, G. P.; Meirer, F.; Keurentjes, J. T. F.; Weckhuysen, B. M. Beyond mechanical recycling: giving new life to plastic waste. *Angewandte Chemie (International ed. in English), 59* (36), 15402-15423, 2020.

24. Zhao, Y. B.; Lv, X. D.; Ni, H. G. Solvent-based separation and recycling of waste plastics: A review. *Chemosphere, 209*, 707-720, 2018.

25. Cervantes-Reyes, A.; Nunez-Pineda, A.; Barrera-Diaz, C.; Varela-Guerrero, V.; Martinez-Barrera, G.; Cuevas-Yanez, E. Solvent effect in the polyethylene recovery from multilayer post-consumer aseptic packaging. *Waste Management, 38*, 61-64, 2015.

26. Poulakis, J. G.; Papaspyrides, C. D. The dissolution reprecipitation technique applied on high-density polyethylene .1. Model recycling experiments. *Advances in Polymer Technology, 14* (3), 237-242, 1995.

27. Achilias, D. S.; Antonakou, E.; Roupakias, C.; Megalokonomos, P.; Lappas, A. Recycling techniques of polyolefins from plastic wastes. *Global Nest Journal, 10* (1), 114-122, 2008.

28. Poulakis, J. G.; Papaspyrides, C. D. Recycling of polypropylene by the dissolution/reprecipitation technique .1. A model study. *Resources Conservation and Recycling, 20* (1), 31-41, 1997.

29. Hadi, A. J.; Najmuldeen, G. F.; Ahmed, I. Potential solvent for reconditioning polyolefin waste materials. *Journal of Polymer Engineering, 32* (8-9), 585-591, 2012.

30. Leray, C. Waxes. In *Kirk-Othmer Encyclopedia of Chemical Technology*, 2006.

31. Globenewswire.   https://www.globenewswire.com/news-release/2019/08/05/1897015/0/en/Polyethylene-Wax-PE-Market-To-Reach-USD-1-47-Billion-By-2026-Reports-And-Data.html (accessed 04.03.2020).

32. Hsia, H. T. Ethylene-Vinylacetate Copolymer Waxes. US 6,623,855 B2, 2003.

33. Henderson, A. M. Ethylene-vinyl acetate (EVA) copolymers: a general review. *IEEE Electrical Insulation Magazine, 9* (1), 30-38, 1993.

34. Rahimi, A.; García, J. M. Chemical recycling of waste plastics for new materials production. *Nature Reviews Chemistry, 1* (6), 0046, 2017.

35. Hájeková, E.; Bajus, M. Recycling of low-density polyethylene and polypropylene via copyrolysis of polyalkene oil/waxes with naphtha: product distribution and coke formation. *Journal of Analytical and Applied Pyrolysis, 74* (1), 270-281, 2005.

Part 7

INNOVATIONS IN REMANUFACTURING

39

Image-Based Machine Learning in Automotive Used Parts Identification for Remanufacturing

Abu Islam[1]*, Suvrat Jain[1], Nenad G. Nenadic[1], Michael G. Thurston[1], Justin Greenberg[2] and Brad Moss[2]

[1]*Rochester Institute of Technology, Rochester, NY, USA*
[2]*Dieselcore, Katy, TX, USA*

Abstract

Remanufacturing of durable goods has the potential to prevent them from entering landfills or being melted down for recycling. Manufacturing and remanufacturing both require raw material to create a finished product. In remanufacturing this material called core, originates from previously used manufactured products. These products are rescued from landfills and scrap processors by specialized core suppliers or returned by consumers when purchasing replacement products. Due to the nature of cores being previously used or broken, identification and sorting of this material is challenging. Often core is damaged, without proper identification markings and often shows signs of rust, carbon deposits and wear. Most of the current identification, sorting and handling processes are very labor intensive, error prone and can have poor ergonomics. An automated part identification and sorting process is a potential solution for human induced error given its ability to effectively sort and inspect cores that have similar geometries or set of features. To automate the part identification, a neural network has been trained with images of different automotive parts at multiple orientations. Inception Transfer Learning has been used to reduce the number of training data requirements, speeding up training and higher accuracy. In order to keep the resolution of feature detection insensitive to parts sizes, a YOLO (You Only Look Once) algorithm has been used to create bounding box around the area of the image being analyzed. An automated sorting system has been developed that consists of a smart conveyor, multiple cameras, and laser line scanners. Once on the conveyor, the automated sorting system coordinates the movement and imaging of the core at multiple stations in order to capture a near-360 degree view of the core part. The algorithm detects the part types and models from the images captures from the vision system. It has been demonstrated that the vision based sorting system can classify parts with better than 95% accuracy. This paper will present the results in detail.

Keywords: Image based, machine learning, remanufacturing, parts identification, core identification, computer vision

Corresponding author: asigis@rit.edu

Nabil Nasr (ed.) Technology Innovation for the Circular Economy: Recycling, Remanufacturing, Design, Systems Analysis and Logistics, (509–526) © 2024 Scrivener Publishing LLC

39.1   Introduction

Remanufacturing of durable goods has the potential to prevent them from entering land-fills or being melted down for recycling. Manufacturing and remanufacturing both require raw material to create a finished product. In remanufacturing, this material called core, originates from previously used manufactured products. These products are rescued from landfills and scrap processors by specialized core suppliers or returned by consumers when purchasing replacement products. Due to the nature of cores being previously used or bro-ken, identification and sorting of this material is challenging. Often core is damaged, with-out proper identification markings and often shows signs of rust, carbon deposits and wear. Most of the current identification, sorting and handling processes are very labor intensive, error prone and can have poor ergonomics. An automated part identification and sorting process is a potential solution for human induced error given its ability to effectively sort and inspect cores that have similar geometries or set of features.

A remanufacturing facility is a type of manufacturing center that uses previously used or broken finished goods to manufacture sellable finished product that meets or exceeds the original product manufacturing specifications. Each previously used finished product is received into the remanufacturing center, properly identified, and sorted. These products are then completely disassembled, reusable components are compared to original product specifications and consumable components are discarded and replaced with new. Reusable components that do not meet original product specifications are machined to specification or replaced. The product is then assembled and tested to validate meeting original specifica-tions before being packaged and returned to the parts distributors for sale. One of the major problems in this industry is the accuracy of identifying core material and sorting them for future consumption. The identification of core historically has relied on years of experience and intrinsic industry knowledge. This results in labor shortages of employees with the knowledge, skills, and abilities to do the job. If a part is wrongly identified, it could lead to a delay or the complete halt of remanufacturing. For this problem, an object detection and recognition system can be used to automatically identify the different parts and put them into the required category.

39.2   Literature Review

Though there are multiple approaches available for solving this problem, we have chosen deep learning as it works well with large volume of data and scales up well to support real time identification. Due to large volume of data, training models such as convolutional neu-ral networks (CNN) is ideal as they can easily accommodate huge amount of samples while providing accuracy in predictions. There are several approaches proposed by researchers in the field of object detection and recognition.

Although there have been numerous advancements in the field of object detection, chal-lenges still persist. [1] proposed SIFT (Scale Invariant Feature Transform) algorithm which was a promising start but it is known to be computationally expensive and at times inaccu-rate. This resulted in [2] proposing SURF (Speeded Up Robust Features) which improved upon SIFT algorithm by reducing the computational cost but its performance wasn't satis-factory enough to be applicable in a use case like ours dealing with production environment, such as a remanufacturing facility. While various deep learning models such as convolutional neural networks (CNNs) and deep neural networks (DNNs) have been applied to train better object detection systems, the computational cost and speed is still an issue. [3, 4] proposed deep learning approaches of mechanical parts identification, but all the parts data were from CAD images. [5] proposed a means to identify mechanical components in a moving conveyor. Most existing methods [6, 7] are not able to handle real-time constraints for applications like automotive parts detection in remanufacturing that need high-level classification and localization accuracy. In our use case, remanufacturing industry consists of a lot of complications and difficulties like varying lighting conditions, background distortion, and part orientation which calls for a more robust, efficient, and fast object detection approach. From a lot of previous research, it has been found that object detection systems can be implemented in different ways [8]. A lot of them have been developed and tested on several datasets. However, the scope and feasibility of object detection systems for automotive parts in the remanufacturing industry is not yet explored completely. Therefore, in this paper we are proposing an approach using the YOLO (You Only Look Once) algorithm [9, 10] for real-time object detection and recognition. Lightweight and fast, YOLO reduces the complexity of deep neural networks while still maintaining high accuracy. By allowing only one forward pass through the network, it is possible to build an object detector that is capable of processing images in real time. Previous algorithms required many passes through the image to detect objects, which resulted in reduced efficiency. The main motivation behind this project is to have an efficient real time solution to detect and classify objects which are used in the remanufacturing industry like automotive parts.

39.3   Goals

The goal of the project is to first sort between different types of core, brake shoes (Figure 39.1) vs fuel injectors (Figure 39.2). Then sort between make and models within each type

Figure 39.1 Brake shoe cores.

Figure 39.2 Fuel injector cores.

Figure 39.3 Test fixture.

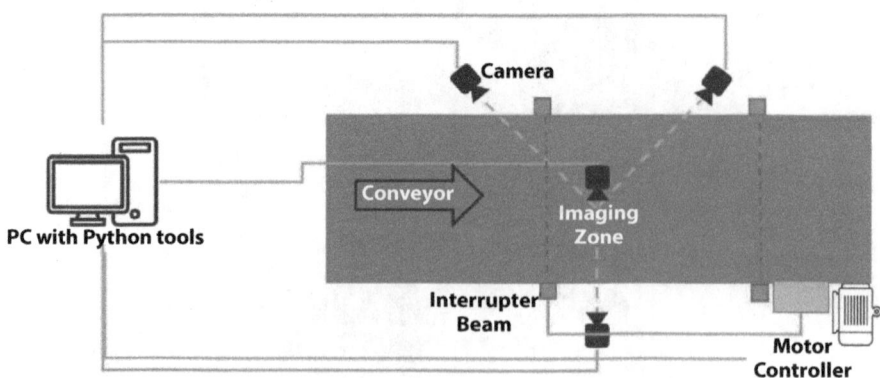

Figure 39.4 Fixture schematic.

of component. It is also necessary to recognize core with accuracy of better than 90% within a cycle time of 10 seconds or less. Automation for identification recognition of components makes the system as robust as possible, flagging any new parts that the system has not been trained on.

39.3.1   Experiment Description

An automated sorting system has been developed that consists of a smart conveyor with multiple cameras as shown in Figure 39.3. Dorner 3200 series conveyor 120" L x 18" W was chosen for its compact footprint. The belt material was chosen to minimize glare and reflections off the belt. The belt is driven by variable speed .75 HP VFD gear motor capable of maintaining full torque throughout belt speed range of 6.9 - 69 ft/min. The motor is controlled by Lenze AC Tech Protech i550 Inverter. The belt also has 2 Photo Eye Kit, 24V DC retro reflective sensor, one for detecting parts at the imaging zone and the other to detect end of travel. Imaging Source cameras with Sony CMOS IMX264LQR, 5MP, 38 fps image sensors were chosen for image capture. Cameras were mounted with Kowa 1" 16.0mm F1.4 Manual Iris C-Mount Lens. Edmunds Optics M35.5X0.5 Linear Glass Polarizer filters were used to minimize ambient light disturbances. Smith-Victor Slim Panel 800W 2-Light Daylight LED Kit was used for illumination. Light diffuser boards placed on the top of the fixture and lights off the illuminator was made to reflect off the diffuser boards to minimize glare. The cameras are connected to the computer via USB cable. The reflective sensors are connected to the PLC digital input ports. The PLC communicates with the PC thru Ethernet cable as shown in Figure 39.4. The Control software in written in Python ecosystem. Once on the conveyor the part moves towards the imaging zone. Once the optical sensor detects the part, the cameras capture images and the part moves on. The image is sent to the control computer thru USB. The algorithm in the control PC detects the part types and models from the images captured by the vision system.

39.3.2   Model Used

To automate the core identification, a neural network has been trained with images of different core at multiple orientations. Inception Transfer Learning as shown in Figure 39.5, has been used to reduce the number of training data requirement, speeding up training and higher accuracy.

As shown in Figure 39.6 the model was trained using 5 different fuel injectors and 6 different brake shoe models supplied by DieselCore. Each part model had multiple replications as shown in Figure 39.7 and Figure 39.8. Multiple pictures were captured from each part and their labels used for ground truth generation.

39.3.3   Results

As the cameras capture new images from the parts in the conveyor, the images are first classified using Brake shoe-Fuel Injector model. Once the type of part is determined,

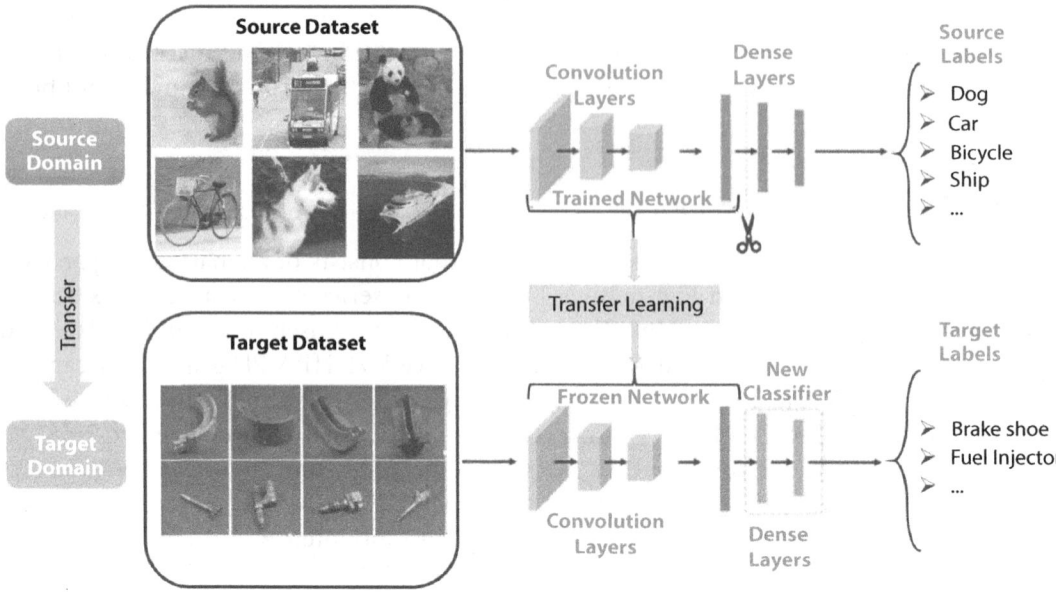

Figure 39.5 Inception transfer learning model.

Type	Part No	OEM	description	Number
Fuel Injectors	445117016	Bosch	6.7 Ford injector	6
	AE	Ford	7.3 Ford injector	5
	OR4528	CAT	3176 injector	5
	445120082	Bosch	LMM injector	5
	445120027	Bosch	LLY injector	5
Brake Shoes	4707Q	Meritor	Drum Brake Shoe	2
	4718Q	Meritor	Drum Brake Shoe	4
	4725E	Eaton	Drum Brake Shoe	4
	4524Q	Meritor	Drum Brake Shoe	4
	4729E	Eaton	Drum Brake Shoe	4
	4515Q	Meritor	Drum Brake Shoe	3

Figure 39.6 Training data.

Figure 39.7 Labeled brake shoe.

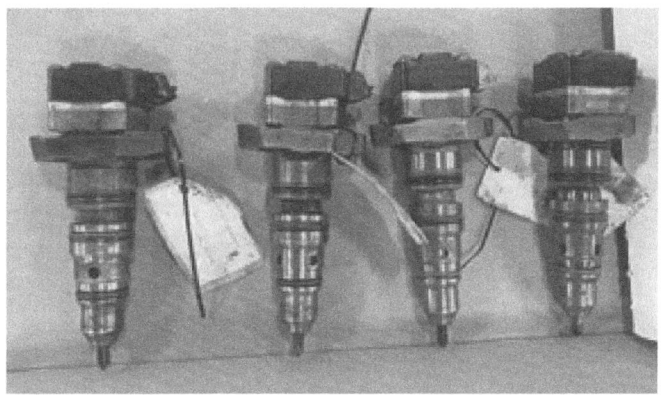

Figure 39.8 Labeled fuel injectors.

the images are further classified using Brake shoe or Fuel Injector model classifier as shown in Figure 39.9.

As can be seen from Figure 39.10, the model is able to distinguish between a fuel injector and a brake shoe with a high degree of accuracy. When it comes to distinguishing between different types of fuel injectors, the model works quite well except for the case of Bosch LLM+LLY vs Bosch LML as showing in Figure 39.11. Similarly, the model works quite well for identifying different models of brake show, except for 4515Q and 4707Q as shown in Figure 39.12. Upon further investigation about causes for these errors, it was found that the reason for the fuel injector errors was, the fuel injector sizes are small in the field of view of the camera, so the feature details are not clean in the region of interests as shown in Figure 39.13. Similarly, as can be seen from Figure 39.14, the main difference

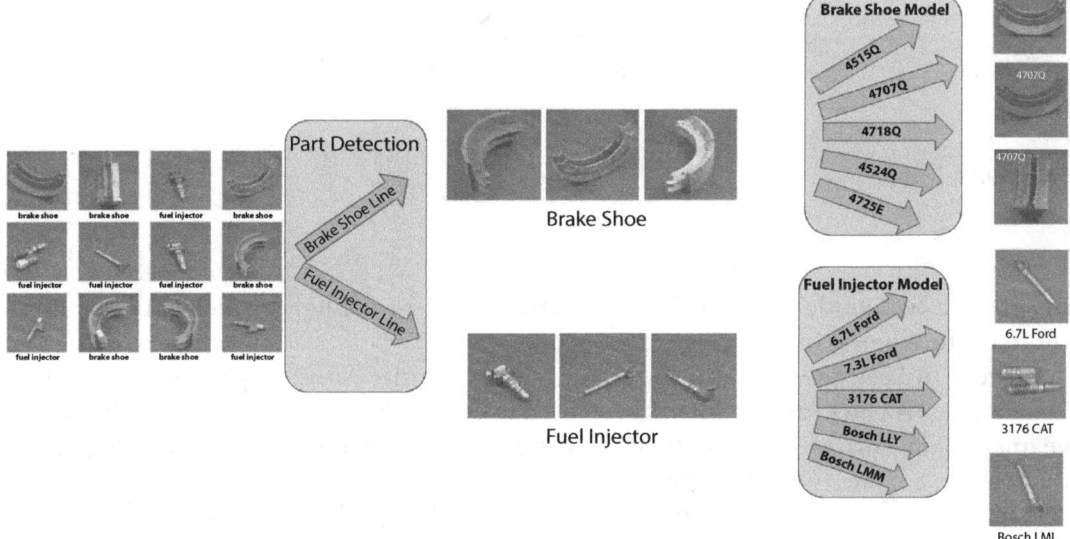

Figure 39.9 Process flow.

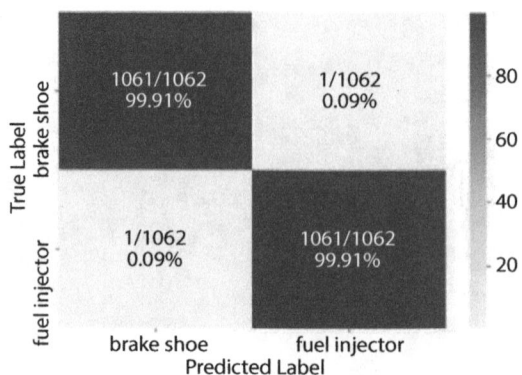

Figure 39.10 Brake shoe vs fuel injector results.

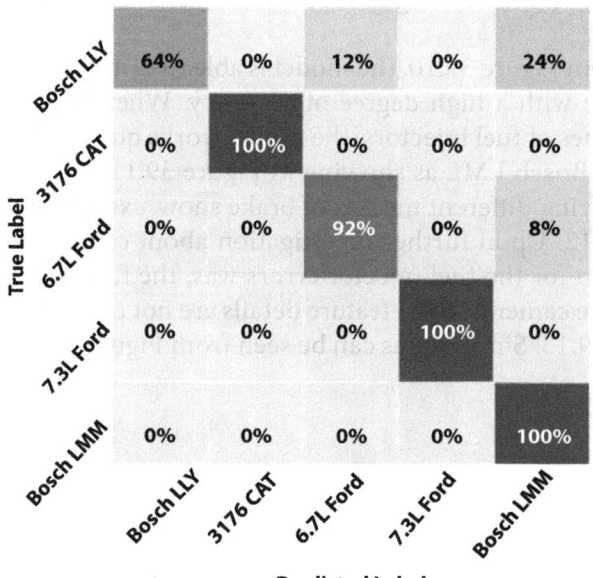

Figure 39.11 Fuel injector results.

between the two models are the distance between the 2 rivet holes which could be difficult to capture in the images (Figure 39.14).

In order to keep the resolution of feature detection insensitive to parts sizes, a YOLO (You Only Look Once) algorithm has been used to create a bounding box around the area of the image being analyzed. YOLO treats object detection as a regression problem to predict what objects are present and where they are.

YOLO uses a single neural network that predicts class probabilities and bounding box coordinates from an entire image in one pass. The YOLO model divides the image into S x S grid, shown in Figure 39.15. Each of these grid cells predicts a number of a bounding

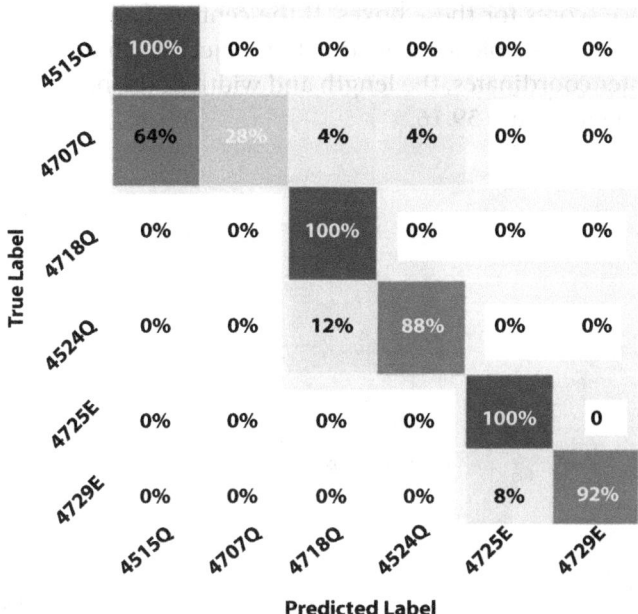

Figure 39.12 Brake shoe results.

Figure 39.13 Different part size issues.

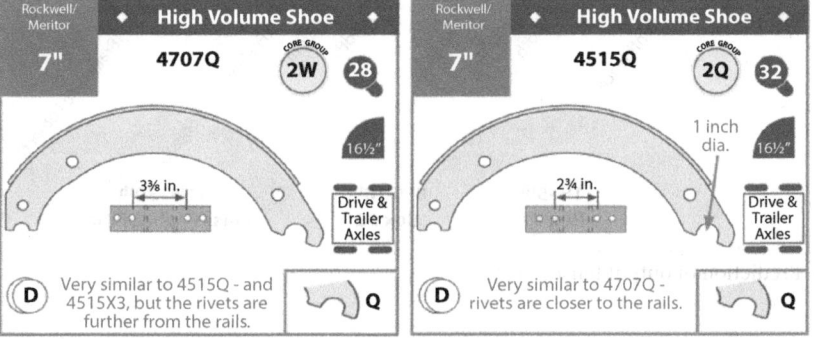

Figure 39.14 Hidden features issue.

boxes and confidence scores for these boxes. If the center of an object falls into one of the grids, then that cell is responsible for detecting that object. Each bounding box outputs five predictions: the center coordinates, the length and width of the bounding box and the class prediction as showing in Figure 39.16.

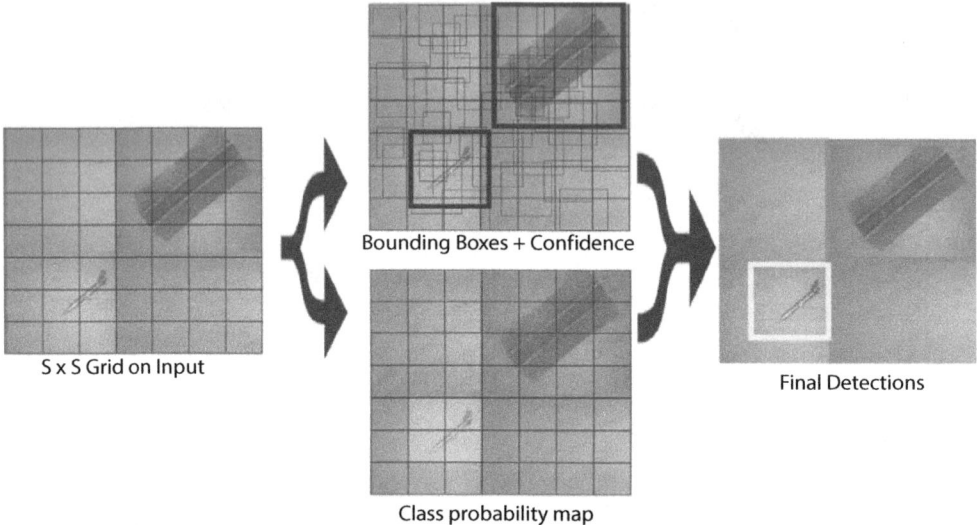

Figure 39.15 YOLO grid architecture [17].

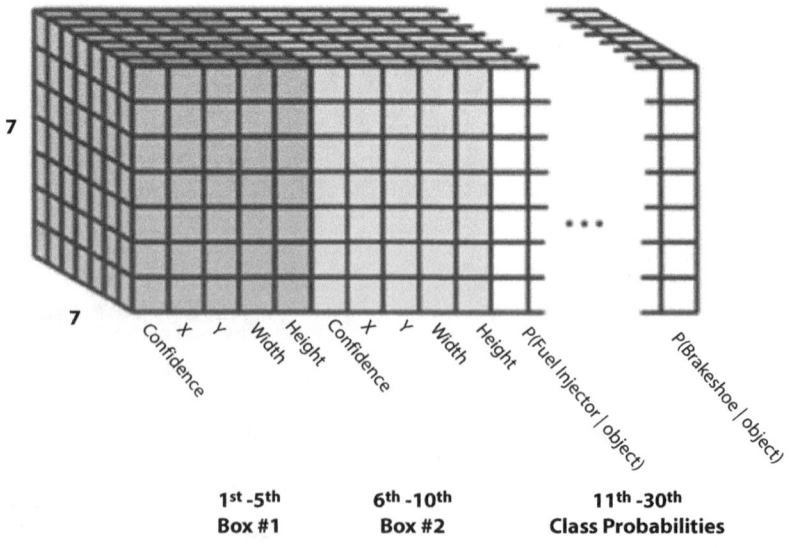

Figure 39.16 Prediction of output tensor [17].

The loss function [17] defined in YOLO incorporates the errors in determining the bounding [1] box size and locations, confidence in the presence of the object and the confidence in classification.

$$Loss_{yolo} = \lambda_{coord} \sum_{i=0}^{S^2} \sum_{j=0}^{B} 1_{ij}^{obj} \left[\left(x_i - \hat{x}_i\right)^2 + \left(y_i - \hat{y}_i\right)^2 \right] + \lambda_{coord}$$

$$\sum_{i=0}^{S^2} \sum_{j=0}^{B} 1_{ij}^{obj} \left[\left(\sqrt{w_i} - \sqrt{\hat{w}_i}\right)^2 + \left(\sqrt{h_i} - \sqrt{\hat{h}_i}\right)^2 \right] + \longrightarrow \text{Bounding Box coord}$$

$$\sum_{i=0}^{S^2} \sum_{j=0}^{B} 1_{ij}^{obj} \left[\left(C_i - \hat{C}_i\right)^2 + \lambda_{noobj} \sum_{i=0}^{S^2} \sum_{j=0}^{B} 1_{ij}^{noobj} \left[\left(C_i - \hat{C}_i\right)^2 + \longrightarrow \text{Confidence}$$

$$\sum_{i=0}^{S^2} 1_{ij}^{noobj} \sum_{C \in classes} \left[\left(p_i(C) - \hat{p}_i(C)\right)^2 \right] \longrightarrow \text{Classification}$$

where,

xi,yi denote the location of the centroid of the anchor box

wi,hi, denote the width and height of the anchor box

Ci, denote is the confidence score of whether there is an object or not, and

pi(c) denote the classification loss.

1_{ij}^{obj} denotes if object is present in cell *i*.

1_{ij}^{obj} denotes j_{th} bounding box responsible for prediction of object in the cell *i*.

λ_{coord} = **5** and λ_{noobj} = **.5** are regularization parameter required to balance the loss function

All losses are *mean-squared* errors, except classification loss, which uses *cross-entropy* function.

The confidence score indicates how sure the model is that the box contains an object and also how accurate it thinks the box is that predicts.

*confidence_score = Pr(object) * IoU*

IoU: Intersection over union is used to ensure that the predicted bounding boxes are equal to the real boxes of the objects. This phenomenon eliminates unnecessary bounding boxes that do not meet the characteristics of the objects (like height and width). The final detection will consist of unique bounding boxes that fit the objects perfectly.

$$Pr(Class\ i|Object)*Pr(Object)*IoU = Pr(Class\ i)*IoU.$$

The final predictions are encoded as an S x S x (B*5 + C) tensor.

YOLO base architecture has overall 24 convolutional layers, four max-pooling layers, and two fully connected layers as shown in Figure 39.17.

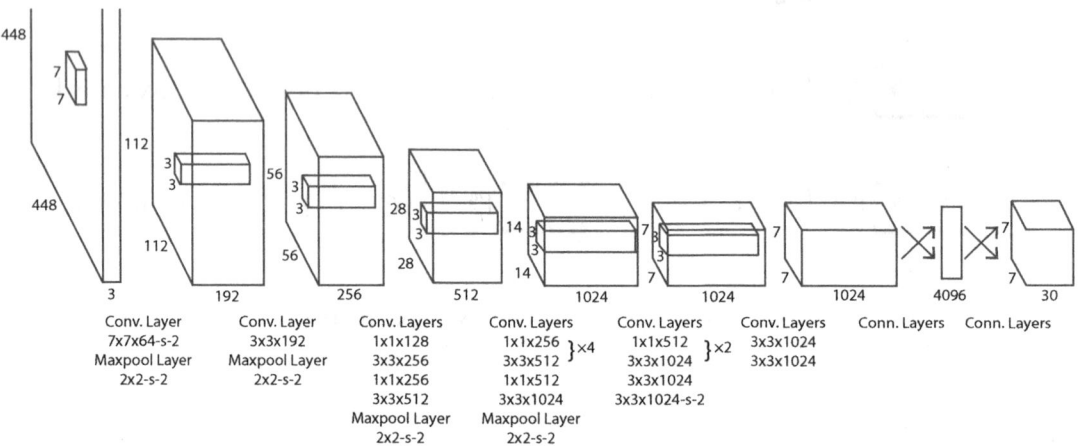

Figure 39.17 YOLO network architecture [17].

Network Architecture

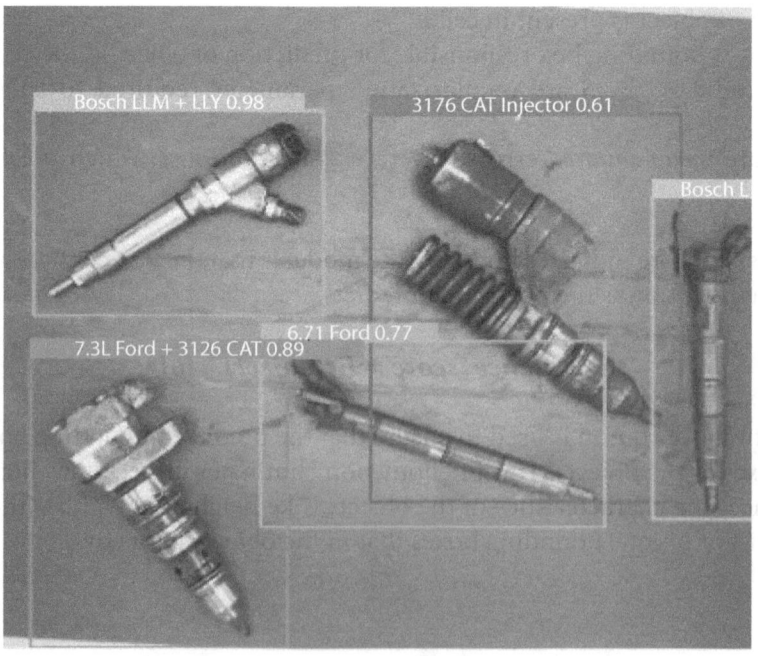

Figure 39.18 Fuel injector detection.

As can be seen from Figure 39.18 that YOLO algorithm can not only detect, localize with bounding box, and classify small objects in the field of view, but it can also detect multiple objects successfully simultaneously. Figure 39.19 shows the confusion matrix for Fuel injector classification and it completely eliminated the errors between Bosch LML and Bosch LML+LLY parts. Figure 39.20 shows corresponding results for Brake shoe, as the Brake

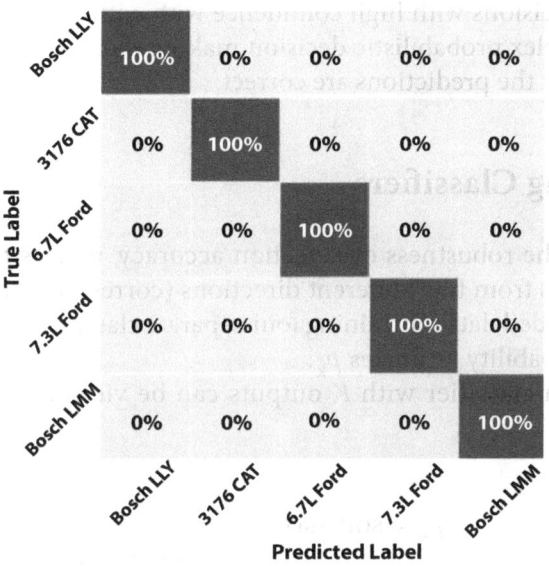

Figure 39.19 Fuel injector results.

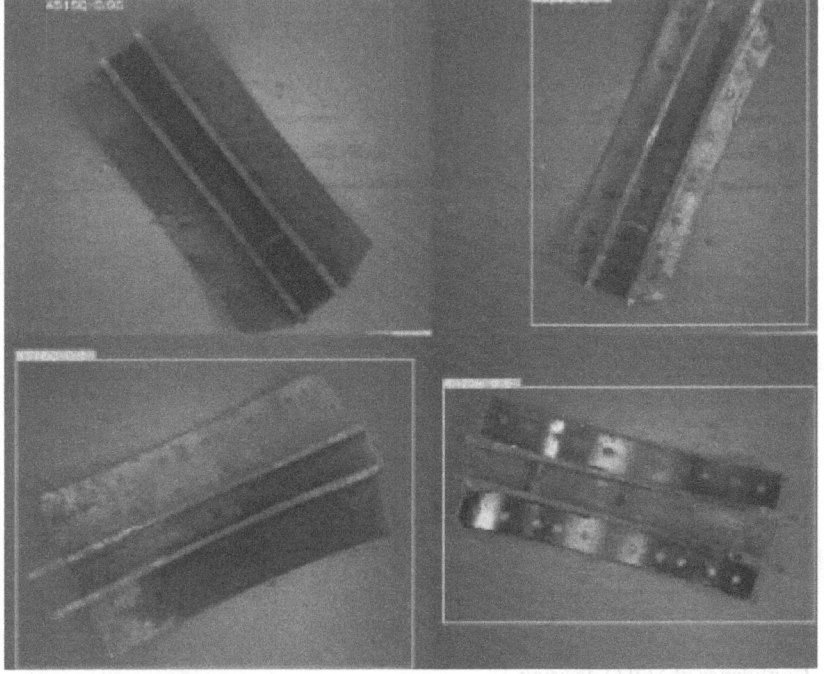

Figure 39.20 Brake shoe detection.

shoes are large compared to the field of view, it didn't show any improvement in accuracy from previous approach.

Figure 39.21 shows the Confidence of YOLO decision. Although the confidence associated with most correct decisions (blue curve) are quite high (greater than 90%) there are some correct decisions with confidence as low as 40%. The model was tested with parts it is not trained on to test the failure modes under extreme conditions. It can be seen there are some incorrect decisions with high confidence with confidence greater than 90%. This warrants a more complex probabilistic decision making than choosing a single confidence threshold to know that the predictions are correct.

39.4 Combining Classifiers

In order to improve the robustness of detection accuracy, images from multiple cameras can be utilized. Images from four different directions (corresponding to the different camera orientations) provided data for training four separate classifiers. The four classifiers provide independent probability estimates $p_{\hat{k}}$.

The kth output of a classifier with K outputs can be viewed as the estimated probability [11]

$$p_{\hat{k}} = \text{softmax}(z_k) = \frac{e^{z_k}}{\sum_{i=1}^{K} e^{z_i}} \tag{39.1}$$

Where, the condition $\sum_{k=1}^{K} p_{\hat{k}} = 1$ is automatically satisfied. YOLO also provides confidence associated with its predictions [12]

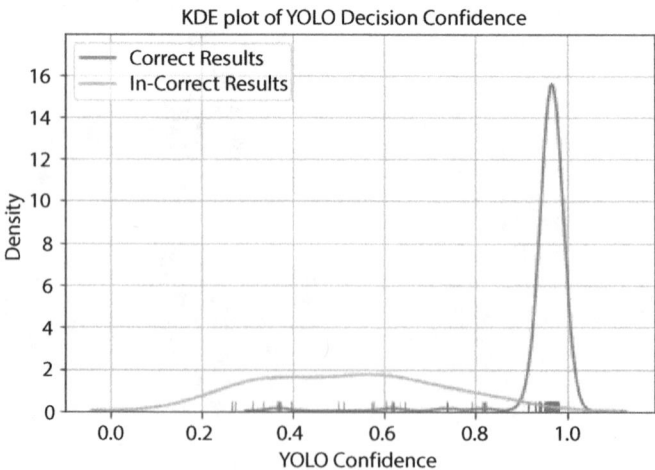

Figure 39.21 Confidence of YOLO decision.

Because the probability estimates can be mapped readily onto a multinomial distribution, they can be efficiently fused. The multinomial distribution is given by [13]

$$p_{\text{Mult}[k]} = \prod_{i=1}^{K} p_k^{\text{oi},k}, \quad k \in \{1,2,\ldots,K\} \tag{39.2}$$

where the $\delta_{i,k}$ is the Kronecker function, which is 1 if $i = k$ and 0 if $i = tk$.

The multinomial distributions are probabilistically combined in the Bayesian sense using the Dirichlet distribution. This approach is consistent with coherent reasoning [14, 15] and is viewed as an extension to logic in physical sciences [16]. The Dirichlet distribution is the conjugate prior to the multinomial distribution given by

$$p_{\text{Dirichlet}}(x_1,x_2,\ldots,x_K|\alpha_1,\alpha_2,\ldots,\alpha_K) = \frac{\Gamma(\sum_{k=1}^{K}\alpha)}{\prod_{k=1}^{K}\Gamma(\alpha_k)} \prod_{k=1}^{K} x_k^{a_k-1}, \tag{39.3}$$

$$0 \text{-.5 } x_k \text{-.5 1 } \forall k \text{ and } \sum_{k=1}^{N} x_k = 1$$

And $\Gamma(x)$ is the gamma function

$$\Gamma(x) = \int_0^{CX)} t^{x-1}e^{-t}dt, x > \tag{39.4}$$

The updates using the conjugate priors are straightforward and fast (in contrast to the general Bayesian updating, which are involved and require computationally expensive evaluation of integrals).

In some cases, multiple estimated probabilities have comparable values. When these multiples include the maximum, the confidence of the decision is relatively low. Classifiers often employ the reject option to avoid making low confidence decisions by setting the minimum level of the estimated probability θ_{min} [13]

$$\max(p_{\hat{k}}) < \theta_{\text{min}} \tag{39.5}$$

If the condition given by Eq. (39.1) is not satisfied, the automated system rejects to make the decision and involves the human into the decision loop. If θ_{min} is set to 1, the system will never make the decision; if it is set to $1/K$ it will never reject to make the decision. The user and designer set a sensible threshold based on the track record of the performance.

In addition to setting the minimum value for the maximum estimate, we can specify the minimum difference between the two highest estimates

$$\max(p_{\hat{k}}) \ \max_{(k=\text{targmax } p\square k)}(p_{\hat{k}}) < (\Delta\theta)_{\text{min}} \tag{39.6}$$

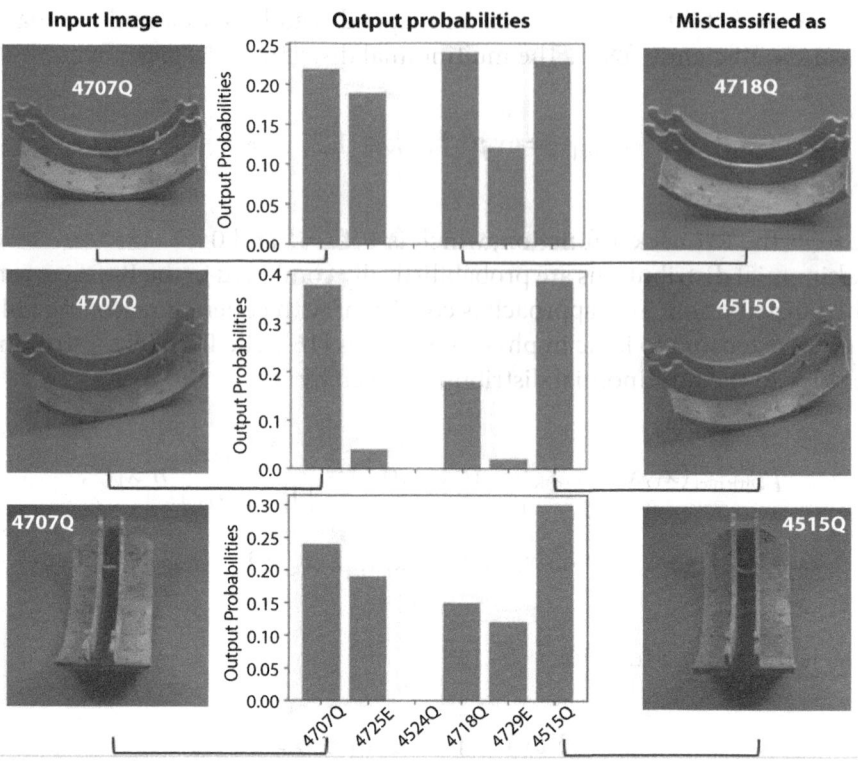

Figure 39.22 Difficulty with similar parts.

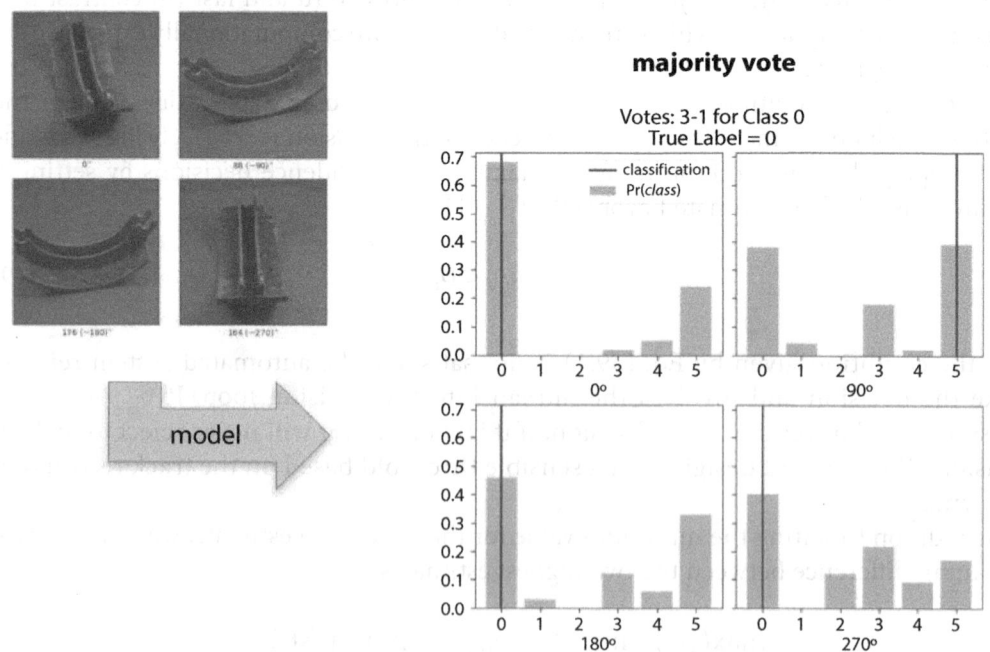

Figure 39.23 Voting for better classification.

Figures 39.22 and 39.23 are important in demonstrating the challenges faced in identifying similar looking objects in computer vision and how the use of multiple cameras can help overcome these challenges. In Figure 39.22, we can see examples of two brake shoe parts, 4707Q and 4718Q, whose output probabilities are too similar to make a decision as to which part it actually is. This is a common problem in computer vision where similar looking objects can lead to confusion in classification. To address this issue, Figure 39.23 demonstrates the use of a probabilistic voting system based on multiple cameras to aid in better decision making. In this approach, multiple cameras capture images of the same part from different angles, and the detections are aggregated to make a final decision. If a given part's probability or confidence of detection is higher in majority of the detections with different cameras, we can conclude that the part majorly detected is the actual part in the image. The significance of these figures lies in their ability to showcase the limitations of single-camera detection and the potential benefits of using multiple cameras in computer vision applications. By demonstrating how a probabilistic voting system can improve detection accuracy, these figures can inform the development of more robust and reliable computer vision systems

39.5   Conclusions and Recommendations

The remanufacturing of used parts can be a viable alternative to purchasing new parts. Like manufacturing, the process begins with correctly identified raw materials for production. Due to the source of acquisition of this raw material known as core, an accurate and efficient identification process is essential. This process can be overseen manually, but an automated system has the potential to provide more accurate, consistent, and efficient results. This paper presents the implementation of an automated visual recognition system for automating the identification and sorting of various core parts. The system uses computer vision and object detection techniques to identify and separate different types of core based on their characteristics. The algorithm detects the part type and model from the image. The part is then identified and sorted accordingly. This process effectively provides identification and sorting of material with greater than 95% accuracy within the constraints of 10 seconds or less cycle time. This solution is reproducible and new products can be added to the learning model as requirements emerge. Proper implementation of this solution reduces reliance on manual error prone human candidates with the knowledge, skills, and abilities to complete the task. As a future work more rigorous statistical decision making based on the theoretical approach presented in this paper will be conducted.

Acknowledgements

Funding for this work was provided by the New York State Department of Economic Development under Grant #AC118. Any opinions, results, findings and/or interpretations of data contained herein are the responsibility of Rochester Institute of Technology and do not necessarily represent the opinions, interpretation or policy of New York State.

References

1. Lowe, D. G. Distinctive Image Features from Scale-Invariant Keypoints. In *International Journal of Computer Vision* 60, 2, 2004.
2. Bay, H., Ess, A., Tuytelaars, T., & Van Gool, L. Speeded-Up Robust Features (SURF). *Computer Vision and Image Understanding, 110*(3), 346–359, 2008. https://doi.org/10.1016/J.CVIU.2007.09.014
3. Kim, S., Chi, H. gun, Hu, X., Huang, Q., & Ramani, K. A Large-Scale Annotated Mechanical Components Benchmark for Classification and Retrieval Tasks with Deep Neural Networks. *Lecture Notes in Computer Science (Including Subseries Lecture Notes in Artificial Intelligence and Lecture Notes in Bioinformatics), 12363 LNCS,* 175–191, 2020. https://doi.org/10.1007/978-3-030-58523-5_11
4. Rucco, M., Giannini, F., Lupinetti, K., & Monti, M. A methodology for part classification with supervised machine learning. *Artificial Intelligence for Engineering Design, Analysis and Manufacturing: AIEDAM, 33*(1), 100–113, 2019. https://doi.org/10.1017/S0890060418000197
5. Borrelly, J. J., & Laurgeau, C. *Recognition of Mechanical Parts on a Moving Conveyor.* 577–581, 1980. https://doi.org/10.1016/s1474-6670(17)64763-3
6. Yanagisawa, H., Yamashita, T., & Watanabe, H. A study on object detection method from manga images using CNN. *2018 International Workshop on Advanced Image Technology, IWAIT 2018,* 1–4, 2018. https://doi.org/10.1109/IWAIT.2018.8369633
7. Erhan, D., Szegedy, C., Toshev, A., & Anguelov, D. Scalable object detection using deep neural networks. *Proceedings of the IEEE Computer Society Conference on Computer Vision and Pattern Recognition,* 2155–2162, 2014. https://doi.org/10.1109/CVPR.2014.276
8. Zou, Z., Shi, Z., Guo, Y., & Ye, J. *Object Detection in 20 Years: A Survey.* 1–39, 2019. http://arxiv.org/abs/1905.05055
9. Ahmad, T., Ma, Y., Yahya, M., Ahmad, B., Nazir, S., Haq, A. U., & Ali, R. Object Detection through Modified YOLO Neural Network. *Scientific Programming, 2020,* 1–10, 2020. https://doi.org/10.1155/2020/8403262
10. Bochkovskiy, A., Wang, C.-Y., & Liao, H.-Y. M. *YOLOv4: Optimal Speed and Accuracy of Object Detection,* 2020. http://arxiv.org/abs/2004.10934
11. Goodfellow, I., Bengio, Y., & Courville, A. *Deep learning.* MIT Press, 2016.
12. Redmon, J., Divvala, S., Girshick, R., & Farhadi, A. You only look once: Unified, real-time object detection. *Proceedings of the IEEE Computer Society Conference on Computer Vision and Pattern Recognition, 2016-December,* 779–788, 2016. https://doi.org/10.1109/CVPR.2016.91
13. Bishop, C. M. Pattern recognition and machine learning. In *Information science and statistics.* Springer, 2006.
14. Koller, D., & Friedman, N. *Probabilistic graphical models: principles and techniques.* MIT press, 2009.
15. Pearl, J. *Probabilistic reasoning in intelligent systems: networks of plausible inference.* Morgan Kaufmann, 2014.
16. Jaynes, E. T., Bretthorst, G. L., & Inc., ebrary. Probability theory the logic of science. In *ebrary Electronic Books.* (pp. xxix, 727 p.). Cambridge University Press, 2003.
17. Joseph Redmon, Santosh Divvala, Ross Girshick, and Ali Farhadi. You only look once: Unified, real-time object detection. In *Proceedings of the IEEE Conference on Computer Vision and Pattern Recognition (CVPR),* pages 779–788, 2016.

40

Image-Based Methods for Inspection of Printed Circuit Boards

Nicholas Gardner[1], Cooper Linsky[1], Everardo FriasRios[2] and Nenad Nenadic[1]*

[1]Golisano Institute for Sustainability, Rochester Institute of Technology, Rochester, NY, USA
[2]CoreCentric Solutions, Carol Stream, IL, USA

Abstract

The remanufacturing and reusing of printed circuit boards (PCBs) is an important component of the circular economy. Current practices on the remanufacturing floor employ several manual processing steps, including identification of the PCB type, localization and degradation assessment of various components, and keying data entries into the system. The repetitive nature of the manual processing steps places a heavy burden on technicians, who tire, make mistakes, and introduce subjectivity into the assessment process. Furthermore, these tedious tasks make employees unhappy, which leads to high turnover. Machine learning has become state-of-the-art for automating inspection tasks but typically requires a large amount of labeled data. We describe the process of introducing machine learning and computer vision for two tasks associated with the remanufacturing process: 1) part number identification and 2) localization of components with an assessment of their degradation. The components selected for the visual inspection were light-emitting diodes (LEDs) because the traditional assessment was based on manual visual inspection. The solution incorporated commercially-available solutions, newly-trained models, and novel approaches for relaxing requirements for machine learning development, all integrated into one development environment. Specifically, the part-number identification solution leveraged the Google Cloud Vision API for extracting character strings from images. The solution for the degradation assessment involved two steps, localizing components and classifying their health. The localization of components used a novel approach that employed classical, deterministic image processing and machine learning. The localized LED sub-images were classified using a custom-trained deep-learning model. Because labeling can also be time-consuming and expensive, we propose a localization scheme that leverages the efficacy of deep learning and significantly reduces the time required to label a dataset. LED localization and assessment performance showed a better than 97% detection rate on the validation data for a specific PCB when the false detection rate was held below 5%. In addition to software development, we explored and discussed trade-offs related to different options for image captures, industrial cameras, and smart devices relative to the use of the captured images.

Keywords: Remanufacturing, printed circuit boards, light emitting diode, degradation assessment, computer vision, machine learning, text identification

**Corresponding author*: nxnasp@rit.edu

Nabil Nasr (ed.) Technology Innovation for the Circular Economy: Recycling, Remanufacturing, Design, Systems Analysis and Logistics, (527–540) © 2024 Scrivener Publishing LLC

40.1   Introduction

Cores and *product returns* arrive on the manufacturing floor randomly, under unknown conditions. The sorting and handling of these cores require flexible processes, which currently depend on manual labor. Moreover, the variation of products is high, often featuring tens of thousands of classes of parts that must be identified and grouped by different applications and manufacturers. The receiving process starts with a manual teardown identifying the core with a *part number* identifier. With a large inbound of core, up to thousands of pieces per week, the sorting and receiving require speed, which often sacrifices the accuracy of the manual condition assessment. Errors in condition assessment create waste downstream because they failed to sort out cores cost-prohibitive to remanufacture. Moreover, failure to identify unusable cores wastes handling and storage capacity.

Training technicians takes 4-8 weeks and includes imparting the requisite domain knowledge, internalizing the knowledge, and practicing to attain the necessary efficiency. However, monotonous tasks lead to low employee retention in core receiving. Business motivation to adopt machine-learning-based automation includes efficiency improvement, increased accuracy, and a path to expanded capabilities.

Analysis of *light-emitting diode* (LED) degradation from the *printed circuit board* (PCB) images required two steps: 1) the localization of LED sub-images and 2) modeling degradation using the data of the sub-images. Two widely used data-driven localization models are *you-only-look-once* (YOLO) and U-Net. YOLO is a single-stage detector: it detects a bounding box and classifies the associated object in one shot [1]. YOLO has been continuously improving over the last six years [2–6] and has been successfully applied to detect electronic components on PCBs [7]. However, YOLO's bounding boxes are aligned with the image axes, which is a limitation for LED detection because PCB layouts employ both aligned and oblique LEDs relative to the PCB. A YOLO-based solution for slanted bounding boxes requires a post-processing step to standardize the diode layout for the assessment model. Initially developed for image segmentation of medical images, the U-Net model [8] and its variants [9, 10] have been successfully deployed in other fields, e.g., crack detection [11–13]. Machine learning models like U-Net typically require large training datasets to learn. To train U-Net for diode localization, every diode on the PCBs must be labeled with bounding box locations. This is highly time-consuming process, as there can be over a hundred diodes on each board and dozens of boards in the training set.

In addition to purely data-driven methods, classical image-processing approaches [14] provide significant value in developing practical computer-vision solutions. For example, image registration is the task of transforming one image's coordinate system into another. It is necessary for computer vision tasks such as *simultaneous localization and mapping* (SLAM) [15, 16], panoramic stitching [17], and image alignment [18].

Feature detection is the task of localizing features in an image. A feature can be thought of as a point of interest. Features that let us recognize human faces are eyes, skin, hair, etc. After features are localized, a feature description is required to assign a numerical representation. Good features should not change under perspective or illumination changes and should be well localized. Several open-source or expired-patented algorithms are successful in doing so. SIFT [19] relies on difference-of-Gaussian for detection and histogram of oriented gradients for description. ORB [20] adds an orientation component to FAST corner detection [21] and modifies [22] to be rotation invariant for description. BRISK [23] also modifies FAST by searching for maxima in scale space in addition to the image plane. KAZE [24] detects and describes features in nonlinear scale space. Regarding the computational time to register two images, ORB is the fastest, followed by BRISK, with KAZE and SIFT significantly slower.

The degradation assessment can be treated either as a classification, with two (healthy and degraded) or more states (healthy and few specified discrete levels of degradation), or a regression where the output maps healthy to failed onto, e.g., the 0-1 range. Regardless if the supervised learner is a classifier or a regressor, the dominant approach for practical image-based machine learning is *transfer learning* [25], which adapts a pre-trained model, e.g., Inception [26] to the desired task, using a major deep learning framework, such as Tensorflow [27] with Keras [28] or PyTorch [29]. Some highly-specialized images, e.g., maps, microstructures, or biological data, work better from custom models [30].

Computer vision approaches found applications in solving manufacturing problems related to LEDs [31], leveraging heuristics to detect manufacturing issues such as mouse bites, missing components, or incorrect packing orientation. More recently, a deep learning model based on CNN layers was proposed for LEDs inspection to detect line blemishes and scratch marks [32]. To our knowledge, the degradation of LED performance has not been studied.

40.2   System-Level Approach to Introducing Machine Learning-Based Automation

This section describes the system-level development and implementation of computer vision and machine learning solutions for the remanufacturing floor. Specifically, the solution was developed for CoreCentric Solutions, a third-party after-market solutions provider, including returns management, in-warranty, out-of-warranty repairs, full product remanufacturing, and materials recycling. While the solution was developed for the specific remanufacturer, the methodology, components, and aspects should apply to similar problem settings commonly encountered by other organizations.

40.2.1   System-Level Description

The system-level approach is depicted in Figure 40.1. It consisted of image capture, part number identification, and one or more detection algorithms, including LED degradation assessment. Part number identification and detection can operate on the same image, but only sometimes. For example, some PCBs containing LEDs have part number information on the opposite side of the PCB relative to the side which contains LEDs.

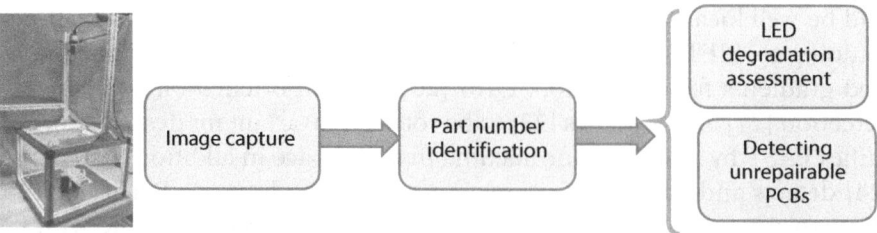

Figure 40.1 System block diagram.

40.2.2   Image Capture

Image captures were mainly based on an industrial camera, as shown in Figure 40.1. However, images can also be taken by a smart device, a phone, or a tablet. This option is further explored in the section on part number identification.

In the context of the development and deployment of machine vision systems, image captures have multiple purposes: they are inputs for machine learning models described below, and they can also be used as information gathering that can unlock future opportunities. For example, Figure 40.2a shows the user interface for capturing images with ground truth in the form of the bounding box and failure mode description that the user specifies.

The information is stored in the accompanying data storage structure, which consists of a file with the ground truth table and a folder with the associated images. The ground truth

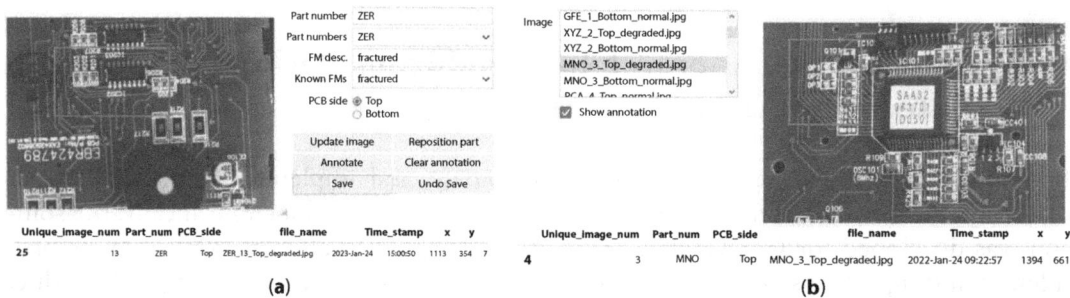

Figure 40.2 (a) User interface for capturing images with ground truth information on failures (b) Companion user interface for reviewing the annotated image captures.

Table 40.1   Ground truth information associated with an image.

Image #	Part #	PCB side	File name	Timestamp	x	y	w	h	note
1	XYZ	Top	XYZ_2_Top_degraded.jpg	2022-Jan-21 17:03:27	1273	1575	640	466	burned
2	XYZ	Bottom	XYZ_2_Bottom_OK.jpg	2022-Jan-21 17:03:56	-1	-1	0	0	NaN
3	MNO	Top	MNO_3_Top_degraded.jpg	2022-Jan-24 09:22:57	1394	661	546	660	crack

table contains part numbers to allow the user to populate the table quickly and accurately and an extendible file that contains known failure modes, as shown in Table 40.1. The data in the table allows the reconstruction of the user's image captures with annotation, as illustrated in Figure 40.2b, which shows a companion user interface for reviewing image captures. This information can be further edited by domain experts or used by the developers of machine learning solutions.

40.2.3   Part Number Identification

Part number identification was the second main block in Figure 40.1. The solution leveraged the Google Cloud Vision API. The block diagram of the system is displayed in Figure 40.3. It consists of four main blocks: 1) image update, 2) Google AI, 3) comparison of text and existing part numbers, and 4) display and visualization. The blocks are described in turn.

The image update block can either read a still image from the disk (still images) or the camera (live images). Live images can be captured using an industrial camera or smart devices via third-party apps. The system can decimate the captured image to accelerate cloud processing using the Google Cloud Vision API. The API takes the updated image and detects all text boxes it can find. It returns a structure that contains a collection of text strings and the associated bounding boxes in pixels.

The next block attempts to find a match for any of the text strings extracted from the image to an existing part number. The processing employs a custom implementation of the Trie data structure. The initial algorithm testing identified two leading causes of possible failure match: 1) on occasion, the Google AI platform breaks the part number into substrings, and 2) sometimes, the Google AI platform misclassifies one or more characters within the part number string. Two solutions were developed to overcome the text identification issues. First, the substrings were concatenated using the information of the associated bounding boxes (they have to be aligned and adjacent). Second, the character misclassification was addressed with partial text string matching. In partial string matching, the existing stored

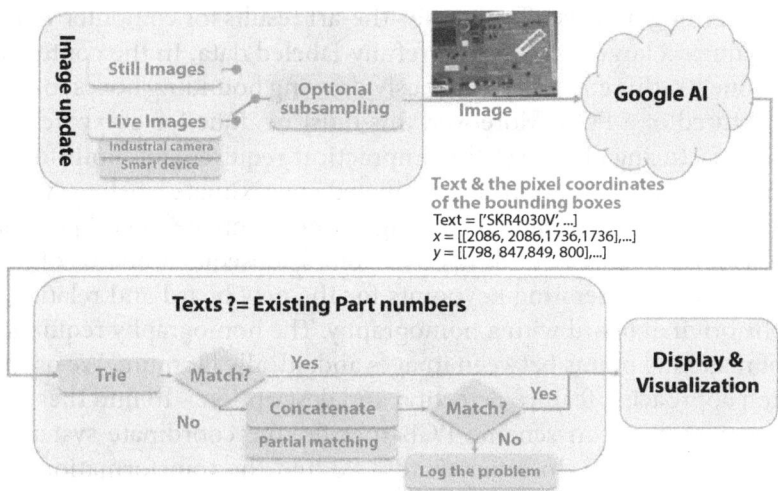

Figure 40.3 Block diagram of the part number identification algorithm.

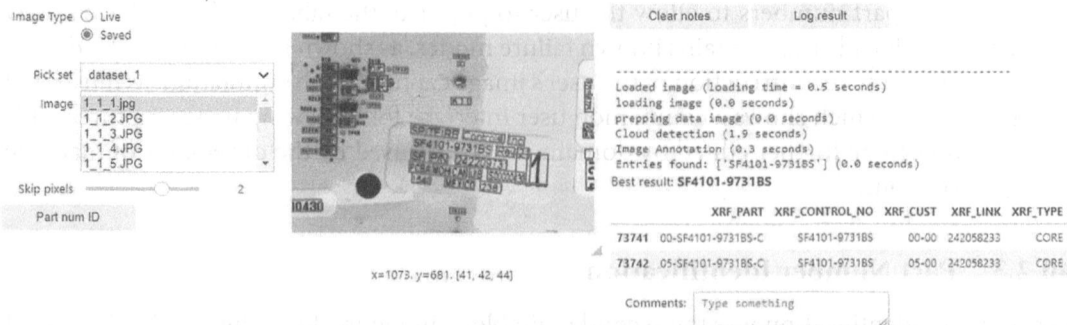

Figure 40.4 Integrated system of part number identification.

part numbers were compared to text strings extracted from the PCB image by measuring the ratio of character agreements. The ratios and the associated strings were then sorted, and the top matches were shown to the user in descending order. If no suitable match is found, the problem is logged to enable future improvements. The logging consists of capturing the image of the PCB and an optional comment from the operator.

The final block serves to display the results. It shows the rows of the reference table with identified parts.

If the part is still undetected, the user can log the event for future troubleshooting by saving the associated image and providing an optional text message.

Figure 40.4 shows the user interface for the integrated part number identification. The interface can operate in *live mode* for operations on the remanufacturing floor or in *saved mode* for training purposes. The table on the right of the image shows the successfully identified part (the orange rectangle in the image). The interface also allows the user to log results and add notes when encountering problematic edge cases, enabling continuous improvement.

40.2.4   LED Degradation Assessment

While machine learning models offer state-of-the-art results for computer vision problems, they typically require a large amount of carefully-labeled data. In the context of localizing electrical components, this entails meticulously drawing bounding boxes for each diode on every image captured of a PCB. Moreover, this must be done for every new PCB scheme requiring remanufacturing. To offset this impractical requirement, both for research and for potential on-site implementation, an automated approximate labeling system was leveraged. This system still requires manual labeling of one "template" board per board type but yields automatic, high-quality bounding boxes for all subsequent boards of the same type. This is accomplished by generating keypoints for the new board and relating them to the keypoints of the original board with a homography. The homography requires a minimum of four pairs of matching points between images and, ideally far more. We used the classical computer vision approach – feature detection and description – to find these points.

By making our localization scheme PCB-specific, the coordinate system of two PCB images can always be related. In other words, we find the transformation that maps the points of a PCB to the corresponding points of another. We model this transformation with a homography. Assuming a pin-hole camera model, two planes captured by a stationary camera can be related by a homography. Enforcing a stationary camera ensures that the two planes share the same projection center. Our capture system enforces a stationary camera, and even if the camera is rotated, two images can still be related. Therefore, no matter how a PCB is placed on the capture bed, whether translated or rotated, we can register its image to another. Specifically, the localization was based on deterministic image processing techniques that require a small manual setup associated with a new part number.

The localization process is depicted in Figure 40.5. It starts with a one-time initialization that must be repeated for each new PCB design. The initial setup consists of manual labeling of LED bounding boxes on a template PCB, then generation of keypoints for that template PCB. The one-time initialization steps are grouped in the black rectangle on the left side of Figure 40.5. Once the initialization is complete, the system is ready for online localization, the localization of LEDs on new PCBs of the same type. Online localization starts with keypoint generation and matching, followed by transforming the bounding boxes of LEDs from the template board (see the top row in Figure 40.5). These steps represent the automated registration system. This system can also be utilized to extract the diode sub-images themselves. The final row represents the steps necessary to extend the automated registration system for this purpose. First, minor adjustments are performed to standardize diode orientation. Then, sub-images associated with individual LEDs are extracted and stored for assessment.

With the automated registration system complete and the dataset approximately labeled, our exploration could move on to machine learning models. With a training dataset consisting of three different PCBs and ten or more duplicate boards for each, training was conducted for U-Net. The model displayed high bounding-box precision on validation boards but would routinely yield a false positive or false negative on each board. Tuning the confidence parameters was attempted but this effort quickly reached diminishing returns.

Additionally, generalization was quite limited. These mixed results are likely due in part to the limited dataset. As discussed earlier, machine learning models benefit from very large datasets. Due to practical considerations, it can be challenging to gather and image a sufficiently large dataset even with automated labeling. Despite its occasional performance flaws, the U-Net model was able to localize components very quickly. For an application

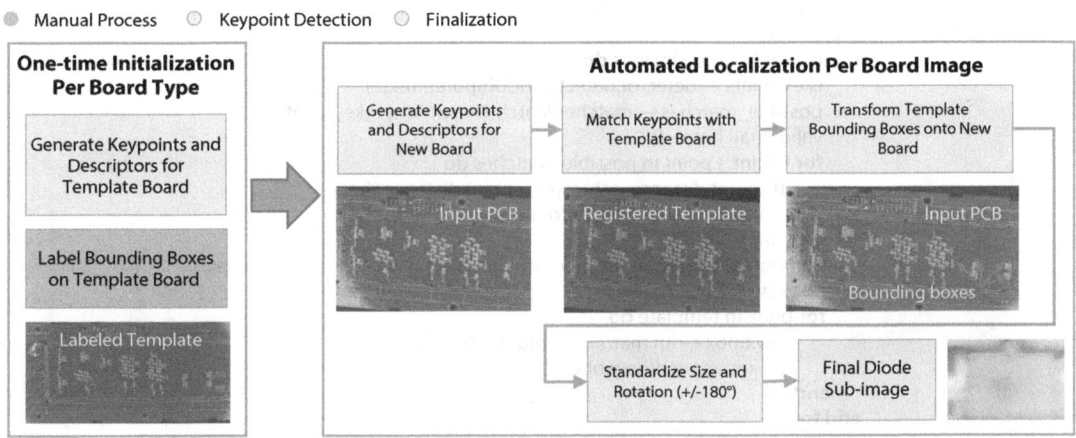

Figure 40.5 Localization approach.

with a larger amount of data and or a small number of unique board types to operate on, this could be the right solution.

Another path investigated localization using YOLO, specifically yolov5. Similarly to U-Net, performance was swift and effective on trained board types, but no meaningful generalization to new board types was observed. Different data splits and data augmentation hyperparameters were attempted, but none led to a general solution. Like U-Net, this model is an excellent solution for evaluating quickly and accurately on a small number of consistent board types. Additionally, with a sufficiently large dataset, it is possible that generalization to diode detection on any board would be possible.

Observing the lack of generalization from YOLO and U-Net, it was decided to examine the possibility of using the automated approximate labeling system for localization. This system generates almost perfect bounding boxes, with no missed or extra predicted diodes, and requires only labeling a single template board for each board type. This labeling was required for the machine learning models as well, as currently they must be trained on each new board type. This solution is not perfect either. For keypoint detectors, performance is dataset dependent. In this dataset, it is important that the algorithms are scale and rotation-invariant. We find that KAZE is most consistent at finding sufficient, high-quality keypoints on all boards in the dataset. However, unlike BRISK and ORB, it is quite slow. This method's low overhead and simplicity makes it a good choice for general purpose applications and datasets containing many unique board types. Additionally, for datasets where faster keypoint detectors, like BRISK and ORB, are effective, this solution is likely the best one. Figure 40.6 shows a pseudocode that summarizes localization steps.

LED assessment was based on sub-images of individual LEDs. The initial attempt to develop the model was based on transfer learning of pre-trained models. Inception-ResNet-v2 was selected for its versatility, accuracy, and availability. The weights of this

---

Algorithm 1 Automated Localization (For Multiple Board Images)

Require: Board type template exists
    template ← load(board_type)
    detector ← cv.KAZE.create()
    t_keypoints ← detector.detectAndCompute(template.image)
    matcher ← cv.Matcher(params)
    thresh ← a
    for images **in** board.images **do**
        i.keypoints ← detector.detectAndCompute(image)
        possible_matches ← matcher.Match(i.keypoints, t_keypoints)
        valid_matches ← []
        for i.point, t.point **in** possible_matches **do**
            if i.point.distance ≤ thresh * t_point.distance **then**
                valid.matches.append(match)
        end for
        homography ← cv.findHomography(t.valid_matches, i.valid.matches)
        bboxes ← []
        for bbox **in** template **do**
            new.bbox ← np.matmul(homography, bbox)
            bboxes.append(new.bbox)
        end for
    end for

Figure 40.6 Automated localization pseudocode.

model were frozen, and a single dense layer was attached and trained to predict diode health based on the features extracted. Multiple tests were conducted, with poor performance when evaluated on the validation data. The lack of success suggested that diode sub-images differed considerably from the images that Inception-ResNet-v2 had been trained on for naïve transfer learning to be effective. A future investigation into transfer learning for this application could examine the efficacy of partially unfreezing weights (progressively moving from the last to the first layers) in the pre-trained model and training with a more extensive set of individual LED sub-images.

The second, more successful approach was based on a custom-based convolutional neural network, trained on examples of sub-images of healthy and degraded LEDs. The general topology was based on standard neural network convolutional models (see, e.g., [33, 34]), as depicted in Figure 40.7. The first part of the model consisted of two convolutional layers, with kernel size 5, separated by 2x2 max pooling and 25% dropout [35]. After these layers, the output was flattened and sent through three decreasing linear layers, each separated with dropout and a ReLU activation, defined as $\text{ReLU}(z) = \max(z, 0)$. Finally, because the model was designed as a binary classification, the sigmoid $\sigma(z) = 1/(1 + e^{-z})$ was applied to the output, leaving a value to be compared to the health threshold chosen for the model. The necessary ground-truth data on the state of degradation was obtained by applying voltage on individual LEDs and observing the potential full or partial loss of brightness. The loss of brightness is not dramatic, the ground truth for assessment was subjective.

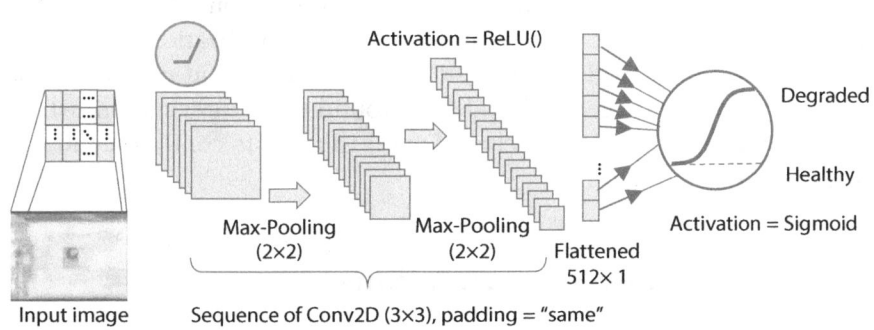

Figure 40.7 Topology of the LED assessment neural network with convolutional layers.

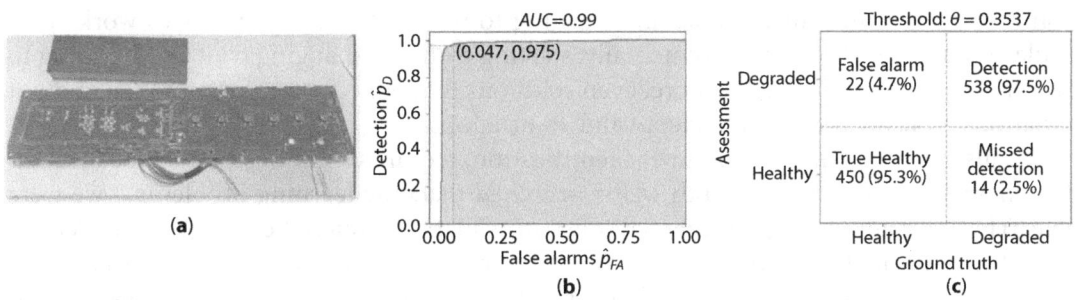

Figure 40.8 Performance of the LED degradation assessment for a PCB (a) image of the PCB (b) receivers operating characteristic (c) confusion matrix.

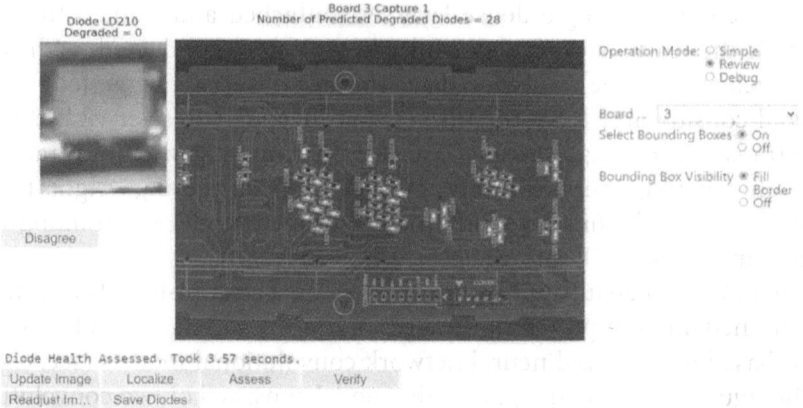

Figure 40.9 Integrated solution for LED assessment: image loading, localization, and assessment.

A typical performance on the validation data for a specific PCB (Figure 40.8a) was indicated in the *receiver operating characteristic* (ROC), which plots the estimated detection rate $\hat{p}_D$ or true positive rate vs. estimated false alarm rate $\hat{p}_{FA}$ (Figure 40.8b), and the confusion matrix, which summarizes a specific point on the ROC curve in a tabular form (Figure 40.8c). The specific ROC point corresponded to the initial requirement that false alarm rate $\hat{p}_{FA}$ did not to exceed 5%. With this requirement, the model attained the estimated detection rate $\hat{p}_D$ of 97.5%.

Figure 40.9 shows the integrated solution the LED localization and assessment. Because some levels of degradation can be subjective, in addition to localization and assessment, the tool allows a subject-matter expert to review the assessments and provide feedback. The feedback consists of saved images and updated ground truth information, which enables the retraining of machine learning models and paves the path to continuous adaptive learning of the deployed system.

40.3   Conclusions & Recommendations

Machine learning and computer vision provide a clear path to automate and eliminate tedious, error-prone tasks on the remanufacturing floor. This article illustrated developing solutions for two carefully-selected tasks: part number identification and LED degradation assessment. Introducing machine learning to remanufacturing processes works particularly well when brought incrementally, starting from low-hanging fruit and building to higher-value solutions. Furthermore, even solutions that only cover a partial range of input variations typically provide sufficient value to be adopted.

Because the first task, part number identification, required solutions for text localization and character classification – both well-researched machine learning problems – we were able to leverage an existing commercial solution available through the Google Cloud Vision API. It is important to emphasize that the commercial solution, while very cost-effective, was not a turn-key solution and needed some data preprocessing and post-processing. For example, down-sampling images of high resolution accelerated the part number identification. The post-processing steps included parsing the responses from the API and mapping them onto the data from previously registered part numbers. Sometimes detection broke part number strings into sub-strings that had to be concatenated during post-processing.

Moreover, occasionally, matching the character strings to the existing part had to be probabilistic because of the occasional misclassification of individual characters. Because of the large number of different part numbers, it was not practical to address all edge cases. Instead, the integrating solution was equipped with a mechanism to capture and store critical information when a problem arises to enable continuous improvement.

Images of LEDs are not commonly used for training large machine-learning models, and no commercial solutions exist. Moreover, they are significantly different from typical images used for training large machine learning models that even transfer learning for degradation assessment did not work well. Future work can investigate improving transfer learning results by gradually unfreezing weights from the based model and training the model on a more extensive dataset of sub-images of LEDs.

Furthermore, popular deep-learning models for localization (YOLO and U-Net) showed limited generalization, including occasional missed LED detection or occasional false positives. In our experiments, the best solution for degradation assessment employed a CNN model trained from scratch. The most effective localization was primarily based on classical image-processing techniques. However, we combined classical methods and leveraged a more recent, popular segmentation model, U-Net, to accelerate the requisite localization ground truth labeling.

The integrated solution for LEDs was equipped with the tools for careful inspection, which enabled the expert to overwrite individual assessments and create data for adaptive learning. This capability was especially useful because the assessment of mild loss of brightness was subjective. The models were developed for three different types of LEDs, but only one was presented in this manuscript. The next development step is to integrate LED degradation detection and localization.

The solution structure presented in this article can be extended to another component failure. The part identification system requires an updated part number database to be repurposed. The automated registration system drastically lowers the manual labor cost of building a relevant dataset. The assessment model can be trained to solve a different component failure problem with a new, labeled dataset. It is important to note: the best methods for localization and assessment will change depending on the dataset. However, this article has demonstrated a modular process flow that can be modified to suit new health assessment tasks. In addition to developing specific solutions, an organization significantly benefits from adopting machine-learning philosophy by capturing data associated with critical processes. This data represents an asset that can unlock additional solutions down the road, be used for training, and better characterize the inventory.

Acknowledgments

The authors gratefully acknowledge our colleagues: Scott Nichols for developing the image capturing fixtures, Neel Surendra Sancheti for capturing the initial ground truth data on LED degradation, and Zane Kitchen Lipski for the initial development of the methods for concatenated bounding boxes.

This material is based upon work supported by the U.S. Department of Energy's Office of Energy Efficiency and Renewable Energy (EERE) under the Advanced Manufacturing

Office Award Number DE-EE0007897 awarded to the REMADE Institute, a division of Sustainable Manufacturing Innovation Alliance Corp. In addition, RIT received funding, in part, from New York State Empire State Development under Grant #AC118 that supported the work on this project.

The views expressed herein do not necessarily represent the views of the U.S. Department of Energy or the United States Government or New York State.

References

1. Diwan, T., Anirudh, G., & Tembhurne, J. V. Object detection using YOLO: Challenges, architectural successors, datasets and applications. *Multimedia Tools and Applications*, 1–33, 2022.

2. Redmon, J., Divvala, S., Girshick, R., & Farhadi, A. *You Only Look Once: Unified, Real-Time Object Detection,* 2016. http://pjreddie.com/yolo/

3. Redmon, J., & Farhadi, A. *YOLO9000: Better, Faster, Stronger* (pp. 7263–7271), 2017. http://pjreddie.com/yolo9000/

4. Redmon, J., & Farhadi, A. *YOLOv3: An Incremental Improvement,* 2018. https://doi.org/10.48550/arxiv.1804.02767

5. Wang, C.-Y., Bochkovskiy, A., & Liao, H.-Y. M. *YOLOv7: Trainable bag-of-freebies sets new state-of-the-art for real-time object detectors,* 2022. https://doi.org/10.48550/arxiv.2207.02696

6. Wang, C. Y., Bochkovskiy, A., & Liao, H. Y. M. Scaled-YOLOv4: Scaling Cross Stage Partial Network. *Proceedings of the IEEE Computer Society Conference on Computer Vision and Pattern Recognition*, 13024–13033, 2020. https://doi.org/10.48550/arxiv.2011.08036

7. Li, J., Gu, J., Huang, Z., & Wen, J. Application Research of Improved YOLO V3 Algorithm in PCB Electronic Component Detection. *Applied Sciences 2019, Vol. 9, Page 3750*, 9(18), 3750, 2019. https://doi.org/10.3390/APP9183750

8. Ronneberger, O., Fischer, P., & Brox, T. U-Net: Convolutional Networks for Biomedical Image Segmentation. In N. Navab, J. Hornegger, W. M. Wells, & A. F. Frangi (Eds.), *Medical Image Computing and Computer-Assisted Intervention – MICCAI 2015* (pp. 234–241), 2015. Springer International Publishing.

9. Du, G., Cao, X., Liang, J., Chen, X., & Zhan, Y. Medical image segmentation based on u-net: A review. *Journal of Imaging Science and Technology*, 64, 1–12, 2020.

10. Siddique, N., Paheding, S., Elkin, C. P., & Devabhaktuni, V. U-net and its variants for medical image segmentation: A review of theory and applications. *Ieee Access*, 9, 82031–82057, 2021.

11. Cheng, J., Xiong, W., Chen, W., Gu, Y., & Li, Y. Pixel-level crack detection using U-net. *TENCON 2018-2018 IEEE Region 10 Conference*, 462–466, 2018.

12. Hsieh, Y.-A., & Tsai, Y. J. Machine learning for crack detection: Review and model performance comparison. *Journal of Computing in Civil Engineering*, 34(5), 4020038, 2020.

13. Liu, Z., Cao, Y., Wang, Y., & Wang, W. Computer vision-based concrete crack detection using U-net fully convolutional networks. *Automation in Construction*, 104, 129–139, 2019.

14. Gonzalez, R., & Wood, R. *Digital Image Processing* (4th ed.). Pearson, 2017.

15. Bailey, T., & Durrant-Whyte, H. Simultaneous localization and mapping (SLAM): Part II. *IEEE Robotics and Automation Magazine*, 13(3), 108–117, 2006. https://doi.org/10.1109/MRA.2006.1678144

16. Durrant-Whyte, H., & Bailey, T. Simultaneous localization and mapping: Part I. *IEEE Robotics and Automation Magazine*, 13(2), 99–108, 2006. https://doi.org/10.1109/MRA.2006.1638022

17. Brown, M., & Lowe, D. G. Automatic panoramic image stitching using invariant features. *International Journal of Computer Vision, 74*(1), 59–73, 2007. https://doi.org/10.1007/S11263-006-0002-3/METRICS

18. Szeliski, R. Image Alignment and Stitching: A Tutorial. *Foundations and Trends® in Computer Graphics and Vision, 2*(1), 1–104, 2007. https://doi.org/10.1561/0600000009

19. Lowe, D. G. Object recognition from local scale-invariant features. *Proceedings of the Seventh IEEE International Conference on Computer Vision, 2*, 1150–1157, 1999.

20. Rublee, E., Rabaud, V., Konolige, K., & Bradski, G. ORB: An efficient alternative to SIFT or SURF. *2011 International Conference on Computer Vision*, 2564–2571, 2011.

21. Rosten, E., & Drummond, T. Machine learning for high-speed corner detection. *European Conference on Computer Vision*, 430–443, 2006.

22. Calonder, M., Lepetit, V., Strecha, C., & Fua, P. Brief: Binary robust independent elementary features. *European Conference on Computer Vision*, 778–792, 2010.

23. Leutenegger, S., Chli, M., & Siegwart, R. Y. BRISK: Binary Robust invariant scalable keypoints. *Proceedings of the IEEE International Conference on Computer Vision*, 2548–2555, 2011. https://doi.org/10.1109/ICCV.2011.6126542

24. Alcantarilla, P. F., Bartoli, A., & Davison, A. J. KAZE features. *European Conference on Computer Vision*, 214–227, 2012.

25. Pan, S. J., & Yang, Q. A Survey on Transfer Learning. *IEEE Transactions on Knowledge and Data Engineering, 22*(10), 1345–1359, 2010. https://doi.org/10.1109/TKDE.2009.191

26. Szegedy, C., Ioffe, S., Vanhoucke, V., & Alemi, A. A. *Inception-v4, inception-resnet and the impact of residual connections on learning*, (n.d.).

27. Abadi, M., Barham, P., Chen, J., Chen, Z., Davis, A., Dean, J., Devin, M., Ghemawat, S., Irving, G., Isard, M., & others. Tensorflow: A system for large-scale machine learning. *12th Symposium on Operating Systems Design and Implementation (16)*, 265–283, 2016.

28. Chollet, F. *Deep Learning with Python*, (2017).

29. Paszke, A., Gross, S., Massa, F., Lerer, A., Bradbury Google, J., Chanan, G., Killeen, T., Lin, Z., Gimelshein, N., Antiga, L., Desmaison, A., Xamla, A. K., Yang, E., Devito, Z., Raison Nabla, M., Tejani, A., Chilamkurthy, S., Ai, Q., Steiner, B., … Chintala, S. PyTorch: An Imperative Style, High-Performance Deep Learning Library. *Advances in Neural Information Processing Systems, 32*, 2019.

30. Goodfellow, I., Bengio, Y., & Courville, A. *Deep learning*. MIT Press, 2016.

31. Perng, D. B., Liu, H. W., & Chang, C. C. Automated SMD LED inspection using machine vision. *International Journal of Advanced Manufacturing Technology, 57*(9–12), 1065–1077, 2011. https://doi.org/10.1007/S00170-011-3338-Y/METRICS

32. Lin, H., Li, B., Wang, X., Shu, Y., & Niu, S. Automated defect inspection of LED chip using deep convolutional neural network. *Journal of Intelligent Manufacturing, 30*(6), 2525–2534, 2019. https://doi.org/10.1007/S10845-018-1415-X/TABLES/6

33. Geron, A. *Hands-on Machine Learning with Scikit-Learn, Keras & Tensorflow* (2nd ed.). O'Reilly, 2019.

34. LeCun, Y., Bengio, Y., & Hinton, G. Deep learning. *Nature, 521*(7553), 436–444, 2015.

35. Srivastava, N., Hinton, G. E., Krizhevsky, A., Sutskever, I., & Salakhutdinov, R. Dropout: a simple way to prevent neural networks from overfitting. *Journal of Machine Learning Research, 15*(1), 1929–1958, 2014.

41

Effects of Ultrasonic Impact Treatment on the Fatigue Performance of the High Strength Alloy Steel

Joha Shamsujjoha[1]*, Shirley Garcia Ruano[1], Mark Walluk[1], Michael Thurston[1]
and M. Ravi Shankar[2]

[1]*Golisano Institute for Sustainability (GIS), Rochester Institute of Technology (RIT), Rochester, NY, USA*
[2]*Industrial Engineering Department (IED), University of Pittsburg (PITT), Pittsburgh, PA, USA*

Abstract

In this study, the feasibility of Ultrasonic Impact Treatment (UIT) as a repair method to recover the surface properties and fatigue life of surface-hardened steel components degraded during operation by corrosion, wear, or fatigue is investigated. UIT introduces severe plastic deformation in the materials through multiple sliding impacts at a high frequency (~40 kHz). The study examines the role of UIT processing variables, such as traverse speed, static force, pin size, and the number of passes, on the surface topography (roughness and waviness), microstructures, and the depth distributions of hardness using a design of experiment. Cross-sectional microscopic analysis of the UIT- processed specimen reveals gradient microstructures with dynamically recrystallized (DRX) nanocrystalline grains close to the surface, followed by the elongated grains in the deformed zone. The size of the DRX and deformed zone depends on the UIT processing conditions and increases with decreasing traverse speed and an increasing number of passes. However, reducing the traverse speed and/or increasing the number of passes can cause near-surface damage to the impacted material, which can adversely affect the performance of UIT-treated components. After UIT, a 55% increase in the hardness value close to the surface area was observed. The depth of the hardened zone is about ~1mm. Samples treated with UIT that exhibit high hardness, hardening depth, and low near-surface damage are selected for fatigue performance evaluation. Finally, rotational bend fatigue testing (RBF) confirms that materials subjected to UIT treatment exhibit significantly better fatigue performance compared to untreated samples.

Keywords: Remanufacturing, ultrasonic impact treatment, surface topography, strain hardening, rotational bend fatigue

41.1 Introduction

Large volumes of surface-hardened steel components are currently being scrapped, leading to unnecessary primary material consumption and associated energy emissions. Remanufacturing (reman) could alleviate this problem. However, the remanufacturing of

---

Corresponding author: mdsgis@rit.edu

Nabil Nasr (ed.) Technology Innovation for the Circular Economy: Recycling, Remanufacturing, Design, Systems Analysis and Logistics, (541–554) © 2024 Scrivener Publishing LLC

these high-value components is currently limited because available repair technologies cannot reliably compensate for the degradation of surface properties caused by operational wear, fatigue, and/or the repair process itself. These high-value components are used for applications that impose high loads and/or low weight allowances, and they are often manufactured with thermochemical surface treatments to enhance fatigue resistance [1–3]. While these surface treatments are critical to fatigue performance, they restrict remanufacturing in several ways. Firstly, the performance improvements are degraded by material removal and/or thermal annealing. Secondly, duplicating the original surface properties can be uneconomical and may require treating the entire component to achieve a local repair. Lastly, non-thermal treatments such as Ultrasonic Impact Treatment (UIT) are not yet fully understood at the mechanistic level to provide confidence in reproducing the original surface properties. The current repair strategy often involves grinding out damaged material, which removes the enhanced surface material necessary to meet performance specifications or adding material via welding or other high-temperature processes, which thermally degrade (anneal) the surface properties [4]. As a result, the performance of the repaired components is uncertain, their value decreases, and their reuse cycles are limited. Therefore, there are significant economic and sustainability benefits to researching new repair techniques that can recover the nominal 'as-new' fatigue life of used or worn surface-hardened steel components [5].

Ultrasonic Impact Treatment (UIT) is a widely used surface processing technique for post-weld treatment, which aims to reduce tensile residual stresses, increase resistance to fatigue, stress corrosion cracking, and wear, and extend the useful life of materials [6–9]. Essentially, UIT is a type of needle peening that employs ultrasonic vibrations to deform the material surface, resulting in novel properties by introducing a nanocrystalline surface microstructure. During UIT, low-amplitude ultrasonic oscillations drive one or more spherical tips to impact the material surface up to 40,000 times per second. At the same time, a static axial load is applied through the tip, and the tip is moved across the surface [10]. This micro-cold-forging process introduces severe plastic deformation to the surface layers and forges closed superficial micro-cracks. UIT also produces a surface layer (>200 µm deep) of nanocrystalline grains while inducing higher compressive residual stresses (~2x) to deeper depths (0.5 mm - <1.0 mm) than conventional shot peening. In a recent study [11], Maleki *et al.* compared the effects of different severe plastic deformation techniques on the surface modification behavior of Inconel 718 superalloy, including shot peening, laser peening, and ultrasonic impact treatment. Their investigations revealed that the compressive residual stress layer was approximately 400µm, with the highest magnitude of residual stress at ~1000MPa, which is significantly higher than that produced by shot peening (with a depth of compressive stress after shot peening reported to be around 250µm, and a stress value at the surface of ~400 MPa). Additionally, they reported a 44% increase in surface hardness. Their study concluded that ultrasonic impact had the most significant influence on improving fatigue life, as was also reported in [12] when compared to shot peening and tungsten inert gas (TIG)-dressing.

While it is evident that UIT can extend the fatigue life under optimal conditions, previous studies did not examine the role of UIT processing conditions. In addition to producing nanocrystalline and hardened microstructures, as well as compressive residual stress, UIT can increase surface roughness and cause damage to components, especially with high kinetic energy impacts (as reported in [13]). The overall effect of UIT on the fatigue life of treated components depends on the balancing effects of beneficial strain hardening, grain refinement, and compressive residual stress, as well as detrimental stress concentrations. However, the interaction between UIT-induced microstructural changes and ultimate component performance is complex. Furthermore, uncertainties in the final performance of UIT-treated components have limited its widespread application in improving surface properties of high-valued components. Therefore, designers need to consider both the beneficial and detrimental aspects of UIT processing to maximize the fatigue life of treated components. However, a systematic study to quantify the effects of UIT parameters on the fatigue life of materials is lacking. Thus, the main objective of this study is to investigate the role of UIT parameters on the rotating bend fatigue performance of high-strength micro-alloyed steels.

41.2   Materials and Methods

41.2.1   Materials

A microalloyed steel, typically used for crankshafts, is chosen as the baseline material. It contains 0.35-0.40 wt. percent carbon, 1.30-1.50 wt. percent manganese, and other minor alloying elements such as V and Nb. The microstructure of the baseline microalloyed steel consists of ferrite and pearlite, as shown in the optical micrograph in Figure 41.1.

41.2.2   Ultrasonic Impact Treatment (UIT)

Progress Rail's UIT-6000 series ultrasonic impact testing (UIT) equipment was used in this study. The equipment includes an ultrasonic generator with a frequency of 27 kHz and an output voltage of up to 80.6 V, as well as Progress Rail's proprietary magnetostrictive

Figure 41.1 Optical micrograph of the microalloyed steel showing ferrite-pearlite microstructure.

transducer that converts electrical energy into mechanical vibration. A waveguide (ultrasonic horn) is used to amplify the vibration output, and a holder containing high wear-resistant pins is installed at the end of the waveguide to strike the workpiece. The pins can oscillate freely between the waveguide and the workpiece. The UIT equipment offers both single and multiple pin configurations, with up to four pins placed side by side inside the pin holder. Three pin diameters are available: 3mm (0.12 inch), 4.8 mm (0.19 inch), and 6.35 mm (0.25 inch).

For the UIT peening process, the UIT tool was securely placed inside a custom-made fixture designed to maintain a level of preload on the workpiece being treated. The tool and the sample were mounted to the tool post and chuck of the lathe machine, respectively, to perform the peening on the cylindrical sample. The adaptation made for the tool fixing was stable enough to withstand the vibrations produced by the impact of the needle/pin on the material . The experimental setup for the UIT peening process is shown in Figure 41.2.

41.2.3   Microstructure Evaluation

For microstructural analysis, samples were sectioned using the conventional band saw and slow-speed diamond saw methods, mounted in Polyfast and mechanically ground up to 1200 grit silicon carbide paper, and polished with diamond pastes of 3 and 1 µm. The cross-sections were examined using a Zeiss Optical Microscope. The surface topography of the samples was analyzed using Keyence VHX700 digital microscope.

41.2.4   Mechanical Evaluation-Hardness & Fatigue Testing

The Vickers microhardness test was performed on the cross-sectional samples to analyze the depth-resolved hardness distributions using the "LECO LM248 AT" machine with an indentation load of 100 gf and a dwell time of 13 seconds. At least 10 readings were taken at each depth to obtain satisfactory statistics. Samples for microhardness measurement were prepared in a similar manner to that described for microstructure analysis in the previous section.

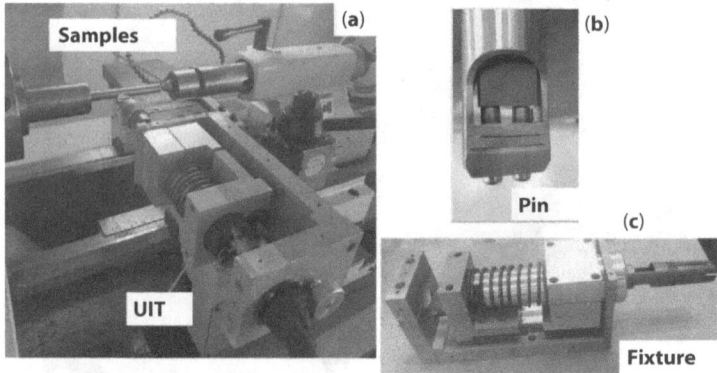

Figure 41.2 Experimental setup for UIT treatment, where the UIT tool and workpiece are mounted to a lathe machine, is shown in (a). Pictures of the pin holder and custom-made fixture are shown in (b) and (c), respectively.

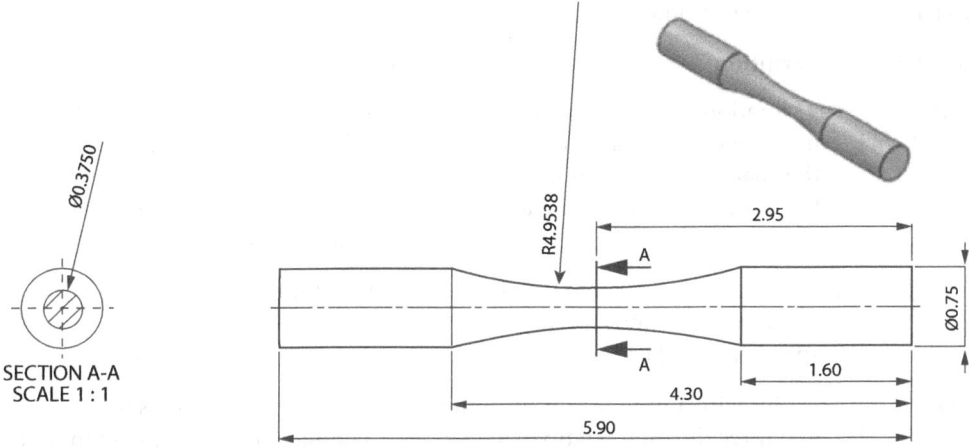

Figure 41.3 Geometry and dimensions of the hourglass samples prepared according to the ISO 1143-210 standard [14].

The rotating beam fatigue (RBF) tests of the as-fabricated and UIT-treated samples were performed using ADMET's expert 9300 fatigue tester in accordance with the ISO 1143-2010 standard [14] at room temperature using the hourglass sample geometry. The geometry and dimensions of the hourglass samples are shown in Figure 41.3. All specimens were fabricated in-house using a lathe machine. After fabrication, the specimens were carefully polished to remove any machining marks using 1200 grit silicon carbide paper, and these specimens will be referred to as "machined" throughout the manuscript. The load-controlled fatigue tests were performed at a stress ratio of R=-1 and a frequency of 50 Hz. The tests were stopped when specimens broke or after 1×107 cycles (referred to as "runout"), whichever came first. The bending stress value corresponding to a given load was calculated using the formula given below:

$$\sigma = \frac{32FL}{\pi d^3},$$ where F is the applied force, L is the force arm length, d is the specimen diameter.

41.2.5   Design of Experiments

A significant amount of screening experiments was performed to identify the critical UIT processing parameters that have a major impact on the surface integrity, evaluated from the surface morphology, microstructures and microhardness distributions. Table 41.1 lists the parameters identified as crucial on the screening experiment.

Screening experiments revealed that surface hardness increases with an increasing power level, which is related to increasing ultrasonic vibrational amplitude. However, following the manufacturer's recommendation, the power was set to the highest level (100%) for all experiments. Both mono-pin and multi-pin experiments were conducted, and the results showed that the surface hardness and depth of the deformation zone increase with an increasing number of pins in the pin holder. However, obtaining consistent UIT treatment of hourglass-shaped fatigue samples, especially in the curved area, was challenging.

Table 41.1 UIT processing parameters.

Parameters	Description	Ranges
Amplitude	The vibration amplitude of the ultrasonic horn, can be adjusted by the changing the power output of the ultrasonic power generator.	Fixed at 100% power level (~1.5 kV), which corresponds to a frequency of ~25 kHz and an amplitude of ~32 μm, based on manufacturer recommendation
Scanning Speed	The resultant speed is calculated from the sample rotation and traverse speed of the UIT tool.	0.16 in/min to 0.48 in/min
Pin size	UIT pins with a fixed length of 25mm and three different pin tip diameters (3mm, 4.8 mm and 6mm) were available.	Most of the experiments were conducted using the 4.8 mm pin.
Multi-pin assembly	UIT tool holders are designed to contain different combinations of pins.	Due to the geometrical constraint of the hourglass-shaped fatigue samples used in the study, the mono-pin UIT treatment was primarily performed.
Number of UIT passes	Number of times the surface was processed. With increasing number of passes, the impact duration per unit area increases.	1-3 passes
Pressing Force	Static force can be adjusted by compressing the spring to a certain distance; i.e., changing the spring gap.	75 N to 130 N

Hence, only the mono-pin experiment was performed in this study. In the case of mono-pin experiments, the material undergoes more severe plastic deformation when treated with smaller diameter pins. However, medium diameter pins were selected because the smallest diameter pins experienced severe wear. Both scanning speed (resultant speed calculated from the tool's traverse speed and sample rotational speed) and the number of passes significantly impact the surface properties of the UIT-treated specimens. A scanning speed that is too slow increases the hardness and depth of the deformation zone but also causes surface damage and significant wear of the UIT pins. Conversely, the deformation is minimized when the scanning speed is too fast. Therefore, a scanning speed between 0.16 inch/min and 0.48 inch/min was selected as the optimal range. Similarly, the degree of deformation increases with an increasing number of UIT passes; however, more than three passes result in significant surface damage. Additionally, the pressing force, which is the nominal force between the UIT pin and the workpiece, significantly impacts the surface properties and is considered an independent factor for setting up the Design of Experiment (DOE).

Table 41.2 Showing sets of UIT processing conditions based on the Taguchi L_9 array method.

ID	Scanning speed (in/min)	Number of passes	Static force, N
UIT7	0.16	1	130
UIT9	0.16	2	105
UIT8	0.16	3	75
UIT4	0.32	1	105
UIT3	0.32	2	75
UIT2	0.32	3	130
UIT5	0.48	1	75
UIT1	0.48	2	130
UIT6	0.48	3	105

The Taguchi L9 orthogonal array method was employed in setting up the Design of Experiment to comprehend the interplay between independent factors. A 3-factor, 3-level DOE was created using Minitab software. Table 41.2 displays the 9 unique UIT processing conditions obtained from the Taguchi L9 array method.

It should be noted that the kinetic energy per impact, which results from the frequency and amplitude of the waveguide tip vibrations, remains the same across all conditions. However, changes in scanning speed and number of passes affect the number of impacts per unit area. When these parameters are combined with pressing force, the total impact energy per unit area caused by UIT pin impacts varies. For instance, the impact energy per unit area is highest with the slowest scanning speed, the highest number of UIT passes, and the largest static force, whereas it is lowest with the fastest scanning speed, the lowest number of UIT passes, and the lowest pressing force.

41.3 Results and Discussions

41.3.1 Surface Morphology

Figure 41.4 shows the surface morphologies of different UIT-treated samples, collected using a Keyence VHX700 digital microscope. This qualitative assessment reveals distinct changes in surface morphologies for various UIT processing conditions. Typically, a surface treated with higher impact energy per unit area using UIT shows a rougher surface appearance than a surface treated with lower impact energy per unit area. For instance, UIT9 samples treated with higher impact energy (the lowest speed, a medium pressing force, and 2 UIT passes) exhibit a rougher surface appearance compared to UIT5, treated with lower impact energy per unit area (highest speed, the lowest pressing force, and 1 pass).

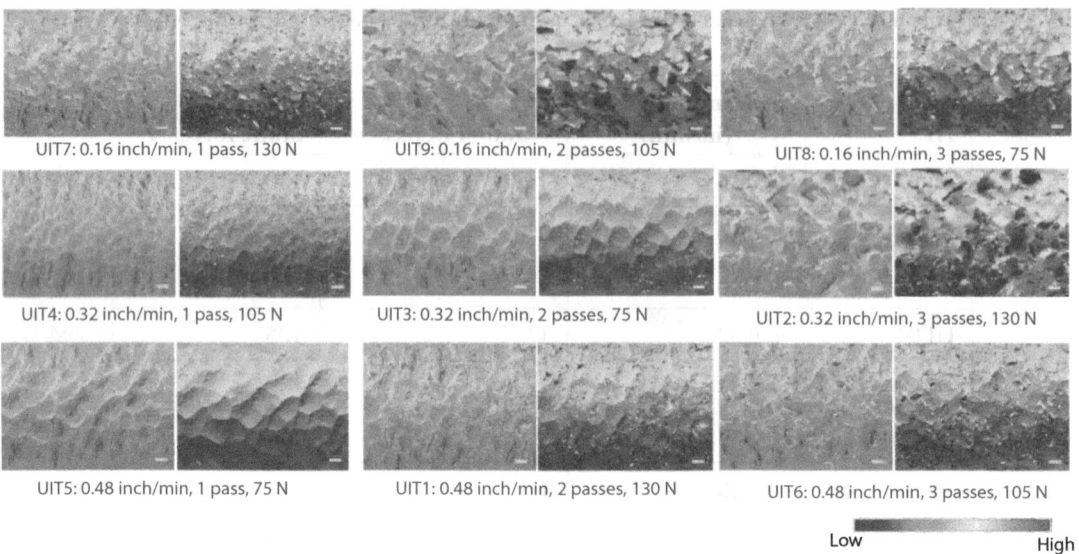

UIT7: 0.16 inch/min, 1 pass, 130 N   UIT9: 0.16 inch/min, 2 passes, 105 N   UIT8: 0.16 inch/min, 3 passes, 75 N

UIT4: 0.32 inch/min, 1 pass, 105 N   UIT3: 0.32 inch/min, 2 passes, 75 N   UIT2: 0.32 inch/min, 3 passes, 130 N

UIT5: 0.48 inch/min, 1 pass, 75 N   UIT1: 0.48 inch/min, 2 passes, 130 N   UIT6: 0.48 inch/min, 3 passes, 105 N

Low   High

Figure 41.4 Showing the surface morphologies of the samples treated with different UIT processing conditions.

41.3.2   Microstructures

Figure 41.5 shows cross-sectional optical micrographs of samples subjected to UIT using different processing conditions. The elongated grains observed indicate severe plastic deformation in the surface and subsurface layers. The extent of deformation is dependent on the UIT process parameters, specifically the impact energy per unit area. For instance, the most substantial deformation zone thickness (~700 µm) was observed when the specimen was treated with the lowest UIT speed, a medium pressing force, and 2 passes. Conversely, mild plastic deformation was observed in the specimen treated with a low impact energy, such as in the case of specimen UIT 5 (0.48 inch/min scanning speed, 1 pass, and 75 N pressing force). However, higher impact energy during UIT processing leads to significant damage to underlying microstructures, such as the formation of micro-notches (as depicted in Figure 41.6). These notches increase the local elastic stress concentration (K_t) and can have adverse effects on fatigue performance.

41.3.3   Microhardness

The microhardness distributions of the samples treated with UIT are depicted in Figure 41.7. Regardless of the processing conditions, an increase in hardness value is observed for all the samples after UIT treatment. The highest hardness is observed close to the surface, followed by a gradual decrease to a depth of approximately 2.5mm from the surface. Consistent with the microstructural analysis, the samples treated with higher impact energy per unit area exhibit higher hardness values. UIT9 (0.16 inch/min scanning speed, 2 passes, and 105 N pressing force) shows the highest hardness value of approximately 455 HV, which is 55% higher than the original value of 290 HV for the untreated baseline material. Only a modest

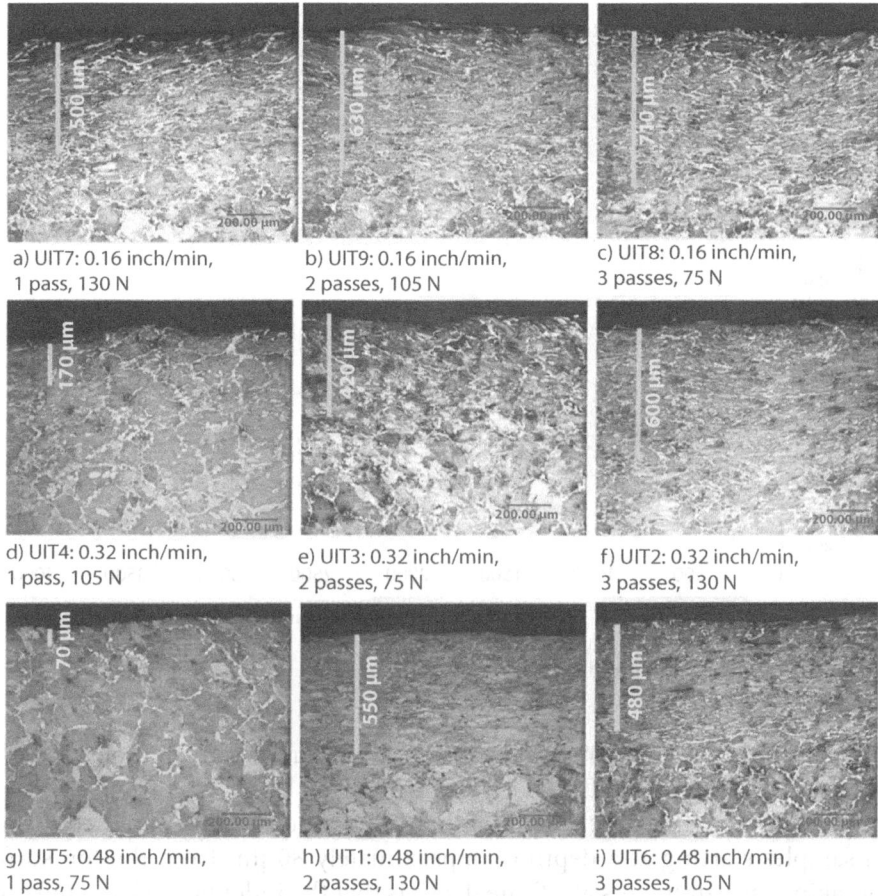

a) UIT7: 0.16 inch/min,
1 pass, 130 N

b) UIT9: 0.16 inch/min,
2 passes, 105 N

c) UIT8: 0.16 inch/min,
3 passes, 75 N

d) UIT4: 0.32 inch/min,
1 pass, 105 N

e) UIT3: 0.32 inch/min,
2 passes, 75 N

f) UIT2: 0.32 inch/min,
3 passes, 130 N

g) UIT5: 0.48 inch/min,
1 pass, 75 N

h) UIT1: 0.48 inch/min,
2 passes, 130 N

i) UIT6: 0.48 inch/min,
3 passes, 105 N

Figure 41.5 Showing the microstructures of the samples treated with various UIT processing conditions.

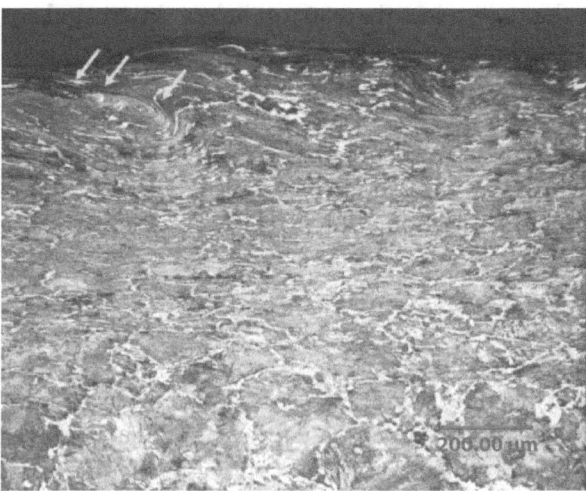

Figure 41.6 Evidence of surface damaged (yellow arrows) for the specimen processed at 0.16 inch/min scanning speed, 2 passes and 105 N pressing force.

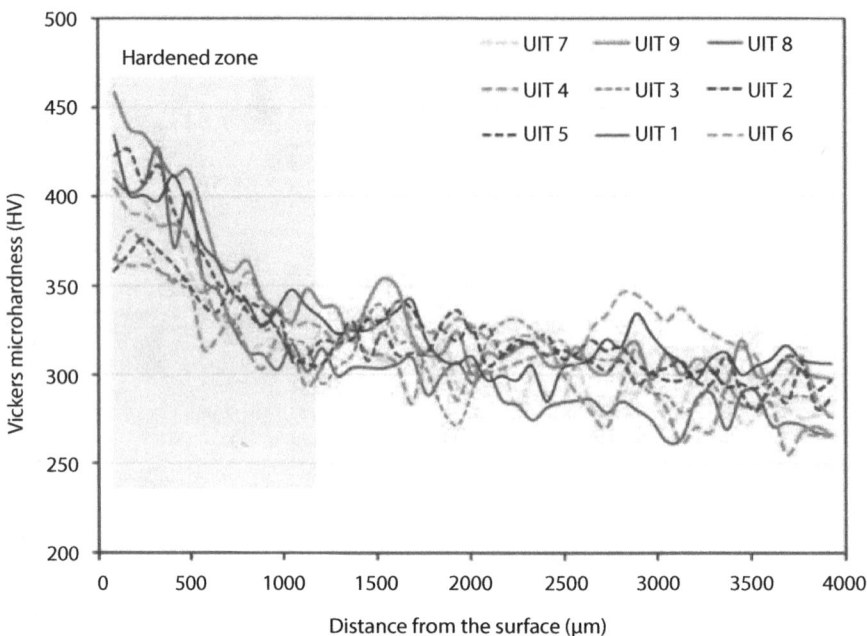

Figure 41.7 Microhardness depth profile of UIT-treated specimens.

increase in hardness value (~20%) was observed for samples treated at a lower impact energy per unit area, such as UIT3, UIT4, and UIT5.

It should be noted that Vickers microhardness values were obtained from cross-sectioned samples starting at a depth of approximately 80 μm from the top surface. The reason for taking measurements at a finite depth is that the indentation has a finite size, and the volume of material interacting with the indenter is even larger. To avoid edge effects, it is standard practice to obtain hardness data at a distance no less than 2½ times the linear dimension of the indentation from the surface. Hence, assuming that the hardness value will be even higher in the region less than 80 μm below the treated surface is reasonable.

41.3.4 Rotational Bend Fatigue

The Wöhler (S-N) curve of the as-machined microalloyed steel is shown in Figure 41.8. The curve in the plot is the best fit to the experimental data obtained by least squares non-linear regression of the Stromeyer equation, which relates N_f, the cycles to failure, to the stress amplitude, S, as follows:

$$S = SL + A(N_f)^m$$

where S_L, A, and m are fitting parameters. S_L is the endurance limit, approximately 374 MPa for microalloyed steel. It should be noted that employing this relation with an endurance limit, S_L, does not represent an endorsement of a particular fatigue design philosophy Rather, it is provided as a means to visualize and compare the obtained fatigue data.

Variations in fatigue lives for different UIT processing conditions tested at a constant stress amplitude of approximately 620 MPa are presented in Figure 41.9. The plot also includes the fatigue life of as-machined (AM) samples tested at the same stress amplitude (~620 MPa) for comparison purposes. While a few UIT-treated samples were tested at a lower stress amplitude (~550 MPa), they all resulted in run-out tests (> 106 cycles). Therefore, the high- stress amplitude was selected to differentiate the fatigue lives of the different UIT-treated samples.

UIT3 specimens, with the third-lowest hardness enhancement and medium depth of the deformed layer, exhibit the most significant improvement in fatigue life. Conversely, UIT1 specimens, with one of the highest hardness increments and a high depth of the deformation layer, have the worst fatigue life among all UIT-treated samples. Notably, UIT2 specimens, with moderate hardness enhancement and deformation depth, have the second-highest fatigue life. These findings highlight the complex interaction of UIT-induced changes in material properties, such as strain hardening, grain refinement, surface roughness/damage, and residual stresses, on fatigue performance [15, 16]. While the importance of UIT-induced compressive residual stresses on fatigue life needs to be addressed, the residual stresses were not measured in this investigation and will be the focus of a subsequent study. However, it is well documented that UIT produces a sub-surface layer of compressive residual stresses [10–12].

Furthermore, it is known that, under a constant residual stress field, fatigue crack initiation life decreases with increasing surface roughness [16]. Qualitative analysis indicates that surface roughness/damage increases with increasing impact energy (low scanning speed, high pressing force, and a high number of UIT passes) of the UIT. However, the impact of elastic stress concentration associated with surface roughness (damage) is somewhat counteracted by the presence of a deeper deformation layer, compared to conventional shot peening, in the UIT-processed samples. The enhancement of local mechanical properties

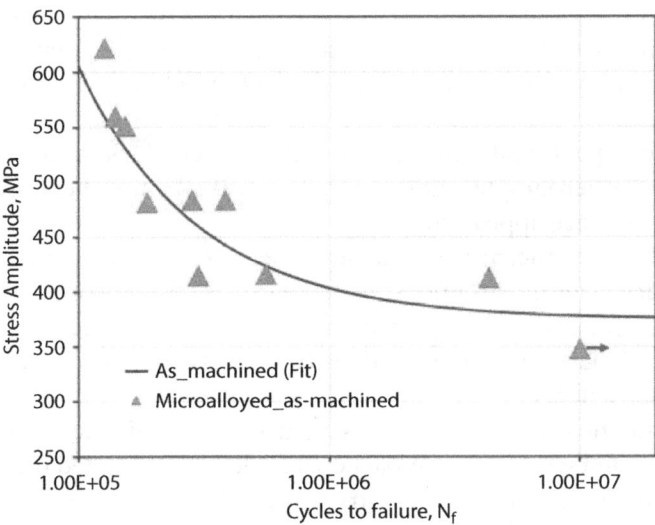

Figure 41.8 Fatigue performance of as-fabricated and UIT-treated microalloyed steel. The blue arrow in the figure indicates a run-out test.

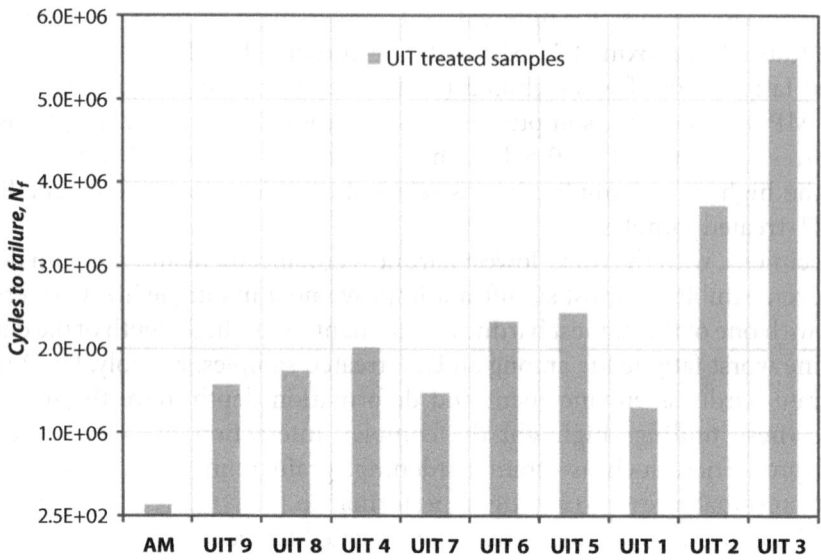

Figure 41.9 Fatigue performance of UIT-treated microalloyed steel.

in the deformed zone resists both fatigue crack nucleation and propagation. Due to this balancing effect of beneficial strain hardening and compressive residual and detrimental surface roughness/damage, UIT2, the sample with moderate hardness enhancement and deformation depth, exhibits one of the highest fatigue lives. Nevertheless, it is worth noting that UIT-processed samples show a significant improvement in fatigue lives, regardless of the processing conditions (see Figure 41.8 and Figure 41.9).

41.4   Conclusions

In this study, the impact of ultrasonic impact treatment on the fatigue performance of a high-strength microalloyed steel was evaluated, and the following findings were obtained:

1. Ultrasonic impact treatment caused noticeable changes in surface morphology. Surface micrographs indicated that increasing impact energy density led to rougher surface appearances.
2. Cross-sectional microstructural analysis revealed a heavily deformed near-surface layer in specimens treated with UIT. The depth of the deformation zone increased with higher impact energy per unit area.
3. UIT with higher kinetic energy density impacts may result in micro-notch type surface defects.
4. Surface and near-surface hardness values increased after UIT treatment. Samples treated with higher impact energy density exhibited higher hardness values, consistent with the microstructural analysis.
5. Rotational bending fatigue testing showed a significant improvement in fatigue performance after UIT treatment. However, a detailed analysis of the

microstructural and residual stress evolution is needed to determine the relative contributions of UIT-induced factors such as strain hardening, residual stress, and surface defects on the fatigue performance, which will be the focus of a follow-up study.

Acknowledgements

This material is based upon work supported by the U.S. Department of Energy's Office of Energy Efficiency and Renewable Energy (EERE) under the Advanced Manufacturing Office Award Number DE-EE0007897 awarded to the REMADE Institute, a division of Sustainable Manufacturing Innovation Alliance Corp. This report was prepared as an account of work sponsored by an agency of the United States Government. Neither the United States Government nor any agency thereof, nor any of their employees, makes any warranty, express or implied, or assumes any legal liability or responsibility for the accuracy, completeness, or usefulness of any information, apparatus, product, or process disclosed, or represents that its use would not infringe privately owned rights.

Reference herein to any specific commercial product, process, or service by trade name, trademark, manufacturer, or otherwise does not necessarily constitute or imply its endorsement, recommendation, or favoring by the United States Government or any agency thereof. The views and opinions of authors expressed herein do not necessarily state or reflect those of the United States Government or any agency thereof.

In addition, RIT received funding, in part, from New York State Empire State Development under Grant #AC118 that supported the work on this project. The views expressed herein do not necessarily represent the views of New York State.

The authors would also like to express their gratitude to Chris Kinney and Curtis Graham (Remanufacturing Division, Caterpillar Inc.) for providing the materials for the study.

References

1. Vencl, A., Rac, A., Diesel engine crankshaft journal bearings failures: Case study. *Eng. Fail. Anal.*, vol. 44, pp. 217–228, Sep. 2014.
2. Torims, T., Pikurs, G., Ratkus, A., Logins, A., Vilcans, J., Sklariks, S., Development of Technological Equipment to Laboratory Test *In-situ* Laser Cladding for Marine Engine Crankshaft Renovation. *Procedia Eng.*, vol. 100, no. January, pp. 559–568, Jan. 2015.
3. Torims, T., Bruckner, F., Ratkus, A., Fokejevs, A., Logins, A., The Application of Laser Cladding to Marine Crankshaft Journal Repair and Renovation. *ASME 2014 12th Bienn. Conf. Eng. Syst. Des. Anal. ESDA 2014*, vol. 1, Oct. 2014.
4. Cai, Z., Zhang, P., Liang, Z., Novel Remanufacturing Technology as an Alternative to Maintenance for Diesel Engine and its Benefit Analysis. *Adv. Mater. Res.*, vol. 216, pp. 435–439, 2011.
5. Liu, C., Cai, W., Dinolov, O., Zhang, C., Rao, W., Jia, S., Li, L., Emergy based sustainability evaluation of remanufacturing machining systems. *Energy*, vol. 150, pp. 670–680, May 2018.
6. Roy, S., Fisher, J.W., Yen, B.T., Fatigue resistance of welded details enhanced by ultrasonic impact treatment (UIT). *Int. J. Fatigue*, vol. 25, no. 9–11, pp. 1239–1247, Sep. 2003.

7. Liu, Y., Wang, D., Deng, C., Xia, L., Huo, L., Wang, L., Gong, B., Influence of re-ultrasonic impact treatment on fatigue behaviors of S690QL welded joints. *Int. J. Fatigue*, vol. 66, pp. 155–160, Sep. 2014.

8. Suzuki, T., Okawa, T., Shimanuki, H., Nose, T., Ohta, N., Suzuki, H., Moriai, A., Effect of Ultrasonic Impact Treatment (UIT) on Fatigue Strength of Welded Joints. *Adv. Mater. Res.*, vol. 996, pp. 736–742, 2014.

9. Mori, T., Shimanuki, H., Tanaka, M., Effect of UIT on fatigue strength of web-gusset welded joints considering service condition of steel structures. *Weld. World*, vol. 56, no. 9–10, pp. 141–149, Mar. 2012.

10. Statnikov, E.S., Korolkov, O.V., Vityazev, V.N., Physics and mechanism of ultrasonic impact. *Ultrasonics*, vol. 44, no. SUPPL., pp. e533–e538, Dec. 2006.

11. Maleki, E., Unal, O., Guagliano, M., Bagherifard, S., The effects of shot peening, laser shock peening and ultrasonic nanocrystal surface modification on the fatigue strength of Inconel 718. *Mater. Sci. Eng. A*, vol. 810, p. 141029, Apr. 2021.

12. Statnikov, E.S., Muktepavel, V.O. and Blomqvist, A., Comparison of ultrasonic impact treatment (UIT) and other fatigue life improvement methods. *Weld. World*, vol. 46, no. 3–4, pp. 20–32, Feb. 2002.

13. Lesyk, D.A., Martinez, S., Mordyuk, B.N., Dzhemelinskyi, V.V., Lamikiz, A., Prokopenko, G.I., Grinkevych, K.E., Tkachenko, I.V., Laser-Hardened and Ultrasonically Peened Surface Layers on Tool Steel AISI D2: Correlation of the Bearing Curves' Parameters, Hardness and Wear. *J. Mater. Eng. Perform.*, vol. 27, no. 2, pp. 764–776, Feb. 2018.

14. ISO 1143:2010, Metallic Materials—Rotating Bar Bending Fatigue Testing (International Organization for Standardization, 2010).

15. Shamsujjoha, M., Agnew, S.R., Melia, M.A., Brooks, J.R., Tyler, T.J., Fitz-Gerald, J.M., Effects of laser ablation coating removal (LACR) on a steel substrate: Part 1: Surface profile, microstructure, hardness, and adhesion. *Surf. Coatings Technol.*, vol. 281, pp. 193–205, Nov. 2015.

16. Shamsujjoha, M., Agnew, S.R., Melia, M.A., Brooks, J.R., Tyler, T.J., Fitz-Gerald, J.M., Effects of laser ablation coating removal (LACR) on a steel substrate: Part 2: Residual stress and fatigue. *Surf. Coatings Technol.*, vol. 281, pp. 206–214, Nov. 2015.

42

Mechanical Properties of High Carbon Steel Coatings on Gray Cast Iron Formed by Twin Wire ARC

K. DePalma, M. Walluk* and L. P. Martin

Rochester Institute of Technology, Rochester, NY, USA

Abstract

Twin wire arc is a thermal spray technology commonly used for applying coatings to cast iron components, however, studies of properties such as the in-plane strength and elastic modulus of coatings are rare. This is because the measurement of these properties is complicated by the maximum achievable TWA coating thickness. This study relates the microstructure and mechanical properties to the spray parameters used during deposition of high carbon steel coatings on gray cast iron substrates. Directional relationships are established between the spray parameters, including air pressure, traverse speed, standoff distance, and arc current, and the coating properties, such as porosity, hardness, and strength. Bend testing was used to establish the in-plane elastic modulus and strength of the coatings, and adhesion testing was used to determine the out-of-plane (normal) strength. The flexural strength was found to be an order of magnitude greater than the adhesive strength, suggesting orthotropic behavior associated with the lamellar structure of the coatings. The resistance of the coatings to both abrasive and sliding wear was investigated and found to be on the same order of magnitude as cast iron substrate. This report is an initial evaluation of twin wire arc repair of cast iron components damaged by wear or corrosion, with relevance to a wide range of automotive and heavy industrial applications.

Keywords: Twin wire arc, elastic modulus, mechanical properties, remanufacturing

42.1 Introduction

A wide range of thermal spray processes exist to apply metallic coatings or rebuild worn surfaces. Twin wire arc (TWA) is among the more common thermal spray processes because it provides the capability to deposit most relevant metallic materials, including steels [1], aluminums [2, 3], and nickel alloys [4, 5]. Additionally, TWA operates at relatively high deposition rates and low cost relative to other thermal spray methods. In the TWA process, a pair of wires is fed into a spray nozzle and a voltage difference is applied between the wires. This produces an arc between the wire tips, melting the feedstock. A stream of compressed

---

**Corresponding author*: mrwasp@rit.edu

Nabil Nasr (ed.) Technology Innovation for the Circular Economy: Recycling, Remanufacturing, Design, Systems Analysis and Logistics, (555–572) © 2024 Scrivener Publishing LLC

gas atomizes the molten metal and accelerates it toward the substrate surface. The resultant droplets are formed at temperatures upwards of 5000°C and can be accelerated to velocities up to 300 m/s [6]. As the droplets travel from the spray nozzle to the substrate they cool, slow, and begin to oxidize, all of which have a strong effect on the coating microstructure. The extent of these changes depends upon the distance between the nozzle and substrate, referred to as standoff distance (SOD), the atomizing gas pressure, and the atomizing gas composition. Finally, the spray gun is translated across the substrate surface to provide even coverage of the area to be coated or repaired.

Optimization of the TWA process relies on modification of five key process parameters: SOD, traverse speed, atomizing gas pressure, arc voltage, and arc current. Understanding the relationship between these parameters and the microstructure and mechanical properties of the coating is critical for designing the coating process. The mechanical properties of interest include the strength and hardness of the coating, and the bond strength at the interface with the substrate. Optimization of coating properties remains an area of active research, and further development of test methods to evaluate the composite substrate-coating system has been identified to be a "critical aspect" of leveraging all thermal spray methods [7].

A wide selection of prior works have investigated the mechanisms of the formation of TWA coatings, and the relationship between these mechanisms and the resultant microstructure [6, 8, 9]. TWA coatings consist of successive impacts of molten droplets upon the substrate surface, often referred to as "splats." During each splat, the droplet flattens and splashes outwards. The proportion of these two mechanisms relative to one another is governed by the droplet size, temperature, and velocity at impact, and it has important implications for the microstructure of the coating [10]. For example, a detailed investigation of the relationship between spray parameters, spray droplet morphology, and resultant coating microstructure is given by Johnston, *et al.* which concludes that higher spray pressure leads to smaller droplets and therefore reduced porosity, at the expense of reduced deposition efficiency [11].

Various prior works on mechanical properties of TWA coatings have investigated hardness and adhesion strength, and the role of oxide content and porosity. For example, a comprehensive study by Sampath, *et al.* reported on the microstructural and mechanical properties of Ni-5Al TWA coatings, including investigation of elastic modulus by bend testing [12]. A parametric study by Jandin, *et al.* related the coating process parameters to coating hardness and microstructure, finding that hardness and oxide content of steel coatings both increase when compressed air is used as the high pressure accelerating gas, but decrease when nitrogen is used [13]. A similar study by Gedzevicus and Valiulis produced results consistent with Jandin, *et al.*, with the addition of an investigation of adhesion strength which reported strengths in the range of 60 MPa [6]. Another parametric study by Fitriyana, *et al.* measured the adhesive strength and hardness of iron-chrome coatings on stainless steel substrates. The key findings of that study were that the adhesive strength was strongly correlated to SOD, with stronger coating bonds generally being formed at smaller SODs and lower air pressure. The increased adhesion strength at lower SODs are attributed to reduction in residual stresses. This stems from the finding that lower SOD produces thinner coating layers, with thinner coating layers exhibiting lower residual stresses associated with thermal expansion effects [8].

Studies of other properties such as the in-plane strength and elastic modulus of coatings are rare. This is likely because the measurement of these properties is complicated by the maximum achievable TWA coating thickness. TWA coating thickness is limited by residual stresses caused by thermal expansion mismatch between the coating and substrate. Coatings generally cannot exceed approximately 6 mm in thickness without active cooling methods to mitigate the residual stresses. Measurement of in-plane strength and elastic modulus typically cannot be performed directly, but rather rely on measurements performed on a substrate-coating bilayer. Numerical analysis must then be performed to differentiate the contribution of each layer of the system. Jandin, *et al.* provides one example of this, using a coated cantilever beam method to measure the elastic modulus of a 1080 steel coating to be 100±10 GPa [13]. You, *et al.* used a three-point bending methodology to measure the elastic modulus of a tungsten based coating. The mathematical model used by You, *et al.* allows direct calculation of the elastic modulus using the load-displacement curve from three-point bending [14]. That study found the elastic modulus of the tungsten thermal spray coatings to be 14% of bulk tungsten, indicating that the thermal spray microstructure has a potentially significant effect on coating modulus. The elastic modulus has particular importance for the design of thermally sprayed components because the ratio of substrate modulus to coating modulus determines the distribution of stresses in the system.

The wear performance is another important field of study when seeking to understand the performance of a coating in a broader system. Rodriguez, *et al.* note that understanding the relationships between coating microstructure and wear resistance is critical when designing a coating process [15]. Their study considered the abrasive wear properties of 420 stainless steel coatings, finding a combination of both rolling and sliding wear mechanisms. A similar study by Cooke, *et al.* investigated TWA coatings when subjected to abrasive wear and found that the wear results were "contrary to the general law that wear decreases with increase in hardness" due to overhanging brittle particles that break off during testing [16].

This paper reports results from a Design of Experiments on the study of the impact of standoff distance, air pressure, traverse speed, and arc current on TWA coatings of 1080 carbon steel on gray cast iron substrates. The effects of these parameters on the microstructure, hardness, and adhesive strength of the coatings are presented and discussed. Additionally, bend testing was performed on selected samples to determine the in-plane strength and elastic modulus, and the failure modes were assessed by metallographic methods. Abrasive and sliding wear tests were also performed on selected samples to compare the coating wear resistance to that of the substrate material. These materials were selected because of their applicability to large cast iron components such as cylinder heads and housings, but the results are broadly applicable to a wide range of heavy industrial applications.

42.2   Main Content of Chapter

42.2.1   Materials

Pearlitic Class 40 gray cast iron, provided by Dura-Bar under the designation G2, was coated with AISI 1080 carbon steel provided by Polymet under the tradename PMET 714. The nominal compositions of the alloys is specified in ASTM A48 [17] and A830 [18]. Round test coupons with a diameter of 54 mm and thickness of 13 mm were used as substrates for adhesion, hardness, and wear testing. Rectangular bars with a width of 30 mm, length of 125 mm, and thickness 9.5 mm, were used as substrates for 3-point bend testing.

All test coupons were saw cut from bar stock and milled flat prior to coating. The edges of the samples were filed to remove sharp edges. The diameter of the 1080 steel wire was 1.6 mm.

42.2.2   Coating Procedure

Prior to coating, substrate surfaces were grit-blasted with 36 grit Al_2O_3 to achieve a roughness of at least 6 μm Ra, as measured with a Starrett SR-100 portable profilometer. Dust from the grit blasting was blown off using compressed air, and the surfaces were wiped with isopropyl alcohol to remove any residual cutting fluid or other contaminants. Coating was performed within one hour after grit blasting to minimize the formation of an oxide layer on the surface [19]. All coatings were applied with a Thermach AT-400 TWA system using compressed air as the accelerating gas. This system allows for control of five spray parameters: arc voltage, arc current (which also dictates wire feed rate), atomizing gas pressure, SOD, and traverse speed. In this study, the arc voltage was held constant at 30 V, while the other parameters were varied per the 2^{4-1} fractional factorial design of experiments (DOE), Table 42.1. The DOE defines a total of nine parameter groups, where the midpoint parameters, Group 9, represent the manufacturer-recommended spray parameters.

In order to maintain consistent SOD and traverse rate, the spray gun was attached to an ABB IRB-1600 industrial robot. Because both the arc current and traverse speed impact the layer thickness, the number of layers deposited on each substrate was adjusted, based on the process parameters, to yield a consistent coating thickness of approximately 1.2 mm. The number of layers required to achieve the target coating thickness is given in Table 42.1 for each parameter set. Each consecutive layer was deposited immediately following the preceding layer. In order to better ascertain the contribution of the coating during 3-point bend testing, the bending samples were prepared with slightly thicker coatings of approximately 1.5 mm. This was achieved by increasing the number of spray layers by 25% from the values indicated in Table 42.1.

Table 42.1 2^{4-1} Fractional factorial experimental design.

Group	Pressure [kPa]	SOD [mm]	Arc current [A]	Traverse speed [mm/s]	Spray layers
1	276	75	75	250	26
2	276	75	125	100	7
3	276	150	75	100	11
4	276	150	125	250	18
5	483	75	75	250	26
6	483	75	125	100	7
7	483	150	75	100	11
8	483	150	125	250	18
9	379	115	100	175	14

42.2.3   Test Methods

Adhesion testing was performed in accordance with ASTM D4541 [20] using a Defelsko PosiTest device to measure the load to failure. To improve bonding of the epoxy to the coating, the top surface of the coatings was sanded to 220 grit to remove the majority of the roughness. A 4 mm wide channel was milled around the test sites to control the failure area. The test dollies were bonded in place using Solvay FM1000 epoxy. Each coupon was sized for two adhesion tests, and two coupons were fabricated at each test condition. After testing, each dolly was inspected to determine if the failure occurred at the coating/substrate interface (adhesive failure), within the coating (cohesive failure) or at the epoxy/dolly interface (epoxy failure). The diameter of the pulled-off coating was also measured in order to accurately calculate the stress at failure.

Samples from each parameter group were cross-sectioned, mounted, and polished for metallograhic inspection. The samples were ground with 220 and 600 grit SiC paper, then polished with 9, 3, and 1 μm diamond suspensions using a Buehler AutoMet 250 polishing system. Microstructures were imaged using light optical microscopy (LOM). Porosity was calculated by image analysis of the micrographs following ASTM E2109 Test Method B [21]. For the calculation of average porosity, eight micrographs were evaluated for each parameter group. Quantitative measurements of oxide content were not pursued, but oxide content was evaluated qualitatively. Coating hardness was evaluated using both Vickers and Rockwell B methods. Vickers micro-indentation hardness testing was performed on the metallographic cross-sections at three depths: near the substrate/coating interface, in the middle of the coating, and near the surface. Vickers testing used a 200 g load and was performed on a LECO LM-248 AT micro-hardness tester. Each specimen was tested a total of 24 times, eight tests at each depth. Rockwell B hardness was measured normal to the coating on the sanded top surfaces, and followed testing parameters from ASTM E18 [22]. Each specimen was tested five times, yielding a total of ten Rockwell B hardness measurements per spray condition.

Bend testing was performed using a 3-point bend fixture on an Instron universal test system equipped with an Instron 2527 load cell. The test geometry placed the coatings in tension and the uncoated side of the sample in compression. Bend tests were load controlled with a constant load rate of 500 N/min. In order to validate the strain measurements, and to control for local deformation at the supports, selected samples were instrumented with a strain gauge on the coating surface. After the coatings were fractured, the samples were sectioned and their fracture surfaces were imaged. A loop abrasion test in accordance with ASTM G174 Option C [23] was performed on selected parameter groups. Wear specimens of 32 mm x 8 mm x 4 mm were cut from larger test coupons using wire EDM. The coated face was then polished to approximately 0.2 μm Ra. The abrasive media was a 30 μm aluminum oxide adhered to a polyester tape. The applied load was 100 g and the test was run for a total of 75 belt passes at 100 RPM. Four replicate samples were tested of the uncoated substrate and spray conditions 4, 5, and 9, leading to a total of 16 tests. This test generates a semi-cylindrical wear scar with a radius equal to the radius of the driven shaft. Calculation of the wear volume is possible by measuring only the scar width, the sample length (equal to the length of the scar), and the shaft radius. The wear volume is calculated by:

$$V_{scar} = \frac{D^2 t}{8}[2 \sin^{-1}\frac{(b)}{D} - \sin(2 \sin^{-1}(\frac{b}{D}))] \tag{42.1}$$

where D is the shaft diameter (17 mm), t is the sample length (8 mm), and b is the scar width measured using LOM [23]. Because wear resistance is inversely related to the scar volume, this test allows a direct comparison between the abrasive wear resistance of the uncoated and coated samples.

Sliding wear testing was performed on a custom reciprocating wear fixture. The wear fixture utilized a horizontal dowel pin as the mating body in order to produce a linear contact that is more representative of several industrial applications than the high Hertzian stresses that can occur in the standard ball on plate testing. The wear test was guided by ASTM G133 [24], although the use of a pin rather than a ball is nonstandard. The pin was 4140 steel, hardened to 50 HRC, and had a diameter of 3.2 mm and a length of 5.5 mm. The pin was loaded with 170 N, lubricated with 15W40 oil, and held stationary while the sample was reciprocated with a stroke length of 9 mm at a frequency of approximately 2 Hz. The coated samples were polished to a roughness of approximately 0.2 μm prior to testing. The wear scars were evaluated by measuring the depth of the wear scar using a Starrett SR-400 profilometer and integrating the depth curve over the length of the scar. Testing was performed on three replicate samples from two spray groups, and the uncoated substrate, for a total of nine wear tests.

42.3   Results

42.3.1   Adhesion and Hardness Testing

The results of the adhesion tests are given in Figure 42.1. All of the failures were adhesive in nature. Note that the data in Figure 42.1 represent the average value for each group while the error bars represent the range of observed values. Group 4 produced coatings with the highest average adhesion strength and relatively low scatter in the data. The manufacturer recommended settings, Group 9, produced coatings with intermediate adhesion strength.

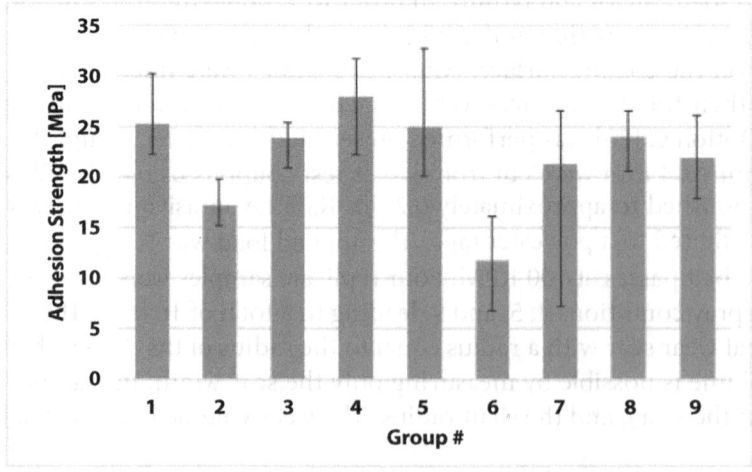

Figure 42.1 Adhesion results by spray parameter group. Error bars represent the minimum and maximum observed values.

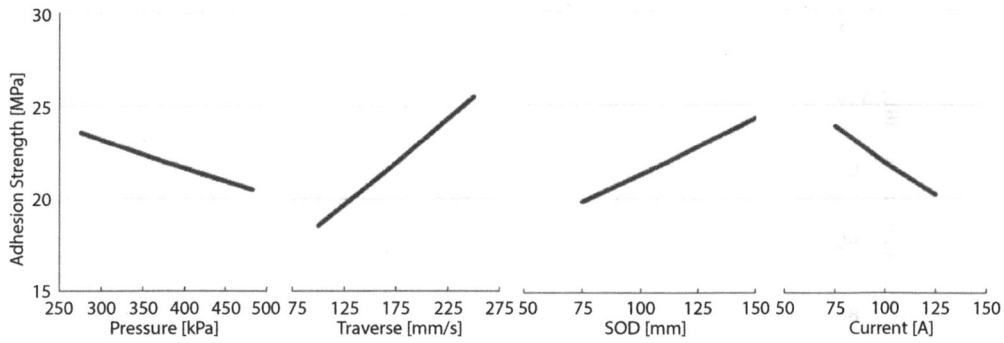

Figure 42.2 Adhesion main effects plot for 2^{4-1} experiment.

The main effects plot showing the impact of each parameter, Figure 42.2, indicates that traverse speed is the parameter with the greatest impact on adhesion strength, with faster traverse speed linked to stronger coatings. Additionally, the directionality of the impact of arc current (which is tied to wire feed rate) on adhesion also indicates that a coating made up of many thin layers will have greater adhesive strength than one with few thick layers. SOD is also correlated to adhesion strength, with larger SOD leading to improved adhesive strength. Two key trends may be found from these main effects plots. First, parameters which decrease the oxide content of the coating lead to higher adhesive strengths. Second, parameters which reduce layer thickness, leading to a coating constituted of many thin layers, lead to higher adhesion strengths.

The trend linking lower oxide content to improved adhesion strength concerns the accelerating gas pressure. Higher gas pressure produces finer spray droplets with a greater surface area than larger droplets produced by lower gas pressures. When the accelerating gas is air, this additional surface area allows for more oxidation, ultimately leading to higher oxide content in the coating. The oxide is far more brittle than the other constituents in the coating, and the presence of large amounts of oxide at the coating/substrate interface appears to detrimentally affect adhesive strength.

There is a trend linking thinner spray layers to improved adhesion for both the arc current and traverse speed parameters. These parameters are both directly related to layer thickness, with faster traverse speeds and lower arc currents (and therefore, material feed rates) producing thinner layers. These relationships, combined with Figure 42.2, indicate that coatings with many thin layers will generally have better adhesion strength than coatings with a few thick layers. This effect was also noted by Fitriyana, *et al.* who link this effect to a reduction in residual stresses within the coating [8]. This reduction in residual stress with thinner coating layers is attributed to thermal effects, since deposition of thinner coating layers imparts less heat into the system [25].

Micro-indentation hardness testing revealed relatively small differences in coating hardness across the parameter groups. The range of measured Vickers hardness values was approximately 320 – 360 HV, all slightly harder than bulk 1080 steel, which was measured to have Vickers hardness of 309. Rockwell B hardness measurements, see Figure 42.3, revealed larger differences which were analyzed using a main effects plot. The coatings measured between 90 and 100 HRB, which is comparable to bulk 1080 steel which was measured to have Rockwell B hardness of 99. The main effects plot, Figure 42.4, indicated that air

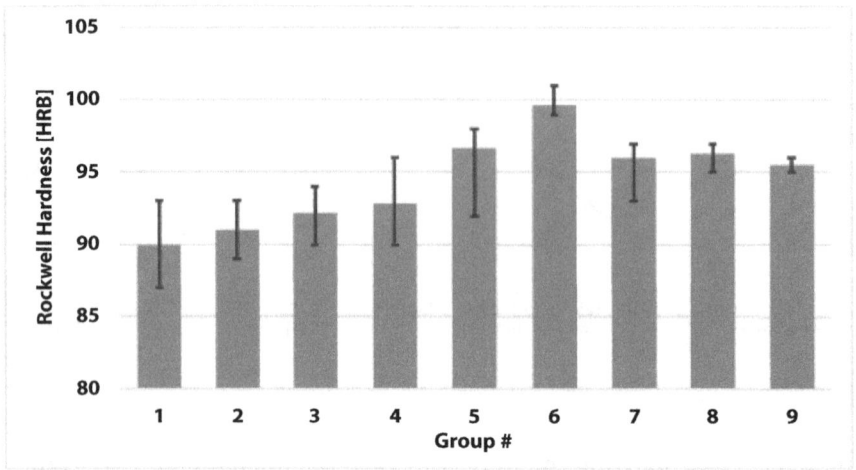

Figure 42.3 Rockwell B hardness measurements. Error bars represent the standard deviation of the measurements.

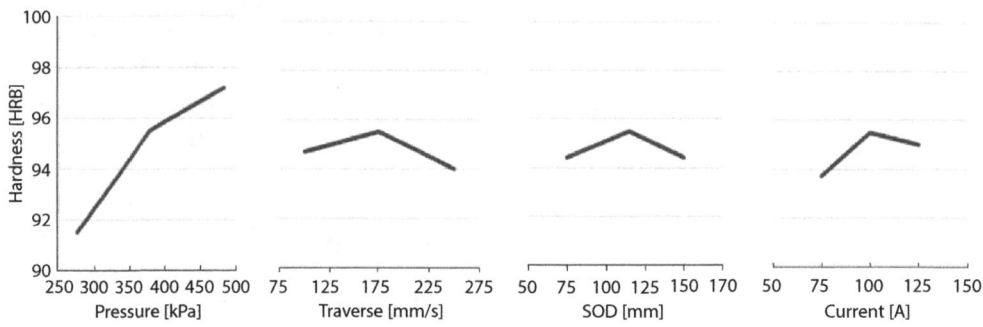

Figure 42.4 Rockwell B hardness main effects plot.

pressure was the most influential parameter for hardness. This result was expected based on the discussion above where increased air pressure is linked to the formation of smaller molten droplets which are more susceptible to oxidation. The oxide phase is significantly harder than the metallic matrix, leading to higher hardness in coatings with a high oxide content. The effects of the other three parameters studied: current, SOD, and traverse, were small.

42.3.2 Metallography

Specimens were cross-sectioned, mounted, polished, and imaged to evaluate the microstructural differences caused by the differing spray parameters. Porosity, voids, oxides, and pre-solidified droplets (PSD's) were all features of interest. Microstructures differed significantly among the spray groups, ranging from: Group 9 with few defects, Group 4 with significant large porosity, and Group 5 with large oxide colonies. In addition, interface quality varied from Group 4, with the fewest interfacial defects, to Group 6 with the most.

Figure 42.5 shows the microstructure from Group 9 (midpoint parameters), with a uniformly distributed layered structure of metallic alloy (light phase) and oxide (gray phase),

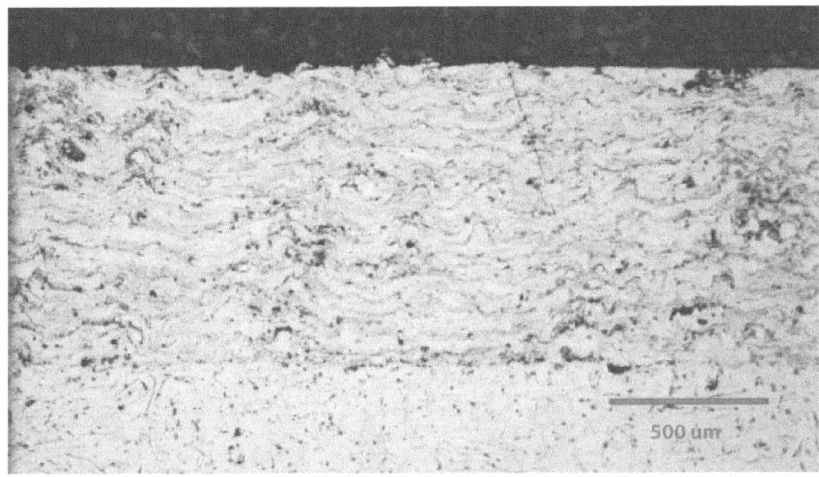

Figure 42.5 Microstructure of sample 9-1.

and intermittent isolated residual porosity (black phase). This microstructure shows no major defects and is typical of a desirable TWA microstructure. Small colonies of oxide are present, but do not cascade through the entire coating thickness. Because Group 9 used manufacturer recommended coating parameters, the microstructure of Group 9 was used as a baseline against which to compare the other coating microstructures.

Shown in Figure 42.6 is the microstructure from Group 4 (low pressure, high traverse, high SOD, high current) which is noteworthy for having both an elevated proportion of large pores and the highest adhesive strength of any group. The coating interface shows almost no defects, which is expected to be a factor in providing a strong adhesive bond. The large pores are explained by the combination of low pressure and high SOD. Low accelerating gas pressure leads to larger spray droplets, combined with lower impact velocity, which in turn leads to reduced flattening of the droplets at impact and therefore increased porosity. While oxide is still present in the coating, along with many pores, the high adhesive

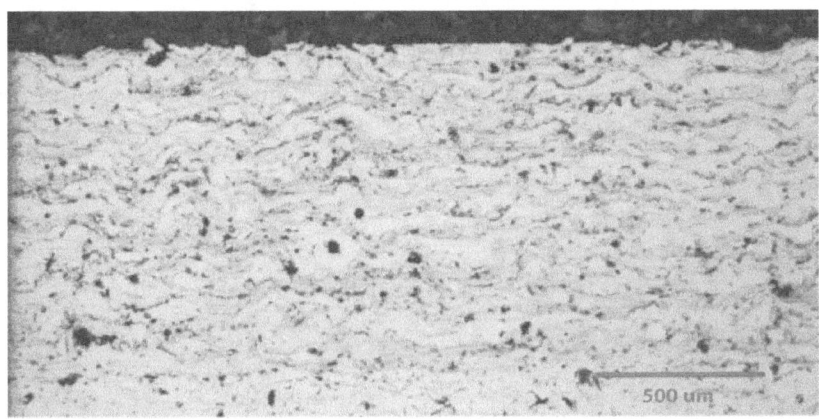

Figure 42.6 Microstructure from sample 4-2.

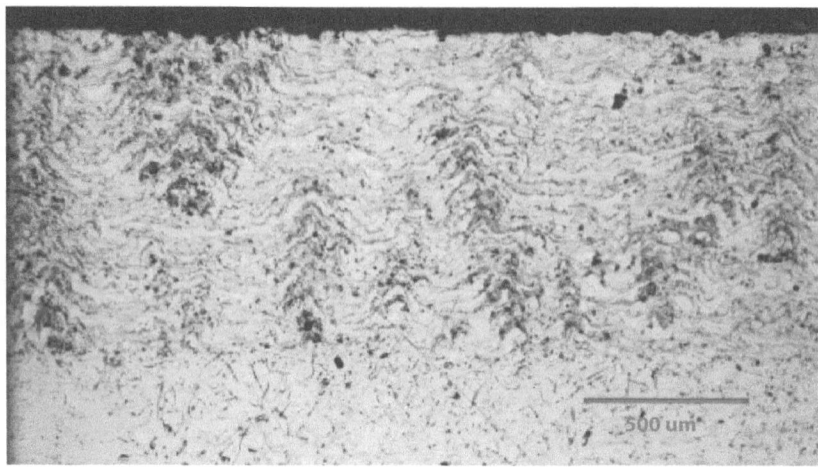

Figure 42.7 Microstructure from sample 5-2.

strength leads to the conclusion that strength is more closely tied to the microstructure at the interface rather than the microstructure elsewhere in the coating.

Figure 42.7 shows a microstructure from Group 5 (high pressure, high traverse, low SOD, low current) with an extremely high oxide content. A cascade effect of oxide formation shows oxide colonies spanning many spray layers. The porosity content of Group 5 is lower than the other groups. As discussed above, coatings created at high pressure have greater oxide content due to the larger surface area of smaller spray droplets. The high spray pressure, combined with the low SOD, also implies high impact velocity of the spray droplets, leading to improved compaction of the droplets and overall low porosity. Figure 42.8 shows higher-magnification images of the microstructures from the same samples shown in Figure 42.5-Figure 42.7. Figure 42.8a shows a representative section of the Group 9 microstructure, with uniformly distributed oxide, some small colonies of oxide, and intermittent porosity. Figure 42.8b highlights the high porosity and low oxide contents of Group 4. Figure 42.8c shows a large oxide concentration in Group 5. These large oxide-rich features, which extend across many spray layers, are attributed to the entrapment of liquid oxide splatter by the surface perturbation caused by the oxide content of the previous layer.

Porosity measurements, see Figure 42.9, indicate that all of the coatings have relatively low porosity, ranging from 0.7 to 2.6%. The main effects plot for the porosity, see Figure 42.10, concludes that air pressure and SOD have significant impacts on porosity, while

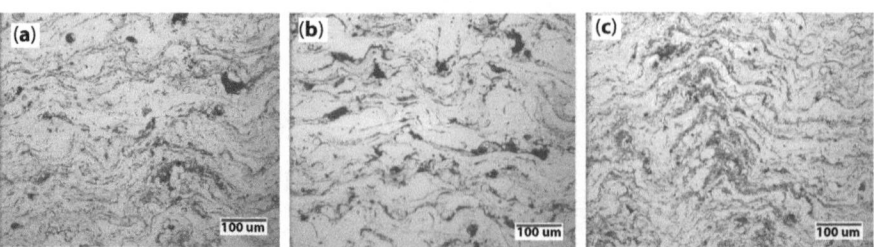

Figure 42.8 200x microstructure from (a) Group 9, (b) Group 4, (c) Group 5.

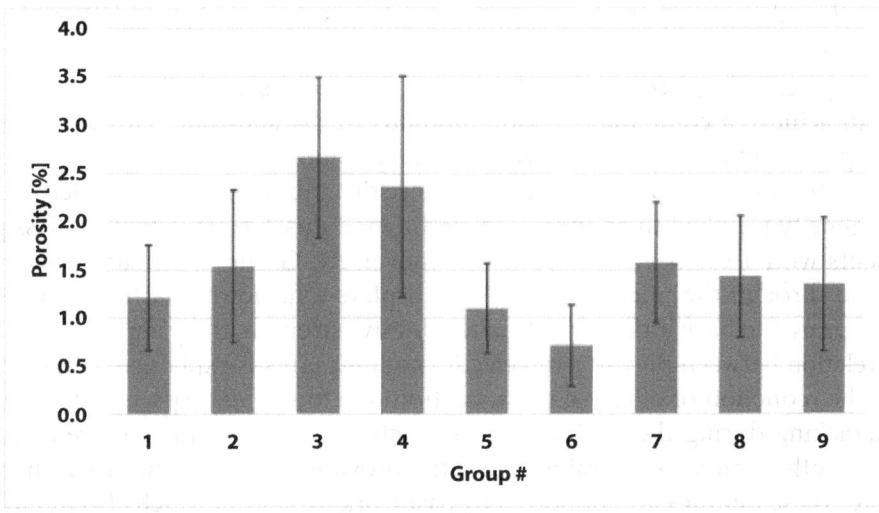

Figure 42.9 Coating porosity measurements. Error bars represent standard deviation.

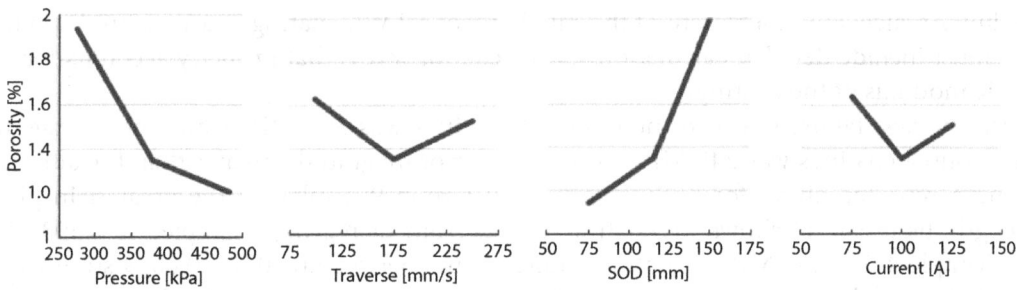

Figure 42.10 Main effects plot of porosity measurements.

the impacts of traverse speed and current are small within the range of conditions tested. Increased air pressure has the effect of reducing porosity, which is attributed to both the formation of smaller droplets and increased flattening of droplets on impact. The reduction in porosity associated with high air pressure is coupled with the increase in oxide content exemplified by Figure 42.7. Similarly, the effect of SOD is believed to be caused by increased flattening of particles at lower SOD, due to the relatively higher particle velocities at impact.

42.3.3 Bend Testing

Bend testing was used to calculate the elastic modulus, ultimate strength, and elongation at break of the coatings. Similar methodologies have been previously applied to calculate the elastic moduli of thermal spray coatings [14], [26], [27]. For calculation of the elastic modulus, the bending strains were limited to 0.002. The samples were comprised of substrates 9.5 mm thick that were coated using two parameter sets with high adhesion, Groups 4 and 9. These parameter sets were chosen to compare coatings known to have good adhesive properties, but possessing different microstructures. Uncoated substrates were tested first to confirm the substrate elastic modulus, E_a, which was found to be 112 GPa and in agreement with published values.

The average calculated coating modulus for Group 9 was 145 GPa, indicating that the coating can achieve a comparable elastic modulus to the substrate. However, the results from Group 4 are significantly lower, with an average modulus of only 67 GPa. Comparison of the microstructures discussed above and the elastic moduli indicate that increased material homogeneity is linked to an increased elastic modulus. Group 9 contains less porosity and presents with a stiffer elastic modulus. Conversely, Group 4 contains more porosity interspersed through the microstructure and displays a far lower elastic modulus. While porosity appeared to have little correlation to adhesive strength, there appears to be a significant correlation between porosity and elastic modulus. This is explained by the combined effects of the reduction of effective cross- section due to the void space and the initiation of microcracking during the testing. The latter effect becomes more severe as entrained porosity and other microstructural irregularities provide stress concentrators where cracks can initiate. These findings are supported by past work by Jandin, which also found that the elastic modulus increases with increased material homogeneity [13].

These are important findings with critical design implications because the distribution of stress between the coating and substrate is dependent on the ratio of the respective elastic moduli. As such, any assessment of the suitability of a TWA coating in a structural application must include detailed calculations of the coating stress that properly accounts for the elastic modulus of the coating.

The average failure stress in the Group 9 coatings was 396 MPa while the average for the Group 4 coatings was 245 MPa, both an order of magnitude greater than the adhesive strength reported above for both groups, with Group 9 displaying the greatest in-plane strength. Figure 42.11 shows stress-strain curves from each group (Sample 9-2 = 393 MPa and Sample 4-2 = 261 MPa). Coating failure occurs between 0.0031 - 0.0039 strain, with final failure of the substrate around 0.0058 strain. Cross-sections of the fracture surfaces from these samples are presented in Figure 42.12, where it can be seen that the microstructure around the failure in Sample 4-2 contains a comparatively low fraction of oxide, coarse layers of alloy, and minimal secondary cracking. The secondary cracking that is present

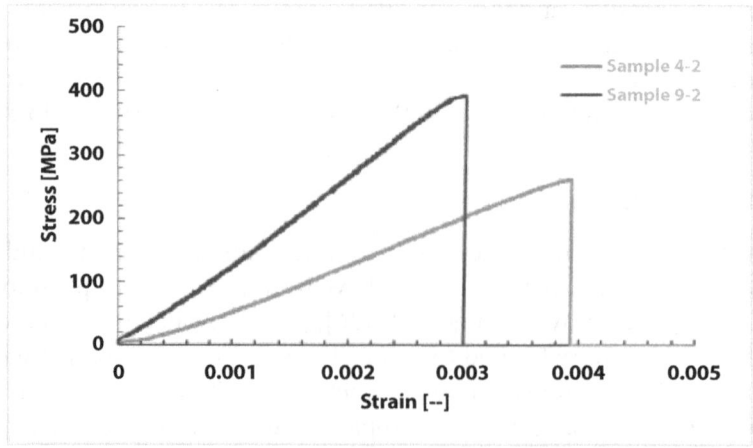

Figure 42.11 Calculated maximum stress/strain in the coatings from samples 4-2 and 9-2.

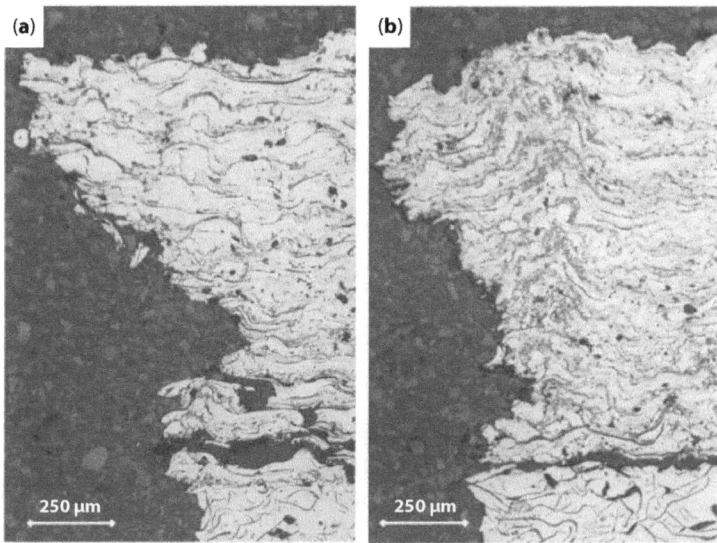

Figure 42.12 Metallurgical cross-sections of the fracture surfaces from (a) group 4 and (b) group 9.

near the interface is generally limited to in-plane (parallel to the interface). In Sample 9-2, the failure occurred at, or adjacent to, a columnar feature of high oxide fraction. In addition, the overall microstructure is significantly finer, with a higher fraction of

oxide, and there is significant secondary cracking near the interface, some of which is oriented vertically and follows the contours of the oxide layers. The conclusion to be drawn is that the oxide layers, especially when oriented vertically, make the coating significantly more brittle, decreasing the in-plane strain to failure.

It should be noted that once the coating fails, the maximum stress in the substrate abruptly increases because the fractured coating no longer bares any load. The significance of this effect depends on the ratio of coating to substrate thickness and is more pronounced in systems where the coating consists of a large portion of the overall system thickness. In this case, the ultimate strength of the substrate was not significantly higher than that of the coating, leading to the failure of the substrate shortly after failure of the coating. This is another critical aspect to consider when designing a coated system to ensure that coating failure does not cascade into further failure of the substrate material.

42.3.4 Wear

Results from the sliding and abrasive wear tests are plotted in Figure 42.13. The uncoated cast iron performed consistently better than the steel coatings in both the abrasive and sliding wear tests. In abrasive wear, Groups 4 and 9 wore approximately 20% and 30% more than the base material, respectively. In sliding wear, Groups 4 and 9 wore approximately 55% and 20% more than the base material, respectively. As the wear resistance is inversely related to the wear volume, this result indicates that both coatings are more susceptible to both types of wear relative to the base material.

Analysis of the wear scars suggests that the two coating groups behaved substantially differently in both types of wear test. The abrasive wear scars in Figure 42.14 show some

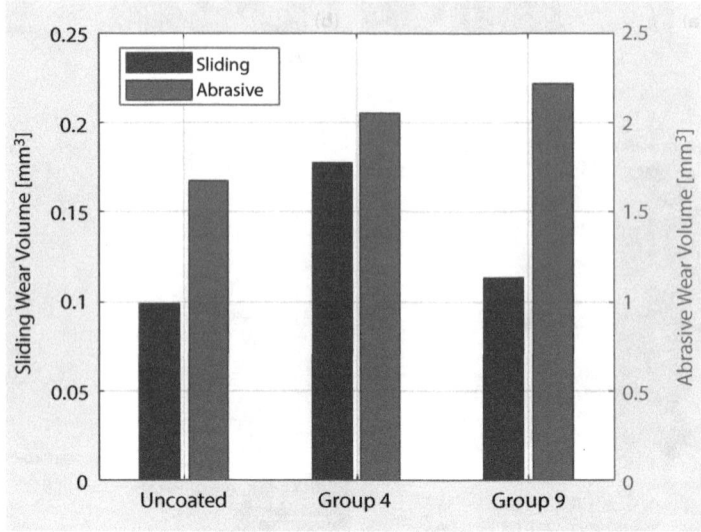

Figure 42.13 Average wear scar volume results from ASTM G174 abrasive and sliding wear tests.

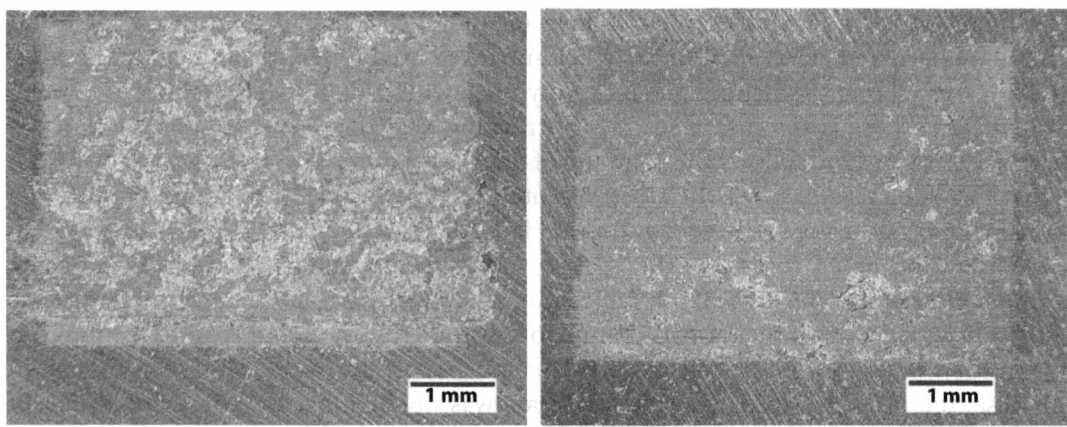

Figure 42.14 Sliding wear scar example. Group 4 (left) and group 9 (right).

material pullout in Group 4, but not in Group 9. This is attributed to the higher porosity content of Group 4. As the wear begins, the higher proportions of soft ferrite and porosity in Group 4 will allow the coating material to reorient

into a high hardness configuration. This is not possible in Group 9, where the lower amount of porosity and higher amount of oxide will not allow significant relocation and reorientation of coating particles. Additionally, the increased oxide content is likely to cause increased incidence of particles breaking off. These phenomena lead to less overall abrasive wear in Group 4, despite the lower overall hardness. This result is consistent with previous findings that harder TWA coatings tend to wear more in abrasive applications [16]. The base cast iron exhibited no evidence of material pullout and lacks oxide particles, leading to superior performance than either Group 4 or Group 9.

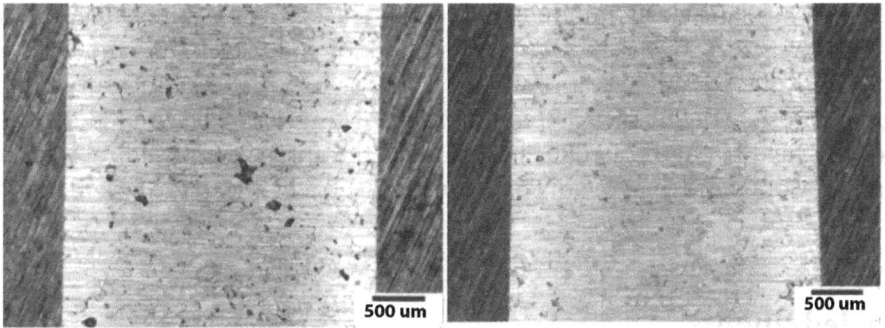

Figure 42.15 Abrasive wear scar images, group 4 (left) and group 9 (right).

The sliding wear scars also indicate that microstructural differences between the groups significantly impacted the wear mechanism. Figure 42.15 shows large differences between Group 4 and Group 9, where Group 4 displays large pits not present in Group 9. The cause of this is the reciprocating action of the sliding wear test used. As discussed above, the microstructure of Group 4 allows for more movement of hard particles within the coating when loaded. Contrary to abrasive wear, where this effect was beneficial, the reciprocating nature of the test causes material to work harden during a stroke in one direction, then to fracture during a subsequent stroke in the opposite direction. This leads to the large amount of pull-out clear in Group 4. Additionally, the vertical crack-like features present in Group 4 are believed to be locations in which the material has begun the process of relocating and work hardening, but has yet to fracture. The lower porosity and higher oxide present in Group 9 leads to a different mechanism, in which the mating dowel pin will wear away ferrite until the dowel is predominantly supported by hard oxide particles, at which point wear will continue slowly. This is evidenced by the presence of wear grooves in Figure 42.15. The base cast iron exhibited none of the pitting seen in the coatings, indicating no material pullout, leading to an overall higher wear resistance.

42.4 Conclusion

The results presented above indicate that the mechanical properties of TWA coatings of 1080 steel on cast iron substrates are significantly affected by the spray parameters: air pressure, SOD, traverse speed, and arc current. Increased air pressure leads to reduced coating porosity and increased coating hardness, but with the tradeoff of reduced adhesion. Increased traverse speed leads to stronger adhesion and slightly increased hardness, without any discernable tradeoff within the range of values studied. Increased SOD leads to increased adhesion for the distances applied, although it also leads to a large increase in porosity and decrease in hardness. Arc current was the least influential parameter and followed the same trends as modification to traverse speed, as both primarily influence the layer thickness. Coatings with preferential spray parameters were able to achieve adhesion strengths up to 30 MPa, with average values of approximately 25 MPa. Further study of the mechanical properties of a subset of the sample groups provided further information on the bending behavior and wear properties of the coatings. Bend testing found that the coatings can achieve an in-plane elastic modulus up to 145 GPa, which is stiffer than the substrate and approximately 73% of the modulus of the bulk of 1080 steel. This modulus is highly dependent on spray parameters, and values as low as 67 GPa were measured in coatings with higher porosity. The in-plane strength of the coatings was determined to be 396 MPa or 68% of the yield strength of 1080 steel. In both the abrasive and sliding wear tests, the TWA samples wore more than the uncoated substrate, likely due to the lack of homogeneity in the coating.

Acknowledgments

The authors would like to thank Domenic Maiola, Jason Purvee, and Mitch Campbell for their important contributions to this research. This material is based upon work supported by the Office of Naval Research under Award No. N00014-18-1-2339. Disclaimer - Any opinions, findings, and conclusions or recommendations expressed in this material are those of the author(s) and do not necessarily reflect the views of the Office of Naval Research. © 2023 Rochester Institute of Technology.

References

1. D. Poirier *et al.*, Performance Assessment of Protective Thermal Spray Coatings for Lightweight Al Brake Rotor Disks, *Journal of Thermal Spray Technology*, vol. 28, pp. 291–304, 2019.
2. E. A. Esfahani, H. Salimijazi, M. Golozar, J. Mostaghimi, and L. Pershin, Study of Corrosion Behavior of Arc Spraying Aluminum Coating on Mild Steel, *Journal of Thermal Spray Technology*, vol. 21, no. 6, pp. 1195–1202, 2012.
3. G. Burkle, H. J. Fecht, A. Sagel, and C. Wanke, Dynamic mechanical analysis of the mechanical properties of Al- and Fe- based thermal spray coatings, in *Thermal Spray: New Surfaces for a New Millenium*, C. C. Berndt, K. A. Khor, and E. F. Lugscheider, Eds. ASM International, 2001, pp. 999–1002.
4. J. Wang, J. Liu, L. Zhang, J. Sun, and Z. Wang, Microstructure and mechanical properties of twin-wire arc sprayed Ni-Al composite coatings on 6061-T6 aluminum alloy sheet, *International Journal of Minerals, Metallurgy and Materials*, vol. 21, no. 5, pp. 469–478, 2014.
5. T. C. Chen, C. C. Chou, T. Y. Yung, K. C. Tsai, and J. Y. Huang, Wear behavior of thermally sprayed Zn/15Al, Al and Inconel 625 coatings on carbon steel, *Surface and Coatings Technology*, vol. 303, pp. 78–85, 2016.
6. I. Gedzevicius and A. v. Valiulis, Analysis of wire arc spraying process variables on coatings properties, *Journal of Materials Processing Technology*, vol. 175, no. 1–3, pp. 206–211, 2006.
7. G. M. Smith and S. Sampath, Sustainability of Metal Structures via Spray-Clad Remanufacturing, *Jom*, vol. 70, no. 4, pp. 512–520, 2018.
8. D. F. Fitriyana *et al.*, The Effect of Compressed Air Pressure and Stand-off Distance on the Twin Wire Arc Spray (TWAS) Coating for Pump Impeller from AISI 304 Stainless Steel, in *NAC 2019*, 2020, pp. 119–130.
9. A. P. Newbery and P. S. Grant, Oxidation during electric arc spray forming of steel, *Journal of Materials Processing Technology*, vol. 178, no. 1–3, pp. 259–269, 2006.

10. M. P. Planche, H. Lioa, and C. Coddet, Relationships between in-flight particle characteristics and coating microstructure with a twin wire arc spray process and different working conditions, *Surface & Coatings Technology2*, vol. 182, pp. 215–226, 2004.

11. A. L. Johnston, A. C. Hall, and J. F. McCloskey, Effect of process inputs on coating properties in the twin-wire Arc zinc process, *Journal of Thermal Spray Technology*, vol. 22, no. 6, pp. 856–863, 2013.

12. S. Sampath, X. Y. Jiang, J. Matejicek, L. Prchlik, A. Kulkarni, and A. Vaidya, Role of thermal spray processing method on the microstructure, residual stress and properties of coatings: An integrated study of Ni-5 wt. % Al bond coats, *Materials Science and Engineering A*, vol. 364, no. 1–2, pp. 216–231, 2004.

13. G. Jandin, H. Liao, Z. Q. Feng, and C. Coddet, Correlations between operating conditions, microstructure and mechanical properties of twin wire arc sprayed steel coatings, *Materials Science and Engineering A*, vol. 349, no. 1–2, pp. 298–305, 2003.

14. J. H. You, T. Höschen, and S. Lindig, Determination of elastic modulus and residual stress of plasma-sprayed tungsten coating on steel substrate, *Journal of Nuclear Materials*, vol. 348, no. 1–2, pp. 94– 101, 2006.

15. E. Rodriguez, M. A. González, H. R. Monjardín, O. Jimenez, M. Flores, and J. Ibarra, Heat treated twin wire arc spray AISI 420 coatings under dry and wet abrasive wear, *Metals and Materials International*, vol. 23, no. 6, pp. 1121–1132, 2017.

16. K. Cooke, G. Oliver, V. Buchanan, and N. Palmer, Optimisation of the electric wire arc-spraying process for improved wear resistance of sugar mill roller shells, *Surface & Coatings Technology*, vol. 202, pp. 185–188, 2007.

17. ASTM, A-48 Standard Specification for Gray Iron Castings. ASTM International, West Conshohocken, PA, 2021.

18. ASTM, A-830 Standard Specification for Plates, Carbon Steel, Structural Quality, Furnished to Chemical Composition Requirements. ASTM International, West Conshohocken, PA, 2018.

19. F. N. Longo, Coating Processing, in *Handbook of Thermal Spray Technology*, J. R. Davis, Ed. ASM International, 2004.

20. ASTM, D-4541-17: Standard Test Method for Pull-Off Strength of Coatings Using Portable Adhesion Testers. ASTM International, West Conshohocken, PA, pp. 1–16, 2017.

21. ASTM, E-2109-01 Standard Test Methods for Determining Area Percentage Porosity in Thermal Sprayed Coatings, vol. 01, no. Reapproved. ASTM International, West Conshohocken, PA, 2014.

22. ASTM, E-18-20: Standard Test Methods for Rockwell Hardness of Metallic Materials. ASTM International, West Conshohocken, PA, pp. 1–38, 2020.

23. ASTM, G-174-17: Standard Test Method for Measuring Abrasion Resistance of Materials by Abrasive Loop Contact, vol. i, no. Reapproved. ASTM International, West Conshohocken, PA, pp. 1–7, 2017.

24. ASTM, G-133-16 Standard Test Method for Linearly Reciporcating Ball-on-Flat Sliding Wear, vol. 05. ASTM International, West Conshohocken, PA, pp. 1–10, 2016.

25. P. Araujo, D. Chicot, M. Staia, and J. Lesage, Residual stresses and adhesion of thermal spray coatings, *Surface Engineering*, vol. 21, no. 1, pp. 35–40, Feb. 2005.

26. C. Leither, J. Risan, M. Bashirzadeh, and F. Azarmi, Determination of the elastic modulus of wire arc sprayed alloy 625 using experimental, analytical, and numerical simulations, *Surface and Coatings Technology*, vol. 235, pp. 611–619, 2013.

27. K. DePalma, M. Walluk, L. P. Martin, and K. Sisak, Investigation of Mechanical Properties of Twin Wire Arc Repair of Cast Iron Components, *Journal of Thermal Spray Technology*, 2022.

43

Towards Development of Additive Manufacturing Material and Process Technologies to Improve the Re-Manufacturing Efficiency of Commercial Vehicle Tires

Yiqun Fu[1], Tadek Kosmal[2], Ren Bean[3], Robert Radulescu[4], Timothy E. Long[3] and Christopher B. Williams[5]*

[1]Macromolecules Innovation Institute, Virginia Tech, Blacksburg, Virginia, USA
[2]Department of Mechanical Engineering, Virginia Tech, Blacksburg, Virginia, USA
[3]Biodesign Center for Sustainable Macromolecular Materials and Manufacturing, Arizona State University, Tempe, AZ, USA
[4]Michelin North America, Greenville, SC, USA
[5]Macromolecules Innovation Institute & Department of Mechanical Engineering, Virginia Tech, Blacksburg, Virginia, USA

Abstract

In the US, tire retreading is applied mainly to commercial vehicle tires and involves about 14.5 million units for end-users in the transportation industry. During retreading, the incoming casings must be buffed to a specific contour to accommodate a new tread band, which leads to material loss of about 4kgs, or ~8% of the original tire weight. Furthermore, the resulting remanufactured tire has ~2kgs additional weight compared to a new tire, which increases the tire rolling resistance by ~0.1Kgs/T and contributes to increased vehicle energy consumption and CO_2 emissions during the tire usage cycle. This project seeks to improve the re-manufacturing efficiency of the tire retreading process through the concurrent development of additive manufacturing (AM) technologies and printable elastomers. The primary aim is to develop a novel AM technology and material such that tires can be directly retreaded to reduce by ~10% the use of primary feedstock to manufacture new tires without decreasing the tire longevity performance or increasing its re-manufacturing cost versus the conventional retread process.

In this paper, the authors present progress towards this aim through developments in two key project objectives: (1) advancement of 3D scanning technologies to automatically generate multi-axis robotic deposition toolpaths for conformal printing directly onto worn tires, and (2) development of a 3D printable rubber latex emulsion that provides exceptional elastomeric properties and can serve as a bonding material to ensure cohesion between the tire casing and new tire tread bands.

Corresponding author: cbwill@vt.edu

Nabil Nasr (ed.) Technology Innovation for the Circular Economy: Recycling, Remanufacturing, Design, Systems Analysis and Logistics, (573–584) © 2024 Scrivener Publishing LLC

Keywords: Additive manufacturing, elastomer, latex, tire retreading, 3D scanning

43.1 Introduction

According to the January 2020 edition of the Modern Tire Dealer Magazine [1], 14.5 million tires are retreaded each year in the US for end users in the transportation industry. This process is applied mainly to commercial vehicle tires and consists of replacing the tread of worn tires while preserving their casings and, in the process, saving about 70% of raw materials that are required to make new tires. During retreading however, the incoming casings must be buffed to a specific contour that can accommodate a new tread band that will cover the whole width of the tire (Figure 43.1: Traditional Process). Consequently, about 3 millimeters of unused tread and about 2 millimeters of under tread are removed as an average from each casing. This removal leads to a material loss of about 4 kgs, or about 8% of the original tire weight. Furthermore, the remanufactured tire ends up with about 2mm more under tread than the corresponding new tire since the bands that make up the new tread are held together by a thin strip of rubber about 2mm thick. This results not only in an additional material loss of ~2 kgs for each tire that is remanufactured, but also in an increase in tire rolling resistance of about 0.1 kgs/T that will increase the vehicle energy consumption and CO_2 emissions during the tire usage cycle. Finally, since the buffing process has about 0.5% scrap rate, additional material losses can be attributed to the traditional retreading process.

Solutions to the technical barriers described above can lead to an increase in the efficiency of the retreading process and hence a reduction in the use of primary feedstock, energy consumption and CO_2 emissions to manufacture retreaded tires. Additive manufacturing (AM) processes, also referred to as 3D printing, present an opportunity to achieve these efficiency improvements through the selective deposition of elastomeric material to retread partially worn tires. However, current AM processes neither (1) enable the selective deposition of material onto non-planar surfaces, nor can they (2) deposit rubber-type materials that can provide a good level of cohesion between the tire casing and a new tread band.

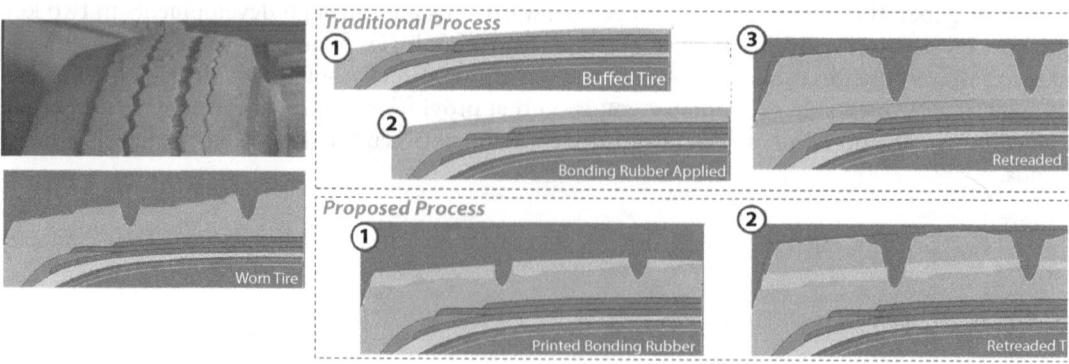

Figure 43.1 Schematic of traditional retreading process (above) and proposed changes (below).

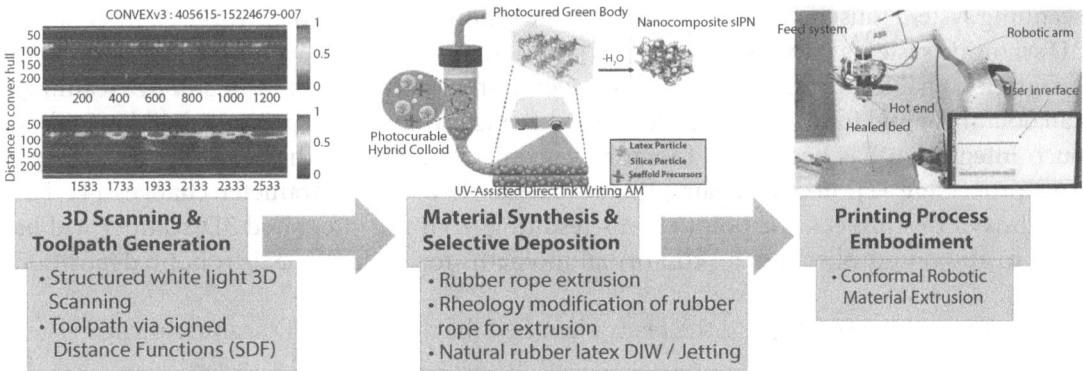

Figure 43.2 Three core project objectives and corresponding technology/material options to be evaluated.

To address these issues, the authors are developing a suite of new complementary technologies that can lay bonding rubbers on uneven non-axisymmetric surfaces through an AM process to create even surfaces:

1. A *3D scanning system* capable of identifying regions in need of repair and of generating conformal AM toolpaths for selective repair;
2. A *3D-printable elastomeric material system* capable of selective material extrusion and of meeting the use specifications of bonding tire bands onto commercial vehicle tire carcasses;
3. A *multi-axis robotic AM extrusion system* capable of selectively depositing elastomeric materials conformally onto the circumference of a worn tire.

Through the concurrent development of these three technologies (Figure 43.2), bonding rubber will be able to be selectively dispensed onto a tire carcass to create the surface on which individual strips of tread will be added, before the tire is cured using a traditional cold-retreading process (Figure 43.1: Proposed Changes). It is estimated that this advancement in tire repair technologies will reduce the consumption in primary feedstock to manufacture new tires by 10% without decreasing the tire longevity performance or increasing its re-manufacturing cost, versus the conventional retread process. Additionally, a 0.1 kgs/T improvement in tire rolling resistance is expected when using the new re-manufacturing process.

The overall aim of this manuscript is to present advances towards two of these integral technologies: evaluation and development of a high-resolution 3D scanning system capable of imaging a worn tire (Section 43.2) and the development of a suitable 3D-printable elastomeric material (Section 43.3).

43.2   3D Scanning of Worn Tires

Enabling selective printing of a bonding rubber onto a worn tire requires (i) a digital three-dimensional representation of the tire's surface topography from which (ii) a conformal printing toolpath can be generated. Thus, to achieve the overall project aims, a 3D scanning system must be first integrated into the robotic AM tire repair workcell. Integrating 3D scanning systems into AM processes is an emerging field of research, as such technology integration enables documentation of the evolution of part geometry. *In-situ* 3D scanning can document and inform responsive process changes that amend previous build mistakes. Such integration has been used to address geometric errors in the printed part through post-processing [2], *in-situ* repair [3], and on-the-fly process parameter changes [4]. For the aims of this project, the point clouds resulting from the integrated 3D scanner will be used to autonomously generate conformal toolpaths for the multi-axis robotic deposition system.

To achieve the project goals, the team identified that a suitable 3D scanner must:

- be able to quickly scan the worn tire surface, and thus provide a sufficient field of view (FOV) in each scan (ideally the width of the tire carcass);
- provide accurate and precise scan data sufficient for identifying relevant tire tread features (< 0.5 mm);
- and scan the tire in an autonomous fashion, and thus automatically conduct and patch separate scans while rotating the tire into the scanner's FOV.

43.2.1   Structured Light Scanning

From this list of requirements, the team identified Structured Light Scanning (SLS) as an ideal technology candidate. Compared to other modes of 3D scanning, SLS offers fast, passive sensing ability with high resolution (~0.5mm) within a suitable scanning area (0.5m x0.5m), which matches a tire's defect and overall size, respectively.

With SLS, a series of fringe patterns can be projected (encoded) onto the tire carcass and imaged using a stereo- camera pair. The projected fringe patterns display spatial correspondence between the two camera's image sets. With spatial correspondence documented, matching pixels between two camera views can be accurately identified, or decoded, as seen in Figure 43.3a, from the projected fringe pattern images. The decoding process is susceptible to error from illumination inconsistencies (reflection and over-exposure), but with low/medium albedo polymers and proper SLS hardware selection, errors can be mitigated. Finally, with a known spatial transformation between two cameras, matching pixels can be triangulated across image sets to form point clouds (Figure 43.3c).

43.2.2   Structured Light Scanning for Tire Repair

To achieve the project goals, the team developed a custom SLS system capable of conducting quick scans at sufficient FOV and resolution. By self-developing a custom scanner, 3D scan results can be directly integrated into the proposed digital geometric framework for generating multi-axis deposition toolpaths, which increases autonomy and speed.

The custom SLS system is comprised of a Texas Instruments (TI) Digital Light Processing (DLP) Projector and two 5.4-megapixel Lucid Triton High Dynamic Range (HDR) cameras (Figure 43.4). The TI projector allows rapid cycling of fringe patterns at 20Hz, which enables conducting scans in only 2 seconds. In addition, the TI projector's DLP technology reduces unnecessary light exposure, thereby increasing scanner precision and decreasing noise. Furthermore, the integrated stereo HDR cameras also mitigate excess exposure while

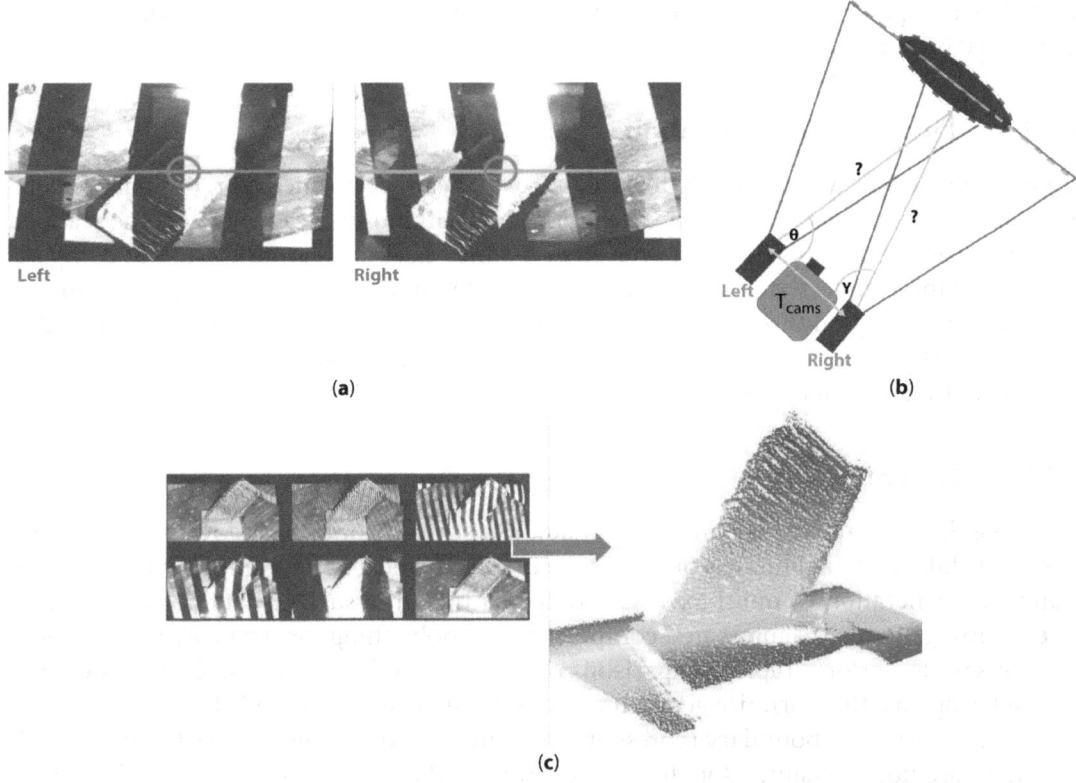

Figure 43.3 Structured Light Scanning: (a) overall schematic, (b) set of images from cameras, (c) sequence of scans and images resulting in 3D point cloud.

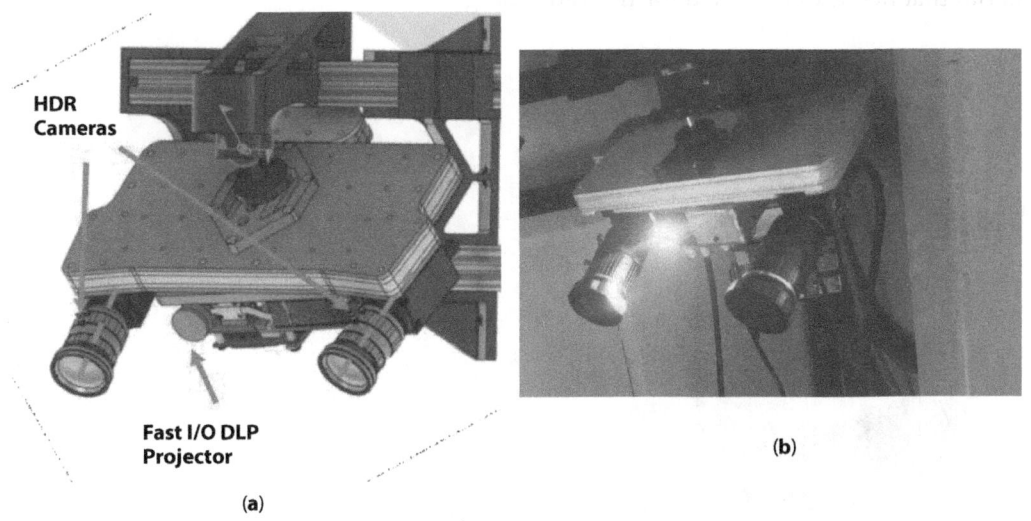

Figure 43.4 Custom Structured Light Scanning system: (a) schematic and (b) actual system.

their 5.4-megapixel sensors provide a scan resolution of <0.1mm when scanning a tire section in a 0.65m x 0.42m FOV.

43.2.3   Tire Scanning Validation

To validate the SLS system's ability to scan a tire, the team performed a preliminary experimental scan on a vehicle tire, as seen in Figure 43.5. With no surface treatment, the scanned tired precision should be <0.5 mm or even less (note: there is no ground truth to compare, yet). Since the scanners are positioned less than 600mm away from the tire, the scan features low noise (found through statistical outlier analysis) with <4% of points identified as outliers. Furthermore, the predicted resolution of 0.1mm is confirmed, with the average spacing between points computed at 0.12mm.

43.2.4   Next Steps

With the SLS system created, and early validation scans demonstrating the ability to capture high-fidelity digital models of worn tire surface, the next key technical challenge is the automatic generation of multi-axis AM toolpaths from the surface data acquired during tire scanning. To realize automated computation of toolpathing for repair AM processes, it is necessary to perform rapid comparisons of the digital solid models of the specified part geometry against the worn tire geometry that is constructed from the 3D scan. Traditional parametric surface or boundary representation schema (explicit modeling), typical of CAD software, are not well-suited for the computation of the intersections of multiple, curved surfaces. As such, the team looks to the use of implicit models, or signed distance functions (SDFs), which store an analytical function representing distance to the part's surface. 3D scan points can be rapidly inputted into the SDF to gather distance from the digital model of the desired surface, forming a basis for evaluating error between as-scanned and as-designed solid models. Using SDF's, this comparison between the worn tire surface and the desired surface can occur in near real-time, thus enabling rapid rendering of the bonding material that needs to be added for tire retreading.

Figure 43.5 3D scanning worn tire surface; (a) SLS projection onto worn tire, (b) resulting 3D point cloud.

43.3   Additive Manufacturing of Elastomeric Materials via Photopolymerization of Latex Resins

43.3.1   3D Printable Latex Rubber

A core challenge in the proposed approach lies in the concurrent design of a bonding rubber material that can be selectively, and precisely, deposited onto the worn tire surface. To enable printing, the material must have a suitable viscosity and shear thinning profiles for extrusion, while still maintaining the prerequisite molecular weight and latent reactivity for curing needed to provide the necessary adhesive strength and performance. Decoupling the inherent relationship between melt viscosity and molecular weight of the bonding rubber is therefore a core technical challenge in order to provide precise dispensing of the material.

In general, high molecular weight polymers are not suitable for AM extrusion, as a greater amount of chain entanglements induces melt and solution viscosities above the suitable maximum for the processes (up to approximately 10^5 Pa·s with appropriate shear-thinning behavior). To address this, the VT/ASU research groups have developed a novel means of 3D printing rubber latex systems in which a photoactive monomer scaffold is integrated into a colloidal suspension of high molecular weight polymer nanoparticles [5, 6]. These polymer colloids mitigate the molecular weight-viscosity relationship through preventing long-range entanglement as chains are sequestered into discrete particles. Selective photo-curing of the scaffold results in a solid, structural green body embedded with high molecular weight polymer. A subsequent post-processing drying step allows the polymer particles to coalesce and further entangle, resulting in a high-performance elastomer.

Virginia Tech and Arizona State University researchers have successfully printed these rubber latex photocurable resins via both vat photopolymerization (VP) and ultraviolet-assisted material extrusion (UV-MEX) AM processes. In VP, a vat of latex resin is selectively cured in a layer-wise fashion via a UV projector. In UV-MEX, the latex resin is selectively extruded through a nozzle onto the desired substrate and then photocured via UV irradiation. As it allows high-resolution, conformal deposition of rubber latex onto curved substrates, UV-MEX is appropriate for the project aims. To enable processing via UV-MEX, the latex resin's viscosity must be tuned by increasing the colloid concentration and/or inclusion of nanoscale fillers such as silica, nanoclay, etc. A schematic of the overall proposed reinforced-latex resin and UV-MEX system is provided in Figure 43.6.

Silica is widely used for reinforcement of rubber thanks to its polar nature. The inorganic nanoparticles allow access to high-performance nanocomposites. However, when introduced to latexes, silica's polar surface leads to undesired aggregation and flocculation of the latex particles. To prepare a stable photocurable resin with a processable viscosity, the surface of silica must be modified chemically, and the concentration of silica must be finely tuned to avoid formation of colloidal gel. Herein, the team reports on progress on developing a photocurable silica-reinforced styrene butadiene rubber (SBR) elastomeric composite.

43.3.2   SBR Latex Resin Synthesis, Printing, and Characterization

The surface functionalization of silica is carried out as per Scheme 43.1. 25 g silica (Sigma-Aldrich, 15-20 nm) was combined with 100 mL of tetrahydrofuran (THF) and purged 20

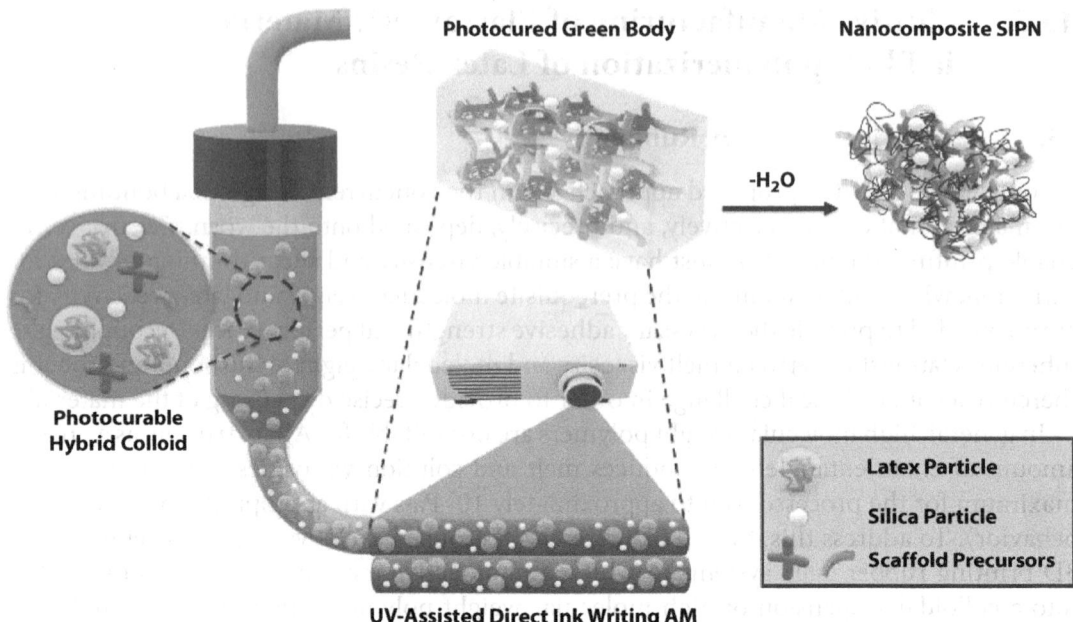

Figure 43.6 UV-MEX of a photocurable latex colloid reinforced with silica nanoparticle.

min with argon while stirring vigorously. 3.258 g (14.7 mmol) of (3-aminopropyl)triethoxysilane was added and reacted under 60 °C for 16 h. After the reaction, the dispersion was diluted by hexanes and precipitated by centrifuge. The precipitate was redispersed in THF and went through another two rounds of purification process. After repeats, the sample was dispersed in dimethylformamide (DMF) and purged with argon for 20 min while stirring. 4.22 g (42.2 mmol) of succinic anhydride was dissolved in 10 mL DMF and added via syringe and the reaction was allowed to proceed for 12 h at room temperature. The resultant carboxylic acid (COOH) functional particles were precipitated from diethyl ether, centrifuged three times. The final product was dried under vacuum and stored in a vial.

To prepare the resin, functionalized silica was first dispersed in water. Then the scaffold, composed of N-Vinylpyrrolidone (NVP) and poly(ethylene glycol) diacrylate (PEGDA; average molecular weight = 575), and a photo-initiator (TPO-L), were added to the dispersion and vortexed until fully mixed. Lastly, the mixture was added to the SBR latex dropwise while stirring. The final formulation is listed in Table 43.1.

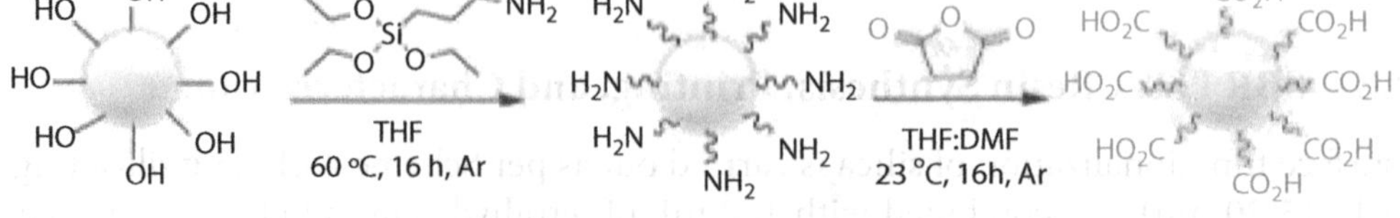

Scheme 43.1 Carboxylate functionalization of silica nanoparticles.

Table 43.1 Formulation of photo-curable SBR-silica resin.

Ingredient	Weight (g)	Percentage
silica-COOH	2.4836	5.10%
H20	13.4	27.53%
NVP	3.2435	6.66%
PEGDA	1.254	2.58%
TPO-L	0.2863	0.59%
SBR	28	57.53%
Total	48.6674	100.00%

The resulting resin had suitable viscosity for processing via VP, and thus provides a quick path to evaluating both the resin's photocuring behaviour and the resultant printed parts' mechancial properties. To obtain the appropriate VP printing parameters, a working curve is first prepared (Figure 43.6). A test layer is projected on the resin surface with intensity of 14.2 mW/cm^2 at varying exposure times (2, 2.25, 2.5, 2.75, and 3 s). The thickness of the resultant cured layers from the varied exposure times were measured with a microscope. The cured depths are plotted against the corresponding applied exposure, as shown in Figure 43.7. The values of the resin's intrinsic curing properties, depth of penetration (Dp) and critical energy (Ec), are measured from this working curve as 129.29 mm and 13.17 mJ, respectively.

Five scaled ASTM D-412 C tensile specimens (gauge width ~1.1 mm) were printed with the printing parameters generated from the working curve. In addition, a set of additional

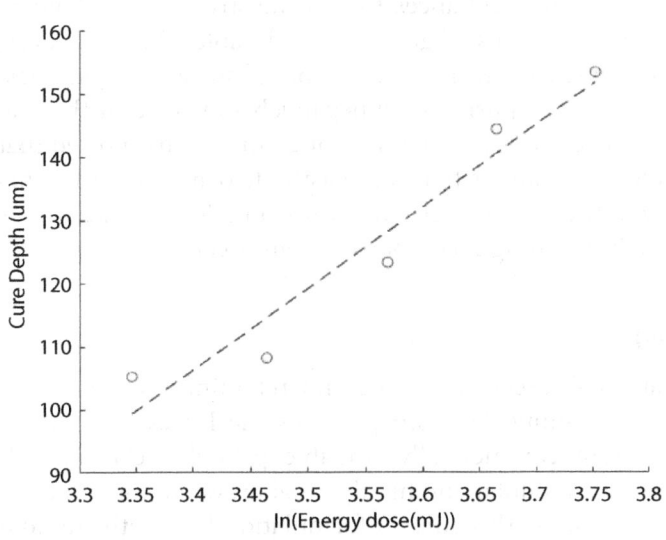

Figure 43.7 Working curve of photocurable composite SBR-silica resin.

Table 43.2 Summary of the Tensile Test Results of Silica-reinforced SBR Resin

Metric	Silica-reinforced SBR	Neat SBR
Elongation at break (%)	432.07 ± 44.59	331.88 ± 18.52
Tensile strength (MPa)	8.80 ± 0.77	5.87 ± 0.48
Young's Modulus (MPa)	3.83 ± 0.14	2.37 ± 0.06

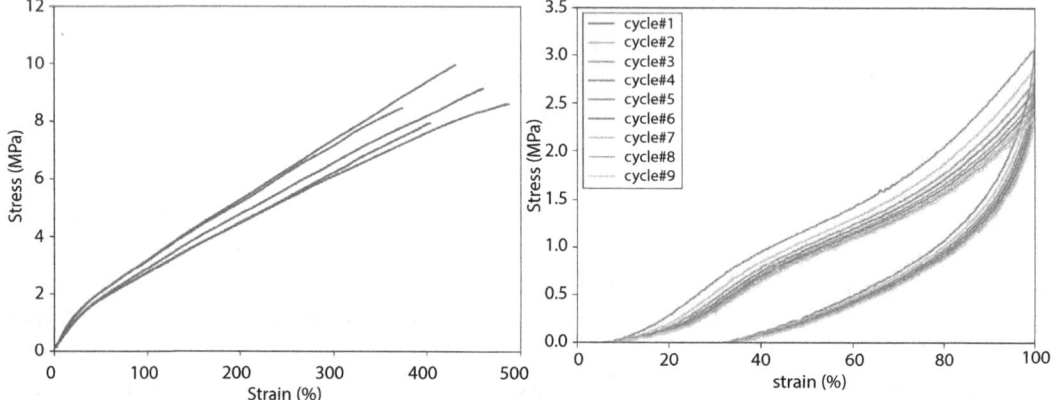

Figure 43.8 Uniaxial tensile test (Left) and cyclic tensile test (Right) of SBR-silica composites.

specimens were printed from a neat SBR photoresin that did not include silica nanoparticles. The printed green bodies were post-cured for three minutes on each side and then dried under 60 °C for 36 hours in air environment.

The subsequent tensile tests (conducted at a strain rate of 25 mm/min) showed that the addition of silica significantly enhances the tensile strength and Young's modulus, when compared to the neat SBR resins (Figure 43.8 and Table 43.2). The average of elongation at break increased to 432.07% with the addition of silica nanoparticles, which confirms that the elongation of the composite is not negatively impacted by the silica-reinforcement. Cyclic tensile tests (0% to 100% strain at rate of 25 mm/min) showed that, due to the existence of the scaffold, the composite has a plastic deformation at the first stretch (Figure 43.8). Compared with the neat SBR elastomer formulation, the addition of silica improves the hysteresis stress drop through nanoparticle reinforcement.

43.3.3   Next Steps

With elongation at break exceeding 400% and retaining a tensile strength greater than 8 MPa, this SBR-silica composite resin provides the largest elongation, toughness, and strength of any current commercially-available printable elastomer [7]. Photocurable hybrid colloids present a modular and highly tunable system for additive manufacturing of elastomeric nanocomposites. This current formulation falls in the middle of commercially available SBR's 3.5-20 MPa tensile strength range and 450-600% elongation [8]. Changing the morphology and concentration of the inorganic reinforcing filler will allow us to further tune the mechanical properties and rheology for UV-MEX processability. Next steps will include further exploration of reinforcing materials and other rubber latex resins, including natural rubber and styrenic block copolymers.

43.4   Conclusions & Recommendations

To increase the re-manufacturing efficiency of the tire retreading operation, the authors aim to co-create a multi- axis robotic additive manufacturing workcell capable of selectively depositing a tailored bonding rubber material conformally onto a tire carcass. Achieving this goal requires the concurrent development of (i) a 3D scanning technology capable of acquiring digital models of a worn tire, (ii) a printable elastomeric material featuring both the mechanical properties for end-use tire application and the requisite rheology for selective printing via extrusion, and (iii) a multi-axis printing system capable of depositing the material system conformally onto the worn tire.

In this manuscript, the authors report progress in (i) creating a custom 3D scanner capable of scanning a tire carcass at the requisite resolution for detecting tread damage, and (ii) developing a printable rubber latex nanocomposite material. The team has created a custom Structured Light Scanning (SLS) system that provides a scan resolution of < 0.1mm over 0.65m x 0.42m field of view, which is sufficient for scanning damaged features across the entire width of a commercial truck tire. In addition, the team has created a photocurable rubber latex system that can be selectively patterned via extrusion and UV irradiation. The use of a colloidal latex suspension provides a path for printing high molecular weight elastomers within processable viscosity ranges. The addition of silica nanoparticles has increased the ultimate tensile strength to 8.8 MPa, while still providing an elongation at break in excess of 430%. Thus, this 3D printable polymer exceeds the reported toughness and elongation of any commercially-available elastomer.

With the 3D scanner complete, the team now looks to using the resultant data to plan multi-axis toolpaths for the robotic extrusion system. The team is working on an implicit solid modeling kernel that uses signed distance functions (SDF) to represent the 3D point cloud from the SLS system. This will enable rapid Boolean comparison from the as- scanned worn tire and the original tire geometry specification. This will then be used to generate the toolpaths required to repair the worn tire. Concurrently, the team will look to continue to tailor the formulation and rheology of the printable latex rubber to further enhance performance and processability.

Acknowledgements

This material is based upon work supported by the U.S. Department of Energy's Office of Energy Efficiency and Renewable Energy (EERE) under the Advanced Manufacturing Office Award Number DE-EE0007897 awarded to the REMADE Institute, a division of Sustainable Manufacturing Innovation Alliance Corp. This report was prepared as an account of work sponsored by an agency of the United States Government. Neither the United States Government nor any agency thereof, nor any of their employees, makes any warranty, express or implied, or assumes any legal liability or responsibility for the accuracy, completeness, or usefulness of any information, apparatus, product, or process disclosed, or represents that its use would not infringe privately owned rights. Reference herein to any specific commercial product, process, or service by trade name, trademark, manufacturer, or otherwise does not necessarily constitute or imply its endorsement, recommendation, or favoring by the United States Government or any agency thereof. The views and opinions of authors expressed herein do not necessarily state or reflect those of the United States Government or any agency thereof.

References

1. MTD Facts Section – Commercial Tire Market, *Modern Tire Dealer Magazine*, January 2020 edition, pg. 40.
2. J. Dvorak *et al.*, A machining digital twin for hybrid manufacturing, *Manufacturing Letters*, vol. 33. Elsevier BV, pp. 786–793, Sep. 2022.
3. M. Perini, P. Bosetti, and N. Balc, Additive manufacturing for repairing: from damage identification and modeling to DLD, *Rapid Prototyping Journal*, vol. 26, no. 5. Emerald, pp. 929–940, Jan. 02, 2020.
4. I. Garmendia, J. Leunda, J. Pujana, and A. Lamikiz, In-process height control during laser metal deposition based on structured light 3D scanning, *Procedia CIRP*, vol. 68. Elsevier BV, pp. 375–380, 2018.
5. P. J. Scott, V. Meenakshisundaram, M. Hegde, C. R. Kasprzak, C. R. Winkler, K. D. Feller, C. B. Williams, T. E. Long, 2020, 3D Printing Latex: A Route to Complex Geometries of High Molecular Weight Polymers, *ACS Applied Materials and Interfaces*, 12 (9): 10918-10928
6. P. J. Scott, D. A. Rau, J. Wen, M. Nguyen, C. R. Kasprzak, C. B. Williams, T. E. Long, Polymer-inorganic hybrid colloids for ultraviolet-assisted direct ink write of polymer nanocomposites, *Additive Manufacturing*, 35, 101393
7. Proto3000, Elastomers, Rubber, and Foam 3D Printing Materials, https://proto3000.com/materials/elastomer-rubber-foam-3d-printing/, accessed January 2022.
8. Rahco Rubber, SBR – Styrene Butadiene Rubber, https://rahco-rubber.com/materials/sbr-styrene- butadiene-rubber/

Part 8

TIRE RECYCLING AND REMANUFACTURING

44

Crumb Rubber From End-of-Life Tires to Reduce the Environmental Impact and Material Intensity of Road Pavements

Angela Farina*, Annick Anctil and M. Emin Kutay

Department of Civil and Environmental Engineering, Michigan State University, East Lansing, Michigan, United States

Abstract

Flexible pavements cover 95% of the roads in the USA and include approximately 400 million tons of asphalt mixtures composed of natural aggregates and asphalt binder. Polymer modifications improve the durability of flexible pavements and are proven to prolong the service life. Crumb rubber (CR) from end-of-life tires might substitute the synthetic polymer in the asphalt mixtures, reducing tires disposed of in landfills. In the USA, the recycling of end-of- life tires contributed to lowering landfill disposal from 75% in the 1990s to 17% today. In 2021, in the USA, the use of CR in asphalt mixtures accounted for only 10% of the total CR market. This percentage could be increased by demonstrating the benefit of using end-of-life tires in road pavements. In this work, an integrated approach between life cycle assessment and mechanistic-empirical pavement design was used to evaluate long-term mechanical performance, environmental impact, and quantity of materials saved over the service life of road pavements. Laboratory-prepared asphalt mixtures modified with polymer coated rubber (PCR, dry and wet technology) and devulcanized rubber (DVR) were compared to reference blends (control - unmodified, and SBS - modified with synthetic polymer styrene-butadiene-styrene). Global warming potential (GWP), fossil depletion, and cumulative energy demand were quantified for a 1-mile surface layer over the pavement service life in Michigan. Environmental impacts depended on the long-term mechanical response of each asphalt mixture because the major contribution of the total environmental impacts was attributed to the number of cycles of reconstruction obtained from the mechanist- empirical pavement design. The PCR dry mixture had the lowest environmental impact compared to the other mixtures containing CR. The GWP for the PCR dry layer was 45.7% lower than the carbon footprint of the control layer and 8.0% lower than the SBS layer. Using asphalt mixtures modified with SBS or PCR dry can save up to 58.0% of materials compared to the unmodified control mixture over 50 years. Tires recycling contributes to the circularity of materials by minimizing the use of virgin materials and reducing waste and lowers the environmental impact of road pavements.

Keywords: Crumb rubber, polymer coated rubber, devulcanized rubber, end-of-life tires, asphalt pavements, life cycle assessment, environmental impact, material intensity

Corresponding author: farinaan@msu.edu

Nabil Nasr (ed.) Technology Innovation for the Circular Economy: Recycling, Remanufacturing, Design, Systems Analysis and Logistics, (587–598) © 2024 Scrivener Publishing LLC

44.1　Introduction

The construction of flexible pavements requires the use of large volumes of materials. Flexible pavements cover 95% of the paved roads in the USA. Every year, approximately 400 million tons of asphalt mixture (natural aggregates and asphalt binder) is required for road construction and maintenance [17]. The million tons of materials currently in use will be potentially available for recycling in new infrastructures. Recycling contributes to materials circularity by minimizing the use of virgin materials and reducing waste. Recycled materials, such as crumb rubber (CR) from end-of-life tires, can improve the durability of asphalt mixtures, reduce the use of virgin materials, and their disposal in landfills. CR might substitute the synthetic polymer (e.g., Styrene- Butadiene-Styrene, SBS) in the asphalt mixtures.

The CR has been used as an additive in asphalt pavements since the 1960s in the USA [1]. One significant advantage of using CR is that it can improve the mechanical performance of traditional asphalt mixtures if designed and blended properly [18, 19]. The CR is incorporated in the asphalt mixture through two broad methods: wet and dry techniques [2]. In the wet process, CR is blended with asphalt binder before mixing with aggregates. In dry technology, CR particles are added as a partial replacement of the aggregates during the mixing phase of the asphalt mixture. Recently, enhanced CR- based products have been designed. Some examples of these enhanced materials are polymer coated rubber (PCR) and devulcanized rubber (DVR). The PCR is a chemically enhanced rubber where a polymer film partially covers the surface area of the rubber particles [20]. The advantage of PCR is that it can be used as dry technology. The DVR is produced through the devulcanization process and is used to modify the asphalt binder. Methods for rubber devulcanization involve chemical, thermal, mechanical, and thermomechanical processes [3]. The environmental evaluation of materials such as PCR and DVR is essential to assess the benefits of recycled materials in asphalt pavements. Life Cycle Assessment (LCA) is a comprehensive methodology used to evaluate and quantify the environmental impacts of products and service [4–6]. This methodology can help support paving solutions decisions when applied to road pavements by adding environmental factors to economic and mechanical performance. LCA is particularly useful to evaluate the benefits of asphalt mixtures containing recycled materials such as reclaimed asphalt pavement (RAP, a recycled material obtained from milling operations of old pavements used in partial substitution of the natural aggregates) and CR. Multiple studies using LCA have found that the global warming potential and cumulative energy demand were lower for mixes containing RAP compared to a traditional blend using virgin material only [7–9]. The carbon footprint and the cumulative energy demand of an asphalt mixture were, respectively, 8% and 19% lower than a traditional mixture by using 30-40% by weight of RAP and decreasing the mixing temperature (150°C instead of 160-170°C) [10]. Some studies showed that thickness and service life play an important role in the LCA of road pavement. The global warming and cumulative energy demand reduction varied from 36 to 45% when using CR instead of the traditional mix due to higher durability and reduced thickness of the surface layer [8, 9]. Predicting the service life of road pavements is essential to perform a reliable cradle-to-grave LCA. The most accurate approach to predict long-term performance is the mechanistic-empirical pavement design guide (MEPDG) [11, 21]. The MEPDG considers the effect of the climate, traffic, and materials properties to calculate the damage accumulation and the distress prediction over time (NCHRP Project 1-37A, 2004). By predicting the distresses, it is possible to predict for how many years, after the construction, the pavement will last. The MEPDG approach allows having more realistic simulations and can reduce the number of unjustified assumptions about the performance and lifetime, which are commonly found in LCA studies. In this work, an integrated approach between life cycle assessment and mechanistic- empirical pavement design was used to evaluate the environmental impact of a 1-mile single-lane road pavement by using results of the long-term performance prediction. The goal of this study was the evaluation of the environmental impact of different asphalt mixtures containing CR used in the surface layer of road pavement. We also calculate the quantity of materials saved over the service life of road pavements. Laboratory-prepared asphalt mixtures modified with polymer coated rubber (PCR, dry and wet technology) and devulcanized rubber (DVR) were compared to reference blends (control - unmodified, and SBS - modified with synthetic polymer styrene-butadiene-styrene).

44.2   Materials and Methods

Figure 44.1 shows the methodology applied in this work. We performed the mix design of the mixtures and performed mechanical performance testing. The mix design was used to determine the quantity of each material in the life cycle inventory. Results of the mechanical performance (dynamic modulus, repeated load permanent deformation, fatigue cracking testing indirect tensile strength) were then used to calibrate the models for the long-term prediction performance using the MEPDG software (AASHTOWare Pavement ME). From the MEPDG results, we calculated the number of reconstructions necessary over the service life of the pavement for each asphalt mixture by dividing the number of years between the construction (year 0) and the time necessary for the pavement structure to pass the failing threshold, by the total service life of the pavement (50 years). We used this information in the LCA reconstruction phase to calculate the road pavement's material intensity over 50 years.

44.2.1   Asphalt Mixtures Preparation

At the Michigan State University asphalt laboratory, we designed five asphalt mixtures based on the Superpave specifications [12]: control, SBS, PCR dry, PCR wet, and DRV. The control mixture was unmodified (made of aggregates and bitumen). Mixtures with similar aggregate gradation with a nominal maximum size of 12.5 mm were designed with target air voids of 3±0.5% for surface layers. All mixtures had an equivalent single axle load (ESAL) of 30 million, except the DVR mixture designed for lower-level traffic (3 million) because this mix was part of a previous project. The RAP content in the DVR mix was 20% by weight of the total mix. Instead, in the other mixtures, RAP content was 15% by weight. Table 44.1 reports the details of the mixtures and the material content for each blend. After preparing the mixtures, dynamic modulus, repeated load permanent deformation, and fatigue cracking tests were performed, while the indirect tensile strength was calculated. Data collected were used to calibrate the models of the MEPDG.

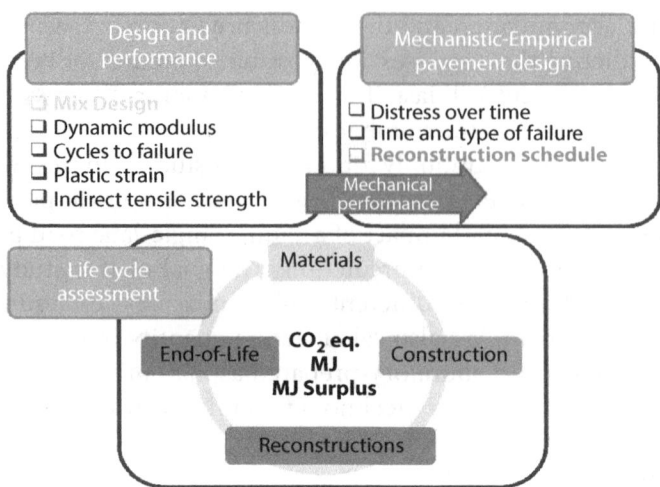

Figure 44.1 Scheme of the study.

Table 44.1 Material content and mixtures characteristics.

	Control	**SBS**	**PCR dry**	**PCR wet**	**DVR**
Mix type	unmodified	wet	dry	wet	wet
ESALS (million)	30	30	30	30	3
Asphalt PG†	58-28	70-28	58-28	82-28	70-28
Materials	(%)				
Aggregates	80.34	80.05	79.75	80.04	80.15
RAP	14.96	14.90	14.86	14.91	14.93
Bitumen	4.70	4.86	4.89	4.38	4.46
SBS	-	0.17	-	0.15	0.10
PCR	-	-	0.50	0.35	-
DVR	-	-	-	-	0.34
Sasobit	-	-	-	0.15	-
Cross-linker	-	0.02	-	0.02	0.02

†PG: performance grade, range of optimal pavement temperatures for bitumen

44.2.2   Mechanistic-Empirical Pavement Design

The analyses conducted with the AASHTOWare Pavement ME software were performed with the highest level of accuracy (Level 1) (NCHRP Project 1-37A, 2004). Level 1 requires site-specific input data determined from actual measurements in laboratory testing. A high-traffic level structure was selected to run the analyses among the road pavements existing in Michigan. Table 44.2 reports data regarding the pavement structure implemented in the software.

The three asphalt concrete layers (top, leveling, base) were assumed to be constructed with each of the five asphalt mixtures considered in this work. The design life of the pavement structures was set to 20 years, and the results obtained were used to project the number of reconstructions over a service life of 50 years. The analyses were performed for the wet freeze climate conditions of Michigan. The main input data implemented in the AASHTOWare Pavement ME software were: i) resilient modulus and thickness for unbound layers, ii) volumetric parameters (air voids, binder content), iii) dynamic shear modulus and phase angle of asphalt binders, iv) dynamic modulus of mixtures at various frequencies and temperatures, v) creep compliance, vi) indirect tensile strength for each layer, vii) traffic data (e.g., vehicles distribution, axles per truck), and, finally, viii) climate data for Michigan. The MEPDG analysis was used to calculate the reconstruction schedule

Table 44.2 Pavement structures data.

High traffic structure	
ESALs (million)	30
Asphalt concrete layers	
Thickness (in) - Top	1.5
Thickness (in) - Leveling	2
Thickness (in) - Base	3
Modulus (E* - psi)	from master curves
Unbound base layer	
Thickness (in)	6
Resilient modulus M_R (psi)	33,000
Unbound subbase layer	
Thickness (in)	18
Resilient modulus M_R (psi)	20,000
Subgrade	
Thickness	Semi-infinite
Resilient modulus M_R (psi)	7,000

over the lifetime of the pavement using the five different asphalt mixtures. The reconstruction schedule was used as input data in LCA.

44.2.3   Life Cycle Assessment

The objective of the LCA was to compare global warming potential (GWP), fossil depletion (FD), and cumulative energy demand (CED) of pavement structures built with asphalt mixes modified with CR from scrap tires (PCR dry, PCR wet, and DVR) compared to the two reference mixtures (unmodified control and SBS mixtures). An attributional approach was used to carry out the LCA. The analysis was conducted for the United States, and all data regarding electricity production, chemicals, and transportation were adjusted to represent the average conditions. Material transportation was included based on the commodity flow survey for the United States [13]. In this work, the boundaries of the cradle-to-grave system included material acquisition (e.g., aggregates, bitumen, CR, SBS, etc.), material production (asphalt mixtures), construction, use phase (reconstruction schedule), and the end of life of the surface layers (Figure 44.2). The use phase included the reconstruction of the surface layers based on the response of the pavement structure to the distresses under the level of traffic and climate condition. The use phase did not include the emissions from vehicles traveling on the roads over the service life. The functional unit was a 1-mile single-lane pavement structure with a fixed thickness over the service life of the road pavement (50 years). The reference flow reported in Table 44.3 was the amount, in metric tons (t), of the different asphalt mixtures used to build 1 mile of pavement. The mixtures had different weights due to the different specific gravity. TRACI 2.1 method [14] was used to evaluate GWP and FD impacts and the Cumulative Energy Demand impact category for the energy

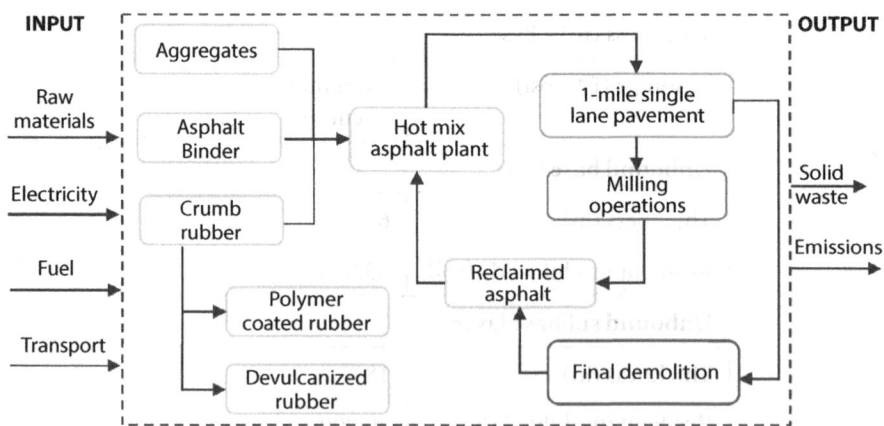

Figure 44.2 System boundary for the cradle-to-grave analysis of the study.

Table 44.3 Reference flow (t/1-mile single lane).

Control	SBS	PCR dry	PCR wet	DVR
2,246.40	2,252.30	2,198.30	2,277.70	2,251.00

demand [22]. SimaPro 9.1 [15] software was used for the assessment. The life cycle inventory was performed based on [16].

44.3   Results

Results obtained from the MEPDG analysis were in terms of pavement distress predictions over the design life compared to threshold values. These thresholds represent the maximum allowable distress magnitude to ensure that the pavement structure performs satisfactorily over its design life (NCHRP Project 1-37A, 2004). As expected, each asphalt mixture had a different behavior under the simulated loads and weather condition. The number of reconstructions over the service life was calculated by dividing the number of years between the construction (year 0) and the time needed to reach the distress threshold, reported in Table 44.4, by the total service life of the pavement (50 years), as shown in Figure 44.3. The PCR dry, PCR wet, and DVR performed better than the control mix, and both mixtures with PCR had similar long-term performance as the SBS-modified mix.

Table 44.5 reports the total GWP, FD, and CED for 1-mile single-lane road pavement constructed with all five mixtures over 50 years in Michigan. The PCR dry mixture had the lowest environmental impact compared to the other mixtures containing CR. The GWP for the PCR dry layer was 45.72% lower than the carbon footprint of the control layer and 8.04% lower than the SBS layer (Figure 45.4). As shown in Figure 44.5, environmental impacts depended on the long-term mechanical response of each asphalt mixture because the major contribution of the total environmental impacts was attributed to the number of cycles of reconstruction obtained from the mechanist-empirical pavement design.

Table 44.4 Number of years before distress reach the threshold.

	Control	SBS	PCR dry	PCR wet	DVR
years	9.83	23.42	22.92	22.58	16.08

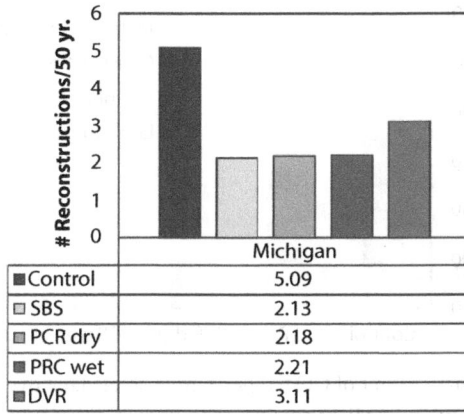

Michigan	
■Control	5.09
□ SBS	2.13
▨ PCR dry	2.18
■ PRC wet	2.21
▨DVR	3.11

Figure 44.3 Number of reconstructions over 50 years for the high traffic level structure in Michigan.

Table 44.5 Environmental impact of each mixture per 1 mile lane over 50 years.

	GWP	FD	CED
	t CO$_2$eq	Surplus TJ	TJ
Control	927.00	3.82	29.04
SBS	547.19	2.21	16.85
PCR dry	503.18	2.06	15.66
PCR wet	550.86	2.05	16.01
DVR	623.11	2.23	17.35

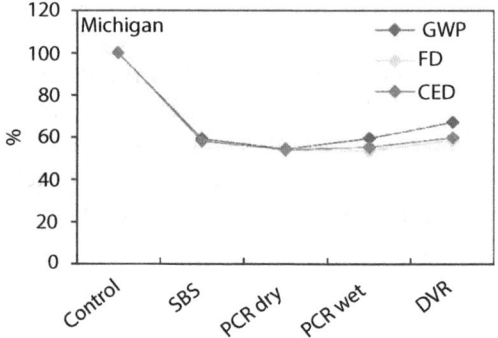

Figure 44.4 Global warming potential (GWP), fossil depletion (FD), and cumulative energy demand (CED) from cradle-to-grave LCA of a high traffic pavement using reference and rubberized mixtures in Michigan.

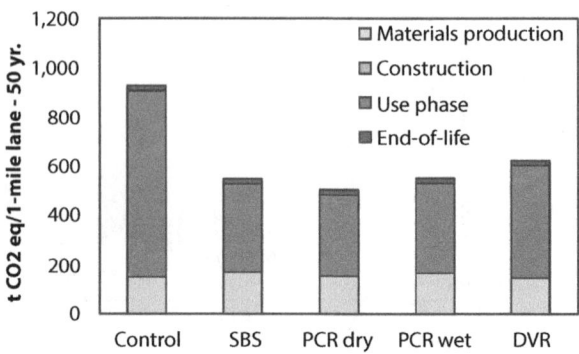

Figure 44.5 Global warming potential in t of CO2eq per 1-mile single lane over 50 years: contribution of each LCA phase for a high traffic pavement in Michigan.

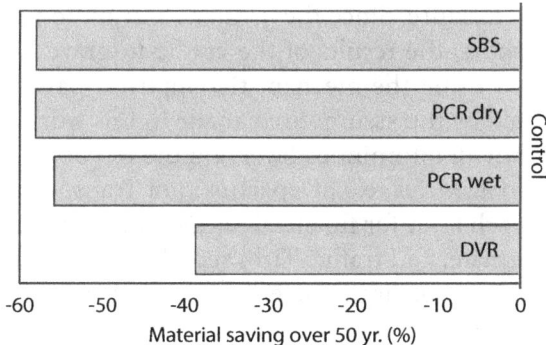

Figure 44.6 Percentage of material saving over 50 years for each asphalt mixture compared to the control reference mixture.

Based on the amount of asphalt mixtures used for the single-lane pavement structure in Michigan and the number of reconstructions, we calculated the total material usage for each mixture over 50 years. Then, we quantified the percentage difference of materials saved with respect to the control blend. SBS and PCR dry asphalt mixtures can save up to 58.0% of materials compared to the unmodified control mixture over the 50 years of service life (Figure 44.6). Similar results were reported for the PCR wet mixtures compared to the control mix. Among the modified mixtures, the PCR dry blend saved the same quantity of materials over 50 years as the SBS mixture. PCR dry mixture can reduce the use of virgin materials up to 2.3 times compared to the control mix for a high-traffic level structure. PCR dry uses 271.7 t/1-mile single lane of asphalt binder and approximately 3,830 t/1-mile single lane of natural aggregates over the service life. By comparison, the control mix uses 626.6 ton/1-mile single lane of asphalt binder and approximately 9,140 t/1-mile single lane of natural aggregates in 50 years.

Results for the DVR mixture were higher than those reported for PCR (both technologies). The mixture modified with DVR was designed slightly differently from the other mixtures. It was designed for 3 million ESALs instead of 30 million, with different aggregates (per type and gradation) and a percentage of RAP higher than that used in the others (20% instead of 15% by weight of the total mix). For these reasons, the DVR mix performed differently in terms of long-term pavement response.

44.4 Conclusions

The goal of this work was to evaluate whether asphalt mixtures containing recycled materials such as polymer coated rubber (PCR) and devulcanized rubber (DVR) might perform as good as or better than mixes modified with synthetic polymers (SBS) and unmodified mixtures from the environmental perspectives. Overall, in Michigan, the mixtures containing crumb rubber (PCR dry, PCR wet, DVR) performed as good as and better than the SBS mix, especially the PCR dry mixture, having the lowest environmental impact compared to the PCR wet and DVR. In terms of material saving, using asphalt mixtures modified with SBS or PCR (dry technology) could save up to approximately 58% of materials compared to the unmodified control mixture. Since the greater contribution to the total impacts can be attributed to the use phase, the results of the cradle-to-grave LCA were driven by the distress response predicted using the mechanistic-empirical pavement design methodology. Results may be affected by the assumptions made in this work. For instance, the same asphalt mixture was used in all bituminous layers on the top of the unbound materials. In real field applications, road agencies (e.g., Department of Transportation) prescribe the use of different mixtures per each layer for financial reasons and to tune mechanical properties on the project location, climate, and traffic. Tires recycling contributes to the circularity of materials by minimizing the use of virgin materials and waste and reduces the environmental impact of road pavements.

References

1. United States Environmental Protection Agency. Markets for Scrap Tires -EPA/530-SW-90-074A October 1991
2. Epps JA. NCHRP 198. *Uses of Recycled Rubber Tires in Highways.* 1994.
3. Garcia, P,S,, de Sousa, F.D.B., de Lima, J.A., Cruz, S.A., Scuracchio, C.H., Devulcanization of ground tire rubber: Physical and chemical changes after different microwave exposure times. *Express Polym Lett* 9(11):1015-26, 2015.
4. EC JRC, ILCD Handbook: General Guide for Life Cycle Assessment – Detailed Guidance. European Commission, Joint Research Centre, Institute for Environment and Sustainability, Luxembourg. 2010.
5. ISO 14040, Environmental management-Life cycle assessment-Principles and Framework. International Organization for Standardization, Geneva, Switzerland, 2006.
6. ISO 14044, Environmental management – Life cycle assessment – Requirements and guidelines. International Organization for Standardization, Geneva, Switzerland, 2006.
7. Chiu, C-T., Hsu, T-H., Yang, W-F., Life cycle assessment on using recycled materials for rehabilitating asphalt pavements. *Resour Conserv Recycl,* 2008:
8. Farina, A., Zanetti, M.C., Santagata, E., Blengini, G.A., Life cycle assessment applied to bituminous mixtures containing recycled materials: Crumb rubber and reclaimed asphalt pavement. *Res, Cons & Rec* 117:204-12, 2017.
9. Bressi, S., Santos, J., Orešković, M., Losa, M., A comparative environmental impact analysis of asphalt mixtures containing crumb rubber and reclaimed asphalt pavement using life cycle assessment. *Int J Pavement Eng* 1-15. 2019.
10. Giunta, M., Mistretta, M., Pratico', FG., Environmental Sustainability and Energy Assessment of Bituminous Pavements Made with Unconventional Materials. p 1-503, 2020.
11. Lanotte, M., Kutay, M.E., Haider, S.W., Musunuru, G.K., New calibration approach to improve Pavement ME Design thermal cracking prediction: mixture-specific coefficients–the Michigan case study. *Road Mater Pavement Des;*21(7):1859-71, 2019.
12. Asphalt Institute, MS-2, *Asphalt Mix Design Methods,* 2014.
13. US Census Bureau, *Commodity Flow Survey,* 2012.
14. Environmental Protection Agency, Tool for the Reduction and Assessment of Chemical and other Environmental Impacts (TRACI) - USER ' S MANUAL, 2012.
15. PRé Sustainability, SimaPro, https://simapro.com/about/, 2022.
16. Farina, A., *Environmental Impact of Civil Engineering Infrastructures: Flexible Pavements Using End-of-Life Tires and Material Intensity for Wind Turbines.* PhD Dissertation. 2021.

17. National Asphalt Pavement Association (NAPA), NAPA FAST FACTS, https://www.asphalt-pavement.org/uploads/documents/NAPA_Fast_Facts_Sept_2020_Final_Version.pdf. 2020.

18. Santagata, E., Baglieri, O., Riviera, PP., Lanotte,M., Alam, M., Influence of lateral confining pressure on flow number tests. Bear Capacit Roads, Railw Airfields - Proc 10th Int Conf Bear Capacit Roads, Railw Airfields, BCRRA 2017 2018:237-42, eBook ISBN9781315100333, 2017.

18. Kocak, S., Kutay, M.E., Use of crumb rubber in lieu of binder grade bumping for mixtures with high percentage of reclaimed asphalt pavement. *Road Mater Pavement Des* 0629:1-14, 2016. https://doi.org/10.1080/14680629.2016.1142466, 2016.

20. Kurgan, G., Dongre, RN., International Trends in Low-Carbon/Low-Energy Pavement Construction. Workshop on the Adoption of Innovative Technologies and Materials for Road Construction in India. 2015. http://www.iahe.org.in/5.1.pdf

21. NCHRP Project 1-37A, Implementation of the AASHTO Mechanistic-Empirical Pavement Design Guide and Software guide, http://onlinepubs.trb.org/onlinepubs/archive/mepdg/guide.htm, DOI: 10.17226/22406, 2014.

22. Frischknecht, R., Jungbluth, N., Implementation of Life Cycle Impact Assessment Methods, Data v2.0 (2007), Ecoinvent report No.3, https://inis.iaea.org/collection/NCLCollectionStore/_Public/41/028/41028089.pdf, 2007.

45

Tire Life Assessment for Increasing Re-Manufacturing of Commercial Vehicle Tires

Vispi Karkaria[1], Jie Chen[1], Chase Siuta[2], Damien Lim[2], Robert Radelescu[2] and Wei Chen[1]*

[1]*Department of Mechanical Engineering, Northwestern University, Evanston, IL*
[2]*Michelin North America, Greenville, SC*

Abstract

Re-manufacturing at commercial freight carriers includes the management of the life of freight tires. Freight carriers have limits on the age, number of retreads, and number of repairs a tire can have before it must be removed from service. When a tire tread is worn out, the freight carrier can decide to send the tire to the retreader and if it falls within the freight carriers limits, the tire can be retreaded and returned to the fleet for another life. This decision is made with little knowledge of the tire's ability to withstand another service life. Lacking this knowledge, the carriers tend to be conservative and truck tires are, on average, only retread once. This practice leads to increased operating costs for the fleets and wasted raw materials that often go directly to the landfill. To reduce the rate at which tires are discarded with usable casing life remaining, we propose novel machine learning and data science techniques that can be combined to determine tire retreadability by estimating the remaining casing health. If applied over the full perimeter of the US commercial vehicle tire industry, it is estimated that this work could result in an extra retread on average, achieving a yearly reduction of 580 metric kilotons of primary feedstock used to manufacture new tires, saving 37.1PJ of energy required to obtain the raw materials needed to fabricate new tires and a reduction of 1476 metric kilotons of eCO2 emissions. The objective of this paper is to present the preliminary building blocks of this methodology. A casing health model is developed, combining physical models and machine-learning models to differentiate the aggregate behavior between commercial freight fleets and different wheel positions based on their measured or simulated usage parameters. With the aid of these building blocks, the importance and global sensitivity of different tire usage parameters are obtained, and the tires can be used by the fleets in a way to maximize the expected tire life before the casing is removed from service.

Keywords: Tire life prediction, tire life comparison, random forest model, data balancing, tire energy, fleets, wear, retread

Corresponding author: weichen@northwestern.edu

Nabil Nasr (ed.) Technology Innovation for the Circular Economy: Recycling, Remanufacturing, Design, Systems Analysis and Logistics, (599–612) © 2024 Scrivener Publishing LLC

45.1   Introduction

A circular economy is a crucial component in achieving sustainable and responsible utilization of natural resources. The concept of a circular economy encompasses the reuse and regeneration of materials and products, promoting sustainable production and consumption practices [1]. This system prioritizes environmental conservation and resource efficiency. Inefficient use of the full potential of the tire is a threat to the circular economy because it represents a loss of resources [2]. Tires are made from a mixture of natural and synthetic materials, including rubber, steel, and textile fibers. When they are discarded to a landfill, these materials are often lost and are not easily reclaimed or reused, and new tire casings are required to be manufactured, requiring more raw materials and energy for their fabrication [3]. Inefficient use of the tire tread and casing undermines this principle, and it is important to find ways to optimize their use to support a circular economy.

In order to identify the best approach to promoting a circular economy in the commercial freight industry, it is crucial to have a thorough understanding of the construction and composition of tires [4]. A tire is composed of two primary components: the tire casing and the tire tread [5]. The tire casing, the structural layer of a tire that provides support for the tread and other layers, is composed of rubber and fabric reinforced with steel cables [6]. The tread, the part of the tire that comes into contact with the road and provides traction, is made of rubber and is designed to wear down gradually over time [5]. Tread patterns and compounds are optimized depending on the targeted application. Tread designed for short distance delivery trucks are different than line haul cross country types of usage [6]. Retreading is the process of applying a new tread layer to a tire that has worn down, in order to extend the lifespan of the tire casing and improve its performance [7]. Retreading is a cost-effective way to extend the life of tires and reduce the overall environmental impact [8]. By accurately predicting the remaining casing potential of a tire, fleet operators can schedule retreading at the optimal time, maximizing the number of retreads a tire can have throughout its life, ultimately reducing the amount of primary feedstock required. Additionally, using a tire life prediction framework can help fleet operators identify the usage parameters that are causing their tires to wear out more quickly and make adjustments that prolong the life of their tires.

In the field of tire life prediction, a variety of techniques such as machine learning methods, physics-based modeling and analytical methods have been used to accurately forecast the remaining lifespan of a tire. Machine learning methods like Decision Tree [9], Support Vector Machines [10], and Neural Networks [11] have been used for predicting tire life by analyzing historical data. The challenge with these machine learning models is handling imbalanced data, which can negatively impact model performance and efficiency. Physics-based modeling uses mathematical equations and simulations to predict tire wear and durability based on variables such as load, pressure, and temperature. Finite Element Analysis (FEA) can simulate the behavior of a tire under different loading conditions and predict its performance [12] [13], however, it may not be able to accurately predict tire life under a wide range of operating conditions [14]. A unified tire life prediction framework is needed to integrate the advantages of all these methods and balance the input dataset to predict tire life accurately by taking into account the real usages of different fleets.

To this end, we propose a novel tire life prediction framework, which can predict remaining casing potential, and compare tire life in different scenarios to give the fleet the tools necessary for maximizing tire casing life. By accurately predicting the remaining lifespan of a tire, retreading can be planned and executed before the tire reaches the end of its usable life.

The tire life prediction framework is trained using data obtained from various sources such as historical telematics, historical inspection, and FEM model. The data includes a response variable, which is the Remaining casing potential (RCP) and is represented by binary values of 0 and 1. An RCP value of 0 implies that the tire casing is unlikely to pass retread inspection, whereas an RCP value of 1 indicates that the tire has no damage accumulated. RCP is derived from endurance damage of the tire which is collected during tire inspection. Endurance damage in tire casing refers to the gradual weakening of the tire's structure due to prolonged exposure to stress, such as heat, friction, and impacts, leading to eventual failure or breakdown of the tire. This type of damage can be caused by factors such as overloading, rough road conditions, and improper maintenance, and can significantly reduce the lifespan of the tire.

The framework includes global sensitivity analysis to identify the input parameters that have the greatest impact on the tire life. By understanding which variables have the most influence, steps can be taken to optimize or adjust those variables to prolong the life of the tire. Due to the sparsity of the damage data compared to the undamaged data, we balance the data using the proposed Variance-reduction Synthetic Minority Oversampling Technique (VR-SMOTE). Then we utilize the random forest model to predict the remaining casing potential (RCP) of the tire. RCP refers to the amount of additional tread life that a tire can provide after the current tread is worn out. It is a measure of the quality and durability of the tire's casing to assess the feasibility of retreading a tire. the number of retreads. The tire life prediction framework shows high predictive accuracy for damage occurrence and is then tested on different scenarios, such as different fleet types, tire locations, and usage conditions. Results show that framework can learn the change of RCP in different scenarios where the truck can be driven.

The structure of the paper is organized as follows. In Section 45.2, the description of the framework is presented. In Section 45.3, the framework is validated using different identified scenarios where tire life is measured. In the first scenario tire life of two truck fleets are compared, and in the second scenario tire life in different tire locations are studied. In section 45.4, the closing remarks on the proposed framework are given, and future work is given.

45.2   Method

This section presents an overview of the tire life prediction framework, which utilizes data science and machine learning techniques to predict the remaining casing potential (RCP) of a tire. The subsequent subsection provides detailed information on the various components of the tire life prediction framework.

45.2.1   Description About the Tire Life Prediction Framework

The tire life is influenced by many factors such as the type of vehicle, type of tire, the usage condition, the road conditions, and the climate conditions. In Figure 45.1 we describe the cause effect relationship of different tire variables with respect to their impact on damage. It is observed that damage, represented by the remaining casing potential, is influenced by usage conditions which impact the usage state of tire. The usage state of tire is also influenced by truck design and tire design (represented by their IDs) and "time in service", captured by the total vehicle mileage and casing age. In Figure 45.1, different data sources associated with each of the attributes are indicated, namely (A) Historical inspection input data, (B) Historical telematics input data, (C) Historical telematics data serving as inputs to FEA model, and (D) input data obtained from FEA model.

Our proposed tire life prediction framework utilizes a combination of machine learning and data science techniques to predict the RCP of a tire. RCP is a unitless term which ranges from 0 to 1, where if RCP reaches 0 it indicates that the tire is about to fail. The RCP of a tire is directly influenced by the endurance defect, as the more damage that is inflicted on the tire's casing, the lower the RCP will be. Inputs of the prediction model include attributes denoted by letters A, B, and D in Figure 45.1.

Our proposed life prediction framework includes several major modules (steps) shown in Figure 45.2, following the sequence of data fusion, data reduction, data balancing, and machine learning. Initially, all data from various sources are fused with respect to time. Data fusion with respect to time refers to the process of combining and synchronizing data from multiple sources at different points in time to form a more complete and accurate representation of a given phenomenon. Then global sensitivity analysis is conducted for

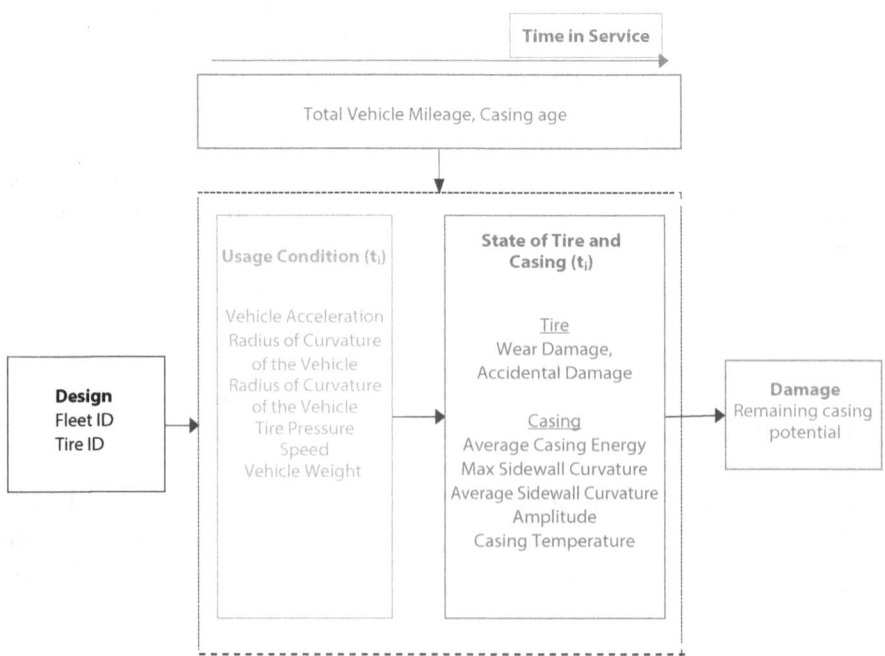

Figure 45.1 Cause-effect diagram for damage prediction.

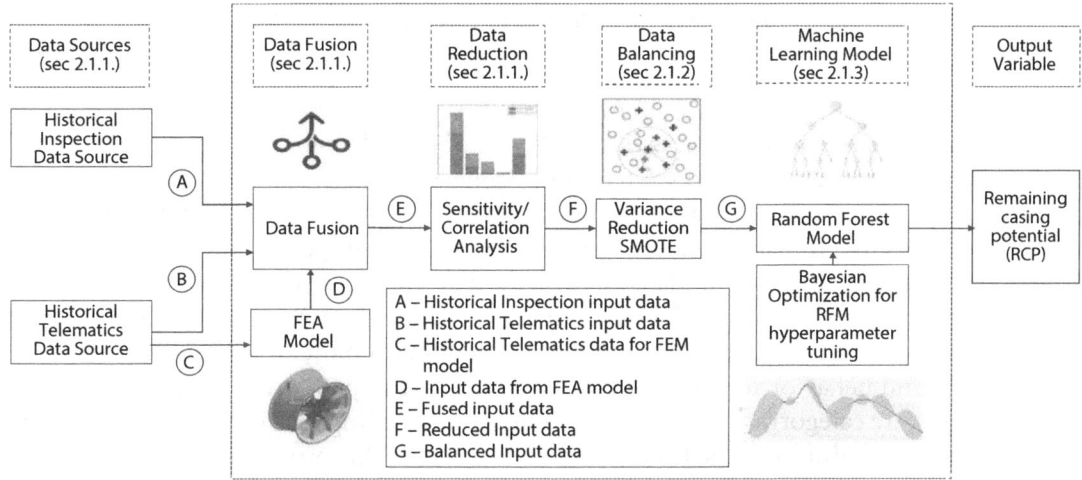

Figure 45.2 Flowchart of tire life prediction framework. The input variables denoted by (A), (B), (C) and (D) are given in Figure 45.1.

data reduction to evaluate the importance of the input variables with respect to tire life, and insignificant parameters are removed from the analysis. Global sensitivity analysis is critical in this analysis, as data associated with 31 input variables are available, however not all of them are needed in modeling. After global sensitivity analysis, eleven significant input variables are identified as the inputs for prediction. To obtain a machine learning model with good prediction using the random forest model in this study, data balancing technique such as the variance reduction SMOTE method proposed in this work is first applied, with details introduced next.

The data received from tires is imbalanced because most tires do not reach the end of their usable life, therefore data points of tire failure are sparse. To address this imbalance, a newly proposed variance-reduction SMOTE method is applied to the data for data balancing. This method increases the number of minority class samples to match the number of majority class samples, thus balancing the data. Using the balanced data, random forest model is created for tire life prediction. Random forest method is chosen because it can handle imbalanced data and non-linear relationships between input variables and tire wear better than neural networks or regression models. Random forest method predicts probability distribution of remaining casing potential (RCP) even if it is trained with a binary prediction of remaining casing potential. This is because probability calibration techniques, such as Platt scaling is used to adjust the predicted probabilities output by the random forest model. These techniques can ensure that the predicted probabilities more accurately reflect the true probability distribution of the remaining casing potential. The objective of the random forest model is to predict RCP with the highest F-1 score, which is a commonly used accuracy metric. To optimize the model performance, Bayesian optimization is used to identify the optimal parameters of the random forest model. This is accomplished by identifying the optimal set of hyperparameters, by balancing the trade- off between overfitting and underfitting, while considering the inherent uncertainty in the data. The final output of the tire life framework is the probabilistic predictions of RCP. In the subsequent subsections we will discuss each step in the tire prediction framework in detail.

45.2.1.1 Data Sources, Data Fusion and Data Reduction for the Tire Life Prediction Framework

The proposed tire life prediction framework in this study utilizes data from three sources to train the model: historical inspection data, historical telematics data, and FEA data. Historical inspection data for tires refer to the collected data from past inspections and assessments of tire performance and condition. Historical telematics data refer to information that is collected, transmitted, and analyzed remotely using telecommunication technologies and devices. This type of data can be collected from a variety of sources, including GPS, sensors, cameras, and other devices. Data collected from FEA for tire includes energy data such as average casing energy, which provides insight into the tire's internal and external structure and behavior under different loads, speed, tire pressure and conditions. The input variables are categorized and listed in Table 45.1 according to their sources. The input data are fused with respect to the time they were collected. This process is done by aligning the data based on the time it was collected, and then combining it in a way that preserves the temporal relationships between the data points. The total number of datapoints are 1713 out of which 85% datapoints have RCP response as 1 and 15% datapoints have RCP response as 0.

Global sensitivity analysis is conducted next to finalize 11 variables from 31, and it is observed in Figure 45.3 that all the 11 input variables have a significant impact on the prediction of tire life. Global sensitivity analysis is a method used to determine the degree to which the output of a system or model is affected by changes in each input variable, across the entire range of possible input values. In the next subsection, we will discuss the data balancing step of the framework.

45.2.1.2 Data Balancing for the Tire Life Prediction Framework

Data imbalance is created when there is an uneven distribution of response variable, and one class label has very high number of observations than others [15]. Initially, the tire life prediction framework was trained on imbalanced dataset and an F-1 score of 0.48 was observed. The reason behind the low F-1 score is that the datapoints with RCP value as 1 defect class is very high as compared to datapoints with RCP value as 0.

Minority data over sampling, majority data down sampling, and synthetic data creation are methods used to solve the data imbalance problem [16]. Synthetic Minority Oversampling Technique (SMOTE) is a popular oversampling method that is used to address the problem

Table 45.1 Variables for the tire life prediction framework.

	Historical inspection data	Historical telematics data	FEA datasource			
Vehicle	Total Vehicle Mileage (A)	Vehicle Acceleration (B)		Time in Service		
Tire	Tread Depth (A) Wear Damage (A) Accidental Damage (A)	Radius of Curvature of the Road (B)		Usage Condition (ti)		
Casing	Casing Age (A) Endurance Damage	Casing Temperature (B)	Average Casing Energy (D) Max Sidewall Curvature (D)	Usage State of Tire and Casing (ti)		
			Average Sidewall Curvature Amplitude(D)	Damage		

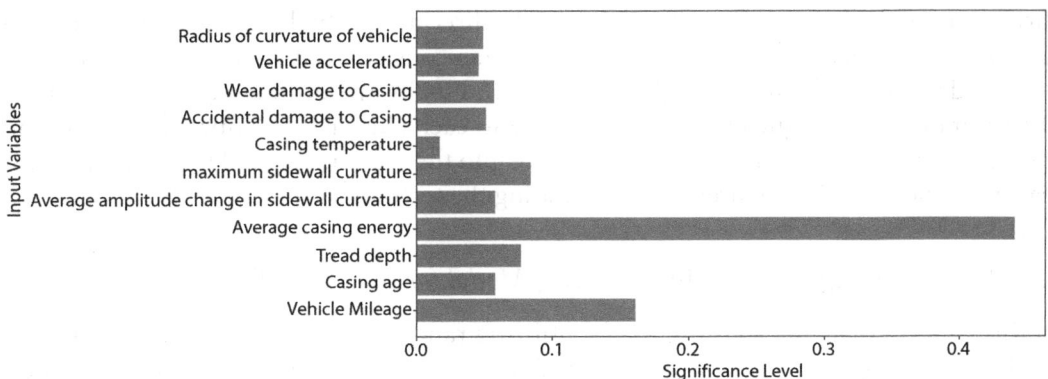

Figure 45.3 Significance level of input variables of the tire life prediction framework.

of imbalanced data sets in machine learning. Disadvantage of the SMOTE method is that it can lead to over-sampling of the minority class, which can negatively impact the performance of the model. Additionally, SMOTE may not always reduce uncertainty, which is related to variance, in the synthetic samples, which can lead to overfitting or poor generalization of the model [17]. We propose a new Variance Reduction-SMOTE (VR-SMOTE) method, which is a variation of the ordinary SMOTE method, in this paper which works on the principle of reducing variance of the minority data created. It is specifically designed to reduce the variance of the synthetic samples generated by SMOTE. In VR-SMOTE, we initialize variables as depicted in the Figure 45.4, and proceed to cluster the minority dataset using K-Nearest Neighbors (KNN) clustering. Subsequently, we compute the relative

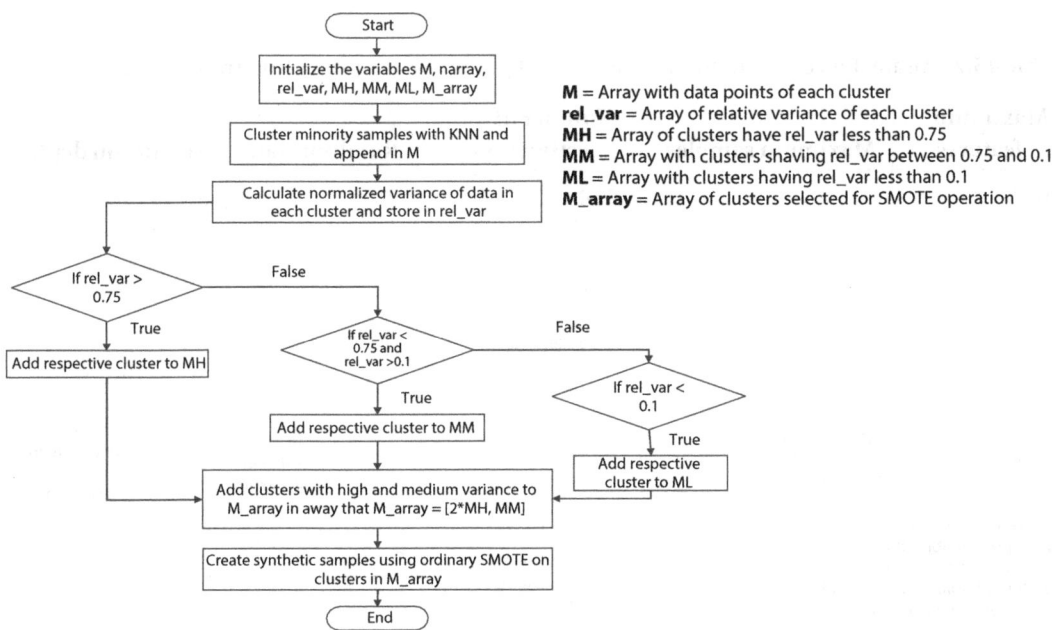

Figure 45.4 Flowchart of the Variance Reduction SMOTE method (VR-SMOTE).

variance of each cluster and separate them into three arrays: high variance, medium variance, and low variance clusters. A final array is then constructed by selecting clusters based on a predefined criterion $M_{array} = [2 * MH, MM]$. Finally, the ordinary Synthetic Minority Over-sampling Technique (SMOTE) is applied to each cluster in the final array to balance the dataset. In the subsequent section, we delve into the methodology of training a machine learning model for tire life prediction, including the techniques and algorithms used.

45.2.1.3   *Training the Machine Learning Model for the Tire Life Prediction*

In this paper, a machine learning model utilizing random forest algorithm is proposed to predict tire life. Figure 45.2 shows input and output variables to the random forest model. Random Forest was selected for its ability to handle high dimensionality, unbalanced data, and non-linear outliers. We use Bayesian optimization to optimize the parameters of the random forest model, aiming to maximize the F-1 score, particularly when working with synthetic data. Results of the Bayesian optimization, including number of estimators, maximum depth, maximum features, and maximum samples, are presented in Table 45.2. These techniques result in a robust tire life prediction framework that can accurately predict tire life.

45.2.2   Tire Life Comparison

The tire life prediction framework can be deployed to compare tire life in different scenarios. It is useful for predicting future tire life and understanding the rate at which a tire may need to be removed from a vehicle. For comparing tire life in different scenarios, the input variables need to be divided into two categories: those that depend on vehicle mileage and those that are independent of vehicle mileage. The quantitative relationship between these

Table 45.2 Parameters of the random forest model of the tire life prediction framework.

Maximum features	Maximum samples	Number of estimators	Random state	Maximum depth
0.55	0.644	154	91	3

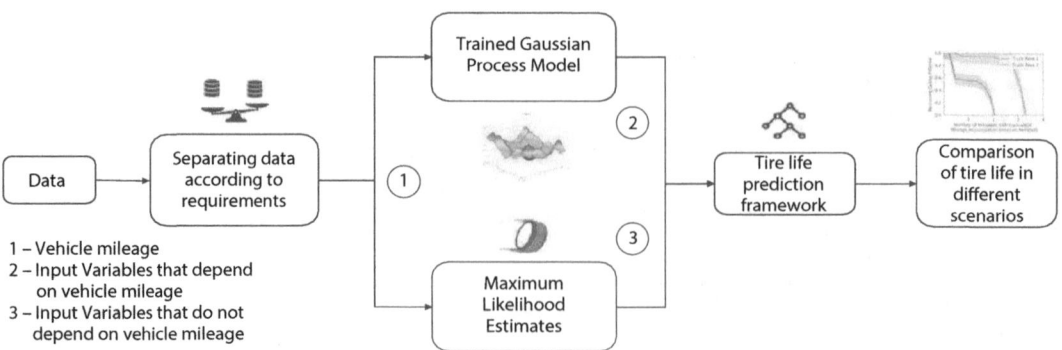

1 – Vehicle mileage
2 – Input Variables that depend on vehicle mileage
3 – Input Variables that do not depend on vehicle mileage

Figure 45.5 Flowchart of comparing tire life in different scenarios.

input variables with vehicle mileage is then used to predict remaining casing potential with the aid of tire life prediction framework as shown in Figure 45.5. The tire life prediction framework utilized in this step is explained in detail in Figure 45.2.

In this study, a Gaussian Process (GP) model is employed to predict the change of usage parameters with respect to vehicle mileage in different scenarios. The GP model is trained on input variables such as Average Casing Energy, Tread Depth, Casing Age, Total Vehicle Mileage, Max Sidewall Curvature, and Average Sidewall Curvature Amplitude, which are dependent on vehicle mileage. By analyzing patterns and trends in the validation data, the GP model is able to predict the rate of change of these input variables in different scenarios, such as when the truck fleets are changed. The predictions made by the GP model are then utilized as input to the tire life prediction framework for tire life comparison. Independent parameters, such as Radius of Curvature, Acceleration, Wear, Damage and Temperature, are analyzed using MLE. These parameters are used in tire life prediction for comparison across different scenarios. Results are presented in the results section, with more information on input tire variables.

45.3 Results

The proposed tire life prediction framework was validated using a set of validation data. The validation results indicate that the framework is capable of accurately predicting the remaining casing potential (RCP) of a tire. To further demonstrate the effectiveness of the tire life prediction framework, a tire life comparison analysis was conducted in different scenarios.

45.3.1 Validation of the Tire Life Prediction Framework

The performance of the Random Forest model, which has been fine-tuned using Bayesian optimization, is evaluated using the confusion matrix and F-1 score metrics. These accuracy measures are also employed to identify the most suitable data balancing method for the tire life prediction framework. The confusion matrix, a two-dimensional table that compares the predicted values of the model to the actual values, is utilized to assess the model's accuracy and identify any errors that may be present. A higher F-1 score, approaching 1, indicates a better balance between precision and recall. The table below illustrates the F-1 scores obtained by applying different data balancing methods.

From Table 45.3, it is inferred that the F-1 score obtained by SMOTE method is satisfactory, but this method has a big drawback which is multiclass imbalance. In multiclass imbalance data, the minority data space is overwhelmed by majority data space. Applying SMOTE in a multiclass imbalance problem creates synthetic minority data in the majority data space which creates a wrong representation of the data space [18]. For solving the problem of multiclass imbalance, we use the Clustered SMOTE method which creates clusters to identify minority data space, and then applies the SMOTE method in those minority data clustered regions [19]. The results achieved by applying Clustered SMOTE are much better than SMOTE. But it is observed that the synthetic minority data created had large variance, which reduces the predictability of the random forest model. Hence, a new method named Variance reduction SMOTE is proposed, in which clusters are initially created, and then

Table 45.3 Details of the data balancing methods used in the tire life prediction framework.

	Balancing method	F-1 score	Analysis
1	None	0.48	Random forest model trained on highly imbalanced data gives low F-1 score
2	Increasing minority data	0.56	The disadvantage of using this method as compared to Variance reduction SMOTE is that we are repeating minority data which makes the random forest model overfit
3	Decreasing majority data	0.52	The disadvantage of using this method as compared to Variance reduction SMOTE is that there is loss of information as some datapoints are deleted
4	Regular SMOTE	0.6	The disadvantage of this method as compared to Variance reduction SMOTE is that synthetic minority data might invade majority data space
5	Clustered SMOTE	0.75	The disadvantage of using this method as compared Variance reduction SMOTE is that this method does not reduce the variance in data while creating synthetic minority data
6	Variance reduction SMOTE	0.88	This method utilizes the concepts of Clustered SMOTE with a novel method of reducing variance in minority data class

clusters having variance more than a given value are selected again to create synthetic data. VR-SMOTE aids the random forest model to achieve the highest F-1 score. In Figure 45.6 the confusion matrix achieved after applying VR-SMOTE is given, in which it is observed that correct prediction is higher than incorrect predictions. This shows that the tire life prediction framework is accurate and robust to uncertainty.

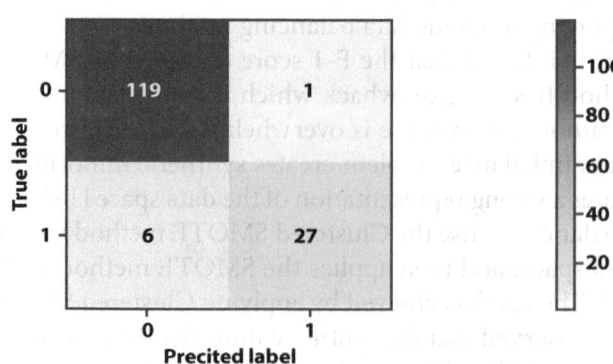

Figure 45.6 Confusion matrix of tire life prediction model achieved from testing data.

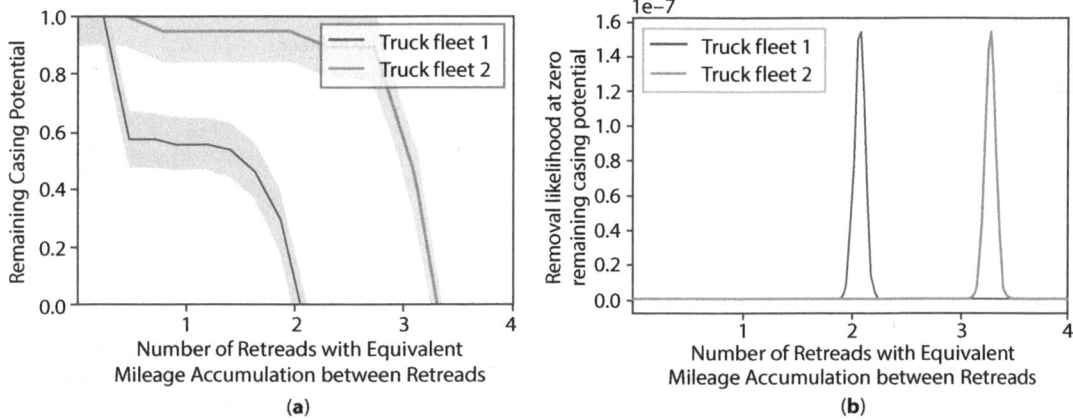

Figure 45.7 (a) Tire life comparison of tire at different truck fleets. (b) Probability density of tire life of tire at different truck fleets.

45.3.2   Comparison for Tires at Different Truck Fleets

In this study, we compare the tire life of two different fleets, named fleet 1 and fleet 2. These fleets differ in terms of usage conditions, such as climatic conditions, road conditions, and driving behaviors of the operators. The tire life prediction framework incorporates data from both fleets, utilizing the three data sources described in section 45.2.1.1, and the outcomes are visualized in Figure 45.7a. The random forest model provides an uncertainty estimate in the form of a 95% confidence interval. This interval represents the range of values within which the true value of the parameter being estimated is likely to fall, with a probability of 95%. The x-axis represents the number of tread life, and the y- axis represents the remaining casing potential. It can be observed that the tire life in fleet 1 is shorter than that of fleet 2, as the remaining casing potential of fleet 1 decreases at a faster rate than that of fleet 2. Additionally, it can be inferred that the tires in fleet 1 can be retreaded once, while tires in fleet 2 can be retreaded twice. The results are verified by comparing it with the real-life data where it is observed that 33% more tires of fleet 1 fail before tires of fleet 2.

In Figure 45.7 b, we present the results of our analysis comparing the likelihood of removal of tires at zero remaining casing potential for truck fleets 1 and 2. The data is represented by normal distributions with their means indicating the number of treads. Our analysis shows that the tires in truck fleet 1 have a higher likelihood of failure before reaching zero remaining casing potential, as compared to truck fleet 2, with a 95% level of confidence. In the following subsection, we apply the tire life prediction framework to analyze tire life at different tire locations. The results are verified by comparing it with the real-life data where it is observed that 35% more tires at steer location fail before tires at drive location.

45.3.3   Comparison for Tires at Different Tire Location

In this study, we compare the tire life expectations at different tire positions using our tire life prediction framework. Our analysis, as shown in Figure 45.8 a, reveals that the remaining casing potential at the steer axle position decreases faster compared to the remaining casing potential at the drive axle position. This can be attributed to the fact that a tire on

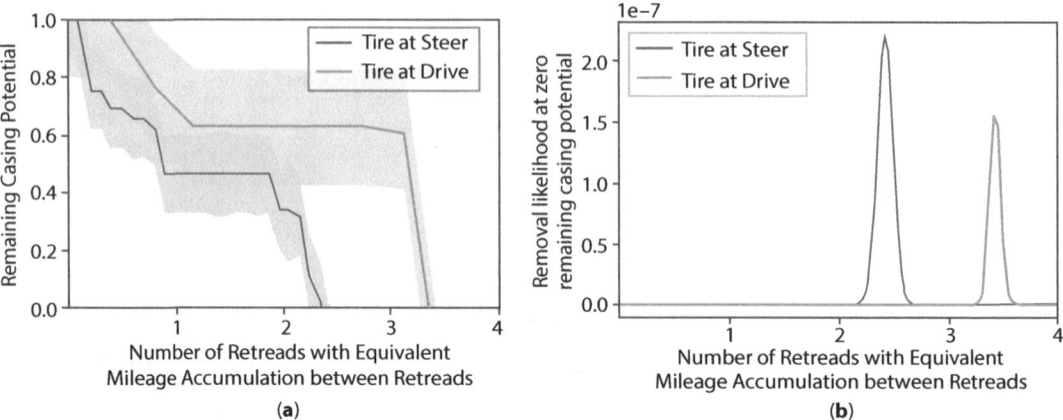

Figure 45.8 (a) Tire life comparison of tire at different tire position, (b) Probability density of tire at different tire position.

the steer axle position carries more load at a higher pressure on average and is thus subject to higher stress, resulting in a shorter expected life before the removal of the casing. As a result, tires on the steer axle are often moved to the drive or trailer axle after the first retread.

In Figure 45.8 b, we present the probability densities of the remaining casing potential of tires at the steer and drive locations when the remaining casing potential reaches 0. From the data, it can be inferred that the tire at the drive location has a longer life expectancy than the tire at the steer location. Specifically, it is observed that the tire at the steer location can be used for a maximum of one retreading, while the tire at the drive location can be used for up to two retreading. The results are verified by comparing it with the real-life data where the chances of tire at steer location failing is 33% higher than tire at drive location.

45.4 Conclusion

In this chapter, we proposed a tire life prediction framework aimed at enhancing the longevity of truck tires through informed rethreading decisions. Our approach incorporates data from both historical inspection and telematics sources and leverages a Finite Element Analysis (FEA) model to incorporate critical physical information about the tire's internal and external structure. The incorporation of physical aspects from the FEA model is crucial as it allows for a more accurate prediction of tire behavior under varying loads and conditions, and aids in identifying key factors that impact tire life, ultimately enabling tire design optimization for increased lifespan.

Eleven significant variables which impact the tire life are identified by utilizing global sensitivity analysis, and the imbalance in the dataset is reduced by applying a new VR-SMOTE method. It is observed that VR-SMOTE gives better F-1 score as compared to its earlier variant cluster-SMOTE method, as it focuses on reducing the variance while generating synthetic data. The balanced input variables are used to train a Random Forest model, whose parameters are optimized using Bayesian optimization. The resulting F-1 score of 0.88 demonstrates the high accuracy of the proposed model in predicting tire life in terms of remaining casing potential. Additionally, the model is be used for tire life comparison, enabled us to differentiate tire life in various scenarios such as fleet type and tire position.

The results of our study indicate that the proposed framework effectively differentiates fleets and tire locations and provides insight into the potential for retreading based on usage and design characteristics. Overall, the proposed framework provides valuable tools for truck fleet operators to improve tire longevity and make informed decisions about tire usage. Furthermore, the tire life prediction framework allows for the differentiation of tire life between different fleets and tire locations, enabling users to identify specific usage characteristics that may be negatively impacting their tire's life. To further enhance the framework, future work will involve training it on real-time data from sensors and utilizing machine learning methods for time-series data analysis, with the ultimate goal of implementing a digital twin concept for tire life prediction.

Acknowledgements

Grant support from the REMADE institute research program (DE-EE0007897) is greatly acknowledged.

References

1. G. Moraga *et al.*, Circular economy indicators: What do they measure?, *Resour. Conserv. Recycl.*, vol. 146, pp. 452–461, Jul. 2019.
2. N. Ferronato and V. Torretta, Waste Mismanagement in Developing Countries: A Review of Global Issues, *Int. J. Environ. Res. Public. Health*, vol. 16, no. 6, p. 1060, Mar. 2019.
3. H. Pakdel, D. M. Pantea, and C. Roy, Production of dl-limonene by vacuum pyrolysis of used tires, *J. Anal. Appl. Pyrolysis*, vol. 57, no. 1, pp. 91–107, Jan. 2001.
4. B. Rodgers and W. Waddell, Tire Engineering, in *Science and Technology of Rubber*, Elsevier, 2005, pp. 619–II.
5. B. Lorenz, B. N. J. Persson, G. Fortunato, M. Giustiniano, and F. Baldoni, Rubber friction for tire tread compound on road surfaces, *J. Phys. Condens. Matter*, vol. 25, no. 9, p. 095007, Mar. 2013.
6. L. A. P. Barbosa and P. S. G. Magalhães, Tire tread pattern design trigger on the stress distribution over rigid surfaces and soil compaction, *J. Terramechanics*, vol. 58, pp. 27–38, Apr. 2015.
7. B. Lebreton and A. Tuma, A quantitative approach to assessing the profitability of car and truck tire remanufacturing, *Int. J. Prod. Econ.*, vol. 104, no. 2, pp. 639–652, Dec. 2006.
8. V. Simic and S. Dabic-Ostojic, Interval-parameter chance-constrained programming model for uncertainty- based decision making in tire retreading industry, *J. Clean. Prod.*, vol. 167, pp. 1490–1498, Nov. 2017.
9. N. Sharma and M. Kalra, Predictive Maintenance for Commercial Vehicles Tyres Using Machine Learning, in *2022 13th International Conference on Computing Communication and Networking Technologies (ICCCNT)*, Oct. 2022, pp. 1–6.
10. J. Zhu, K. Han, and S. Wang, Automobile tire life prediction based on image processing and machine learning technology, *Adv. Mech. Eng.*, vol. 13, no. 3, p. 168781402110027, Mar. 2021.
11. C. Sivamani *et al.*, Tyre Inspection through Multi-State Convolutional Neural Networks, *Intell. Autom. Soft Comput.*, vol. 27, no. 1, Art. no. 1, 20215.

12. D. Lee, S. Kim, K. Sung, J. Park, T. Lee, and S. Huh, A study on the fatigue life prediction of tire belt-layers using probabilistic method, *J. Mech. Sci. Technol.*, vol. 27, no. 3, pp. 673–678, Mar. 2013.

13. W. Nyaaba, E. O. Bolarinwa, and S. Frimpong, Durability prediction of an ultra-large mining truck tire using an enhanced finite element method, *Proc. Inst. Mech. Eng. Part J. Automob. Eng.*, vol. 233, no. 1, pp. 161–169, Jan. 2019.

14. K. M. Jeong, Prediction of Burst Pressure of a Radial Truck Tire Using Finite Element Analysis, *World J. Eng. Technol.*, vol. 04, no. 02, Art. no. 02, 2016.

15. T. Zhu, Y. Lin, and Y. Liu, Synthetic minority oversampling technique for multiclass imbalance problems, *Pattern Recognit.*, vol. 72, pp. 327–340, Dec. 2017.

16. I. Domingues, J. P. Amorim, P. H. Abreu, H. Duarte, and J. Santos, Evaluation of Oversampling Data Balancing Techniques in the Context of Ordinal Classification, in *2018 International Joint Conference on Neural Networks (IJCNN)*, Jul. 2018, pp. 1–8.

17. N. V. Chawla, K. W. Bowyer, L. O. Hall, and W. P. Kegelmeyer, SMOTE: Synthetic Minority Over-sampling Technique, *J. Artif. Intell. Res.*, vol. 16, pp. 321–357, Jun. 2002.

18. S. Barua, Md. M. Islam, and K. Murase, A Novel Synthetic Minority Oversampling Technique for Imbalanced Data Set Learning, in *Neural Information Processing*, Berlin, Heidelberg, 2011, pp. 735–744.

19. Z. Xiang, Y. Su, J. Lan, D. Li, Y. Hu, and Z. Li, An Improved SMOTE Algorithm Using Clustering, in *2020 Chinese Automation Congress (CAC)*, Nov. 2020, pp. 1986–1991.

Appendix

Definition of tire input parameters which depend on vehicle mileage:

Average Casing Energy: the cumulative elastic energy that the casing sees throughout its life, calculated by finite element models and known conditions of pressure, load and speed, then summed over each revolution of the tire.

Max Sidewall Curvature: The maximum curvature of the sidewall during a revolution, related to the load and pressure settings of the tire. Higher load and lower pressure can lead to higher sidewall curvature and result in lower fatigue lives of the cables.

Average Sidewall Curvature Amplitude: The average difference between the maximum and minimum curvature in the sidewall during a revolution, related to the amplitude of the bending stresses seen during each cycle and leading to cyclic fatigue of the cables.

46

Recycling Waste Tire Rubber in Asphalt Pavement Design and Construction

Dongzhao Jin and Zhanping You*

Department of Civil, Environmental, and Geospatial Engineering, Michigan Technological University, Townsend Drive, Houghton, MI, USA

Abstract

Non-biodegradable solid tire wastes create a major environmental and public health risk. It is imperative to discover an effective method to recycle tires. The objective of this research is to study recycled waste tire rubber in asphalt pavement design and construction in Michigan's cold and wet regions. Different construction and design technologies of recycled waste tire rubber have been studied in asphalt pavement, including: 1). Recycled waste tire rubber as rubber modified asphalt binder by dry process in asphalt overlay, 2). Recycled waste tire rubber as hot rubber chip seal in pavement maintenance, 3). Recycled waste tire rubber as stress absorbing membrane interlayer in pavement pre-maintenance, and 4). Recycled waste tire rubber as tire-derived aggregate in the subgrade. The field performance and lab performance were evaluated for recycled waste tire rubber construction and design and conventional asphalt pavement construction and design. The results showed that recycled tire rubber in asphalt overlay will improve the high temperature rutting and low temperature cracking performance and exhibit adequate performance compared with the polymer (styrene-butadiene-styrene) asphalt overlay in high traffic conditions. Also, rubber-modified asphalt pavement mitigated the noise level by 2-3 dB on the road at different vehicle speeds. Recycled tire rubber as hot rubber chip seal can have 56.5% higher pull-off strength compared with conventional chip seal. The fracture energy of the recycled tire rubber as a stress absorbing membrane interlayer increased 8.17-19.2% for various material combinations compared with the control asphalt mixture. Recycled tire rubber as tire-derived aggregate can exhibit equal elastic modulus compared with the soil subgrade. In conclusion, recycled waste tire rubber in asphalt pavement can improve pavement performance in the overlay, chip seal layer, tire-derived aggregate layer, and stress absorbing membrane interlayer. In addition, it reduces waste tire accumulation in the environment and supports sustainable waste management practices.

Keywords: Eecycled waste tire rubber, asphalt overlay, tire-derived aggregate, stress absorbing membrane interlayer, hot rubber chip seal

---

*Corresponding author: zyou@mtu.edu

Nabil Nasr (ed.) Technology Innovation for the Circular Economy: Recycling, Remanufacturing, Design, Systems Analysis and Logistics, (613–624) © 2024 Scrivener Publishing LLC

46.1 Introduction

Waste rubber imposes a huge burden on the environment. Present estimates suggest that 2.9 billion tons of end-of-life tires are annually produced worldwide [1]. Most accumulate in landfills or in the natural environment. Therefore, our environments are inundated with rubber. Adding rubber into asphalt mixtures is an opportunity to dramatically reduce these landfilled rubbers.

Since the 1960s, asphalt pavements in the United States have been infused with waste tire rubber particles. The wet process, dry process, and terminal blend process are the three methods used to include rubber particles in asphalt mixes that are widely accepted. Among these three methods, there are significant differences in how rubber particles interact with other combinations of ingredients. In the first scenario, rubber particles are utilized as a polymeric modifier of the basic bitumen, where they can be digested totally or partly. The aggregate structure is partially replaced by rubber particles in the dry technology, and their contact with the asphalt binder is low. The terminal blend has a stable storage life and is made in a manner comparable to polymer-modified asphalt. According to reports, wet process products greatly increase the asphalt mixes' resistance to rutting and cracking [2–5]. High blending temperatures (around 180°C) combined with mechanical action during the synthesis of rubber-modified asphalt encourage the diffusion of the oily portion of the asphalt into the cross-linked polymeric matrix of the tire [6]. As a consequence, rubber particles inflate up to 2-3 times their original size, and the original rubber's core is assumed to be surrounded by a layer of gel-like substance [7]. Wet technology has the benefit of swelling rubber particles as well as a considerable disadvantage. On the one hand, the final combination benefits from the bitumen's very elastic qualities [8, 9]. Additionally, because some of the rubber's oily portion has been absorbed, the remaining binder is less susceptible to thermo-oxidative aging [10]. On the other hand, swelling results in a simultaneous decrease in the space between particles and the hardness of the base bitumen, which results in an overall rise in viscosity [11]. Given that the ingredients must be handled at temperatures greater than typical, this poses a serious problem for effective pumping, shipping, mixing, and lay-down stages. Additionally, the degree of interaction phenomenon varies and is strongly influenced by the properties of the raw materials [12]. Simply adding the rubber to the heated aggregates will yield mixtures created utilizing the dry technique without requiring any changes to conventional asphalt plants. The above-mentioned phenomena of interactions for the wet process start to occur when rubber particles come into contact with bitumen [13, 14]. But the development happens under uncontrolled circumstances. As a result, while dry technology mixes are simple to manage in the asphalt factory, their performance in the field is uneven [15–17]. The terminal blend method, as opposed to the dry or wet process, does not require a blending unit at the project site because it is supplied by oil refineries, which lowers the related expenses. Due to the complete digestion of the rubber particle into the asphalt binder, the terminal blend method has better storage stability, and it also performs rather well on paved surfaces [18, 19]. Much research has focused on the potential effects of TDA on surrounding water quality [20–22]. TDA does not pose a significant threat to the outside environment, while the TDA fill below the groundwater table should have negligible off-site effects on the water quality. Copper, iron, manganese, and zinc may be released from the TDA. Finally, Asphalt Pavement Stress Absorbing Membrane

Interlayer (SAMI) is a layer placed between the asphalt concrete layer and the underlying layer in road construction [23, 24]. Its purpose is to absorb the stress generated by traffic loads, reducing the stress transmitted to the underlying layer, and thus extending the overall pavement life [25].

Based on the limitations mentioned in the above description, the function of waste tire rubber when applied in each layer of asphalt pavement is still unclear. There is a need to summarize the recycled waste tire rubber in different layers of asphalt pavement structure and evaluate the performance of the recycled waste tire rubber in each layer. The objective of this research is to study recycled waste tire rubber in asphalt pavement design and construction in Michigan's cold and wet region. The lab and field performance of the recycled tire rubber and conventional asphalt pavement are evaluated and compared.

46.1.1 Waste Tire Rubber Materials

The scrap tire process and crumb rubber used in this study are shown in Figure 46.1. Companies buy and collect used tires from individuals, gas stations, tire shops, etc. Stage I is shredding the used tires into chips. It can be used as TDA in subgrade. Stage II is removing the steel wire. The wire has to be removed beforehand to recycle automobile tires. The revived steel is then used to produce other steel goods, and the rubber moves on to the next stage. Stage III is granulation and fiber removal, and Stage IV is final milling and screening. The screening stage is necessary to ensure there are no wires or other contaminants left that can affect the further usage of the rubber. It can be used in asphalt overlay, hot rubber chip seal, and the SAMI layer.

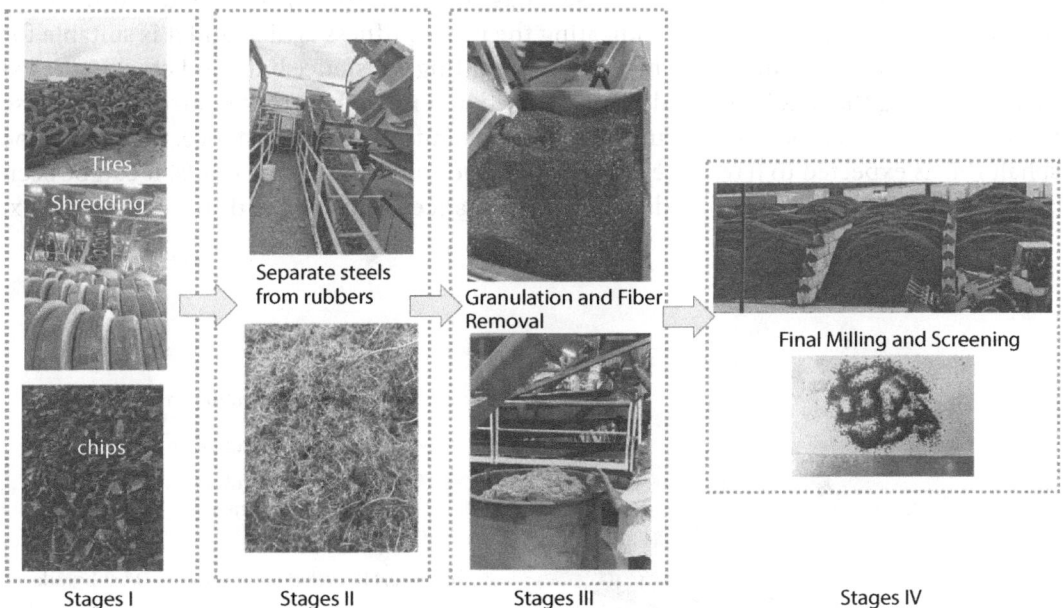

Figure 46.1 Scrap tire process and crumb tire rubber applied in this research.

46.2   Recycled Tire Rubber in Asphalt Pavement

In this section, different construction and design technologies of recycled waste tire rubber in asphalt pavement were studied, including recycled waste tire rubber as rubber modified asphalt binder in asphalt overlay (section 46.2.1), recycled waste tire rubber as hot rubber chip seal in pavement maintenance (section 46.2.2), recycled waste tire rubber as stress absorbing membrane interlayer in pavement pre-maintenance (section 46.2.3), and recycled waste tire rubber as tire-derived aggregate in subgrade (section 46.2.4) as seen in Figure 46.2. The field performance and lab performance were evaluated for recycled waste tire rubber construction and design and conventional asphalt pavement construction and design. The details are shown below.

46.2.1   Recycled Tire Rubber in Asphalt Overlay and Performance Evaluation

Recycled tire rubber in asphalt overlay is an effective method to utilize waste tire rubber. The demonstration construction project of recycled tire rubber in asphalt overlay was conducted by Kent County Road Commission (KCRC) on Cascade Road, Michigan, in June 2021. The condition before and after construction is shown in Figure 46.3. The current average daily traffic (ADT) is 16,500, which makes it a highly trafficked main road. The 10-year projected ADT is over 20,000. Commercial ADT is around 7%. Approximately two inches of the existing asphalt layer was cold-milled, and two inches of HMA was placed. The most commonly used equivalent load in pavement design in the U.S. is 18,000 pounds (80 kN). That means the new design will carry up to 10,000,000 standard 18,000 pound trucks.

The total number of passenger tires to be used in this project is about 2,270. The rutting–cracking performance space diagram results are shown in Figure 46.4. Phase I means the mix has both poor rutting resistance and poor cracking resistance as seen in Figure 46.5a, and it could be used only for temporary use. Phase II means the mix has good rutting resistance but poor cracking resistance, indicating the mix's stiffness is high and it is suitable for bottom layers. Phase III means the mix has poor rutting resistance but good cracking resistance, indicating the mix is soft and could be used as a reflective crack control layer. Phase IV means the mix has both good rutting resistance and cracking resistance, indicating the asphalt mix is expected to have excellent performance on heavy traffic roads. It can be seen that the rubber mix will exhibit adequate performance compared with the polymer mix.

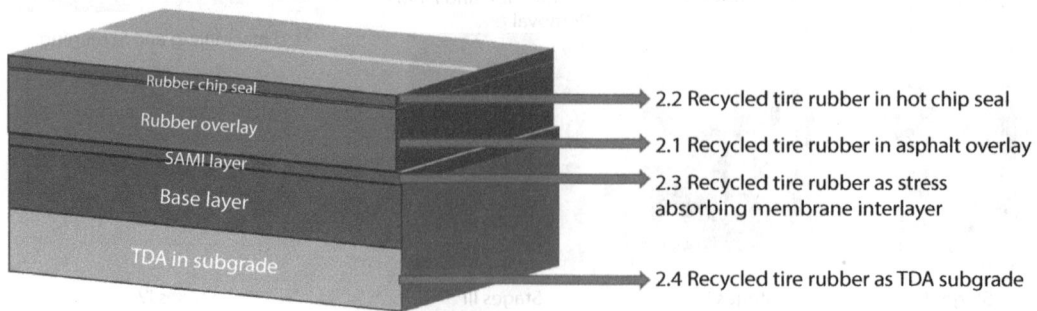

Figure 46.2 Technologies of recycled waste tire rubber in asphalt pavement used in this study.

Figure 46.3 Cascade Road before and after construction provided by Kent County Road Commission [26].

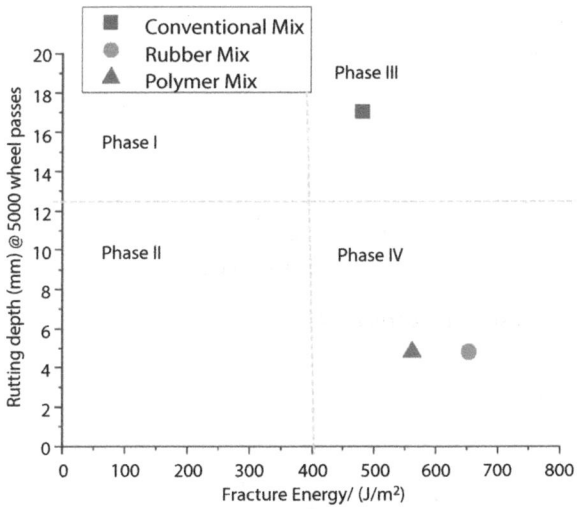

Figure 46.4 Rutting–cracking performance space diagram.

The results of the noise test demonstrated that the rubber-modified asphalt pavement mitigated the noise level by 2-3 dB on the road at different vehicle speeds. as seen in Figure 46.5.

46.2.2 Recycled Tire Rubber in Hot Rubber Chip Seal and Performance Evaluation

Recycled tire rubber in a hot rubber chip seal is also a promising technology. The demonstration project of rubber asphalt chip seal was constructed in Michigan in 2021. The hot rubber asphalt at 200 °C was sprayed from the asphalt distributor in the construction section at a controlled, constant speed to prevent bleeding or aggregate loss in the future. After the asphalt was applied, the aggregate chip was quickly applied to the construction lane, and a uniform chip seal layer was applied across the test section, as shown in Figure 46.5 (a). The field pull-off test results are shown in Figure 46.6 (b). Two sections with the same average daily traffic (AADT=500) on the road were selected to compare after one year of service. The rubber chip seal showed 56.5% higher pull-off strength compared with the conventional chip seal.

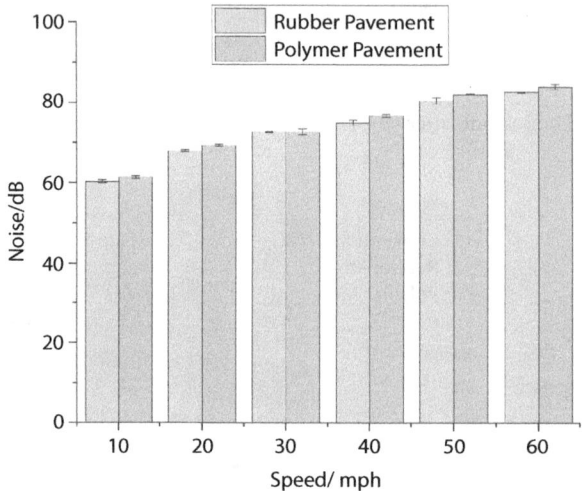

Figure 46.5 Field noise level outside of the truck [26].

(a)   (b)

Figure 46.6 Hot rubber asphalt chip seal construction process and field pull-off test results.

46.2.3   Recycled Tire Rubber as Stress Absorbing Membrane Interlayer and Performance Evaluation

Recycled tire rubber as SAMI layer is an effective way to eliminate surface distress and improve the pavement service life [27]. The demonstration project of utilizing recycled tire rubber in the SAMI layer is located on 7-Mile Road between Midland Road and E Beaver Road in Michigan. Two field test sections with a total of 8.85 km for this project were demonstrated and tested: 1). 7.24 km of asphalt pavement with SAMI layer; 2). 1.6 km of asphalt pavement without SAMI layer, as shown in Figure 46.7. The two sections are located within the same climatic and traffic volume conditions since they are both on the same road. This creates very valuable conditions for comparing the pavement performance with or without the SAMI layer. The DCT test result showed that the rubber-modified asphalt mixture showed better cracking resistance than the conventional asphalt mixture, and the Trans-Polyoctylene Rubber (TOR) additive would have a slightly negative effect on the low-temperature properties. The fracture of the conventional asphalt mixture with the SAMI layer is 13.06 % higher than the conventional asphalt mixture without the SAMI layer. The fracture of the rubber mix with the SAMI layer is 8.17 % higher than the rubber mix without the SAMI layer. The fracture of the rubber mix with TOR and SAMI layer is 19.2 % higher than the rubber mix without the SAMI layer.

46.2.4   Recycled Tire Rubber as TDA Subgrade and Performance Evaluation

Tire-Derived Aggregate (TDA) is a material created from shredded scrap tires that can be used as a substitute for traditional aggregate materials in construction and engineering applications. TDA is produced by shredding waste tires into small pieces that are then screened to a uniform size. The construction in Clare county is shown in Figure 46.8. The Field Light Weight Deflectometer (LWD) test results are shown in Figure 14.9. A total of 35 points of the TDA subgrade were tested during the construction. The total test length of

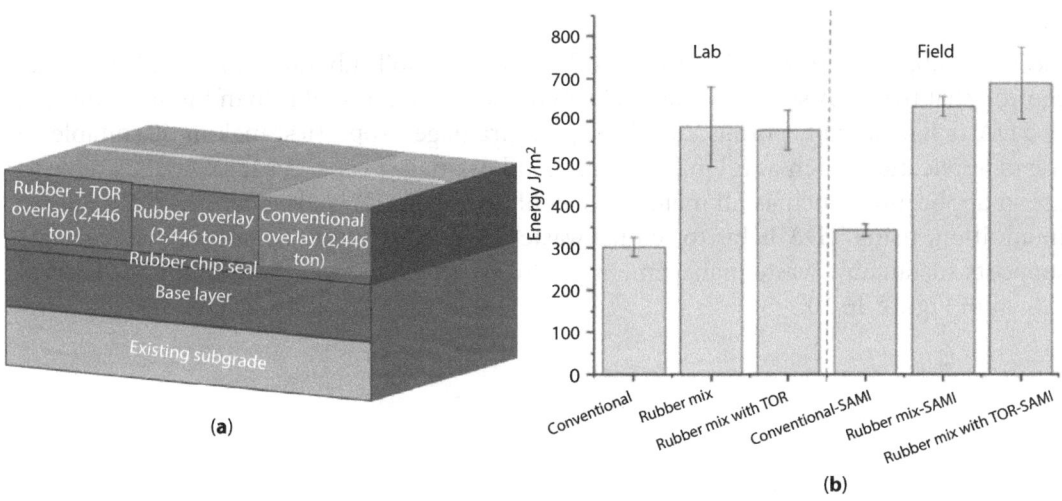

Figure 46.7 Recycled tire rubber as stress absorbing membrane interlayer structure and lab low temperature performance.

Figure 46.8 6 inches of tire-derived aggregate in Clare County construction.

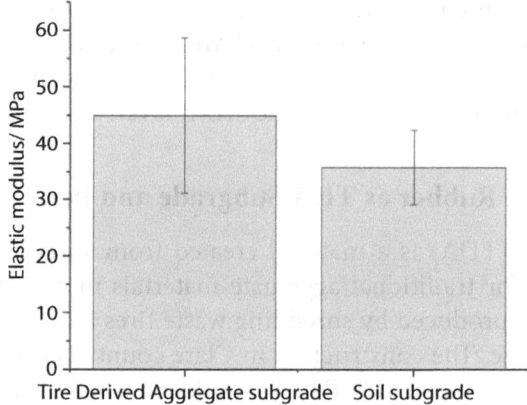

Figure 46.9 Field LWD test results in Clare County construction. Note: (elastic modulus data of the soil subgrade is displayed based on US41 subgrade data).

the TDA subgrade is 600 ft. The total test length of the soil subgrade is 1 mile. LWD result showed that the TDA subgrade has a 25.6 % higher elastic modulus than the soil subgrade.

TDA is lightweight, porous, and has good drainage properties, making it suitable for use in applications such as fill material has good drainage properties making it suitable for use in applications such as fill material, backfill for retaining walls, and drainage material. In addition, using TDA helps to reduce waste tire accumulation in the environment and supports sustainable waste management practices. The construction in Ingham County is shown in Figure 46.10.

Figure 46.10 Tire-derived aggregate in Ingham County construction.

46.3   Conclusions & Recommendations

This research studied recycled waste tire rubber in asphalt pavement design and construction in Michigan's cold and wet regions. The lab and field performance of the recycled tire rubber and conventional asphalt pavement were evaluated and compared. Some findings can be summarized as follows:

(1) Recycled tire rubber in asphalt overlay will improve the high temperature rutting and low temperature cracking performance and exhibit adequate performance compared with the polymer asphalt overlay in high traffic conditions.

(2) Recycled tire rubber as hot rubber chip seal can have 56.5% higher pull-off strength compared with conventional chip seal.

(3) Recycled tire rubber as stress absorbing membrane interlayer can increase 8.17-19.2% fracture energy compared with the control asphalt mixture.

(4) Recycled tire rubber as tire derived aggregate can exhibit equal elastic modulus compared with the soil subgrade.

In conclusion, the recycled waste tire rubber in asphalt pavement can improve the pavement performance in the overlay, chip seal layer, tire derived aggregate layer, and stress absorbing membrane interlayer. In addition, it reduces waste tire accumulation in the environment and supports sustainable waste management practices.

Acknowledgements

These projects were completed in collaboration with Clare, Bay, Antrim, Ingham, and Kent Counties, and the projects were funded by The Scrap Tire Management Program of the Michigan Department of Environment, Great Lakes, and Energy (EGLE). We appreciate Dr. Bora Cetin of Michigan State University for conducting the LWD test on TDA subgrade. The research team declares no conflict of interest.

References

1. Shoul, Behnaz, Yousef Marfavi, Banafsheh Sadeghi, Elaheh Kowsari, Peyman Sadeghi, and Seeram Ramakrishna. Investigating the potential of sustainable use of green silica in the green tire industry: A review. *Environmental Science and Pollution Research* 29, no. 34: 51298-51317, 2022.
2. de Mello, L.G.R., M.M. de Farias, and K.E. Kaloush, Effect of temperature on fatigue tests parameters for conventional and asphalt rubber mixes. *Road Materials and Pavement Design*, 19(2): p. 417-430, 2018.
3. Harvey, John, Manuel Bejarano, and Lorina Popescu. Accelerated pavement testing of rutting and cracking performance of asphalt-rubber and conventional asphalt concrete overlay strategies. *Road Materials and Pavement Design* 2, no. 3: 229-262, 2001.
4. Chiu, Chui-Te, and Li-Cheng Lu. A laboratory study on stone matrix asphalt using ground tire rubber. *Construction and Building Materials* 21, no. 5: 1027-1033, 2007.
5. Kök, Baha Vural, and Hakan Çolak. Laboratory comparison of the crumb-rubber and SBS modified bitumen and hot mix asphalt. *Construction and Building Materials* 25, no. 8: 3204-3212, 2011.
6. Artamendi, Ignacio, H. Khalid, G. C. Page, P. G. Redelius, L. J. Ebels, and I. Negulescu. Diffusion kinetics of bitumen into waste tyre rubber. *Asphalt Paving Technology* 75: 133, 2006.
7. Jamrah, Anas, M. Emin Kutay, and Sudhir Varma. Backcalculation of swollen crumb rubber modulus in asphalt rubber binder and its relation to performance. *Transportation Research Record* 2505, no. 1: 99-107, 2015.
8. Gogoi, Rupam, Krishna Prapoorna Biligiri, and Narayan Chandra Das. Performance prediction analyses of styrene-butadiene rubber and crumb rubber materials in asphalt road applications. *Materials and Structures* 49: 3479-3493, 2016.
9. Venudharan, Veena, and Krishna Prapoorna Biligiri. Effect of crumb rubber gradation on asphalt binder modification: rheological evaluation, optimization and selection. *Materials and Structures* 50: 1-14, 2017.
10. Ibrahim, Mohd Rasdan, Herda Yati Katman, Mohamed Rehan Karim, Suhana Koting, and Nuha S. Mashaan. A review on the effect of crumb rubber addition to the rheology of crumb rubber modified bitumen. *Advances in Materials Science and Engineering* 2013, 2013.
11. Peralta, Joana, Hugo MRD Silva, Loic Hilliou, Ana V. Machado, Jorge Pais, and R. Christopher Williams. Mutual changes in bitumen and rubber related to the production of asphalt rubber binders. *Construction and Building Materials* 36: 557-565, 2012.
12. Airey, Gordon, Mujibur Rahman, and Andrew C. Collop. Crumb rubber and bitumen interaction as a function of crude source and bitumen viscosity. *Road Materials and Pavement Design* 5, no. 4: 453-475, 2004.

13. Chen, Siyu, Dongdong Ge, Dongzhao Jin, Xiaodong Zhou, Chaochao Liu, Songtao Lv, and Zhanping You. Investigation of hot mixture asphalt with high ground tire rubber content. *Journal of Cleaner Production* 277: 124037, 2020.

14. Jin, Dongzhao, Dongdong Ge, Xiaodong Zhou, and Zhanping You. Asphalt Mixture with Scrap Tire Rubber and Nylon Fiber from Waste Tires: Laboratory Performance and Preliminary ME Design Analysis. *Buildings* 12, no. 2: 160, 2022.

15. Amirkhanian, Serji N. Utilization of crumb rubber in asphaltic concrete mixtures–south carolina's experience. See ref 3: 163-174, 2001.

16. Farina, Angela, Maria Chiara Zanetti, Ezio Santagata, and Gian Andrea Blengini. Life cycle assessment applied to bituminous mixtures containing recycled materials: Crumb rubber and reclaimed asphalt pavement. *Resources, Conservation and Recycling* 117: 204-212, 2017.

17. Santagata, Ezio, Michele Lanotte, Davide Dalmazzo, and Maria Chiara Zanetti. Potential performance-related properties of rubberized bituminous mixtures produced with dry technology. In *Proceedings, 12th International Conference on Sustainable Construction Materials, Pavement Engineering and Infrastructure*, Liverpool, UK, pp. 27-28. 2013.

18. Tang, Naipeng, Weidong Huang, and Feipeng Xiao. Chemical and rheological investigation of high-cured crumb rubber-modified asphalt. *Construction and Building Materials* 123: 847-854, 2016.

19. Gandhi, Tejash, Trey Wurst, Courtney Rice, and Brandon Milar. Laboratory and field compaction of warm rubberized mixes. *Construction and Building Materials* 67: 285-290, 2014.

20. Downs, Lisa A., Dana N. Humphrey, Lynn E. Katz, and Chet A. Rock. Water quality effects of using tire chips below the groundwater table. Master's Thesis, in Civil Engineering, University of Maine, 1996.

21. Selbes, Meric. Leaching of dissolved organic carbon and selected inorganic constituents from scrap tires. PhD Diss., Clemson University, 2009.

22. Maeda, Richela K. Water quality evaluation of tire derived aggregate. 2016.

23. Abe, Nagato, Hironobu Maehara, and Teruhiko Maruyama. The research on reflective cracking inhibition effect using stress absorbing membrane interlayer. *Journal of Pavement Engineering, JSCE* 4: 95-102, 1999.

24. de Souza Gaspa, Matheus, Kamilla L. Vasconcelos, and Liedi Légi Bariani Bernucci. Adhesion Between Asphalt Layers Through the Leutner Shear Test. In *8th RILEM International Conference on Mechanisms of Cracking and Debonding in Pavements*, pp. 495-500. Springer Netherlands, 2016.

25. Roy, Satyajit, and Mahabir Dixit. Use of Glass Grid and SAMI as Reinforced Interlayer System in Runway. In *Geotechnics for Transportation Infrastructure: Recent Developments, Upcoming Technologies and New Concepts*, Volume 2, pp. 283-294. Springer Singapore, 2019.

26. Jin, Dongzhao, Kwadwo Ampadu Boateng, Dongdong Ge, Tiankai Che, Lei Yin, Wayne Harrall, and Zhanping You. A Case Study of the Comparison between Rubberized and Polymer Modified Asphalt on Heavy Traffic Pavement in Wet and Freeze Environment. *Case Studies in Construction Materials*: e01847, 2023.

27. Ogundipe, O. M., N. H. Thom, and A. C. Collop. Evaluation of performance of stress-absorbing membrane interlayer (SAMI) using accelerated pavement testing. *International Journal of Pavement Engineering* 14, no. 6: 569-578, 2013.

47

Chemical Pre-Treatment of Tire Rubbers for Froth Flotation Separation of Butyl and Non-Butyl Rubbers

Haruka Pinegar and Jeffrey Spangenberger*

Argonne National Laboratory, Lemont, Illinois, USA

Abstract

Recycling of tire rubbers is an emerging issue for the reduction of greenhouse gas emissions in manufacturing industries, considering tremendous energy consumption in tire manufacturing and waste generation. Current technologies allow shredding of used tire rubber compounds into a fine powder mixture, Micronized Rubber Powders (MRPs). MRPs are used for additives in various rubber products and cement filler. Reincorporation of MRPs into new tire production will significantly reduce the environmental impact of tire manufacturing. However, MRPs are a mixture of butyl and non-butyl rubbers from different tire parts. The current reincorporation rate of MRPs into the new tread production is limited to 10% to ensure tire performance. Separation of butyl rubber from MRPs to reduce butyl rubber content down to 2% will make it possible to double the current reincorporation rate of MRPs, corresponding to the yearly reduction of primary stock requirement by 110,000 metric tons. However, robust method for butyl and non-butyl rubbers separation has yet to come.

This research work investigated froth flotation as a low-cost method to separate butyl and non-butyl rubber materials. Experiments were conducted on cut pieces of representative butyl and non-butyl tire rubber plaques, which had close texture and chemical composition to real tire rubbers, and shreds of real used tires. Tire rubbers were chemically treated with acidic or basic solutions at various operating conditions before froth flotation of tire rubbers. Effective separation of representative butyl and non-butyl rubber pieces was demonstrated by froth flotation after the 10 minutes of pre-treatment in hydrochloric acid solution at pH2 and 60°C, achieving 97% purity butyl rubber in the sinking fraction and 100% purity non-butyl rubber in the floating fraction. However, the same acidic pre-treatment and froth flotation of the real used tire shreds did not separate butyl and non-butyl rubbers. This is likely due to the additives in the real used tires that were not contained in the representative tire plaques. Further investigation is required for the effective separation of butyl and non-butyl rubbers from a mixture of real used tire shreds.

Keywords: Tire recycling, froth flotation, surface treatment

Corresponding author: JSpangenberger@anl.gov

Nabil Nasr (ed.) Technology Innovation for the Circular Economy: Recycling, Remanufacturing, Design, Systems Analysis and Logistics, (625–638) © 2024 Scrivener Publishing LLC

47.1   Introduction

Vehicle tires are composed of various materials including several types of rubbers, fibers, steel, carbon black, silica and special oil. Manufacturing of vehicle tires is highly energy intensive as it involves extraction of raw materials, transportation, manufacturing of components, and assembling. The tire manufacturing process is estimated to require approximately 74 million BTU of energy per short ton of tire produced [1]. Despite the high energy consumption in manufacturing, tremendous amounts of used tires are discarded to landfills, which adversely affects the environment. When used tires are recycled, they are shredded into crumbs, and then further ground into micronized rubber powders (MRPs) by cryogenic grinding. MRPs are currently used for additives in various rubber products and cement filler [1]. However, the usage of MRPs in tire manufacturing is still limited. With current technology, tire-to-tire tread compound recycling of MRPs is about 10% to maintain tire performance, as MRPs are a mixture of butyl and non-butyl rubbers if the feedstock of MRPs is the entire used tire. Butyl rubber contamination in MRPs (up to about 15%) diminishes the performance of tire tread when present at high loading levels due to incompatibility with non-butyl rubber [2]. Therefore, butyl rubber in MRPs needs to be separated from non-butyl rubbers so that MRPs can be used for the new tire tread production at a higher recycling rate. Tire inner liner is composed of halo-butyl rubber, and traditionally it can be stripped manually. However, automation of this method is necessary to process tons of used tire rubbers at an industrial scale. It requires new technologies that can separate butyl and non-butyl rubbers from MRPs continuously, which improves MRP's recyclability into new tire production at lower cost. If the butyl content in the separated non-butyl fraction is reduced to less than 2%, it will make it possible to double the current MRP's tire-to-tire recycling rate into the tread production. This is corresponding to the yearly reduction of primary stock requirement by 110,000 metric tons, the reduction of energy consumption by 9.7 trillion BTU, and the reduction of about 189,000 metric tons of CO_2 produced by light-duty and commercial vehicle tire manufacturers in the U.S.

Froth flotation is a commercially established material separation method, which separates hydrophilic and hydrophobic materials. In this process, a layer of small bubbles (froth) is created on top of the solution containing organic surfactant by mechanical agitation or air injection. Hydrophobic materials attach to the froth with the assist of "collector" chemical, while hydrophilic materials sink to the bottom of the solution. Froth flotation is a low-cost and industrially established method, which will be beneficial to apply to the tire recycling process. As tire rubbers are hydrophobic as they are, surface treatment is required to modify the difference of hydrophobicity/hydrophilicity between butyl and non-butyl tire rubbers.

This research aimed to conduct proof-of-concept experiments for the butyl/non-butyl rubber separation by froth flotation after the chemical treatment. First, the experiments were conducted using millimeter-size tire rubber plaques. These plaques were colored differently between butyl and non-butyl rubbers so that the difference in the flotation behavior was clearly observed. After determining the process condition, it was applied to the real used tire rubber. With a success of butyl/non-butyl tire rubbers separation in millimeter-size, these separated pieces can be further ground to be MRPs with higher recyclability.

47.2   Research Work

47.2.1   Materials

Experiments were conducted on the four kinds of tire rubber plaques, representing tire tread, sidewall and inner liner, and the shreds of real used tire rubbers. Characterization of tire rubber samples were conducted by X-ray Fluorescence Spectrometer (Niton XL5 XRF analyzer), Energy Dispersive Spectroscopy (Phenom XL Benchtop SEM Microscope) with applying gold coating on the surface of samples, and Fourier Transform Infrared Spectroscopy (Nicolet iS20 FTIR spectrometer) using a Germanium Attenuated Total Reflection (ATR) crystal. Chemical bonds in materials were determined using the IR spectrum table and chart provided by Sigma-Aldrich [3] and other literatures.

Representative tire rubber plaques were made specifically for the flotation separation experiments. Butyl rubber plaques were colored in white or beige and non-butyl tire rubbers were colored in black so that the separation of butyl and non-butyl rubbers would be visually observed. Before the experiments, each tire rubber plaque was cut into 3 mm – 5 mm size. Figure 47.1(a) shows the picture of tire rubber plaques pieces. The top black pieces are non-butyl rubbers representing tread (top-left) and sidewall (top-right), and the bottom pieces are butyl rubbers representing an inner liner. The shreds of real used tires were sieved, and the fraction larger than 5.6 mm was separated from the other smaller fractions. Several pieces were manually picked up from the >5.6 mm fraction, and analyzed to determine whether they were butyl or non-butyl rubbers. After verifying the types, butyl and non-butyl rubber pieces were separated, and cut into 3 – 5 mm size for the chemical treatment and froth flotation experiments. Figure 47.1(b) shows that real used tire rubbers are black due to the carbon black filling, and butyl (top in Figure 47.1(b)) and non-butyl (bottom in Figure 47.1(b)) rubbers would hardly be distinguished visually.

Chemical structures of non-butyl tire rubbers and butyl tire rubbers are quite different as shown in Figure 47.1(c). Non- butyl tire rubbers, such as natural rubber, butadiene rubber and styrene butadiene rubber, contain carbon-carbon double bonds (C=C) in each monomer. Halobutyl rubber is composed of ~98% poly isobutylene (-C_4H_8-) and ~2% of polyisoprene (-C_5H_8-) with small quantity of H atoms replaced with halogen. Thus, compared to non-butyl tire rubbers, butyl tire rubbers are mostly composed of single carbon-carbon

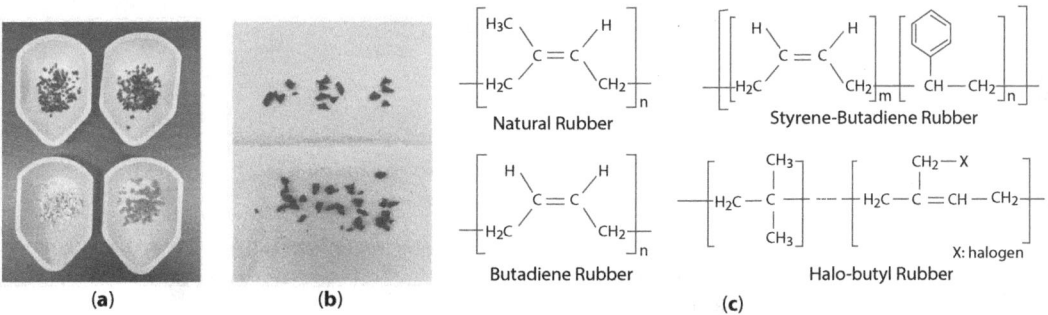

Figure 47.1 Millimeter-size pieces of (a) representative tire rubber plaques and (b) real used tire rubber shreds used for the chemical treatment and froth flotation experiments and (c) chemical structures of non-butyl and butyl tire rubbers.

bonds (C-C). Figure 47.2 shows FTIR peaks of non-butyl tire rubbers (Figure 47.2(a)) and butyl tire rubbers (Figure 47.2(b)), comparing the peaks of representative tire rubber plaques and real used tire rubbers. The graphs are also enlarged at wavelength 600 – 1800 cm^{-1}, where various peaks are shown. In Figure 47.2(a), peaks around 3010cm^{-1}, 1600-1700cm^{-1}, 950-1000cm^{-1}, 800-850cm^{-1} and 700-750cm^{-1} indicates that non-butyl rubber materials contain C=C bonds. From 700-750cm^{-1} peaks, which indicate presence of 1,2-disubstituted alkene, these non-butyl tire rubbers likely contain Butadiene rubbers. In Figure 47.2(b), strong peaks in 1000-1100cm^{-1} suggests butyl rubber contains silica [4]. It also shows stronger peaks for C-H bonds in alkane in 1400- 1500cm^{-1} and 1350-1400cm^{-1}. The peaks in 650-700cm^{-1} are likely for C-Br bonds. Real used tire rubber materials contain various additives such as antioxidants, which may attribute to the peaks of amine (N-H bonds: 1580-1650cm^{-1}, C-N bonds: 1266-1342cm^{-1} and 1020-1250cm^{-1}) and aromatic bonds (C=C bonds: 1500-1700cm^{-1}, C-H bonds: 680- 860cm^{-1}).

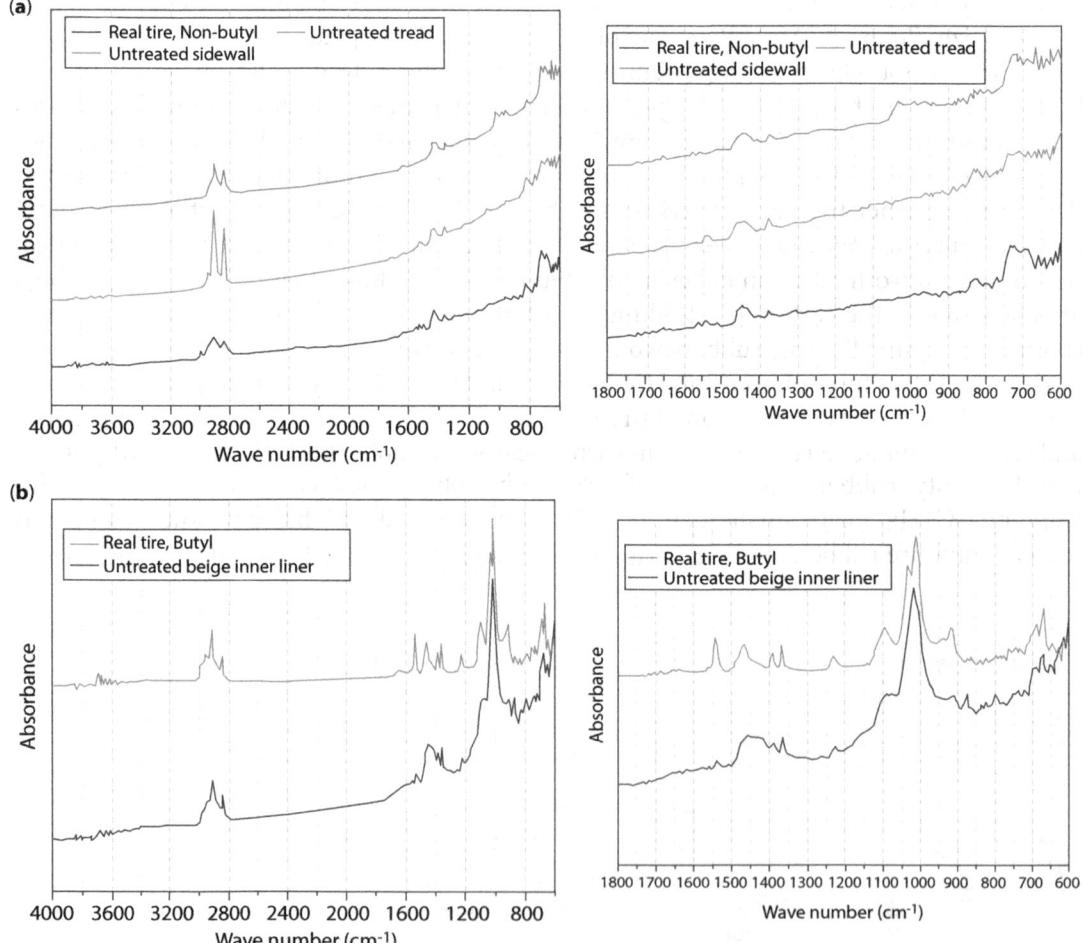

Figure 47.2 FTIR peaks of (a) non-butyl tire rubbers (representative tread and sidewall plaques and real tire non-butyl rubber) and (b) butyl tire rubbers (representative beige-colored inner liner and real tire butyl rubber).

Table 47.1 Result (mass percentage) of elemental analysis by XRF using the plastic analysis mode.

Element	Representative tire rubber plaques				Real used tire rubber	
	Tread	Sidewall	Inner liner (white)	Inner liner (beige)	Butyl rubber	Non-butyl rubber
Balance	97.39	97.78	89.72	97.15	95.88	96.93
Zinc	2.56	2.18	1.25	1.08	2.15	2.67
Bromine	0.01	< 0.01	1.47	1.29	1.476	<0.01
Iron	0.01	< 0.01	0.08	0.21	0.18	0.168
Selenium	0.012	< 0.01	0	0	0	<0.01
Titanium	0	0	7.31	0.17	0.021	0.02
Vanadium	0	0	0.07	0	<0.01	<0.01
Calcium	0	0	0.06	<0.01	0.09	0.182
Chlorine	0	0	0	0	0.184	0

Table 47.1 is the XRF results of tire rubber plaques and real used tire rubbers, showing elemental contents above 0.01 wt% detected using the plastic analysis mode setting. Butyl rubber was verified by the amount of bromine content. It must be noted that XRF analysis is not effective for the elements with atomic number less than 16, such as oxygen, magnesium, silicon and sulfur. Therefore, EDS analysis was also conducted to detect other elements that were not detected by XRF. EDS analysis of the samples detected that real non-butyl tire rubbers contained more oxygen than the non-butyl tire rubber plaques. XRF and EDS analysis suggested that white and beige butyl tire rubber plaques were in their colors due to their titanium and magnesium contents. Compared to non-butyl tire rubbers, butyl tire rubbers contained more silicon in both tire rubber plaques and real used tire rubbers.

47.2.2 Experimental Methods

Three grams total of butyl rubber and non-butyl tire rubber plaque pieces were mixed at the mass ratio of 15% butyl and 85% non-butyl tire rubber (2:3 mass ratio of tread: sidewall), considering the approximate composition of MRPs (15% inner liner, 20% tread, 35% sidewall and 30% body ply/tread ply/bead area). Chemical treatment of the tire rubber plaque mixture in the conical flask was performed using pH2 hydrochloric acid (HCl) solution or pH12 sodium hydroxide (NaOH) solution. 300 ml of prepared solution was heated up to various testing temperatures on the magnetic hotplate stirrer before adding the tire rubber plaque mixture in the solution. Chemical treatment was performed at 300 RPM stirring speed, and after the treatment, the solution was immediately cooled down to room temperature by soaking the conical flask in a water bath. After the chemical treatment, tire rubber pieces were separated from the process solution by vacuum filtration, and rinsed with de-ionized water.

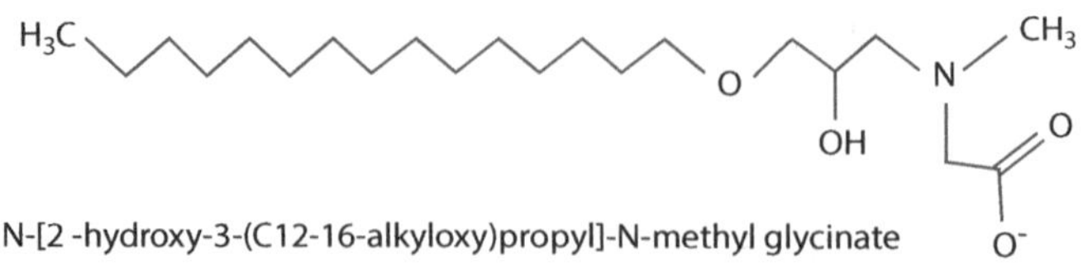

N-[2 -hydroxy-3-(C12-16-alkyloxy)propyl]-N-methyl glycinate

Figure 47.3 Chemical structure of N-[2 -hydroxy-3-(C12-16-alkyloxy) propyl]-N-methyl glycinate.

Simple froth flotation tests were performed using Atrac 922 collector. Collector is a surfactant chemical used to make the surface of targeted materials hydrophobic so that they are selectively attached to a froth layer. Atrac 922 is composed of 50-70 % N-[2-hydroxy-3-(C12-16-alkyloxy)propyl]-N-methyl glycinate, 10 – 20% Propylene glycol and 1 – 5 % Ethyl diglycol. The anionic head of the main chemical, N-[2-hydroxy-3-(C12-16-alkyloxy) propyl]-N-methyl glycinate (Figure 47.3), attaches to the surface of targeted materials by dipole-dipole attraction, while the long hydrocarbon chain on the other end attaches to bubbles.

Froth flotation test was set up using a 1 L beaker. The tire rubbers were placed in 1000 ml de-ionized water, and manually stirred to verify that tire rubber pieces sink before starting the flotation process. The water was stirred at 300 RPM with the 5 drops (~20 ppm) of frother, Methyl Isobutyl Carbinol (MIBC) for 3 minutes. Afterwards, diluted Atrac922 solution was added to the water and stirred for another 3 minutes. Air was provided through a plastic tube connected to the in-house air supply line to generate a froth layer. Airflow was adjusted so that the flotation solution would not splash out of the beaker. The rubber pieces floating on the froth layer and those sinking at the bottom of the beaker were collected separately, and they were dried in a vacuum oven at 80°C. Mass of floating and sinking fractions were measured to determine the butyl/non-butyl tire rubber separation performance.

To verify the floatability of rubbers after the chemical treatment, the experiments of butyl and non-butyl pieces collected from the real used tire rubber shreds were conducted separately. 0.16 grams of butyl rubber pieces and 0.85 grams of non- butyl rubber pieces, respectively, were used in the chemical treatment test. The chemical treatment was performed with a 100 mL solution in a 100 mL conical flask. Flotation test was conducted in the same manner as the experiment with tire rubber plaques.

After the experiments, tire rubber samples were analyzed by FTIR, XRF, and X-ray Photoelectron Spectroscopy (PHOIBOS 150 hemispherical energy analyzer).

47.2.3   Experimental Results

Table 47.2 shows the froth flotation separation results of the mixture of representative tire rubber plaques after the chemical treatment with various conditions. Almost 100% of non-butyl tire rubber plaques were collected in the floating fraction regardless of the testing condition. Majority of butyl rubber plaques were collected with non-butyl tire rubber plaques in the floating fraction if no chemical pretreatment was conducted. pH2 HCl treatment was effective to increase the butyl rubber recovery in the sinking fraction. By increasing the HCl solution temperature to 60°C from 20°C, butyl rubber recovery was improved to 100% from 49%, compared with the same the chemical treatment time and froth flotation condition. On the other hand, increasing the chemical treatment time and Atrac 922

Table 47.2 Butyl/non-butyl rubber separation results and conditions of chemical treatment and froth flotation.

| Operating condition | | | | Results | | | |
Solution	Chemical treatment temperature	Chemical treatment time	Atrac 922 concentration	Butyl rubber collected in the sinking fraction	Non-butyl rubber collected in the floating fraction	Butyl purity in the sinking fraction	Non-butyl purity in the floating fraction
No treatment	N/A	N/A	0.5 ppm	27%	100%	100%	89%
pH 2 HCl	20°C	10 min	0.5 ppm	49%	100%	98%	92%
pH 2 HCl	60°C	10 min	0.5 ppm	100%	99%	97%	100%
pH 2 HCl	60°C	30 min	1 ppm	32%	100%	100%	90%
pH 12 NaOH	60°C	10 min	0.5 ppm	72%	99%	92%	95%

concentration worsened the separation efficiency. After the 10 minutes of chemical treatment with 60°C NaOH solution, 72% butyl rubber was collected in the sinking fraction, although the separation efficiency was not as high as the treatment with 60°C HCl solution. The best separation result was achieved with 97% purity butyl tire rubber and 100% purity non-butyl tire rubber after the 10 minutes of pH2 HCl treatment at 60°C followed by froth flotation with 20 ppm MIBC frother and 0.5 ppm Atrac 922 collector.

Figure 47.4(a) shows the photo of the non-butyl tire rubber plaque pieces floating on top of the froth layer and the butyl tire rubber plaque pieces sinking at the bottom of the beaker during the froth flotation experiment. Figure 47.4(b) and (c) shows the obvious difference in butyl tire rubber collected in the sinking fraction.

Figure 47.5 shows the result of FTIR analysis comparing the representative tire rubber plaques before and after the 10 minutes pH2 HCl solution treatment at 60°C. In Figure 47.5(a) and (b), the strength of peaks in 700-750cm^{-1} reduced. Figure 47.5(b) also shows the reduction of peaks in 3010cm^{-1} and 950 – 1000 cm^{-1}, while new peaks are shown in 1450- 1500cm^{-1}. These changes in FTIR peaks indicate that C=C bonds in non-butyl rubbers were cracked by the HCl treatment. Figure 47.5(c) shows small reduction of peaks in 950-1000cm^{-1} and 650-700cm^{-1}, indicating the reduction of C=C and C-Br bonds. The cracking of C=C bonds would be much smaller considering small quantity of polyisoprene in butyl rubber. Peaks shown in 1500-1550cm^{-1} after the HCl treatment are possibly due to new formation of N-O bonds from the small quantity of N-containing additives.

Figure 47.6 compares the XPS peaks of C 1s and O 1s before and after the 10 minutes pH2 HCl treatment of representative tire rubber plaques at 60°C. XPS peaks of butyl tire rubber plaques were calibrated based on the location of Si 2p peak of SiO2. XPS peaks of non-butyl rubber plaques were not calibrated due to the lack of knowledge of carbon condition and no existence of SiO2, although they were most likely shifted to the lower binding energy (typical C-H/C-C binding energy range: 284.8-285.5). The results shows O/C peak intensity ratio of tread plaque increased from 0.015 to 0.11 after the treatment, while those

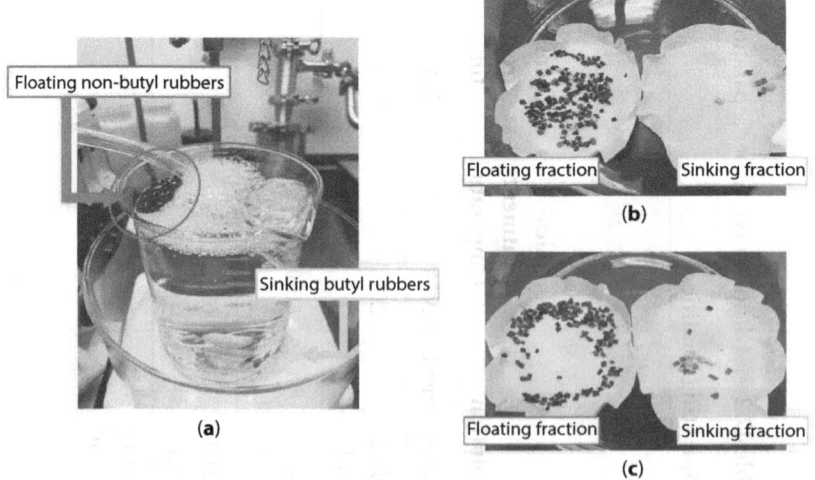

Figure 47.4 Photos of (a) floating non-butyl rubber plaque pieces and sinking butyl rubber plaque pieces during the froth flotation experiment, separated floating and sinking fractions after the flotation test of a tire rubber plaque mixture treated by pH2 HCl solution for 10 minutes at (b) 20°C and (c) 60°C.

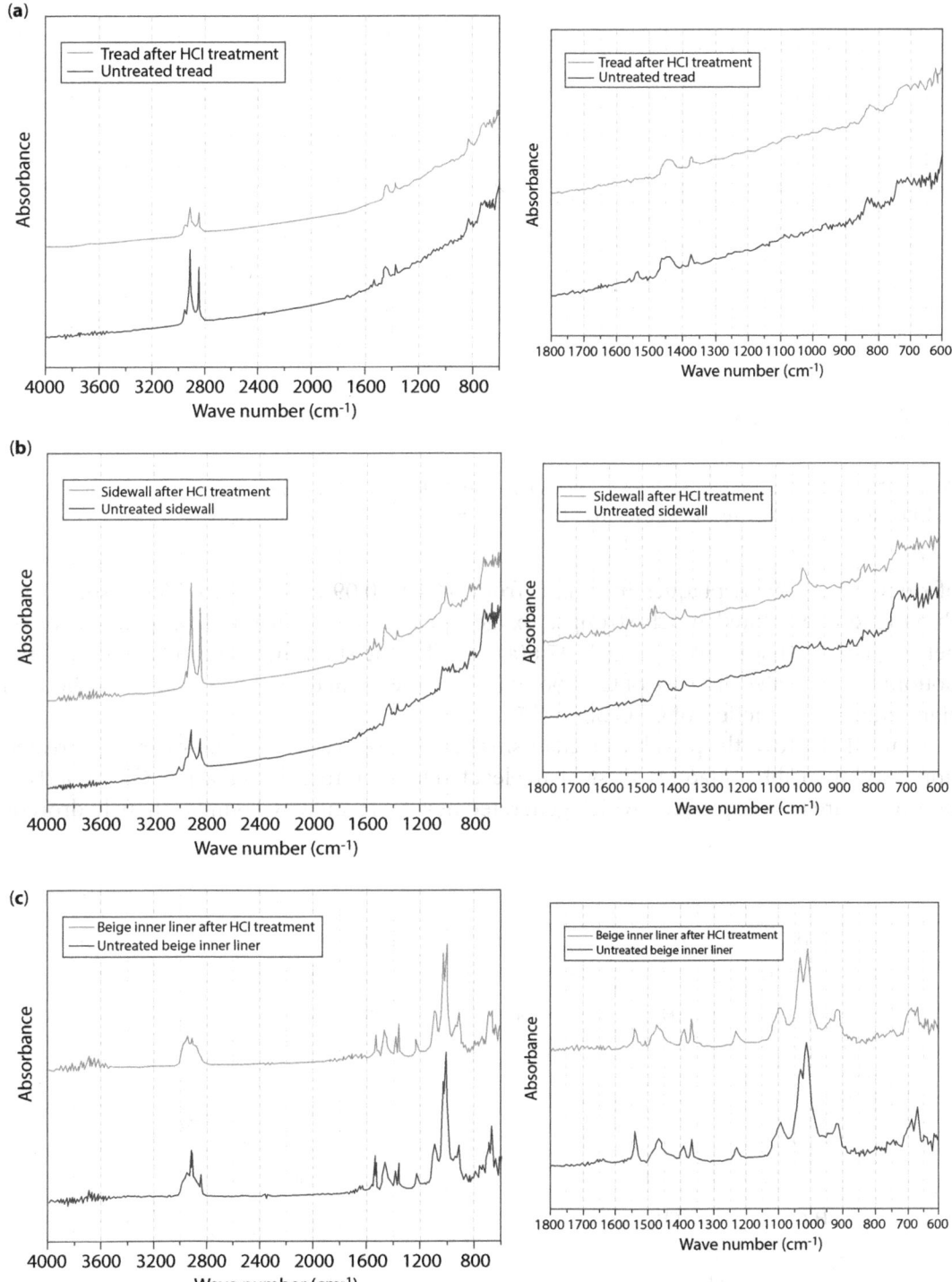

Figure 47.5 FTIR peaks of tire rubber plaques before and after the treatment with pH2 HCl solution at 60°C for 10 minutes: (a) tread, (b) sidewall and (c) beige inner liner.

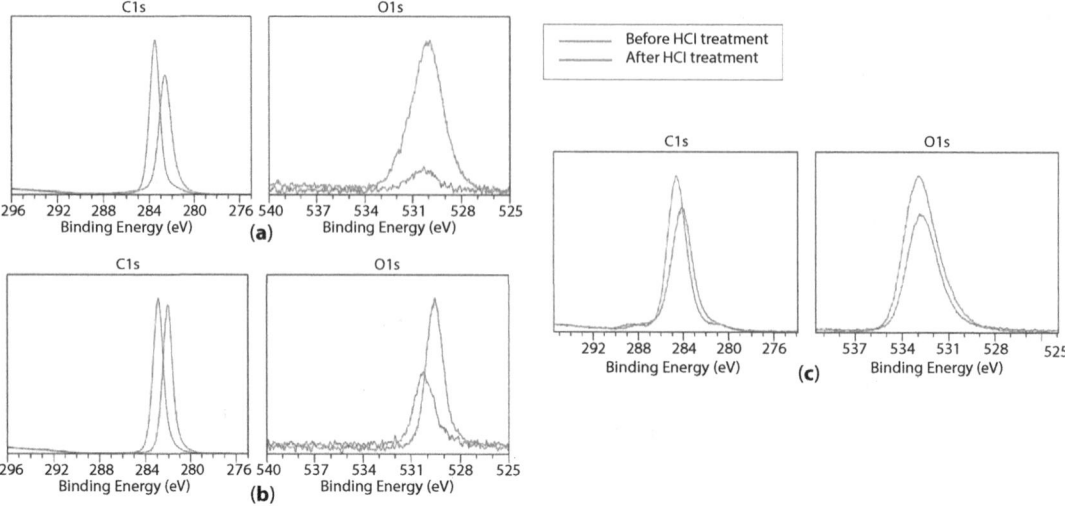

Figure 47.6 XPS peaks of representative (a) tread, (b) sidewall and (c) beige inner liner rubber plaques before and after the 10 minutes treatment with pH2 HCl solution at 60°C.

of sidewall and butyl plaques increased from 0.04 to 0.09 and 0.95 to 1.59, respectively. Butyl plaque exhibited much stronger oxygen peak due to silica and magnesium oxide before the HCl treatment. In Figure 47.6(a) and (b), the slight increase in C1s peak at the bottom left indicates increase of C-O bonds, while the reduction of C1s peak at the bottom right indicates reduction of C=C bonds [5].

Figure 47.7 shows the possible mechanisms that allowed selective flotation of HCl-treated non-butyl tire rubber plaques. First, at an elevated temperature, HCl cracked C=C bonds in non-butyl tire rubber plaques, which generated new C-H and C-Cl bonds. Then, hydrolysis

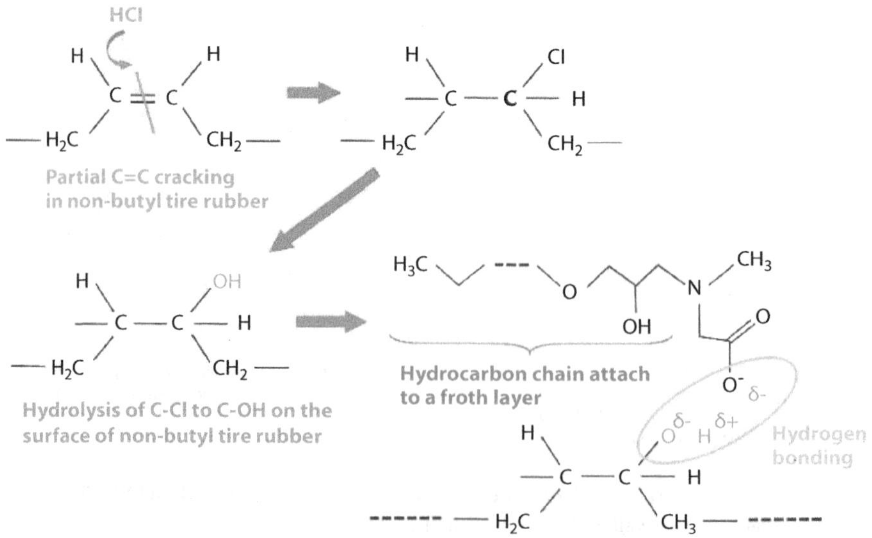

Figure 47.7 Possible mechanism of selective flotation of HCl-treated non-butyl tire rubber.

of C-Cl bonds in water generates C-OH bonds. Due to hydrogen bonding between anionic head of N-[2-hydroxy-3-(C12-16-alkyloxy)propyl]-N-methyl glycinate in Atrac 922 and electropositive H atom in hydroxyl, non-butyl rubbers attach to a froth layer and are collected in the floating fraction. Although this C=C cracking and hydrolysis may happened in butyl rubber, a quantity of hydroxyl that would be attracted to the collector agent was much smaller compared to non-butyl rubber plaques. Also, metal oxides such as silica and magnesium oxide in butyl rubber likely attributed to the higher hydrophilicity of butyl rubber plaques.

The 10 minutes of chemical treatment with 60°C pH2 HCl solution was applied to the butyl and non-butyl tire rubbers from the real tire rubber shreds separately, and the froth flotation was conducted with 20 ppm MIBC frother and 0.5 ppm Atrac 922 collector mixed in 1000 mL de-ionized water. Using the real tire rubber shreds in this experiment, all of butyl and non-butyl tire rubbers were collected in the sinking fraction. Even with Atrac 922, non-butyl rubbers were repelled from the froth layer and was not attached to the froth. Figure 47.8 shows the FTIR peaks of non-butyl tire rubber from the real tire rubber shreds, comparing before and after the HCl treatment. The clear peak at 3010cm^{-1} and peaks in 1600-1700cm^{-1}, which respectively represent C-H bond and C=C bond of alkene, are shown after the HCl treatment. These results show that cracking of C=C bonds on non-butyl tire rubber did not occur. The characterization results shown above suggests that non-butyl tire rubbers from the real used tire rubber shreds contained various additives that were not in the non-butyl tire rubber plaques. Such additives, including various organic compounds, likely prevented the cracking of C=C bonds by HCl and hydrolysis of C-Cl to C-OH that would make the surface of non-butyl rubber attach to the collector agent. In addition, surface contamination of tread and sidewall due to their usage over the long period of time may affect the chemistry of non-butyl tire rubbers.

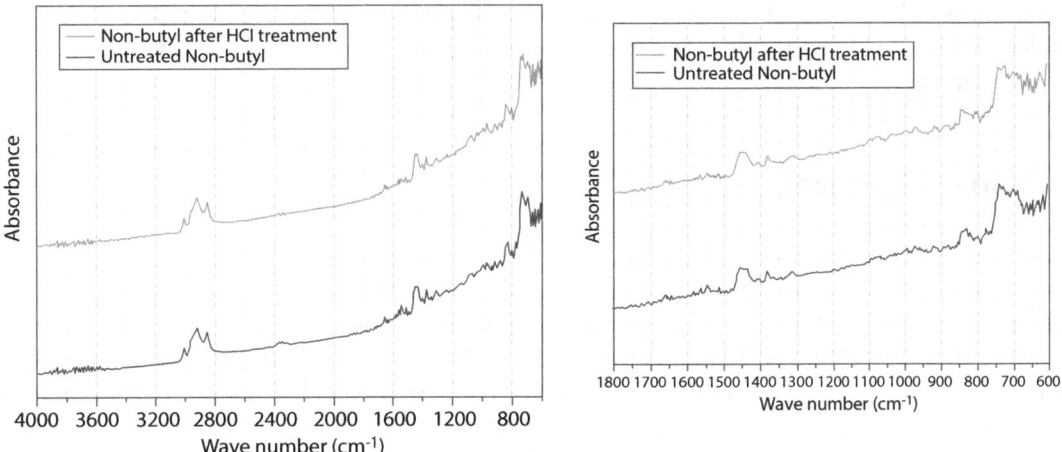

Figure 47.8 FTIR peaks of non-butyl tire rubber from the real used tire rubber shreds, comparing before and after the treatment with pH2 HCl solution at 60°C for 10 minutes.

47.3　Conclusions and Recommendations

In this study, froth flotation separation of butyl and non-butyl tire rubbers after the chemical pre-treatment was investigated. Millimeter size tire rubber pieces were used for the experiments, assuming that separated butyl and non- butyl tire rubber pieces will be further ground to MRPs with a higher recyclability. After the 10 minutes of chemical treatment with pH2 HCl solution at 60°C, representative butyl and non-butyl tire rubber plaques were successfully separated by the froth flotation method using 20 ppm and Methyl Isobutyl Carbinol and 0.5 ppm Atrac 922. The best result achieved 97%-purity butyl tire rubber in the sinking fraction and 100%-purity non-butyl tire rubber in the floating fraction. FTIR results suggested the reduction of C=C bonds, and XPS results exhibited increase in oxygen in both tire rubber plaques, which would make them hydrophilic, although these reactions happened more on non-butyl tire rubber plaques. These results suggested the possible mechanism of selective flotation of non-butyl rubber. HCl cracked C=C bonds on the surface of non-butyl tire rubber and generated new C-H and C-Cl bonds on each carbon atom in the remaining C-C bonds. Subsequently, hydrolysis of C-Cl bonds in water replaced Cl with hydroxyl (-OH), and electropositive H atom on hydroxyl attracted anionic head of N-[2-hydroxy-3-(C12-16-alkyloxy)propyl]-N- methyl glycinate in Atrac 922. The long hydrocarbon portion of the collector on the other side attached to the floating bubbles, which separated non-butyl tire rubbers to the floating fraction and butyl tire rubbers into the sinking fraction.

However, the HCl treatment was not effective to separate butyl and non-butyl tire rubbers from the real used tire rubber shreds, as both butyl and non-butyl tire rubbers sunk during the flotation process. It is likely that real non-butyl tire rubbers did not float due to the additives that were not contained in the representative non-butyl tire rubber plaques. These additives possibly prevented the cracking of C=C bonding on non-butyl tire rubbers by HCl and generation of hydroxyl, and thus, the collector agent did not attach to the surface of non-butyl rubbers.

More investigation is required to successfully separate butyl and non-butyl tire rubbers from the real used tire shreds. That will include further investigation of the changes in surface chemistry of butyl and non-butyl rubbers from the real tire rubber shreds, effective collector for non-butyl rubbers and reduction of the impurity's impacts on froth flotation separation performance.

Acknowledgement

This research work was funded by Laboratory Directed Research and Development (LDRD) Seed funding at Argonne National Laboratory and performed at Materials Engineering Research Facility (MERF). The tire rubber plaques and real used tire rubber shreds were kindly supplied by Robert Radulescu at Michelin. He also estimated the energy consumption reduction in tire manufacturing industry as a potential outcome of the success of butyl/non-butyl tire rubber separation by the proposed method. XPS analysis was performed by Justin Connell at Joint Center for Energy Storage Research (JCESR) facility at Argonne National Laboratory.

References

[1] U.S. Environmental Protection Agency Office of Resource Conservation and Recovery, Documentation for Greenhouse Gas Emission and Energy Factors Used in the Waste Reduction Model (WARM) Tires, https://www.epa.gov/sites/default/files/2020-12/documents/warm_tires_v15_10-29-2020.pdf , 2020

[2] M. Theusner, Method for processing rubber granules or rubber powders, US Patent 2013/0225766 A1, assigned to Continental Reifen Deutschland GmbH, 2013

[3] Sigma-Aldrich, IR Spectrum Table & Chart, https://www.sigmaaldrich.com/US/en/technical-documents/technical-article/analytical-chemistry/photometry-and-reflectometry/ir-spectrum-table, 2023

[4] Kralevich, M.L., Koenig, J. L., FTIR Analysis of Silica-Filled Natural Rubber. *Rubb. Chem. Tech.*, 71(2), 300, 1998

[5] He, M., Gu, K., Wang, Y., Li, Z., Shen, Z., Liu, S., Wei, J., Development of high-performance thermoplastic composites based on polyurethane and ground tire rubber by *in-situ* synthesis. *Res. Cons. Rec.*, 173, 105713, 2021

48

Development of Manufacturing Technologies to Increase Scrap Steel Recycling Into New Tires

Seetharaman Sridhar[1]*, Subramaniam Rajan[1], Robert Radulescu[2] and Narayanan Neithalath[1]

¹Arizona State University, Tempe, AZ, USA
²Michelin North America, Greenville, USA

Abstract

This paper describes our projects efforts to develop innovative processing technologies that can enable a greater rate of recycling of steel scrap into the manufacturing of new tires since it is known that recycling steel scrap has the potential to reduce up to 90% of embedded energy when compared to carbon-based reduction from iron-ore. To resist harsh conditions during use, tires are built with steel cables constructed by assembling together fine wires of diameters as low as 0.2 mm. To obtain these fine wires, steelmakers first hot roll square billets to steel cords of 5.5 mm in diameter. Steel wire makers or tire manufacturers then cold-draw these cords to the finer diameters mentioned. During the cold drawing process, the fine wires are subjected to high levels of tensile and torsional stresses that are known to generate wire breakage when impurity levels are too high. For this reason, since most steel scrap has about 0.23% Cu originating from wires and motors in cars, current use of recycled steel scrap for producing new tires is limited to 20%, with the remaining coming from virgin iron sources, primarily pig iron. The goal of this work is to increase the scrap recycling rate from 20% to 80% while maintaining the industrial performance of the tire manufacturing process and the product quality. We will identify the location of Cu-containing phases, and develop, validate and optimize a processing step to mitigate the detrimental role of copper.

Keywords: Steel, tire cord, tires, recycling, copper, wire drawing, oxidation, cracking

48.1 Introduction

Life cycle analysis (LCA) of light duty and commercial vehicle tires indicates that about 90% of their environmental impact is generated during product use, as demonstrated in Figure 48.1 [1]. For this reason, tires are designed to be long lasting, energy efficient, and reused as often as possible. Unfortunately, recycling of secondary feedstock into new tires is today limited to small reincorporation rates so as to not impact the product performance and manufacturing process. Seeking solutions that can increase tire recycling content while preserving the product and industrial performance will have significant environmental

---

Corresponding author: seetharaman@asu.edu

Nabil Nasr (ed.) Technology Innovation for the Circular Economy: Recycling, Remanufacturing, Design, Systems Analysis and Logistics, (639–650) © 2024 Scrivener Publishing LLC

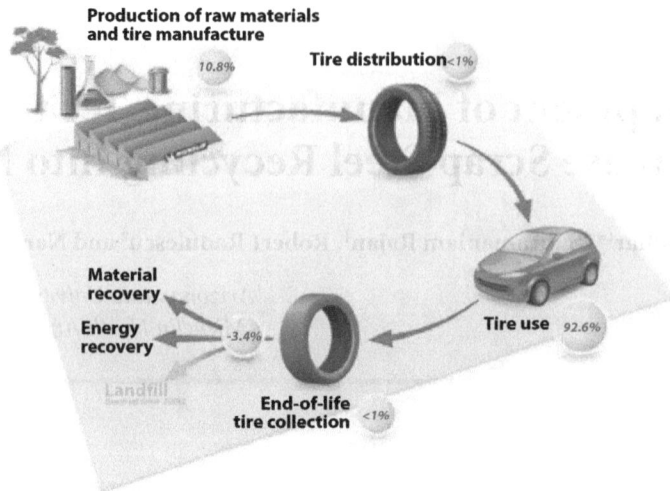

Figure 48.1 Life cycle assessment of a passenger car tire [1].

benefits since the US tire industry requires a yearly consumption of more than 3.6 million tons of materials to manufacture 300 million tires for the transportation industry.

Steel is among the world's most recycled material, with end-of-life recovery rates as high as 90% [2]. This is quite appropriate since steel production accounts for approximately 25% of industrial emissions [3]. To provide long-lasting performance and ensure energy efficiency, tires are built with steel cables constructed by assembling together fine wires of diameters as low as 0.2 mm (Figure 48.2(a)). To obtain these fine wires, steelmakers first hot roll square billets to steel cords of 5.5 mm in diameter. Steel wire makers or tire manufacturers then cold draw these cords to finer diameters. During the cold drawing process, the fine wires are subjected to high levels of stresses. However, when the end-of-life steel scrap is contaminated with other metals, as often is the case, they result in reduced mechanical properties of the material, such as premature wire breakage and low tire performance,

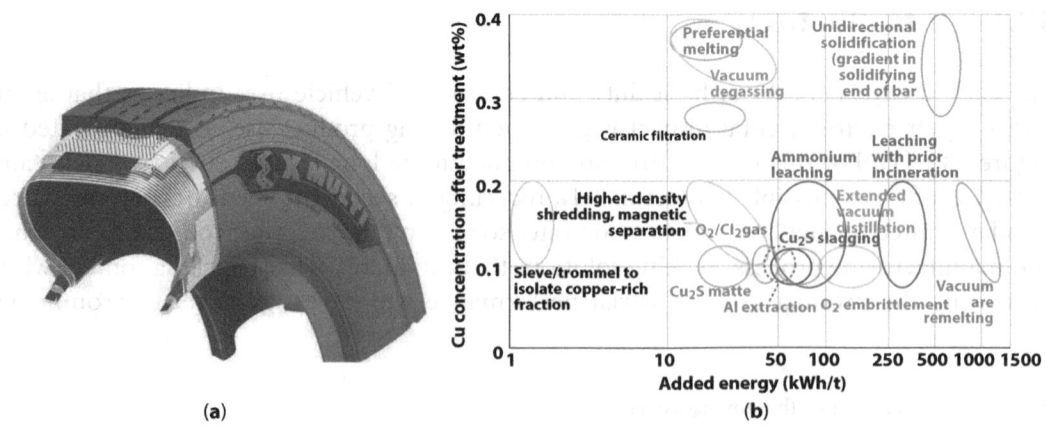

Figure 48.2 (a) Internal construction of the tires, showing the steel cords; and (b) Energy consumption in reducing copper concentration in steel, through the use of several methods [7].

Table 48.1 Allowable maximum copper content (wt.%) [6].

Product	IF steel	DDQ	Drawing	Commercial	Structural	Fine wire	Rebar
Cu	0.030	0.040	0.060	0.100	0.120	0.070	0.400

restricting their use in tire manufacture. Copper, as an impurity in steel, is pervasive in end-of-life scrap, known to cause metallurgical problems, and cannot currently be removed commercially once in the melt. Copper is very commonly present in end-of-life steel scrap, originating mostly from copper wires and motors in automobiles, appliances, and machinery. Concentrations of copper > 0.1 wt % cause hot shortness, a phenomenon leading to surface cracking in hot rolling and forming [4], [5]. Hot shortness causes embrittlement through the formation of liquid Cu-enriched phases that infiltrates the grain boundaries. Embrittlement can occur during casting, re-heating, and hot rolling in the steel plant, when the surface of the steel is oxidized. The tolerances to copper are dependent on the targeted product, and the more thin gauged the product is, the lower the tolerance, as shown in Table 48.1. Economically, thin gauged products are more valued per ton, and therefore steel recycling in general results in a down grade in terms of value. Thus copper limits the recyclability of steel.

Thermodynamically, removing copper from scrap is a viable option, but in reality, the impurities and the included copper in the melt of steel scrap are difficult to remove by conventional methods. This is an energy intensive process as well, as is shown in Figure 48.2(b) [6]. For this reason, since most steel scrap has about 0.23% Cu [7], current use of recycled steel scrap for producing new tires is limited to 20%, or 0.046% Cu. In fact, the actual Cu impurity threshold where the fine wires begin to break during the cold drawing process is not precisely known.

This necessitates innovative processing technologies that can enable a greater rate of recycling of steel scrap into the manufacturing of new tires since it is known that recycling steel scrap has the potential to reduce approximately 70% of the energy consumption and carbon emissions when compared to carbon-based reduction from iron-ore [8]. When considering the energy and C intensity of agglomeration and sintering the savings become even more. Alternative methods for impurity removal that can be applied upstream of the steel cord production are not viable today. For example, it is known that shredding plants can maintain the final Cu content in steel scraps between 0.20-0.30 wt%, and that through handpicking after shredding or through magnetic separation, the final Cu content could reach 0.10 wt% [9]. These techniques are however not very cost-effective or are at very early stages of development. Additionally, leaching processes that could be applied during steel making are not industrially scalable and numerous studies have shown that impurity removal in molten state incurs high operating cost and is energy intensive.

The objectives of the current work (ongoing, and thus detailed results are yet unavailable), thus, are to address the knowledge gaps identified in REMADE's manufacturing material optimization road map, specifically the fact that manufacturing processes (hot rolling and cold drawing) developed for primary feedstock (virgin steel) are unable to tolerate chemistry or performance variations (% Cu) frequently seen in secondary feedstock (scrap steel). We intend to increase the recycling rate of scrapped steel back into new tires from 20% to 80% through development of technologies that can allow the wire drawing

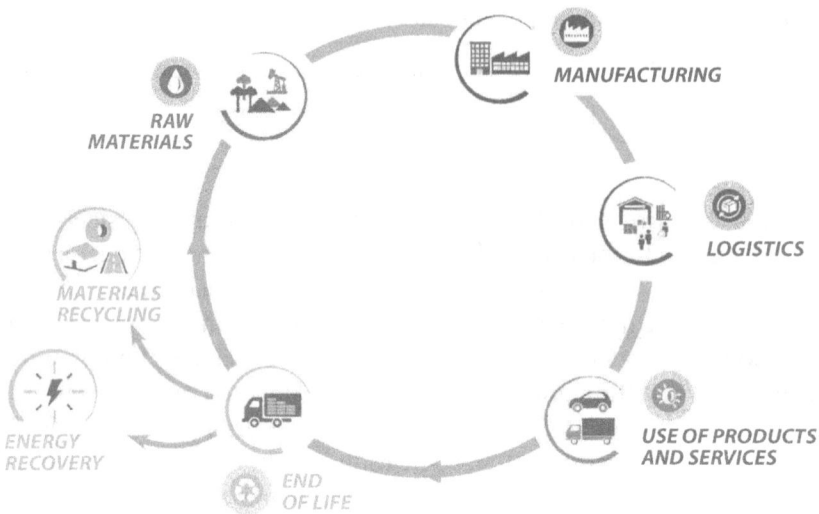

Figure 48.3 Michelin's roadmap for sustainable materials use in tire manufacturing.

manufacturing process to tolerate higher degrees of impurities from scrap steel. Reaching an 80% steel recycling rate is a very ambitious target that is well aligned with Michelin's road map for 100% use of sustainable materials when manufacturing new tires by 2050, as shown in Figure 48.3. The specific objective of the project is to push the allowable limits of copper from 0.046% to 0.184%, while maintaining other impurities, including Sn at a constant level. This would increase the scrap recycling rate from 20% to 80%, thereby increasing secondary feedstock consumption by 60%.

48.2   Technical Approach

The technical approach adopted for this work, shown in Figure 48.4 consists of material characterization and testing of strained and unstrained samples containing different copper contents, followed by thermal and chemical mitigation of sensitized features, followed by the commercialization strategy. The first phase will allow for a determination of where and in what forms Cu is concentrated in the steel cord microstructure and the consequent

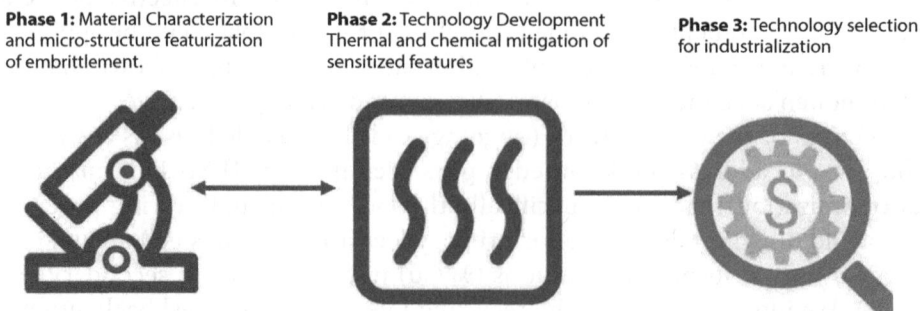

Figure 48.4 Technical approach adopted to enhance recycled steel content in tire manufacturing.

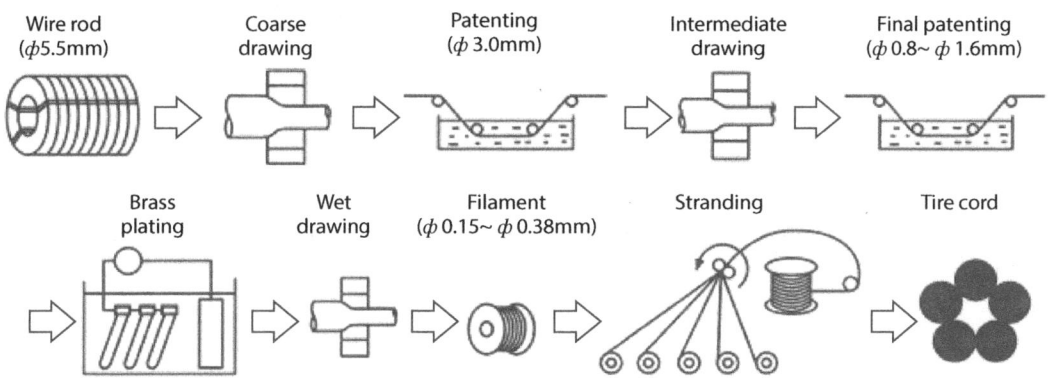

Figure 48.5 Illustration of the cold drawing process, reducing a 5.5 mm diameter rod to a ~0.2 mm wire, and then developing steel cords out of the strands.

sensitivity of the wire to cracking. A correlation between Cu content and the sample micro-structure features will then provide the basis for prediction of wire breakage during the cold drawing process based on Cu content and wire microstructure. The second phase will consist of developing thermal and chemical processing techniques that are necessary to ameliorate the micro-structure to tolerate higher Cu contents. These treatments will be applied ahead of the cold-drawing process and will be developed through lab-scale annealing and surface reduction experiments. We will identify at least one pre-treatment solution to eliminate the sensitization of wire samples with the targeted Cu level. Specifically, the mechanical testing will need to yield the same mechanical resistance for the treated sample at Cu levels of at least 0.184% Cu as those of baseline samples with 0.046% Cu. We will use steel cord samples with 0.046% Cu and mechanical tests that have been designed to represent the physical process of cold drawing These tests will rely on the use of existing finite element models that simulate the cold drawing process and yields the same stress fields in the steel cords as those obtained during cold drawing (Figure 48.5).

48.2.1 Evaluation of Material Characteristics and Microstructure Features of Embrittlement

Based on the authors' experience [10], [11], the cause of impurity-induced embrittlement depends on the scrap chemistry and processing history of the steel, i.e. time, temperature and the oxidizing environments, and cycles of austenite decomposition, formation and grain evolution. It has been reported that Ni forms Cu-Ni-Fe solution in the enrichment zone when the melting temperature is increased. Furthermore, Si promotes a rougher interface morphology and occlusion through internal oxidation, and slows the diffusion of Fe^{2+} by forming Fayalite, thereby reducing Fe oxidation. Electron microscopy images of metal-oxide interface of 0.2 wt% Cu-bearing steel with varying Si content [11] is shown in Figure 48.6.

Oxidizing conditions also influence cracking, with the most severe cracking observed at temperatures between 1100 and 1150°C. Fe oxidation rate is countered by diffusion of Cu from grain boundaries. Below 1093°C, Cu does not melt, and thus hot shortness is not observed. The oxygen concentration is also responsible for the scale morphology – inert

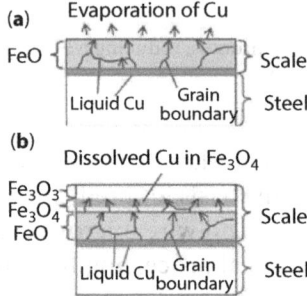

Figure 48.6 Electron microscopy images of metal-oxide interface of 0.2 wt% Cu-bearing steel with varying Si content [11].

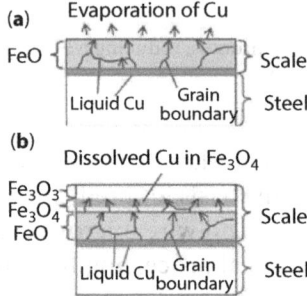

Figure 48.7 Steel-scale interface morphology and Cu behavior for: (a) low O_2 and (b) high O_2 oxidizing environments [12].

atmospheres suppress hot shortness, as shown in Figure 48.7 [12]. Past work has also shown that crack sensitivity is dependent upon the degree of separated Cu-rich phases along grain-boundaries and that this sensitivity increases in the presence of embrittling elements such as Sn and Sb in this phase [12]–[15].

A critical knowledge gap here is to identify the extent of microstructural features responsible for mechanical failure, i.e. separated Cu, as a function of copper content for a defined processing history from the steel plant. The localization, shape and composition of the Cu-rich phases will be observed, as well as any evident cracks that could be present. It is important to note that, at times when Sn is present, the Cu could have diffused away from the formed cracks. Examples from past work of expected results are shown in Figure 48.8.

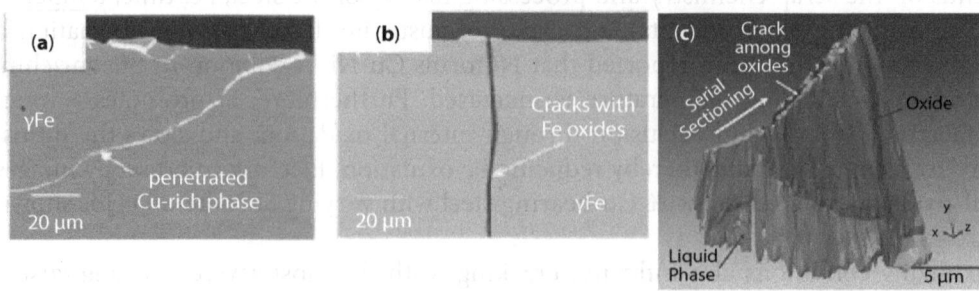

Figure 48.8 (a) Imageable Cu in prior austenite grain-boundaries in a Fe-Cu sample; (b) cracks left after Cu has diffused away in a Fe-Cu+Sn sample; (c) FIB re-construction of a partially Cu filled cracked grain boundary.

From the material characterization studies, a correlation can be developed between Cu content and the embrittling features. The results will be mathematically correlated in terms of number of identified features vs. Cu content in the wire. The key microstructure feature will be sensitization, i.e. Cu-penetrated in prior austenite grain boundaries (e.g. Figure 48.8a), and the occurrence of such features may be defined in this case as the number of features and/or an acceptable size/thickness of such features. We will draw from previous experience on data informatics when seeking to correlate alloy chemistries to microstructural features [16].

48.2.2 Mechanical Property Tests to Elucidate the Influence of Straining

It has been reported that the presence of Cu can result in significant reductions in mechanical properties steel samples. A recent study showed that 0.14% and 0.28% of Cu as impurity in steel reduces the toughness although the yield stress and ultimate tensile stress are increased (Figure 48.9(a)). To quantify such effects, and to design mitigation measures, mechanical tests will be carried out on the steel rods and the drawn out wires. The 5.5 mm diameter steel cords will also be subjected to mechanical tests – specifically tension tests - that simulate the cold drawing process. The mechanical tests will then rank the different 5.5 mm diameter steel cord samples with respect to tensile (yield and ultimate) strength, ultimate tensile strain, elastic modulus, and ductility of the virgin steel cords (before drawing them into wires). Because the highest strain is of particular interest for the cold-drawing process, the lateral strains in the steel cord will also be monitored. To address the mechanical test procedure variability, a statistical approach will be adopted. Assuming that the Cu content is the only variable with respect to the composition, its influence on the mechanical properties will be examined using these tests. SEM observations of the surface as well as metallographic observations of the longitudinal cross-section of the specimen after the test will be carried out.

To increase the reliability of the analysis, the same mechanical and material characterization tests will be carried out for cords of lower diameter. Specifically, the different 5.5 mm

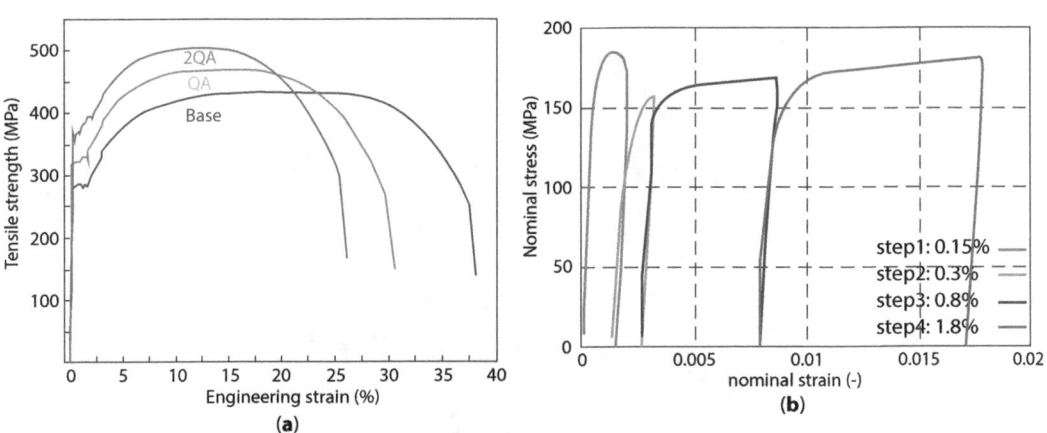

Figure 48.9 (a) Stress-strain relationships of steel containing 0.14% (denoted as QA) and 0.28% (denoted as 2QA) Cu [17], and (b) illustration of an interrupted tensile test [18].

diameter steel cord cords with Cu contents ranging from 0 to 0.30% will be cold drawn to intermediary diameters between 1.0-2.0 mm. After the cold drawing transformation, the 1.0-2.0 mm diameter samples will be characterized with the same material characterization and mechanical testing protocol used for the 5.5 mm diameter steel cords (optical and SEM observations of the microstructure, tensile tests and observations of the fracture area as well as internal damage). However, two additional characterizations will be carried out here. The first one is a torsion test. The second additional characterization will be interrupted tensile tests. In this case, tensile tests will be stopped at 50%, 75%, and 90% of the ultimate strains. For these two additional tests, the internal damage will be systematically observed in order to identify the evolution of the micro-structure damage and the localization of the initiation of the final crack.

In order to test the mechanical properties of wires drawn from the 5.5 mm diameter steel cords, the procedures described in ASTM D4975 with requisite modifications will be employed. The tension tests on the 1.0-2.0 mm diameter wires will be carried out on a universal testing machine with fixtures and load cell sensitivity to test small diameter specimens (Figure 48.10). As described above, the yield strength, ultimate tensile strength, yield and ultimate strains, and the elongation will be determined and related to the Cu content as well as the micro-structure. The wire properties will also be related to the properties of the parent steel cords from which the wires were drawn to evaluate the influence of micro-structural changes due to wire drawing. The torsional resistance of the wires, which

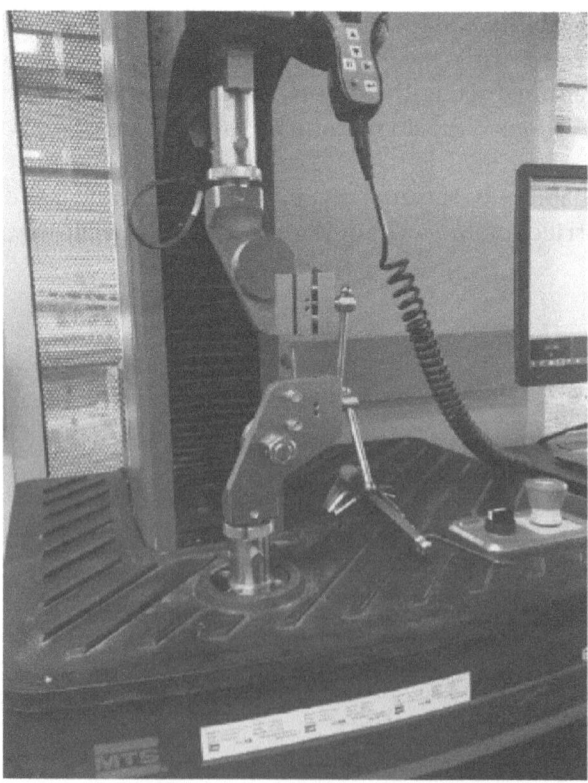

Figure 48.10 Low force tension testing machine.

signifies the metallurgical soundness and surface quality of a drawn wire, will be determined based on the number of full rotational turns of the wire to fracture. The team will also convert the torque-number of turns (cycle) curves into shear stress-shear strain curves to extract parameters that can be related to the compositional and micro-structural features. For example, utilizing a power law relationship between shear stress and shear strain that gives a strain-hardening index, the strain-hardening behavior of the wire during torsion will be obtained. This behavior will be a function of the wire composition, processing, and its micro-structure. Moreover, an analysis of the wire sample failure mode during the torsion test will be carried out to determine whether the failure is ductile or brittle. The failure mode will be related to the wire sample micro-structure and composition.

We will also develop quantitative correlations between the sample Cu content and its mechanical properties and sensitized microstructures. This correlation will be obtained for steel cords of 5.5 mm diameter and cold-drawn wires of 1.0-2.0 mm diameter. The correlations should allow the determination of microstructural features responsible for sample failure during mechanical testing, as well as a more precise Cu reincorporation limit beyond which the current cold-drawing process will generate wire breakage.

48.2.3   Thermal and Chemical Mitigation of Sensitized Features

Based on aforementioned results and the confirmation that Cu does not pose an embrittlement risk when dissolved in the interior of the grains (Figure 48.11 which shows Cu filled grains through as observed through a confocal laser scanning microscope and how the Cu rich darker groves are re-distributed during austenite re-formation), thermal and chemical treatments that can dissolve and re-distribute the separated Cu without degrading the steel microstructure can be developed. Our approaches include:

(i)   If the sensitization is primarily near the surface, then chemical etching or mechanical grinding at sufficient but controlled depths will be pursued since these techniques could be sufficient to remove the sensitized features.

(ii) If sensitization is found deeper in the material and is associated with the presence of Cu, then thermal annealing under inert atmospheres will be pursued since this technique can help to re-distribute the impurity.

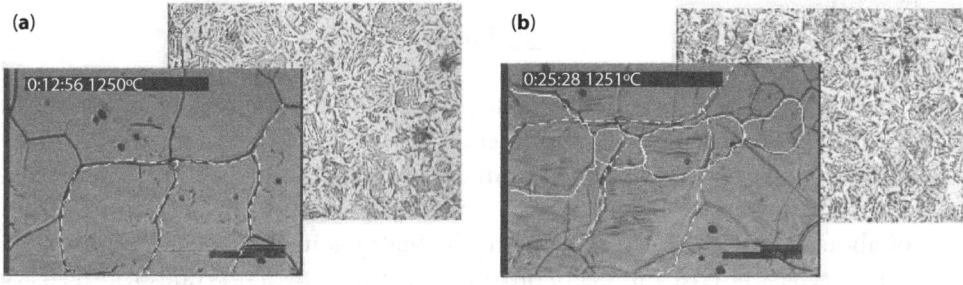

Figure 48.11 Front CSLM images obtained at temperature and back SEM image after cooling (a) Sensitized austenite appear as groves filled with Cu. (b) Liquid phase no longer as concentrated at austenite boundaries and is dissipated amidst the finer grain structure.

A similar approach was applied by the PI for Zn induced Liquid Metal Embrittlement (LME) during direct hot forming of Zn coated B-steels [19].

(iii) Finally, a thermal cycling through austenite and austenite decomposition could be an alternative path since it can re-form grain boundaries and decomposition structures and hence reduce sensitization. Such an approach was employed by the PI for direct hot-charging of continuously cast steels [20].

48.3   Conclusions and Recommendations

A key knowledge gap identified in REMADE's manufacturing and material optimization roadmap is that the manufacturing processes (hot rolling and cold drawing) developed for primary feedstock (virgin steel) are unable to tolerate chemistry or performance variations (% Cu) frequently seen in secondary feedstock (scrap steel). The scientific investigations and processing technologies described here are designed to address this knowledge gap, and have the potential to allow for an increase in the Cu content from scrap steel from 0.046% to 0.184% in tire manufacturing. The thermal (and potentially chemical) treatments to improve the hot shortness of the steel cords with Cu content above 0.046% is expected to be of significant consequence to the industry looking for unique ways to enhance the recycling rate of scrap steel in their processes and products. The novelty of this approach comes from the development of technologies that render Cu and other impurities harmless to the wire drawing process and distinguishes itself from energy- and cost-intensive processes for Cu removal from the melt, than nascent scrap sorting techniques.

The solutions developed here will seek to increase the reincorporation of recycled steel back into new tires from the current limit of 20% to 80%. As a consequence, it is estimated that these solutions can be categorized as having the potential to reduce the primary feedstock consumption and increase secondary feedstock consumption by 60%, as well as enabling cross-industry reuse of recycled feedstock, specifically the use of scrap steel collected from the automotive industry. This work, when fully implemented, is expected to result in a yearly reduction of about:

- 682 metric kilotons of virgin steel material needed to make new tires
- 10.7 PJ of energy required to obtain the raw materials needed to fabricate new tires
- 1,190 metric kilotons of CO_2e emissions

For the tire industry eco-system, the value proposition for the new technologies developed comes from the reduction in environmental impact and raw material cost. The reduction in environmental impact is obtained from the decrease in the use of primary feedstock, energy use and associated CO_2 emissions, while the reduction in cost comes from the decrease of about 20% (15-25% range due to fluctuations in the price of pig iron) in the raw material purchasing cost for steel cord manufacturing when crude iron feedstock is replaced with scrap steel.

References

[1] Performance - Environment | Michelin The tire digest, https://thetiredigest.michelin.com/performance- environment, 2023

[2] C. Broadbent, Steel's recyclability: demonstrating the benefits of recycling steel to achieve a circular economy. Int. J. Life Cycle Assess., 21, 11, 2016.

[3] Iron and Steel Technology Roadmap – Analysis, IEA, https://www.iea.org/reports/iron-and-steel-technology-roadmap, 2020

[4] K. E. Daehn, A. Cabrera Serrenho, and J. M. Allwood, How Will Copper Contamination Constrain Future Global Steel Recycling?. Environ. Sci. Technol., 51, 11, 2017.

[5] L. Wang, H. Gu, and Q. Li, Susceptibility to Surface Hot Shortness of a Cu-containing Steel. J. Phys. Conf. Ser., 1965, 1, 2021.

[6] H. Jin and B. Mishra, Minimization of Copper Contamination in Steel Scrap, in: Energy Technology 2020: Recycling, Carbon Dioxide Management, and Other Technologies, Cham, pp. 357–364, 2020.

[7] David, Ferrous Scrap Steel," Metallurgist & Mineral Processing Engineer, https://www.911metallurgist.com/ferrous-scrap-steel, 2017

[8] R. J. Fruehan, O. Fortini, H. W. Paxton, and R. Brindle, Theoretical Minimum Energies to Produce Steel for Selected Conditions, https://www.energy.gov/sites/prod/files/2013/11/f4/theoretical_minimum_energies.pdf, 2000

[9] Z. Gao, S. Sridhar, D. E. Spiller, and P. R. Taylor, Applying Improved Optical Recognition with Machine Learning on Sorting Cu Impurities in Steel Scrap. J. Sustain. Metall., 6, 4, 2020.

[10] E. Sampson and S. Sridhar, Effect of Water Vapor During Secondary Cooling on Hot Shortness in Fe-Cu-Ni-Sn-Si Alloys. Metall. Mater. Trans. B Process Metall. Mater. Process. Sci., 45, 5, 2014.

[11] E. Sampson and S. Sridhar, Effect of Silicon on Hot Shortness in Fe-Cu-Ni-Sn-Si Alloys During Isothermal Oxidation in Air. Metall. Mater. Trans. B, 44, 5, 2013.

[12] Y. Kondo and H. Tanei, Effect of Oxygen Concentration on Surface Hot Shortness of Steel Induced by Copper, ISIJ Int., 55, 5, 2015.

[13] L. Yin, E. Sampson, J. Nakano, and S. Sridhar, The Effects of Nickel/Tin Ratio on Cu Induced Surface Hot Shortness in Fe. Oxid. Met., 76, 5, 2011.

[14] L. Yin and S. Sridhar, Effects of Residual Elements Arsenic, Antimony, and Tin on Surface Hot Shortness. Metall. Mater. Trans. B, 42, 5, 2011.

[15] L. Yin, S. Balaji, and S. Sridhar, Effects of Nickel on the Oxide/Metal Interface Morphology and Oxidation Rate During High-Temperature Oxidation of Fe–Cu–Ni Alloys. Metall. Mater. Trans. B, 41, 3, 2010.

[16] A. Rahnama, S. Clark, and S. Sridhar, Machine learning for predicting occurrence of interphase precipitation in HSLA steels. Comput. Mater. Sci., 154, 2018.

[17] J. Duan, D. Farrugia, C. Davis, and Z. Li, Effect of impurities on the microstructure and mechanical properties of a low carbon steel. Ironmak. Steelmak., 49, 2, 2022.

[18] D. Colas et al., Strain field measurements in polycrystalline tantalum, 2013.

[19] V. Janik, P. Beentjes, D. Norman, G. Hensen, and S. Sridhar, Role of heating conditions on microcrack formation in zinc coated 22mnb5, in: Materials Science & Technology Conference and Exhibition 2014 (MS&T'14) : Proceedings, 2014.

[20] B. A. Webler, E.-M. Nick, R. O'Malley, and S. Sridhar, "Influence of cooling and reheating on the evolution of copper rich liquid in high residual low carbon steels," Ironmak. Steelmak., vol. 35, no. 6, pp. 473–480, Aug. 2008.

Part 9
E-SCRAP RECYCLING

49

Selective Leaching and Electrochemical Purification for the Recovery of Tantalum from Tantalum Capacitors

R. Adcock*, T. Chen, N. Click, M.-F. Tseng and M. Tao

Arizona State University, Tempe, AZ, USA

Abstract

Tantalum (Ta) is a critical element that is used in many different electronic products. One of such products is epoxy-coated tantalum capacitors. They are composed of a tantalum metal anode, a tantalum oxide (Ta_2O_5) dielectric and a manganese dioxide (MnO_2) cathode. These capacitors are incredibly useful due to their capacitance per unit volume, frequency characteristics and stability. As the supply of tantalum that is economical to extract dwindles, it becomes important to develop a recycling technology for its recovery from tantalum capacitors. Our process uses pyrolysis, sequential leaching and electrochemical purification before the precipitation of a tantalum salt that feeds into the current tantalum production feedstock. First, the capacitors are pyrolyzed to remove the epoxy coating, where time and temperature were investigated. The resulting mass was milled to create a homogenous feedstock. In the leaching step, the manganese dioxide cathode is leached in hydrochloric acid (HCl) with hydrogen peroxide (H_2O_2), and the silver (Ag), tantalum and tantalum oxide are leached in hydrofluoric acid (HF), again with hydrogen peroxide. Tantalum and silver do not leach into hydrochloric acid. The leaching rate was determined as a function of acid concentration and temperature. The hydrofluoric acid leachate is then electrochemically purified by the electrowinning of silver. The final tantalum-containing solution is treated with potassium fluoride (KF) to produce potassium heptafluorotantalate (K_2TaF_7), the feedstock from which most current tantalum metal is produced through sodium (Na) reduction in molten salt. The total recovery rate and purity of the tantalum and silver are then confirmed with mass measurements as well as chemical analysis. This method provides an improvement over current methods due to its simplicity and reduction in the number and quantity of chemicals involved, only requiring 4 chemicals. It also demonstrates a new method for the separation of manganese from tantalum, a challenge in current recycling processes due to its similar properties to tantalum. Our process also produces an important byproduct in pure metallic silver.

Keywords: Tantalum, silver, e-waste, critical metals, metal separation, acid leaching, electrochemical recovery

Corresponding author: rjadcock@asu.edu

Nabil Nasr (ed.) Technology Innovation for the Circular Economy: Recycling, Remanufacturing, Design, Systems Analysis and Logistics, (653–664) © 2024 Scrivener Publishing LLC

49.1   Introduction

With the growing amount of electronic waste across the world as well as the diminishing of natural resources, it is important to investigate recycling of target elements from waste sources for a circular economy. One of these elements is tantalum (Ta), a hard and corrosion resistant metal that is commonly used in capacitors. Its unique properties allow it to be used in a number of industries, including electronics, biomedical, aerospace and chemical [1]. Tantalum has high, temperature insensitive volumetric capacitance that allows miniaturization of electronic devices and circuits [1]. It also has very high strength and resistance to high temperatures that allow it to be used in the aerospace industry, and it is bioinert which allows it to be used in the biomedical industry [1]. It is clear to see that tantalum has many important applications, and its utilization will only increase as these industries grow.

As the utilization of tantalum increases, it shifts the balance of tantalum in the form of ore to be mined and as a resource in waste streams. The United States is already recognizing the supply risk of tantalum. Currently, there is no mining of tantalum in the United States, so the country is completely reliant on imports and recycling to meet the needs [2]. Considering that currently less than 1% of the tantalum used is recycled [3], it is important to improve the current recycling technologies. Some of the current methods to recycle tantalum from capacitors include chloride metallurgy, hydrometallurgy with solvent extraction, and automated dismantling [3–5]. These methods, while innovative, use large numbers of chemicals, require large energy inputs or are complex to carry out. The chloride metallurgy method uses temperatures of nearly 500°C that requires significant energy input. In the hydrometallurgical method [5], harsh conditions and significant pretreatment steps make the process unsafe and complex. In addition, both of these chemical methods are only able to recover tantalum as tantalum pentoxide (Ta_2O_5) rather than tantalum metal. The recovered tantalum pentoxide requires further processing to obtain tantalum metal. The automated dismantling of printed circuit boards utilizes complex technology and is only able to separate the capacitors for reuse or further recycling and not pure tantalum metal. Therefore, research should focus on 1) reduce the number of chemicals involved, 2) reduce the energy required, and 3) reduce the overall complexity.

A diagram of a tantalum capacitor is shown in Figure 49.1, which contains silver (Ag), manganese oxide (MnO_2), tantalum, tantalum pentoxide, among other materials. In general, the epoxy molding compounds (EMC's) are halogenated hydrocarbons and occasionally contain silica [5]. The lead frames are generally iron (Fe) or nickel (Ni) metal [5]. A common technique in the recovery of tantalum is to leach the capacitors in acid, isolate the tantalum from other compounds, and precipitate it out. Other researchers have found that the leaching effectiveness of hydrofluoric acid (HF) is much higher than other acids like hydrochloric acid (HCl), sulfuric acid (H_2SO_4) or nitric acid (HNO_3), and it is common to do so at high concentrations, temperatures, and pressures, sometimes reaching 49% v/v, 220°C, and 23 atm [5]. While these conditions are very effective at leaching tantalum, the harsh conditions make the leaching process energy intensive, complex and risky for the workers. Another challenge in recovery of tantalum from capacitors is to isolate the tantalum from the manganese found in the anode (Figure 49.1). Manganese has very similar properties to tantalum and is very electrochemically stable, reducing from Mn^{2+} to Mn(s) at −1.17V vs standard hydrogen electrode (SHE) [7], so it cannot be separated from tantalum

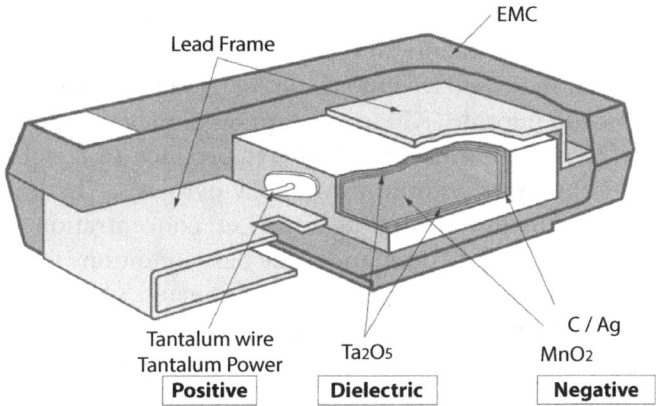

Figure 49.1 Tantalum capacitor diagram [6].

electrochemically in an aqueous solution. They are often separated with solvent extraction; however this increases the number of chemicals involved in the process [5]. In the precipitation step, tantalum is generally recovered as tantalum pentoxide, which is not as desirable as tantalum metal. This is reflected in the price difference and the fact that Ta can be more easily oxidized to Ta_2O_5 than reduced the other way around [8–10].

Proposed Process for Tantalum Recovery

The goal of this work is to recover tantalum through a simple process, at safe and energy efficient temperatures, and in a form that can be used to obtain pure tantalum metal. To balance safety, low energy input, minimizing the number of chemicals and creating a product that can be readily converted to tantalum metal, our proposed process includes 5 main steps: pretreatment, leaching in hydrochloric acid, leaching in hydrofluoric acid, electrochemical purification of silver and precipitation of tantalum. The steps are summarized in Figure 49.2.

The pretreatment step involves thermal and mechanical methods to remove the epoxy and increase the surface area, as well as expose the MnO_2 for the HCl leaching. The next step

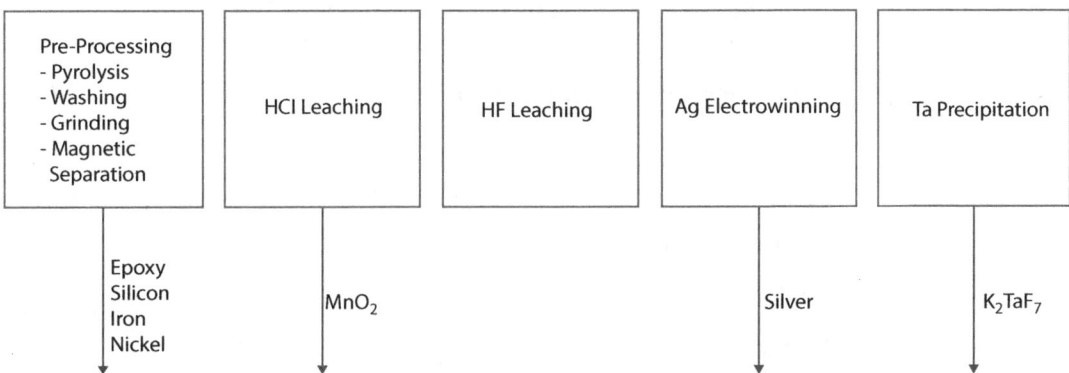

Figure 49.2 Proposed process for Ta recovery from capacitors.

is hydrogen peroxide (H_2O_2) assisted leaching in HCl to remove Fe, Ni and MnO_2. Then, H_2O_2-assisted leaching is performed in 10% HF at 50°C to dissolve the Ta, Ag and Ta_2O_5. Electrochemical purification is used to remove the Ag from the HF leachate, and precipitation is done with potassium fluoride (KF) to create potassium heptafluorotantalate (K_2TaF_7) which can be readily fed into the current industry to produce Ta metal. Previous research was used to determine the best conditions for epoxy pyrolysis. The leaching of MnO_2 in HCl was determined as a function of stirring and H_2O_2 concentration. The Ta leaching in HF was determined as a function of time under the best conditions found, although lower concentrations of HF and temperatures were also investigated. Our own research was used to determine the best conditions for Ag removal by electrowinning, and finally the recovery rate of K_2TaF_7 was determined as a function of KF concentration.

49.2   Materials and Methods

The tantalum capacitors used in this study were model TCKOJ226AT from Cal-Chip Electronics. A detailed diagram of the capacitor is shown in Figure 49.1. It contains a tantalum metal cathode, a tantalum pentoxide dielectric, a manganese dioxide/carbon/silver anode. Also important are the nickel and iron metal leads, and the epoxy coating. The as-received capacitors were milled with a ball mill (8000D, Spex-Certiprep) and analyzed using X-ray fluorescence spectroscopy (XRF, S2 PUMA, Bruker). Table 49.1 shows the physical breakdown and Table 49.2 shows the chemical breakdown of the capacitors obtained from XRF.

Once the capacitors were quantified, the capacitors were pyrolyzed at 600°C for 30 minutes, as based on previous research this was found to be effective at removing the epoxy coating [5].

Table 49.1 Capacitor physical composition.

Epoxy resin	Electrodes and dielectric
57.36%	42.64%

Table 49.2 Capacitor atomic breakdown. †This includes silver as the XRF machine cannot quantify silver because it is used as part of the detector mechanism.

Capacitor elemental composition	
Not Measurable†	57.36%
Tantalum (Ta)	30.30%
Manganese (Mn)	9.75%
Tin (Sn)	1.55%
Iron (Fe)	0.63%
Nickel (Ni)	0.40%

Next, the pyrolyzed capacitors were milled with a ball mill (8000D, Spex-Certiprep) to reduce the particle size and expose any sealed metal and/or dielectric of the capacitors.

The chemicals used in this study include hydrochloric acid (Sigma-Aldrich, 37%), manganese dioxide (Loudwolf, 99%), hydrogen peroxide (Alfa Aesar, ACS 29-32% w/w aqueous solution), hydrofluoric acid (JT Baker, 49% w/w), tantalum wire (Thermo Scientific, 1 mm diameter, 99.95% metals basis), potassium fluoride (Alfa Aesar, ACS 99%), and tantalum fluoride (TaF_5, Alfa Aesar, 99.9% metals basis). The leaching of MnO_2 was determined as a function of H_2O_2 concentration, both with and without stirring at 350 rpm. The time that it took to completely dissolve 0.5 g of MnO_2 was used to determine the optimal concentration of H_2O_2 and stirring. In the HF leaching, the leach rate of Ta wire was determined as a function of time. The Ta wire was weighed every 3 hours until it had completely dissolved. For the actual source material, i.e. tantalum capacitors, the Ag was removed using conditions of our own research [11]. For the precipitation, KF was used to precipitate TaF_5 dissolved in aqueous solutions of 1% v/v HF. Five different molar ratios of KF to TaF_5 were tested. Once the precipitate was filtered, it was weighed to determine the recovery rate. The precipitate was analyzed using X-ray diffraction (XRD, Aeris, Malvern PANalytical).

49.3   Results and Discussion

Leaching in Hydrochloric Acid

One of the main difficulties in recovering Ta from capacitors is the significant amount of MnO_2 that is present. Mn has similar properties to Ta and is also able to be dissolved in HF under the conditions used in our process. Rather than separating the Mg from the Ta in the HF solution, our process employs selective leaching of the metals. MnO_2 in the presence of H_2O_2 can be leached into a HCl solution. MnO_2 is not soluble in HCl alone under normal conditions, but it reacts with HCl violently and exothermically when oxidized with H_2O_2,

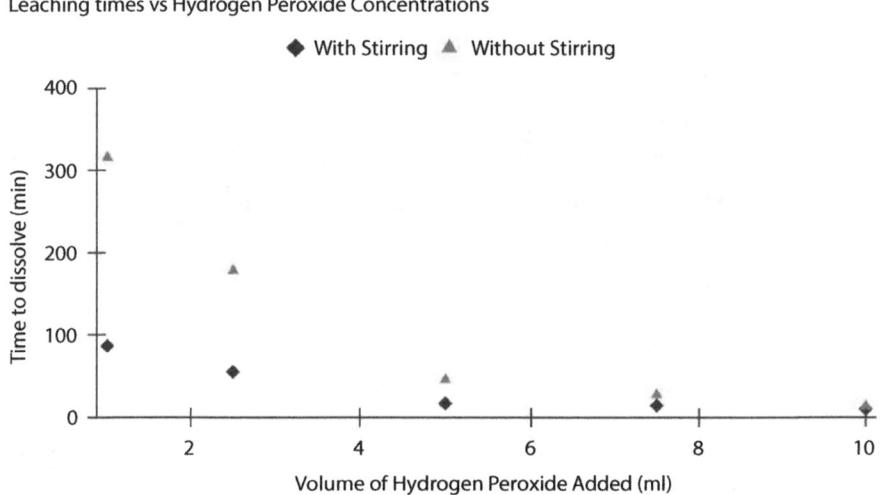

Figure 49.3 Time to complete leaching of MnO_2 as a function of H_2O_2 concentration.

Tantalum Leaching in Hydrochloric Acid (HCl)

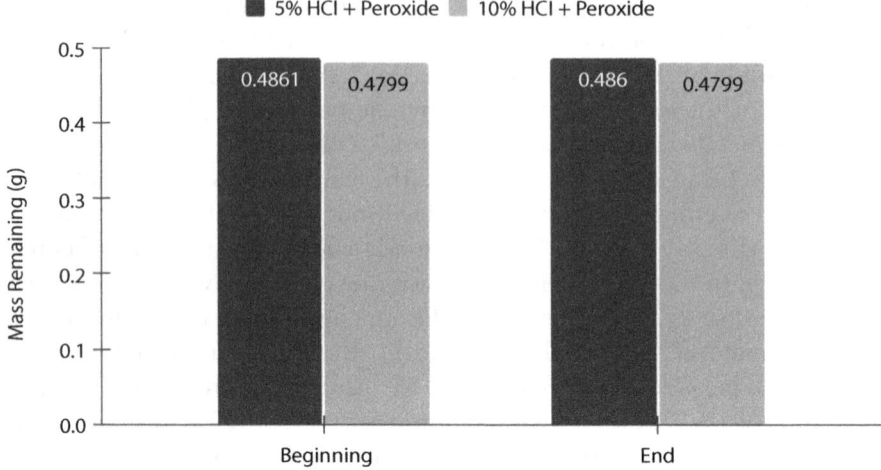

Figure 49.4 Ta leaching in HCl with H_2O_2 for 6 hours.

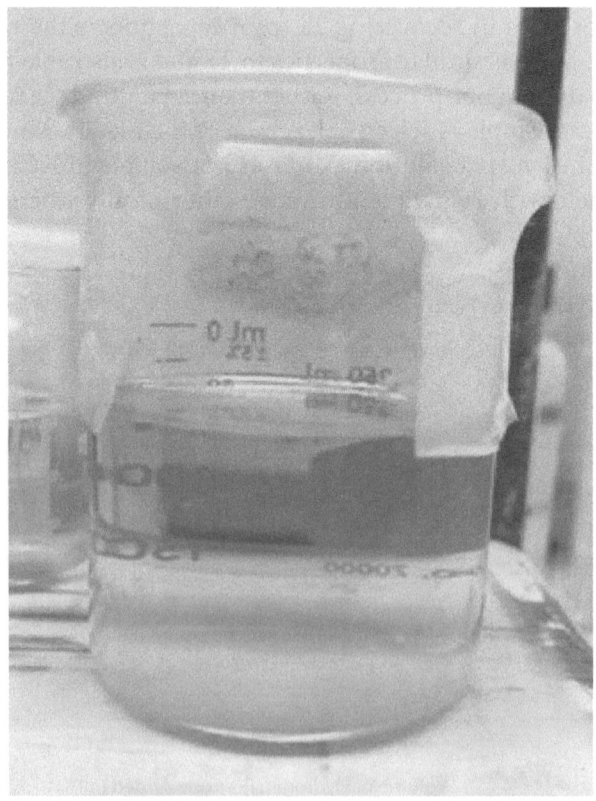

Figure 49.5 Filtered HCl leachate after 3 hours of leaching capacitors.

thus leached. In a control experiment, no MnO_2 was able to leach into a HCl solution in 72 hours without the addition of H_2O_2. On the other hand, with the addition of H_2O_2, it was able to be leached within hours. The effect of stirring was also investigated. This is summarized in Figure 49.3. All the H_2O_2 was added at once to a 50 mL solution of 1 M HCl. For the trials that had stirring, it was performed with a frequency of 350 rpm. With a concentration of about 0.5 M H_2O_2 or greater and the addition of stirring, the leaching of MnO_2 is completed in less than an hour.

The primary goal of our process is to recover Ta from the HF leachate. Therefore, it needs to be shown that the HCl leaching, while being able to leach MnO_2, is unable to leach Ta. Experiments were performed with and without H_2O_2, and in both 5% v/v HCl and 10% v/v HCl for 6 hours. There were no leaching conditions where HCl was able to dissolve Ta at all. The mass measurements are shown in Figure 49.4.

Based on Figures 49.3 and 49.4, HCl plus H_2O_2 is able to leach MnO_2 but is unable to leach Ta into the solution. Also contained in the tantalum capacitors are Fe and Ni. While they were not investigated in detail in this study, both Fe and Ni are known to leach into HCl with the presence of H_2O_2. It is also possible to magnetically separate Fe and Ni before the HCl leaching. When this HCl plus H_2O_2 chemistry was applied to actual capacitors, the leachate changes color to green. Figure 49.5 shows the HCl leachate after 3 hours of leaching capacitors. The green color suggests Fe and Ni dissolved in the solution.

Leaching in Hydrofluoric Acid

Once all the Fe, Ni and Mn that would have an impact on Ta recovery are removed, the Ta, Ta_2O_5 and Ag are leached in HF with the presence of H_2O_2. To determine the optimum leaching conditions, a Ta wire was submerged in HF with H_2O_2. First, trials were run at room temperature and with low concentrations of HF (around 1%). These conditions were unable to significantly leach Ta. Other researchers were able to quickly dissolve Ta at very high pressures and temperatures under normal HF concentrations [5]. High temperatures

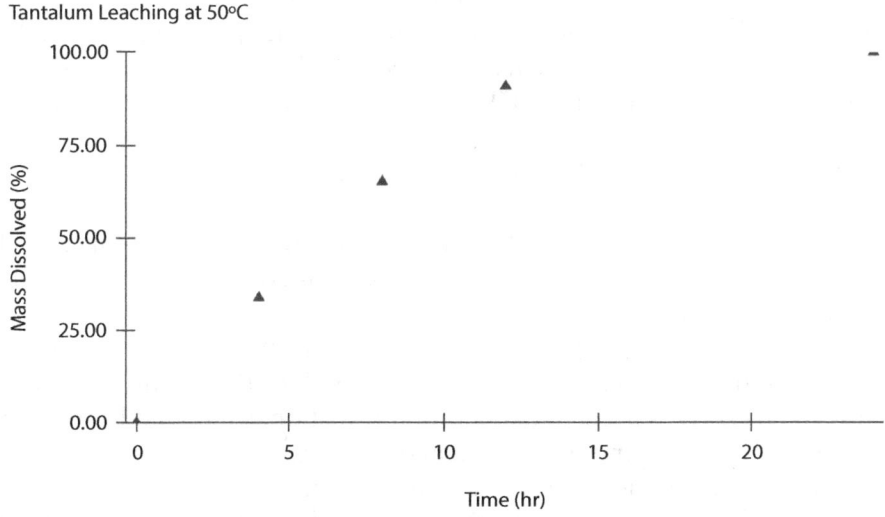

Figure 49.6 Percent of Ta leached in 10% w/w HF with H_2O_2 at 50°C.

and pressures are not only dangerous if performed without extreme care, but also energy intensive and require specialty equipment to carry out in the presence of hazardous HF. To balance safety, energy intensity and the ability to use general lab equipment, a concentration of 10% HF with H_2O_2 at 50°C was chosen. The leaching results are summarized in Figure 49.6. After 12 hours of leaching under those conditions, 91% of the Ta had been leached, and after 24 hours it had leached completely. This method is slower than the 3 hours achieved at 220°C, 23 atm and 5% concentration, however the conditions are much more manageable.

The capacitor powders filtered from the HCl leachate, after being washed and dried, were added to a HF solution with H_2O_2 under the conditions identified above for 12 hours and 31% of the powders by weight were dissolved. Using our own research, it has also been shown that Ag in the presence of H_2O_2 is able to be leached by HF [11]. As a byproduct of the process, Ag can then be recovered through electrowinning.

Electrochemical Purification

Once leached in HF, the Ta had to be separated from the Ag in the HF leachate. Ta is very electrochemically stable and cannot be electrowon from an aqueous solution. The standard reduction potential of Ta(V) is far lower than the hydrogen evolution reaction [7]. Ag, on the other hand, can be electrowon from an aqueous solution. Based on our own research [11], Ag electrowinning was carried out in the HF leachate with graphite working electrode and counter electrode. –0.8 V vs a silver/silver chloride (Ag/AgCl) reference electrode was applied for 24 hours and 19.2 mg of Ag was recovered. Once no more Ag was plated onto a fresh graphite electrode and no Ag reduction peak was observed during a cyclic voltammetry, it was determined that there was no more Ag in the HF leachate and the process could continue with precipitation.

Precipitation

At this point, only Ta should have remained in the HF leachate. As stated previously, our goal was to recover Ta in a form that could be used to produce pure Ta metal. In the industry, K_2TaF_7 is fed into a molten salt system and recovered as pure Ta metal, and K_2TaF_7 is relatively simple to produce from Ta in HF. The precipitation reaction is shown below:

$$TaF_5 + 2KF \rightarrow K_2TaF_7 \tag{49.1}$$

To understand the reaction more and estimate the recovery rate from the actual capacitors, 0.1 M TaF_5 was dissolved in 1% v/v HF, and KF was added in different molar ratios. This is shown in Figure 49.7.

A recovery rate greater than 90% can be achieved with the KF concentration only 1.5 times the concentration of TaF_5. This knowledge will reduce the chemical waste from the process since no excess KF is needed. Using this information, KF was added to the post-Ag electrowinning HF leachate, and 4.4 g of material was recovered. The powder was analyzed using XRF and determined to be 7.7% Ta by weight. Pure K_2TaF_7 should contain 46% Ta by weight. Therefore the recovered precipitate was not pure K_2TaF_7. Assuming that the

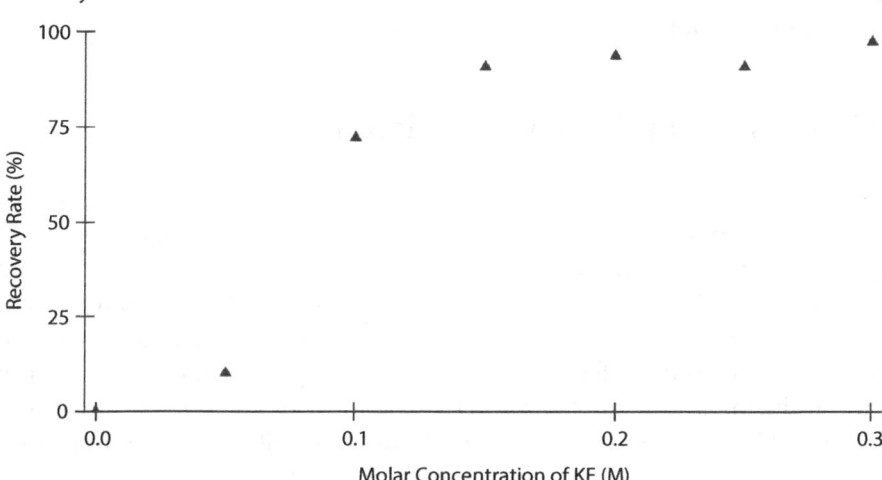

Figure 49.7 Recovery rate of K_2TaF_7 as a function of KF concentration.

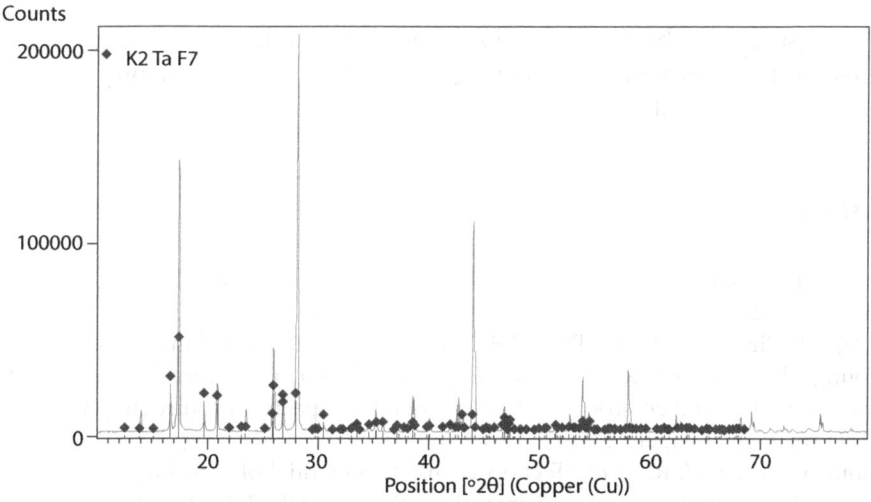

Figure 49.8 XRD pattern of recovered K_2TaF_7 precipitate. All the peaks are associated with K_2TaF_7.

capacitors have a concentration of 30% Ta based on the XRF data (Table 49.2), the overall Ta recovery rate by our process is 28%. The silicon (Si) in the original capacitors was unable to be quantified by XRF so it was not accounted for in the process, however more than 99.5% of Si can be removed during a washing and grinding step [5]. The recovery rate is also likely dependent on the concentration of HF. Since 10% v/v HF was used during the leaching the precipitation had to be performed in the same concentration. In a batch process where the concentration of TaF_5 is allowed to build significantly, this would not be an issue.

To ensure that the recovered precipitate from pure TaF_5 was K_2TaF_7, powder x-ray diffraction was used. The XRD pattern from the precipitate with the peak locations from literature are shown in Figure 49.8. All the peaks were associated with K_2TaF_7. The precipitate seemed to be pure with K_2TaF_7, showing that when all impurities are removed through the real process, a pure precipitate could be recovered.

49.4   Conclusions and Recommendations

Tantalum is a critical element that will need to be recycled in the future to meet demand and reduce supply chain risks. A process was developed and investigated to recover tantalum in a form that can be used to produce pure tantalum metal in the current industrial process. Compared to other processes used to recover tantalum from capacitors, this process reduces the number of chemicals, utilizes a new method to separate manganese from tantalum, improves the safety and simplify the complexity of the leaching process, in addition to producing a valuable byproduct in the form of pure silver metal. A tantalum recovery rate of 97.5% was accomplished from tantalum salts, and a recovery rate of 28% was achieved from applying the process to actual tantalum capacitors.

Acknowledgement

This work was supported by the Defense Advanced Research Project Agency. We acknowledge the use of facilities within the Eyring Materials Center at Arizona State University supported in part by NNCI-ECCS-1542160.

References

1. Tercero Espinoza, L. A. Case study: Tantalum in the world economy: History, uses and demand. Polinares, 2012.
2. Schulz, K.J., Piatak, N.M., and Papp, J.F., 2017, Niobium and tantalum, chap. M of Schulz, K.J., DeYoung, J.H., Jr., Seal, R.R., II, and Bradley, D.C., eds., Critical mineral resources of the United States—Economic and environmental geology and prospects for future supply: U.S. Geological Survey Professional Paper 1802, p. M1–M34.
3. Ramon, H., Peeters, J. R., *et al.*, Techno-economic potential of recycling Tantalum containing capacitors by automated selective dismantling. Procedia CIRP, 90, 421, 2020.
4. Niu, B., Chen, Z., and Xu, Z., An integrated and environmental-friendly technology for recovering valuable materials from waste tantalum capacitors. Journal of Cleaner Production, 512, 2017.
5. Chen., W-S., Ho, H-J., and Lin., K-Y., Hydrometallurgical Process for Tantalum Recovery from Epoxy- Coated Solid Electrolyte Tantalum Capacitors. Materials (Basel), 1220, 2019.
6. Cal-Chip, Standard Tantalum Capacitors - TC Series,
7. https://secureservercdn.net/198.12.145.38/88c.e57.myftpupload.com/wp-content/uploads/2022/06/tc_series.pdf, 2023
8. Bard, A. J., Parsons, B., Standard Potentials in Aqueous Solutions. J. Jordon (Ed.) Dekker 1985.
9. SMM, Tantalum Pentoxide (Ta2O5≥99.95), RMB/Kg, https://www.metal.com/Niobium-Tantalum/202107020002, 2023
10. SMM, Tantalum Ingot (Ta≥99.95%), RMB/Kg, https://www.metal.com/Niobium-Tantalum/202211090002, 2023

11. Pérez-Pérez, D., Acosta-Vera, R., *et al.*, Synthesis of Irregular Tantalum Pentoxide (Ta2O5) Microparticles by Direct Thermal Oxidation of Ta Foils in Atmospheric Oxygen. IOP Conf. Series: Materials Science and Engineering, 897, 2020.
12. Chen, T., Silver Recovery through a Fluoride Chemistry for Solar Module Recycling. Submitted to PVSC, 2023.

50

Recovery of Lead in Silicon Solar Modules

Natalie Click, Randy Adcock and Meng Tao*

Arizona State University, Tempe, Arizona, USA

Abstract

Silicon solar modules contain toxic lead (Pb) in the solder. To prevent contamination of the environment, it is imperative all the Pb is recovered from end-of-life silicon solar modules. In this study, acetic acid (AcOH) is shown as a successful leaching agent for Pb from solder and end-of-life silicon solar cells. Hydrogen peroxide (H_2O_2) is used to facilitate Pb leaching. To quantify the effect of H_2O_2, 3 g of virgin solder ribbon was leached for 24 hours using 16.5 mL of 10% v/v AcOH and different mole ratios of H_2O_2 to Pb with and without stirring. It was observed that the amount of mass loss from the solder increases with H_2O_2 amount. The leachate with a mole ratio of 1.5:1 H_2O_2:Pb and no stirring was electrowon using two half cells with a potassium chloride (KCl) salt bridge. Gray-colored dendrites were formed on a graphite working electrode by applying a potential of –0.8 V vs the silver/silver chloride (Ag/AgCl) reference electrode. Scanning electron microscopy (SEM) imaging shows large, flat Pb crystallites. Energy dispersive X-ray spectroscopy (EDS) of the dendrites confirms that they are metallic Pb. No tin (Sn) is found in the dendrites, and leaching studies of Sn in AcOH + H_2O_2 confirm Sn does not leach into solution. Finally, Pb leaching from an end-of-life silicon solar cell was tested. 25 cm^2 of solar cell was milled in liquid nitrogen, then leached in 178 mL of 10% v/v AcOH and 35 mL of H_2O_2 for 80 hours with no stirring or heating. Inductively coupled plasma optical emission spectroscopy (ICP-OES) confirms 0.001 M Pb in the leachate, which suggests near complete leaching.

Keywords: Silicon solar cell, silicon photovoltaic, recycling, electrochemistry, lead recovery, lead leaching, acetic acid, green chemistry

50.1 Introduction

The transition from a linear to circular economy is imperative for meeting global sustainability goals. The materials stream for green technologies like solar energy is no exception. Presently, there exists no sustainable solution to managing crystalline silicon photovoltaic (PV) waste, which makes up the majority of the PV market. Silicon PV modules contain many elements, including Si, Al, Ag, Cu, Sn, and toxic Pb. Pb has long been known for its toxicity [1], yet there is no national recycling program and module waste disposal is left up to individual state regulations [2]. The ethylene vinyl acetate (EVA) encapsulant used in the

---

Corresponding author: meng.tao@asu.edu

Nabil Nasr (ed.) Technology Innovation for the Circular Economy: Recycling, Remanufacturing, Design, Systems Analysis and Logistics, (665–676) © 2024 Scrivener Publishing LLC

majority of silicon PV modules has been shown to decompose into acetic acid (AcOH) [3, 4], which is the same chemical used for Pb leaching in this study. EVA degradation could pose dangers of Pb leaching into the environment from improper disposal of PV modules. The Environmental Protection Agency has set the Maximum Contaminate Level Goal (MCLG) for Pb in drinking water equal to zero mg/L [5]. Therefore, it is imperative that all the Pb from waste silicon PV modules be collected and managed safely.

Some work has been done to investigate silicon PV recycling, with focuses on Si [6–9] and Ag [9–11]. Despite Pb's toxicity, there is relatively little literature studying its recovery from waste silicon PV. Huang *et al.* demonstrated nitric acid (HNO_3) leaching and sequential electrowinning for Pb recovery from a synthetic leachate. They achieved 99% Pb recovery. Jung *et al.* investigated sequential precipitation to recovery Al, Cu, Pb, Si, and Ag from real solar cells, and achieved 93% Pb removal. Their chemistry includes sequential HNO_3, sodium hydroxide (NaOH), and sodium sulfide (NaS). In this paper, we propose the use of a low-concentration AcOH + hydrogen peroxide (H_2O_2) solution to leach Pb from solder on silicon solar cells. One advantage of AcOH is that it is milder than other chemicals, such as HNO_3. In fact, household vinegar is roughly 7% AcOH. The American Chemical Society lists 12 Principles of Green Chemistry as a way to promote enviro-mindful thinking in the chemistry space. Successful leaching of Pb in AcOH aligns with the 'Designing Safer Chemicals' principle. Currently, AcOH has been demonstrated for Pb leaching in battery recycling technology [14, 15]. We predict it will also work as a leaching agent for Pb recovery from end-of-life silicon solar modules. In this paper, we discuss progress made using AcOH for leaching of Pb from virgin solder and silicon solar cells.

50.2   Methodology and Materials

The following chemicals and materials were used: H_2O_2 (Alfa Aesar, 29-33% w/w), AcOH (Sigma Aldrich, glacial), solder (63%Sn 37%Pb 1.8% flux), Sn powder (Alfa Aesar, 325 mesh), potassium chloride (KCl) (Alfa Aesar), agar (Alfa Aesar), and graphite electrodes (McMaster Carr, 2.3 mm and 3 mm diameters). A Jeol JXA-8530F or Zeiss Auriga scanning electron microscope (SEM), both equipped with energy dispersive x-ray (EDS), was used for imaging and elemental analysis. An Agilent 5900 inductively coupled plasma optical emission spectrometer (ICP-OES) was used for elemental analysis of the leachates.

3 g of virgin leaded solder was leached in 16.5 mL 10% v/v acetic acid with varying amounts of H_2O_2 for 24 hours. Three different mole ratios of Pb:H_2O_2 were tested: 1:1.5, 1:2, and 1:2.5. These conditions were selected to ensure an oversupply of H_2O_2 for complete Pb leaching. For each mole ratio, the effect of stirring and no stirring was investigated. Stirring was performed using a magnetic stir pill rotating at 700 rpm. To facilitate more complete Pb leaching, a different experiment was performed where 3 g solder was first chopped up into small pieces and then leached for 72 hours in 16.5 mL of 10% v/v AcOH with a 1:2.5 Pb:H_2O_2 ratio and aggressive 700 rpm stirring. Stirring was made more aggressive by increasing the stir bar size.

To electrowin Pb from the AcOH leachate, chronoamperometry was performed. Two half cells with a 2 M KCl salt bridge were used to prevent Pb oxidation and deposition on the counter electrode. A potential of –0.8 V vs Ag/AgCl was applied to a graphite working electrode submerged in the solution from leaching with a 1:1.5 Pb:H_2O_2 mole ratio and no stirring. After approximately 5.5 hours, small gray colored crystals had formed in the beaker. These crystals were collected and analyzed using EDS.

Finally, AcOH leaching and electrowinning was demonstrated on a silicon solar cell. Two small pieces of silicon module, which were previously removed from the Al frame and cut into strips, were used in the experiment. A third, unleached strip of solar cell was set aside and used for elemental analysis. The total area of the two strips used for leaching was 25 cm². These strips were milled in liquid nitrogen to form a fine powder. 35 g of this powder was leached in 178 mL of 10% v/v AcOH with 35 mL H_2O_2 for 80 hours with no stirring. The glass container was covered with parafilm to prevent evaporation during leaching. After 80 hours, the leachate was filtered, dried, and converted to a 2% HNO_3 matrix, and then analyzed using ICP-OES. To account for error in the ICP-OES analysis, a Pb standard was volumetrically diluted in 10% v/v AcOH matrix to 1 ppm, then dried, converted to 2% HNO_3, and measured.

50.3 Results and Discussion

50.3.1 Virgin Solder Leaching

SEM/EDS analysis of the solder before and after AcOH leaching is shown in Figures 50.1a and 50.1b. Although the manufacturer lists 37% Pb, EDS analysis of the virgin solder surface shows high Pb content with some Sn and 1% O by mass. The C detected is likely from contamination. The virgin solder has a smooth and homogeneous surface structure. Post leaching with no stirring and a Pb:H_2O_2 ratio of 1:1.5 shows a nonuniform surface covered with etch pits. This is indicative of surface penetration from the acid. The average pit diameter is 10.8 μm. The Pb concentration has been reduced by over 40% and there is a higher percentage of O by mass, indicating electrochemical oxidation of the solder wire. The Sn concentration has increased by over 35%. This indicates that Sn is not leached into AcOH and matches results reported previously [16]. To further confirm Sn does not leach,

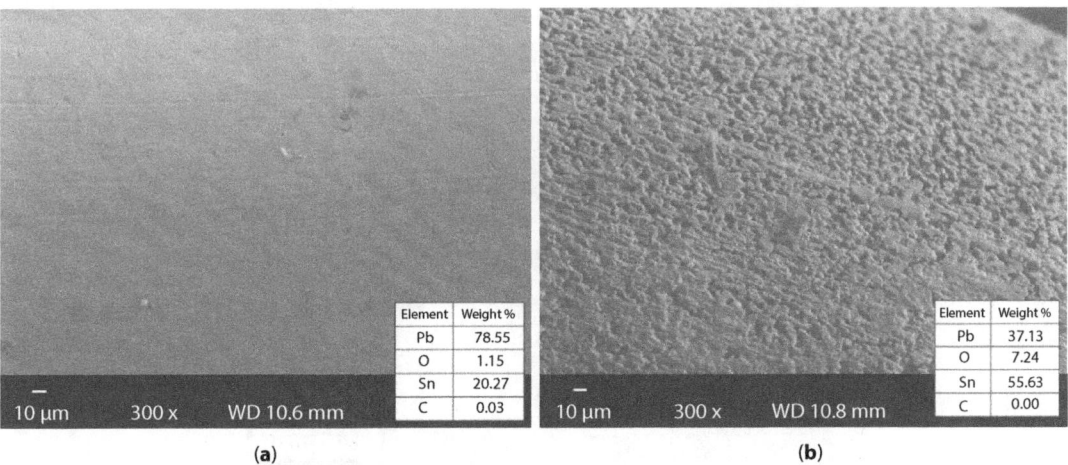

Element	Weight %
Pb	78.55
O	1.15
Sn	20.27
C	0.03

10 μm   300 x   WD 10.6 mm

(a)

Element	Weight %
Pb	37.13
O	7.24
Sn	55.63
C	0.00

10 μm   300 x   WD 10.8 mm

(b)

Figure 50.1 (a) Secondary electron image of virgin solder ribbon with wt. % EDS. (b) Secondary electron image of leached solder ribbon in 10% v/v AcOH, 1:1.5 Pb:H_2O_2 mole ratio, no agitation, with wt. % EDS.

a separate experiment was conducted in which 1 g Sn powder was leached in 16.5 mL of 10% v/v AcOH with H_2O_2 (1:2.5 Sn:H_2O_2 mole ratio) and aggressive stirring for 72 hours. The mass recovered post-leaching was over 99%.

To electrowin Pb from the leachate, chronoamperometry was performed. Small gray dendrites were observed after applying a potential of –0.8 V vs Ag/AgCl for approximately 20,000 sec to the leachate obtained with parameters: 1:1.5 Pb:H_2O_2 mole ratio and no stirring. EDS confirms these dendrites are Pb with mild oxidation (Figure 50.2). The dendrites are flat and smooth, with a fan-like morphology which appears to grow outward from a central vertex. Small deposits on top of the fan show a more characteristic box shape with an average diameter of 13.8 μm. Further optimization of the electrowinning process to recover all the Pb from the solder leachate is needed. Most of the experiments overloaded before 24 hours, which stopped the electrowinning process. This could be due to a current density limitation, drifting compliance voltage, low conductivity of the leachate, and/or complexation issues from the solder resin.

Pb leaching with varying Pb:H_2O_2 mole ratios with and without stirring was investigated. The results are shown in Figure 50.3. There is a clear linear correlation between solder weight loss and Pb:H_2O_2 mole ratio without stirring. The correlation by fitting is $y = 0.9095x - 0.7438$, with an R^2 of 0.9998. Moreover, stirring produces a super-linear correlation. In general, as the H_2O_2 amount increases, more mass of the solder is leached out. When stirring is present, there is also more mass loss for two out of the three conditions. The mass losses for the 1:2 ratio are very similar with and without stirring. However, mass loss may not be indicative of the amount of Pb leached, for a crumbly powder is observed in the beaker after each leaching. EDS of one collected powder sample shows this powder contains mostly Pb, with some O and Sn (Figure 50.4).

In an attempt to leach all the Pb from the solder into AcOH, 3 g of virgin solder was cut into tiny pieces and leached in 16.5 mL of 10% v/v AcOH with 1:2.5 Pb:H_2O_2 ratio for 72 hours with aggressive stirring. The leachate turned gray and formed a frothy white film around the edges of the beaker. After leaching, a sample of the leachate was collected, filtered, and analyzed using ICP-OES. The Pb concentration in the leachate was 133,850 ppm.

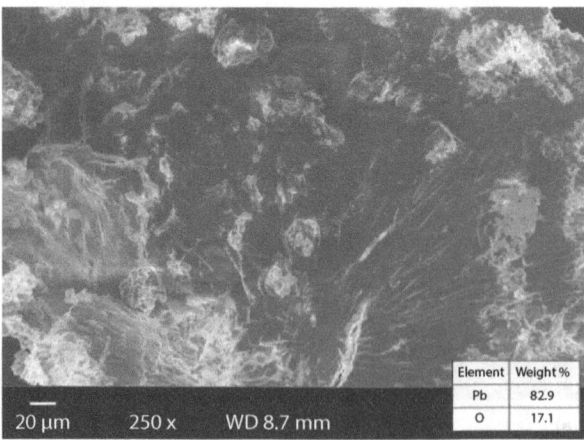

Element	Weight %
Pb	82.9
O	17.1

Figure 50.2 Secondary electron image and EDS data for Pb dendrites electrowon at –0.8 V vs Ag/AgCl in a leachate obtained with 1:1.5 Pb:H_2O_2 ration and no stirring.

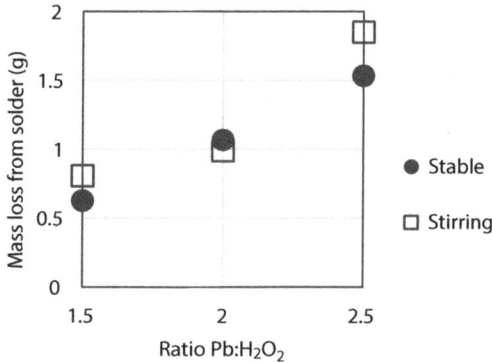

Figure 50.3 Mass loss from solder ribbon vs Pb:H$_2$O$_2$ mole ratio with and without stirring.

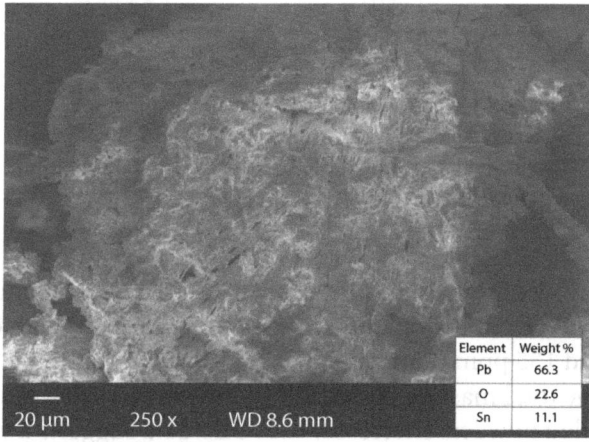

Figure 50.4 Secondary electron image and EDS data for the powder in the beaker after leaching with 10% v/v AcOH, 1:1.5 Pb:H$_2$O$_2$, no stirring.

With a total leaching volume of 17.9 mL (16.5 mL AcOH + 1.4 mL H$_2$O$_2$), this corresponds to 2.4 g Pb, which is more characteristic of an 80:20 Pb:Sn original solder ratio. EDS data of the unleached, virgin solder also suggests a starting Pb concentration of 80% (Figure 50.1a). These results conflict with the manufacturer's reported 37% Pb concentration in the solder.

If the solder is actually an 80:20 alloy, then ICP-OES data suggests 99.86% of Pb has leached from the aggressive stirring process. However, EDS of the powder leftover from this leaching reported 21.12 % wt. Pb. Complete leaching of Pb from Sn has not occurred, and further investigation is needed. One theory is the Sn could be acting as a protective coating around small particles of Pb, preventing the acid and H$_2$O$_2$ from fully attacking the Pb. AcOH concentration could also have an effect on leaching, as An *et al.* [17] observed in the dealloying of Mn-Cu using AcOH.

50.3.2 Leached Solar Cell

Table 50.1 shows the results after leaching 35 g of milled solar cell for 80 hours in 10% v/v AcOH with 35 mL H$_2$O$_2$. A total of 8 elements were detected. In addition to the 8 elements

Table 50.1 Elemental concentration by ICP-OES and corresponding redox potential for each element.

Element	Concentration (M)	E (V)
Ag	0.001	0.627
Al	0.008	-1.703
Ba	0.0001	-3.038
Ca	0.011	-2.926
Cu	0.040	0.300
Fe	0.0001	-0.564
Na	0.018	-2.814
Pb	0.001	-0.219

E is the reduction potential calculated using the Nernst equation and the concentration detected for each element using ICP-OES. The concentration is in mol/L.

detected, trace Sn may also be present in the leachate. It was not included in the ICP-OES analysis because Sn was not in the off-the-shelf standard used. The Pb concentration in the final leachate is around 155 ppm (33.1 mg Pb in solution). Assuming 0.2% of the silicon cell by weight is Pb and the total mass of silicon cell in the powder is 1.3 g, we estimate 26 mg of Pb in the original milled powder. Therefore, ICP-OES data suggests a very high Pb dissolution, as was seen in the aggressive solder leaching experiment. Although this is very promising, complete Pb dissolution cannot be claimed because there is still the possibility that lingering polymer or Sn could protect small particles of Pb from being leached. Because Pb is toxic, it needs to be recovered completely from PV waste. Additionally, 80 hours is too long for an industrial-scale recycling process. Further investigation to speed up the leaching and ensure 100% Pb removal is needed.

Surprisingly, Al leached into AcOH at over 200 ppm. Ca, Fe, and Na are likely contaminants. Cu has been shown to leach into AcOH [18] and can be explained as the electrical wire on the cell. Cu leached into solution the most at over 2500 ppm. The most surprising result is the detection of Ba at a concentration of 7.5 ppm. Two wavelengths were used for Ba detection in ICP-OES: 455.403 nm and 493.408 nm. Both detected a concentration of 7.5 ppm Ba. EDS of a solar cell cut in half detected Ba peaks at 1.00 keV and 4.50 keV (Figure 50.5). The peak at 4.50 keV does not overlap with anything. The peak at 1.00 keV does overlap with Na, and there is no expected Ba peak at 4.00 keV. Still, this data, in conjunction with the ICP-OES results, does suggest there is a small amount of Ba contamination in the silicon solar cell. Because the Ba was detected with Al on the silicon solar cell (Figure 50.6), it is hypothesized that it is a contaminant from the Al back contact. Knowing that Ba(II) acetate is soluble in water, it is reasonable to believe it could leach into AcOH.

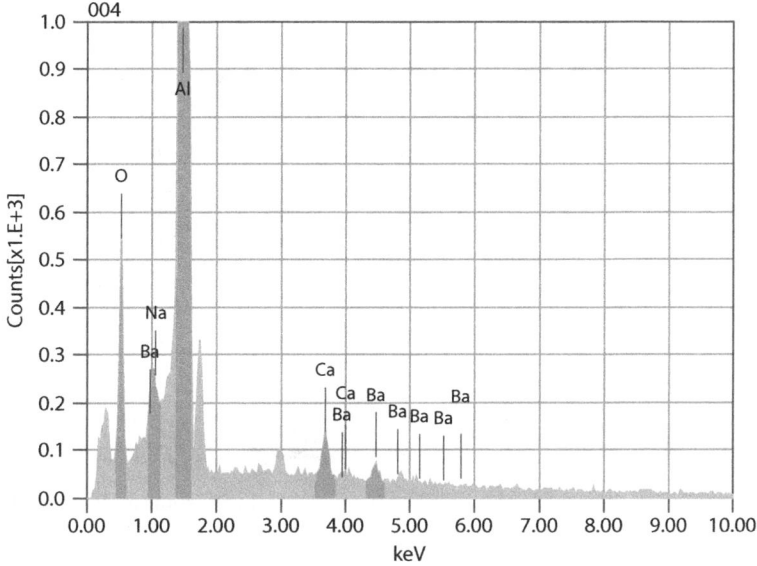

Figure 50.5 ICP-OES of Ba detected in the leachate from silicon solar cell.

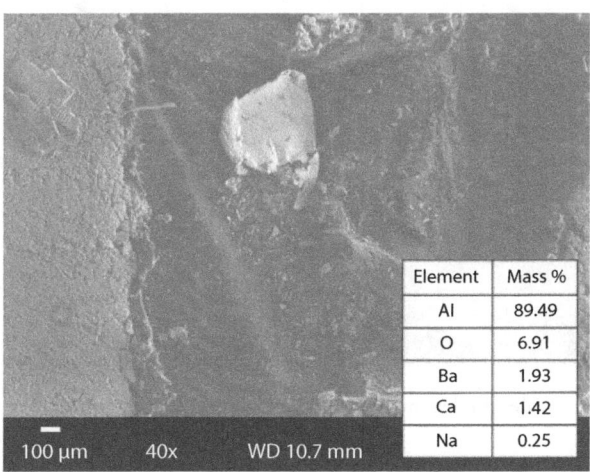

Element	Mass %
Al	89.49
O	6.91
Ba	1.93
Ca	1.42
Na	0.25

Figure 50.6 Secondary electron image and EDS data of the backside of an unleached silicon solar cell, where Ba was detected.

EDS confirmed other elements detected by ICP-OES as well. Figure 50.7 shows a nugget containing mostly Sn with some Si, Pb, and mild oxidation. This is likely a piece from the electrical contact in the cell. Figure 50.8 contains Al from the back sheet and Si from the wafer. Figure 50.9 shows the solar cell strip cut in half, separating the Si wafer from the Al back sheet. The Si wafer portion (gold) is what was coated in carbon then placed in the SEM.

For recovering the metals from solution, electrowinning is preferred over chemical precipitation. This is because little secondary processing is needed after electrowinning and the pure metals can be sold back to a circular economy. Unfortunately in a laboratory setting,

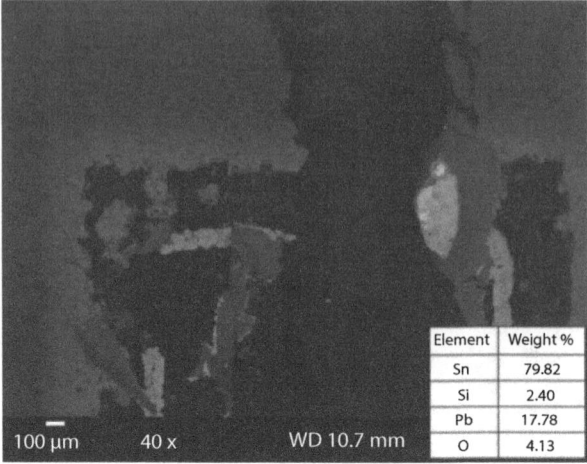

Element	Weight %
Sn	79.82
Si	2.40
Pb	17.78
O	4.13

Figure 50.7 Point analysis of unleached solar cell showing mostly Sn with some Si, Pb, and O.

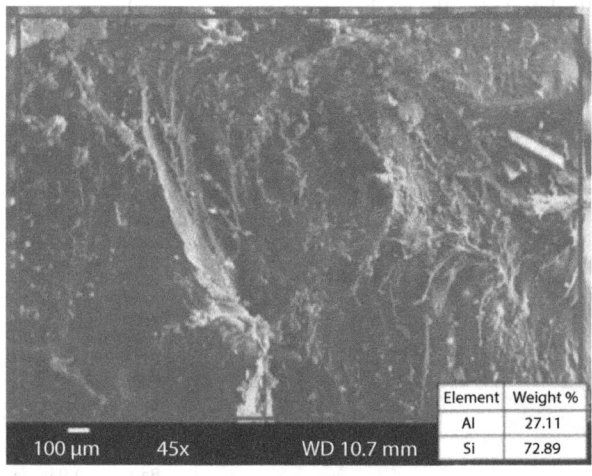

Element	Weight %
Al	27.11
Si	72.89

Figure 50.8 Area analysis of unleached solar cell showing Al and Si are present.

mass processing of solar cells is not possible, and the concentration of Pb in our leachate was too low. Because of this, sequential electrowinning was not attempted for this paper. Of the elements which leached the most, Ag, Cu, and Pb could be recovered via electrowinning. As seen by Table 50.1, Ca and Na are too negative and would need to be recovered via precipitation chemistry. Al is also too negative for electrowinning, but could be recovered using a caustic pre-treatment [13].

Jung *et al.* proposed sequential leaching and precipitation for silicon solar cell recycling, and Huang *et al.* proposed sequential electrowinning to recover Ag, Pb, and Cu from one HNO_3 leachate. We propose using sequential leaching followed by sequential electrowinning for recycling of solar cells. A schematic of this proposed process is shown below in Figure 50.10. The region marked by a dashed black line indicates the area this paper helps expand knowledge in. Ag, Cu, Pb, and Sn all have reasonable electrochemical potentials.

Figure 50.9 Solar cell cut in half, exposing the back of the Si wafer (right) and Al back sheet (left).

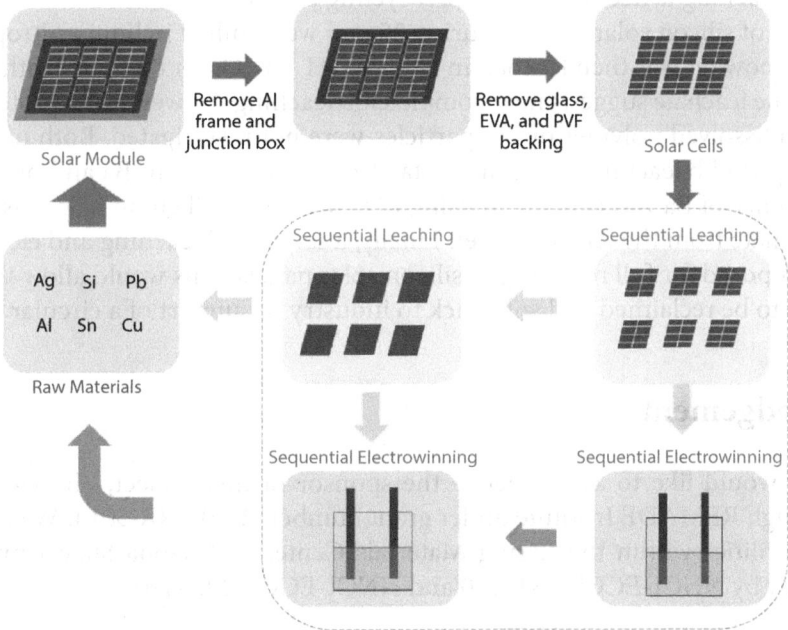

Figure 50.10 Overview of proposed sequential leaching and electrowinning process for recycling of silicon solar modules.

Al would need to be recovered via chemical precipitation, as its electrochemical potential is too negative. Ideally, the solar-grade silicon wafer will also be reclaimed, such as through a process proposed by Klugmann-Radziemska *et al*. All these reclaimed materials would then enter the feedstock for new PV fabrication. In this way, silicon PV would be made circular. One of the biggest challenges for any method of silicon solar cell recycling is the fluorinated back sheet. If burned, this material releases toxic fluorinated gases [12].

We avoided burning and broke apart the stiff polymer by milling in liquid nitrogen. However, this is not practical on an industrial scale. The undissolved polymer encapsulants could also hinder complete leaching. Because of this, further investigation into safely removing the polymers is needed.

50.4   Conclusion

Leaching and electrowinning of Pb into AcOH is demonstrated for virgin solder and silicon solar cells. SEM captured surface etch pits from the H_2O_2+AcOH attack on the solder ribbon. A strong correlation between the amount of H_2O_2 added and the mass loss on solder was observed. The effect of stirring on leaching rate was also investigated. Under all leaching conditions, a crumbly powder was formed in the beaker. EDS showed this powder contained mostly Pb, so a more aggressive leaching process was explored. After exposing 3 g virgin solder to vigorous stirring for 72 hours, ICP-OES showed very high Pb dissolution and the amount of Pb in the remaining powder was lower. It is hypothesized that Sn is acting as a protective coating, preventing complete Pb leaching. This needs more attention, as complete Pb leaching is necessary for safe recycling of silicon PV.

Two pieces of silicon solar cell measuring 25 cm^2 were milled in liquid nitrogen to a fine powder. This powder was then leached in 10% AcOH + H_2O_2 for 80 hours with no stirring. ICP-OES of the leachate suggests near complete Pb leaching. However, the protective effects of Sn and undissolved polymer on Pb particles were not investigated. Both of these could prevent complete Pb leaching. Leaching data also showed 7.5 ppm Ba and over 2500 ppm Cu. The presence of Ba was confirmed using EDS of a solar cell cut in half. It is assumed to be a contaminate from the Al back sheet. Finally, a sequential leaching and electrowinning process is proposed for full recycling of silicon solar panels. This would allow the materials in solar cells to be reclaimed and sold back to industry in support of a circular economy.

Acknowledgement

The authors would like to acknowledge the sponsor of this project, The Department of Energy through REMADE Institute under grant number: 21-01-RR-5014. We acknowledge the use of facilities within the Eyring Materials Center at Arizona State University supported in part by NNCI-ECCS-1542160 and NNCI-ECCS-2025490.

References

[1] Riva, M. A., Lafranconi, A., D'Orso, M. I., and Cesana, G., Lead poisoning: Historical aspects of a paradigmatic "occupational and environmental disease." *Saf. Health Work*, 3, 11, 2012.

[2] End-of-life solar panels: regulations and management, Environmental Protection Agency, https://www.epa.gov/hw/end-life-solar-panels-regulations-and-management#:~:text=The%20 discarded%20solar%20panel%2C%20which,meeting%20the%20characteristic%20of%20 toxicity.

[3] Kempe, M. D., Jorgensen, G. J., Terwilliger, K. T., McMahon, T. J., Kennedy, C. E., and Bore, T. T., Acetic acid production and glass transition concerns with ethylene-vinyl acetate used in photovoltaic devices. *Solar Energy Materials and Solar Cells*, 91, 315, 2007.

[4] Kraft, A., Labusch, L., Ensslen, T., Dürr, I., Bartsch, J., Glatthaar, M., Glunz, S., and Reinecke, H., Investigation of acetic acid corrosion impact on printed solar cell contacts. *IEEE J. Photovolt.*, 5, 736, 2015.

[5] National Primary Drinking Water Regulations, Environmental Protection Agency, https://www.epa.gov/ground-water-and-drinking-water/national-primary-drinking-water-regulations#Radionuclides

[6] Klugmann-Radziemska, E., and Ostrowski, P., Chemical treatment of crystalline silicon solar cells as a method of recovering pure silicon from photovoltaic modules. *Renew. Energy*, 35, 1751, 2010.

[7] Punathil, L., Mohanasundaram, K., Tamilselavan, K. S., Sathyamurthy, R., and Chamkha, A. J., Recovery of pure silicon and other materials from disposed solar cells. *Int. J. Photoenergy*, 2021, 1, 2021.

[8] Shin, J., Park, J., and Park, N., A method to recycle silicon wafer from end-of-life photovoltaic module and solar panels by using recycled silicon wafers. *Sol. Energy Mater. Sol. Cells*, 162, 1, 2017.

[9] Yousef, S., Tatariants, M., Denafas, J., Makarevicius, V., Lukošiūtė, S.-I., and Kruopienė, J., Sustainable industrial technology for recovery of Al nanocrystals, Si micro-particles, and Ag from solar cell wafer production waste. *Sol. Energy Mater. Sol. Cells*, 191, 493, 2019.

[10] Kuczyńska-Łażewska, A., Klugmann-Radziemska, E., Sobczar, Z., and Klimczuk, T., Recovery of silver metallization from damaged silicon cells. *Sol. Energy Mater. Sol. Cells*, 176, 190, 2018.

[11] Zhang, C., Jiang, J., Ma, E., Zhang, L., Bai, J., Wang, J., Bu, Y., Fan, G., and Wang, R., Recovery of silver from crystal silicon solar panels in self-synthesized choline chloride-urea solvents system. *Waste Manage.*, 150, 280, 2022.

[12] Huang, W.-H., Shin, W. J., Wang, L., Sun, W.-C., and Tao, M., Strategy and technology to recycle wafer-silicon solar modules. *Solar Energy*, 144, 22, 2017.

[13] Jung, B., Park, J., Seo, D., and Park, N., Sustainable system for raw-metal recovery from crystalline silicon solar panels: from noble-metal extraction to lead removal. *ACS Sustainable Chem. Eng.*, 4, 4079, 2016.

[14] Zhu, X., He, X., Yang, J., Gao, L., Liu, J., Yang, D., Sun, X., Zhang, W., Wang, Q., and Kumar, V. R., Leaching of spent lead acid battery paste components by sodium citrate and acetic acid. *J. Haz. Mater.* 250-251, 387, 2013.

[15] Hu, G., Zhang, P., Yang, J., Li, Z., Liang, S., Yu, W., Li, M., Tong, Y., Hu, J., Hou, H., Yuan, S., and Kumar, V. R., A closed-loop acetic acid system for recovery of PbO@C composite derived from spent lead-acid battery. *Resour. Conserv. Recycl.*, 184, 1, 2022.

[16] Click, N. and Tao, M., Selective lead recovery from a mixture of lead and tin for silicon solar module recycling. Electrochemical Society, 242nd Meeting, 832, 2022.

[17] An, S., Zhang, S., Liu, W., Fang, H., Zhang, M., and Yu, Y., Dealloying behavior of Mn-30Cu alloy in acetic acid solution. *Corros. Sci.*, 75, 256, 2013.

[18] Elomma, H., Seisko, S., Lehtola, J., and Lundström, M., A study on selective leaching of heavy metals vs. iron from fly ash. *J. Mater. Cycles Waste Manag.*, 21, 1004, 2019.

51

Thermolysis Processing of Waste Printed Circuit Boards: Char-Metals Mixture Characterization for Recovery of Base and Precious Metals

Mohammad Rezaee[1]*, Joelson P. M. Alves[1], Sarma V. Pisupati[1], Charles Ludwig[2], Henry Brandhorst[2] and Ernest Zavoral[2]

[1]*John and Willie Leone Family Department of Energy and Mineral Engineering, Center for Critical Minerals, EMS Energy Institute, College of Earth and Mineral Science, Pennsylvania State University, University Park, PA 16802, USA*
[2]*CHZ Technologies LLC., Auburn, AL 36830, USA*

Abstract

Waste printed Circuit boards (WPCBs) encompasses 3% of electronic waste, one of the fastest-growing categories of global waste. These waste materials contain base and precious metals with concentrations higher than those of primary resources. Therefore, recovery of the metals (especially precious metals) is the driving economic factor for recycling WPCBs. However, this effort requires the liberation of metals from non-metal fractions and each other. While comminution can effectively liberate metals from non-metals, it requires significant energy due to the high hardness and tenacity of WPCBs. Although the non-metal fraction has a low economic value, it could provide high energy recovery instead of being landfilled if the release of toxic flame retardants can be controlled. In this study, pilot scale testing of a patented thermolysis process was performed to demonstrate the capability of safely converting the plastic components of WPCBs into clean fuel gas substantially free of halogenated and volatile organic compounds, thereby liberating the metals from non-metal fractions. The char-metal mixture obtained from the process was then comprehensively characterized for the liberation and recovery of base and precious metals. The dioxin content analysis proved that dioxins were not generated, and the existing dioxins in the feedstock were cracked under chemical-reducing conditions of the process. The characterization of size and density fractions of the char-metal mixture revealed that most metals including Cu, Al, and Au are moderately and well liberated in particle size fractions smaller than 0.5 mm, and 0.149 mm, correspondingly. Based on these findings, a process flowsheet was proposed to liberate and recover base and precious metals from WPCBs. The process involves shredding and thermolysis processing of the WPCBs to convert the plastic components to fuel gas. The remaining char-metal mixture is then subjected to sizing to fractionate the sample to +0.595 mm, -0.595 mm+0.149 mm, and -0.149 mm size fractions (i.e., coarse, medium, and fine). Only the coarse fraction will require comminution for the metal liberation. The medium and fine fractions could be directly treated through physical separation to recover and separate base

---

*Corresponding author: m.rezaee@psu.edu

Nabil Nasr (ed.) Technology Innovation for the Circular Economy: Recycling, Remanufacturing, Design, Systems Analysis and Logistics, (677–696) © 2024 Scrivener Publishing LLC

and precious metals. Through the proposed process, the feed to the comminution process, the most energy-consuming stage in recycling, is reduced by 65%.

Keywords: E-waste, waste printed circuit boards (WPCBs), thermolysis, base metals, precious metals, liberation

51.1 Introduction

Electronic waste (e-waste) is one of the fastest-growing categories of global waste, constituting approximately 8% of municipal waste [1, 2]. Waste printed circuit boards (WPCBs) encompass about 3% of nearly 45 Mt/year of global e-waste generation [3, 4]. WPCBs consist of a number of metallic components, including base metals (Cu, Al) and precious metals (Au, Ag, Pd), with a concentration higher than those of primary resources, making them economically attractive for recycling [5]. The process of recovering metals from WPCBs involves multiple stages, i.e., pretreatment/liberation, physical separation, and chemical separation, with liberation being the most critical and challenging stage [4–7].

The goal of the liberation is to disassociate metals from non-metals and each other, leading to more efficient downstream physical and chemical separations [8, 9]. WPCBs are made of metallic and non-metallic components, glass fibers, and reinforced resin, which are challenging to efficiently liberate using conventional comminution methods [10, 11]. Due to the low economic value, the recycling of the non-metal fraction is rarely practiced and considered as the by-product of the WPCBs recycling. Liberation of WPCBs is typically performed through two main steps: pretreatment (disassembly, cutting, heat treatment) and size reduction through comminution [12]. Pretreatment of WPCBs is necessary to dismantle, separate, and classify the components according to their chemical compositions and hazardous nature and has been reviewed in detail by Ghosh *et al.* [3]. Dissociation of metals from the non-metal fraction could be obtained through comminution (e.g., by pulverizing to less than 0.6 mm) [13]. However, it requires significant energy due to the high hardness and tenacity of WPCBs. Furthermore, local temperatures of WPCBs tend to increase during crushing, forming metals and plastics agglomeration, further complicating the liberation process. It also makes the resins undergo partial pyrolysis, producing brominated hydrocarbons, benzynes, and phenols [11]. Uncontrolled grinding WPCBs also produces a large amount of fine particles of less than 150 microns, resulting in metal loss during grinding [14, 15]. It is crucial to address all these challenges encountered during the liberation process to recover metals successfully.

Liberation/recycling of the non-metal from metal fraction through various chemical methods, such as thermolysis (also called pyrolysis), gasification, supercritical fluid depolymerization, and hydrogenolytic degradation, for their conversion to the polymers or fuel, have been investigated mostly in laboratory scale by researchers (e.g., [16]). However, WPCB combustion requires close monitoring to avoid the release of toxic flame retardants present in these materials [17, 18]. Among the chemical methods, thermolysis (with advantages of less temperature, pressure, and cost) offers a viable alternative for comminution to liberate the metals from the non-metal fraction. This method decomposes the organics and enriches the metal content [8, 19–23]. To address the environmental and health effects of the process (i.e., dioxin precursors in the thermolysis oil and toxic gasses such as hydrogen bromide and organobrominated compounds), debromination could be performed by introducing $CaCO_3$ or Fe_2O_3, which control brominated and other organic compounds like benzene [3, 24, 25]. This process generates salts such as $CaBr_2$, which can be used in various applications such as extinguishing agents, petroleum drilling, and refrigeration. However, the industrial application of this process for recycling WPCBs and the characterization of the remaining char-metal mixture have rarely been reported.

In this study, a pilot scale testing of a patented thermolysis process, CHZ Technologies, LLC. was performed for converting the plastic components of WPCBs to useable fuel gas sources substantially free of halogenated organic compounds (including volatile organic compounds, i.e., VOCs) [26]. The remaining char-metal mixture was then comprehensively analyzed for liberation and recycling base (Cu and Al) and precious metals (Au, Ag, and Pd).

51.2   Material and Methods

250 kg WPCBs were collected and shredded to 50.8 mm to be processed by the thermolysis process. The feedstock was characterized for major elemental content, flame retardant, heating value, and dioxin/furans. The base metal composition of WPCBs consists mostly of Cu (6% - 26%) used in circuitry, Al (1% - 7%) used in capacitors, Fe (1% - 8%), followed by Pb and Sn (4%-6%) used in lead frames and Pb/Sn solder solders, and Zn and Ni [27]. The metal content of the sample (Table 51.1) aligned well with the typical base metal content

Table 51.1 Feedstock analysis.

Parameter (Unit)	Value
Dry residue (Mass %)	99.2
Iron (mg/kg)	5860
Aluminum (mg/kg)	22500
Copper (mg/kg)	352000
Zinc (mg/kg)	778
Tin (mg/kg)	92100
Cadmium (mg/kg)	0.18
Chromium (mg/kg)	21.4
Nickel (mg/kg)	5630
Lead (mg/kg)	24000
Bromine (mg/kg)	20000
Heating value (kJ/kg)	8185
Chlorine (Mass %)	11.3

Table 51.2 Analysis of dioxin/furans (in ng/kg) in WPCBs and the char-metal mixture obtained by the thermolysis process.

Analysis	Feedstock	Char-metal mixture
2,3,7,8- TetraCDD	< 1	<1
1,2,3,7,8-PentaCDD	< 1	<1
1,2,3,4,7,8-HexaCDD	<2	<1
1,2,3,6,7,8-HexaCDD	<3	<1
1,2,3,7,8,9-HexaCDD	<5	<1
1,2,3,4,6,7,8-HeptaCDD	< 15	<5
OctaCDD	< 200	<10
2,3,7,8-TetraCDF	< 1	<2
1,2,3,7,8-PentaCDF	<2	<1
2,3,4,7,8-PentaCDF	<2	<1
1,2,3,4,7,8-HexaCDF	<2	<1
1,2,3,6,7,8-HexaCDF	< 1	<1
1,2,3,7,8,9-HexaCDF	< 1	<1
2,3,4,6,7,8-HexaCDF	< 1	<1
1,2,3,4,6,7,8-HeptaCDF	<5	<3
1,2,3,4,7,8,9-HeptaCDF	<3	<3
OctaCDF	< 15	<10

of WPCBs. It should be noted that the data reported in Table 51.1 was obtained by digestion of the sample using hydrofluoric acid, and may underestimate the elemental content. The heating value of the feedstock is well aligned with that of the non-metallic fraction of WPCBs (11 to 16 GJ/t reported by [17], and [18]). This value is comparable with the calorific value of wood and peat coal and can provide high energy recovery if the release of toxic flame retardants can be controlled. The flame retardants in the feedstock contain bromine, which was analyzed as expected with 2%. The analysis of the dioxin/furans in the sample (Table 51.2) shows the elevated levels of dioxins present in the feedstock.

51.2.1 Thermolysis Process

The objectives of the pilot scale testing of the patented thermolysis process (described in detail by [26]) for processing WPCBs were to (i) avoid creating new dioxin/furans and reduce existing dioxin/furans by cracking, (ii) utilize the thermolysis method to generate clean, useable fuel gas sources substantially-free or free of halogenated organic compounds (including volatile organic compounds, i.e., VOCs), and (iii) generate char containing valuable metals, precious metals, glass reinforcement, and other materials, all of which are substantially-free or free of halogenated organic compounds. The schematic of the process is shown in Figure 51.1. The shredded feedstock was continuously fed to the process in short intervals, steam was injected, condensates were removed from scrubbers 1 and 2, and the char was removed from the char screw conveying unit. The plant throughput was 170 kg of material, averaging 35 kg/hr. The total average gas volume was 15 m³/hr. The plant was operated with an elevated temperature range in the reactor to secure a complete chemical conversion of the test material. The plant operated at 700 to 800 °C for the reactor, where the temperature of the reactor head was measured to be between 600 to 650 °C. The pressure in the reactor was, on average, about five mbar. Despite the material's high reactivity and rapid gasification, significant pressure spikes were not recorded. The generated gas volumes had significant spikes due to the high reactivity of the material; a more constant infeed volume could equalize these spikes. The gas cleaning was performed at 75 °C for scrubber 1, and at 30 °C for scrubbers 2 and 3. The resulting materials and media were sampled throughout the process for characterization. The mass balance of the thermolysis process is provided in Table 51.3. The data shows that the char encompasses 63% of the input WPCBs. The process reduced the feedstock mass by 37% by converting the plastic components to fuel gas, a portion of which is used to drive the thermolysis process. Therefore, the process concentrates the metals in the remaining char-metal mixture.

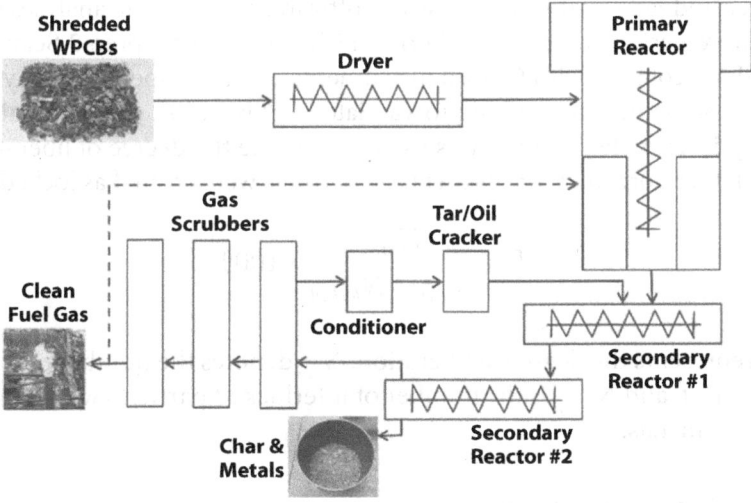

Figure 51.1 Schematic of the thermolysis process.

Table 51.3 Mass [kg] of the input and output streams.

Input		Output			
Material	Steam	Gas	Oil/Oligomers	Char	Water
170	23	64	3	108	15

51.2.2   Characterization

The product gas and solid char-metal mixture were comprehensively characterized. The product gas was characterized by gas chromatography to confirm that the gas output contains suitable compositions with high methane, hydrogen, and carbon monoxide content for further usage, and hydrogen bromide or hydrogen chloride neutralized in the gas scrubbers with sodium hydroxide. Dioxin and VOC contents of the char were characterized to confirm the efficacy of the process for the destruction of these components.

The char-metal mixture was further characterized for elemental content and liberation study to recover base and precious metals. The materials were first analyzed for size distribution, ash and elemental contents, and composition phases. The size analysis was performed using dry and wet screening for size fractions larger and smaller than 0.149 mm, respectively, sizing the sample into eight different size fractions, +25.4 mm, -25.4+12.7 mm, -12.7+4.76 mm, -4.76+1.19 mm, -1.19+0.595 mm, -0.595+0.297 mm, - 0.297+0.149 mm, -0.149 mm. Representative samples of each size fraction were pulverized in the RETSCH ZM 200 pulverizer for characterization. The pulverized samples were analyzed for proximate analysis using thermogravimetric (TGA) analysis, and elemental content using Thermo Fisher Scientific XSeries 2 Inductivity Coupled Plasma-Mass Spectrometry (ICP-MS) with a detection limit of pg/L. The samples were digested using hydrofluoric acid and aqua regia, according to the ASTM D6357-11, before ICP-MS analysis. Triplicate samples were analyzed, and ICP-MS standard solutions with known concentrations of elements were used for quality control. The XRD analysis was performed to determine the phase characterization of the samples using a Malvern Panalythical Empyrean X-ray powder diffraction (XRD) with Cu (K-α) radiation source and Jade® pattern processing software. Microscopic analyses of the various size fractions were performed using a Thermo Fisher Scientific Apreo 2 Scanning Electron Microscope (SEM) coupled with Oxford instruments energy-dispersive X-ray spectroscopy (EDS) detector for elemental mapping to validate the liberation results obtained from the elemental analysis. Particle counting was used to evaluate the degree of liberation of metals (equation 51.1); therefore, metal-non-metal composites were treated as locked particles.

$$DL = \frac{N_{free}}{N_{free} + N_{locked}} \times 100\% \tag{51.1}$$

Where DL represents the degree of liberation, N_{free} denotes the number of free particles of the metal of interest, and N_{locked} is the number of interlocked particles with either non-metal particles or other metals.

51.2.3   Size-Density Fractionation

The size-density fractionation, shown in Figure 51.2, was performed for the liberation analysis. Upon size classification of the char-metal mixture, a representative sample of each size fraction (except for -0.149 mm) was subjected to density fractionation (i.e., float-sink analysis) at specific gravity (SG) values of 1.8, 2.5, and 2.95 using lithium metatungstate (LMT) solution. LMT is a heavy liquid with an SG of 2.95 and can be mixed with water to obtain lower SG values. The SG values were selected based on the density of heavy metals, light metals, and organic matter present in the sample. For the density fractionation, each size fraction was submerged and mixed in a 2-L beaker containing the lowest SG solution.

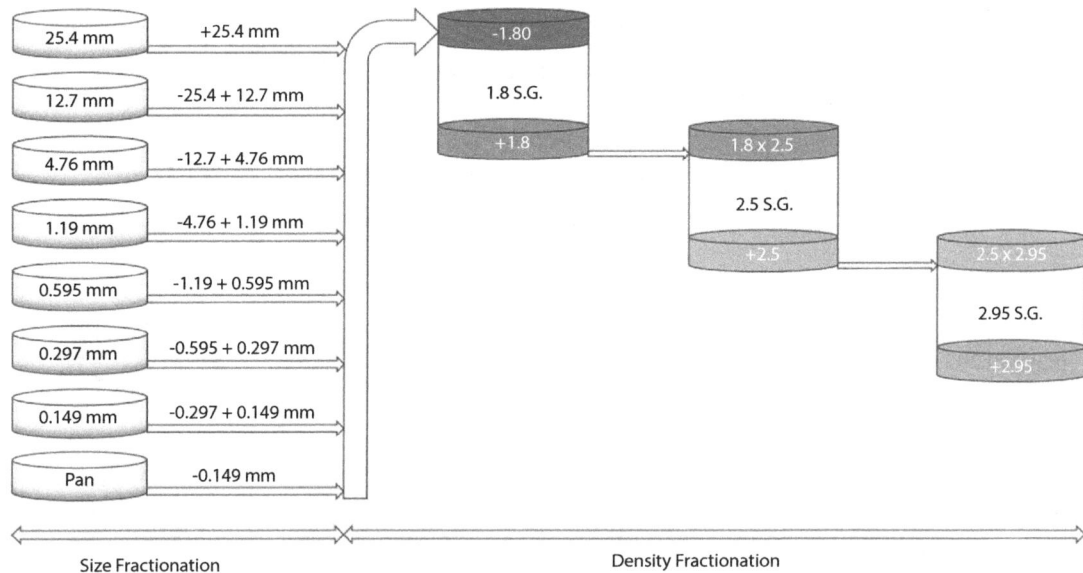

Figure 51.2 Schematic of the size-density fractionation process.

Enough time was then provided for the materials to be segregated in the float and sink fractions. The float fraction was removed using a screening tool, allowing the drainage of the excess medium, and washed multiple times with DI water. The remaining medium was then filtered to recover the particle that sank at the bottom of the beaker. The sink fraction was then rinsed and subjected to the next denser medium, and the process was repeated. As a result, each size fraction was separated into four density fractions, i.e., -1.8 SG, 1.8x2.5 SG, 2.5 x 2.95 SG, and +2.95 SG. Table 51.4 shows the pictures of the density fractions obtained from each size class. Those vivid illustrations provide indirect evidence (i.e., color changes)

Table 51.4 Size-density fractions.

Size Fraction (mm)	Specific Gravity Fractions			
	+2.95	2.95x2.5	2.5x1.8	-1.8
-25.4+12.7				
-12.7+4.76				
-4.76+1.19				

(Continued)

Table 51.4 Size-density fractions. (*Continued*)

Size Fraction (mm)	Specific Gravity Fractions			
	+2.95	2.95x2.5	2.5x1.8	-1.8
-1.19+0.595				
-0.595+0.297				
-0.297+0.149				

of the materials' segregation through size and density fractionation. For instance, the reddish-brown color of the +2.95 SG fractions might indicate heavy-weight metal content such as Cu, while the silvery-white particles in the 2.5x2.95 SG could indicate the presence of light-weight metals such as Al. The dark color of the 1.8 SG float fraction could be attributed to the presence of carbon-rich (i.e., char) particles obtained from the thermolysis processing of the plastic components in WPCBs. The size-density fractions were fully characterized to provide insights into how the target metals (Cu, Al, Au, AG, and Pd) are liberated.

51.3 Results and Discussions

51.3.1 Analysis of the Gas Product

The analysis of the gas product (Table 51.5) reflects homogeneous operating conditions. The gas analysis shows relatively high hydrogen content and low paraffin content. The ratio of CO and CO_2 is characteristic of the temperature profile of the process. Oxygen and nitrogen are not components of the original gas compositions but intruders into the gas

Table 51.5 Gas product analysis.

Gas composition [Volume %]	Gas sample 1	Gas sample 2
H_2	36.4	34.1
O_2	0.30	0.31
N_2	1.70	1.9

(*Continued*)

Table 51.5 Gas product analysis. (*Continued*)

Gas composition [Volume %]	Gas sample 1	Gas sample 2
CH_4	22.2	22.5
CO_2	11.2	10.3
CO	20.8	21.5
Ethane	0.66	1.0
Ethene	2.5	2.6
Propane	<0.01	0.01
Propene	0.06	0.15
i-Butane	<0.01	<0.01
n-Butane	<0.01	<0.01
Heating value [kWh/m^3]	4.8	4.9

sample during the transport of the glass vessels. The molecular structure of the feedstock with characteristic phenol components does not generate significant amounts of long-chain molecules like paraffin and olefins. The low amount of generated oily components could be pumped at average temperatures and converted with the cracking reactor.

51.3.2 Char-Metal Mixture Characterization

51.3.2.1 Dioxin Content Analysis

The char was analyzed for dioxin/furan content (Table 51.2). Most measured data were below the detection limit. The analysis proves that dioxins were not generated and that even existing dioxins in the feedstock were cracked during this test under chemical-reducing conditions. Comparing the dioxin/furan values in the feedstock and the char proves the destruction of the dioxin/furan molecules. This result is based upon reactions without oxygen under chemically reducing conditions. The process can be optimized for this feedstock and be improved by additional reactors to secure complete cracking by follow-up reactions [26]. Furthermore, no VOCs was detected from the char.

51.3.2.2 Elemental Content

The elemental content analysis of the char-metal mixture revealed that the sample contains mainly Cu (19.5%), followed by Al (4.3%). Regarding the precious metals, the sample contains 32.2 ppm of Au, 10 ppm of Ag, and 3 ppm of Pd. The concentration of the rest of the precious metals (i.e., Rh, Ru, Pt, Re) was below 1 ppm. The major impurities in the sample were identified as Fe (3.7%) and Si (0.6%).

Table 51.6 Size and proximate analysis.

Size fraction (mm)	Ash (%)	Volatile (%)	Fixed carbon (%)	Weight (%)	Cumulative passing (%)	Cumulative retained (%)
+25.4	81.5	9.5	9.0	0.5	100	0.5
-25.4+12.7	88.3	7.0	4.7	4.2	95.3	4.7
-12.7+4.76	91.2	2.6	6.1	14.6	80.7	19.3
-4.76+1.19	84.8	5.7	9.5	22.5	58.2	41.8
-1.19+0.595	69.5	13.1	17.4	12.5	45.7	54.3
-0.595+0.297	71.9	10.3	17.8	8.5	37.2	62.8
-0.297+0.149	71.7	8.6	19.7	14.7	22.5	77.5
-0.149	77.2	8.7	14.1	22.5	0.0	100.0
Total	79.2	7.7	13.0	100.0		

51.3.2.3   Characterization of Size Fractions

The size fractionation results show that the majority of the sample, approximately 60%, is present in size fractions smaller than 1.19 mm (Table 51.6). The higher ash content in the coarse fractions, +1.19 mm, is attributed to base metals and ceramic content WPCBs, which are mainly present in coarse sizes in WPCB components. The carbon content is less than 10% in the coarse fractions (+1.19 mm) and 10-20% in the finer fractions. The carbon and volatile content in the char could be associated with plastic resins in WPCBs. Removing carbon content by physical separation could further increase the grades of the metals.

Elemental analysis of size fractions was conducted to evaluate the grade and distribution of base and precious metals and major impurities in various size fractions. The data revealed that most (i.e., 60-80%) of the base metals (Cu and Al) are reported in the coarse fractions (+1.19 mm). The concentration of Cu in the coarse fractions varies in the range of 20 to 30% and is less than 20% in the finer fractions (Figure 51.3). Al has higher concentrations in the fractions coarser than 0.149 mm, ranging from 4-6%. Its concentration decreases to less than 1% in the finest size fractions, -0.149 mm (Figure 51.3). The majority (i.e., 69%) of the Au presents in the fine fractions (-1.19 mm), out of which approximately 34% by weight with a concentration of about 50 ppm is reported in the finest size fraction (i.e., -0.149 mm) (Figure 51.3). Therefore, efforts should be concentrated on the recovery of base metals from coarse fractions (+1.19 mm), and the Au recovery from the fine fractions (-1.19 mm), upon liberation. Ag, Pd, and the remaining Au are equally distributed in all other fractions with grades ranging from 5-12 ppm and 1-4 ppm, respectively (Figure 51.3). Fe follows the same trends as base metals. Majority of this element is distributed in coarse fraction (+0.19mm), with more than 30% reported to the -4.76+1.19 mm fraction (Figure 51.3). About 15% of Fe was also found in the finest fraction (-0.149 mm), as one of the major impurities for this fraction, which contains the majority of Au.

The size fractions were further characterized for the phase composition using X-ray diffraction (XRD) analysis. The XRD results were consistent with the elemental content analyzed by ICP-MS. The most significant peaks in the XRD patterns of each size fraction were Cu and Si, representing metal and non-metal contents, respectively. The XRD patterns

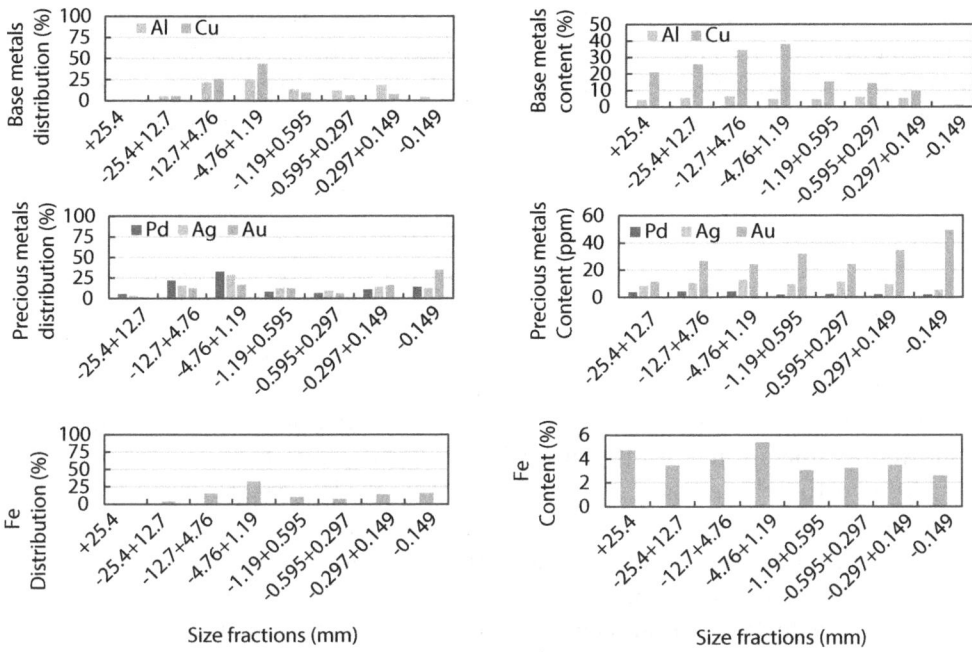

Figure 51.3 Base (Cu, Al) and precious (Au, Ag, Pd) metals and Fe distribution and content in various size fractions.

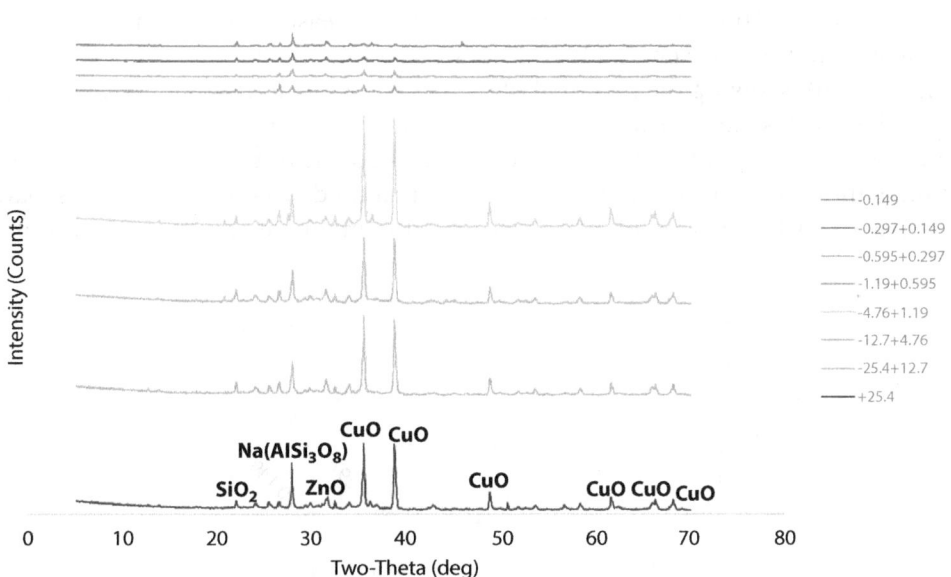

Figure 51.4 XRD analysis of various size fractions.

showed that the highest intensity diffraction peaks related to CuO occur in the four size fractions coarser than 1.16 mm, verifying the ICP-MS elemental analysis results. The samples were ashed at 750 °C for eight hours in a muffle furnace to remove the organic matter before XRD analysis. The ashing process might have caused the conversion of metals to their oxides, e.g., Cu to CuO (Figure 51.4).

51.3.2.4   Characterization of Size-Density Fractions

The weight distribution results from the size-density fractions (shown in Table 51.4) revealed that, for most of the size fractions, the majority of the weight is reported in the 1.80 SG float and 2.50 x 1.80 SG fractions, indicating that most of the heavy metals are interlocked with light metals and non-metal (e.g., ceramic) particles (Figure 51.5). The distribution and concentration of Cu, the primary element in WPCBs, as a function of size and density (Figure 51.6) showed its liberation from the other metals increases as the particle size decreases, resulting in reporting this element (with SG of 8.93) to the heaviest fraction (i.e., +2.95 SG). As shown in Figure 51.6, more than 75% of Cu is recovered in +2.95 SG fraction of size fractions smaller than 1.19 mm. The Cu grade in these fractions is more than 50%. Therefore, this element can be potentially concentrated using gravity separation in these size fractions without any size reduction. Upon recovery of this element, the remaining materials can be further liberated, as needed, for effective separation of other elements. However, its separation in the coarser size fraction (i.e., + 1.19 mm) requires size reduction/liberation.

Al was the second dominant element in the char-metal mixture (i.e., 4.3%). This element was partially liberated in size fractions smaller than 0.595 mm, as more than 60% of this element (with SG of 2.70) was reported in lower density fractions (Figure 51.6). Reporting this element to the density fraction of 1.8 x 2.5 SG, instead of 2.5 x 2.95 SG, was attributed to partial liberation of this element. Its separation with high purity needs further liberation from the other components, mainly Si and carbon content.

The distribution and concentration of Au in size density fractions showed that, similar to Cu, the liberation of Au increases as the particle size decreases. More than 60% of Au with a grade of 90 ppm was recovered in the +2.95 SG fraction of size fractions finer than 0.595 mm (Figure 51.6), showing the potential of using gravity separation for concentrating Au. The liberation of this element in coarser size fractions requires liberation/size reduction. The data also revealed that Pd and Ag do not follow the same trend as Au; therefore, further liberation of these elements in the fine-size fractions should be considered upon separation of Au. Analysis of the density of alloys and metal components of WPCBs, summarized

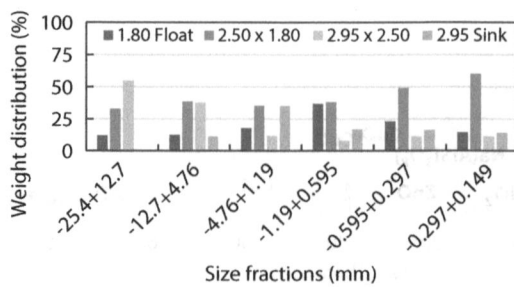

Figure 51.5 Weight distribution of size-density fractions.

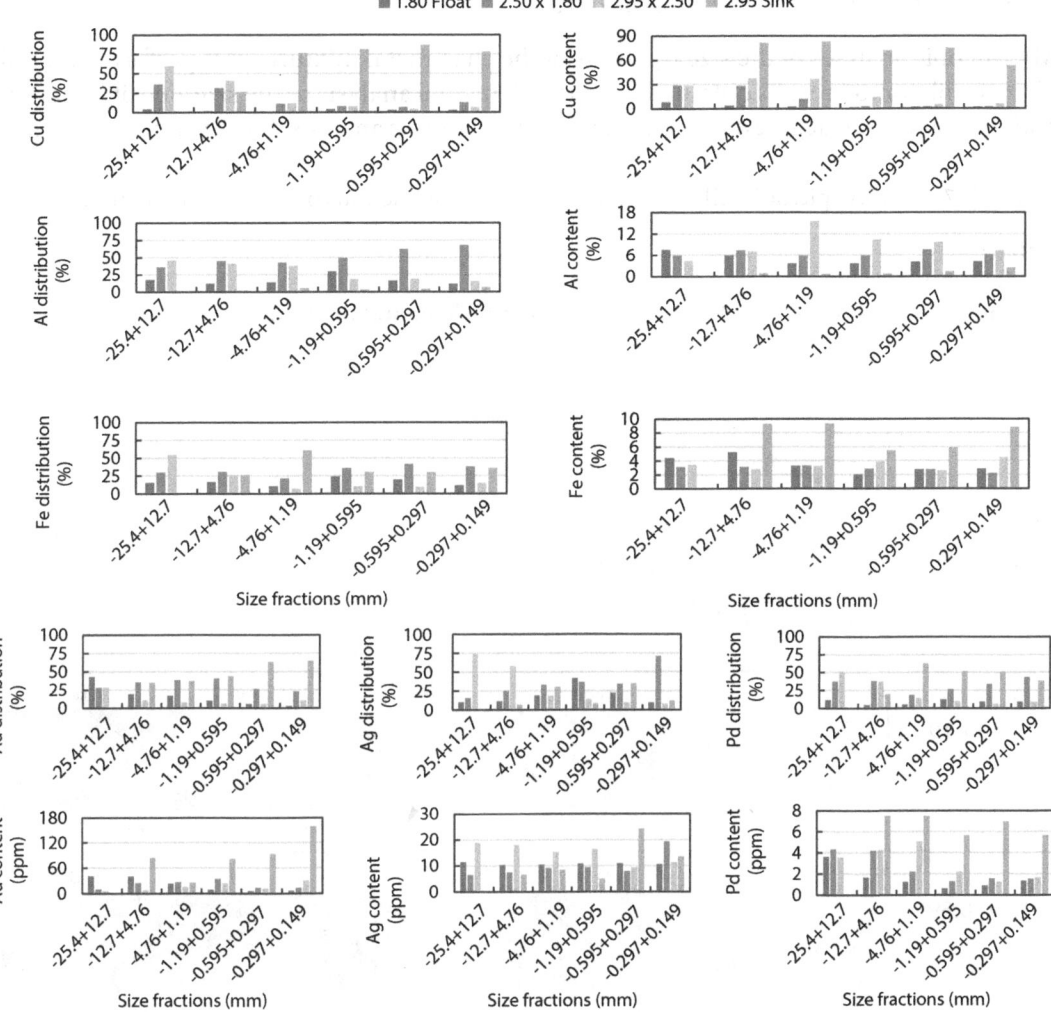

Figure 51.6 Distribution and content of Cu, Al, Fe, and precious metals (Au, Ag, and Pd) in the size-density fractions.

by Cui and Forssberg [28], reveals that the majority of alloys and base metals in WPCBs have SG < 10, while precious metals have +10 SG (i.e., Au: 19.32, Pd: 12, and Ag: 10.49). Therefore, gravity separation could potentially separate precious metals from base metals.

As one of the major impurities in recycling the target metals, Fe could be removed by utilizing the differences in its magnetic susceptibility and density with the other metals. However, the distribution of this element did not follow the same trend as Cu and Au, and no liberation was found for this element. Based on its distribution among all density fractions, its separation from the concentrated products will likely require further liberation (Figure 51.6). It should be noted that this element also presents in various alloys contained in WPCBs [28]. In this case, its separation will require utilizing chemical separation techniques.

51.3.2.5 Degree of Liberation Analyses

Microscopic analyses of the size fractions of the char-metal mixture were performed to validate the liberation results obtained from the elemental analysis of the size-density fractions and to measure the degree of liberation of metals. For this analysis, the size fractions coarser

Table 51.7 Microscopic/SEM-EDS analysis of various size fractions of char-metal mixture.

Size fractions (mm)	Microscopic/SEM-EDS images and elemental mapping
(a) -25.4+12.7	
(b) -12.7+4.76	
(c) -4.76+1.19	
(d) -1.19+0.595	
(e) -0.595+0.297	
(f) -0.297+0.149	
(g) -0.149	

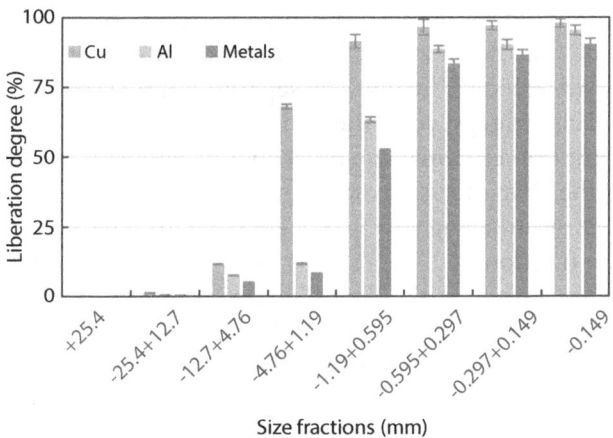

Figure 51.7 Calculated the degree of liberation of Cu, Al, and total metals.

than 4.76 mm were analyzed by a microscope, and the finer size fractions were analyzed using SEM-EDS for elemental mapping. The images confirmed that, in general, more metal liberation is achieved with a decrease in particle sizes (Table 51.7).

The images in Table 51.7, rows (a) and (b), show that the metals are interlocked and not liberated from each other in size fractions coarser than 4.76 mm. The rows (d)-(g) images confirm that Cu is well-liberated in fractions finer than 1.19 mm. Although other elements such as Al and Si follow the same trend as Cu, it is evident from the SEM images and EDS mapping results that they are not as liberated as Cu. The analysis of the -0.149 mm size fraction, row (g), revealed that most of the elements in this fraction are well-liberated, providing an opportunity for the separation of target metals in this fraction using enhanced gravity separators or flotation process.

The analysis of the degree of liberation confirmed that, in general, more metal liberation is achieved with a decrease in particle sizes (Figure 51.7). The results show that the metals are not liberated from each other in the coarse fractions, + 4.76 mm, in which Cu, Al, and total metals are liberated less than 12%, 8%, and 6%, respectively. However, Cu is more than 91% liberated in size fractions smaller than 1.19 mm. Although Al follows the same trend as Cu, it is less well-liberated than Cu. Al reaches a degree of liberation of more than 88% at size fractions smaller than 0.595 mm, in which total metals have a liberation degree of more than 80%. The degree of liberation of all metals reaches more than 90% in the finest size fraction (i.e., - 0.149 mm).

51.4   Conceptual Process Flowsheet for Liberation and Recovery of Base and Precious Metals

The elemental analysis of size and size-density fractions along with the analysis of the degree of liberation showed that Cu is well liberated in size fractions finer than 1.19 mm (with a degree of liberation of more than 90%). However, Al, Au, and total metals reach a high liberation degree (i.e., more than 80%) in size fractions smaller than 0.595 mm. The majority of metals were found to be well liberated in the size fraction smaller than 0.149 mm, with a more than 90% liberation degree for total metals. As a result, only the +0.595mm fraction, which encompasses 54% of the char-metal mixture, should be liberated, while the -0.595 mm could be directly subjected to the physical separation processes for the recovery of base and precious metals. This strategy, combined with the thermolysis process, reduces 65% of the feed to the comminution circuitry, the most energy extensive process in recycling.

It is proposed that the char-metal mixture obtained from the thermolysis process is first subjected to size classification using a two-deck screen with opening sizes of 0.595 mm and 0.149 mm. The coarse fraction (+0.595 mm) is then subjected to liberation. For this purpose, we recommend using a rod mill followed by size classification to utilize the differences in the original size of the metals present in WPCB components and their malleability for the selective liberation of metals in various size fractions. The malleable base metals (such as Cu and Al) which present in coarse size in WPCBs components (such as circuitry and capacitors) are expected to liberate in their original coarse sizes, while the precious metals (used as plating layers and contact materials) and the brittle components (such as Si present in ceramic components) are expected to liberated in fine size fractions (-0.595 mm). Upon liberation, the coarse size fraction (+0.595 mm) could be subjected to gravity separators (such as jig) for the separation of the dense metals (such as Cu) from light metals (such as Al).

The liberated metals in medium and fine size fractions, i.e., -0.595+0.149 mm, and -0.149 mm, respectively, obtained from size classification of the char-metal mixture and rod mill product could be treated with appropriate physical separation units. For example, the medium fraction could be treated by shaking table or spiral units, and the fine fraction could be treated by enhanced gravity separation or flotation. Depending on the performance of these units and the target recovery and grade values, the product streams could be subjected to further liberation to enhance the recovery or grade. Magnetic separation

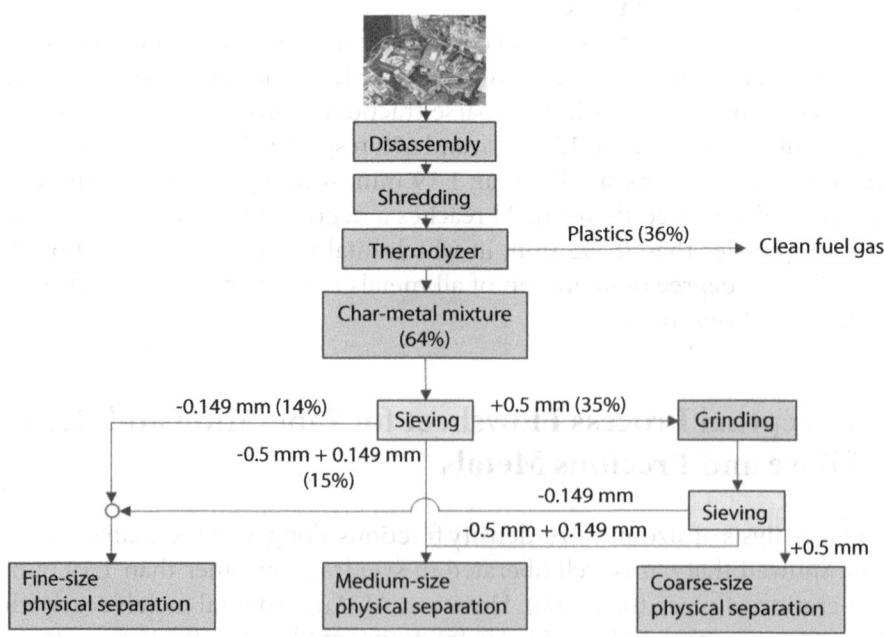

Figure 51.8 Schematic of the proposed recycling process.

could be subsequently used for all fractions to potentially remove ferrous metals and further concentrate the target metals (Al, Cu, Au, Ag, and Pd) in the non-ferrous fractions. The schematic of this proposed conceptual process is presented in Figure 51.8, which will be validated in our future study. The modification of this process (patent pending) could be also directly applied for recycling original WPCBs, rather than char-metal mixture [29].

51.5   Conclusions

In this study, pilot scale testing of a thermolysis process was conducted to safely convert the plastic components of WPCBs to clean fuel gas sources substantially free of halogenated organic compounds (including volatile organic compounds, i.e., VOCs), thereby liberating metals from the non-metal fraction. The char-metal mixture obtained from the process (encompassing 64% of the WPCB feedstock) was then comprehensively characterized for potential recovery of base and precious metals. The analysis of the gas product showed relatively high hydrogen content and low paraffin content, with a CO to CO_2 ratio of 2:1. The dioxin/furan content of the char were below the detection limit, proving that dioxins were not generated, and the existing dioxins in the feedstock were cracked under the chemical-reducing conditions. Furthermore, no VOCs were detected from the char. The elemental content, phase, and microscopic characterizations of size and density fractions revealed that Cu is well liberated in size fractions smaller than 1.19 mm. However, the other metals including Al, and Au were liberated in size fractions smaller than 0.5 mm, and more than 90% degree of liberation for metals was observed in the -0.149 mm size fraction. The density fractions results revealed that the base and precious metals could be processed and separated from the liberated size fractions using gravity separation units. Based on the obtained results, a process flowsheet for the liberation and physical separation of the base and precious metals was formulated. In the proposed process, upon disassembling and shredding the WPCBs to a top size of 50.8 mm, the materials are processed through the thermolyzer, where 36% of the materials (plastic components) are converted to clean fuel gas. The remaining char-metal mixture is then sized to fine, medium, and coarse size fractions (i.e., -0.149 mm, +0.149 mm-0.595 mm, and +0.959 mm). Based on the liberation study, only the coarse fraction requires liberation through comminution, which will be conducted using a rod mill to liberate the metal components in various size fractions by utilizing the differences in the malleability of WPCB components. Therefore, only 54% of the char will be subjected to the comminution process. This strategy, combined with the thermolyzer, reduces the feed to the mill unit by 65%, thereby a significant energy reduction in the recycling process. Upon liberation, each size fraction could be subjected to appropriate physical separation (including gravity separation) units for the separation of base and precious metals.

Acknowledgment

This material is based upon work supported by the U.S. Department of Energy's Office of Energy Efficiency and Renewable Energy (EERE) under the Advanced Manufacturing Office award number DE-EE0007897 awarded to the Remade Institute, a Division of

Sustainable Manufacturing Innovation Alliance Corp. The views expressed herein do not necessarily present the views of the U.S. Department of Energy or the United States government. Appreciation is also extended to Penn State EMS Energy Institute (EI), Center for Critical Minerals (C²M), Energy and Mineral Engineering (EME) Department, and Material Research Institute (MRI) for providing technical facilities.

References

1. Widmer, R., Oswald-Krapf, H., Sinha-Khetriwal, D., Schnellmann, M., Böni, H., Global perspectives on e-waste. Environmental Impact Assessment Review, 25(5 SPEC. ISS.), 436–458, 2005.
2. Awasthi, A.K., Cucchiella, F., D'Adamo, I., Li, J., Rosa, P., Terzi, S., Wei, G., Zeng, X., Modelling the correlations of e-waste quantity with economic increase. Science of the Total Environment, 613, 46-53, 2018.
3. Ghosh, B., Ghosh, M. K., Parhi, P., Mukherjee, P. S., Mishra, B. K., Waste printed circuit boards recycling: an extensive assessment of current status. Journal of cleaner production, 94, 5-19, 2015.
4. Otsuki, A., Mensbruge, L. D. la, King, A., Serranti, S., Fiore, L., Bonifazi, G., Non-destructive characterization of mechanically processed waste printed circuit boards - particle liberation analysis. Waste Management, 102, 510–519, 2020.
5. Kaya, M., Recovery of metals and non-metals from electronic waste by physical and chemical recycling processes. Waste Management, 57, 64–90, 2016.
6. Duan, C., Wen, X., Shi, C., Zhao, Y., Wen, B., He, Y. Recovery of metals from waste printed circuit boards by a mechanical method using a water medium. Journal of Hazardous Materials, 166(1), 478–482, 2009.
7. Veit, H. M., Diehl, T. R., Salami, A. P., Rodrigues, J. S., Bernardes, A. M., Tenório, J. A. S., Utilization of magnetic and electrostatic separation in the recycling of printed circuit boards scrap. Waste Management, 25(1), 67–74, 2005.
8. Hao, J., Wang, Y., Wu, Y., Guo, F., Metal recovery from waste printed circuit boards: A review for current status and perspectives. Resources, Conservation and Recycling, 157, 104787, 2020.
9. Das, A., A. Vidyadhar, and S. P. Mehrotra, A Novel Flowsheet for the Recovery of Metal Values from Waste Printed Circuit Boards. Resources, Conservation and Recycling 53(8), 464–69, 2009.
10. Yoo, J. M., Jeong, J., Yoo, K., Lee, J. chun, Kim, W. Enrichment of the metallic components from waste printed circuit boards by a mechanical separation process using a stamp mill. Waste Management, 29(3), 1132–1137, 2009.
11. Qiu, R., Lin, M., Ruan, J., Fu, Y., Hu, J., Deng, M., Tang, Y., Qiu, R., Recovering full metallic resources from waste printed circuit boards: A refined review. Journal of Cleaner Production, 244, 118690, 2020.
12. Kaya, M., Recovery of metals and non-metals from waste printed circuit boards (PCBs) by physical recycling techniques. Minerals, Metals and Materials Series, 9783319521916, 433–451, 2017.
13. Quan, C., Li, A., Gao, N., Characterization of products recycling from PCB waste pyrolysis. Journal of Analytical and Applied Pyrolysis, 89(1), 102-106, 2010.
14. Barnwal, A., Mir, S., Dhawan, N., Processing of Discarded Printed Circuit Board Fines via Flotation. Journal of Sustainable Metallurgy, 6(4), 631–642, 2020.
15. Marra, A., Cesaro, A., Belgiorno, V., Separation efficiency of valuable and critical metals in WEEE mechanical treatments. Journal of Cleaner Production, 186, 490–498, 2018.

16. Guo, J., Guo, J., Xu, Z., Recycling of non-metallic fractions from waste printed circuit boards: A review. Journal of Hazardous Materials, 168(2–3), 567–590, 2009.

17. Bizzo, W. A., Figueiredo, R. A., de Andrade, V. F., Characterization of printed circuit boards for metal and energy recovery after milling and mechanical separation. Materials, 7(6), 4555–4566, 2014.

18. Kumar, A., Holuszko, M. E., Janke, T., Characterization of the non-metal fraction of the processed waste printed circuit boards. Waste Management, 75, 94–102, 2018.

19. Cunliffe, A. M., Jones, N., Williams, P. T., Recycling of fibre-reinforced polymeric waste by pyrolysis: Thermo-gravimetric and bench-scale investigations. Journal of Analytical and Applied Pyrolysis, 70(2), 315–338, 2003.

20. Zhou, Y., Wu, W. B., Qiu, K., Recycling of organic materials and solder from waste printed circuit boards by vacuum pyrolysis-centrifugation coupling technology. Waste Management, 31(12), 2569–2576, 2011.

21. Undri, A., Rosi, L., Frediani, M., Frediani, P., Efficient disposal of waste polyolefins through microwave assisted pyrolysis. Fuel, 116, 662–671, 2014.

22. Pivato, A., Vanin, S., Raga, R., Lavagnolo, M. C., Barausse, A., Rieple, A., Laurent, A., Cossu, R. Use of digestate from a decentralized on-farm biogas plant as fertilizer in soils: An ecotoxicological study for future indicators in risk and life cycle assessment. Waste Management, 49, 378–389, 2016.

23. Shen, Y., Chen, X., Ge, X., Chen, M., Chemical pyrolysis of E-waste plastics: Char characterization. Journal of environmental management, 214, 94-103, 2018.

24. Luyima, A., Zhang, L., Kers, J., Laurmaa, V., Recovery of metallic materials from printed wiring boards by green pyrolysis process. Medziagotyra, 18(3), 238–242, 2012.

25. Sun, J., Wang, W., Liu, Z., Ma, C., Recycling of waste printed circuit boards by microwave-induced pyrolysis and featured mechanical processing. Industrial and Engineering Chemistry Research, 50(20), 11763–11769, 2011.

26. Bradhorst, H. W., Engel, U. H., Ludwig, C. T., Zavoral, E., Multistafe thermolysis method for safe and efficient conversion of e-waste materials, International patent WO 2017/116750 A1, 2017.

27. Duan, H., Hou, K., Li, J., Zhu, X., Examining the technology acceptance for dismantling of waste printed circuit boards in light of recycling and environmental concerns. Journal of environmental management, 92(3), 392-399, 2011.

28. Cui, J., Forssberg, E., Mechanical recycling of waste electric and electronic equipment: A review. Journal of Hazardous Materials, 99(3), 243–263, 2003.

29. Rezaee, M. Apparatus and process for selective liberation and mechanical recycling of electronic waste to recover valuable elements. U.S. Provisional Pat. App. No. 63/519,376, 2023.

52

Circular Economy and the Digital Divide: Assessing Opportunity for Value Retention Processes in the Consumer Electronics Sector

Kyle Parnell*, Constanza Berrón, Chelsea Gulliver, Michael Thurston and Nabil Nasr

Golisano Institute for Sustainability, Rochester Institute of Technology, Rochester, NY, USA

Abstract

End-of-use (EOU) and end-of-life (EOL) consumer electronic products (CEPs) are commonly discarded or sent to recycling well before their functional life is exhausted, forfeiting substantial embodied economic value and environmental impact. In many cases, product flows end in the Global South, where informal recycling processes create harmful environmental, health, and safety impacts. Extant efforts to address these impacts focus on restricting the trade of EOU and EOL products to the Global South, stifling the supply chain of recycling industries that make nontrivial contributions to developing economies. This paper explores whether a more sustainable system is possible through the circular economy principle of value retention processes (VRPs), namely, remanufacturing, repair, and direct reuse. We posit that current product flows support the use of EOU and EOL CEPs as inputs to VRP industries in both developed and developing economies, and suggest that increasing the prevalence of VRPs in this sector can benefit developing economies more than recycling alone by increasing access to technology and thus standards of living. To that end, we study the movement of CEPs within and between case nations representing developed and developing economies—the United States (US) and Ghana (GH), respectively. Product-level material flow analysis quantifies trends in CEP use, and a new model to estimate product disposition is proposed. Results suggest product flows that already exist support an increase in VRPs and technology access in developing nations. We discuss how a shift toward VRPs may improve the resource efficiency of the CEP sector, and consider barriers to such a shift.

Keywords: Material flow analysis, value retention processes, consumer electronics, digital divide

Corresponding author: kxp6496@rit.edu

Nabil Nasr (ed.) Technology Innovation for the Circular Economy: Recycling, Remanufacturing, Design, Systems Analysis and Logistics, (697–712) © 2024 Scrivener Publishing LLC

52.1   Introduction

52.1.1   Consumer Electronics & The Circular Economy

Consumer electronic products (CEP) manufacturing and retail contribute markedly to global industrial productivity; worldwide estimates suggest revenue growth from $1 trillion in 2020 to $1.5 trillion by 2026 [1]. Consequently, the production, distribution, use, and end-of-life (EOL) disposition of these products are responsible for considerable social and environmental impacts, including landfill waste and toxicity (35 MMT/year) [2], greenhouse gas (GHG) emissions (793 MMT CO_2e/year and growing) [3], and adverse effects on human health and development. CEPs also often require critical materials [4], high energy use in manufacturing [5, 6], and complex treatment of hazardous waste streams [7]. Further, in the age of information, access to CEPs is critical to technological, economic, and social evolution in developing societies [8]. Serving these growing consumer needs while mitigating environmental and human health impacts is a considerable challenge. The circular economy is broadly emerging as a means to address these challenges; specifically, value retention processes (VRPs) including remanufacturing, refurbishing, repair, and direct reuse are gaining market share and acceptance as practical applications of circular economy principles [9]. In some industry sectors—e.g., automotive and commercial machinery—VRPs have been found to be more economically efficient than and environmentally preferable to incumbent linear business models, offering a means to decouple economic growth from increasing environmental impact [10]. Accelerating demand for and an accordingly growing waste problem in CEPs highlights the necessity of such decoupling to sustainable development in this sector as well [1].

52.1.2   The Digital Divide

Access to CEPs is also linked to many consumer benefits. Refrigeration and air conditioning, for example, are vital to food security and human quality of life as global mean temperature continues to rise [11]. Similarly, access to information and communication technology (ICT) products—e.g., mobile phones and computers—is a key enabler in educational and employment opportunities and a viable healthcare tool through telemedicine [8,12]. Increasing access to CEPs in developing economies, then, can be seen as a valuable contributor to socio-economic development. Growth in consumption and ownership of CEPs is, however, concentrated mostly in the Global North, dominated by the wealth of developed economies. Though a valuable element of global trade, growing consumption in populations with nearly ubiquitous basic access is a luxury, yielding diminishing returns in user benefits [13]. In contrast, large portions of developing and transitional economy populations in the Global South lack basic access to CEPs. A review of consumer technology ownership rates between the US and GH highlights this disparity, commonly termed the "digital divide" (Table 52.1).

This disparity is due largely to a lack of affordability, but is also influenced by underdeveloped utility infrastructure and traditional cultural perspectives [18,19]. This divide is partially addressed by CEP products specifically designed to fit the economic, environmental, and infrastructural criteria of consumer populations in Global South economies. Inexpensive mobile phones, small household refrigerators, and other market-specific CEPs are available through online and brick-and-mortar retailers; however, performance, energy

Table 52.1 Household penetration of consumer electronic products in United States and Ghana, estimated 2020.

Product	US (%households)	US source	GH (%households)	GH source
washing machine	84%	[14]	3%	[15]
refrigerator	99%	[14]	49%	[15]
air conditioner	88%	[14]	20%	[16]
television	97%	[17]	70%	[18]
computer, tablet	60%	[17]	2%	[18]
computer, laptop	75%	[17]	5%	[18]
computer, desktop	41%	[17]	1%	[18]
smartphone	97%	[17]	50%	[18]

efficiency, and reliability are often sacrificed in the name of cost, further contributing to material, energy, and waste challenges in the CEP sector.

Despite this disparity, the Global South often bears the burden increased consumption in the Global North creates, as its nations are in many cases the final destination of discarded CEPs [20]. Ghana, for example, is one of the largest waste electronic and electrical equipment (WEEE) processors in the world, receiving millions of tonnes of used CEPs each year, primarily from the Global North [21]. The majority of these flows are subjected to informal (i.e., improper) recycling and disposal, where processes such as destructive manual disassembly, uncontrolled acid leaching, and open pit burning are used to recover components and trace amounts of valuable materials [22,23,24]. These practices are inefficient in material recovery, damage fragile ecosystems, poison land and water resources, and harm laborers [25,26,27]. Several policies (e.g., Basel, Rotterdam, and Bamako Conventions) aim to mitigate the harm from improper CEP management by restricting the flow of WEEE from Global North to Global South nations. These mechanisms have been broadly ineffective [28] due to Global North WEEE generators exploiting regulatory loopholes and enforcement weaknesses to avoid high labor and disposal costs (e.g., mislabeling or disguising WEEE to deceive customs officials [25,29,30]) and importers smuggling WEEE in pursuit of recycling value recovery by labeling it as old stock, retail rejects, or diplomatic material [31].

52.1.3 Circular Opportunity in CEPs

The flow of EOU and EOL CEPs both within and between nations in the Global North and South is thus likely to continue, if slightly hindered. Low policy efficacy, inefficient and harmful incumbent recycling systems, and room for growth in access and ownership in turn support the need to strengthen the circular economy in CEPs through VRPs. A potential means to both increase access (i.e., bridge the digital divide) and advance circularity in the CEP sector at large is thus to leverage existing CEP flows at EOU or EOL as a feedstock for consumer technology industries in developing economies where used, repaired, or refurbished products are more broadly accepted. Although the aggregate flow of EOU/

EOL products from the Global North to the Global South is well known, little product-level detail is available on how they might be used more valuably. Research in this space generally focuses on recycling processes, both in technology and effects on environmental and human health [22-27,32]. Although both [33] and [34] explore the circular economy in Africa specifically, the primary aim is to highlight the characteristic suitability of VRPs for developmental foci of job creation and resource efficiency, rather than to quantify their physical feasibility. Others [35] address the feasibility question, but focus on the automotive industry, where VRPs have long been established industrial practice with demonstrated success. Finally, [36] explore the potential for CEP remanufacturing in developing economies; however, they provide mostly qualitative evidence that VRPs can benefit both OEMs and local economies, supported by case studies on products for which technology and policy have evolved considerably since the time of publication. Consequently, characterization of modern CEP flows and estimation of their potential for use in VRPs is not well established.

52.2 Methods

52.2.1 Key Product Identification

To ensure tractability, we first identify a subset of CEP categories as key products whose flows, use, and ownership may serve as a sound indicator for basic technology access in both developed and developing economies. Critical products are selected from a high-level population of product types from the Consumer Technology Association [17,37] and organized by product categorization criteria under the European Union's WEEE Directive [38]. Key products are selected based on qualitative assessment of the degree to which they serve as indicators of basic technology access.

52.2.2 Domestic Material Flow Analysis (MFA)

Product-level material flow analysis (MFA) is applied to selected products in both the US and Ghana as examples of developed and developing economies in the Global North and South, respectively. The baseline system boundary for both US and Ghana extends from domestic sales to EOU or EOL at the consumer level from 2010-2019. This system is illustrated in Figure 52.1.

The baseline system is intended only to characterize the flow of products through the domestic economy of each case nation. In accordance with the root data sources, inputs are assumed to include both new OEM products and used products; although we can infer significant flows of both new and used products into Ghana, the relative contribution of each product type to the total value is not available. Although many organizations estimate aggregate WEEE generation at a national level, product-level data for EOU and EOL disposition is not widely available. Therefore we use reported year-to-year data for product shipments as well as household ownership to impute flows, estimating outputs (i.e., EOU/EOL products discarded by the user) using the basic MFA model [39]:

Σ *inputs* $= \Sigma$ *outputs* $+ \Delta$ *stock* (1), and solve for outputs with: Σ *outputs* $= \Sigma$ *inputs* $-$ Δ *stock* (2)

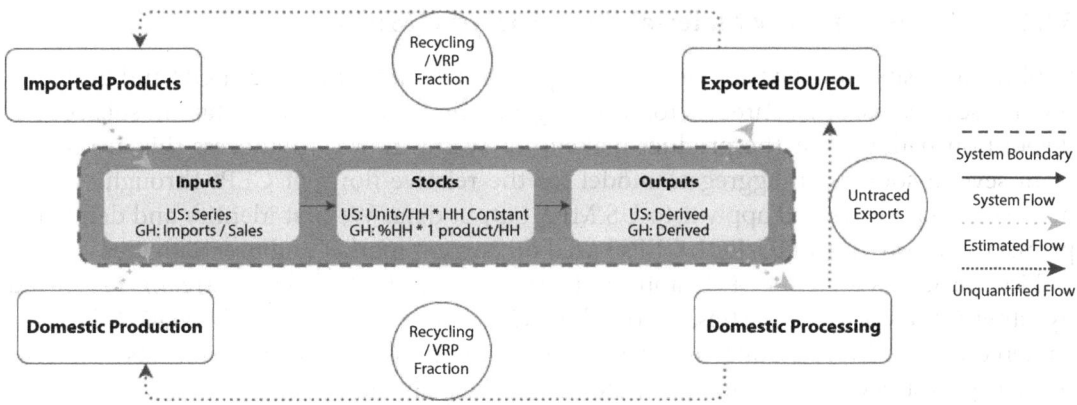

Figure 52.1 Product-level Material Flow Analysis system boundary for both United States and Ghanaian markets.

For the US market, ICT product inputs are derived from CTA estimates on year-to-year product shipments; appliance product inputs are derived from US EPA reports on ENERGYSTAR rated product shipments and relative market share [37,40]. GH product shipments are derived from the UN Comtrade Database, World Integrated Trade Solution (WITS) web tool, and the Green Cooling Initiative inventory [41,42,43]. US stock estimates are derived from CTA and US Energy Information Administration (EIA) Residential Energy Consumption Survey (RECS) data on technology access (in percentage of households that own a product) and consumption intensity (number of units per household), as well as household population growth values from the US Census Bureau [14,44]. GH stock values are derived from technology access data published by Ghana Statistical Service and the Global Data Lab, and an assumed consumption intensity of one product per household [15,18]. Although there are likely cases where a single Ghanaian household owns more than one of a given product, this assumption is based on the relative lack of basic access in Global South nations and the ownership of product multiples as a relative and thus uncommon luxury. Ghana household data is derived from Ghana Statistical Service Living Standards Survey, Population and Housing Census, and Labour Force reports, using logistic regression to estimate number of households for years in which data is not available [66,67,68,69]. Where specific values for unit shipments are not available, the historic mean ratio of trade value (in USD) to import weight (in kg) per unit imported is used as a basis for estimation. This method is used to account for the variability of trade intensity, which the data suggests is not always growing or linear. However, available data does suggest that consumer ownership of CEPs grows linearly; thus, where specific values for product stocks are not available, linear regression of historical ownership rates is used to estimate technology access and, in turn, stock values. To estimate mass flows, average unit mass values across 2010-2019 are applied to unit flows. Mass values between US and GH cases are assumed to be the same. These baseline models illustrate domestic consumer metabolism for CEPs in both the US and Ghana. To understand whether developed EOU/EOU product flows can support developing VRP economies, we further characterize these flows.

52.2.3   Transboundary Material Flow Analysis (MFA)

Publicly accessible information on what happens to product flows after EOU and EOL disposal is scarce; existing estimates for WEEE generation provide little detail on subsequent disposition pathways or the product makeup of WEEE flows. To address this, we derive from several sources an aggregate model for the relative flows of CEPs through various disposition pathways and apply it to US MFA output data. We first identify and define the primary pathways for EOU/EOL CEPs based on current market and user activities.

First, products are either (1) captured in WEEE-Specific Collection streams organized by either OEMs or third parties, destined for either reuse or recycling [2,46,47,48]; or (2) gathered in Non-WEEE Collection, i.e., disposed in general municipal solid waste streams primarily destined for landfill [49], stored, or illegally dumped. From WEEE-Specific Collection, products may then be either (1.1) Reused, i.e., reintroduced into markets and use via direct reuse, repair, or remanufacturing [50]; or (1.2) Recycled, i.e., sent to bulk processors for material separation and recovery through either formal or informal means. Recycled products are either (1.2.1) Properly Recycled, i.e., formally processed by means of controlled hazard removal and disposal, shredding, material separation (e.g., magnetic ferrous, eddy current, or spectroscopic detection), and commodity collection [51]; or (1.2.2) Improperly Recycled, i.e., informally processed by means of manual disassembly and separation, uncontrolled acid leaching and open pit burning for precious metal recovery, and residual waste dumping [52]. We then apply geographic assumptions to each disposition pathway. Non-WEEE Collection flows are assumed to be treated domestically, as their primary destination is landfills. Reused flows are assumed to represent best-case EOU products (e.g., user returns with low age, high function, little required processing, and high potential profit margin) that are consequently retained domestically. Properly Recycled flows are also assumed to be treated domestically, given that the necessary infrastructure requires high economic development and known US EOU/EOL CEP exports to developed trade partners are minimal [53]. Improperly Recycled flows are assumed to be exported to the Global South, the primary destinations for which are China, India, and West Africa, with Nigeria and Ghana leading the latter [54]. Adapting mass balance ratios from [54] for the distribution of imports across these economies, we estimate approximately 10% of US improper recycling flows may be available to Ghana.

Although the labeling of flows for recycling suggests a lack of function and thus unsuitability for VRPs, failure rate data for US product types in these categories suggests that irrecoverable failure is relatively uncommon [55]. Further, [56] suggests that users in developed economies often discard EOU CEPs in landfill or recycling streams despite remaining function. Likewise, at least one major CEP OEM reports that its product takeback program—which in many cases captures fully- and partially-functional products—includes "no reuse…parts harvesting…[or] resale" and instead ends in recycling only [57]. These conditions imply that much of the Recycling (and even Non-WEEE Collection) flows may still have function, or at least be repairable, and that US Improper Recycling flows may thus be largely suitable for VRPs despite their current designation. Although we label US exports to developing economies as "improper recycling" because that is their most likely disposition, GH treats these imports as EOU/EOL CEPs with multiple possible ends. Assessments of used CEP imports into Ghana that suggest approximately 15% of those imports are truly non-functioning and non-repairable, and thus sent directly to recycling or landfill [58,59].

The remaining 85% of used imports have usable potential, but extant tracking efforts do not estimate what portion of this amount are actually returned to service, and indeed suggest that much of the potentially usable flow is sent to recycling or landfill anyway. For the purpose of this analysis, we thus consider up to 85% of US Improper Recycling flows to GH (8.5% of total Improper Recycling flow) to be potentially harvestable for VRPs (i.e., 15% fallout rate).

This model is then generally applied across product types to characterize US output flows in greater detail. This application yields some uncertainty, as differences in product attributes between main product types (i.e., appliances and ICT) engenders variable suitability for recycling and VRP applications that may affect actual flows. However, in the absence of better product-level post-EOU/EOL tracking, we posit that this model is sufficient for the stated purpose. Using these models, we analyze the potential for EOL/EOU CEP flows of case study products from the US to support CEP economies in GH as a feedstock for VRPs. We then discuss the potential for VRP products to support growth in the Ghanaian CEP sector, and by extension increase public access to key consumer technologies.

52.3 Results

52.3.1 Key Technologies

Product-level assessment of CEP types highlights a subset products for further analysis, parenthetically identified by their respective WEEE categories: refrigerators, representing temperature exchange equipment (I) and large appliances (IV) central to food security; air conditioner, representing temperature exchange equipment (I) that has high use-phase energy impacts and growing functional importance in light of global temperature rise; LCD/LED Television, representing screen-containing equipment (II) that reflects quality of life via entertainment and social satisfaction; laptop computer, representing screen-containing equipment (II) that enables communication and information connectivity and their potential benefits; and smartphone, representing small IT equipment (VI) that is valuable in basic access to communication, financial services, employment, healthcare, and civil systems participation.

52.3.2 Baseline MFA Models

Results of baseline material flow data analysis are documented in Table 52.2. Case products are abbreviated as: air conditioner (AC), refrigerator (RF), laptop computer (LT), LCD/LED television (TV), and smartphone (SP). Stock Growth reflects the 2010-2019 differential relative to 2010 for volume and mass.

Table 52.2 Product flows by type and relative share of total, in million units and thousand tonnes, 2010-2019.

		AC	RF	LT	TV	SP	Total
United States	Volume (million units)						
	Inputs (mil. units)	68.9	101	407	974	1,434	2,985
	Share of Inputs	2.3%	3.4%	13.6%	32.6%	48.0%	100%
	ΔStock (mil. units)	22.1	26.6	25.9	33.3	88.4	194
	Outputs (mil. units)	46.8	74.6	381	940	1,346	2,790
	Share of Outputs	1.7%	2.7%	13.7%	33.7%	48.3%	100%
	Mass (thousand tonnes)						
	Inputs (th. MT)	2,141	6,576	895	9,737	186	19,800
	Share of Inputs	12.2%	33.2%	4.5%	49.2%	0.94%	100%
	ΔStock (th. MT)	772	1,730	57	333	11	2,900
	Outputs (th. MT)	1,639	4,849	838	9,404	175	16,900
	Share of Outputs	12.2%	33.2%	4.5%	49.2%	1.0%	100%
	Stock Growth 2010-19	56%	19%	37%	50%	283%	-
Ghana	Volume (million units)						
	Inputs (mil. units)	1.21	5.98	0.36	2.02	3.25	12.4
	Share of Inputs	9.5%	46.2%	2.8%	15.9%	25.6%	100%
	ΔStock (mil. units)	0.26	5.87	0.15	0.69	0.70	4.13
	Outputs (mil. units)	0.95	3.55	0.21	1.33	2.55	8.29
	Share of Outputs	11.5%	42.7%	2.5%	16.1%	30.7%	100%
	Mass (thousand tonnes)						
	Inputs (th. MT)	42.4	382	0.78	20.3	0.42	446
	Share of Inputs	9.5%	85.7%	0.18%	4.6%	0.1%	100%
	ΔStock (th. MT)	9.1	151	0.33	6.9	0.29	168
	Outputs (th. MT)	33.3	231	0.46	13.3	0.33	278
	Share of Outputs	12.0%	83.0%	0.16%	4.8%	0.12%	100%
	Stock Growth 2010-19	19%	48%	67%	19%	29%	-

Tracking these products across sale, ownership, and disposal, we find a nearly inverse relationship between product size and flow volume. Smartphones account for nearly half of all US unit outputs, but less than one percent by mass. In contrast, refrigerator outputs are far fewer (< 3% by units), but these flows account for nearly 30% of output mass. Stock growth in the US for all products except for smartphone is driven mostly by market population growth (i.e., number of households) and an increase in consumption intensity. US smartphone stock growth is driven primarily by increasing technology access; modern smartphone technology first emerged around the start of the study period, and US consumer uptake increased rapidly. In contrast, Ghanaian consumption intensity is held constant in this model. Instead, we can draw inferences on changes in technology access by comparing stock growth over the study period to household growth over the same time. We find that the number of households in GH increased by 42% from 2010-2019. Considered against stock growth values, we can suggest that relative to population, access to refrigerators remained largely stable, while access to laptop computers increased, and access to air conditioners, televisions, and smartphones may have decreased. We thus look to US outputs to bridge these gaps.

52.3.3   US Output Disposition MFA Models

To estimate whether US outputs are sufficient to support VRPs in Ghana, we construct a consolidated EOU/EOL CEP mass flow model with more specific disposition estimates (Figure 52.2). Mass share values for the share of products (1) collected in WEEE-specific streams, (2) reused (wherein collection is implicit), and (3) properly recycled are used to differentially estimate (1a) the share not collected in WEEEE-specific streams, (2a) the share of WEEE-specific collection that is recycled, and (3a) the share of recycled WEEE that is improperly managed. We assume the relative makeup of each flow follows the product-level distribution of US outputs from the baseline MFA model (Figure 52.3).

In Figures 52.3 and 52.4 below, disposition pathway labels follow the same architecture as those in Figure 52.2; in particular, "Recycled" represents the sum of "Properly Recyceld" and "Improperly Recycled" flows.

This architecture estimates nearly 65% of US EOU/EOL CEPs are improperly managed; either not collected in WEEE-specific streams (i.e., landfill, storage, dumping) or sent to improper recycling streams. The modeled ratio of improper recycling to total generation (21%), which we assume is exported [53], aligns well with other models for US WEEE export (23%) [54]. Converting mass flows to units according to average product mass values, we get Figure 52.4.

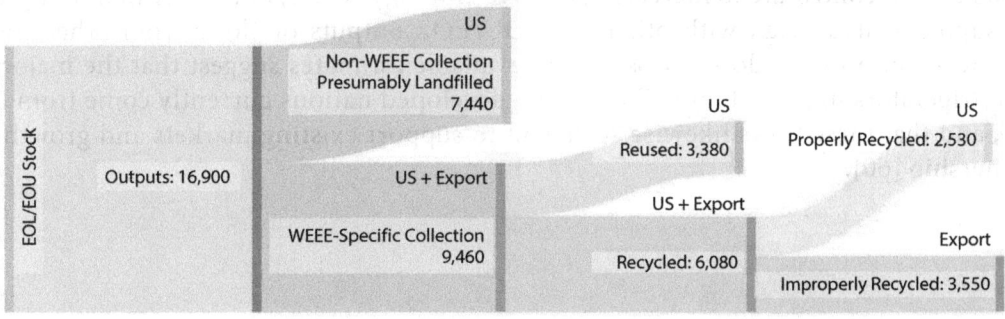

Figure 52.2 United States case study product output and disposition 2010-2019, thousand tonnes.

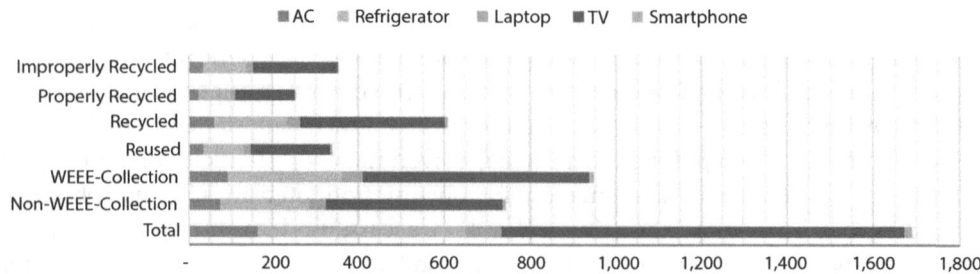

Figure 52.3 United States case study product output flows by type and pathway 2010-2019, ten thousand tonnes.

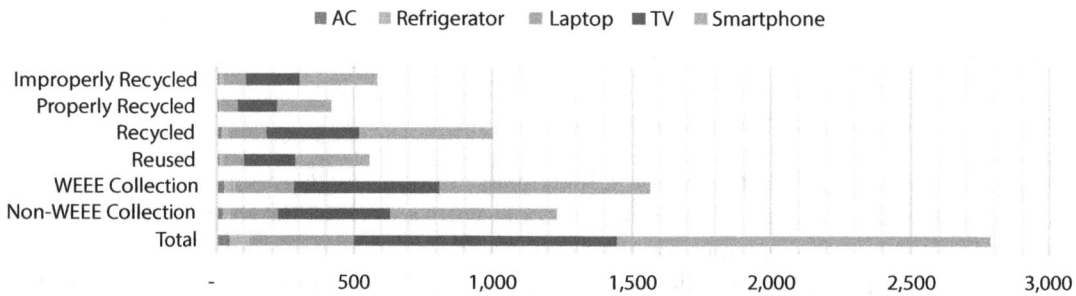

Figure 52.4 United States Consumer Electronic Product output flows by type and pathway 2010-2019, million units.

52.3.4 Transboundary Analysis

Using the disposition model in Figure 52.2, and considering the Improperly Recycled flow properties outlined in Section 52.2.3, we estimate that usable US EOL/EOU Improper Recycling flows are more than sufficient to meet the inflow demand of the Ghanaian CEP economy for all products except refrigerators and air conditioners (Figure 52.5).

In this sense, US EOL/EOU CEP outputs provide enough products to serve existing ICT product markets in Ghana and support considerable growth in CEP ownership—sufficient to help bridge the digital divide. However, US EOL/EOU flows of air conditioners and refrigerators do not meet existing market needs. This is not to say that such flows cannot contribute to increasing basic technology access, but rather that they must be supplemented—e.g., with other US EOU/EOL outputs or flows from other trade partners—in order to do so. Most recent available estimates suggest that the majority of refrigerators imported into Ghana from developed nations currently come from the EU, and that these flows likewise sufficient to support existing markets and growth in ownership [60].

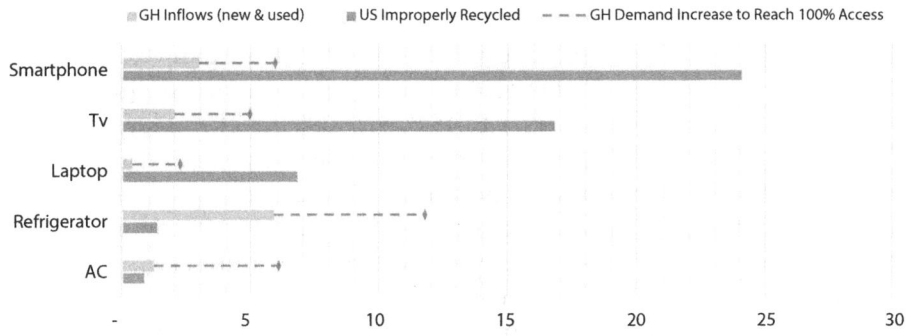

Figure 52.5 Usable United States exports available in Ghana vs. gross demand in Ghana 2010-2019, million units.

52.4   Conclusions & Recommendations

52.4.1   The Case for VRPs

Under current conditions, large portions of EOU/EOL CEP imports into Ghana are sent directly to informal (improper) recycling, despite considerable residual value. Concurrently, Ghanaian consumers continue to struggle with a lack of access to technology, driven primarily by a lack of affordability. VRPs retain a greater share of imported EOU/EOL products' embodied functional and material value that would be lost in recycling, and are well known to reduce the costs of providing functional value; savings that can be passed on to consumers through discounting—refurbished CEPs are routinely found in Agbogbloshie markets at 70% to 80% discount rates [61,62]. Together, these conditions suggest that there is opportunity to leverage existing exported US EOU/EOL flows as feedstocks for VRP pathways to meet technology needs in GH. By shifting to a more VRP-inclusive model, Ghana may reduce the cost and increase the availability of CEPs at scale, thereby increasing opportunity for growth in access. Further, in reducing the cost of products from the Global North, a VRP-inclusive CEP economy may draw consumer attention away from newly-manufactured but low-durability products designed specifically to be inexpensive, which are in many cases preferred based on price but commonly discarded in landfill streams when they fail [19]. Across these benefits, a VRP CEP economy that leverages US EOL/EOU product flows as a low-cost, high-value feedstock stands to increase technology access while decoupling such growth from the increase in environmental impacts typically associated with doing so.

52.4.2   Capacity Limitations & Uncertainty Analysis

The central claim from modeling results is that existing transboundary flows support opportunity to increase VRPs in GH with an associated increase in technology access. The model does not analyze GH's VRP infrastructure capacity, and thus cannot make a claim regarding whether GH industry is equipped to capture this opportunity in full. However, market data demonstrates that industrial capacity for VRPs in GH exists and has been growing, suggesting that while barriers exist, some level of capture is possible. For example, 2010 estimates of 2,000 CEP repair businesses and 10,000-15,000 direct employees in GH [63] have by 2021 grown to an estimated 3,000 refurbishers with 15,000-21,000 direct employees [64]. At scale, 2012-2017 data suggest the GH EEE refurbishment and WEEE management sector generated on average $416 million USD annually, supporting 200,000 domestic jobs through direct and downstream industries [23,65]. However, to advance VRPs in the CEP economy, further infrastructural advancements are a likely necessity. Common CEP failures, e.g., refrigerator compressor faults and ICT touchscreens, often require moderate to high technical knowledge and infrastructure to effectively repair. Unfortunately, CEP OEMs show low to moderate support for service knowledge and replacement parts sharing, even in Global North markets; there is little evidence of any interest in such support for developing economies. A VRP-inclusive economy in Ghana may thus require the development of intrinsic service knowledge, tools, and parts supply infrastructure, which can be difficult to achieve at scale.

Interpretation of results must also consider uncertainties raised by the age of some data sources. In particular, the value representing the portion of imported products that are found to be unusable and unrepairable after they arrive (i.e., 15% fallout rate) is from a 2011 assessment of EEE product flows [58]. A reduction in this rate (e.g., through policy measures that stabilize incoming product condition) would increase the portion of flows potentially usable for VRPs; assuming stable flow magnitude, this correlates to a net increase in usable product supply. In this case, even a 0% fallout rate would not materially change modelling conclusions; allocated US flows would still well exceed total GH market demand for ICT products, and still fall short for appliances, though this gap would close slightly. Increasing the fallout rate, e.g., as a result of lower product durability, would likewise have no material effect on outcomes; even if half of EOU/EOL CEPs imported from the US were unusable and unrepairable, the remainder would still exceed total GH ICT market demand by an order of magnitude, and still fall short for appliances, though by a wider margin. More recent fallout rate data is not available, and there is little evidence to suggest change in either direction.

Conclusions about the suitability of US improper recycling flows to meet GH CEP demand via increased VRP are most sensitive to how much EOU/EOL flow the US generates and how much GH accepts. US domestic MFA data is current and demonstrates consistent and growing generation; further, nonalignment with the Basel Convention and weak internal WEEE management standards suggest this is unlikely to change. In contrast, GH has an active history of efforts to restrict WEEE flow through policy, but the recent effects of these efforts on import volumes are unknown.

We estimate that 10% of US improper recycling flows are available to Ghana based on 2014 mass balance estimates [54]. More recent import ratios are not available; however, even with significant restriction of imports beyond modeled conditions, we suggest increased VRP in GH is still possible. To illustrate this, we apply Generalized Reduced Gradient (GRG) local optimization to identify the balance of import and fallout rates necessary for 2010-2019 US improper recycling flows to drop below 100% fulfillment of 2010-2019 GH market demand in any ICT product category, and find that import must be restricted to 1.3% with a 13% fallout rate to do so. There is little evidence that GH is pursuing restrictions of such magnitude or that current enforcement resources would be able to achieve them.

52.4.3   Challenges & Future Work

It is important to note that leveraging imported EOL/EOU CEPs for VRP feedstocks would make such an economy in part dependent on the mix of product types and conditions that developed nations export. These are supply chain characteristics that are highly variable and over which developing economies in the Global South have little control or ability to predict. Import policies may influence the supply stream by defining acceptable product condition criteria for import, but enforcement of such restrictions has been historically difficult and inconsistently successful. It must also be acknowledged that VRPs are, at their root, lifecycle extension strategies. Extended product lifecycles will still eventually end, at which point EOL products must still be managed. In this sense, WEEE management systems are an inevitable necessity; VRPs only offer a means to shift the point of WEEE generation from predominantly imports to domestic use in exchange for creating additional value for the economies in which VRPs and subsequent product use occur.

Ghana's domestic generation of WEEE is not insignificant, though in most cases still outpaced by the volume of EOU/EOL products imported from the US that are neither functioning nor repairable (and thus considered direct waste). Improvements in WEEE management infrastructure are thus a critical component of VRP advancement—both to reduce environmental, health, and safety hazards of incumbent processes, and because improved collection and recycling methods can serve as a supplemental material feedstock to VRPs at a high value-to-cost ratio. Substantial investment is required to achieve this end, from both Global North economies in which product flows originate and Global South economies who leverage them; presently, however, policy-based prevention seems favored over VRP capacity building.

Finally, although the physical viability of circularization in the CEP sector is evident, we must reconcile that CEPs exhibit many different characteristics than classic VRP case products—e.g., variable consumer preferences, shorter lifecycles, rapid technology change rates, and lack of repair technologies—all of which create technical, economic, and logistical challenges to the adoption of VRPs. These conditions suggest that circularity in consumer product sectors may require considerable change in product strategies, business models, and infrastructure. Additional research is required to specifically identify and characterize these barriers and the potential impacts of their alleviation.

References

1. Global Market Insights. Consumer Electronics Market Size by Product & Application: Industry Analysis Report, Regional Outlook, Growth potential, Competitive Market Share & Forecast 2021-2027, 2020.
2. Baldé, C. *et al.*, *Global Transboundary E-waste Flows Monitor 2022*, UNITAR, 2022.
3. CTA, Global Year-to-Year Emissions Summary, 2019 Industry Report on GHG Emissions, 2020.
4. European Commission, Communication from the Commission to the European Parliament—Critical Raw Materials Resilience: Charting A Path Towards Greater Security And Sustainability, 2020.
5. Yu, J., Williams, E., Ju, M., Analysis of material and energy consumption of mobile phones in China, *Energy Policy*, 38(8), 4135-4141, 2010.

6. Ryen, E., Babbitt, C., Williams, E., Consumption-Weighted Life Cycle Assessment of a Consumer Electronic Product Community, *Envrion. Sci. Technol.*, 49(4), 2549-2559, 2015.

7. Boudier, F., Faouzi B., Hazardous waste management and corporate social responsibility: illegal trade of electrical and electronic waste, *Business and Society Review* 116(1), 29-53, 2011.

8. Warschauer, M., Matuchniak, T., New Technology and Digital Worlds: Analyzing Evidence of Equity in Access, Use, and Outcomes, *Rev. of Res. in Education*, 34(1), 179-225, 2010.

9. ERN, Remanufacturing Market Study For Horizon 2020, grant agreement No 645984, a report for the European Commission by the partners of European Remanufacturing Network, Brussels, 2015.

10. Nasr, N. *et al.*, Re-defining Value—The Manufacturing Revolution: Remanufacturing, Refurbishment, Repair and Direct Reuse in the Circular Economy, *International Resource Panel*, 2018.

11. Mc Carthy, U., Uysal, I., Badia-Melis, R., Mercier, S., O'Donnell, C. and Ktenioudaki, A., Global food security–Issues, challenges and technological solutions, *Trends in Food Sci. & Technol*, 77, 11-20, 2018.

12. Shiferaw, F., Zolfo, M., The role of information communication technology (ICT) towards universal health cover-age: the first steps of a telemedicine project in Ethiopia, *Global Health Action*, 5(1), 15638, 2012.

13. Ahmed, Z., Phong Le, H., Linking Information Communication Technology, trade, global-ization index, and CO2 emissions: evidence from advanced panel techniques, *Environ. Sci. Pollution Res.*, 28, 8770-8781, 2021.

14. US EIA, Appliances by Housing Unit Type (HC3.1), Residential Energy Consumption Surveys 2009/15/20; 2021.

15. Global Data Lab, Wealth, Poverty, & Assets for selected indicators in Ghana, Area Database v4.2, 2020.

16. Centre for Energy, Environment, and Sustainable Development, Domestic Refrigerating Appliance and Room Air Conditioner Market And Feasibility Assessment: ECOWAS Refrigerators and ACs Initiative in Ghana, 2020.

17. Consumer Technology Association, 24th U.S. Consumer Technology Ownership & Market Potential Study, 2022.

18. Ghana Statistical Service, Household Ownership and Usage of ICT Products and Services, Household Survey on ICT in Ghana Abridged Report, pp.6-35, 2022.

19. Sovacool, B. K., Toxic transitions in the lifecycle externalities of a digital society: The complex afterlives of electronic waste in Ghana, *Resources Policy*, 64, 101459, 2019.

20. Peyton F., Big Chill: How Africa is Moving to Battle 'Zombie' Appliances, *Reuters*, 2020.

21. Ghana Energy Commission, Enforcement of Energy Efficiency Legislative Instruments (1815, 1932, and 1958) at Ports Of Entry (With Relevant Indicators/Statistics), 2020.

22. Maphosa, V., & Maphosa, M., E-waste management in Sub-Saharan Africa: A systematic liter-ature review, *Cogent Business & Management*, 7(1), 2020.

23. Daum, K., Stoler, J., & Grant, R.J., Toward a More Sustainable Trajectory for E-Waste Policy: A Review of a Decade of E-Waste Research in Accra, Ghana, *Int. J. Environ. Res. Public Health*, 14(2), 135, 2017.

24. Acquah, A.A., D'Souza, C., Martin, B., Arko-Mensah, J., Nti, A.A., Kwarteng, L., Takyi, S.A., Quakyi, I.A., Robins, T.G., & Fobil, J.N., Processes and challenges associated with informal electronic waste recycling at Agbogbloshie, a suburb of Accra, Ghana, *Proc. Human Factors and Ergon. Soc. Annu, Meet.*, 63, 938-942, 2019.

25. Perkins, D., Bruné Drisse, M., Nxele, T., E-waste: a global hazard, *Annals of Global Health*, 80(4), 286-295, 2014.

26. Yu, E. *et al.*, Informal processing of electronic waste at Agbogbloshie, Ghana: workers' knowledge about associated health hazards and alternative livelihoods, *Glob. Health Promot.*, 24, 90-98, 2017.

27. Wittsiepe, J. *et al.*, Pilot study on the internal exposure to heavy metals of informal-level electronic waste workers in Agbogbloshie, Accra, Ghana, *Environ. Sci. Pollut. Res.*, 24(3), 3097-3107, 2017.

28. Agyarko, K. *et al.*, The Importance of Stopping Environmental Dumping in Ghana: the Case of Inefficient New and Used Cooling Appliances with Obsolete Refrigerants, *Duke Env. Law & Policy Forum*, pp. 51-106, 2022.

29. Clasp & Institute For Governance & Sustainable Development, Environmentally Harmful Dumping of Inefficient and Obsolete Air Conditioners in Africa, 2020.

30. Andoh, D., Second-Hand Fridges Still Flood Markets; Ban Not Enforced Since 2008, *Graphic Comm.Group*, 2019.

31. Gyamfri, S. *et al.*, The Energy Efficiency Situation in Ghana, *Renew. and Sust. Energy Rev.*,1415-1423, 2018.

32. Williams, E., Kahhat, R., Allenby, B.R., Kavazanjian, E., Kim, J., & Xu, M., Environmental, social, & economic implications of global reuse & recycling of personal computers, *Environ. Sci. Tech*, 42(17), 6446-54, 2008.

33. Desmond, P. and Asamba, M., Accelerating the transition to a circular economy in Africa: Case studies from Kenya and South Africa, *Chapter in The Circular Economy and the Global South*, ISBN: 9780429434006, 2019.

34. Ohiomah, I., Sukdeo, N., Challenges of the South African economy to transition to a circular economy: a case of remanufacturing, *J. Remanufacturing* 12(2):213-225, 2022.

35. Fiagbe, Y., Asafo-Adjaye, M.K., Potential of remanufacturing industry in Ghana, GhIE Conference, pp.1-14, 2013.

36. Brent, A., Steinhilper, R., Opportunities for remanufactured electronic products from developing countries: hypotheses to characterise the perspectives of a global remanufacturing industry, *IEEE Africon*, 2:891-896, 2004.

37. CTA, Historical Sales Data, U.S. Consumer Technology Sales & Forecasts, 2022.

38. European Parliament, Annex III Categories of EEE Covered by This Directive, Directive 2012/19/EU of the European Parliament and of the Council on Waste Electrical and Electronic Equipment (WEEE), 2018.

39. Brunner, P.H., & Rechberger, H., *Handbook of material flow analysis: For environmental, resource, and waste engineers*, p.59, CRC press, 2016.

40. US EPA, ENERGY STAR Unit Shipment and Market Penetration Reports, Years 2010-2021, 2022.

41. United Nations Department of Economic and Social Affairs, Imports of selected product codes into Ghana in Recent Periods, UN Comtrade Database, 2022.

42. World Integrated Trade Solution, Ghana Product Imports from World, selected categories under product codes 761, 851, 7752, 74151, World Bank, 2022.

43. Green Cooling Initiative, Ghana's Greenhouse Gas Inventory and Technology Gap Analysis for the Refrigeration and Air Conditioning Sector, p.38, 2018.

44. US Census Bureau, Table HH-1: Households by Type: 1940-Present, Historical Household Tables, 2022.

45. World Bank, World Development Indicators, Dataset 0037712.

46. Duan, H. *et al.*, Quantitative Characterization of Domestic and Transboundary Flows of Used Electronics: Analysis of Generation, Collection, and Export in the United States, StEP Initiative, 2013.

47. PACE, A New Circular Vision for Electronics: Time for a Global Reboot, World Economic Forum, 2019.

48. Miller, T.R., Duan, H., Gregory, J., Kahhat, R., Kirchain, R., Quantifying Domestic Used Electronics Flows using a Combination of Material Flow Methodologies: A US Case Study, *Eniron. Sci. Technol.*, 50, 5711-5719, 2016.

49. US EPA, Understanding E-Waste in Cleaning Up Electronic Waste, 2022.

50. Jorgensen, B., Why the Refurbished Electronics Market is Thriving, *Elec. Eng. Times*, AspenCore Network, 2020.

51. Kang, H., Schoenung, J., Electronic waste recycling: A review of US infrastructure and technology options, *Res., Cons. Recy.* 45(4):368-400, 2005.

52. Thongkaow, P., T. Prueksasit, and W. Siriwong. "Material flow of informal electronic waste dismantling in rural area of Northeastern Thailand." Osaka: *Int. Conf. on Nat. Sci. and Env.*, 2017.

53. USITC, Used Electronics Products: An Examination of U.S. Exports, Invest. No. 332-528, Publication 4379, 2013.

54. Breivik, K., Armitage, J., Wania, F., Jones, K., Tracking the Global Generation and Exports of e-Waste: Do Existing Estimates Add Up?, *Environ. Sci. Technol.*, 48(15), 8735-8743, 2014.

55. Blancco, Trend Report: Q3 2018, State of Mobile Device Repair & Security, 2019.

56. Tiep, H., *et al.* E-waste management practices of households in Melaka, *Int. J. Env. Sci. Dev.* 6(11):811-817, 2015.

57. Michigan Dept. of Environment, Great Lakes, and Energy, Takeback Program Report in State Registration Data FOIA Request 1099-20, p.58, 2020.

58. Amoyaw-Osei, Y., Agyekum, O. O., Pwamang, J. A., Mueller, E., Fasko, R., Schluep, M., Ghana e-Waste Country Assessment, Secretariat of the Basel Convention, 2011.

59. Basel Convention, EEE Imports by Condition, WEEE Generation by Origin, in Where are WEEE in Africa? Find- ings from the Basel Convention E-Waste Africa Programme, 2011.

60. World Bank. Technical Report on the Sustainable Management of E-Waste in Ghana, 2015.

61. Oteng-Ababio, M, E-waste: an emerging challenge to solid waste management in Ghana, *Int. Devel. Planning Rev.*, 32(2), 191-206, 2010.

62. Oteng-Ababio, M, Electronic Waste Management in Ghana–Issues and Practices, in: Sustainable Development- Authoritative and Leading Edge Content for Environmental Management, S. Curkovic (Ed.), InTech, 2012.

63. Prakash, S., Manhart, A., Agyekum, O. O., Amoyaw-Osei, Y., Schluep, M., Müller, E., & Fasko, R., Informal e-Waste Recycling Sector in Ghana: an In-Depth Socio-Economic Study," *Proc. Going Green Care Innov.*, 2010.

64. Chasant, M. Agbogbloshie Demolition: The End of an Era or an Injustice?, Muntaka.com, 2021.

65. Grant, R., Oteng-Ababio, M. Mapping the invisible and real 'African' economy: urban e-waste circuitry, *Urban Geography*, 33(1):1-21, 2012.

66. Ghana Statistical Service, 2010 Population & Housing Census: National Analytical Report, 2010.

67. Ghana Statistical Service, Ghana Living Standards Survey Round 6 (GLSS 6) Main Report, 2014.

68. Ghana Statistical Service, Ghana Living Standards Survey Round 7 (GLSS 7) Main Report, 2019.

69. Ghana Statistical Service, 2021 Population & Housing Census Press Release on Provisional Results, 2021.

Part 10
PATHWAYS TO NET ZERO EMISSIONS

53

Emission Reduction for an Imflux® Constant Pressure Injection Molding Process

Birchmeier, Brandon, Lawless III, William F. and Santini, Kelly*

iMFLUX, Hamilton, OH, USA

Abstract

The goal of achieving a circular economy and net zero emissions is being embraced by many manufacturers of plastic materials, products, and machinery. There are many aspects of plastics manufacturing in which these goals should be addressed, such as raw plastic material consumption, reclamation and reuse of plastic materials and reduction of emissions from manufacturing processes. Emission generation and energy consumption around the world has been raising concerns for decades due to possible depletion of finite resources, overall environmental impacts as well as supply chain issues such as maintaining power grid integrity. Population growth, as well as an increased conversion from glass and metal products over to plastic products, has significantly increased the overall energy consumption and emissions of the plastics industry. Therefore, emission and energy reduction should be the objective during any step of the plastics manufacturing process. This paper analyzes the energy consumption reduction of an iMFLUX constant pressure molding process when compared to a conventional plastic injection molding process. Three different machine and mold pairings are evaluated, the molds ranging from a single cavity 5-gallon bucket to a 16-cavity deodorant cap. The iMFLUX constant pressure molding process makes it possible to manufacture a plastic part at lower pressures and temperatures as well as manufacturing the parts at a faster cycle time. Due to the plastic injection pressure reduction of the iMFLUX process, the clamping force required to keep the mold closed during plastic injection can be reduced. For each machine and mold pairing, a Design of Experiment (DOE) is performed and the following results for energy consumption reduction compared to a conventional injection molding process are reported: 1) iMFLUX baseline - no melt temperature reduction, no clamping force reduction and no cycle time reduction 2) iMFLUX with melt temperature reduction only 3) iMFLUX with clamping force reduction only 4) iMFLUX with cycle time reduction only 5) iMFLUX with melt temperature reduction and clamping force reduction only 6) iMFLUX with melt temperature reduction and cycle time reduction only 7) iMFLUX with clamping force reduction and cycle time reduction only 8) iMFLUX with melt temperature reduction, clamping force reduction and cycle time reduction. The energy consumption reduction is extrapolated to show emission savings for typical annual production.

Keywords: Injection molding, iMFLUX process, CO_2 emission reduction, energy consumption design

*Corresponding author: santini.kl@imflux.com

Nabil Nasr (ed.) Technology Innovation for the Circular Economy: Recycling, Remanufacturing, Design, Systems Analysis and Logistics, (715–724) © 2024 Scrivener Publishing LLC

53.1   Introduction

Injection molding is a widely used manufacturing process for producing a variety of plastic products. However, the high energy consumption of the injection molding machines and operations is a concern for manufacturers and their customers. The energy efficiency of injection molding has a significant impact on the overall cost and environmental footprint of the manufacturing process. Over the years energy efficiency has been tackled by transitioning from hydraulic machines to hybrid machines and in some cases completely electric machines. Many machine manufacturers have made efforts to make their machines more energy efficient but when the conventional injection molding process uses the same methodology that it has been using for decades, there is only so much efficiency that can be achieved by changing from a hydraulic machine to a hybrid or electric machine.

When developing a conventional injection molding process according to the most widely used methodology, a screw velocity set point is selected to inject the material at a constant fill rate relative to the screw's position. This constant velocity set point is selected from a relative viscosity curve developed by injecting at various injection velocities and recording the resulting melt pressures and fill times and calculating apparent viscosities. As the plastic material is injected into the mold at a constant screw velocity, the melt pressure will increase as the resistance to flow through the cavity increases. This constant climb in melt pressure can limit the process window by driving the machine to operate close to its pressure limit or maximum alarm pressure limit. Furthermore, as melt pressure increases for a given part geometry the machine size may potentially need to be upgraded to account for the higher pressures.

Increasing the machine size with all other factors such as machine type and part type being equal will always increase the energy consumption due to the fact that there are larger components to heat and move in the molding machine.

After a velocity profile for conventionally injecting plastic part has been established using the current industry methodology, the process then goes from a velocity controlled portion of the process to a pack and/or hold pressure controlled portion of the process in which an abrupt change in plastic flow front velocity inside the mold/part cavity occurs. This transfer typically takes place as the part is anywhere between 90%-98% of the part being full. This can additionally increase the energy consumption required because any hesitation in the plastic flow front velocity that far away from the injection point (gate) of the part needs to be overcome by increasing the packing/holding pressure. The packing/holding pressure is then maintained for a period of time until no more plastic can be introduced into the mold cavity. The higher the packing/holding pressure and the longer the packing/holding time, the higher the energy consumption will be with all other factors being equal.

With iMFLUX® the way an injection molding machine operates is flipped on its head. Rather than control the majority of filling (90%-98%) with a velocity controlled injection of plastic material, the velocity is variably controlled to maintain a constant pressure. The final filling of the part (2%-10%) is also pressure controlled, so there is no abrupt change in control modes (from velocity control to pressure control) which reduces plastic flow front hesitation. Reducing the hesitation in the final portion of the filling of the part also reduces the amount of time to complete the complete filling of the part, which further reduces energy consumption. The iMFLUX methodology allows for a low constant pressure throughout the injection molding process which is a significantly reduced pressure (up to 50%) than what is required by a conventional injection molding process. Finally, due to reducing injection pressures significantly, the iMFLUX process gives manufacturers the ability to manufacture plastic parts at lower plastic melt temperatures, which can also help to reduce energy consumption [1].

The overall reduction in the pressure required to fill the part gives manufacturers the ability to make energy efficient choices when developing an injection molding process. As recently validated by UL [2], the simple establishment of an iMFLUX process can yield up to a 15% energy savings when comparing a conventional injection molding process to an iMFLUX developed process (kW/kg). This paper explores how to maximize the energy savings with an iMFLUX process and considers what other settings can be optimized.

53.2   Experimental Method and Results

53.2.1   Experimental Approach

This experiment was developed to explore how to maximize the energy savings of the iMFLUX process and understand what aspects of an iMFLUX process have the greatest impact. There can be many variables when injection molding: size of the machine, mold size, cavities, melt temperatures, clamp force, cycle time (which is driven by overall injection time as well as cooling time) and much more.

For this experiment, three different machine and mold pairings are evaluated. The three pairings are as follows: 1) Single cavity (mold makes one part at a time) five-gallon bucket mold was run on a 450 ton (clamping force) KTEC Ferromatik injection molding machine. 2) 16-cavity deodorant cap mold was run on a 250 ton DeMag injection molding machine. 3) Single cavity ASTM tensile bar mold was run on a 180 ton Nissei injection molding machine. Energy was monitored by a Fluke 1738 "Power Logger" Energy Meter with 30 minutes of run time for each portion of the experiment.

For each mold and machine pairings, a Design of Experiment (DOE) was performed and the following results for energy consumption reduction compared to a conventional injection molding process are reported: 1) iMFLUX baseline process - no melt temperature reduction, no clamping force reduction and no cycle time reduction 2) iMFLUX with melt temperature reduction only 3) iMFLUX with clamping force reduction only 4) iMFLUX with cycle time reduction only 5) iMFLUX with melt temperature and clamping force reduction only 6) iMFLUX with melt temperature and cycle time reduction only 7) iMFLUX with clamping force and cycle time reduction only 8) iMFLUX with melt temperature reduction, clamping force reduction and cycle time reduction. The energy consumption reduction is extrapolated to show emission savings for typical annual production.

53.2.2   Single Cavity 4.5 Gallon Bucket Energy Consumption Experiment

Experimental Method

A baseline conventional process was developed according to industry standard methodology. An iMFLUX® constant pressure process was developed, and the Design of Experiment was performed using upper and lower limits for the Melt Temperature (F), Clamping

Table 53.1 Energy consumption design of experiment data results for bucket.

Process	Isolated test target	Melt temp (F)	Clamp tons (kN)	Cooling time (s)	Cycle time (s)	Weight (kg)	Cycles	kW	kW/kg
Conv	Baseline	450	460	20	35	30.86	51	84.99	27.54
iMFLUX	iMFLUX Only	438	448	17.5	32.5	32.23	55	83.02	25.76
iMFLUX	iMFLUX*Melt	425	460	20	35	29.89	51	73.81	24.69
iMFLUX	iMFLUX*Tons	450	435	20	35	30.08	51	76.75	25.51
iMFLUX	iMFLUX*Cycle	450	460	15	30	35.1	60	84.1	23.96
iMFLUX	iMFLUX*Melt *Tons	425	435	20	35	29.99	51	73.41	24.48
iMFLUX	iMFLUX*Melt *Cycle	425	460	15	30	34.94	60	80.2	22.95
iMFLUX	iMFLUX*Tons *Cycle	450	435	15	30	34.92	60	82.8	23.71
iMFLUX	iMFLUX*All	425	435	15	30	35.07	60	80.2	22.87

Tonnage (Tons) and Cooling Time (s) inputs. Outputs of Cycle Time (s), Part Weight of all parts during 30 minute run (kg), Number of Cycles during 30 minute run, Kilowatts (kW) during 30 minute run and Kilowatts per Kilogram (kW/kg) during 30 minute run were all observed and recorded, see Table 53.1 - Energy Consumption Design of Experiment Data Results for Bucket.

Results

The iMFLUX processes showed an energy consumption reduction of 6.5%-17.0% for the parameters that were tested using a single cavity bucket mold in a 450 ton KTEC Ferromatik injection molding machine. A summary of the energy savings potential is demonstrated in Figure 53.1 - Total Energy Savings Potential for iMFLUX Processes vs. Conventional Process for Bucket.

53.2.3 16 Cavity Deodorant Cap Energy Consumption Experiment

Experimental Method

A baseline conventional process was developed according to industry standard methodology. An iMFLUX® constant pressure process was developed, and the Design of Experiment was performed using upper and lower limits for the Melt Temperature (F), Clamping Tonnage (kN) and Cooling Time (s) inputs. Outputs of Cycle Time (s), Part Weight of all parts during 30 minute run (kg), Number of Cycles during 30 minute run, Kilowatts (kW) during 30 minute run and Kilowatts per Kilogram (kW/kg) during 30 minute run were all

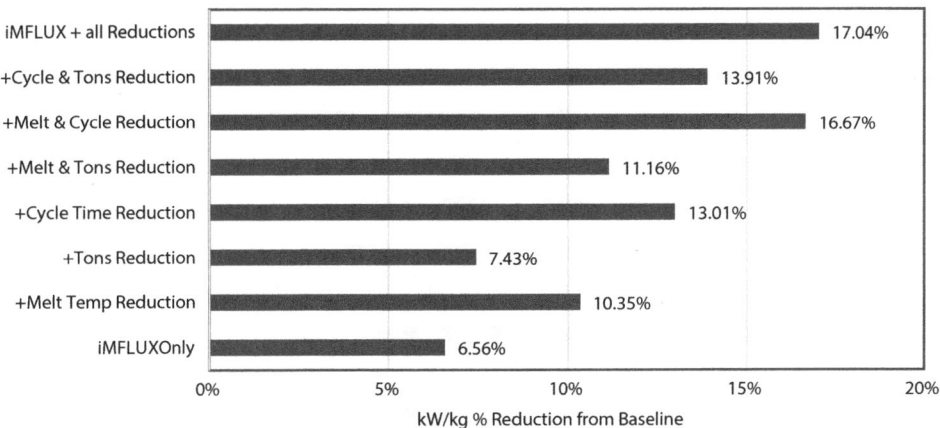

Figure 53.1 Total energy savings potential for iMFLUX processes vs. conventional process for bucket.

Table 53.2 Energy consumption design of experiment data results for deodorant cap.

Process	Isolated test target	Melt temp (F)	Clamp tons (kN)	Cooling time (s)	Cycle time (s)	Weight (kg)	Cycles	kW	kW/kg
Conv	Baseline	420	1700	5	35	6.77	121	58.96	49.24
iMFLUX	iMFLUX Only	410	1050	3.5	32.5	5.86	140	56.65	42.98
iMFLUX	iMFLUX*Melt	400	1700	5	35	6.55	125	55.06	44.86
iMFLUX	iMFLUX*Tons	420	400	5	35	6.5	126	52.65	41.36
iMFLUX	iMFLUX*Cycle	420	1700	2	30	5.18	157	62.35	41.57
iMFLUX	iMFLUX*Melt *Tons	400	400	5	35	6.5	126	53.79	42.27
iMFLUX	iMFLUX*Melt *Cycle	400	1700	2	30	5.23	157	58.55	39.03
iMFLUX	iMFLUX*Tons *Cycle	420	400	2	30	5.18	158	62.3	39.16
iMFLUX	iMFLUX*All	400	400	2	30	5.18	158	62.44	38.16

observed and recorded, see Table 53.2 - Energy Consumption Design of Experiment Data Results for Deodorant Cap.

Results

The iMFLUX processes showed an energy consumption reduction of 8.9%-22.5% for the parameters that were tested using a 16 cavity deodorant cap mold in a 250 ton (2500kN) Demag injection molding machine. A summary of the energy savings potential is

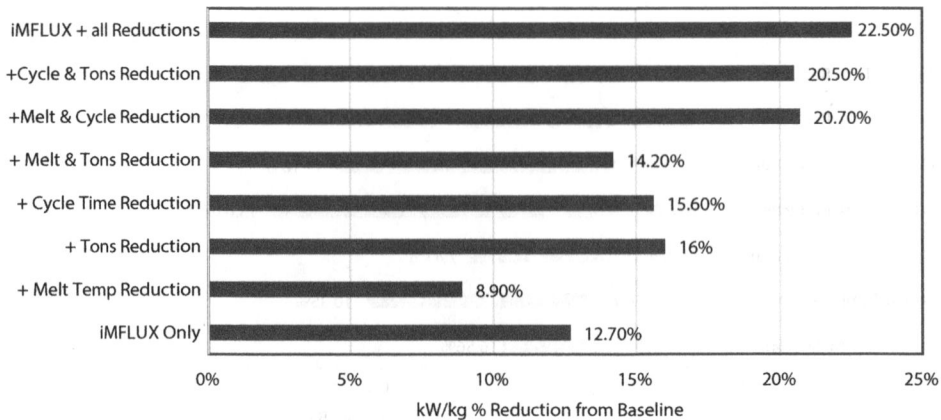

Figure 53.2 Total energy savings potential for iMFLUX processes vs. conventional process for deodorant cap.

demonstrated in Figure 53.2 - Total Energy Savings Potential for iMFLUX Processes vs. Conventional Process for Deodorant Cap.

53.2.4 Single Cavity Tensile Bar Energy Consumption Experiment

Experimental Method

A baseline conventional process was developed according to industry standard methodology. An iMFLUX® constant pressure process was developed, and the Design of Experiment was performed using upper and lower limits for the Melt Temperature (F), Clamping Tonnage (Tons) and Cooling Time (s) inputs. Outputs of Cycle Time (s), Part Weight of all parts during 30 minute run (kg), Number of Cycles during 30 minute run, Kilowatts (kW) during 30 minute run and Kilowatts per Kilogram (kW/kg) during 30 minute run were all observed and recorded, see Table 53.3 - Energy Consumption Design of Experiment Data Results for Tensile Bar.

Results

The iMFLUX processes showed an energy consumption reduction of 6.5%-14.1% for the parameters that were tested using a single cavity tensile bar mold in a 180 ton Nissei injection molding machine. A summary of the energy savings potential is demonstrated in Figure 53.3 - Total Energy Savings Potential for iMFLUX Processes vs. Conventional Process for Tensile Bar.

53.3 Conclusions and Recommendations

53.3.1 Summary of Results for All Experiments

When comparing the energy consumption of a conventional injection molding process to an iMFLUX® process without modifying melt temperature, clamping tonnage or cycle time, there is a minimum energy percentage reduction of at least 6.5% for all of the mold and

Table 53.3 Energy consumption design of experiment data results for tensile bar.

Process	Isolated test target	Melt temp (F)	Clamp tons (kN)	Cooling time (s)	Cycle time (s)	Weight (kg)	Cycles	kW	kW/kg
Conv	Baseline	450	40	19.5	36	1.22	50	1.4	1.15
iMFLUX	iMFLUX Only	435	30	18	35	1.38	52	1.48	1.07
iMFLUX	iMFLUX*Melt	420	40	19.5	36	1.38	50	1.42	1.03
iMFLUX	iMFLUX*Tons	450	20	19.5	36	1.38	50	1.47	1.07
iMFLUX	iMFLUX*Cycle	450	40	17.5	34	1.46	53	1.51	1.03
iMFLUX	iMFLUX*Melt *Tons	420	20	19.5	36	1.39	50	1.41	1.01
iMFLUX	iMFLUX*Melt *Cycle	420	40	17.5	34	1.47	53	1.45	0.99
iMFLUX	iMFLUX*Tons *Cycle	450	20	17.5	34	1.47	53	1.49	1.01
iMFLUX	iMFLUX*All	420	20	17.5	34	1.44	53	1.42	0.99

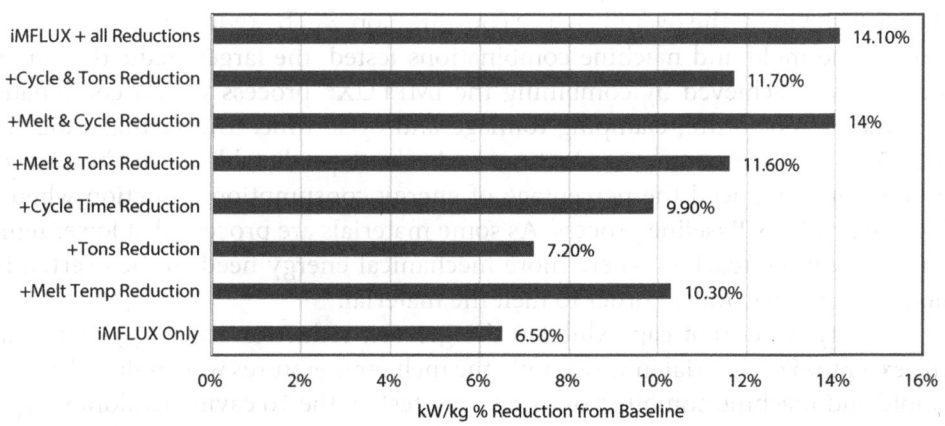

Figure 53.3 Total energy savings potential for iMFLUX processes vs. conventional process for tensile bar.

machine combinations that were tested. This is due to the fact that the machine is using lower pressures which results in lower energy consumption. All energy reduction percentage results for all three mold and material combinations can be found in Table 53.4 - Energy Reduction Percentage Results for Bucket, Deodorant Cap and Tensile Bar.

Table 53.4 Energy reduction percentage results for bucket, deodorant cap and tensile bar.

	Bucket	Deodorant cap	Tensile bar
iMFLUX Only	6.56%	12.70%	6.50%
+ Melt Temp Reduction	10.35%	8.90%	10.30%
+ Tons Reduction	7.43%	16%	7.20%
+ Cycle Time Reduction	13.01%	15.60%	9.90%
+ Melt & Tons Reduction	11.16%	14.20%	11.60%
+ Melt & Cycle Reduction	16.67%	20.70%	14%
+ Cycle & Tons Reduction	13.91%	20.50%	11.70%
iMFLUX + all Reductions	17.04%	22.50%	14.10%

As mentioned previously, the lower pressures allow for a reduction in machine clamping force that is needed to overcome the injected plastic pressure in order to keep the mold closed. In addition, the melt temperatures can also be reduced compared to conventional injection molding. Finally, since the part is being filled under a constant pressure throughout the entire filling of the part, there is more optimal plastic contact with the steel molding surface which improves the cooling efficiency and allows for a faster cycle time. Since even an idle machine will consume energy, the more machine time that a machine can be utilized to make parts the more the overall energy consumption will be reduced.

For all of the mold and machine combinations tested, the largest reduction in energy consumption was achieved by combining the iMFLUX® process with a combination of reducing melt temperature, clamping tonnage and cycle time. The savings ranged from 14.1%-22.5%. Melt temperature reduction is the least predictable as in the case of the Deodorant Cap it reduced the percentage of energy consumption reduction when compared to the iMFLUX Baseline process. As some materials are processed at lower temperatures, a point can be reached where more mechanical energy needs to be exerted by the injection molding machine in order to melt the material.

The 16 cavity deodorant cap exhibited the greatest reduction in energy consumption, with the exception of the trial in which only the melt temperatures were reduced. Out of the three mold and machine combinations that were tested, the 16 cavity deodorant cap mold is the best representation of the majority of high production injection molding processes that are used in manufacturing plastic parts. The single cavity bucket had the next highest energy consumption, with the exception of combining melt temperature and clamping tonnage reductio. Finally, the tensile bar showed the lowest reduction in energy consumption. It should be noted that tensile bars are not a typical production part for consumer use but are rather used in industry and research for material properties testing.

Table 53.5 CO_2 Emission reduction results for 1 million parts each of bucket, deodorant cap and tensile bar.

	Conventional baseline kWh	iMFLUX optimized kWh	Savings kWh	CO_2 emission reduction (Tons)
Bucket	16066	11138	4928	3.800
Deodorant Cap	126	78	48	0.037
Tensile Bar	280	253	27	0.021

53.3.2   Summary of CO_2 Emissions Reduction

Using the data that was collected during the trials that were performed, the amount of energy that was required to manufacture 1 million parts for each of the bucket, deodorant cap and tensile bar was calculated. The number of kilowatt hours (kWh) required to make 1 million parts manufactured by a conventional injection molding process was compared to the kWh required to make 1 million parts manufactured by an optimized iMFLUX process. It should be noted that the iMFLUX process also allows for more machine utilization because it manufactures the parts in a shorter period of time in addition to the CO_2 emissions reduction. The EPA website calculator was used for the CO_2 Emission Reduction calculations [3]. CO_2 emissions reduction results for producing 1 million parts for each of the mold and material combinations can be found in Table 53.5 - CO_2 Emission Reduction Results for 1 million parts each of Bucket, Deodorant Cap and Tensile Bar.

For the 1 million part calculations it can be seen that the bucket would show the most CO_2 emission reduction, mostly due to its size in comparison to the other parts. It should be noted that it would take over a year to produce 1 million buckets at the cycle times that were achieved. As mentioned earlier, tensile bars are not a typical production part for consumer use but are used for the testing of plastic material physical properties in labs. However, if you put the deodorant CO_2 emission reduction into context, there are hundreds of millions of this type of cap produced every year. For every billion deodorant cap parts that are produced, there is a potential 37 tons of CO_2 emission reduction possible. This doesn't even consider the other components of the deodorant package.

References

1. iMFLUX, 'Education - Plastic Show Demonstrations - 2020 Plastec West Show', https://youtu.be/oYT_WkWcYEw, 2020
2. Underwriters Laboratories, Marketing Claim Verification ID: V653137, https://verify.ul.com/verifications/1051, 2022
3. Environmental Protection Agency, Greenhouse Gas Equivalencies Calculator, https://www.epa.gov/energy/greenhouse-gas-equivalencies-calculator, Updated March 2022

54

Circular Economy Contributions to Decarbonizing the US Steel Sector

Julien Walzberg[1]* and Alberta Carpenter[1,2]

[1]*National Renewable Energy Laboratory, 15013 Denver West Parkway, Golden, CO, USA*
[2]*Renewable and Sustainable Energy Institute, University of Colorado, Boulder, CO, USA*

Abstract

The potential benefits of the circular economy (CE) for decarbonization have recently attracted much attention in the academic and grey literature. The department of energy (DOE) industrial decarbonization roadmap highlight that in addition to the four pillars (energy efficiency, electrification, low-carbon fuels, feedstocks, and energy sources, and carbon capture, utilization, and storage (CCUS)) identified in the roadmap, "scenarios [...] will need to be developed that incorporate materials efficiency and circular economy strategies". Moreover, the roadmap identifies hard-to-abate CO_2 emissions across the five industrial subsectors included in the analysis (iron & steel, chemical, food & beverage, refining, and cement manufacturing). Thus, additional options to decarbonize the industrial sector, like the CE, could prove beneficial to address those hard-to-abate emissions. Moreover, they may be less costly, require less R&D or incentives, and be more readily adopted than other strategies such as energy efficiency and CCUS. While the Industrial Decarbonization Roadmap includes some CE strategies (e.g., the increasing market share of steel from electric arc furnaces – which incorporate steel scrap), a deeper dive into what role the CE could play in the United States (US) industrial decarbonization is needed. In this work, we present several industrial decarbonization scenarios that incorporate materials efficiency and circular economy strategies for the iron & steel subsector. The scenarios identify barriers (including technical limits and constraints), opportunities, and R&D needs.

Moreover, we estimate the contribution of those scenarios to reducing the industrial sector's CO_2 emissions and their potential synergies with the Industrial Decarbonization Roadmap's four pillars. Iron & steel manufacturing was responsible for 90 million tons of industrial CO_2 emissions in 2020 – 7% of the industrial sector's total process- related emissions. Besides recycling – which presents technical limitations due to trace contaminants – CE strategies on the demand side could lower iron & steel manufacturing emissions (e.g., the development of lighter products or business models that encourage a more intensive use). This study reviews barriers, opportunities, and trade-offs for 5 material efficiency strategies. Many topics explored in the review call for further research.

Keywords: Circular economy, industrial decarbonization, iron & steel, material efficiency, barriers, opportunities, trade-offs

**Corresponding author*: julien.walzberg@nrel.gov

Nabil Nasr (ed.) Technology Innovation for the Circular Economy: Recycling, Remanufacturing, Design, Systems Analysis and Logistics, (725–738) © 2024 Scrivener Publishing LLC

54.1 Introduction

The circular economy (CE) is an alternative economic model aiming to solve environmental issues stemming from the exploitation of earth's natural resources. The CE narrows, slows, regenerates, and closes material flows while keeping materials' value within the economy. Through strategies such as redesigning, reusing, repurposing, or recycling, the CE minimizes waste, providing economic and environmental benefits the economy while preserving the environment. CE measures could contribute to decarbonization by reducing global greenhouse gas (GHG) emissions from material use, thereby complementing the mitigation measures outlined in the Nationally Determined Contributions [1]. Human societies are consuming growing quantities of raw materials – about 100 billion metric tons in 2020 – some with high embodied GHG emissions such as concrete and steel [1]. The extraction of those materials and subsequent processing account for about 50% of global GHG emissions [2]. Thus, the CE has a high potential to contribute to decarbonization.

As Figure 54.1 illustrates, the contribution of CE to decarbonization is increasingly investigated by scholars. This is consistent with the growing interest from various stakeholders, as evidenced by recent reports [1, 3–5]. Wang *et al.* [5], for instance, argue that to reach the Paris climate goals, products need to deliver the same function or wellbeing benefits with fewer materials. They propose nine actions that CE stakeholders such as governments, businesses, NGOs, and researchers need to take to make circularity contribute most effectively to decarbonization: shift consumption patterns, spark circularity from the design phase, incorporate circularity within clean energy technologies, integrate CE into climate policies, incentivize cross-border GHG emissions reduction, connect circularity metrics with climate change indicators, increase transparency and comparability in modeling frameworks, apply systemic impact assessment, and further research the role of the CE in climate change adaptation. And while the latest report by the intergovernmental panel on climate change has highlighted the role of the CE to mitigate climate change [6], global circularity is wilting, rather than improving [7].

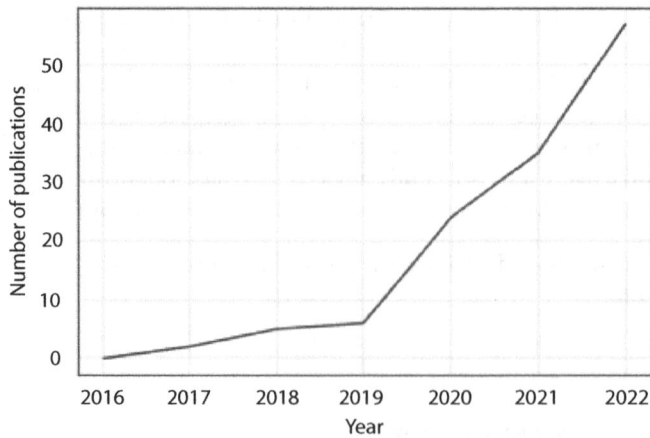

Figure 54.1 A Scopus search was performed using the keyword combination "circular economy" and "decarbonization". The figure shows the number of articles published each year according to this query.

CE is often viewed as an umbrella concept [8]. In a recent literature review, Calisto Friant, Vermeulen, and Salomone [9] identified 72 concepts that have been influencing CE discourses, some dating back the 1940's. Among them, material efficiency (ME) has been particularly used in decarbonization studies [10–12]. ME seeks to achieve similar outcomes with the use of less materials or less detrimental materials [13]. For decarbonization purposes, ME strategies will, therefore, seek to achieve similar outcomes with the use of less materials or less GHG emissions-intensive materials [4]. While other concepts that have influenced CE discourses – such as the symbiotic economy or the natural step – can contribute to decarbonization, this work focuses on ME. Finally, as a concept contributing to the CE, strategies improving ME can be considered CE strategies as well.

While ME research is sometime inconclusive regarding benefits to decarbonization in certain economic sectors (e.g., electronics [4], there is consensus that ME can reduce GHG emissions of materials and products with high embodied energy [1, 3, 14]. For instance, ME could reduce vehicles' and buildings' emissions by 35% to 40% and 35% to 60%, respectively [4].

The department of energy (DOE) industrial decarbonization roadmap highlight that in addition to energy efficiency, electrification, low-carbon fuels, feedstocks, and energy sources, and carbon capture, utilization, and storage (CCUS), solutions "[…] will need to be developed that incorporate materials efficiency and circular economy strategies" [2]. Moreover, the roadmap identifies hard-to-abate CO_2 emissions across the five industrial subsectors included in the analysis (iron & steel, chemical, food & beverage, refining, and cement manufacturing). Thus, additional options to decarbonize the industrial sector, like the ME, could prove beneficial to address those hard-to-abate emissions. Those solutions could also be less costly, require less R&D or incentives, and be more readily adopted than other strategies such as energy efficiency and CCUS.

Iron & steel manufacturing was responsible for 90 million tons of the United States (US) industrial CO_2 emissions in 2020 – 7% of the industrial sector's total process-related emissions [2]. Besides recycling – which presents technical limitations due to trace contaminants such as Cu, Sn, and Ni [15] – ME strategies on the demand side could lower this industrial sector's GHG emissions (e.g., through the development of lighter products or business models that encourage a more intensive use). However, several research gaps exist in the current literature. Specific US estimates are lacking, and most studies focus on products rather than materials. Moreover, while barriers, opportunities and trade-offs are crucial to the effective implementation of ME strategies, there are not comprehensively discussed in the literature. Thus, the objective of this study is to fill those research gaps by conducting a critical literature review for the iron & steel sector.

54.2 Method

To collect the articles in this review, a Scopus search was performed using keyword combinations of the expressions "material efficiency" or "circular economy" and "decarbonization" (or synonyms) and "iron and steel". The selection of keywords for the Scopus search follows from a preliminary review (after reading highly cited publications such as Hertwich *et al.* [4] and Wijkman and Skånberg [16]). More articles were then found using the "snowballing" technique [17] and after reading their abstracts, 13 out of 37 articles were kept.

Publications that were not focusing on the iron and steel sector (e.g., articles and reports discussing ME strategies from an end use perspective such as Hertwich *et al.* [10] and Ellen MacArthur Foundation [3]) were not included.

During the literature review the authors, title and publication year were documented. Moreover, the relationships between the publications were visualized with a graph (based on mutual citations). Besides bibliographic information, the articles were analyzed to extract ME strategies information. The study's context, and the strategies' barriers, opportunities, trade-offs, and estimates of GHG emission reductions were summarized. This step enables the comparison of studies and identification of policy recommendations. Finally, other sources were used to discuss the insights from the literature review on the and next steps for the research were highlighted.

54.3   Results

54.3.1   Bibliographic Summary

Table 54.1 summarizes the bibliographic information of the articles included in this literature review. As seen in the table, more than half the publications selected were published in the last three years, confirming that the contribution of ME and CE to decarbonization is a trending topic. While scholars have mostly been leading the research, NGOs and one steel manufacturer have also published reports. The international energy agency (IEA), for instance, produced an iron and steel technology roadmap that analyzes the decarbonization potential of ME in the global iron and steel sector. Unsurprisingly given its scope, the environmental science & technology journal has published 5 out of the 13 publications included in this review.

Figure 54.2 shows how the publications relate with each other's. With an average node degree of 2, the different studies seem to fairly influence one another. Unsurprisingly, academic publications share more edges (i.e., mutual citations) than reports from NGOs and the steel industry. Zhu, Syndergaard, and Cooper [18] is the publication that is citing the most other studies from our literature review. The authors use a network-based material flow analysis to map the flow of steel in the US and, based on their results and the literature (e.g., Allwood [12]), propose policies to reduce the US steel stocks per capita.

In addition to post-consumer recycling, five ME strategies were identified: direct reuse, fewer steel products, lifespan extension, lightweighting, and process yields improvements. Post-consumer recycling is excluded from this analysis because it only reduces demand for primary steel and it is a strategy that is reaching saturation in the US (with a recycling rate of 79% in 2020) [19, 20]. Moreover, the availability of high grade scrap is limiting post-consumer recycling making it a less attractive ME strategy in the long term [15, 21].

Direct reuse of steel may take several forms [22]. In relocation a steel component in good conditions from one product is reused in a similar product (e.g., reuse of steel components in cars). Alternatively, the component can be reused in a different application. For instance, steel from dismantled buildings can be used in agriculture sheds [22]. When the condition of the steel component is low, it can be remanufactured to be used in either the same or a different application. Ship recycling, for example, can be reformed and used as construction rebars [23]. The fewer steel products strategy mainly relies on the sharing economy and

Table 54.1 Bibliographic information of the articles in this literature review.

Authors	Title	Publication year	Publication type
AISC	More than Recycled Content: The Sustainable Characteristics of Structural Steel	2017	White paper
Allwood et al.	Sustainable materials: with both eyes open	2012	Book
Cooper et al.	Reusing Steel and Aluminum Components at End of Product Life	2012	Journal article (Environmental Science & Technology)
Cooper et al.	The potential for material circularity and independence in the U.S. steel sector	2020	Journal article (Journal of Industrial Ecology)
IEA	Iron and steel technology roadmap: Towards more sustainable steelmaking	2020	Report
Milford et al.	Assessing the potential of yield improvements, through process scrap reduction, for energy and CO_2 abatement in the steel and aluminium sectors	2011	Journal article (Resources, Conservation and Recycling)
Milford et al.	The Roles of Energy and Material Efficiency in Meeting Steel Industry CO_2 Targets	2013	Journal article (Environmental Science & Technology)
Miller & Grubert	US industrial sector decoupling of energy use and greenhouse gas emissions under COVID: durability and decarbonization	2021	Journal article (Environmental Research Communications)
Ryan et al.	Reducing CO_2 Emissions from U.S. Steel Consumption by 70% by 2050	2020	Journal article (Environmental Science & Technology)
Skelton et al.	Comparing energy and material efficiency rebound effects: an exploration of scenarios in the GEM-E3 macroeconomic model	2020	Journal article (Ecological Economics)
USS	Sustainability report	2021	Report
Worrell & Boyd	Bottom-up estimates of deep decarbonization of U.S. manufacturing in 2050	2022	Journal article (Journal of Cleaner Production)
Zhu et al.	Mapping the Annual Flow of Steel in the United States	2019	Journal article (Environmental Science & Technology)

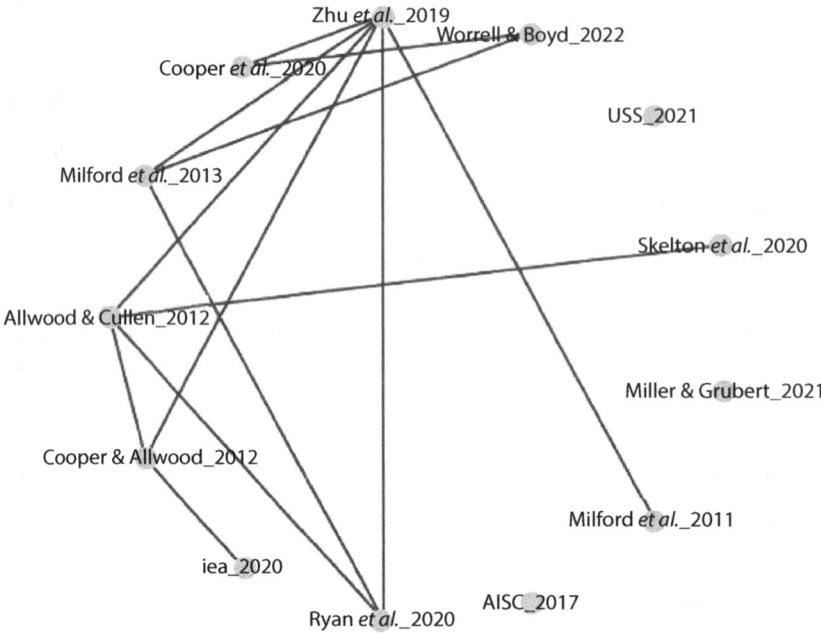

Figure 54.2 Relationships between the publications in this literature review.

dematerialization technologies. Teleworking and ridesharing are two examples of the application of this strategy [20]. Lifespan extension seeks to holds steel in in-use stocks for longer [11]. As an example Cai *et al.* [24] showed that doubling a building's lifetime could save about 400 million metric tons (Mt) of GHG emissions per year. Lightweighting consists in making products that contain less steel while providing the same function. As an example, building the Golden Gate Bridge today would require only half of the amount of steel that were used during its construction in 1937, for the same structure and strength [25]. Finally, improving process yields, for instance by reducing the amount of process scrap, also contributes to ME [19].

54.3.2 Review of Material Efficiency Strategies in the Steel Sector

Table 54.2 summarizes barriers, opportunities, and trade-offs of material efficiency in the steel sector. Overall, some barriers and opportunities are common to several ME strategies while others are more specific. The lack of support from steel users may hinder the adoption of all ME strategies besides process yield improvements. Indeed, ME requires steel users to develop new lightweight and modular designs, develop new business models based on accessing a service rather than owning a product, and increase data sharing – initiatives that they may judge too costly or risky [26]. Regarding the iron & steel sector, ME could lead to job losses, especially for the fewer steel products, lifespan extension, and lightweighting strategies [20].

One opportunities all ME strategies could benefits from are the design of performance-based standards – as opposed to prescriptive requirements – for instance in the construction industry [11, 27]. This type of standards would spark the design of more

Table 54.2 Barriers, opportunities, and possible trade-off of the reviewed material efficiency strategies.

Material efficiency strategy	Barriers	Opportunities
Direct reuse	• Labor costs • Demolition practices • Retrievable? • Low degree of standardization • Lack of support from steel users • Data availability about product	• New construction regulations • Maintain value • More durable and standardized products • Easier disassembly, knowledge conservation • Reinforcement steel • Lighten load on technological shift • Emissions requirements in regulations • Performance based standards • Testing standards
Fewer steel products	• Lack of support from steel users • Rural vs urban areas • Loss in living standards • Job losses • Population density • Partial substitution	• Lighten load on technological shift • Emissions requirements in regulations • Performance based standards • Value vs quantity • Cultural changes • Less is more? • Cost reductions
Lifespan extension	• Lack of support from steel users • Job losses • Long term	• Lighten load on technological shift • Emissions requirements in regulations • Performance based standards • Repair incentives • Failure risks • Servicing
Lightweighting	• Lack of support from steel users • Perceived risks • Manufacturing processes • Performance losses • Job losses	• Lighten load on technological shift • Emissions requirements in regulations • Performance based standards • Value vs quantity • Standards • Green procurement
Process yields improvements	• Costs • Data availability	• Lighten load on technological shift • Emissions requirements in regulations • Performance based standards • Waste handling, testing • Machining > Stamping and blanking > others • Tessellation • Best practices

durable and less wasteful designs of steel components that can be reused in different applications. However, one downside of such policies could be the increase in labor costs to ensure the components integrity through an extended lifetime or multiple reuse cycles. Similarly, incorporating life-cycle emissions requirements into regulations for products and constructions would stir end-user into preferring more durable and light steel products and encourage them to reuse steel [11]. It would also prompt steel manufacturers and end users to reduce process scraps [28].

One benefit that all ME strategies provide is to lighten the load on the process technology shifts that need to occur to decarbonize the steel sector in the next 30 years [2]. This could avoid the lock-in of polluting steel manufacturing capacities (e.g., in India) and leave time for the industry to overhaul [11]. At the moment, avoiding the production of 1 ton of steel emits less GHG emissions than producing it even if it is recycled later [11]. Moreover, GHG emission reductions from mature technological solutions such as energy efficiency [2] roughly amount to the process yields improvements strategy [19], meaning there is much room for other ME strategies to further reduce emissions. By contrast, less mature technological solutions (e.g., electrolysis or natural gas/hydrogen direct reduced iron steelmaking) may not be able to contribute to decarbonization in the near term [11]. While ME enables overhauling of the steel industry, the drawback is that it could divert necessary investments in cleaner technologies [20]. Such trade-offs could limit the contribution of both ME and technological improvements to decarbonizing the US steel sector. Other trade-offs shared by the ME strategies are the diversion of scrap for recycling (i.e., the competition between recycling and other ME strategies).

Additional barriers to direct reuse include the increased labor costs related to disassemble products and verify the steel component properties [12]. In the construction sector, reuse is also limited by unsuited demolition practices [12] and the difficulty of retrieving steel components such as reinforcing steel [22]. Once steel components are retrieved, the lack of standardization between products providing the same function (e.g., the door designs in different car models) and knowledge sharing between manufacturers may still hinder reuse [11, 22]. Conversely, better standardization and increase conservation of knowledge (e.g., regarding physical properties and disassembly) represents opportunities to improve steel reuse [22]. While relocating and remanufacturing steel products are usually preferable as they maintain the steel component values, the choice of the reuse option depends on both the demand for the used component and the component performance once disassembled [22]. One of the biggest opportunities for steel reuse is in the construction sector. For instance, reinforcement steel used in concrete represents a fifth of the steel demand [22]. Therefore, new construction regulations – that mandate or set up targets for steel reuse – could promote ME [12]. Improving waste-handling infrastructure by creating material inventories and robust testing standards and developing business models based on product value rather than quantity could push also the adoption of reuse practices [11]. Finally, research and development activities focusing on re-certification standards, automated disassembly, and machines to validate steel component properties are needed [12, 22].

The sharing economy and dematerialization technologies may reduce the need for steel products. One barrier to this ME strategy is the households' desire to live in more spacious, rural communities rather than in dense and urban areas, thereby limiting the business case for sharing business models, such as ridesharing [29, 30]. Moreover, such service-based economies may not affect everyone equally and lead to losses in living standards [5, 20].

Perhaps even more crucially than for steel reuse, replacing business models based on quantity of product sold by ones based on product value (focusing on the service that is delivered by the product rather than the product itself) represents a great opportunity for this ME strategy [11]. Recent increases in online shopping and remote working (especially during the COVID-19 pandemic) seem to – at least partially – indicate that such sharing economy and dematerialization phenomena are well under way [20, 29]. Regarding feasibility, the 30% decline in production (between Q2- 2019 and Q2-2020) due to lower demand sparked by COVID-19 did not seem to affect the steel sector too negatively [29] – although a sustained reduction in demand would without doubts be more impactful. On the consumption side, lower steel stock per capita in other developed countries such as France or the United Kingdom indicate that it is possible to maintain high standards of living with fewer steel products [20]. Moreover, Skelton, Paroussos, and Allwood [28] showed that switching to ridesharing would reduce transportation costs by 85%, which could be a strong motivator to behavioral change. However, the sharing economy and dematerialization technologies may not always fully substitute to traditional product-based functions. For instance, ridesharing may not directly displace new car demand on a 1:1 basis (as it may depend on user of sharing services selling their car on the secondary market) – a phenomenon known as substitution rebound effect [31]. Because ME can lead to lower costs, an income rebound effect could also occur. Skelton, Paroussos, and Allwood [28], for instance, showed that reducing demand for vehicles by 10% could lead to a 85% rebound effect, meaning that 85% of the GHG emission reductions due to ME would be offset by households' re-spending in other GHG emitting products and services.

Due to long lifetime of steel products (in average), applying the lifetime extension strategy to new steel products would have a limited contribution to the US steel sector decarbonization during the 2020-2050 period [20]. However, including current products (e.g., by retrofitting building and improving cars' maintenance and repair services) could make this strategy more interesting in the near-medium terms [20]. Lifetime extension is especially relevant for industries where interruptions are costly such as train and electricity generators [12]. Other opportunities to enhance the adoption of this ME strategy include creating business models centered around servicing, maintaining, and upgrading [12], and developing business models based on product value rather than quantity [11]. For instance, a steel product that is more durable could be sold at a premium, favoring ME without necessarily creating disadvantageous situations for either end-users or steel manufacturers. Creating incentives or reduced taxes (e.g., reduces sales & use taxes) for repair and maintenance services would also create opportunities for a wider adoption of the lifetime extension strategy [20]. Lastly, a specificity of the lifetime extension strategy warrant caution: the contribution of this ME strategy to decarbonization heavily depends on the trade-off between higher operational emissions due to the use of older products (e.g., older less insulated buildings) and lower embodied emissions due to longer lifetimes [4, 12].

Lightweighting consist in lowering the amount of steel used in an application. An unexpected barrier to this strategy is that installation load requirements may constraint steel requirements [12]. Thus, even if in-use steel requirements are low, installation loads may require higher amount of steel (e.g., this is the case for steel pipes).

Moreover, perceived risks associated with component failure prompt to use extra steel [12]. This is not surprising considering that the lack of trust in CE solutions has been showed to be a limiting factor to its development [32]. Further complicating matters, it is possible that lightweighting decreases products' performances [20]. In addition, current manufacturing processes favor designs that use more steel [12]. Fortunately, lightweighting also presents several opportunities. First, manufacturing costs may decrease as less raw materials are needed [12]. Flexible metal casting, forming, and fabrication process along with businesses centered around value may also prove to be win-win situations for both end-users and manufacturers [11, 12]. AISC [33] argues that the inclusion of a structural steel manufacturer in buildings' design discussions (during the project's conceptual phase) provides major benefits from a material, costs, and environmental perspective, benefiting both parties. The white paper cites the construction of a medical center in Ohio as an example: the involvement of a steel manufacturer decreased material inputs, costs and GG emissions by about 15%. Writing mandatory or voluntary standards (e.g., steel utilization targets into green building rating systems such as LEED) and green procurement at the public level (e.g., construction of new lightweight government buildings) could further encourage and strengthen such win-win business models [20, 11]. Interestingly, lightweighting is greatly discussed by the US steel industry [25, 33], showing its interest for that ME strategy.

Specific barriers to the process yields improvement strategy are the potential high costs of process improvements [28] and the dispersed nature of the steel supply chain which makes it difficult to identify where yield losses occur [19]. While high costs may hinder steel manufacturers to adopt less wasteful processes, it would also reduce the risks of rebound effect [28]. Opportunities for improvements could focus in priority on machining processes because they have the highest yield losses [19]. Other processes with high yield losses in car manufacturing (between 10% and 32%) are stamping and blanking. Tessellating could be a possible strategy to reduce yield losses. Given the high variability in yield losses in car manufacturing, another opportunity would be to share best practices within that industry [28]. According to Skelton, Paroussos, and Allwood [28], improving process yields could lead to a rebound effect due to the decreased costs of steel products. However, a carbon price could mitigate this unintended consequence, as long as governments spend surpluses in a responsible manner (e.g., to reduce debt).

54.3.3   Estimations of Greenhouse Gas Emission Reductions from Material Efficiency

Figure 54.3 shows the estimated GHG emission reductions (in Mt/year) in the iron & steel sector due to ME strategies from the publications reported in Table 54.1. While the US GHG emissions represent less than 5% of global emissions, Figure 54.3 reveals that estimated reductions from ME are close (e.g., for the fewer steel products or lifespan extension strategies) or even higher (lightweighting) than global estimated reductions. This is explained by the more ambitious steel demand reduction that ME strategies were assumed to cause in the US (e.g., a 30% versus a 3% reduction in the US and global demands, respectively, for the lightweighting strategy), leading to greater GHG emission reductions. Results from Figure 54.3 also show that depending on the assumptions and data used, similar demand reduction scenarios can lead to divergent estimates – highlighting the uncertainty that come with the estimation of ME contributions to decarbonization.

Several publications focused on global steel demand and estimated global GHG emission reductions from ME strategies. Allwood [12] assumed that 1% of steel components were reused, leading to a 3 Mt reduction in current GHG emissions. In their decarbonization

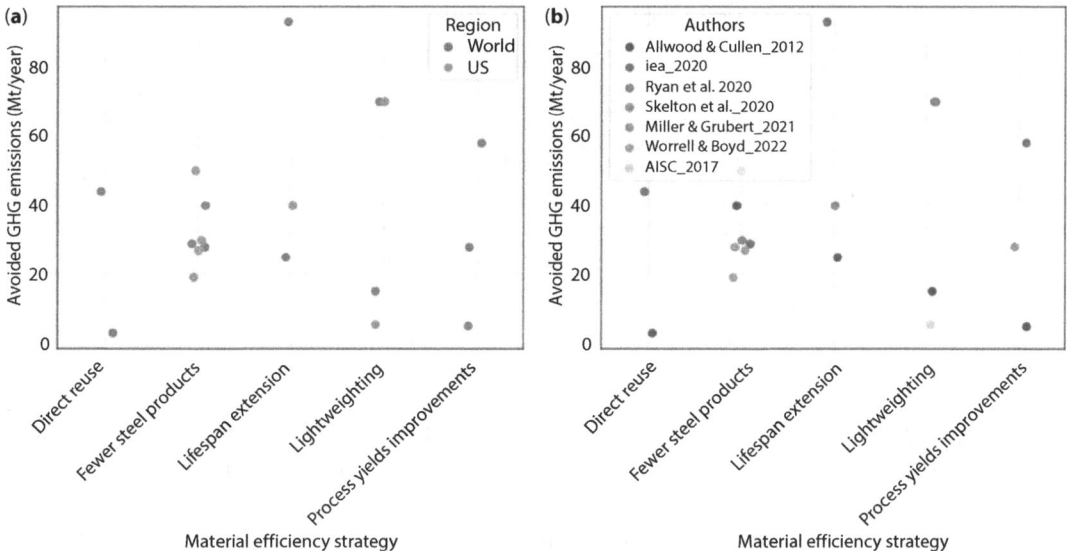

Figure 54.3 Estimates of avoided greenhouse gas emissions (million metric tons per year) from the reviewed material efficiency strategies by (a) geographic locations and (b) authors. Estimates are either directly taken from publications or, if reported as demand reductions, converted into GHG emission reductions using DOE's roadmap [2].

roadmap, IEA [11] estimated that a 4% reduction in global steel demand due to reuse could cause a reduction in emissions of 44 Mt/year. By decreasing the steel demand by 1%, using fewer steel products could lead to a 40 Mt reduction in current GHG emissions [12]. By contrast, IEA [11] calculated that a 2% reduction in steel demand (assumed to be due to a reduced use of vehicles) could lead to GHG emission reductions of 29 Mt/year. In another study, Skelton, Paroussos, and Allwood [28] showed that a 10% reduction in demand for vehicles (e.g., through car-sharing) could avoid 28 Mt/year of GHG emissions. A 1% increase in lifespan could save 25 Mt/year of GHG emissions [12]. The IEA [11] estimated that an 8% reduction in steel demand (assumed to be due to buildings lifespan extension) could lead 93 Mt/year – the highest estimation reported in Figure 54.3. Regarding lightweighting, making products 1% lighter would help save 15 Mt/year of GHG emissions globally. A 3% reduction in steel demand due to better building design (using less steel) combined with a 3% demand reduction due to vehicles lightweighting would save 70 Mt/year [11].

Improving process yields by 1% could reduce GHG emissions by about 5 Mt/year [12]. The IEA [11] found that a 2% and 3% reduction in steel demand due to improved yields during semi-manufacturing (intermediate steel products) and during product manufacturing, respectively would save 58 Mt/year. Finally, another study estimated a 28 Mt/year reduction in GHG emissions due to a 10% reduction in the amount of scrap generated when making cars [28].

Other works made the same type estimations but using US specific data. Taking an original approach based on time series that covers the COVID-19 pandemic, Miller and Grubert [29] calculated that a 30% decline in production (between Q2-2019 and Q2-2020) could lead to GHG emission reductions of 27 Mt/year in the US. In another study, a 50% reduction in steel demand for office & commercial buildings and transport led to about reduction of about 50 Mt/year [20]. Reducing steel stock per capita by 1 ton/capita (from about 11 ton/capita) could reduce GHG emissions by 30 Mt/year. Through the combination of a slightly higher recycling rate (80%) and a 33% reduction in iron production Worrell and Boyd [34] calculated 19 Mt/year reduction in GHG emissions. If new and existing steel products saw a 30% lifetime extension, 40 Mt/year of GHG emissions could be avoided. Using less steel in buildings could reduce steel demand by 15% in the construction sector, leading to a reduction in GHG emissions of about 5 Mt/year [33]. Finally, Ryan *et al.* [20] calculated that a 30% weight reduction for steel product could reduce emissions by 70 Mt/year.

54.4   Conclusions & Recommendations

This study strived to analyze the literature on ME and its application to decarbonization. The review showed that academic and non-academics publications do no tend to leverage much each other. Thus, future research activities (from scholars and non-scholars) could probably benefit from a better utilization of all stakeholders' work. The review also summarized barriers, opportunities, and trade-offs for 5 ME strategies. Many topics explored in this literature review call for further research. For instance, blockchain technology emerges especially promising to improve trust among stakeholder, enhance knowledge sharing and track material flows [35] – all topics that were identified as barriers to ME in this review. Next steps for this research are to study the decarbonization potential of ME in other industrial sectors (e.g., cement) and in other countries and apply text analysis techniques to draw more insights from the literature.

Acknowledgements

This work was authored by the National Renewable Energy Laboratory, operated by Alliance for Sustainable Energy, LLC, for the US Department of Energy (DOE) under Contract No. DE-AC36-08GO28308. Funding provided by the US Department of Energy Office of Energy Efficiency and Renewable Energy Advanced Manufacturing Office. The views expressed in the article do not necessarily represent the views of the DOE or the US Government. The US Government retains and the publisher, by accepting the article for publication, acknowledges that the US Government retains a nonexclusive, paid-up, irrevocable, worldwide license to publish or reproduce the published form of this work, or allow others to do so, for US Government purposes.

References

1. Circle Economy. 2021. Circularity Gap Report 2021.
2. DOE. Industrial Decarbonization Roadmap, 2022.
3. Ellen MacArthur Foundation. Completing the Picture: How the Circular Economy Tackles Climate Change edited by Material Economics, 2019.
4. Hertwich, Edgar G., Saleem Ali, Luca Ciacci, Tomer Fishman, Niko Heeren, Eric Masanet, Farnaz Nojavan Asghari, Elsa Olivetti, StefanPauliuk, Qingshi Tu, and Paul Wolfram. Material

efficiency strategies to reducing greenhouse gas emissions associated with buildings, vehicles, and electronics—a review. *Environmental Research Letters* 14 (4):043004, 2019.

5. Wang, Ke, Milo Costanza-van den Belt, Garvin Heath, Julien Walzberg, Taylor Curtis, Jack Berrie, Patrick Schroder, Leah Lazer, and Juan Carlos Altamirano. *Circular Economy as a Climate Strategy: Current Knowledge and Calls-to-Action.* United States: Washington, D.C.: World Resources Institute, 2022.

6. IPCC. Climate Change 2022: Mitigation of Climate Change. Contribution of Working Group III to the Sixth Assessment Report of the Intergovernmental Panel on Climate Change [P.R. Shukla, J. Skea, R. Slade, A. Al Khourdajie, R. van Diemen, D. McCollum, M. Pathak, S. Some, P. Vyas, R. Fradera, M. Belkacemi, A. Hasija, G. Lisboa, S. Luz, J. Malley, (eds.)], 2022.

7. Circle Economy. 2022. Circularity Gap Report 2022.

8. CIRAIG. Circular economy: a critical literature review of concepts. Polytechnique Montréal, 2015.

9. Calisto Friant, Martin, Walter J. V. Vermeulen, and Roberta Salomone. A typology of circular economy discourses: Navigating the diverse visions of a contested paradigm. *Resources, Conservation and Recycling* 161:104917, 2020. https://doi.org/10.1016/j.resconrec.2020.104917.

10. Hertwich, Edgar, Reid Lifset, Stefan Pauliuk, Niko Heeren, Saleem Ali, Qingshi Tu, Fulvio Ardente, Peter Berrill, Tomer Fishman, and Koichi Kanaoka. Resource Efficiency and Climate Change: Material Efficiency Strategies for a Low-Carbon Future. 2020.

11. IEA. 2020. *Iron and steel technology roadmap: Towards more sustainable steelmaking*: OECD Publishing.

12. Allwood, Jullian. *Sustainable materials – with both eyes open.* 2012. accessed 22/05/2015. http://www.cisl.cam.ac.uk/publications/julian- allwood-sustainable-materials.

13. Allwood, Julian M., Michael F. Ashby, Timothy G. Gutowski, and Ernst Worrell. Material efficiency: providing material services with less material production. *Philosophical Transactions of the Royal Society A: Mathematical, Physical and Engineering Sciences* 371 (1986):20120496, 2013.

14. Cantzler, Jasmin, Felix Creutzig, Eva Ayargarnchanakul, Aneeque Javaid, Liwah Wong, and Willi Haas. Saving resources and the climate? A systematic review of the circular economy and its mitigation potential. *Environmental Research Letters*, 2020.

15. Cooper, Daniel R., Nicole A. Ryan, Kyle Syndergaard, and Yongxian Zhu. The potential for material circularity and independence in the U.S. steel sector. *Journal of Industrial Ecology* 24 (4):748-762, 2020. https://doi.org/10.1111/jiec.12971.

16. Wijkman, Anders, and Kristian Skånberg. *The circular economy and benefits for society: Jobs and climate clear winners in an economy based on renewable energy and resource efficiency: A study pertaining to Finland, France, the Netherlands, Spain and Sweden*: Club of Rome, 2015.

17. Badampudi, Deepika, Claes Wohlin, and Kai Petersen. Experiences from using snowballing and database searches in systematic literature studies. *Proceedings of the 19th International Conference on Evaluation and Assessment in Software Engineering*, Nanjing, China, 2015.

18. Zhu, Yongxian, Kyle Syndergaard, and Daniel R. Cooper. Mapping the Annual Flow of Steel in the United States. *Environmental Science & Technology* 53 (19):11260-11268, 2019.

19. Milford, Rachel L., Julian M. Allwood, and Jonathan M. Cullen. Assessing the potential of yield improvements, through process scrap reduction, for energy and CO2 abatement in the steel and aluminium sectors. *Resources, Conservation and Recycling* 55 (12):1185- 1195, 2011. https://doi.org/10.1016/j.resconrec.2011.05.021.

20. Ryan, Nicole A., Shelie A. Miller, Steven J. Skerlos, and Daniel R. Cooper. Reducing CO2 Emissions from U.S. Steel Consumption by 70% by 2050. *Environmental Science & Technology* 54 (22):14598-14608, 2020.

21. Worldsteel association. Climate change and the production of iron and steel, 2021.

22. Cooper, Daniel R., and Julian M. Allwood. Reusing Steel and Aluminum Components at End of Product Life. *Environmental Science & Technology* 46 (18):10334-10340, 2012.

23. Rahman, S. M. Mizanur, Robert M. Handler, and Audrey L. Mayer. Life cycle assessment of steel in the ship recycling industry in Bangladesh. *Journal of Cleaner Production* 135:963-971, 2016. https://doi.org/10.1016/j.jclepro.2016.07.014.

24. Cai, Wenjia, Liyang Wan, Yongkai Jiang, Can Wang, and Lishen Lin. Short-Lived Buildings in China: Impacts on Water, Energy, and Carbon Emissions. *Environmental Science & Technology* 49 (24):13921-13928, 2015.

25. USS. Sustainability report. 2021.

26. International Energy Agency (IEA). Technology Roadmap: Solar Photovoltaic Energy, 2014.

27. Allwood, Julian M, Jonathan M Cullen, Mark A Carruth, Daniel R Cooper, Martin McBrien, Rachel L Milford, Muiris C Moynihan, and Alexandra CH Patel. *Sustainable materials: with both eyes open*. Vol. 2012: UIT Cambridge Limited Cambridge, UK, 2012.

28. Skelton, Alexandra C. H., Leonidas Paroussos, and Julian M. Allwood. Comparing energy and material efficiency rebound effects: an exploration of scenarios in the GEM-E3 macroeconomic model. *Ecological Economics* 173:106544, 2020. https://doi.org/10.1016/j.ecolecon.2019.106544.

29. Miller, Sabbie A., and Emily Grubert. US industrial sector decoupling of energy use and greenhouse gas emissions under COVID: durability and decarbonization. *Environmental Research Communications* 3 (3):031003, 2021.

30. Jensen, J. P., and K. Skelton. Wind turbine blade recycling: Experiences, challenges and possibilities in a circular economy. *Renewable and Sustainable Energy Reviews* 97:165-176, 2018. https://doi.org/10.1016/j.rser.2018.08.041.

31. Makov, Tamar, and David Font Vivanco. Does the Circular Economy Grow the Pie? The Case of Rebound Effects From Smartphone Reuse. *Frontiers in Energy Research* 6 (39), 2018.

32. Walzberg, Julien, Robin Burton, Fu Zhao, Kali Frost, Stéphanie Muller, Alberta Carpenter, and Garvin Heath. An investigation of hard- disk drive circularity accounting for socio-technical dynamics and data uncertainty. *Resources, Conservation and Recycling* 178:106102, 2022. https://doi.org/10.1016/j.resconrec.2021.106102.

33. AISC. 2017. More than Recycled Content: The Sustainable Characteristics of Structural Steel.

34. Worrell, Ernst, and Gale Boyd. Bottom-up estimates of deep decarbonization of U.S. manufacturing in 2050. *Journal of Cleaner Production* 330:129758, 2022. https://doi.org/10.1016/j.jclepro.2021.129758.

35. Rejeb, Abderahman, Andrea Appolloni, Karim Rejeb, Horst Treiblmaier, Mohammad Iranmanesh, and John G. Keogh. The role of blockchain technology in the transition toward the circular economy: Findings from a systematic literature review. *Resources, Conservation & Recycling Advances* 17:200126, 2023. https://doi.org/10.1016/j.rcradv.2022.200126.

55

Environmentally Extended Input-Output (EEIO) Modeling for Industrial Decarbonization Opportunity Assessment: A Circular Economy Case Study

Samuel Gause[1]*, Heather Liddell[1], Caroline Dollinger[1], Jordan Steen[1] and Joe Cresko[2]

[1]*Energetics, Columbia, Maryland, USA*
[2]*U.S. Department of Energy, Washington, D.C., USA*

Abstract

In increasingly complex supply chains, technology changes in one sector can have cascading impacts in other sectors that are not fully considered in most emissions analyses. Environmentally extended input-output (EEIO) models enable top-down analysis of the interactions between sectors and across the entire economy, including characterization of Scope 1, 2, and 3 emissions. EEIO models utilize matrix-based techniques to infer flows of material resources and emissions between economic sectors. Here we describe a new Excel-based EEIO model and scenario analysis tool designed for rapid "what-if" opportunity assessment. Baseline data in the model are drawn from publicly available sources, including recently released data from the U.S. Energy Information Administration's 2018 Manufacturing Energy Consumption Survey and the Bureau of Economic Analysis's economic input-output accounts data for the same year. A user interface allows for modification of fuel, electricity, and demand assumptions in each industrial sector—facilitating rapid assessment of the potential impacts of sustainability interventions, such as reduced demand, increased use of recycled material, or a change in energy sourcing to low-carbon options. Scenario analyses such as these can inform policymakers on priorities for decarbonization and improved resource utilization, while also enabling researchers and manufacturers to understand the drivers of cumulative emissions impacts in different products and sectors. We will demonstrate capabilities of this unique tool through a pair of illustrative case studies focused on the circular economy.

Keywords: EEIO, supply chain emissions, industrial decarbonization

55.1 Introduction

One of the most powerful available tools for supply-chain analysis is economic input-output (EIO) modeling – a comprehensive, top-down technique that can be used to analyze interactions across entire economies. Environmentally extended input-output (EEIO)

Corresponding author: sbgause@energetics.com

Nabil Nasr (ed.) Technology Innovation for the Circular Economy: Recycling, Remanufacturing, Design, Systems Analysis and Logistics, (739–754) © 2024 Scrivener Publishing LLC

models leverage EIO techniques to examine environmental impacts involving energy, emissions, and other environmental factors. These models can provide robust data for holistic analysis of decarbonization and supply chain opportunities, including opportunities to increase circularity and resource efficiency. Major EEIO models for the United States, including the Environmental Protection Agency's (EPA's) USEEIO model [1, 2] and Carnegie Mellon University's EIO-LCA webtool [3, 4], have demonstrated the powerful capabilities of EEIO to reveal systemic opportunities for emissions reduction in supply chains through top-down analysis of resource flows. Input-output methods are mature, and have been deployed recently to explore many aspects of industrial sustainability, including emissions hotspots [5–7], supply chain risk [8], social impacts of manufacturing [9], and effects of recycling/demand reduction [5, 8, 10, 11]. However, a limitation of EEIO models in general is that the underlying economic, energy, and environmental data are typically static (and therefore provide only a "snapshot" of the economic, technology, and energy conditions of the model base year). Data recency and lack of adjustability can constrain the utility of these models for forward-looking analysis, especially for the rapidly evolving industries, supply chains, and energy systems that characterize the circular economy. Researchers have made recent strides in adjustable input-output tools for specific use-cases, such as the multiregional RaMa-Scene webtool for the circular economy [5]. Nonetheless, facile adjustability remains rare in EEIO models and tools, and no existing EEIO tool yet enables rapid scenario modeling of industrial decarbonization strategies through user-defined "what-if" queries.

In this paper, we report on a new "EEIO for Industrial Decarbonization Analysis" (EEIO-IDA) scenario modeling tool that is now under development by the U.S. Department of Energy's (DOE's) Industrial Efficiency and Decarbonization Office. This interactive, Excel-based tool is designed for rapid "what-if" analysis based on user-defined decarbonization assumptions. Grid conditions, energy use, and decarbonization technology status for individual manufacturing sectors are user-adjustable, as summarized in Table 55.1. The scenario-building dashboard in the EEIO-IDA tool is aligned with the industrial decarbonization pillars presented in the DOE *Industrial Decarbonization Roadmap* [12]. For each user scenario, CO_2-equivalent (CO_2e) greenhouse gas emissions (for each sector and for the economy overall) are automatically calculated and summary results are presented. This scenario-modeling functionality is made possible by the fuel-use vectors that comprise the basis of the underlying model dataset. Rather than deploying emissions vectors (such as CO_2 emissions by sector) as the foundational environmental impact vectors of the environmentally extended model, as EPA's USEEIO and other models do, the foundational environmental vectors in the EEIO-IDA model reflect energy consumption (electricity and fuel use) drawn from primary sources such as the U.S. Energy Information Administration (EIA) Manufacturing Energy Consumption Survey (MECS) and other sources. These fuel-use vectors are supplemented by vectors for process emissions in applicable sectors (such as nonmetallic mineral products, where cement manufacturing processes emit significant non-energy-related CO_2), to complete the foundational model accounting. Based on user scenario inputs in the tool, electricity and fuel use breakdowns by sector are automatically adjusted from the base-case (2018) data, and subsequently the emissions for each sector (CO_2, CH_4, N_2O, and fluorinated compounds) are automatically calculated based on standard emissions factors for stationary and/or mobile combustion.

Table 55.1 User-adjustable features in the scenario-building dashboard of the EEIO-IDA tool.

Dashboard element	Description of user-adjustable assumptions in EEIO-IDA tool
U.S. Electric Grid	The user can define the U.S. average electric grid makeup by specifying the fraction of electricity generated from each of the following energy sources: coal, natural gas, petroleum, nuclear, biofuels, and renewable sources other than biofuels. Eight built-in grid scenarios are provided, ranging from the current U.S. grid to a hypothetical net-zero 2050 grid. The user can select one of these scenarios or build a custom scenario.
Energy Mix	For each of the 26 industrial sectors in the model (19 manufacturing sectors plus 7 non-manufacturing industrial sectors), the user can define the energy mix by specifying the fraction of energy supplied by the following energy sources: electricity, coal, natural gas, petroleum, biofuels, and green hydrogen. Reference values for the year 2018 are provided as the default and can be retained or adjusted for each sector.
Energy Requirements	For each of the 26 industrial sectors in the model, the user can specify the total energy requirements of the sector (in trillion Btu). Reference values for the year 2018 are provided as the default and can be retained or adjusted for each sector.
Non-Energy-Related Emissions	For the 8 industrial sectors in the model with significant non-energy-related process emissions, the user can specify the total non-energy-related emissions of CO_2, CH_4, N_2O, and fluorinated compounds resulting from the industrial activity (in million metric tons). Reference values for the year 2018 are provided as the default and can be retained or adjusted for each sector.
Carbon Capture	For each of the 26 industrial sectors in the model, the user can specify the fraction of energy- related CO_2 and (for applicable sectors) the fraction of non-energy-related CO_2 captured using carbon capture, utilization, and storage (CCUS) technologies. The reference value for all sectors is zero (i.e., no carbon capture). This value can be retained or adjusted for each sector.
Waste or Demand Reduction	For each of the 26 industrial sectors in the model, plus 4 additional user-selected sectors (commercial or governmental), the user can specify a percent demand reduction. A reduction in final product demand in one sector will result in upstream impacts to supplier sectors, and this dashboard element captures those impacts. The reference value for all sectors is zero (i.e., product demand is the same as in 2018). This value can be retained or adjusted for each sector.
Biogenic Emissions	User can toggle "on" or "off" inclusion of biogenic emissions from combustion of biofuels in CO_2-equivalent greenhouse gas emissions totals.

With its scenario-building user interface and capabilities for dynamic updating of environmental impact vectors, the EEIO-IDA tool provides a unique sandbox for rapidly exploring the potential impacts of technology changes in the industrial sector. Supply and demand relationships between industrial sectors are integral to the EEIO methodology, and therefore built into the tool. This makes the EEIO-IDA tool well suited for analysis of circular economy strategies and their potential role in reducing economy-wide emissions in the United States. Case studies from the cement and automobile industries are offered as examples in this paper.

55.2 Methods

EEIO-IDA is powered by publicly available data sources for energy use, economic flows, and emissions across the U.S. economy. Table 55.2 summarizes model data sources for electricity and fuel use by sector. These data are used to calculate baseline energy-related greenhouse gas emissions (of CO_2, CH_4, and N_2O), applying standard emissions factors for stationary and mobile combustion from the Intergovernmental Panel on Climate Change (IPCC) [13]. Table 55.2 also shows the data sources for process emissions by sector, which were primarily drawn from EPA's *Inventory of U.S. Greenhouse Gas Emissions and Sinks (1990-2020)* [14], with adjustments made to avoid double-counting with energy-related emissions calculated based on fuel use. Process emissions considered include CO_2, CH_4, N_2O, and fluorinated compounds. For summary reporting, greenhouse gas emissions are converted to a CO_2-equivalent value utilizing 100-year global warming potential (100-year GWP) values from IPCC's Fifth Assessment Report [15]. Equation 55.1 below shows how these data are used to calculate the emissions (e) for each industry k and greenhouse gas t.

$$e_{k,t} = C_{1,k} * \Sigma_i(f_i * x_{i,t}) + C_{2,k} * p_t \tag{55.1}$$

Here, f is the quantity of fuel i, x is the emissions factor for fuel i and greenhouse gas t pulled from the IPCC factors for stationary and mobile combustion [13], p is the process emissions of greenhouse gas t, and C_1 and C_2 are constants which are utilized in the user-defined scenarios (both equal to 1 in the base case scenario). Electricity is treated as a "fuel" in the model with a weighted-average emissions factor defined based on the assumed grid mix. All user-adjustable features (except for demand reductions, which induce a proportional decrease in the final demand vector for the selected industry) will modify portions of this equation, producing a new vector of emissions by industry to use in the EEIO calculation.

The model base year is 2018, corresponding to the most recent available EIA MECS dataset for industrial energy use [16] as well as Bureau of Economic Analysis (BEA) economic input-output data [17] for the same year. For economic data, a 2018 base year is achieved by selecting a mid-resolution representation of the overall economy (73 commodities).

Table 55.2 Data sources for electricity, fuel use, and non-energy-related emissions by sector.

Industries and corresponding NAICS codes	Data sources – electricity and fuel use	Data sources – non-energy-related emissions
Agriculture, Forestry, Fishing, and Hunting (11)	Farm Production Expenditures 2018 Summary [20]; Census of Agriculture - 2017 [21]; Census of Irrigation - 2018 [22]; Census of Horticultural Specialties - 2019 [23]; BEA - IO Tables - 2012 & 2018 [18]; EIA - Electric Power Monthly - 2020 [24]	US EPA - GHG Inventory, Chapter 5 [14]
Mining (21)	Economic Census 2017 [25]	US EPA - GHG Inventory, Chapter 3 [14]
Utilities (22)	BEA - IO Tables - 2012 & 2018 [18]; EIA - Electric Power Monthly - 2020 [24]	US EPA - GHG Inventory, Chapter 3 [14]
Construction (23)	EIA - AEO [26]	n/a
Manufacturing (31-33)	EIA - MECS 2018 [16]; ASM 2018 [27]	US EPA - GHG Inventory Chapter 4 [14]
Wholesale Trade (42)	BEA - IO Tables - 2012 & 2018 [18]; EIA - Electric Power Monthly - 2020 [24]	n/a
Retail Trade (44)	BEA - IO Tables - 2012 & 2018 [18]; EIA - Electric Power Monthly - 2020 [24]; EIA - CBECS 2018 [28] Economic Census 2017 [25]	n/a
Transportation and Warehousing, Excluding Postal Service (48, 49)	Transportation Energy Data Book [29]; BEA - IO Tables - 2012 & 2018 [18]; EIA - Electric Power Monthly - 2020 [24]; CBECS 2018 [28]	US EPA GHG Inventory Chapter 2 [14]; Alternative Fuels Data Center [33]; IPCC 2005 [34]
Other Service-Providing Industries, Except	BEA – IO Tables – 2012 & 2018 [18]; EIA – Electric Power Monthly – 2020 [24]; EIA – CBECS 2018 [28]	US EPA GHG Inventory Chapter 7 [14]; CalRecycle [35]; Bureau of Labor Statistics [36]; U.S. Census [37]
Government (51-81)		
Government (no NAICS code)	Comprehensive Annual Energy Data and Sustainability Performance [30]; Association of American Railroads, Railroad Facts [31]; EIA - CBECS 2018 [28]; FTA Annual Database Energy Consumption 2018 [32]	n/a

BEA releases input-output accounts data annually at this level of resolution.[1] In certain cases, time-based adjustments to datasets for years other than 2018 were made to project data to 2018. To do this, the project team utilized growth factors based on relevant data points from multiple years. An example of this type of growth factor is BEA's Chain Type Quantity Index [18]. Fuel usage data for some sectors were estimated based on energy purchases (of electricity, petroleum, natural gas, and coal) as specified in the Make/Use tables of BEA's 2012 benchmark dataset—following a similar methodology to that of the EIO-LCA model [19] for energy purchase allocation to sectors. These data were then projected to 2018 utilizing the Chain Type Quantity Index values for that sector. While energy consumption and fuel allocations estimated using this method are not as accurate as U.S. Energy Information Administration (EIA) or other governmental data reported explicitly for an industry, this provides a reasonable allocation strategy for those sectors for which no better data sources are available.

55.3   Results

Figure 55.1 shows model results for direct (scope 1) and indirect (scope 2 and 3) emissions for the 26 industrial sectors included in the EEIO-IDA model. The results highlight the significance of supply chains in the embodied emissions of industrial goods. For many sectors, such as the food and beverage sector and the motor vehicles sector, supply chain emissions account for a large majority of the embodied emissions. These scope-3 emissions reflect scope-1 or scope-2 emissions for other industries. For example, a large majority of the scope 3 emissions of the food and beverage and tobacco products sector (grey bar in Figure 55.1) coincide with the scope 1 and 2 emissions of the farms sector (blue bars in Figure 55.1). Importantly, scope 3 emissions should never be summed across multiple sectors due to the inherent (and purposeful) double-counting. However, this visualization of emissions-by-sector provides a valuable view into the final fate of industrial emissions as they accrue through suppy chains and ultimately become embodied into products. In particular, IO-based accounting of cumulative scope 1, 2, and 3 emissions can facilitate identification amd quantification of major industrial opportunities to reduce economy-wide emissions through demand-based strategies such as waste reduction (because such strategies have ripple effects on emissions in upstream sectors). The two case studies that follow provide examples of the impact of material efficiency strategies (reduced material requirements and improved product longevity) on the supply chain emissions of an industry or product.

---

[1]The current EEIO-IDA model (73 commodities) makes use of BEA's annual summary-level input/output accounts datasets. BEA's detail-level benchmark datasets (405 commodities) are released less frequently—only every five years. At the 73-sector level of resolution, most industrial sectors are represented at the level of their three-digit North American Industry Classification System (NAICS) code. Our team plans to develop a full-resolution version of EEIO-IDA (405 commodities) following the anticipated late 2023 release of the 2017 benchmark economic dataset from BEA. This level of aggregation will be consistent with EPA's USEEIO and other major EEIO models for the United States.

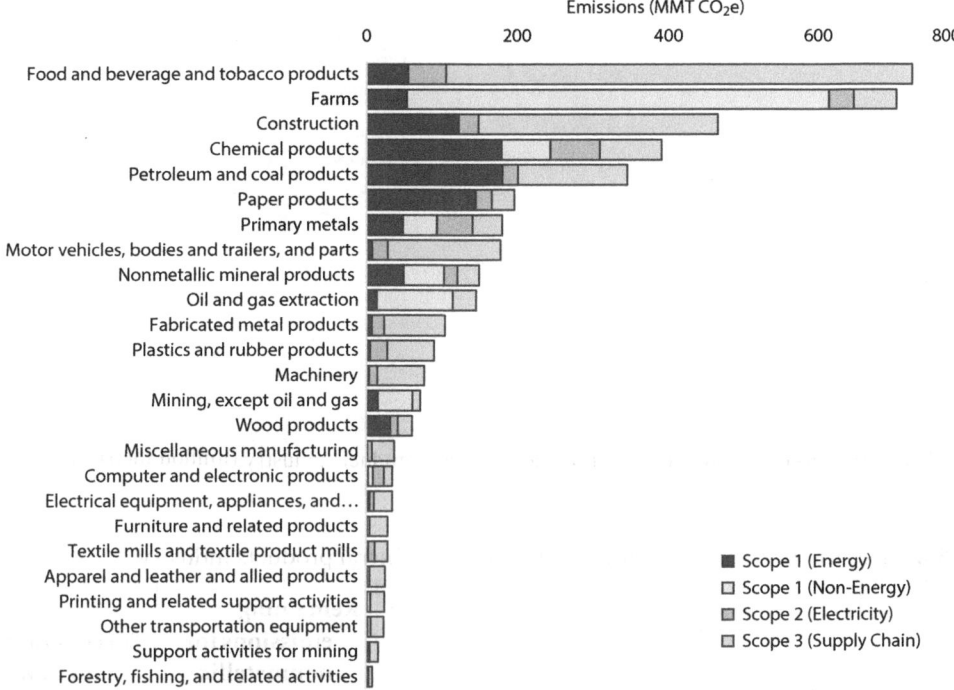

Figure 55.1 Direct sector and indirect supply chain emissions for industrial sectors for base year 2018 in million metric tons (MMT) CO$_2$e.

55.3.1 Case Study 1: Construction Improvements to Reduce Cement Emissions

The first case study looks at the cement industry, contained within the "nonmetallic mineral products" sector (NAICS 327) which also includes other products such as lime, glass, and clay. The cement industry is a major source of greenhouse gas emissions for the United States, producing 67 million metrics tons of CO$_2$e direct onsite emissions annually [38] —about 3.7% of the total U.S. industrial sector emissions [14]. Figure 55.2 shows the emissions for the entire nonmetallic mineral products sector by source. Many of the emissions in this sector are process-related (Scope 1, non-energy) and are considered difficult to abate, with most of these emissions coming from the chemical process of converting limestone (CaCO$_3$) to lime (CaO) which emits CO$_2$. As these emissions are inherent to the process, the only available mitigation strategies are carbon capture, alternative cement chemistries, or alternative routes to cement production [12]. These routes tend to require significant research and development (R&D) and most are capital intensive.

Another option to reduce emissions from the cement industry is to reduce the demand for cement. The EEIO-IDA tool can be utilized to understand the accrual of cement-related emissions in downstream industries. Model results indicate that 40% of the emissions from nonmetallic mineral products are embodied in the construction sector. Therefore, reducing concrete required for construction is a viable strategy for reducing cement industry emissions. Improvements in construction techniques (such as post-tensioning, precasting, and reducing overdesign) can significantly reduce cement demand while maintaining required structural properties. Estimates show a reduction of up to 50% of cement demand may be

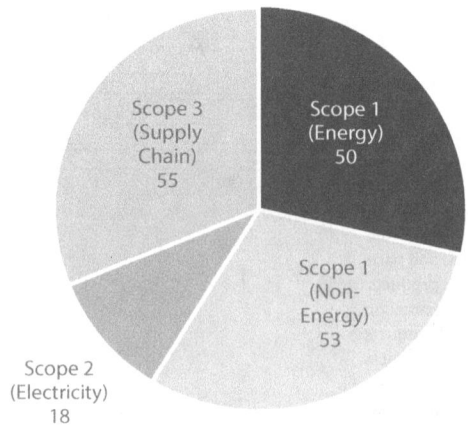

Figure 55.2 Emissions by source for the nonmetallic mineral products industry (million metric tons CO_2e) – 2018 base case.

Table 55.3 Decarbonization scenarios for nonmetallic mineral products industry.

Scenario	Model parameters	Remaining emissions for nonmetallic minerals (MMT CO_2e)	% reduction in emissions from base case
Base Case	• Base case, no emissions reductions	121	
Scenario #1	• 20% Reduction in demand for construction industry	110	9%
Scenario #2	• 50% Reduction in demand for construction industry	94	23%
Scenario #3	• Replacement of 30% of cement with SCM • 25% reduction in energy use for cement production • 90% carbon capture applied to cement industry • IEA's "Net Zero by 2050" electric grid 2050 scenario	41	67%
Scenario #4	• 50% reduction in demand for construction industry • Replacement of 30% of cement with SCM • 25% reduction in energy use for cement production • 90% carbon capture applied to cement industry • IEA's "Net Zero by 2050" electric grid scenario	31	74%

possible with these and other techniques [39]. In the EEIO-IDA tool, this reduction can be modeled by reducing final demand for the construction industry.

Several scenarios were evaluated and compared to the 2018 base case for these industries. The first scenario assumes a moderate increase in the usage of material-efficient construction techniques, for an overall 20% reduction in cement requirements for construction. The second is a more aggressive scenario, assuming 50% reduction. A third scenario adds other decarbonization options for the cement industry, including carbon capture (of 90% of cement industry CO_2 emissions), a net zero electric grid (following the International Energy Agency's (IEA's) "Net Zero by 2050" scenario for year 2050 [40]), replacement of 30% of current cement use with supplementary cementitious material (SCM, low carbon alternative materials such as fly ash) [41], and a 25% improvement in energy efficiency for cement production through implementation of state of the art technologies identified in the DOE Energy Bandwidth analysis for cement [42]. The last scenario combines scenarios #2 and #3.

A summary of the scenarios analyzed and the resulting emission reductions for the non-metallic mineral products industry are shown in Table 55.3. Results show that reducing construction demand for cement can cut emissions in the nonmetallic mineral products industry by up to 23%. Many of the methods for reducing concrete requirements for construction are available now [39], making this an option for near-term emissions reductions. Material efficiency strategies can also offer cascading benefits in other upstream industries, such as truck transportation for carrying cement to construction sites, increasing impacts beyond the construction and cement industries.

55.3.2 Case Study 2: Improving Longevity of Motor Vehicles

The second case study evaluates the motor vehicles, bodies and trailers, and parts sector (also referred to as the "motor vehicle" sector in this paper for brevity). Unlike the cement sector, which is dominated by scope 1 and 2 emissions, 84% of emissions embodied in motor vehicle products are attributed to scope 3 sources in the supply chain (Figure 55.3). Figure 55.4 shows the supply chain industries that contribute most to the emissions embodied in

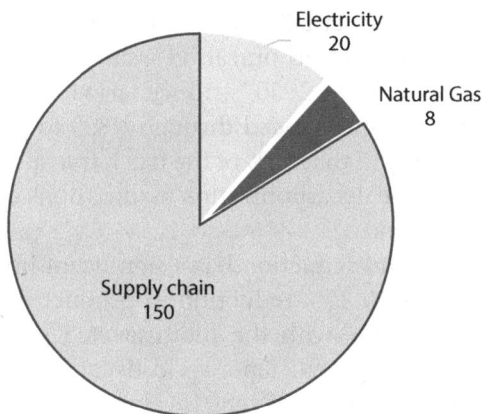

Figure 55.3 Emissions by source for the motor vehicles, bodies and trailers, and parts sector (million metric tons CO_2e) – 2018 base case.

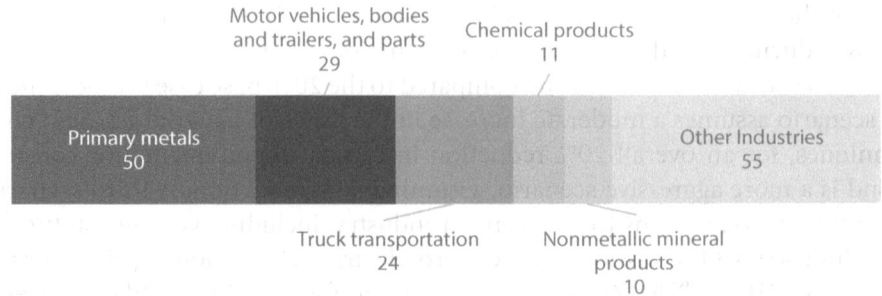

Figure 55.4 Cradle-to-Gate emissions (Scope 1, 2, &3) for the motor vehicles, bodies and trailers, and parts sector (MMT CO_2e) – 2018 base case.

motor vehicle products. As shown, the primary metals sector is the largest contributor to cradle-to-gate emissions, followed by truck transportation.

Because the embodied emissions of motor vehicles are dominated by upstream supply chain effects, there is relatively little room for reducing emissions of motor vehicles through process changes in the industry. Even switching current electricity use for the industry to 100% renewables would only reduce embodied emissions by 11%. Strategies that target the supply chain provide the best opportunity for reducing life-cycle emissions. One way to do this is by increasing the lifespan of motor vehicles to reduce demand for new vehicle production. The average age of vehicles in the U.S. in 2021 was 12 years [43], with estimates of the lifespan of these vehicles being 13-17 years [44]. The average age of the U.S. vehicle fleet has steadily increased over the years, with the average vehicle age increasing by 3 years since the year 2000 [43]. Assuming a similar average yearly increase going forward, the average age of cars would reach 14 years in 2035 and 16 years in 2050. These cases could correspond to a 14% and 25% decrease in demand for new vehicles in 2035 and 2050, respectively.

Table 55.4 shows the potential emission-reduction results for the two decreased-vehicle-demand scenarios (Scenarios #1 and #2), in addition to scenarios with additional decarbonization interventions included for comparison. Scenario #3 assumes that the electric grid follows IEA's "Net Zero by 2050" pathway using the year 2030 values [40]; that primary metals energy requirements are reduced through state-of-the-art steel and aluminum production methods [45, 46]; and that 20% of the truck transportation fleet is electrified. Scenario #4 further intensifies these decarbonization assumptions by assuming that the U.S. electric grid matches IEA's "Net Zero by 2050" grid scenario (for the year 2050); that energy requirements for primary metals are reduced through R&D to practical minimum values for steel and aluminum [45, 46]; and that 50% of the truck transportation fleet is electrified. A final scenario (#5) combines the decarbonization assumptions of Scenario #4 with a 25% reduction in motor vehicle demand.

As shown in Table 55.4, demand reduction has a significant impact on embodied emissions in motor vehicle products. A 25% reduction in product demand (Scenario #2) has a net emissions impact comparable with the multi-faceted decarbonization strategy of Scenario #3, which combines energy efficiency, grid decarbonization, and electrification interventions. While increasing longevity of vehicles alone is not sufficient to reach net zero emissions, this analysis shows the importance of circularity in an overall strategy to reach U.S goals of net zero carbon emissions by 2050.

Table 55.4 Decarbonization scenarios for the motor vehicles, bodies and trailers, and parts sector.

Scenario	Model parameters	Remaining emissions for automotive industry (MMT CO$_2$e)		% Reduction in emissions, scope 1, 2, & 3
		Scope 1 & 2	Scope 1, 2, & 3	
Base Case	• Base case, no emissions reductions	29	179	
Scenario #1	• 14% reduction in demand for motor vehicles	25	156	13%
Scenario #2	• 25% reduction in demand for motor vehicles	22	138	23%
Scenario #3	• Reduce energy demand for primary metals by bringing industry up to state of the art • 20% electrification of truck transportation industry • IEA's "Net Zero by 2050" electric grid year 2030 scenario	19	142	20%
Scenario #4	• Reduce energy demand for primary metals to practical minimum • 50% electrification of truck transportation industry • IEA's "Net Zero by 2050" electric grid year 2050 scenario	12	114	36%
Scenario #5	• 25% reduction in demand for motor vehicles • Reduce energy demand for primary metals to practical minimum • 50% electrification of truck transportation industry • IEA's "Net Zero by 2050" electric grid year 2050 scenario	10	88	51%

55.4 Conclusions and Recommendations

This paper highlighted the capabilities of the new EEIO-IDA tool to identify targets for reducing embodied emissions through demand reduction and material efficiency strategies, using cement and automobile production industries as examples. EEIO-IDA is a unique tool for rapid quantification and comparison of decarbonization opportunities in the U.S. industrial sector. The authors plan to make the present (73-sector) version of the

EEIO-IDA tool available to the public in 2023 for review and use by the community. This tool can help inform decision-makers on the strategies likely to yield the greatest potential emissions reductions—and the scale of implementation required to reach net-zero. Modeled scenario results reflect "big picture" estimated outcomes and can be combined with technology deep-dives to explore the practical opportunities for decarbonization at a more granular level. The case studies presented illustrate the flexibility of the model in handling various decarbonization assumptions. The 2018 model base year of EEIO-IDA provides greater data recency than existing U.S.-focused EEIO models with a 2012 base year (an important attribute for industries and energy systems that are rapidly changing). While a recycling module is not yet available in the tool, researchers have used EEIO modeling previously to evaluate such scenarios [47], and this tool can be modified in the future to include these capabilities. Future updates to the tool are also expected to include an expansion to the 405-sector level once BEA releases the 2017 benchmark input-output data set, expected in late 2023. This update will address the broad sector definitions that represent one of the key limitations of the mid-resolution (73-sector) version of the EEIO-IDA tool, which is the assumption of homogeneity in industries that manufacture diverse products (in terms of economic value and environmental impact) in practice.

Acknowledgments

This work was funded by the U.S. Department of Energy's Industrial Efficiency and Decarbonization Office.

References

1. Yang, Y., Ingwersen, W. W., Hawkins, T. R., Srocka, M., and Meyer, D. E., USEEIO: A New and Transparent United Stated Environmentally-Extended Input-Output Model, *Journal of Cleaner Production*, 158, 308, 2017.
2. Ingwersen, W. W., Li, M., Young, B., Vendries, J., and Birney, C., USEEIO v2.0, The US Environmentally-Extended Input-Output Model v2.0, *Scientific Data*, 9, 194, 2022.
3. Carnegie Mellon University, Economic Input-Output Life Cycle Assessment (EIO-LCA) tool, http://www.eiolca.net/, 2018.
4. Hendrickson, C. T., Lave, L. B., and Matthews, H. S.: *Environmental life cycle assessment of goods and services: an input-output approach*, Routledge, New York, 2006.
5. Donati, F., Niccolson, S., de Koning, A., Daniels, B., Christis, M., Boonen, K., Geerken, T., Rodriges, J. F. D., Tukker, A.: Modeling the circular economy in environmentally extended input-output: A web application, *Journal of Industrial Ecology*, 25, 36, 2020.
6. Chuang, Y., Huang, Z., Ries, R. J., Masanet, E.: The embodied air pollutant emissions and water footprints of buildings in China: a quantification using disaggregated input-output life cycle inventory model, *Journal of Cleaner Production*, 113, 274, 2016.
7. McDowall, W., Rodriguez, B. S., Usabiaga, A., Fernandez, J. A.: Is the optimal decarbonization pathway influenced by indirect emissions? Incorporating indirect life-cycle carbon dioxide emissions into a European TIMES model, *Journal of Cleaner Production*, 170, 260, 2018.

8. Yamamoto, T., Merciai, S., Mogollon, J. M., Tukker, A.: The role of recycling in alleviating supply chain risk–Insights from a stock-flow perspective using a hybrid input-output database, *Resources, Conservation, and Recycling*, 185, 106474, 2022.

9. Richter, J. S., Mendis, G. P., Nies, L., Sutherland, J. W.: A method for economic input-output social impact analysis with application to U.S. advanced manufacturing, *Journal of Cleaner Production*, 212, 302, 2019.

10. Vunnava, V. S. G., Singh, S.: Integrated mechanistic engineering models and macroeconomic input–output approach to model physical economy for evaluating the impact of transition to a circular economy, *Energy & Environmental Science* 14, 5017, 2021.

11. Donati, F., Aguilar-Hernandez, G. A., Siguenza-Sanchez, C. P., de Koning, A., Rodrigues, J. F. D., Tukker, A.: Modeling the circular economy in environmentally extended input-output tables: Methods, software, and case study, *Resources, Conservation, and Recycling* 152, 104508, 2020.

12. Cresko, J., Rightor, E., Carpenter, A., Peretti, K., *et al.*: Industrial Decarbonization Roadmap, U.S. Department of Energy, DOE Technical Report No. DOE/EE-2635, 2022.

13. Gomez, D. R., Watterson, J. D., *et al.*: Stationary Combustion, Volume 2, Chapter 2 in: *2006 IPCC Guidelines for National Greenhouse Gas Inventories.* Published on behalf of IPCC by the Institute for Global Environmental Strategies, Japan, 2006.

14. EPA (2022) Inventory of U.S. Greenhouse Gas Emissions and Sinks: 1990-2020, U.S. Environmental Protection Agency, EPA 430-R-22-003, 2022.

15. Myhre G., Shindell, D., *et al.*: Anthropogenic and Natural Radiative Forcing, Chapter 8 in: *Climate Change 2013: The Physical Science Basis.* Contributions of Working Group I to the Fifth Assessment Report of the Intergovernmental Panel on Climate Change, Cambridge University Press, 2013.

16. U.S. Energy Information Administration, Manufacturing Energy Consumption Survey (MECS), https://www.eia.gov/consumption/manufacturing/, 2021.

17. U.S. Bureau of Economic Analysis, Input-Output Accounts Data, https://www.bea.gov/industry/input-output- accounts-data, 2022.

18. Bureau of Economic Analysis, Input-Output Tables, https://www.bea.gov/itable/input-output, 2022.

19. Weber, C., Matthews, D., Venkatesh, A., Costello, C, and Matthews, H. S., The 2022 US Benchmark Version of the Economic Input-Output Life Cycle Assessment (EIO-LCA) Model (Model Documentation), Carnegie Melon, 2009.

20. U.S. Department of Agriculture, Farm Production Expenditures 2018 Summary, https://www.nass.usda.gov/Publications/Todays_Reports/reports/fpex0819.pdf, 2019.

21. U.S. Department of Agriculture, United States Summary and State Data, https://www.nass.usda.gov/Publications/AgCensus/2017/Full_Report/Volume_1,_Chapter_1_US/usv1.pdf, 2019.

22. U.S. Department of Agriculture, 2018 Irrigation and Water Management Survey, https://www.nass.usda.gov/Publications/AgCensus/2017/Online_Resources/Farm_and_Ranch_Irrigation_Survey/fris.pdf, 2019.

23. U.S. Department of Agriculture, 2019 Census of Horticultural Specialties, https://www.nass.usda.gov/Publications/AgCensus/2017/Online_Resources/Census_of_Horticulture_Specialties/HORTIC.pdf, 2020.

24. Energy Information Administration, Electric Power Monthly, https://www.eia.gov/electricity/monthly, 2022.

25. United States Census Bureau, 2017 Economic Census Data, https://www.census.gov/programs-surveys/economic-census/year/2017/economic-census-2017/data.html, 2022.

26. U.S. Energy Information Administration, Annual Energy Outlook 2022, https://www.eia.gov/outlooks/aeo/, 2022.

27. United States Census Bureau, Annual Survey of Manufacturers (2018), https://www.census.gov/programs-surveys/asm.html, 2022.

28. Energy Information Administration, 2018 Commercial Buildings Energy Consumption Survey, https://www.eia.gov/consumption/commercial, 2022.

29. Oak Ridge National Laboratory, Transportation Energy Data Book: Edition 40, https://tedb.ornl.gov/data, 2022.

30. US Department of Energy, Energy Efficiency and Renewable Energy, Comprehensive Annual Energy Data and Sustainability Performance, https://ctsedwweb.ee.doe.gov/Annual/Report/Report.aspx, 2022.

31. Association of American Railroads, Freight Rail Facts & Figures, https://www.aar.org/facts-figures, 2023.

32. Federal Transit Administration, 2018 Annual Database Energy Consumption, https://www.transit.dot.gov/ntd/data-product/2018-annual-database-energy-consumption, 2022.

33. U.S. Department of Energy, Alternative Fuels Data Center, https://afdc.energy.gov/data/10661, 2021.

34. Devotta, S., Sicars, S. *et al.*, Refrigeration, in: *Safeguarding the ozone layer and the global climate system*, C. Aduardo and I. Elgizouli (Ed.), pp226-268, Intergovernmental Panel on Climate Change, 2005.

35. CalRecyle, Estimated Solid Waste Generation Rates, https://www2.calrecycle.ca.gov/wastecharacterization/general/rates, 2006.

36. U.S. Bureau of Labor Statistics, Employment by major industry sector, https://www.bls.gov/emp/tables/employment-by-major-industry-sector.htm, 2022.

37. U.S. Census Bureau, Quick Facts, https://www.census.gov/quickfacts/fact/table/US/HSD410221, 2022.

38. U.S. Cement Industry Carbon Intensities (2019), U.S. Environmental Protection Agency, EPA 430-F-21-004, 2021.

39. Favier, A., De Wolf, C., Scrivener, K., Habert, G., A sustainable future for the European cement and concrete industry: Technology assessment for full decarbonization of the industry by 2050, ETH Zurich, 2018.

40. Bouckaert, S., Pales, A.F. *et al.*, Net zero by 2050: A roadmap for the global energy sector, *International Energy Agency*, 2021.

41. Hasanbeigi, A., Springer, C., Deep decarbonization roadmap for the cement and concrete industries in California, Global Efficiency Intelligence, 2019.

42. Schwartz, H., Ward, N., Levie, B., Chadwell, B., Brueske, S., Bandwidth study on energy use and potential energy savings opportunities in U.S. cement manufacturing, U.S. Department of Energy, DOE Technical Report No. DOE/EE-1660, 2017.

43. U.S. Bureau of Transportation Statistics, Average age of automobiles and trucks in operation in the United States, https://www.bts.gov/content/average-age-automobiles-and-trucks-operation-united-states, 2022.

44. Automotive News, Scrapping cycle points to 2 years of strong sales, https://www.autonews.com/article/20140120/OEM09/301209932/scrapping-cycle-points-to-2-years-of-strong-sales, 2014.

45. Jamison, K., Kramer, C., Brueske, S., Fisher, A., Bandwidth study on energy use and potential energy savings opportunities in U.S. iron and steel manufacturing, U.S. Department of Energy, DOE Technical Report no. DOE/EE-1231.

46. Das., S., Dollinger, C., Fisher, A., Brueske, S., Bandwidth study on energy use and potential energy savings opportunities in U.S. aluminum manufacturing, U.S. Department of Energy, DOE Technical Report no. DOE/EE-1664, 2017.

47. Aguilar-Hernandez, G.A., Sigüenza-Sanchez, C.P., Donati, F., Rodrigues, J.F.D., Tukker, A., Assessing circularity interventions, a review of EEIOA-based studies, *Journal of Economic Structures*, 7, 2018.

56

Pathways to Net Zero Emissions in Manufacturing and Materials Production- HVAC OEMs Perspective

Deba Maitra*, Swathy Ramaswamy, Cal Krause and Tiffany Waymer

Trane Technologies, Davidson, N.C., USA

Abstract

Increasingly, companies around the world are beginning to recognize reduction in embodied carbon as a means to achieve Net Zero emissions as a growing priority to address global climate change. The Heating, ventilation, and air conditioning (HVAC) original equipment manufacturers (OEMs) are part of one such industry that, utilizes a large volume of metal and alloy components in cast as well as wrought forms, including aluminum alloys, copper alloys, carbon steel, stainless steel and cast iron.

It has been established by the metal industry that replacing primary aluminum with recycled aluminum has much larger impact on reducing embodied carbon compared to any other material. World average CO_2 emission footprint per ton of product of aluminum alloy made from primary aluminum is more than three times of CO_2 emission footprint associated with copper alloys and close to ten times compared to steel [8]. Per The Aluminum Association's LCA (Life cycle analysis) report, 93% of total energy consumed and 94% of total carbon footprint in manufacturing aluminum components is associated with primary aluminum production [8]. OEMs are evaluating possible ways to get to Net Zero emissions for aluminum commodities and are actively engaged with supply chain. Three pathways being explored include i) identifying or developing alternate alloys that can incorporate higher scrap without adversely affecting manufacturing and performance requirements of HVAC systems, ii) forming a closed loop within the supply chain for internal scrap utilization and iii) engaging with smelters for utilizing low carbon aluminum where primary aluminum is currently used in the commodities. This paper proposes several approaches that HVAC OEMs can follow towards achieving aforementioned pathways toward Net Zero emissions through reductions in embodied carbon and any associated challenges and mitigation plans.

In addition to aluminum, steel presents a huge decarbonization opportunity for HVAC OEMs, because steel constitutes a significant portion of their total commodity purchases. Correspondingly, the industry is well equipped for collaboration with steel suppliers to implement a closed loop scrap system. There is also ample opportunity to engage in partnerships and coalitions to expedite the steel industry's transition to net zero. As it is a complex material to ameliorate emissions for, no one solution can be relied on. Instead, similar to aluminum, a multi-faceted approach is required to transform the steel industry.

Keywords: Net zero emission, HVAC OEM, low carbon aluminum, steel and copper, closed-loop recycling, alternate scrap-friendly alloy

---

**Corresponding author:* deba.maitra@tranetechnogies.com

Nabil Nasr (ed.) Technology Innovation for the Circular Economy: Recycling, Remanufacturing, Design, Systems Analysis and Logistics, (755–766) © 2024 Scrivener Publishing LLC

56.1   Introduction

It has been extensively discussed that increased presence of greenhouse gases in the atmosphere has been linked to global warming and changes in the climate. If proper steps are not taken, it is expected that the global temperature could increase to 1.5 degrees above pre-industrial levels by 2050 [1]. For this reason, the Paris Agreement on climate change dictates the global responsibility of reducing greenhouse gas emissions. One of the major contributors towards the greenhouse gas emissions is metal manufacturing processes. In the case of the HVAC industry, significant amounts of two metals for which decarbonization could have a significant impact on reducing embodied carbon: steel and aluminum. Steel is considered the most commonly used metal resource in the world and will be key in the transition to net-zero as it is a crucial ingredient for clean energy applications like wind turbines, solar panels and electric vehicles. The production of steel is associated with around 7% of all global emissions and about a quarter of all industrial emissions [3]. To meet the Paris Agreement, emissions from the steel industry must be reduced by greater than 50% before 2050 [3]. Aluminum is considered as one of the most valuable material resources in the modern world due to its applicability in several industries such as construction, packaging, transportation, electronics, and most importantly heat exchangers for HVAC industry. As per the International Aluminum Institute, the demand for aluminum alloys is expected to increase by 81 per cent by 2050 [2]. Both aluminum and steel are infinitely recyclable. In other words, both the metals and their alloys can be melted down and reused without adversely impacting physical and mechanical properties. Due to such advantage, these metals can play a crucial role in meeting sustainability goals throughout the value chain.

Use of recycled aluminum is gaining more attention in the aluminum manufacturing industry due to several reasons. Based on the Life Cycle Analysis (LCA) reported by The Aluminum Association, 94% of total carbon footprint and 93% of total energy consumed in manufacturing aluminum components is associated with primary aluminum production. The global aluminum industry's total energy-related CO_2 emissions were around 663 million tons in 2019, and 93% or more energy consumption was associated with primary aluminum production comprising of mining, Alumina refining from Bauxite and Electrolysis Steps to produce virgin aluminum [8]. The US industry average of CO_2 emissions for virgin aluminum production is 11.5 tons per ton of aluminum. However, each recycling process is associated with only 0.6 tons of CO_2 emissions which can be further broken down to 0.4 tons for transformation and 0.2 tons for transportation. The importance of circularity for aluminum compared to other materials is very evident in Figure 56.1, showing the carbon footprint associated with per ton of aluminum product made from primary aluminum is approximately 10 times compared to steel and 3 times compared to copper [8]. These are the three most common metals used in HVAC industry.

Similarly, primary production makes up the largest portion of emissions in steel manufacturing so there is also a growing emphasis on recycling in the steel industry. As detailed in Figure 56.1, steel manufacturing is associated with a little under 2 tons of CO_2 for every ton of steel produced on average. This is driven primarily by the most popular and most emissions intensive steel production method, the blast furnace. Secondary scrap-based steel production in an electric arc furnace is only one-eighth as emissions intensive as traditional blast furnace primary production [3]. However, there is unlikely to be enough scrap to

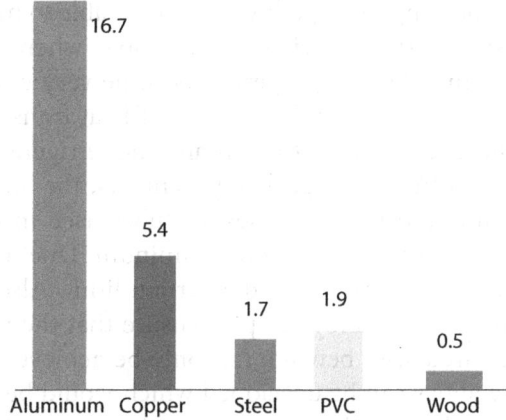

Figure 56.1 World average CO2 footprint of materials (ton/ton of product) [8].

meet total global demand. Therefore, taking a mixed approach by investigating techniques to produce virgin steel with a lower carbon footprint and optimizing scrap collection and recycling in a closed loop process will be essential in achieving global steel emission reduction goals.

Copper is also one of the prime commodities in HVAC Industry and considered as sustainable material because it is infinitely recyclable. Recycling and reuse of copper is happening for decades, with a supporting scrap-collecting infrastructure. More than half of Europe's copper demand is currently being met by recycled materials and, to date, at least 65% of all copper mined remains in circulation, available for use. The mining need for copper is expected to decline with improvement in recycling techniques. Just like Aluminum, use of recycled content in copper can also have a huge impact on embodied carbon and energy consumption reduction. Use of recycled copper reduces the energy consumption by 85% or more compared to use of virgin copper [15].

56.2 Pathways to Net-Zero Solutions

HVAC OEMs are evaluating possible ways to get to net zero emissions for aluminum, steel and copper commodities and this is possible only by actively engaging with the broader supply chain. The literature to date around the overall circularity scope for aluminum alloys discusses three parallel pathways to address all possible aspects in achieving net zero emissions in utilization of aluminum commodities. The three pathways include:

i) Identifying or developing alternate alloys that can incorporate higher scrap without adversely affecting manufacturing and performance requirements of HVAC systems,

ii) Forming a closed loop within the supply chain for internal scrap utilization

iii) Engaging with suppliers about utilizing low carbon aluminum and steel where virgin material is currently used in the commodities.

One challenge of implementing these pathways is being able to meet the product performance achieved today using virgin materials. For example, when aluminum alloy is used in heat exchangers, two of the physical properties become very critical. Thermal conductivity to ensure that coil performance and efficiency of heat transfer is not impacted and the galvanic potential difference between the fins and the refrigerant carrying tubes in the heat exchanger coils. Due to these stringent requirements, the aluminum alloys that are used to manufacture the fins and in some cases for tubes used in residential applications are prime alloys having content more than 98% aluminum. Due to the high purity of the alloys, it is difficult to add more scrap beyond a certain limit. Also, the scrap that can be added is only the same pure prime alloy scrap to ensure that the product meets the alloy chemistry requirements. Circularity benefits can only be achieved in case of HVAC aluminum alloys, if alternate alloys can be developed which would be scrap-based alloys and can replace existing prime alloys without impacting the stringent physical and mechanical requirements. In general, OEMs do not have the capabilities for alloy development. Therefore, progress is possible only when there can be close collaboration with the metal manufacturers and smelters.

56.3   Establishing Pathways for Alternate Alloys

In identifying or developing a new sustainable alloy, the primary goal is to maximize the amount of scrap products that can be added while simultaneously limiting the use of modifiers required to reach the desired composition. Of the three major metals for HVAC applications, establishing pathways for alternate alloys is most challenging in the case of aluminum. Since prime alloys contain >98% aluminum, the maximum scrap input they can have is limited to roughly 10%; in comparison, scrap-based aluminum alloys can take almost 100% scrap due to the high recyclability of aluminum. The types of scrap available can include internal generated scrap, external recycled content (post-industrial), and purchased finished components made from aluminum such as manifolds, distributor tube assemblies, u-bends, etc. [5]. Regardless of the type of scrap used, there should be no negative impact to the product manufacturability or to the product performance.

As alternate alloys are developed, considerations must include the product requirements for the application. As it relates to heat exchangers, the mechanical properties, including tensile strength and elongation, cleanliness, corrosion resistance, galvanic corrosion potential, system performance and thermal conductivity are part of the analysis in the selection of a suitable alloy [6]. The availability of the alloy and alloying elements play an essential role in the selection process as well [4]. 3000 and 8000 series Aluminum alloys are most abundant scraps at Aluminum alloy suppliers' mills serving HVAC industry customers.

Understanding of properties and scrap availability of common aluminum alloys is critical in deciding which alloy chemistry can add value toward HVAC OEM sustainability goals without compromising material requirements. For example 1000-series alloys are readily available and contain limited alloying elements since they are prime alloys; 2000-series alloys from the aircraft industry contain Cu and Zn, but are sensitive to hot cracking during operations such as brazing and soldering; 3000-series alloys from the construction industry contain Mn and Mg, both of which lower thermal conductivity; 4000-series alloys contain Si which lowers strength and these alloys are not heat treatable; 5000-series alloys typically used as automotive sheet contains high amounts of Mg which again leads to low thermal conductivity and may have brazing issues; 6000-series alloys contain Si and Mg and have brazing issues; 7000-series alloys are also used in aircraft contain Zn and Cu and are prime alloys. 8000-series alloys are commonly used in aircraft and radiators contain Fe, Si, and Li, and have thermal conductivity equivalent to 7000-series alloys but low strength. Replacing prime aluminum alloy such as 1000 or 7000 series used in heat exchanger fins and tubes with scrap-friendly alloy in 3000 or 8000 series can substitute 40% or more virgin aluminum with recycled aluminum. Each percentage increase of recycled aluminum in the products will lead to a reduction of cradle-to-gate carbon footprint by as much as 117 kg CO_2e for 1,000 kg of products [8]. Cradle-to-gate refers to life cycle stages from the extraction of raw materials to the completion of products such as cold rolled sheet, extruded tubes, etc. [8].

One option to formulate a sustainable alternate aluminum alloy, if thermal conductivity is compromised, is to consider alternate alloy options for the entire heat exchanger system so that the performance of the coil is not compromised. Simulations can help to understand how much thermal conductivity can be compromised for fins without negatively impacting the thermal performance for all possible combinations of indoor and outdoor HVAC unit configurations. The effect of alloying elements on thermal conductivity can be seen in Figure 56.2. 3000 series alloys are one of the most widely available scraps. However, due to high Mn content, the 3000-series alloys have lower thermal conductivity [10, 11]. If 3000 series is considered as an alternate alloy option for fins due to higher scrap availability, significant modifications are needed in the chemical composition and annealing process to improve the thermal conductivity. Another important factor that needs to be considered for alternate fin alloy is the sacrificial nature of the fins compared to tubes used in heat exchangers. Zn is the most common alloying element added to make an aluminum alloy anodic, when it forms galvanic couple with another aluminum alloy. It is very important to understand the optimum range of the galvanic potential difference between the two alloys, which in this case is the fin and the tube alloy. Another critical factor is ability to braze the

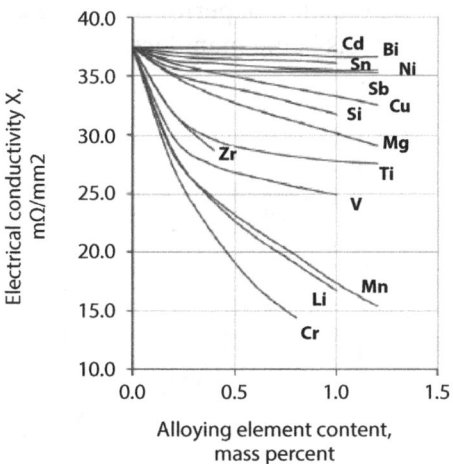

Figure 56.2 Effect of alloying elements in aluminum alloys on electrical conductivity which is directly proportional to thermal conductivity [16].

Aluminum alloy. The filler/base metal interaction should meet the brazing quality requirements. Controlled atmosphere brazing (CAB) is the state-of-the-art mass production technology, and it is very important that any refrigerant leak during production and in service is prevented [9].

56.4   Closing the Loop within the Supply Chain

Aluminum, steel and copper make up a major portion of the commodity materials that HVAC OEMs use in the manufacturing of their products. At the manufacturing facilities, aluminum sheets and tubes are used for the manufacturing of spine fin heat exchangers as well as fin and tube heat exchangers that go into residential and commercial products and applications (Figure 56.3). Steel forms the structure for components and is used in applications like fan grills and the wire ribs of units. A significant quantity of scrap is generated during these manufacturing processes from the cutting, slitting, and bending of metal sheets and tubes.

Currently, most scrap is being sold to third party recyclers with incomplete visibility to downstream processing and final disposition. While the steel copper, and aluminum scrap are being recycled by these third parties due to the inherent high value and infinite recyclability of metals, it is very likely that the scrap is processed and not reused in the HVAC industry. The metal scrap is more likely used in other high-volume manufacturing industries such as automotive or packaging where there is large demand for recycled metal and more tolerance for impurities or contamination that can occur during recycling.

It could be beneficial that the scrap generated during HVAC manufacturing be reused within the HVAC supply chain, particularly for aluminum, due to the nature of the specialized alloys developed primarily for heat exchange applications. HVAC OEMs will need to actively collaborate with raw material suppliers as well as third-party recyclers to effectively close the loop on the scrap generated at their facilities. OEMS should consider redirecting the scrap from their manufacturing facilities back to the suppliers' mills. OEMs should also test to ensure that there is no degradation of performance, when they use that recycled aluminum material within the products being manufactured.

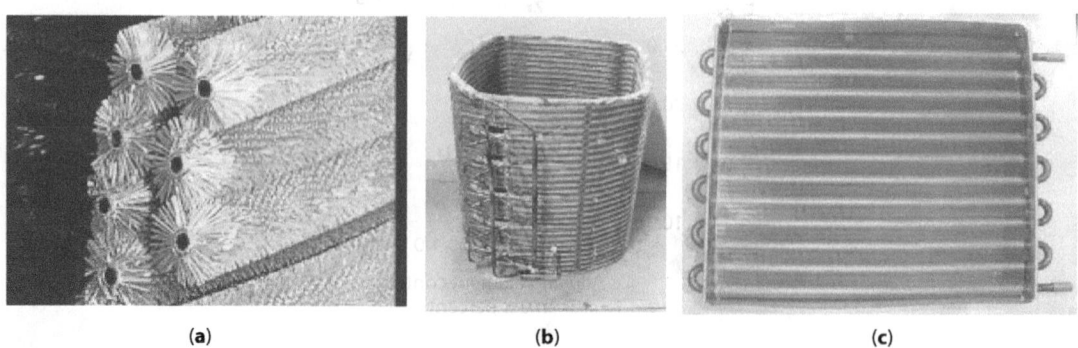

(a)   (b)   (c)

Figure 56.3 a) Spine fins. b) Residential outdoor coils. c) Residential indoor coils (Fin made with spine fins and tube heat exchanger).

In addition to the manufacturing scrap, a significant opportunity exists with the end-of-life waste generated after use of HVAC products. Most of the end-of-life HVAC scrap that ends up in a local or regional metal processing facility is shredded and mixed with automotive, appliance and other post-consumer scrap. This scrap may be shipped overseas and downcycled or recycled to manufacture products for other industries. New technologies and infrastructure are needed to sort and segregate these metal alloy waste streams to ensure that there is sufficient supply of recycled metals to be reused with the HVAC industry and form a truly closed loop within this supply chain.

56.5   The Future of Low Carbon Aluminum

The third approach being worked on to reduce embodied carbon in metals is to develop a low carbon version through greener approaches. In the case of aluminum, it is understood that even though it is 100% recyclable, the quantity of recycled aluminum may not be enough to sustain the growing need of the aluminum alloy utilization in various sectors. Often, due to stringent requirements of physical and mechanical properties of aluminum alloys, especially in HVAC industry, it is necessary to add virgin aluminum to adjust the alloy chemistry. Smelters are willing to collaborate with alloy manufacturers and HVAC OEMs to consider use of low carbon Aluminum as a direct substitute to virgin Aluminum.

Aluminum smelters have already been working to create Low-Carbon aluminum, which is also being called "Green Aluminum" by reducing carbon emissions and minimizing the carbon footprint in the upstream processes of making virgin aluminum through the Scope 1 – 3 approaches. Scope 1 refers to the direct emissions from sources that are owned or controlled by Smelters. Indirect emissions from the consumption of purchased electricity falls under scope 2; for example, electricity produced using renewable energy sources such as geo-thermal or hydroelectricity will significantly reduce the carbon footprint associated with electricity generation. All other indirect emissions that are consequences of the operation of smelters, but not owned or controlled by them is considered as scope 3. An example is the production of purchased goods and materials including mining operations, third-party distribution and logistics.

Smelters are beginning to increase their focus on developing their own versions of low caron virgin aluminum through guidance from the Aluminum Institute. For example, Rio Tinto is developing their version called RenewAL™ [12] Century is developing a product called Natur-Al™ [13] and Hydro's series of low carbon aluminum are called Hydro REDUXA [14], manufactured through use of renewable energy source like hydro power which reduces the carbon footprint to less than fourth of global average.

Two of the smelters, Alcoa and Rio Tinto (Alcan) initiated a project called ELYSIS™ in partnership with OEM Apple Inc [7] with a focus on reducing direct carbon emissions (scope 1) and have a target to achieve less than 4 ton of CO_2 emission per ton of virgin aluminum produced (Figure 56.4). The ELYSIS™ technology is expected to eliminate all direct greenhouse gas (GHG) emissions from the aluminum smelting process and is the first technology ever that emits oxygen as its by-product [4]. In traditional smelting process a carbon anode is used in the electrolysis process for conversion of Al_2O_3 to aluminum; with the ELYSIS™ technology an inert anode will replace the carbon anode releasing oxygen rather than CO_2. Another approach within scope 1 is electrification of manufacturing

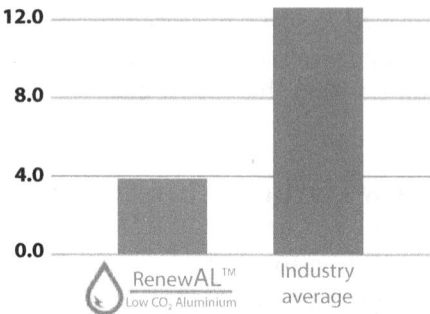

Figure 56.4 Comparison of CO_2 emission in manufacturing RenewAl vs. industry average (Courtesy Rio Tinto [12]).

equipment so that use of fossil fuel such as diesel can be minimized [3]. Most of the smelters are targeting carbon neutrality by 2050. All of the smelters are open for collaboration with OEMs such as Trane Technologies to promote the low carbon versions of virgin aluminum.

56.6   The Future of Low Carbon Steel

Steel, like aluminum, is a material that belongs to an industry where CO_2 emissions are difficult to abate. There is not a readily available golden solution that will unlock the necessary decarbonization. Because of this, OEMs must take a holistic view of the different levers that can be pulled to help improve the sustainability of steel production.

Today, there are three main routes of production in steelmaking: basic oxygen blast furnace (BOF), electric arc furnace (EAF), and direct reduced iron (DRI). Figure 56.5 below shows the scope 1 (direct) and 2 (associated with energy production) emissions associated with these production methods [3]. Of these, BOF is the traditional method and by far

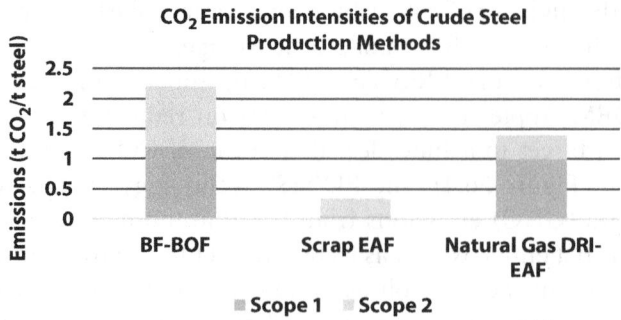

Figure 56.5 CO_2 emission intensities of crude steel production methods [3].

the most popular, accounting for around 70% of 2020 global production [3]. It also the most emissions intensive due to reliance on virgin steel and coal. EAFs are a type of secondary production method that utilizes higher volumes of scrap and is the least emitting option with most of their emissions coming from energy usage. DRI is an alternative to BOF which uses natural gas and is less emission intensive. The direct reduced iron route is interesting as a lower emission alternative for primary steel production, but also because natural gas can be replaced with green hydrogen to produce near zero emission steel. Some major European steelmakers are already piloting this technology, but it is unlikely to be available at scale this decade. Green technologies like carbon capture and bioenergy will also likely have a role to play as 1.8 Gt of additional emissions reduction needed by 2050 heavily relies on breakthrough technology in a Net-Zero Emissions scenario developed by the International Energy Agency [3].

While this new technology provides a long-term solution, in the short-term environmental impact can be lessened through process optimization and demand reduction through improved material efficiency. This can be as simple as avoiding over specification of steel qualities and designing products in a way that requires less steel. Improved manufacturing yields and direct reuse where applicable as opposed to re-melting and recycling steel represent further opportunities for material efficiency gains. Increasing material efficiency may seem sort of contradictory to maximizing scrap usage as it reduces the amount of available scrap, but ultimately these lead to greater emissions and energy savings overall [3]. After optimizing material efficiency, design processes that maximize circularity such as closed loop scrap processes can be implemented to create a more sustainable steel supply chain.

Companies who are downstream customers of steel suppliers, have the opportunity to join coalitions to mobilize demand for greener steel. One example is SteelZero, a Climate Group initiative that brings together organizations across the globe to speed up the transition to a net zero steel industry. With our involvement in SteelZero, organizations commit to 100% net zero steel by 2050 with an interim commitment of 50% by 2030. Another example is First Movers Coalition, a partnership between World Economic Forum and the US State Department that links the public sector and the private sector to accelerate demand for emerging green technologies. Members commit to procuring 10% of their annual steel volume at near-zero emissions by 2030. This is aimed at supporting the early breakthrough of green steel. Manufacturers also have the opportunity to engage in future purchase commitments with steel suppliers to help partner in their transition. This can be as high level as public announcement surrounding future intent to purchase low CO_2 steel, or as granular as direct offtake agreement where volumes and prices are contractually locked in. There is also the possibility of taking on joint investment projects to link the sharing of risks and benefits. Regardless, it is beneficial to engage with steel mills to understand where they are on their sustainability journey and help support their transition to net zero.

56.7 Conclusions

1) It is becoming increasingly important that OEMs address all aspects of material circularity to achieve net zero emissions in the near future.
2) Three pathways to reduce embodied carbon that are being explored by HVAC OEMs together with the supply chain include:

a. Identifying or developing alternate alloys that can incorporate higher scrap without adversely affecting manufacturing and performance requirements of HVAC systems.

b. Forming a closed-loop recycling within the supply chain for internal scrap utilization.

c. Engaging with suppliers on utilizing low carbon aluminum, copper and steel where virgin material is currently used in the commodities.

3) HVAC OEMS should also plan to extend the circularity concept and explore similar solutions for other metals beyond aluminum, steel, and copper.

Acknowledgements

The authors would like to acknowledge discussions with multiple smelters mentioned in the paper and aluminum, copper and steel manufacturers and supply chain team members at Trane Technologies.

References

1. Report of the Intergovernmental Panel on Climate Change, Climate Change 2021-The Physical Science Basis. https://www.ipcc.ch/report/ar6/wg1/downloads/report/IPCC_AR6_WGI_Full_Report.pdf

2. Closing the loop: how to improve aluminum's circularity; Financial Times. https://www.ft.com/partnercontent/international-aluminium-institute/closing-the-loop-how-to-improve-aluminiums-circularity.html

3. Iron and Steel Technology Roadmap Towards More Sustainable Steelmaking, International Energy Agency 2020 https://www.iea.org/reports/iron-and-steel-technology-roadmap

4. Lumley, R. (Ed.), *Fundamentals of aluminum metallurgy: Production, processing and applications,* Woodhead Publishing, Philadelphia, 2011.

5. Das, S.K., Designing Aluminum Alloys for a Recycling Friendly World. *Mat. Sci. Forum.,* 519-521, 1239-1244, 2006.

6. Zhou, B., Liu, B., Zhang, S., Microstructure evolution of recycled 7075 aluminum alloy and its mechanical and corrosion properties. *Journal of Alloys and Compounds*, 879, 2021.

7. WHAT IS ELYSIS? https://www.elysis.com/en/what-is-elysis#carbon-free-smelting

8. Jinlong Wang, The environmental footprint of semi-fabricated Aluminum products in North America- A life cycle assessment report, The Aluminum Association, January 2022.

9. Zhao H, Woods R, Controlled atmosphere brazing of aluminum, Advances in Brazing Science, *Technology and Applications*, 2013, Pages 280-322, 323e.

10. Kim, Cheol-Woo *et al.*, Effect of Alloying Elements on the Thermal Conductivity and Casting Characteristics of Aluminum Alloys in High Pressure Die Casting, *Journal of the Korean Institute of Metals and Materials,* v. 56(11); p. 805-812.

11. Peter Ulintz: Processing Aluminum Stampings, https://metalformingmagazine.com/article/?/materials/aluminum-alloys/processing-aluminum-stampings

12. RenewAl™ A cleaner start to your product lifecycle: Low CO_2 Aluminium; Rio Tinto, file:///C:/Users/irjqxn/Downloads/rt-aluminium-renewal-fact-sheet%20(3).pdf

13. Natur-Al™, Low-Carbon Aluminum for Premium Products; Norðurál- subsidiary of Century Aluminum https://nordural.is/en/natural/

14. Hydro REDUXA low-carbon aluminium; Hydro, https://www.hydro.com/en/aluminium/products/low-carbon-and-recycled-aluminium/low-carbon-aluminium/hydro-reduxa/

15. Andrew Surtees, Copper: The Material Of Choice For Air Conditionaing Systems Important Benefits Of Copper For Air Conditioning Install | HVAC News (hvacinformed.com)

16. Claudio Bunte *et al.*, Proposed Solution for Random Characteristics of Aluminium Alloy Wire Rods Due to the Natural Aging, *Procedia Materials Science* 9, December 2015.

Index

Printed in the USA/Agawam, MA
May 8, 2024

865761.024